The Revolution Continues . . .

The first edition of Stephen Marshak's *Earth: Portrait of a Planet* was warmly embraced by the geology community and practically overnight became the best-selling introductory text. *Earth* has been adopted at over 300 colleges and universities world-wide, including over 240 in North America. Geology instructors have responded to the integration of the theories of plate tectonics and Earth System science; the spectacular art and photographs; the innovative pedagogy, highlighted by "What a Geologist Sees" figures; and the remarkably inviting writing style. Here is just a sampling of some of the comments that we have received:

From Students

"The geology course I'm enrolled in uses your text, and I just wanted to tell you how fantastic it is. The book truly astonishes me. It's very well done, and the illustrations are marvelous. I just thought I would let you know."
—email from a *University of Cincinnati* student to Stephen Marshak

"I originally chose to take a geology class because I had a relatively small interest in Earth Science and it was an interesting way to satisfy my lab credit requirement. Upon starting the class, and reading into your textbook, I have decided to make geology (specifically tectonics) my life's work.

"Without your textbook, I am not so sure I would have found my calling. I look forward to refining my knowledge in geology as my college career moves on."
—email from a student at *Nassau Community College* to Stephen Marshak

"One of the most well-written, informative, and easy-to-read textbooks that I have ever encountered. The diagrams and charts are great. Because of this book, I have a better understanding of the concepts taught in my class."
—email from a *University of Illinois at Urbana-Champaign* student to Stephen Marshak

From Adopters

"The book is extremely well written and very easy to understand. Marshak has a gift for presenting material in a clear, easy-to-understand fashion that I envy. I find I can assign a reading on a topic and count on the students having a pretty good understanding before they come to class. This is definitely not true of most books."
—Tom Juster, *University of South Florida*

"Lucid, thorough writing style and great graphics. . . . Marshak is a fabulous writer. I can't say enough about how good this text is."
—David Osleger, *University of California, Davis*

"I very much like the structure of the book, with plate tectonics as the 'hinge' to understand geology and geological processes."
—Charly Bank, *Colorado College*

"Marshak's book is written from the perspective of the earth as a dynamic system with multiple interrelated components. Each chapter makes sense both from a component perspective and as an integral piece of the overall system."
—Michael Bradley, *Eastern Michigan University*

"*Earth* is a clear and extremely well written textbook. . . . It has excellent illustrations, and my students have praised it for its readability."
—Peter Buseck, *Arizona State University*

"About Marshak—I love it. I found it very complete and understandable."
—James Woodhead, *Occidental College*

"There is simply no competition."
—Torbjorn Tornqvist, *University of Illinois at Chicago*

The next few pages show how the text works and what's been added in the Second Edition.

MODERN ORGANIZATION AND APPROACH

Earth, Second Edition, introduces plate tectonics early—Chapters 3 and 4—after an important chapter on the structure of the earth (Chapter 2). From this point on, the theory of plate tectonics, the unifying theory of the earth sciences, is used to connect all the appropriate subjects, especially volcanoes, earthquakes, mountain formation, and the history of the earth.

Likewise, Earth System science, the framework that stresses the interconnectivity of the Earth's landmasses, bodies of water, and atmosphere, informs the entire narrative. It is particularly evident in the chapters on oceans and coasts, deserts, glaciers, the atmosphere, and global warming. In the Second Edition, an ESS icon appears in the margin next to particularly significant passages.

In short, the text consistently emphasizes the fact that several powerful cycles are at play in nearly every process that drives our Earth.

For the Second Edition, Stephen Marshak has added a great deal of new material on comparative planetary geology, content that is informed by the successful satellite explorations of Mars, Saturn, Venus, and Jupiter.

Of course, all of the chapters are self-contained and can be taught in any order the instructor wants.

Unique chapter on the structure of the Earth; this helps to set up the plate tectonics chapters.

Two chapters on plate tectonics

Full chapter on the history of life as it relates to the history of the Earth

Timely chapter on global warming

BRIEF CONTENTS

STATE-OF-THE-ART ILLUSTRATIONS

One of Stephen Marshak's goals in *Earth* was to develop art that conveys the dynamic way that geologic processes work. Marshak worked directly with a team of top artists to create figures that are clear and simple enough for students to understand but realistic enough to provide a reference framework. These spectacular 3-D illustrations are accompanied by photographs to help students visualize real geology. All of these illustrations are available to instructors as PowerPoint slides on the Norton Media Library CD-ROM that accompanies the text.

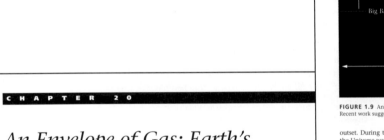

Many new figures have been added for the Second Edition, and many from the First Edition have been enhanced.

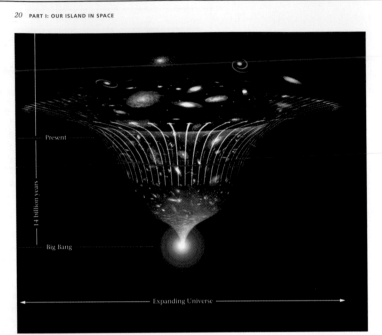

FIGURE 1.9 An artist's rendition of the Big Bang, followed by expansion of the Universe. Recent work suggests that the rate of expansion has changed as time passes.

outset. During the first instant (10^{-43} seconds) of existence, the Universe was so small, so dense, and so hot (10^{28} degrees centigrade) that it consisted entirely of energy—atoms, or even the smallest subatomic particles that make up atoms, could not even exist. (Note that because we use very large and very small numbers in our discussion, we employ scientific notation: $10^1 = 10$, $10^3 = 1,000$, and so on. Similarly, $10^{-3} = 1/1,000$.) But because expansion happened at nearly the speed of light, by the end of 1 second the density (i.e., quantity of mass per unit volume) of the Universe had decreased to "only" about a million billion (i.e., 10^{15}) times the density of water, and under these conditions the first protons and electrons could form (see Appendix A for definitions). Since a single proton constitutes the nucleus of a hydrogen atom (the smallest atom), this statement means that only a second after the big bang occurred, hydrogen atoms—the most abundant atoms—began to populate the Universe.

By the time the Universe reached an age of 3 minutes, its temperature had fallen below the 1 billion degrees, and its diameter had grown to about 100 billion km (60 billion miles). Under these conditions, nuclei of new atoms began to form through the collision and fusion (sticking

---

CHAPTER 20

An Envelope of Gas: Earth's Atmosphere and Climate

20.1 INTRODUCTION

On March 21, 1999, Bertrand Piccard, a Swiss psychiatrist, and Brian Jones, a British balloon instructor, became the first people to circle the globe nonstop in a balloon (▶Fig. 20.1). They began their flight on March 1, following more than twenty unsuccessful attempts by various balloonists during the previous two decades. Their airtight gondola, which could float in case they had to ditch in the sea, contained a heater, bottled air, food and water, and instruments for navigation and communication. The nature of their equipment hints at the challenges the balloonists faced, and why it took so many years before anyone finally succeeded in circling the globe.

Balloons can rise from the Earth only because an **atmosphere,** a layer consisting of a mixture of gases called **air,** surrounds our planet. Any object placed in a fluid feels a buoyancy force, and if the object is less dense than the fluid, then the buoyancy force can lift it off the ground. Balloons rise because the gas (either helium or hot air) in a balloon is less dense than air. Balloonists control their vertical movements by changing either the buoyancy of the balloon or the weight of the payload, but they cannot directly control their horizontal motions—balloons float with the **wind,** the flow of air from one place to another. In order to

Fed by the warm waters of the tropical Atlantic, Hurricane Hugo's spiraling winds flattened buildings, eroded coastal islands, and triggered floods in 1989. Hugo was particularly destructive because it generated a 6-m-high storm surge that arrived on top of very high tides. This image shows the storm striking the southeastern coast of the United States.

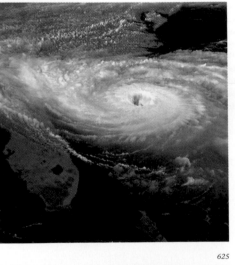

Over a dozen dramatic satellite photographs have been added to the Second Edition.

WHAT A GEOLOGIST SEES

Stephen Marshak has developed a particularly clever device to show students how thoughtful observation can reveal a large amount of information. Using simple drawings, he conveys what a trained geologist sees when viewing a photograph of a landscape. The untrained eye sees only a pretty picture—a geologist sees a page of Earth history.

joints—the water would leak down into the joints. Also, building a road on a steep cliff composed of jointed rock could be risky, for joint-bounded blocks separate easily from bedrock, and the cliff might collapse.

11.5 FAULTS: FRACTURES ON WHICH SLIDING HAS OCCURRED

After the San Francisco earthquake of 1906, geologists found a rupture that ripped across the landscape near the city. Where this rupture crossed orchards, it offset rows of trees,

and where it crossed a fence, it broke the fence in two; the western side of the fence moved northward by about 2 m (Fig. 10.6a). The rupture represented the trace of the San Andreas Fault (▶Fig. 11.13a, b). As we have seen, a "fault" is a fracture on which sliding occurs. Slip events, or "faulting," generate earthquakes. Faults, like joints, are planar structures, so we represent their orientation by strike and dip.

Faults riddle the Earth's crust. Some are currently active (sliding has been occurring on them in recent geologic time), but most are inactive (sliding on them ceased millions of years ago). Some faults, like the San Andreas, intersect the ground surface and thus displace the ground when they move. Others accommodate the sliding of rocks in the

FIGURE 11.13 (a) An oblique air photo showing the San Andreas Fault displacing a creek flowing from the Tremblor Range (background) into the Carizzo Plain, California. (b) What a geologist sees in the previous photo. (c) A road cut in the Rocky Mountains of Colorado, showing a fault offsetting strata in cross section. Note that the fault is actually a band of broken rock about 50 cm wide. (d) What a geologist sees looking at the Rocky Mountain road cut.

(a)

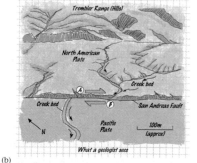
(b)

(c)

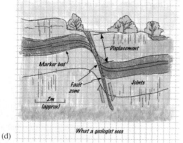
(d)

Eighteen new "What a Geologist Sees" figures have been added to the Second Edition.

TWO-PAGE SYNOPTIC PAINTINGS

These comprehensive paintings by the most respected geology artist in the world—Gary Hincks—are designed to encapsulate a number of topics. Each of the paintings is a complete synopsis of a major part of a chapter. By studying them carefully, students can see interconnections among subtopics covered in the chapter.

Three new paintings were commissioned for the Second Edition. They appear in the chapters on energy, mineral resources, and oceans and coasts.

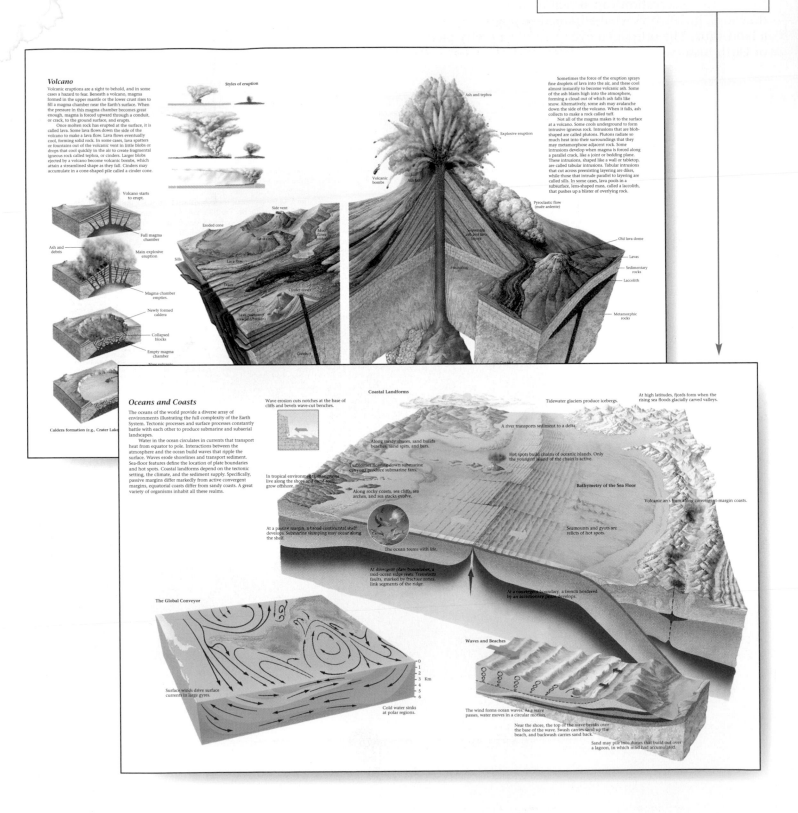

Volcano

Volcanic eruptions are a sight to behold, and in some cases a hazard to fear. Beneath a volcano, magma formed in the upper mantle or the lower crust rises to fill a magma chamber near the Earth's surface. When the pressure in this magma chamber becomes great enough, magma is forced upward through a conduit, or crack, to the ground surface, and erupts.

Once molten rock has erupted at the surface, it is called lava. Some lava flows down the side of the volcano to make a lava flow. Lava flows eventually cool, forming solid rock. In some cases, lava spatters or fountains out of the volcanic vent in little blobs or drops that cool quickly in the air to create fragmental igneous rock called tephra, or cinders. Larger blobs ejected by a volcano become volcanic bombs, which attain a streamlined shape as they fall. Cinders may accumulate in a cone-shaped pile called a cinder cone.

Sometimes the force of the eruption sprays fine droplets of lava into the air, and these cool almost instantly to become volcanic ash. Some of the ash blasts high into the atmosphere, forming a cloud out of which ash falls like snow. Alternatively, some ash may avalanche down the side of the volcano. When it falls, ash collects to make a rock called tuff.

Not all of the magma makes it to the surface at a volcano. Some cools underground to form intrusive igneous rock. Intrusions that are blob-shaped are called plutons. Plutons radiate so much heat into their surroundings that they may metamorphose adjacent rock. Some intrusions develop when magma is forced along a parallel crack, like a joint or bedding plane. These intrusions, shaped like a wall or tabletop, are called tabular intrusions. Tabular intrusions that cut across preexisting layering are dikes, while those that intrude parallel to layering are called sills. In some cases, lava pools in a subsurface, lens-shaped mass, called a laccolith, that pushes up a blister of overlying rock.

Styles of eruption

Ash and tephra

Explosive eruption

Volcano starts to erupt.

Ash and debris

Full magma chamber

Main explosive eruption

Volcanic bombs

Ash fall

Side vent

Eroded cone

Mud flows

Lava flow

Sills

Magma chamber empties.

Dikes

Cinder cones

Newly formed caldera

Collapsed blocks

Empty magma chamber

Pyroclastic flow (nuée ardente)

Old lava dome

Lavas

Sedimentary rocks

Laccolith

Conduit

Metamorphic rocks

Caldera formation (e.g., Crater Lake)

Oceans and Coasts

The oceans of the world provide a diverse array of environments illustrating the full complexity of the Earth System. Tectonic processes and surface processes constantly battle with each other to produce submarine and subaerial landscapes.

Water in the ocean circulates in currents that transport heat from equator to pole. Interactions between the atmosphere and the ocean build waves that ripple the surface. Waves erode shorelines and transport sediment. Sea-floor features define the location of plate boundaries and hot spots. Coastal landforms depend on the tectonic setting, the climate, and the sediment supply. Specifically, passive margins differ markedly from active convergent margins, equatorial coasts differ from sandy coasts. A great variety of organisms inhabit all these realms.

Coastal Landforms

Wave erosion cuts notches at the base of cliffs and bevels wave-cut benches.

Tidewater glaciers produce icebergs.

At high latitudes, fjords form when the rising sea floods glacially carved valleys.

A river transports sediment to a delta.

Along sandy shores, sand builds beaches, sand spits, and bars.

Hot spots build chains of oceanic islands. Only the youngest island of the chain is active.

Turbidities flowing down submarine canyons produce submarine fans.

In tropical environments, mangroves live along the shore and coral reefs grow offshore.

Along rocky coasts, sea cliffs, sea arches, and sea stacks evolve.

Bathymetry of the Sea Floor

At a passive margin, a broad continental shelf develops. Submarine slumping may occur along the shelf.

Volcanic arcs form along convergent-margin coasts.

Seamounts and gyots are relicts of hot spots.

The ocean teems with life.

At divergent plate boundaries, a mid-ocean ridge rises. Transform faults, marked by fracture zones, link segments of the ridge.

At a convergent boundary, a trench bordered by an accretionary prism develops.

The Global Conveyor

Surface winds drive surface currents in large gyres.

0
1
2
3 Km
4
5
6

Cold water sinks at polar regions.

Waves and Beaches

The wind forms ocean waves. As a wave passes, water moves in a circular motion.

Near the shore, the top of the wave breaks over the base of the wave. Swash carries sand up the beach, and backwash carries sand back.

Sand may pile into dunes that build out over a lagoon, in which mud had accumulated.

CAPTURES STUDENT INTEREST

Chapter Opening Story

The Human Angle

Case Studies

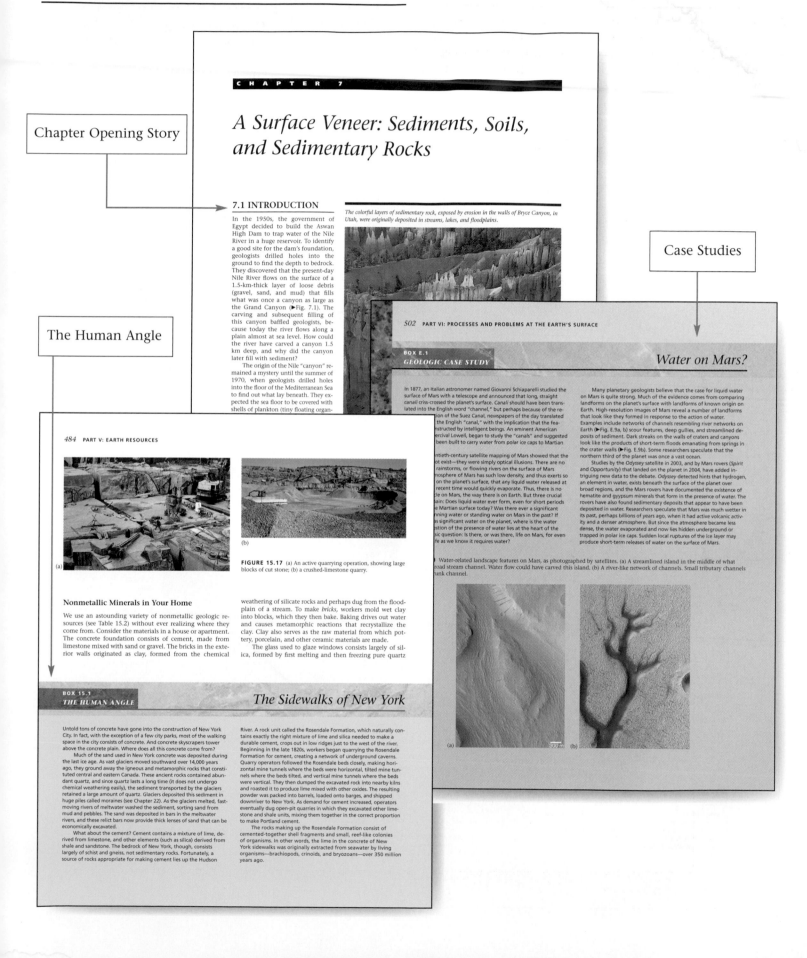

BOXED INSERTS AND SCIENTIFIC APPENDIX

In addition to "The Human Angle" and "Geologic Case Study" boxes, *Earth* has other features that enhance the presentation of important topics and issues. Of particular note are the boxes and appendix that provide background information in physics and chemistry to help students that need it.

BOX 5.2
THE REST OF THE STORY

Where Do Diamonds Come From?

As we saw earlier, diamond consists of the element carbon. Accumulations of carbon develop in a variety of ways: soot (pure carbon) results from burning plants at the surface of the Earth; coal (which consists mostly of carbon) forms from the remains of plants buried to depths of up to 15 km; and graphite develops from coal or other organic matter buried to still greater depths (15–70 km) in the crust during mountain building. Experiments demonstrate that the temperatures and pressures needed to form diamond are so extreme that, in nature, they generally occur only at depths of around 150 km below the Earth—that is, in the mantle. Under these conditions, the carbon curs in carrot-shaped bodies 50–200 m across and at least 1 km deep that are called kimberlite pipes (▶Fig. 5.26a).

Controversial measurements suggest that many of the diamonds that sparkle on engagement rings today were created when subduction carried carbon into the mantle 3.2 billion years ago. The diamonds sat at depths of 150 km in the Earth until two rifting events, one of which took place in the late Precambrian and the other during the late Mesozoic, released them to the surface, like genies out of a bottle. The Mesozoic rifting event led to the breakup of Pangaea.

In places where diamonds occur in solid kimberlite, they can be ...ned only by digging up the kimberlite and crushing it, to separate ...e diamonds (▶Fig. 5.26b). But nature can also break diamonds ...rom the Earth. In places where kimberlite has been exposed at ...round surface for a long time, the rock chemically reacts with ...r and air (a process called weathering; see Chapter 7). These reac-...cause most minerals in kimberlite to disintegrate, creating sedi-...that washes away in rivers. Diamonds are so strong that they ...n as solid grains in river gravel. Thus, many diamonds have been ...ned simply by separating them from recent or ancient river gravel. ...Diamond-bearing kimberlite pipes are found in many places ...nd the world, particularly where very old continental lithosphere ... Southern and central Africa, Siberia, northwestern Canada, ... Brazil, Borneo, Australia, and the U.S. Rocky Mountains all have ... Rivers and glaciers, however, have transported diamond-bearing ...ents great distances from their original sources. In fact, dia-...ds have even been found in farm fields of the midwestern United ...s. Not all natural diamonds are valuable: value depends on color ...larity. Diamonds that contain imperfections (cracks, or specks of ...material), or are dark gray in color, won't be used for jewelry. ...e stones, called *industrial diamonds*, are used instead as abrasives, ...amond powder is so hard (10 on the Mohs hardness scale) that it ...e used to grind away any other substance.

...Gem-quality diamonds come in a range of sizes. Jewelers mea-...diamond size in carats, where one "carat" equals 200 milligrams ...rams)—one ounce equals 142 carats. (Note that a carat mea-...gemstone weight, while a "karat" specifies the purity of gold. ...gold is 24 karat, while 18-karat gold is an alloy containing ...rts of gold and 6 parts of other metals.) The largest diamond ...found, a stone called the Cullinan Diamond, was discovered in ... Africa in 1905. It weighed 3,106 carats (621 grams) before being cut into nine large gems (the largest weighing 516 carats) as well as many smaller ones. By comparison, the diamond on a typical engagement ring weighs less than 1 carat. Diamonds are rare, but not as rare as their price suggests. A worldwide consortium of diamond producers stockpile the stones so as not to flood the market and drive the price down.

• *Generally inorganic:* Most, but not all, minerals are inorganic, by which we mean that most do not contain organic chemicals—there are only about thirty organic chemicals that are classified as minerals. To understand the above statement, we need to examine the meaning of "organic chemical." Organic chemicals are molecules composed of carbon bonded to hydrogen, along with varying amounts of oxygen, nitrogen, and other elements. The name "organic" emphasizes that many such chemicals occur in living organisms. But not all organic materials form in nature—plastic, for example, consists of organic chemicals. In this context, materials such as plastic, plant debris, sugar, fat, and protein are not minerals. It is important to note that the presence of carbon alone does not make a mineral organic. For example, diamond and graphite are minerals composed of pure carbon, and calcite is a mineral composed of carbon bonded to oxygen and calcium—neither are organic. We also emphasize that some inorganic minerals are formed by biologic processes. For example, a clam shell consists mostly of the inorganic mineral calcite, constructed by the life activity of the clam.

With these definitions in mind, we can make an important distinction between a mineral and **glass**. Both minerals and glasses are solids, in that they can retain their shape indefinitely (see Appendix A). But a mineral is crystalline, while glass is not. Whereas atoms, ions, or molecules in a mineral are ordered into a crystal lattice, like soldiers standing in formation, those in a glass are arranged in a semichaotic way, like a crowd of people at a party, in small clusters or chains that are neither oriented in the same way nor spaced at regular intervals (▶Fig. 5.3c, d). Note that the chemical compound silica (SiO_2) forms the mineral quartz when arranged in a crystalline lattice, but forms common window glass when arranged in a semichaotic way.

Some Basic Definitions from Chemistry

BOX 5.1
SCIENCE TOOLBOX

To describe minerals, we need to use several terms from chemistry (for a more in-depth discussion, see Appendix A). To avoid confusion, terms are listed in an order that permits each successive term to utilize previous terms.

• **Element:** a pure substance that cannot be separated into other elements.

• **Atom:** the smallest piece of an element that retains the characteristics of the element. An atom consists of a nucleus surrounded by a cloud of orbiting electrons; the nucleus is made up of protons and neutrons (except in hydrogen, whose nucleus contains only one proton and no neutrons). Electrons have a negative charge, protons have a positive charge, and neutrons have a neutral charge. An atom that has the same number of electrons as protons is said to be "neutral," in that it does not have an overall electrical charge.

• **Atomic number:** the number of protons in an atom of an element.

• **Atomic weight:** the approximate number of protons plus neutrons in an atom of an element.

• **Ion:** an atom that is not neutral. An ion that has an excess negative charge (because it has more electrons than protons) is an **anion**, whereas an ion that has an excess positive charge (because it has more protons than electrons) is a **cation**. We indicate the charge with a superscript. For example, Cl^- (chlorine) has a single excess electron; Fe^{2+} is missing two electrons.

• **Chemical bond:** an attractive force that holds two or more atoms together. For example, *covalent bonds* form when atoms share electrons. *Ionic bonds* form when a cation and anion (ions with opposite charges) get close together and attract each other. In materials with **metallic bonds,** some of the electrons can move freely.

• **Molecule:** two or more atoms bonded together. The atoms may be of the same element or of different elements.

• **Compound:** a pure substance that can be subdivided into two or more elements. The smallest piece of a compound that retains the characteristics of the compound is a molecule.

• **Chemical:** a general name used for a pure substance (either an element or a compound).

• **Chemical formula:** a shorthand recipe that itemizes the various elements in a chemical and specifies their relative proportions. For example, the formula for water, H_2O, indicates that water consists of molecules in which two hydrogens bond to one oxygen.

• **Chemical reaction:** a process that involves the breaking or forming of chemical bonds. Chemical reactions can break molecules apart or create new molecules and/or isolated atoms.

• **Mixture:** a combination of two or more elements or compounds that can be separated without a chemical reaction. For example, a cereal composed of bran flakes and raisins is a mixture—you can separate the raisins from the flakes without destroying either.

• **Solution:** a type of material in which one chemical (the solute) dissolves (becomes completely incorporated) in another (the solvent). In solutions, a solute may separate into ions during the process. For example, when salt (NaCl) dissolves in water, it separates into sodium (Na^+) and chlorine (Cl^-) ions. In a solution, atoms or molecules of the solvent surround atoms, ions, or molecules of the solute.

• **Precipitate:** (noun) a compound that forms when ions in liquid solution join together to create a solid that settles out of the solution; (verb) the process of forming solid grains by separation and settling from a solution. For example, when saltwater evaporates, solid salt crystals precipitate and settle to the bottom of the remaining water.

ANIMATIONS

Developed by Stephen Marshak to illustrate dynamic Earth processes, over sixty original animations emphasize plate tectonics, geologic hazards, and Earth System science concepts. Approximately half of these animations are new to the Second Edition, and many are based on "What a Geologist Sees" figures from the text.

Designed for use by instructors and students, animations can be enlarged to full-screen view for lecture display. VCR-like controls make it easy for instructors to control the pace of the animations during lectures.

Animations are available on the Norton Media Library Instructor's CD-ROM, on the Norton Resource Library Instructor's Website, and on the Student CD-ROM and WebSite at www.wwnorton.com/earth. See pp. *xvii–xviii* for a complete list and descriptions of supplements for students and instructors.

"The Instructor CD contains some of the more imaginative and detailed figures and animations I have found."—Jeff Knott, *California State University, Fullerton*

"The Student CD and Web site were a welcome surprise when I first adopted the book. The animations are very good. . . . I use many of them during class."—Tom Juster, *University of Southern Florida*

Cliff Retreat Along the Coast

Overhangs develop along seacoasts and rivers where the water cuts into a fairly strong slope.

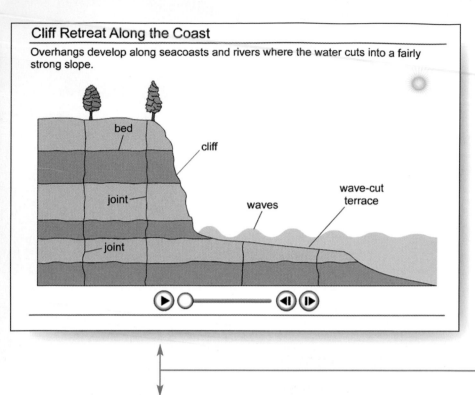

What A Geologist Sees—Thrust Fault

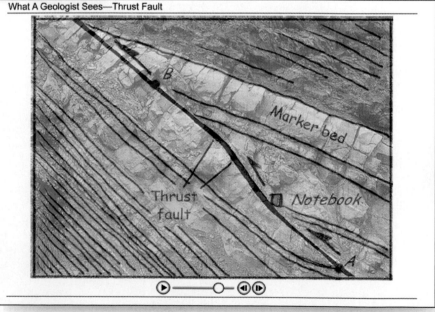

W. W. Norton & Company has been independent since its founding in 1923, when William Warder Norton and Mary D. Herter Norton first published lectures delivered at the People's Institute, the adult education division of New York City's Cooper Union. The Nortons soon expanded their program beyond the Institute, publishing books by celebrated academics from America and abroad. By mid-century, the two major pillars of Norton's publishing program—trade books and college texts—were firmly established. In the 1950s, the Norton family transferred control of the company to its employees, and today—with a staff of four hundred and a comparable number of trade, college, and professional titles published each year—W. W. Norton & Company stands as the largest and oldest publishing house owned wholly by its employees.

Composition by TSI Graphics
Manufacturing by Courier
Illustrations for the Second Edition by Precision Graphics

Editor: Jack Repcheck
Project editor: Thomas Foley
Copy editor: Alice Vigliani
Electronic media editor: April Lange
Director of manufacturing: Diane O'Connor
Photography editors: Neil Ryder Hoos and Stephanie Romeo
Editorial assistant: Sarah Solomon
Book designer: Joan Greenfield
Developmental editor for the First Edition: Susan Gaustad

ISBN 0-393-92502-1

W. W. Norton & Company, Inc., 500 Fifth Avenue, New York, N.Y. 10110
www.wwnorton.com

W. W. Norton & Company Ltd., Castle House, 75/76 Wells Street, London W1T 3QT

2 3 4 5 6 7 8 9 0

DEDICATION

To Kathy, David, and Emma, who helped in (and put up with!) this

endeavor in many ways over many years

BRIEF CONTENTS

PART III
Tectonic Activity of a Dynamic Planet

PART IV
History before History

Imagine a desert canyon at dawn. Stark cliffs of red rock descend like a staircase down to the gravelly bed of a dry stream on the canyon floor. Mice patter among dry shrubs and cactus. Suddenly, the sound of a hammer cracking rock rises from below. Some hours later, a sweating geologist—a scientist who studies the Earth—scales the cliffs, carrying a backpack filled with heavy rock samples that he will eventually take to a lab. Why? By closely examining natural exposures of rocks and sediments in the field (such as those in the canyon just described) as well as studying samples in a laboratory equipped to make sophisticated analyses, use satellite imagery, and complex computer models, geologists can answer a number of profound and fascinating questions about the character and history of our planet: How do rocks form? What do fossils tell us about the evolution of life? Why do earthquakes shake the ground and why do volcanoes erupt? What causes mountains to rise? Has the map of the Earth always looked the same? Does climate change through time? How do landforms develop? Where do we dig or drill to find valuable resources? What kinds of chemical interactions occur among land, air, water, and life? How did the Earth originate? Does our planet resemble others? The modern science of geology (or geoscience), the study of the Earth, addresses these questions and more. Indeed, a look at almost any natural feature leads to a new question, and new questions drive new research. Thus, geology remains as exciting a field of study today as it was when the discipline originated in the eighteenth century.

Before the mid-twentieth century, geologists considered each of the above questions as a separate issue, unrelated to the others. But since 1960, two paradigm-shifting advances have unified thinking about the Earth and its features. The first, *the theory of plate tectonics*, shows that the Earth's outer shell, rather than being static, consists of discrete plates that constantly move very slowly relative to each other, so that the map of our planet constantly changes. We now understand that plate interactions cause earthquakes and volcanoes, build mountains, provide gases for the atmosphere, and affect the distribution of life on Earth. The second advance establishes the concept that our planet is a complex system—*the Earth System*—in which water, land, air, and living inhabitants are dynamically interconnected in ways that allow materials to cycle constantly among various living and nonliving reservoirs. With the Earth System concept in mind, geologists now realize that the history of life links intimately to the physical history of our planet.

Earth: Portrait of a Planet is an introductory geoscience textbook that weaves the theory of plate tectonics and the concept of Earth System science into its narrative from the first page to the last. The book strives to create a modern, coherent image—a portrait—of the very special sphere on which we all live. As such, the book helps students understand the origin of the Earth and its internal structure, the nature of plate movement, the diversity of Earth's landscapes, the character of materials that make up the Earth, the distribution of resources, the structure of the air and water that surround our planet, the evolution of continents during the Earth's long history, and the nature of global change through time. The story of our planet, needless to say, is interesting in its own right. But knowledge of this story has practical applications as well. Students reading *Earth: Portrait of a Planet* will gain insight that can help address practical and political issues too. Is it safe to build a house on a floodplain or beach? How seriously should we take an earthquake prediction? Is global warming for real, and, if so, should we worry about its impacts? Which candidate has a more realistic energy and environmental policy? Should your town sell permits to a corporation that wants to extract huge amounts of water from the ground beneath the town? The list of such issues seems endless.

NARRATIVE THEMES

To understand a subject, students must develop an appreciation of fundamental concepts and by doing so create a mental "peg board" on which to hang and organize observations, ideas, and vocabulary. In the case of *Earth: Portrait of a Planet*, these concepts define "narrative themes" that are carried throughout the book, as discussed more fully in the Prelude.

1. The Earth is a complex system in which rock, oceans, air, and life are interconnected. This system is unique in the solar system.

2. Internal energy (due to the make-up of and processes occurring in our planet's interior) drives the motion of plates, and the interactions among plates, in turn, drive a variety of geologic phenomena, such as the uplift of mountain ranges, the eruption of volcanoes, the vibration of earthquakes, and the drift of continents. But what plate tectonics builds, other Earth phenomena tear down. Specifically, gravity causes materials at the tops of cliffs to slip down to lower elevations. And external energy (provided by the Sun), along with gravity, drives the flow of water, ice, and wind on the Earth's surface—this flow acts like a rasp, capable of eventually grinding away even the highest mountain.

3. The Earth is a planet, formed like other planets from a cloud of dust and gas. Because of its location and history, the Earth differs greatly from its neighbors.

4. Our planet is very old—about 4.57 billion years old. During this time, the map of the planet has changed, surface landscapes have developed and disappeared, and life has evolved.

5. Natural features and processes on Earth can be a hazard—earthquakes, volcanic eruptions, floods, hurricanes, and landslides can devastate societies. But understanding these features can help prevent damage and save lives.

6. Energy and material resources come largely from the Earth. Geologic knowledge can help find them and can help people understand the consequences of using them.

7. Geology ties together ideas from many sciences, and thus the study of geology can increase science literacy in chemistry, physics, and biology.

ORGANIZATION

Topics covered in *Earth: Portrait of a Planet* have been arranged so that students can build their knowledge of geology on a foundation of basic concepts. The parts of the book group chapters so that interrelationships among subjects are clear. Part I introduces the Earth from a planetary perspective. It includes a discussion of cosmology and the formation of the Earth and introduces the architecture and composition of our planet, from surface to center. With this background, students are ready to delve into plate tectonics theory. Plate tectonics theory appears early in this book, a departure from traditional text books, so that students will be able to relate the contents of all subsequent chapters to this theory. Understanding plate tectonics, for example, helps students to understand the chapters of Part II which introduce Earth materials (minerals and rocks). A familiarity with plate tectonics and Earth materials together, in turn, provides a basis for the study of volcanoes, earthquakes, and mountains (Part III). And with this background, students have sufficient preparation to understand the fundamentals of Earth history and the character of natural resources (Parts IV and V).

The final part of this book, Part VI, addresses processes and problems occurring at or near the Earth's surface, from the unstable slopes of hills, down the course of rivers, to the icy walls of glaciers, to the shores of the sea and beyond. This part also includes a summary of atmospheric science and concludes with a topic of growing concern—global change. As we think about the future of the planet, concerns about the warming of the climate and the contamination of the environment loom large.

SPECIAL FEATURES

Broad Coverage

Earth: Portrait of a Planet provides complete coverage of the topics in traditional Physical Geology or Introductory General Geology courses. But increasingly, first-semester courses in geology incorporate aspects of historical geology and of Earth System science. Therefore, this book also provides chapters that address Earth history, the atmosphere, the oceans, and global change. Finally, to reflect the international flavor of geoscience, the book contains examples and illustrations from around the world.

Flexible Organization

Though the sequence of chapters in *Earth: Portrait of a Planet* was chosen for a reason—to make it possible for students to have a complete introduction to plate tectonics early in the course—it has been structured to be flexible, so that instructors can rearrange chapters to fit their own strategies for teaching. Geology is a nonlinear subject—individual topics are so interrelated that there is not a single best way to order them. Thus, each chapter is largely self-contained, repeating background material where necessary for the sake of completeness. Readers will note that the book includes a Prelude and several Interludes. These are, in effect, mini-chapters that treat shorter subjects in a coherent way that would not be possible if they were made into subsections within a larger chapter. Finally, the book includes two Appendices. The first reviews basic physics and chemistry, and as such can be used as an introduction to minerals, if students lack the necessary science background. The second provides full-page versions of important charts and maps.

A Connection to Societal Issues

Geology's practical applications are addressed in chapters on volcanic eruptions, earthquakes, energy resources, mineral resources, global change, and mass wasting. Here, students learn that natural features can be hazardous, but that with a little thought, danger may be lessened. In addition, where relevant, *Earth: Portrait of a Planet* introduces students to some of the ways in which geologic understanding can be applied to environmental issues. Case studies show how geologists have used their knowledge to solve practical problems.

Boxed Inserts

Throughout the text, boxes expand on specific topics. "The Human Angle" boxes introduce links between geologic phenomena and the human experience. "Science Toolboxes" provide background scientific data. And "The Rest of the

Story" boxes give additional interesting, but optional, detail. In this edition, we have also added "*Case Studies."*

Superb Artwork

It's hard to understand features of the Earth system without being able to see them. To help students visualize topics, *Earth: Portrait of a Planet* is lavishly illustrated—the book contains over 200 more illustrations than competing texts! The author has worked closely with the artists to develop an illustration style that conveys a realistic context for geologic features without overwhelming students with extraneous detail. The talented artists who worked on the figures have pushed the envelope of modern computer graphics, and the result is the most realistic pedagogical art ever produced for a geoscience text.

In addition to line art, *Earth* features photographs from all continents. Many of the pictures were taken by the author and provide interesting alternatives to the stock images that have appeared for many years in introductory books. Where appropriate, photographs are accompanied by annotated sketches labeled "What a geologist sees," which help students see the key geologic features in the photos.

In the past, students would need to go to a museum to see bold, colorful paintings of geologic features. Now, they need only flip through the pages of *Earth: Portrait of a Planet.* Famed British painter Gary Hincks has provided spectacular two-page synoptic paintings that illustrate key concepts introduced in the text and visually emphasize the relationships among components of the Earth system.

CHANGES IN THE SECOND EDITION

The Second Edition of *Earth: Portrait of a Planet* is not simply a cosmetic modification of the First Edition. Though the basic organization remains the same, the text has been intelligently modified and thoroughly updated. Key changes include:

1. Incorporation of New Discoveries. Recent research has yielded fundamental new understandings that belong even in introductory books. For example, new computer models now simulate the evolution of Earth's magnetic field, probes to other planets have addressed the issue of whether water existed on Mars, and studies using global positioning satellites (GPS) now allow researchers to measure the drift of continents and even observe vertical movements of continental surfaces. New data clarify patterns of climate change, and new measurements provide insight into where earthquakes might happen. While basic geoscience remains the backbone of *Earth: Portrait of a Planet,* we have incorporated discussions of these new discoveries, and many others, into the text.

2. New Examples. To keep the discussion of natural hazards current, we have updated the treatment to include events that have happened since publication of the First Edition.

3. The View from Space. Satellites orbiting the Earth provide spectacular imagery of our planet's surface. We have added over a dozen satellite images to the Second Edition.

4. New Synoptic Art. Famed geologic artist Gary Hincks has provided new, spectacular two-page paintings that provide an overview of key concepts. The drama of these paintings appeals to students' intuition.

5. New Figures and Photos. Numerous figures have been added to the book, and over 200 figures from the First Edition have been updated and improved. Further, many new photographs have been added, and where possible, photos from the First Edition have been replaced with ones taken by the author that, unlike commercial photographs, can be included on the book's Web page and CD.

6. New WAGS. The "What a Geologist Sees" feature, which provides an annotated sketch of a photograph so that students can better see what the picture shows, have proven to be particularly helpful to students. Numerous new WAGS have been added to this edition.

7. Clarifications. To help ensure that *Earth: Portrait of a Planet* is as accurate and up-to-date as possible, each chapter of the First Edition was sent to an expert reviewer (a world leader on the subject) for comment. From the response, we identified places in the First Edition that did not quite convey the full story and improved them. In addition, we revised the text on a line-by-line basis to ensure that it is as clear as possible and have made many of the more complex discussions even more readable.

8. Key Terms and Case Studies. We have filtered out unnecessary bold-faced terms, so as to make the text visually flow more smoothly, and where relevant, we have added "case studies" to illustrate the practical side of the geosciences, especially as regards environment-related issues.

SUPPLEMENTS

For Instructors:

1. Norton Media Library Instructor's CD-ROM

This instructor CD-ROM offers a wealth of easy-to-use multimedia resources, all structured around the text and designed for use in lecture presentations, including:

• editable PowerPoint lecture outlines by Ron Parker of Earlham College

- all of the art and all photographs from the text
- additional photographs from Stephen Marshak's own archives
- over sixty animations unique to *Earth: Portrait of a Planet,* Second Edition

2. Norton Resource Library Instructor's Website

wwnorton.com/nrl

Maintained as a service to our adopters, this password-protected instructor website offers book-specific materials for use in class or within WebCT, Blackboard, or course websites. Resources include:

- editable PowerPoint lecture outlines by Ron Parker of Earlham College
- over sixty animations unique to *Earth: Portrait of a Planet,* Second Edition
- multiple-choice GeoQuizzes, questions by Stephen Marshak, ideally suited to classes equipped with electronic polling systems
- test-bank questions in MicroTest-, Blackboard-, and WebCT-ready formats
- study guide questions in Blackboard- and WebCT-ready formats
- JPEG versions of all drawn art in the textbook
- instructor's manual in PDF format

3. Instructor's Manual and Test Bank

Prepared by John Werner of Seminole Community College, Terry Engelder of Pennsylvania State University, and Stephen Marshak, this manual offers useful material to help instructors as they prepare their lectures and includes over 1,200 multiple-choice and true-false test questions. The test bank is available both in printed form and electronically in MicroTest III and WebCT- or Blackboard-ready formats.

4. Transparencies

Over 200 figures from the text are available as color acetates.

5. Supplemental Slide Set

This collection of 35mm slides supplements the photographs in the text with additional images from Stephen Marshak's own photo archives. These images are also available as PowerPoint slides on the Norton Media Library CD-ROM.

For Students

1. Student CD-ROM and Web Site

www.wwnorton.com/earth

Developed specifically for *Earth: Portrait of a Planet,* Second Edition, this free online study guide offers materials that re-

inforce core concepts from the text and animations to help students visualize dynamic processes. The CD-ROM version of this resource is automatically packaged with every new book purchased from Norton. It includes:

- **Review Materials**—Chapter overviews, key-term crosswords, and multiple-choice quizzes help students get the most from their reading assignments. A new grade book utility makes it easy for instructors to collect scores from Website quiz assignments.
- **Animations**—Over sixty animations developed by Stephen Marshak, half of them new to this edition, emphasize plate tectonics, geologic hazards, and Earth Systems science. Many of the new animations are based on the "What a Geologist Sees" art from the text.
- **Media Archive**—New to this edition, the media archive augments the collection of original animations with links to video and animation from USGS, NASA, and other outside sources.
- **Earth Science News**—Weekly updates feature news about geologic events, recent advances, and contemporary environmental issues.

2. Art Notebook

Containing all major diagrams from the text, with labels removed and ample space left for notes, the Art Notebook is a valuable resource both for note-taking and for reference.

ACKNOWLEDGMENTS

I am very grateful for the assistance of many people in bringing this book from the concept stage to the shelf in the first place, and for helping to provide the momentum needed to make the Second Edition take shape. First and foremost, I wish to thank my family. My wife, Kathy, has helped throughout in the overwhelming task of keeping track of text and figures and of handling mailings. In addition, she helped edit text, copied drafts, and provided invaluable advice. My daughter, Emma, spent untold days locating and scanning photographs and searching for resources on the Web. My son, David, helped me keep the project in perspective and highlighted places where the writing could be improved. During the initial development of the First Edition, I greatly benefited from discussions with Philip Sandberg, and during later stages in the development of the First Edition, Donald Prothero contributed text, editorial comments, and end-of-chapter material.

The publisher, W. W. Norton & Company, has been incredibly supportive of my work and has been very generous in their investment in this project. Steve Mosberg signed the First Edition, and Rick Mixter put the book on track. Jack Repcheck bulldozed aside numerous obstacles and brought the First Edition to completion. He has continued to be a fountain of sage advice and an understanding friend throughout the development of the Second Edition. Jack has provided numerous innovative ideas that have strengthened the book and brought it to the attention of the geologic community. Under Jack's guidance, the First Edition became one of the most widely used geology texts worldwide.

April Lange has expertly coordinated development of the ancillary materials. She has not only managed their development, but also introduced innovative approaches and wrote part of the material. Her contributions have set a standard of excellence. JoAnn Simony did a superb job of managing production of the First Edition and of doing the page makeup. Thom Foley and Diane O'Connor have continued the tradition by expertly and efficiently handling the task of managing production for the Second Edition. They have calmly handled all the back and forth involved in developing a book and in keeping it on schedule. Susan Gaustad did an outstanding job of copy editing the First Edition. This tradition continued through the efforts of Alice Vigliani on the Second Edition. Neil Hoos and Stephanie Romeo did an excellent job of photo research and of obtaining permissions, and Sarah Solomon, editorial assistant, was a great help in tying up any and all loose ends.

Production of the illustrations has involved many people. I am particularly grateful to Joanne Bales and Stan Maddock, who helped create the overall style of the figures, produced a great many of them for both editions, and have worked closely with me on improvements. I would also like to thank Becky Oles who, along with Joanne and Stan, created all of the new art for this edition. Joanne ably coordinated and supervised the art team at Precision Graphics, Terri Hamer has done an excellent job as production manager, and Jon Prince has creatively programmed the animations. It has been a delight to interact with the artists, production staff, and management of Precision Graphics over the past several years. It has also been great fun to interact with Gary Hincks, who painted the incredible two-page spreads, in part using his own designs and geologic insights. Some of Gary's paintings appeared in *Earth Story* (BBC Worldwide, 1998) and were based on illustrations jointly conceived by Simon Lamb and Felicity Maxwell, working with Gary. Others were developed specifically for *Earth: Portrait of a Planet.* Some of the chapter quotes were found in *Language of the Earth,* compiled by F.T. Rhodes and R.O. Stone (Pergamon, 1981).

In developing the First Edition, I was helped by insightful review and discussions of the manuscript by many geologists. The development of the Second Edition benefited greatly from input by expert reviewers for specific chapters, by new general reviewers of the entire book, and by comments from faculty and students who have used the First Edition and were kind enough to contact me by e-mail. The list of people whose comments were incorporated includes:

Jack C. Allen, Bucknell University
David W. Anderson, San Jose State University
Philip Astwood, University of South Carolina
Eric Baer, Highline University
Victor Baker, University of Arizona
Keith Bell, Carleton University
Mary Lou Bevier, University of British Columbia
Daniel Blake, University of Illinois
Michael Bradley, Eastern Michigan University
Sam Browning, Massachusetts Institute of Technology
Rachel Burks, Towson University
Peter Burns, University of Notre Dame
Katherine Cashman, University of Oregon
George S. Clark, University of Manitoba
Kevin Cole, Grand Valley State University
Patrick M. Colgan, Northeastern University
John W. Creasy, Bates College
Norbert Cygan, Chevron Oil, retired
Peter DeCelles, University of Arizona
Carlos Dengo, ExxonMobil Exploration Company
John Dewey, University of California, Davis
Charles Dimmick, Central Connecticut State University
Robert T. Dodd, State University of New York at Stony Brook
Missy Eppes, University of North Carolina, Charlotte
Eric Essene, University of Michigan
James E. Evans, Bowling Green State University
Leon Follmer, Illinois Geological Survey
Nels Forman, University of North Dakota
Bruce Fouke, University of Illinois
David Furbish, Vanderbilt University
Grant Garvin, John Hopkins University
Christopher Geiss, Trinity College, Connecticut
William D. Gosnold, University of North Dakota
Lisa Greer, William & Mary College
Henry Halls, University of Toronto at Mississuaga
Bryce M. Hand, Syracuse University
Tom Henyey, University of South Carolina
Paul Hoffman, Harvard University
Neal Iverson, Iowa State University
Donna M. Jurdy, Northwestern University
Thomas Juster, University of Southern Florida
Dennis Kent, Lamont Doherty/Rutgers
Jeffrey Knott, California State University, Fullerton
Ulrich Kruse, University of Illinois
Lee Kump, Pennsylvania State University
David R. Lageson, Montana State University
Robert Lawrence, Oregon State University
Craig Lundstrom, University of Illinois
John A. Madsen, University of Delaware
Jerry Magloughlin, Colorado State University
Paul Meijer, Utrecht University (NL)

Alan Mix, Oregon State University
Robert Nowack, Purdue University
Charlie Onasch, Bowling Green State University
David Osleger, University of California, Davis
Lisa M. Pratt, Indiana University
Mark Ragan, University of Iowa
Bob Reynolds, Central Oregon Community College
Joshua J. Roering, University of Oregon
Eric Sandvol, University of Missouri
William E. Sanford, Colorado State University
Doug Shakel, Pima Community College
Angela Speck, University of Missouri
Tim Stark, University of Illinois (CEE)
Kevin G. Stewart, University of North Carolina at Chapel Hill
Don Stierman, University of Toledo
Barbara Tewksbury, Hamilton College
Thomas M. Tharp, Purdue University
Kathryn Thornbjarnarson, San Diego State University
Basil Tickoff, University of Wisconsin
Spencer Titley, University of Arizona
Robert T. Todd, State University of New York at Stony Brook
Torbjörn Törnqvist, University of Illinois, Chicago
Jon Tso, Radford University
Alan Whittington, University of Illinois
Lorraine Wolf, Auburn University
Christopher J. Woltemade, Shippensburg University

I apologize if anyone was inadvertently not included on the list.

ABOUT THE AUTHOR

Stephen Marshak is currently head of the Department of Geology at the University of Illinois, Urbana-Champaign. He holds an A.B. from Cornell University, an M.S. from the University of Arizona, and a Ph.D. from Columbia University. Steve's research interests lie in the fields of structural geology and tectonics. Over the years, he has explored geology in the field on several continents. Since 1983 Steve has been on the faculty of the University of Illinois, where he teaches courses in introductory geology, structural geology, tectonics, and field geology and has won the university's highest teaching award. In addition to research papers and *Earth: Portrait of a Planet*, Steve has authored or co-authored *Essentials of Geology, Earth Structure: An Introduction to Structural Geology and Tectonics,* and *Basic Methods of Structural Geology.*

THANKS!

I am very grateful to the faculty who selected the First Edition for use in their classes and to the students who engaged so energetically with it. The response was very gratifying. I particularly appreciate the comments from readers that helped to make improvements in the Second Edition. I continue to welcome your comments and can be reached at: smarshak@uiuc.edu.

Stephen Marshak

*To see the world in a grain of sand
and heaven in a wild flower.
To hold infinity in the palm of the hand
and eternity in an hour.*

—William Blake (British poet, 1757–1827)

EARTH: PORTRAIT OF A PLANET

Second Edition

And Just What Is Geology?

Civilization exists by geological consent, subject to change without notice.
—WILL DURANT (1885–1981)

We can see the Earth System at a glance near the Maroon Bells, a row of mountains in Colorado. Here, sunlight, air, water, rock, and life all interact.

P.1 IN SEARCH OF IDEAS

In the glare of the midnight sun, our C-130 Hercules transport plane rose from a smooth ice runway on the frozen sea surface at McMurdo Station, Antarctica, and we were off to spend a month studying unusual rocks exposed on a cliff about 250 kilometers (km) away. As we climbed past the smoking summit of Mt. Erebus, Earth's southernmost volcano, we had one nagging thought: no aircraft had ever landed at our destination, so the ground conditions there were unknown; if deep snow covered the landing site, the massive plane might get stuck and would not be able to return to McMurdo. Because of this concern, the flight crew had added a crate of rocket canisters to the pile of snowmobiles, sleds, tents, and food in the plane's cargo hold. "If the turboprops can't lift us, we can clip a few canisters to the tail, light them, and rocket out of the snow," they claimed.

For the next hour, we flew along the Transantarctic Mountains, a ridge of rock that divides the continent into two parts, East Antarctica and West Antarctica (▶Fig. P.1). A vast ice sheet, in places over 3 km thick, covers East Antarctica—the surface of this ice sheet forms a high plain known as the Polar Plateau. Rivers of ice from the Polar Plateau slowly flow down valleys cut through the Trans-

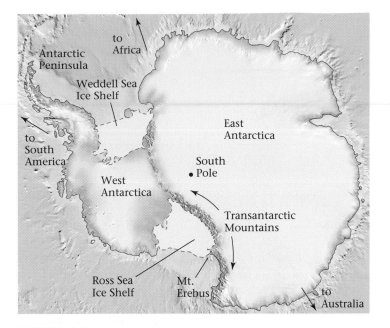

FIGURE P.1 Map of Antarctica.

antarctic Mountains. (Ice sheets and ice rivers are called glaciers.) From the plane's windows, we admired long stripes of rock debris that had been shed from mountains onto the glaciers in these valleys. The stripes move with the ice and highlight the direction of flow. Suddenly, we heard the engines slow.

As the plane descended, it lowered its ski-equipped landing gear. The loadmaster shouted an abbreviated reminder of the emergency alarm code: "If you hear three short blasts of the siren, hold on for dear life!" Roaring toward the ground, the plane touched the surface of our first choice for a landing spot, the ice at the base of the rock cliff

we wanted to study. *Wham, wham, wham, wham!!!!* Sastrugi (frozen snow drifts) rippled the ice surface, and as the plane's skis slammed into them at about 180 km an hour, it seemed as though a fairy-tale giant was shaking the plane. Seconds later, the landing aborted, we were airborne again, looking for a softer runway above the cliff. Finally, we landed in a field of deep snow, unloaded, and bade farewell to the plane (►Fig. P.2). The Hercules trundled for kilometers through the snow before gaining enough speed to take off, but fortunately did not need to use the rocket canisters. When the plane passed beyond the horizon, the silence of Antarctica hit us—no trees rustled, no dogs barked, and no traffic rumbled in this stark land of black rock and white ice. It would take us a day and a half to haul our sleds of food and equipment down to our study site (►Fig. P.3). All this to look at a few dumb rocks?

Geologists, scientists who study the Earth, explore remote regions like Antarctica almost routinely. Such efforts often strike people in other professions as a strange way to make a living. Scottish poet Walter Scott (1771–1832), when describing geologists at work, said: "Some rin uphill and down dale, knappin' the chucky stones to pieces like sa' many roadmakers run daft. They say it is to see how the warld was made!" Indeed—to see how the world was made, to see how it continues to evolve, to find its valuable resources, to prevent contamination of its waters and soils, and to predict its dangerous movements. That is why geologists spend months at sea drilling holes in the ocean floor, why they scale mountains (►Fig. P.4), camp in rain-drenched jungles, and trudge through scorching desert winds. That is why geologists use electron microscopes to examine the atomic structure of minerals, use mass spectrometers to define the composition of rock and water, and use supercomputers to model

FIGURE P.2 Geologists unloading a cargo of tents, sleds, and snowmobiles from the tail of a C-130 Hercules transport plane that has just landed in a snowfield. Note the large skis over the wheels.

FIGURE P.3 Geologists sledding to a field area in Antarctica. The sleds contain a month's worth of food, sample bags, rock hammers, and notebooks as well as tents and clothes (and a case of frozen beer).

FIGURE P.4 A geologist studying exposed rocks on a mountain slope on the desert island of Zabargad, in the Red Sea, off the coast of Egypt.

the paths of earthquake waves. For over two centuries, geologists have pored over the Earth—in search of ideas to explain the processes that form and change our planet.

P.2 THE NATURE OF GEOLOGY

Geology, the study of the Earth, focuses on describing our planet's composition, behavior, and history. As is true with most sciences, geology consists of numerous subdisciplines (Table P.1). In fact, because geology covers such a diversity of subjects, researchers commonly substitute the term "geoscience" for the discipline.

Not only do geologists address academic questions, such as the formation and composition of the Earth, the causes of earthquakes (►Fig. P.5) and ice ages, and the evolution of life, they also address practical problems, such as how to prevent groundwater contamination, how to find oil and minerals, and how to stabilize slopes. And in recent years, geologists have contributed to the study of global climate change, for the long-term record of Earth's past climate lies in layers of

sediment and rock. When news reports begin with "Scientists say . . ." and then continue with "an earthquake occurred today off Japan," or "landslides will threaten the city," or "contaminants from the proposed toxic waste dump are destroying the town's water supply," or "there's only a limited supply of oil left," the scientists referred to are geologists.

The fascination of geology attracts many people to careers in this science. Thousands of geologists work for oil, mining, water, engineering, and environmental companies, while a smaller number work in universities, government geological surveys, and research laboratories. Nevertheless, since the majority of students reading this book will not become professional geologists, it's fair to ask the question "Why should people, in general, study geology?"

First, geology may be one of the most practical subjects you can learn, for geologic phenomena and issues affect our daily lives, sometimes in unexpected ways:

• Landslides, volcanoes, floods, and earthquakes destroy communities.

• The discovery of a new oil field lowers energy costs.

• Concerns about the availability of energy or metallic resources sometimes trigger military conflicts.

• The loss of a town's freshwater supply discourages a factory from locating in the town.

• Your state or province proposes to locate a nuclear-waste-disposal facility near you.

• A storm destroys houses built along a nearby beach, so your insurance rates go up.

FIGURE P.5 Human-made cities cannot withstand the vibrations of a large earthquake. These apartment buildings collapsed during an earthquake in Turkey.

TABLE P.1 Principal Subdisciplines of Geology (Geoscience)

Name	Subject of Study
Engineering geology	The stability of geologic materials at the Earth's surface, for such purposes as controlling landslides and building tunnels.
Environmental geology	Interactions between the environment and geologic materials, and the contamination of geologic materials.
Geochemistry	Chemical compositions of materials in the Earth and chemical reactions in the natural environment.
Geochronology	The age (in years) of geologic materials, the Earth, and extraterrestrial objects.
Geomorphology	Landscape formation and evolution.
Geophysics	Physical characteristics of the whole Earth (such as Earth's magnetic field and gravity field) and of forces in the Earth.
Hydrogeology	Groundwater, its movement, and its reaction with rock and soil.
Mineralogy	The chemistry and physical properties of minerals.
Paleontology	Fossils and the evolution of life as preserved in the rock record.
Petrology	Rocks and their formation.
Sedimentology	Sediments and their deposition.
Seismology	Earthquakes and the Earth's interior as revealed by earthquake waves.
Stratigraphy	The succession of sedimentary rock layers.
Structural geology	Rock deformation in response to the application of force.
Tectonics	Regional geologic features (such as mountain belts) and plate movements and their consequences.

Clearly, all citizens of the twenty-first century, not just professional geologists, will need to make decisions concerning Earth-related issues. And they will be able to make more reasoned decisions if they have a basic understanding of geologic phenomena. History is full of appalling stories of people who ignored geologic insight and paid a horrible price for their ignorance. Your knowledge of geology may help you to avoid building your home on a hazardous floodplain or fault zone, on an unstable slope, or along a rapidly eroding coast. With a basic understanding of groundwater, you may be able to save money when drilling an irrigation well, and with knowledge of the geologic controls on resource distribution, you may be able to invest more wisely in the resource industry.

Second, the study of geology gives you a perspective on the planet that no other field can. As you will see, the Earth is a complicated system; its living organisms, climate, and solid rock interact with one another in a great variety of ways. Geologic study reveals Earth's antiquity (it's about 4.57 billion years old) and demonstrates how the planet has changed profoundly during its existence. What was the center of the Universe to our ancestors becomes, with the development of geologic perspective, our "island in space" today, and what was an unchanging orb originating at the same time as humanity becomes a dynamic planet that existed long before people did.

Third, the study of geology puts human achievements and natural disasters in context. On the one hand, our cities seem to be no match for the power of an earthquake, and a rise in sea level may swamp all major population centers. But on the other hand, we are now changing the face of the land worldwide at rates that far exceed those resulting from natural geologic processes. By studying geology, we develop a frame of reference for judging the extent and impact of changes.

Finally, when you finish reading this book, your view of the world will be forever colored by geologic curiosity. When you walk in the mountains, you will think of the many forces that shape and reshape the Earth's surface. When you hear about a natural disaster, you will have insight into the processes that brought it about. And when you next go on a road trip, the rock exposures next to the highway will no longer be gray, faceless cliffs, but will present complex puzzles of texture and color telling a story of Earth's history.

Here are a few things to do on a highway trip: Play 20 Questions, plug your kids into some sort of electronic anodyne, lose your mind.

Here's another idea: Look for gneisses and amphibolites; seek out scarps, klippes and fault slices. Head for the Silurian boundary. Instead of feeling miserable and confined, feel the bones of the earth as you ride past the exposed evidence of the planet's history.

That's roadside geology, road food for the mind and eye.

—James Gorman (*New York Times,* Nov. 16, 2001, p. E40)

P.3 THEMES OF THIS BOOK

A number of narrative themes appear (and reappear) throughout this text. These themes, listed below, can be viewed as the book's "take home message."

1. *The Earth is a unique, evolving system.* Geologists increasingly recognize that the Earth is a complicated system; its interior, solid surface, oceans, atmosphere, and life forms interact in many ways to yield the landscapes and environ-ment in which we live. Within this **Earth System,** chemical elements pass in cycles between different types of rock, between rock and sea, between sea and air, and between all of these entities and life. Aside from the residue of occasional collisions with fragments of rock or ice (asteroids or comets), all the material involved in these cycles originates in the Earth itself—our planet is truly an island in space.

2.   *Plate tectonics is a unifying idea that explains Earth processes.* Like other planets, Earth is not a homogeneous ball, but rather consists of concentric layers: from center to surface, Earth has a core, mantle, and crust. We live on the surface of the crust, where it meets the atmosphere and the oceans. In the 1960s, geologists recognized that the crust together with the uppermost part of the underlying mantle form a 100- to 150-km-thick semirigid shell. Large cracks separate this shell into discrete pieces, called **plates,** which move very slowly relative to one another (▶Fig. P.6). The theory that describes this movement and its consequences is now known as the **theory of plate tectonics,** and it serves as the foundation for understanding most geologic phenomena. Although plates move very slowly, generally less than 10 centimeters (cm) a year, their movements yield earthquakes, volcanoes, and mountain ranges, and cause the distribution of continents to change over time.

3.   *The Earth is a planet.* The subtitle of this book, *Portrait of a Planet,* highlights the view that despite the uniqueness of Earth's system and inhabitants, Earth fundamentally can be viewed as a planet, formed like the other planets of the solar system from dust and gas that encircled the newborn Sun. Though Earth resembles the other inner planets (Mercury, Venus, and Mars), it differs from them in having plate tectonics, an oxygen-rich atmosphere and liquid-water ocean, and abundant life. Further, because of the dynamic interactions among various aspects of the Earth System, our planet is constantly changing; the other inner planets are static.

4.   *The Earth is very old.* Geologic data indicate that the Earth formed 4.57 billion years ago—plenty of time for geologic processes to generate and destroy features of the Earth's surface, for life forms to evolve, and for the map of the planet to change. Plate-movement rates of only a few centimeters per year, if those movements continue for hundreds of millions of years, can move a continent thousands of kilometers. In geology, we have time enough to build mountains and time enough to grind them down many times over. To define intervals of this time, geologists have invented a time scale, known as the **geologic time scale** (▶Fig. P.7). Geologists call the last 542 million years the **Phanerozoic** Eon, and all time before that the **Precambrian.** They further divide the Precambrian into three main intervals named, from oldest to

FIGURE P.6 Simplified map of the Earth's principal plates. The arrow on each plate indicates the direction the plate moves, and the length of the arrow indicates the plate's velocity (the longer the arrow, the faster the motion). We discuss the types of plate boundaries in Chapter 4.

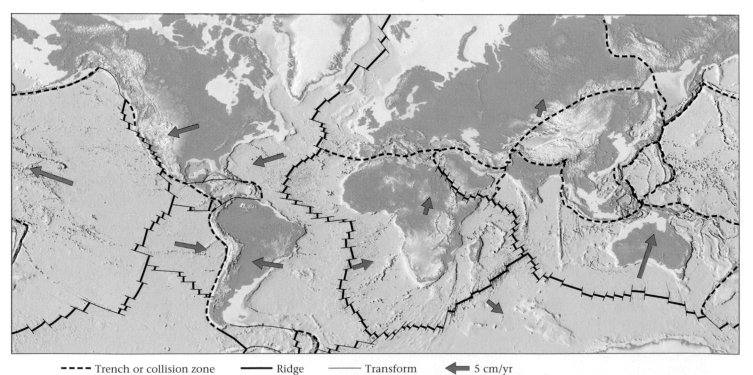

- - - - Trench or collision zone   ——— Ridge   ——— Transform   ⬅ 5 cm/yr

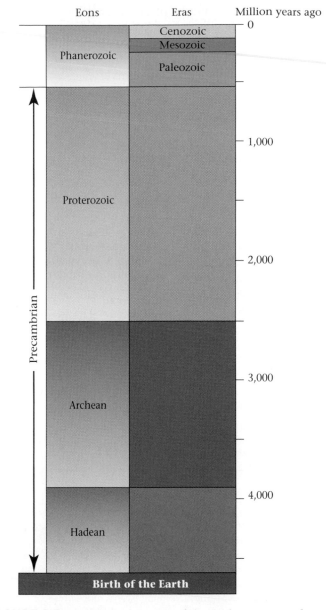

Eons — Eras — Million years ago

FIGURE P.7 The major divisions of the geologic time scale.

TABLE P.2 Abbreviations Used in Specifying Time

C.E.	of the common era. This is equivalent to the familiar Latin abbreviation A.D. For example, World War II ended in 1945 C.E.
B.C.E.	before the common era. This is equivalent to the familiar abbreviation B.C. For example, Julius Caesar was assassinated in 44 B.C.E.
b.p.	before the present day. The year 1950 is often used as a reference point for describing such dates.
k.y.	thousands of years ("k" stands for "kilo")
m.y.	millions of years
b.y.	billions of years
Ka	thousands of years ago ("a" stands for the Latin word "annum," meaning "year"). For example, a date of 15 Ka means 15,000 years ago.
Ma	millions of years ago. For example, the Mesozoic Era ended at 65 Ma.
Ga	billions of years ago ("G" stands for "giga"). For example, at the end of the Archean Eon occurred at 2.5 Ga.

processes are those phenomena that ultimately are driven by heat supplied by radiation coming to the Earth from the Sun. This heat drives the movement of air and water, which grinds and sculpts the Earth's surface and transports the debris to new locations, where it accumulates. The interaction between internal and external processes form the landscapes of our planet.

6. *Geologic phenomena affect our environment.* Volcanoes, earthquakes, landslides, floods, and even more subtle processes such as groundwater flow and contamination or depletion of oil and gas reserves are of vital interest to every inhabitant of this planet. They are often a matter of life and death. Linkages between geology and the environment are therefore stressed throughout the book.

7. *Physical aspects of the Earth system are linked to life processes.* All life on this planet depends on such physical features as the minerals in soil, the temperature, humidity, and composition of the atmosphere, and the flow of surface and subsurface water. And life in turn affects and alters these same physical features. For example, the atmosphere's oxygen comes primarily from plant photosynthesis, a life activity, and it is this oxygen that permits complex animals to survive. The oxygen also affects chemical reactions among air, water, and rock. Without the physical Earth, life could not exist, but without life, this planet's surface might have become a frozen wasteland like that of Mars, or enshrouded in acidic clouds like that of Venus.

youngest, the **Hadean,** the **Archean,** and the **Proterozoic** Eons, and the Phanerozoic Eon into three main intervals named, from oldest to youngest, the **Paleozoic,** the **Mesozoic,** and the **Cenozoic** Eras. (Chapter 13 provides further details about geologic time, and Table P.2 provides abbreviations used for specifying time intervals or dates in the past.)

5. *Internal and external processes interact at the Earth's surface.* Internal processes are those phenomena that ultimately are driven by heat from inside the Earth. Plate movement is an example, and since plate movements cause mountain building, earthquakes, and volcanoes, we call all of these phenomena internal processes as well. External

8. *Science comes from observation, and people make scientific discoveries.* Science is not a subjective guess or an arbitrary dogma, but rather a consistent set of objective statements resulting from the application of the **scientific method** (see Box P.1). Every scientific idea must be constantly subjected to testing and possible refutation, and can be accepted only when supported by documented observations. Further, scientific ideas do not appear out of nowhere, but are the result of human efforts. Wherever possible, this book shows where geologic ideas came from, and tries to answer the question "How do we know that?"

9. *The study of geology can increase general science literacy.* Studying geology provides an ideal opportunity to learn basic concepts of chemistry and physics, because these concepts can be applied directly to understanding tangible phenomena. Thus, in this book, where appropriate, basic concepts of physical science are introduced in boxed features called "Science Toolboxes." Also, Appendix A provides a systematic introduction to matter and energy, for those readers who have not learned this information previously or who need a review.

As you read this book, please keep these themes in mind. Don't view geology as a list of words to memorize, but rather as an interconnected set of concepts to digest. Most of all, enjoy yourself as you learn about what may be the most fascinating planet in the Universe.

KEY TERMS

Archaen (p. 7)
Cenozoic (p. 7)
Earth System (p. 6)
geologic time scale (p. 6)
geologists (p. 3)
geology (p. 4)
Hadean (p. 7)
hypothesis (p. 9)
Mesozoic (p. 7)
Paleozoic (p. 7)

Phanerozoic (p. 6)
plates (p. 6)
Precambrian (p. 6)
Proterozoic (p. 7)
scientific law (p. 9)
scientific method (p. 8)
shatter cones (p. 9)
theory (p. 9)
theory of plate tectonics (p. 6)

BOX P.1
SCIENCE TOOLBOX

The Scientific Method

Sometime during the past 200 million years, a large block of rock or metal, which had previously been orbiting the Sun, crossed the path of Earth's orbit. In seconds, it pierced the atmosphere and slammed into our planet (thereby becoming a meteorite) at a site in what is now the central United States, a landscape of flat cornfields. The impact released more energy than a nuclear bomb—a cloud of shattered rock and dust blasted skyward, and once-horizontal layers of rock from deep below the ground sprang upward and tilted on end in the gaping hole left by the impact. When the dust had settled, a huge crater surrounded by debris marked the surface of the Earth at the impact site. Later in Earth history, running water and blowing wind wore down this jagged scar, and some 15,000 years ago, sediments (sand, gravel, and mud) carried by a vast glacier buried what remained, hiding it entirely from view (▶ Fig. P.8a, b). Wow! So much history beneath a cornfield. Have you ever wondered how geologists come up with such a story? It takes scientific investigation.

The movies often portray science as a dangerous tool, capable of creating Frankenstein's monster, and scientists as warped or nerdy characters with thick glasses and poor taste in clothes. In reality, **science** is simply the use of observation, experiment, and calculation to explain how nature operates, and **scientists** are people who study and try to understand natural phenomena. Scientists carry out their work using the *scientific method,* a sequence of steps for systematically analyzing scientific problems in a way that leads to verifiable results. Let's see how geologists employed the steps of the scientific method to come up with the meteorite-impact story.

1. *Recognizing the problem:* Any scientific project, like any detective story, begins by identifying a mystery. The cornfield mystery came to light when water drillers discovered limestone, a rock typically made of shell fragments, just below the 15,000-year-old glacial sediment. In surrounding regions, the rock at this depth consists of sandstone, made of cemented-together sand grains, which differs greatly in composition from limestone. Since limestone can be used to build roads, make cement, and produce the agricultural lime used in treating soil, workers stripped off the glacial sediment and began excavating the limestone. They were amazed to find that rock layers exposed in the quarry tilted steeply and had been shattered by large cracks. In the surrounding regions, all rock layers are horizontal like the layers in a birthday cake, the limestone layer lies underneath the sandstone, and the rocks contain relatively few cracks. Curious geologists came to investigate, and soon realized that the geologic features of the land just beneath the cornfield presented a problem to be explained: What phenomena had brought limestone up close to the Earth's surface, tilted the layering in the rocks, and shattered the rocks?

2. *Collecting data:* The scientific method proceeds with the collection of observations or clues that point to an answer. Geologists studied the quarry and determined the age of its rocks, measured the orientation of rock layers, and "documented" (made a written or photographic record of) the fractures that broke up the rocks.

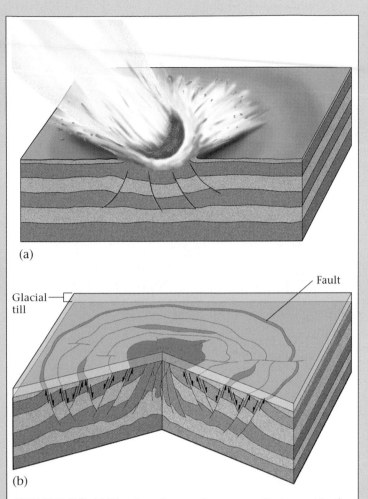

(a)

(b)

FIGURE P.8 (a) The site of an ancient meteorite impact in the American Midwest, before impact. Note the horizontal layers of rock below the ground surface. The thin lines represent boundaries between the successive layers. During impact, a large crater, surrounded by debris, forms. (b) The site of the impact today. The crater and the surface debris were eroded away. Relatively recently, the area was buried by glacial till (sediments). Underground, the impact disrupted layers of rock by tilting them and by generating faults (fractures on which sliding occurs).

3. *Proposing hypotheses:* A scientific **hypothesis** is merely a possible explanation, involving only naturally occurring processes, that can explain a set of observations. Scientists propose hypotheses during or after their initial data collection. The geologists working in the quarry came up with two alternative hypotheses. First, the features in this region could result from a volcanic explosion; and second, they could result from a meteorite impact.

4. *Testing hypotheses:* Since a hypothesis is no more than an idea that can be either right or wrong, scientists must put hypotheses through a series of tests to see if they work. The geologists at the quarry compared their field observations with published observations made at other sites of volcanic explosions and me-

teorite impacts, and studied the results of experiments designed to simulate such events. They learned that if the geologic features visible in the quarry were the result of volcanism, the quarry should contain rocks made of frozen lava (the melt that flows from a volcano). But, no such rocks were found. If, however, the features were the consequence of an impact, the rocks should contain **shatter cones,** small, cone-shaped cracks formed only by meteorite impact (▶Fig. P.9). Shatter cones can easily be overlooked, so the geologists returned to the quarry specifically to search for them, and found them in abundance. The impact hypothesis passed the test!

Theories are scientific ideas supported by an abundance of evidence; they have passed many tests and have failed none. Scientists have much more confidence in a theory than they do in a hypothesis. Continued study in the quarry eventually yielded so much evidence for impact that the impact hypothesis came to be viewed as a theory. Scientists continue to test theories over a long time. Successful theories withstand these tests and are supported by so many observations that they come to be widely accepted. (As you will discover in Chapters 3 and 4, geologists consider the idea that continents drift around the surface of the Earth to be a theory, because so much evidence supports it.) However, some theories may eventually be disproven, to be replaced by better ones.

Some scientific ideas must be considered absolutely correct, for if they were violated, the natural Universe as we know it would not exist. Such ideas are called **scientific laws,** and examples include the law of gravity.

FIGURE P.9 Shatter cones in limestone. These cone-shaped fractures, formed only by severe impact, open up in the direction away from the impact. At this locality, the cones open up downward, indicating that the impact came from above.

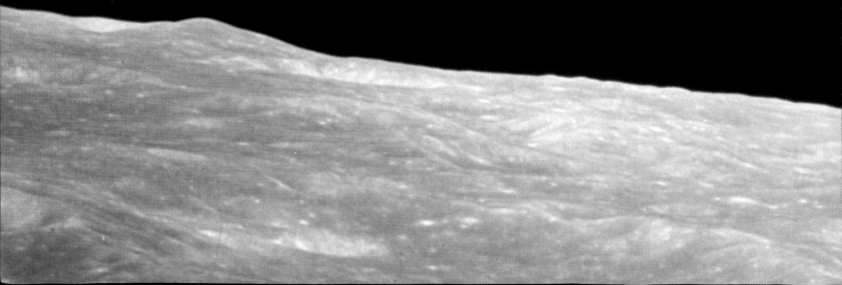

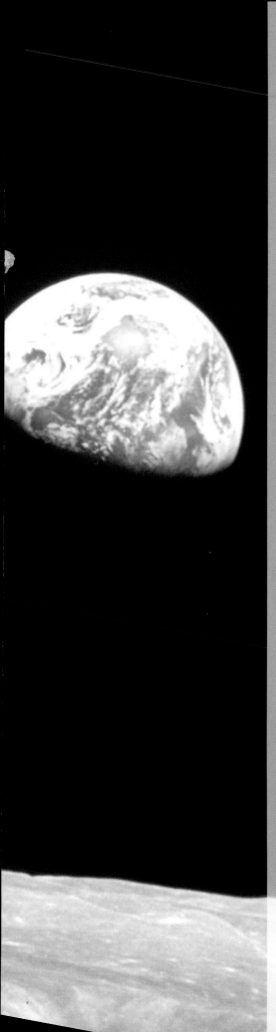

Our Island in Space

*W*hen you look out toward the horizon from a mountain top, the Earth seems endless, and before the modern era, many people thought it was. But to astronauts flying to the Moon, the Earth is merely a small, shining globe—they can see half the planet in a single glance. From the astronauts' perspective, it appears that we are riding on an island in space. This island circles the Sun and, together with the Sun, rushes through space at an immense speed.

Earth may not be endless, but it is a very special planet: its temperature and composition, unlike those of the other planets in the solar system, make it habitable. In this part, we first learn scientific ideas about how the Earth, and the Universe around it, came to be. Then we take a quick tour of the planet, looking particularly at its composition and its various layers. With this background, we're ready to encounter the twentieth-century revolution in geology that yielded the set of ideas we now call the theory of plate tectonics. We'll see that this theory, which proposes that the outer layer of the Earth is divided into "plates" that move with respect to one another, provides a rational explanation for a great variety of geologic features—from the formation of continents to the distribution of fossils. In fact, geologists now realize that plate interactions even lead to the formation of gases from which the atmosphere and oceans formed, and without which life could not exist.

A photograph of the Earth as seen by Apollo 8 astronauts in orbit around the Moon. This image emphasizes that our planet is a sphere with finite limits.

CHAPTER 1

Cosmology and the Birth of Earth

"People like us, who believe in physics, know that the distinction between past, present and future is only a stubbornly persistent illusion."
—ALBERT EINSTEIN (1955)

When the Hubble Space Telescope looks into what, to the naked eye, appears to be the black void of the night sky, it reveals a spectacle of disks and spirals of hazy light. Each of these is a distant galaxy, a cluster of as many as 300 billion stars. This is the fabric of space.

1.1 INTRODUCTION

Sometime in the distant past, perhaps more than 100,000 generations ago, humans developed the capacity for complex, conscious thought. This amazing ability, which distinguishes our species from all others, brought with it the gift of curiosity, an innate desire to understand and explain the workings of ourselves and all that surrounds us—our **Universe.** Questions that we ask about the Universe differ little from questions a child asks of a playmate: Where do you come from? How old are you? Such musings first spawned legends in which heroes, gods, and goddesses used supernatural powers to mold the planets and sculpt the landscape. Increasingly, science, the systematic analysis of natural phenomena, has provided insight into these questions. However, the development of **cosmology,** the study of the overall structure of the Universe, has proven to be a rough one, booby-trapped with tempting but flawed approaches and cluttered with misleading prejudices.

In this chapter, we begin with a brief historical sketch of cosmological thought. For brevity, we restrict the discussion to the Western tradition, though an equally rich history of ideas developed in other cultures. We then look at currently accepted ideas of modern scientific cosmology and the key discoveries that led to scientific ideas about how our planet fits into the fabric of a changing Universe. The chapter concludes with a description of Earth's formation as it may have occurred about 4.57 billion years ago.

1.2 AN EVOLVING IMAGE OF THE EARTH'S POSITION IN SPACE

Three thousand years ago (1000 B.C.E.; "before the Common Era"), the Earth's human population totaled only several million, the pyramids of Egypt had already been weathering in the desert for 1,600 years, and Homer, the great Greek poet, was compiling the *Iliad* and the *Odyssey*. In Homer's day, astronomers of the Mediterranean region knew the difference between stars and planets. They had observed that the positions of stars remained fixed relative to one another but that the whole star field slowly revolved around a fixed point (see Box 1.1), while the planets moved relative to the stars and to each other, etching seemingly complex paths across the night sky. In fact, the word "planet" comes from the Greek word *planēs*, which means "wanderer." Despite their knowledge of the heavens, people of Homer's day did not realize fully that Earth itself is a planet. Some envisioned the Earth to be a flat disk, with land toward the center and water around the margins, that lay at the center of a celestial sphere, a dome to which the stars were attached. This disk supposedly lay above an underworld governed by the fearsome god Hades. Placing the Mediterranean region at the center of the Universe must have made people of that region feel quite important indeed! Philosophers also toyed with numerous explanations for the Sun: to some, it was a burning bowl of oil, and to others, a ball of red-hot iron. Most favored the notion that movements of celestial bodies represented the activities of gods and goddesses.

Around 600 B.C.E., Greek philosophers began to argue about the structure of the Universe. Some advocated the **geocentric Universe concept** (▶Fig. 1.1a), in which the Earth sat motionless in the center of the heavens while other bodies made perfectly circular orbits around it. Notably, most philosophers assumed that orbits had a circular shape, because they considered circles to be the most perfect of geometric forms. A few centuries later, around 250 B.C.E., the **heliocentric Universe concept,** in which all heavenly objects including the Earth orbited the Sun (▶Fig. 1.1b), was proposed, but this idea found little favor.

The argument about the position of the Earth in the Universe seemed unresolved until Ptolemy (100–170 C.E.), an influential Egyptian mathematician, completed calculations that seemed to predict the wanderings of the planets successfully, in the context of the geocentric concept. The church hierarchy of Europe adopted this hypothesis as

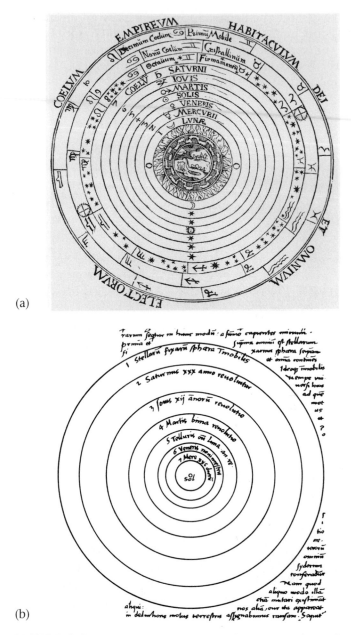

(a)

(b)

FIGURE 1.1 (a) The geocentric image of the Universe. Earth, at the center, is surrounded by air and fire and the Moon, Mercury, Venus, the Sun, Mars, Jupiter, and Saturn. Everything lies within the globe of the stars. (b) The heliocentric view of the Universe, as illustrated in this woodcut from Copernicus's *De revolutionibus*.

dogma, because it certified that the Earth occupied the most important place in the Universe and implied that humans were the Universe's most important creatures. With the fall of Rome in 476 C.E., Europe and the Middle East entered the Middle Ages. For the next millennium, most scientific study in the Western world ceased, and those who disagreed with the Ptolemaic view of the Universe risked charges of heresy.

Then came the Renaissance. The very word means "rebirth" and "revitalization," and in fifteenth-century Europe bold thinkers spawned a new age of exploration and discovery. For example, Nicolaus Copernicus (1473–1543) reintroduced and justified the heliocentric concept in a book called *De revolutionibus* (Concerning the Revolutions)

but, perhaps fearing the wrath of officials, published the book just days before he died (Fig. 1.1b). *De revolutionibus* did indeed spark a bitter battle that pitted astronomers like Johannes Kepler (1571–1630) and Galileo (1564–1642) against the establishment. Kepler showed that the planets follow elliptical, not circular, orbits, and thus demon-

---

BOX 1.1
THE REST OF THE STORY

Earth's Rotation

If you gaze at the night sky for a long time, you'll see that the stars move in a circular path around the North Star. This movement suggests either that the Earth spins on its axis (an imaginary line connecting the North and South Poles) with respect to the stars, or that the stars orbit the Earth (▶Fig. 1.2). Curiously, it was not until the middle of the nineteenth century that Jean-Bernard-Léon Foucault (1819–1868), a French physicist, proved that the Earth spins on its axis. He made this discovery by setting a heavy pendulum, attached to a long string, in motion. As the pendulum continued to swing, Foucault noted that the plane in which it oscillated (a plane perpendicular to the Earth's surface) appeared to rotate around a vertical axis (a line

perpendicular to the Earth's surface). If Newton's first law of motion—objects in motion remain in motion, objects at rest remain at rest—was correct, then this phenomenon required that the Earth rotate under the pendulum while the pendulum continued to swing in the same plane (▶Fig. 1.3a, b). Foucault displayed his discovery beneath the great dome of the Pantheon in Paris, to much acclaim.

We now know that, in fact, the Earth's spin axis is not fixed in orientation; rather, it wobbles. This wobble, known as **precession,** corresponds to the wobble of a toy top as it spins. We'll see later in this book that the precession of the Earth's axis, which takes 23,000 years, may affect the planet's climate.

FIGURE 1.2 Time exposure of the night sky over an observatory. Note that the stars appear to be fixed relative to one another, but that they rotate around a central point, the North Star. This motion is actually due to the rotation of the Earth on its axis.

FIGURE 1.3 Foucault's experiment. (a) An oscillating pendulum at a given time. (b) The same pendulum at a later time. The pendulum stays in the same plane, but the Earth, and hence the frame, rotates.

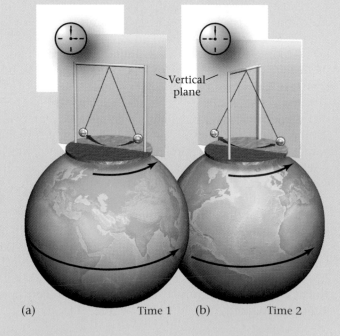

Vertical plane

(a)   Time 1   (b)   Time 2

strated that Ptolemy's calculations, the bulwark of the geocentric hypothesis, were wrong. Galileo, using the newly invented telescope, observed that Venus has phases like our own Moon (a characteristic that could only be possible if Venus orbited the Sun) and that Jupiter has its own moons. His discoveries demonstrated that all heavenly bodies do not revolve around the Earth. So, Galileo's observations also contradicted the geocentric hypothesis. But Europe was not quite ready for Galileo. Officials charged him with heresy, and he spent the last ten years of his life under house arrest.

In the year of Galileo's death, Isaac Newton (1642–1727), perhaps the greatest scientist of all time, was born in England. Newton derived mathematical laws governing gravity and basic mechanics (the movement of objects in space) and thus provided the tools for explaining how physical processes operate in nature. Thus, as the seventeenth century came to a close, people for the first time possessed a clear image of the movements of planets. Unfortunately for human self-esteem, Earth had been demoted from its place of prominence at the center of the Universe, and became merely one of many planets circling the Sun.

1.3 A SENSE OF SHAPE

As we noted earlier, ancient Greeks, like people from many other cultures, originally considered the Earth to be a flat disk. But by the time of Aristotle (c. 257–180 B.C.E.), many philosophers realized that the Earth had to be a sphere, because they could observe sailing ships disappear progressively from base to top as they moved beyond the horizon and they could see that the Earth cast a curved shadow on the Moon during an eclipse. In fact, Ptolemy had developed the concepts of latitude and longitude to define locations on a spherical Earth. Thus, though a few clerics espoused the flat Earth view through the Middle Ages, it is likely that nearly everyone had rejected the idea as the Renaissance began.

The concept of a spherical Earth played a key role in driving European voyages of discovery. Specifically, in the 1480s the Italian geographer and astronomer Paulo Toscanelli (1397–1482) promoted the idea that it would be possible to reach India by sailing westward from Europe. Christopher Columbus made the case for such a voyage to the King and Queen of Spain. The monarchs agreed to fund the voyage, and Columbus set sail. Of course, in 1492 he hit a large roadblock, now known as America, and never made it to India. But two decades later, in 1520, Ferdinand Magellan led a voyage that eventually did succeed in circumnavigating the planet, thereby making it possible to create a globe—a map drawn on a sphere, the first realistic image of our planet.

1.4 A SENSE OF SCALE

We use enormous numbers to describe the size of the Earth, the distance from the Earth to the Sun, the distance between stars, and the distance between galaxies. Who came up with these numbers?

Circumference of the Earth

The Greek astronomer Eratosthenes (c. 276–194 B.C.E.) served as chief of the library in Alexandria, Egypt, one of the great ancient centers of learning in the Mediterranean region. One day while filing papyrus scrolls, he came across a report noting that in the southern Egyptian city of Syene, the Sun lit the base of a deep vertical well precisely at noon on the first day of summer. Eratosthenes deduced that the Sun's rays at noon on this day must be exactly perpendicular to the Earth's surface at Syene, and that if the Earth was spherical, then the Sun's rays could *not* simultaneously be perpendicular to the Earth's surface at Alexandria, 800 km to the north (▶Fig. 1.4). So, on the first day of summer, Eratosthenes measured the shadow

FIGURE 1.4 Eratosthenes discovered that at noon on the first day of summer, the Sun's rays were perpendicular to the Earth's surface (that is, parallel to the radius of the Earth) at Syene, but made an angle of 7.2° with respect to a vertical tower at Alexandria. Thus, the distance between Alexandria and Syene represented 7.2°/360° of the Earth's circumference. Knowing the distance between the two cities, therefore, allowed him to calculate the circumference of the Earth.

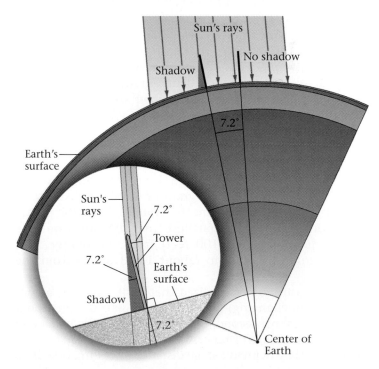

cast by a tower in Alexandria at noon. The angle between the tower and the Sun's rays, as indicated by the shadow's length, proved to be 7.2°. He then commanded a servant to pace out a straight line from Alexandria to Syene. The sore-footed servant found the distance to be 5,000 stadia (1 stadium = 0.1572 km). Knowing that a circle contains 360°, Eratosthenes then calculated the Earth's circumference as follows:

$$\frac{360°}{x} = \frac{7.2°}{5,000 \text{ stadia}}$$

$$x = \frac{360° \times 5,000 \text{ stadia}°}{7.2°} = 250,000 \text{ stadia}$$

$$250,000 \text{ stadia} \times 0.1572 \text{ km/stadium}$$
$$= 39,300 \text{ km} = 24,421 \text{ miles}$$

Thus, twenty-two centuries ago, Eratosthenes determined the circumference of the Earth to within 2% of today's accepted value (40,008 km, or 24,865 miles) without the aid of any sophisticated surveying equipment—a truly amazing feat.

The Distance from Earth to Celestial Objects

Around 200 B.C.E., Greek mathematicians, using ingenious geometric calculations, determined that the distance to the Moon was about thirty times the Earth's diameter, or 382,260 km. This number comes close to the true distance, which on average is 381,555 km (about 237,100 miles). But it wasn't until the seventeenth century that astronomers figured out that the mean distance between the Earth and the Sun is 149,600,000 km (about 93,000,000 miles). As for the stars, the ancient Greeks realized that they must be much farther away than the Sun in order for them to appear as a fixed backdrop behind the Moon and planets, but the Greeks had no way of calculating the actual distance. Our modern documentation of the vastness of the Universe began in 1838, when astronomers found that the nearest star to Earth, Alpha Centauri, lies 40.85 trillion km away.

Since it's hard to fathom the distances to planets and stars without visualizing a more reasonably sized example, imagine that the Sun is the size of an orange. At this scale, the Earth would be a grain of sand at a distance of 10 meters (m) (30 feet) from the orange. Pluto, the planet which on average orbits farthest from the Sun, would lie about 650 m (about 2,100 feet) from the orange, and Alpha Centauri 2,000 km (about 1,243 miles) from the orange.

When astronomers realized that light travels at a constant (i.e., unchanging) speed of about 300,000 km (about 186,000 miles) per second, they realized that they had a way to describe the huge distances between objects in space conveniently. They defined a large distance by stating how long it takes for light to traverse that distance. For example, it takes light about 1.3 seconds to travel from the Earth to the Moon, so we can say that the Moon is about 1.3 light seconds away. Similarly, we can say that the Sun is 8.3 light minutes away. A **light year,** the distance that light travels in one Earth year, equals about 9.5 trillion km (about 6 trillion miles). When you look up at Alpha Centauri, 4.3 light years distant, you see light that started on its journey to Earth about 4.3 years ago.

Astronomers didn't develop techniques for measuring the distance to very distant stars and galaxies until the twentieth century. With these techniques (see an astronomy book for details), they determined that the farthest celestial objects that can be seen with the naked eye are over 2.2 million light years away. Powerful telescopes allow us to see much farther. The edge of the *visible* Universe lies over 13 billion light years away, which means that light traveling to Earth from this location began its journey about 9 billion years before the Earth even existed. When such numbers became available by the middle of the twentieth century, people came to the realization that the dimensions of the Universe are truly staggering.

1.5 THE MODERN IMAGE OF THE UNIVERSE

We've seen that the burst of discovery during the Renaissance forced astronomers to change their view of Earth's central place in the Universe. Eventually, they realized that the Earth is but one of nine planets in the solar system (the Sun, and all objects that travel around it). They also learned that stars are not randomly scattered through the Universe; gravity pulls them together to form immense systems, or groups, called **galaxies.** The Sun is but one of over 300 billion stars that together form the Milky Way galaxy, and the Milky Way is but one of more than 100 billion galaxies constituting the visible Universe (see chapter opening photo). Galaxies are so far away that, to the naked eye, they look like stars in the night sky. The nearest galaxy to ours, Andromeda, lies over 2.2 million light years away.

If we could view the Milky Way from a great distance, it would look like a flattened spiral, 100,000 light years across, with great curving arms gradually swirling around a glowing, disk-like center (▶Fig. 1.5a, b). Presently, our solar system lies near the outer edge of one of these arms and rotates around the center of the galaxy about once every 250 million years. We hurtle through space, relative to an observer standing outside the galaxy, at about 200 km per second.

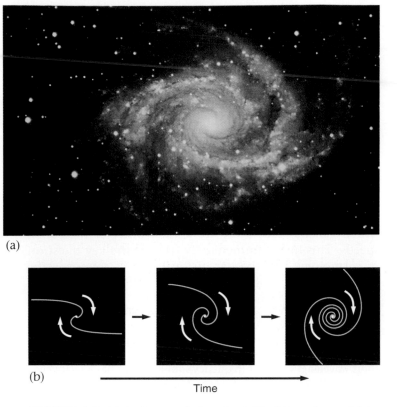

(a)

(b)

Time

FIGURE 1.5 (a) An image of what the Milky Way galaxy might look like if viewed from outside. Note that the galaxy consists of spiral arms around a central cluster. Our Sun lies at the edge of one of these arms. (b) The spiral shape is due to rotation of the galaxy. Here we see three stages in a computer simulation of spiral formation.

1.6 FORMING THE UNIVERSE

Do galaxies move with respect to other galaxies? Does the Universe become larger or smaller with time? Has the Universe always existed? Answers to these fundamental questions came from an understanding of a phenomenon called the Doppler effect. Though the term may be unfamiliar, the phenomenon it describes is an everyday experience. After introducing the Doppler effect, we show how an understanding of it leads to a theory of Universe formation.

The Doppler Effect

When a train whistle screams, the sound you hear has moved through the air from the whistle to your ear in the form of sound waves. (Waves are disturbances that transmit energy from one point to another by causing periodic motions.) As each wave passes, air alternately compresses, then expands. The pitch of the sound, meaning its note in the musical scale, depends on the frequency of the sound

waves, meaning the number of waves that pass a point in a given time interval. Now imagine that as you are standing on the station platform, the train moves toward you. The sound of the whistle gets louder as the train approaches, but its pitch remains the same. Then, the instant the train passes, the pitch abruptly changes; it sounds like a lower note in the musical scale. An Austrian physicist, C. J. Doppler (1803–1853), first interpreted this phenomenon, and thus it is now known as the **Doppler effect.** When the train moves toward you, the sound has a higher frequency (the waves are closer together), because the sound source, the whistle, has moved slightly closer to you between the instant that it emits one wave and the instant that it emits the next (▶Fig. 1.6a, b). When the train moves away from you, the sound has a lower frequency (the waves are farther apart), because the whistle has moved slightly farther from you between the instant it emits one wave and the instant it emits the next.

Light energy also moves in the form of waves. In shape, light waves somewhat resemble water waves. Visible light comes in many colors—the colors of the rainbow. The color of light you see depends on the frequency of the light waves, just as the pitch of a sound you hear depends on the frequency of sound waves. Red light has a longer wavelength (lower frequency) than blue light (▶Fig. 1.7a, b). The Doppler effect also applies to light but is only noticeable if the light source moves very fast (e.g., at least a few percent of the speed of light). If a light source moves away from you, the light you see becomes redder (as the light shifts to lower frequency), and if the source moves toward you, the light you see becomes bluer (as the light shifts to higher frequency). We call these changes the **red shift** and the **blue shift,** respectively (▶Fig. 1.7c).

Red Shifts and the Expanding Universe Theory

In the 1920s, astronomers such as Edwin Hubble (after whom the Hubble Space Telescope was named) braved many a frosty night beneath the open dome of a mountaintop observatory in order to aim telescopes into deep space. These researchers had begun a search for distant galaxies. At first, they documented only the location and shape of newly discovered galaxies. But then, one astronomer began an additional project to study the wavelength of light produced by the distant galaxies. The results yielded a surprise that would forever change humanity's perception of the Universe.

Astronomers found, to their amazement, that the light of distant galaxies displayed red shifts relative to absorption lines in the light spectra of nearby stars. Hubble pondered this mystery and, around 1929, realized that the red shifts must be a consequence of the Doppler effect—and thus that the distant galaxies must be moving away from Earth at an

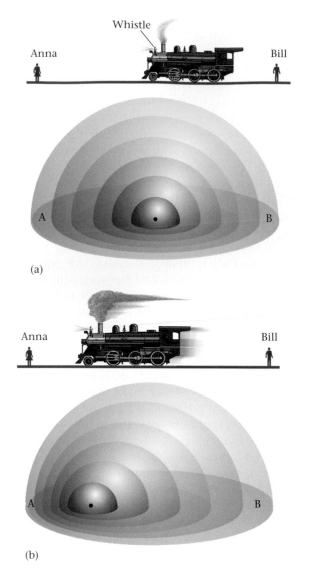

(a)

(b)

FIGURE 1.6 The spacing of waves is wavelength. Wavelength, and, therefore, frequency (the number of waves passing a point in an interval of time) is affected by the speed of the source. Frequency determines pitch. (a) Sound emanating from a stationary source has the same wavelength in all directions (the circular shells represent the waves), and observers at points A and B (Anna and Bill) hear the same pitch. (b) If the source is moving toward Anna, she hears a shorter-wavelength sound than does Bill. Therefore, Anna hears a higher-pitched (higher-frequency) sound than does Bill.

immense velocity. At the time, astronomers thought the Universe had a fixed size, so Hubble initially assumed that if some galaxies were moving away from Earth, others must be moving toward Earth. But this was not the case. On further examination, Hubble concluded that the light from *all* distant galaxies, regardless of their direction from Earth, exhibits a red shift. In other words, *all* distant galaxies are moving rapidly away from us.

How can all galaxies be moving away from us, regardless of which direction we look? Hubble puzzled over this question and finally recognized the solution, one that actually was implied by calculations that Albert Einstein had made. The whole Universe must be expanding! To picture the expanding Universe, imagine a ball of bread dough with raisins scattered throughout. As the dough bakes and expands into a loaf, each raisin moves away from its neighbors, in every direction (▶Fig. 1.8a, b). This idea came to be known as the **expanding Universe theory.**

Hubble's expanding Universe theory marked a revolution in thinking. No longer could we view the Universe as being fixed in dimension, with galaxies locked in position. Now we see the Universe as an expanding bubble, in which galaxies race away from each other at incredible speeds. This image immediately triggers the key question of cosmology: Did the expansion begin at some specific

FIGURE 1.7 Light waves resemble ocean waves in shape, but physically they are quite different. (a) Blue light has a relatively short wavelength (higher frequency). (b) Red light has a relatively long wavelength (lower frequency). (c) The shift in light frequency that an observer sees depends on whether the source is moving toward or away from the observer.

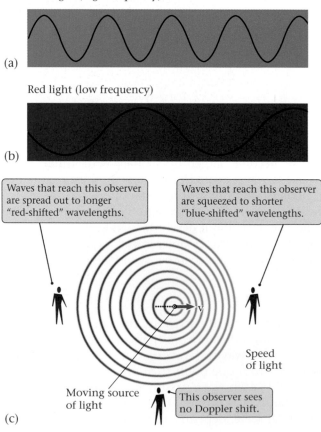

Blue light (high frequency)

(a)

Red light (low frequency)

(b)

Waves that reach this observer are spread out to longer "red-shifted" wavelengths.

Waves that reach this observer are squeezed to shorter "blue-shifted" wavelengths.

Speed of light

Moving source of light

This observer sees no Doppler shift.

(c)

time in the past? If it did, then this instant would mark the beginning of the Universe, the beginning of space and time.

The Big Bang

Most astronomers have concluded that expansion did indeed begin at a specific time, with a cataclysmic explosion called the **big bang.** According to the big bang theory, all matter and energy—everything that now constitutes the Universe—was initially packed into an infinitesimally small point. For reasons that no one fully understands, the point exploded, according to current estimates, 13.7 (± 1%) billion years ago. (To learn what may have happened next, you may first need to review some terms from basic physics and chemistry; please study Appendix A.) Since the big bang, the Universe has been continually expanding (▶Fig. 1.8c).

1.7 MAKING ORDER FROM CHAOS

Aftermath of the Big Bang

Of course, no one was present at the instant of the big bang, so no one actually saw it happen. But by combining clever calculations with careful observations, researchers have developed a consistent model of how the Universe evolved beginning an instant after the "explosion"(▶Fig. 1.9). The calculations come from applying the laws of physics. The observations come from examining the edge of the Universe with large telescopes, for when we look at very distant objects in space we are seeing the distant past. The most distant objects yet observed lie about 13 billion light years away, so looking at them provides a snapshot of the Universe when it was very young indeed.

According to the contemporary model of the big bang, profound change happened at a fast and furious rate at the

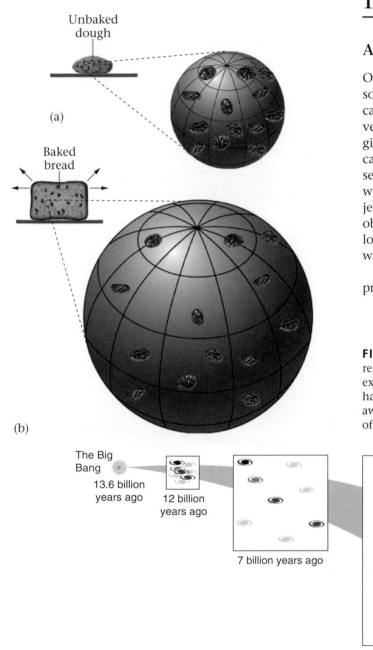

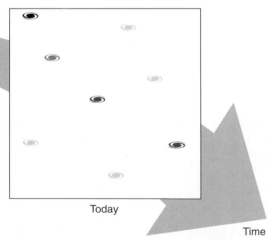

FIGURE 1.8 (a) At the dough stage, raisins in raisin bread are relatively close to one another. (b) During baking the bread expands, and the raisins (like galaxies in the expanding Universe) have all moved away from each other. Notice that *all* raisins move away from their neighbors, regardless of direction. (c) The concept of the expanding Universe; the spirals represent galaxies.

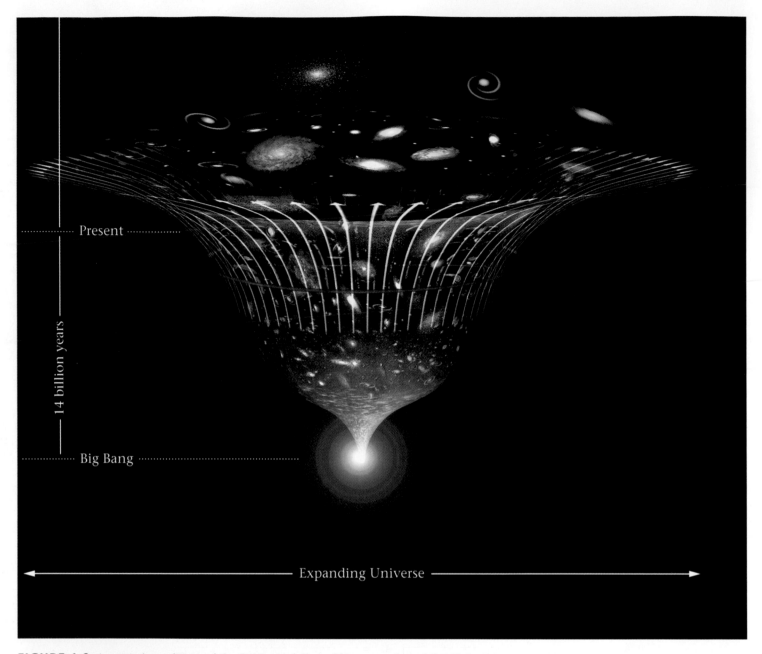

FIGURE 1.9 An artist's rendition of the Big Bang, followed by expansion of the Universe. Recent work suggests that the rate of expansion has changed as time passes.

outset. During the first instant (10^{-43} seconds) of existence, the Universe was so small, so dense, and so hot (10^{28} degrees centigrade) that it consisted entirely of energy—atoms, or even the smallest subatomic particles that make up atoms, could not even exist. (Note that because we use very large and very small numbers in our discussion, we employ scientific notation: $10^1 = 10$, $10^3 = 1,000$, and so on. Similarly, $10^{-3} = 1/1,000$.) But because expansion happened at nearly the speed of light, by the end of 1 second the density (i.e., quantity of mass per unit volume) of the Universe had decreased to "only" about a million billion (i.e., 10^{15}) times the density of

water, and under these conditions the first protons and electrons could form (see Appendix A for definitions). Since a single proton constitutes the nucleus of a hydrogen atom (the smallest atom), this statement means that only a second after the big bang occurred, hydrogen atoms—the most abundant atoms—began to populate the Universe.

By the time the Universe reached an age of 3 minutes, its temperature had fallen below the 1 billion degrees, and its diameter had grown to about 100 billion km (60 billion miles). Under these conditions, nuclei of new atoms began to form through the collision and fusion (sticking together) of hydrogen atoms. Formation of new nuclei by fusion reactions at this time is called "big bang nucleosynthesis" because it happened *before* any stars existed. Big bang nucleosynthesis could only produce small atoms (helium, lithium, beryllium, and boron), meaning ones containing a small number of protons, and it happened very rapidly. In fact, virtually all of the new atomic nuclei that would form by big bang nucleosynthesis had formed by the end of 5 minutes. Why did nucleosynthesis stop? When it reached an age of 5 minutes, the Universe had expanded so much that its average density had decreased to a value one-tenth that of water (i.e., 0.1 kg/m^3), so that atoms were so far apart that they rarely collided. (Note that 0.1 kg/m^3 may not sound like much, but it was vastly greater than the density of today. Now, interstellar space contains almost no atoms, so most of the Universe's volume is almost a perfect vacuum.)

For the next interval of time, the Universe consisted of nuclei dispersed in a turbulent sea of wandering electrons. Physicists refer to such a material as a "plasma." After a few hundred thousand years, temperature dropped below a few thousand degrees. Under these conditions, neutral atoms (in which negatively charged electrons orbit a positively charged nucleus) could form. Eventually, the Universe became cool enough for chemical bonds to bind atoms of certain elements together in molecules—most notably, two hydrogen atoms could join to form molecules of H_2. As the Universe continued to expand and cool further, atoms and molecules slowed down and accumulated into **nebulae,** patchy clouds of gas separated from one another by the vacuum of space. Gas making up earliest nebulae of the Universe consisted entirely of the smallest atoms, namely, hydrogen (98%), helium (2%), and traces of lithium, beryllium, and boron.

Forming the Very First Stars

When the Universe reached its 200 millionth birthday, it contained immense, slowly swirling, dark nebulae separated by vast voids of empty space (▶Fig. 1.10). The universe could not remain this way forever, however, because of the invisible but persistent pull of gravity. Eventually, gravity began to remold the Universe pervasively and permanently.

All matter exerts gravitational pull—a type of force—on its surroundings, and as Isaac Newton first pointed out, the amount of pull depends on the amount of mass. Somewhere in the young Universe, the gravitational pull of an initially denser region of a nebula began to pull in surrounding gases and, in a grand example of "the rich getting richer," grew in mass and, therefore, density. As this denser region sucked in progressively more gas, more matter compacted into a smaller region, and the initial swirling movement of gas transformed into a rotation around an axis that became progressively faster and faster.

FIGURE 1.10 Gases clump to form distinct nebulae, which look like clouds in the sky. In this Hubble Space Telescope picture, new stars are forming at the top of the nebula on the left. Stars that have already formed backlight the nebulae.

(The same phenomenon occurs when a spinning ice skater pulls her arms inward and speeds up.) Because of rotation, the condensing portion of the nebula evolved into a spinning disk-shaped mass of gas called an **accretion disk** (see art spread on pp. 24–25). Eventually, the gravitational pull of the accretion disk became great enough to trigger wholesale inward collapse of the surrounding nebula. Like a house of cards, the inner portion of the cloud fell toward the disk, then the mid-portion fell, and finally the outer portion. With all the additional mass available, gravity aggressively pulled the inner portion of the accretion disk into a dense ball. The energy of motion (kinetic energy) of gas falling into this ball transformed into heat (thermal energy) when it landed on the ball. (The same phenomenon happens when you drop a plate and it shatters—if you measured the pieces immediately after breaking, they would be slightly warmer.) Moreover, the squeezing together of jostling gas atoms and molecules in the ball further increased the gas's temperature. (The same phenomenon happens in the air that you compress in a bicycle pump.) Eventually, the central ball of the accretion disk became hot enough to glow, and at this point it became a **protostar.**

A protostar continues to grow, by pulling in successively more mass, until its core becomes very dense and its temperature reaches about 10 million degrees. Under such conditions, fusion reactions begin to take place; hydrogen nuclei in a protostar join, in a series of steps, to form helium nuclei. Astronomers refer to the process of forming a larger nucleus by combining hydrogen nuclei as "hydrogen burning," because such fusion reactions produce huge amounts of energy and make a star into a fearsome furnace.[1] When the first nuclear fusion reactions began in the first protostar, the body "ignited" and the first true star formed, and when this happened, the first starlight pierced the newborn Universe. This process would soon happen again and again, and many first-generation stars began to burn.

First-generation stars tended to be very massive (e.g., 100 times the mass of the Sun), because compared with the nebulae of today, nebulae of the very young Universe contained much more matter. Astronomers have shown that the larger the star, the hotter it burns and the faster it runs out of fuel and "dies." A huge star may survive only a few million years to a few tens of millions of years before it dies by violently exploding to form a **supernova.**[2] Thus, not long after the first generation of stars formed, the Universe began to be peppered with the first generation of supernova explosions.

1.8 WE ARE ALL MADE OF STARDUST

Element Factories

Nebulae from which the first-generation stars formed consisted entirely of small atoms (elements with atomic numbers smaller than 5), because only these small atoms were generated by big bang nucleosynthesis. In contrast, the Universe of today contains ninety-two naturally occurring elements (see Appendix A). Where do the other eighty-seven elements come from? In other words, how did elements such as carbon, sulfur, silicon, iron, gold, and uranium form? These elements, which are common on Earth, have large atomic numbers (e.g., carbon has an atomic number of 6, and uranium has an atomic number of 92). Physicists now realize that these elements didn't form during or immediately after the big bang. Rather, they formed later, during the life cycle of stars, by the process of "stellar nucleosynthesis."

Because of stellar nucleosynthesis we can consider stars to be element factories, constantly fashioning larger atoms out of smaller atoms. Two processes can be involved: (1) fusion, and (2) neutron capture and decay. In a fusion reaction, smaller nuclei fuse to form larger nuclei. The process of neutron capture and decay involves two steps. First, a neutron sticks to the nucleus of an atom and causes the atomic mass of the nucleus to increase by 1. Then, the neutron "decays," meaning that it transforms into a proton by releasing an electron. When this happens, the atomic mass of the atom stays the same but the atomic number increases by 1 (see Appendix A). Thus, the atom becomes a new element. The specific reactions that take place during stellar nucleosynthesis depend on the mass of the star, because the temperature and density of a more massive star are greater than those of a less massive star; as temperature increases, the velocity of particles increases so larger nuclei can be driven together.

Low mass stars, like our Sun, burn slowly and may survive for 10 billion years. Nuclear reactions in these stars produce elements up to an atomic number of 6 (i.e., carbon). High mass stars (10 to 100 times the mass of the Sun) burn quickly, and may survive for only 20 million years. They mostly produce elements up to an atomic number of 26 (i.e., iron).

Most very large atoms (i.e., atoms with atomic numbers greater than that of iron) require even more violent circumstances to form than can occur within even a high-mass star. These atoms form most efficiently during a supernova explosion. Supernova explosions produce large quantities of neutrons; many collide and attach to nearby atoms and then decay, quickly building very large atoms.

Now you can understand why we call stars and supernova explosions "element factories." They fashion larger atoms—new elements—that had not formed during or immediately after the big bang. What happens to these atoms? Some escape from a star into space during the star's lifetime, simply by moving fast enough to overcome the star's gravitational pull. The stream of atoms emitted from a star during its lifetime is a **stellar wind** (▶Fig. 1.11). Escape also happens upon the death of a star. Specifically, a low-mass star (like our Sun) releases a large shell of gas as its "last gasp" (as we will discuss in Chapter 23), while a high-mass star blasts matter into space during a supernova explosion (▶Fig.1.12). Once in space, atoms form new nebulae or mix back into existing nebulae.

---

1. In this context, the word "burning" refers to fusion reactions. For example, burning of helium means that three helium nuclei fuse to form carbon—these reactions consume atoms of one type to produce atoms of another. Astronomers use the word "burning" because the reactions produce heat and light. But the energy-producing reactions in stars are very different from the reactions that take place when you "burn" wood in your fireplace. The burning of wood is a *chemical reaction* (see Appendix A) during which the chemical bonds holding molecules together break and atoms rearrange into new molecules.

2. The name "supernova" comes from the Latin word *nova*, which means "new"; when the light of a supernova explosion reaches Earth, it looks like a very bright new star in the sky.

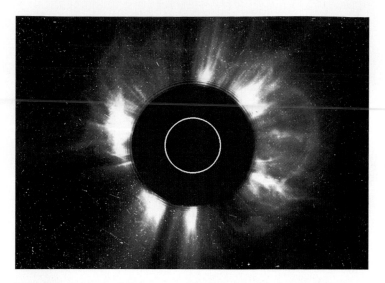

FIGURE 1.11 In this image, a black disk hides the Sun so we can see the stellar wind that the Sun produces.

In effect, when the first generation of stars died, they left a legacy of new elements that mixed with residual gas from the big bang. A second generation of stars and associated planets formed out of the new, compositionally more diverse nebulae. Second-generation stars lived and died, and contributed elements to third-generation stars. Succeeding generations of stars and planets contain a greater proportion of heavier elements. Because different stars live

FIGURE 1.12 Very heavy elements form during supernova explosions. Here, we see the rapidly expanding shell of gas ejected into space that appeared in 1054 C.E. This shell is called the Crab Nebula.

for varied periods of time, at any given moment the Universe contains many different generations of stars, including small stars that have been living for a long time and large stars that have only recently arrived on the scene. The mix of elements we find on Earth includes relics of primordial gas from the big bang as well as the disgorged guts of dead stars. Think of it—the elements that make up your body once resided inside a star!

The relative abundances of the different elements in the solar system reflects the amount of element production and the stability of the atoms. Hydrogen and helium are by far the most abundant elements—together, they constitute 98% of the matter in the solar system. They make up most of the Sun, and the Sun contains most of the matter in the solar system. After hydrogen and helium, the next most abundant elements are carbon and oxygen. Other elements occur in relatively small proportions.

The Nature of Our Solar System

We've just discussed how stars form from nebulae. Can we apply this model to our own solar system—that is, the Sun plus all the objects that orbit it? The answer is yes, but before we do so, let's define the components that make up our solar system.

Our Sun does not sail around the Milky Way in isolation. In its journey it holds, by means of gravitational "glue," many other objects. Of these, the largest are the nine plants: Mercury, Venus, Earth, Mars, Jupiter, Saturn, Uranus, Neptune, and Pluto (named in order from the closest to the Sun to the farthest). A **planet** is a sizable solid object orbiting a star, and it may itself travel with a moon or even many moons. By definition, a **moon** is an object locked in orbit around a planet; all but two of the planets have them. For example, Earth has one moon (the Moon) whereas Jupiter has at least sixteen, of which four are as big as or bigger than the Moon. In the past several years research has documented planets in association with dozens of other stars, and the number keeps growing as more observations become available. In addition to planets and moons, our solar system also includes a belt of asteroids (relatively small chunks of rock and/or metal) between the orbit of Mars and the orbit of Jupiter, and perhaps a trillion comets (relatively small blocks of "ice" orbiting the Sun; ice, in this context, means the solid version of a material that could be gaseous under Earth's surface conditions) in clouds that extend very far beyond the orbit of Pluto (see Box 1.2). Even though there are many objects in the solar system, 99.8% of the solar system's mass resides in the Sun. The next largest object in the Solar system—the planet Jupiter accounts for 99.5% of the non-solar mass of the solar system.

Of the nine planets in our solar system, the inner four (Mercury, Venus, Earth, and Mars) are called the inner

Gravity pulls gas and dust inward to form an accretionary disk. Eventually a glowing ball—the proto-Sun—forms at the center of the disk.

Forming the solar system, according to the nebula hypothesis: A nebula forms from hydrogen and helium left over from the big bang, as well as from heavier elements that were produced by fusion reactions in stars or during explosions of stars.

Gravity reshapes the proto-Earth into a sphere. The interior of the Earth separates into a core and mantle.

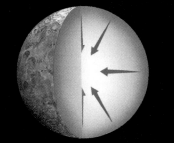

Forming the planets from planetesimals: Planetesimals grow by continuous collisions. Gradually, an irregularly shaped proto-Earth develops. The interior heats up and becomes soft.

Soon after Earth forms, a small planet collides with it, blasting debris that forms a ring around the Earth.

The Moon forms from the ring of debris.

"Dust" (particles of refractory materials) concentrates in the inner rings, while "ice" (particles of volatile materials) concentrates in the outer rings. Eventually, the dense ball of gas at the center of the disk becomes hot enough for fusion reactions to begin. When it ignites, it becomes the Sun.

Dust and ice particles collide and stick together, forming planetesimals.

The Birth of the Earth-Moon System

Eventually, the atmosphere develops from volcanic gases. When the Earth becomes cool enough, moisture condenses and rains to create the oceans. Some gases may be added by passing comets.

FIGURE 1.13 The relative sizes of the planets of our solar system.

planets or the **terrestrial planets** (Earth-like planets) because they consist of a shell of rock surrounding a core of iron alloy, as does the Earth. The next four have been traditionally called the outer planets, the Jovian planets (Jupiter-like planets), or the **gas giant planets.** The adjective "giant" certainly seems appropriate for these planets (▶Fig. 1.13). Jupiter, for example, has a mass that is about 318 times that of Earth and a diameter 11.2 times larger. Notably, the inner two Jovian planets (Jupiter and Saturn) differ significantly from the outer two (Uranus and Neptune). Jupiter and Saturn have an elemental composition similar to the Sun and thus consist predominantly of hydrogen and helium.[3] Uranus and Neptune, in contrast, appear to consist predominantly of "ice" (solid methane, hydrogen sulfide, ammonia, and water). The ninth planet, Pluto, is so small (about two-thirds the size of Earth's moon) that astronomers didn't find it until 1930. It probably consists of a mixture of rock and "ice"—in fact, some astronomers consider Pluto to be a large comet, unworthy of even being called a planet.

Forming the Solar System

Earlier in this chapter we presented ideas about how the first stars formed, early during the history of the Universe, from very light gases (mostly hydrogen) produced during the big bang or by big bang nucleosynthesis. To keep things simple, we didn't mention the formation of planets,

moons, asteroids, or comets in that discussion. Now, let's develop a model to explain where they came from (see art on pp. 24–25).

Our solar system formed at about 4.56 Ga ("Ga" means "giga annum" or a "billion years ago"), over 9 billion years after the big bang, and thus is probably a third- or fourth- or fifth-generation star (no one can say for sure) created from a nebula that, while still predominantly composed of hydrogen, contained all ninety-two elements. Recall that the other elements formed in stars or supernovas.

The materials in nebulae such as the one from which our solar system formed can be divided into two classes. Volatile materials—such as hydrogen, helium, methane, ammonia, water, and carbon monoxide—are ones that could exist as gas at the Earth's surface. In the pressure and temperature conditions of space, some volatile materials remain in gaseous form, but others condense or freeze to form different kinds of "ice." "Refractory materials" are ones that melt only at high temperatures and they condense to form solid soot-sized particles of "dust" in the coldness of space (▶Fig. 1.14a). Thus, when astronomers refer to "dust", they mean specks of solid rocky or metallic material, or clumps of the molecules that make up rock or metal.

We saw earlier that the first step in the formation of a star is the development of an accretion disk. When our solar system formed, this accretion disk contained not only hydrogen and helium gas, but also other gases as well as "ice" and "dust." Such an accretion disk can also be called a **protoplanetary disk,** because it contains the raw materials from which planets form. With time, the central ball of our protoplanetary disk developed into the **proto-Sun,** and the remainder evolved into a series of concentric rings. A protoplanetary disk is hotter toward its center than toward its rim. Thus, the warmer inner rings of the disk ended up with higher concentrations of "dust," whereas

---

3. Notably, even though Jupiter and Saturn have the same composition as the Sun, they did not ignite like the Sun because their masses are so small that their interiors never became hot enough for hydrogen burning to commence. Jupiter would have to be 80 times bigger for it to start to burn.

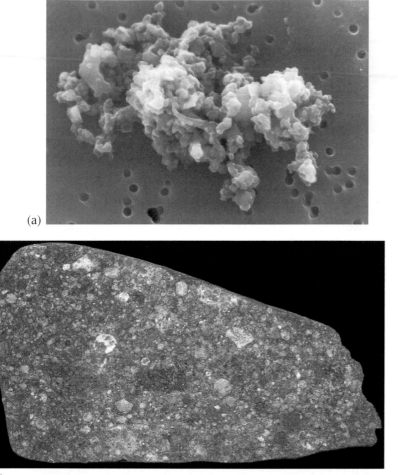

(a)

(b)

FIGURE 1.14 (a) Photograph, taken with a high-powered scanning electron microscope, showing a speck of interplanetary "dust." This speck is 10 (μm is a micron, equal to 1/1,000th of a millimeter). (b) Cross section of a particular type of meteorite (a fragment of solid material that fell from space and landed on Earth) thought to represent the texture of a small planetesimal.

the cooler outer rings ended up with higher concentrations of "ice."

Even before the proto-Sun ignited, the material of the surrounding rings began to coalesce, or accrete. Recall that accretion refers to the process by which smaller pieces clump and bind together to form larger pieces. First, soot-sized particles merged to form sand-sized grains. Then, these grains clumped together to form grainy basketball-sized blocks, which in turn collided. The fate of these blocks depended on the speed of the collision—if the collision was slow, blocks stuck together or simply bounced apart; if the collision was fast, one or both of the blocks shattered, producing smaller fragments that had to recombine all over again.

Eventually, enough blocks accreted to form **planetesimals,** bodies whose diameter exceeded about 1 km (▶Fig. 1.14b). Because of their mass, planetesimals exert enough gravity to attract and pull in other objects that are nearby. Figuratively, planetesimals acted like vacuum cleaners, sucking in debris—small pieces of "dust" and "ice," as well as smaller planetesimals—in their orbit, and in the process they grew progressively larger. Eventually, victors in the competition to attract mass grew into **protoplanets,** bodies almost the size of today's planets. Once the protoplanets had succeeded in incorporating virtually all the debris near their orbits, so that their growth nearly ceased and they were the only inhabitants of their orbits, they became the planets that exist today.

Early stages in the accretion process probably occurred very quickly—some computer models suggest that it may have taken only a few hundred thousand years to go from the "dust" stage to the large planetesimal stage. Estimates for the growth of planets from planetesimals range from 10 million years (m.y.) to 200 m.y. Recent studies favor faster growth and argue that planet formation was essentially completed between 4.568 Ga and 4.558 Ga.

The nature of the planet resulting from accretion depended on distance from the proto-Sun. In the inner orbits, where the protoplanetary disk consisted mostly of "dust," small terrestrial planets composed of rock and metal formed. In the outer part of the protoplanetary disk, where in addition to gas and "dust" significant amounts of "ice" existed, larger protoplanets—as big as 15 times the size of Earth—grew. These, in turn, pulled in so much gas and "ice" that they grew into the gas giant planets. Rings of "dust" and "ice" orbiting these planets became their moons.

When the Sun ignited, toward the end of the time when planets were forming, it generated a strong stellar wind (in this case, the solar wind) of particles traveling rapidly into space. The solar wind blew any remaining gases out of the inner portion of the newborn solar system. But the wind was too weak to blow away the atmospheres from the strong gravitational pull of the gas giant planets.

The overall model that we've just described is called the **nebular theory** of solar system formation. Astronomers like the nebular theory because it explains why the "ecliptic" (the elliptical, or oval, plane traced out by a planet's orbit) of each planet except Pluto is nearly the same, and why all planets orbit the Sun in the same direction (▶Fig. 1.15). These observations make sense if all the planets formed out of a flattened disk of gas revolving in the same direction around a central mass. Pluto, whose ecliptic is inclined relative to other planets' ecliptics, may be an errant comet (Box 1.2) pulled into its present orbit by the tug of another planet's gravity.

THE VIEW FROM SPACE A group of new born stars occurs in a cluster 12 billion light years from Earth, as viewed by the Hubble Space Telescope. The light from the stars makes the gas and dust surrounding them glow. Our solar system may have formed from such a cloud.

beyond a certain critical size, the insides of the planet become warm and the rock becomes soft enough to flow in response to gravity; also, the gravitational force becomes stronger. As a consequence, protrusions are pulled inward toward the center, and the planetesimal re-forms into a shape that permits the force of gravity to be the same at all points on its surface. This special shape is a sphere because in a sphere the distribution of mass around the center has evened out.

CHAPTER SUMMARY

• Most Greek philosophers favored a geocentric Universe concept, which placed the Earth at the center of the Universe, with the planets and Sun orbiting around the Earth within a celestial sphere speckled with stars. The heliocentric (Sun-centered) Universe concept was not widely accepted.

• The Renaissance brought a revolution in scientific thought. The idea of a spherical Earth returned, and Copernicus reintroduced the heliocentric concept, which was then proven by Galileo. Newton introduced physical laws that allowed people to understand motion in the Universe.

• Eratosthenes was able to measure the size of the Earth in ancient times, but it was not until fairly recently that astronomers accurately determined the distance to the Sun, planets, and stars. Distances in the Universe are so large that they must be measured in light years.

• The Earth is one of nine planets orbiting the Sun, and this solar system lies on the outer edge of a slowly revolving galaxy, the Milky Way, which is composed of about 300 billion stars. The Universe contains at least hundreds of billions of galaxies.

• The red shift of light from distant galaxies, a manifestation of the Doppler effect, indicates that all distant galaxies are moving away from the Earth. This observation leads to the expanding Universe theory. Most astronomers agree that this expansion began after the big bang, a cataclysmic explosion about 13.7 billion years ago.

• The first atoms (hydrogen and helium) of the Universe developed about 1 million years after the big bang. These atoms formed vast gas clouds, called nebulae.

• According to the nebular theory of planet formation, gravity caused clumps of gas in the nebulae to coalesce into revolving balls. As these balls of gas collapsed inward, they evolved into flattened disks with bulbous centers. The protostars at the center of these disks eventually became dense and hot enough that fusion reactions began in them. When this happened, they became true stars, emitting heat and light.

• Heavier elements form during fusion reactions in stars; the heaviest are probably made during supernova explosions. The Earth and the life forms on it contain elements that could only have been produced during the life cycle of stars. Thus, we are all made of stardust.

• Planets developed from the rings of gas and "dust," the planetary nebulae, that surrounded protostars. The gas condensed into planetesimals that then clumped together to form protoplanets, and finally true planets. The rocky and metallic balls in the inner part of the solar system did not acquire huge gas coatings; they became the terrestrial planets. Outer rings grew into gas-giant planets.

• The Moon formed from debris blasted free from Earth when our planet collided with a Mars-sized planet early during the history of the solar system.

• A planet the size of the Moon or larger will assume a near-spherical shape because the warm rock inside is so soft that gravity can smooth out irregularities.

KEY TERMS

accretion disk (p. 21)
asteroids (p. 29)
big bang (p. 19)
blue shift (p. 17)
comet (p. 29)
cosmology (p. 12)
differentiation (p. 28)
Doppler effect (p. 17)
expanding Universe theory
 (p. 18)
galaxies (p. 16)
gas giant planets (p. 26)
geocentric Universe concept
 (p. 13)
heliocentric Universe concept
 (p. 13)
light year (p. 16)

meteorite (p. 28)
moon (p. 23)
nebula (p. 21)
nebular theory (p. 27)
planet (p. 23)
planetesimals (p. 27)
precession (p. 14)
protoplanetary disk (p. 26)
protoplanets (p. 27)
protostar (p. 21)
proto-Sun (p. 26)
red shift (p. 17)
stellar wind (p. 22)
supernova (p. 22)
terrestrial planets (p. 26)
Universe (p. 12)

REVIEW QUESTIONS

1. Why do the planets appear to move with respect to the stars?

2. Contrast the geocentric and heliocentric Universe concepts.

3. How did Galileo's observations support the heliocentric Universe concept?

4. Describe how Foucault's pendulum demonstrates that the Earth is rotating on its axis.

5. How did Eratosthenes calculate the circumference of the Earth?

6. Describe how the parallax method can be used to estimate the distance to far objects.

7. Imagine you hear the main character in a cheap science-fiction movie say he will "return ten light years from now." What's wrong with his usage of the term "light year"? What are light years actually a measure of?

8. Describe how the Doppler effect works.

9. What does the red shift of the galaxies tell us about their motion with respect to the Earth?

10. Briefly describe the steps in the formation of the Universe and the solar system.

11. How is a supernova different from a normal star?

12. Why do the inner planets consist mostly of rock and metal, but the outer planets mostly of gas?

13. Why are all the planets in the solar system (except Pluto) orbiting the Sun in the same direction and in the same plane?

14. Describe how the Moon was formed.

15. Why is the Earth round?

SUGGESTED READING

Adams, F., and G. Laughlin. 1999. *The Five Ages of the Universe.* New York: Touchstone.

Ahrens, T. J. 1994. The origin of the Earth. *Physics Today* 47: 38–45.

Allegre, C. 1992. *From Stone to Star.* Cambridge, Mass.: Harvard University Press.

Alves, J. F., and M. J. McCaughrean, eds. 2002. *The Origins of Stars and Planets: The VLT View.* New York: Springer-Verlag.

Beatty, J. K., C. C. Petersen, and A. Chaikin. 1998. *The New Solar System,* 4th ed. Cambridge: Cambridge University Press.

Canup, R. M., and K. Righter, eds. 2000. *Origin of the Earth and Moon.* Tucson: University of Arizona Press.

Clark, S. 1995. *Towards the Edge of the Universe.* New York: Springer-Verlag.

Freedman, R. A., and W. J. Kaufmann, III. 2001. *Universe,* 6th ed. New York: Freeman.

Hawking, S. 1988. *A Brief History of Time.* New York: Bantam.

Hester, J., et al. 2002. *21st Century Astronomy.* New York: W. W. Norton.

Hogan, C. J., and M. J. Rees. 1998. *The Little Book of the Big Bang.* New York: Copernicus Books.

Hoyle, F., G. Burbidge, and J. W. Narlikar. 2000. *A Different Approach to Cosmology: From a Static Universe through the Big Bang towards Reality.* Cambridge: Cambridge University Press.

Kirshner, R. P. 2002. *The Extravagant Universe: Exploding Stars, Dark Energy, and the Accelerating Cosmos.* Princeton: Princeton University Press.

Liddle, A. 2003. *An Introduction to Modern Cosmology,* 2nd ed. New York: John Wiley & Sons.

Silk, J. 1994. *A Brief History of the Universe.* New York: Freeman.

Weinberg, S. 1993. *The First Three Minutes.* New York: Basic Books.

CHAPTER 2

Journey to the Center of the Earth

*The Earth is not a mere fragment of dead history, stratum upon stratum
like the leaves of a book . . . but living poetry like the leaves of a tree.*
—HENRY DAVID THOREAU (1817–62; FROM *WALDEN*)

*The Black Canyon of the Gunnison, in Colorado, is an 829 m (2,722 foot) gash into the
ancient rock comprising North America. The floor of the canyon almost always lies in
shadow. But despite the awesome height of its shear walls, the canyon is a mere scratch on
Earth's surface—its floor is only 0.04% of the way to our planet's center.*

2.1 INTRODUCTION

For most of human history, people perceived the other planets of our solar system to be nothing more than bright points of light that moved in relation to each other and in relation to the stars. In the seventeenth century, when Galileo first aimed his telescope skyward, the planets became hazy spheres, and through the nineteenth and early twentieth centuries, as more powerful telescopes were developed, our image of the planets continued to improve. Now that we have actually sent space probes out to investigate them, we have exquisitely detailed pictures displaying the landscapes of planetary surfaces, as well as basic data about their composition (▶Fig. 2.1). From this information, geoscientists have come up with hypotheses to explain the character of the planets.

What if we turned the tables and became explorers from outside our solar system on a visit to Earth for the first time? What would we see? Even without touching the planet, we could detect its magnetic field and atmosphere, and could characterize its surface. We could certainly distinguish regions of land, sea, and ice. We could also get an idea of the nature of Earth's interior, though we could not see the details. In the first part of this chapter, we imagine rocketing to Earth to study its external characteristics. In

FIGURE 2.1 Satellite studies emphasize that the surfaces of Mercury, Venus, Earth, and Mars differ markedly from one another. All four planets have mountains, valleys, and plains. But only Earth has distinct continents and ocean basins. Surface features provide clues to the nature of geologic processes happening inside a planet.

the second part, we build an image of Earth's interior, based on a variety of data. (Of course, no one can see the interior directly, because high pressures and temperatures would crush and melt any visitor.) This high-speed tour of Earth will provide a frame of reference for the remainder of the book.

2.2 WELCOME TO THE NEIGHBORHOOD

Our journey begins in interstellar space, somewhere between Alpha Centauri and Pluto. Compared with the air we breathe at sea level, interstellar space is a "vacuum," meaning that it contains very little matter in a given volume. In interstellar space, there is only about 1 atom in every 1 to 10 liters. To collect as many atoms as there are in a liter of air at sea level, you would have to compress a cube of interstellar space that is 8,000 to 10,000 km on a side. As we approach the solar system, we enter the Oort Cloud and then the Kuiper Belt of comets. Comets are balls of "ice" and "dust" left over from solar system formation (see Box 1.2). As we pass within the orbit of Pluto, we enter interplanetary space. Here, there are as many as 5,000 (i.e., 5×10^3) atoms per liter, still a vacuum in comparison to air at sea level, where there are 27,000,000,000,000,000,000,000 (i.e., 2.7×10^{22}) atoms per liter. Some of the atoms in interplanetary space are left over from the nebulae out of which the solar system formed, some escaped from the atmospheres of planets, and some were ejected from the Sun (see Chapter 1). The particles ejected from the Sun consist predominantly of protons (positively charged) and electrons (negatively charged) and constitute the **solar wind,** which blows across the full width of the solar system at speeds of up to 400 km/s. ("Charge" is an electrical characteristic that indicates a material's ability to repel or attract other objects.)

As our rocket approaches the Earth, its instruments detect the planet's magnetic field, like a signpost shouting, "Approaching Earth!" A **magnetic field,** in a general sense, is the region affected by the force emanating from a magnet. This force, which grows progressively stronger as you approach the magnet, can attract or repel another magnet and can cause charged particles to move. Earth's magnetic field, like the familiar magnetic field around a bar magnet, is largely a **dipole,** meaning it has a north pole and a south pole. We can portray the magnetic field by drawing **magnetic field lines,** the trajectories or paths along which magnetic particles would align, or charged particles would flow, if placed in the field (▶Fig. 2.2).

The solar wind interacts with Earth's magnetic field, distorting it into a huge teardrop pointing away from the Sun. Fortunately, the magnetic field deflects most of the wind, so that most of the particles in the wind do not reach the Earth's surface. In this way, the magnetic field

like wind separating chaff from wheat. Earth and other terrestrial planets formed from the materials left behind. As a consequence, iron (35%), oxygen (30%), silicon (15%), and magnesium (10%) make up most of Earth's mass (▶Fig. 2.10). The remaining 10% consists of the other eighty-eight naturally occurring elements.

Categories of Earth Materials

The elements making up the Earth combine to form a great variety of materials. We can organize these into several categories.

- *Organic chemicals:* Carbon-containing compounds that either occur in living organisms, or have characteristics that resemble those of living organisms, are called **organic chemicals.** Examples include oil, protein, plastic, fat, and rubber. Certain simple carbon-containing materials, such as pure carbon (C), carbon dioxide (CO_2), carbon monoxide (CO), and calcium carbonate ($CaCO_3$), are *not* considered organic.

- *Minerals:* A solid substance in which atoms are arranged in an orderly pattern is called a **mineral.** Most minerals are not organic. Minerals grow either by the cooling and freezing of a liquid or by precipitation out of a water solution. "Precipitation" occurs when atoms that had been dissolved in water come together and form a solid. For example, solid salt forms by precipitation out of seawater when the water evaporates. A single coherent sample of a mineral that grew to its present shape and has smooth, flat faces is a "crystal," while an irregularly shaped sample, or a fragment derived from a once-larger crystal or group of crystals, is a "grain."

- *Glasses:* A solid in which atoms are not arranged in an orderly pattern is called **glass.** Glass forms when a liquid freezes so fast that atoms do not have time to organize into an orderly pattern.

- *Rocks:* Aggregates of mineral crystals or grains, and masses of natural glass, are called **rocks.** Geologists recognize three main groups of rocks. (1) "Igneous rocks" develop when hot molten (melted) rock cools and freezes solid. (2) "Sedimentary rocks" form from grains that break off preexisting rock and become cemented together, or from minerals that precipitate out of a water solution; an accumulation of loose mineral grains (grains that have not stuck together) is called **sediment.** (3) "Metamorphic rocks" are created when preexisting rocks undergo changes, such as the growth of new minerals in response to heat and pressure.

- *Metals:* Solids composed of metal atoms (such as iron, aluminum, copper, and tin) are called **metals.** In a metal, outer electrons are able to flow freely (see Chapter 15). An **alloy** is a mixture containing more than one type of metal atom (e.g., bronze is a mixture of copper and tin).

- *Melts:* **Melts** form when solid materials become hot and transform into a liquid. Molten rock is a type of melt—geologists distinguish between "magma," which is molten rock beneath the Earth's surface, and "lava," molten rock that has flowed out onto the Earth's surface.

- *Volatiles:* Materials that easily transform into a gas at the relatively low temperatures found at the Earth's surface are called **volatiles.**

Notably, the most common minerals in the Earth contain silicon and oxygen mixed in varying proportions with other elements (typically iron, magnesium, aluminum, calcium, potassium, and sodium). These are called **silicate minerals,** and, no surprise, rocks composed of silicate minerals are silicate rocks. Geologists distinguish four classes of igneous silicate rocks based, in essence, on the proportion of silicon to iron and magnesium. In order, from greatest to least proportion of silicon to iron and magnesium, these classes are *felsic* (or *silicic*), *intermediate, mafic,* and *ultramafic.* As the proportion of silicon in a rock increases, the density (mass per unit volume) decreases. Thus, felsic rocks are less dense than mafic rocks.

Within each class, there are many different rock types, each with a name, that differ from one another in terms of composition (chemical makeup) and crystal size. These will be discussed in detail in Part II, but for now we need to recognize only four rock names: **granite** (a felsic rock with large grains), **basalt** (a mafic rock with small grains), **gabbro** (a mafic rock with large grains), and **peridotite** (an ultramafic rock with large grains).

FIGURE 2.10 The proportions of major elements making up the mass of the whole Earth. Note that iron and oxygen account for most of the mass.

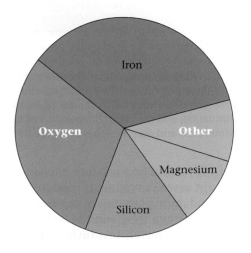

2.6 DISCOVERING THAT THE EARTH HAS LAYERS

The world's deepest mine shaft penetrates gold-bearing rock that lies about 3.5 km (2 miles) beneath South Africa. Though miners seeking this gold must begin their workday by plummeting straight down a vertical shaft for almost ten minutes aboard the world's fastest elevator, the shaft is little more than a pinprick on Earth's surface when compared with the planet's radius (the distance from the center to the surface, 6,371 km). Even the deepest well ever drilled, a 12-km-deep hole in northern Russia, penetrates only the upper 0.03% of the Earth. We literally live on the thin skin of our planet, its interior forever inaccessible to our wanderings.

People have wondered about the Earth's interior since ancient times. What is the source of incandescent lavas spewed from volcanoes, of precious gems and metals, of sparkling spring water, and of the mysterious powers that shake the ground and topple buildings? Without the ability to observe the Earth's interior firsthand, pre-twentieth-century authors dreamed up fanciful images of it. For example, the English poet John Milton (1608–1674) described the underworld as a "dungeon horrible, on all sides round, as one great furnace flamed" (▶Fig. 2.11). Perhaps his image was inspired by volcanoes in the Mediterranean. In the eighteenth and nineteenth centuries, some European writers thought that the Earth's interior resembled a sponge, containing open caverns variously filled with molten rock, water, or air. In this way, the interior could provide both the water that bubbled up at springs *and* the lava that erupted at volcanoes. In fact, in French author Jules Verne's popular 1864 novel *Journey to the Center of the Earth*, three explorers find a route through interconnected caverns to the Earth's center. Our present image of the Earth's interior, one made up of distinct layers, is the end product of many discoveries made during the past two hundred years. The discoveries were made with the help of certain clues.

Clues from Measuring Earth's Density

The first key to understanding the Earth's interior came from studies that provided an estimate of the planet's density (mass per unit volume). To determine Earth's density, one must first determine Earth's mass (the amount of matter making up the Earth). In 1776, the British Royal Astronomer, Nevil Maskelyne provided the first realistic estimate of Earth's mass. Maskelyne postulated that he could weigh the Earth by examining the deflection of a plumb bob attached to a surveying instrument. (A plumb bob is a weight at the end of a string used to determine the orientation of a vertical line.) The angle of deflection (ß) of the plumb bob caused by the gravitational attraction of a mountain indicates the magnitude of gravitational attraction exerted by the mountain's mass relative to the gravitational attraction of the Earth's mass. Maskelyne tested his hypothesis at Schiehallion Mountain in Scotland (▶Fig. 2.12). His results led to an estimate that the Earth's average density is 4.5 times the density of water (i.e., 4.5g/cm³ in the modern metric system). In 1778, another physicist using a different method arrived at a density estimate of

FIGURE 2.11 A literary image of the Earth's insides: *The Fallen Angels Entering Pandemonium, from* [Milton's] *"Paradise Lost," Book 1,* by English painter John Martin (1789–1854).

FIGURE 2.12 A surveyor noticed that the plumb line he was using to level his surveying instrument did not hang exactly vertically near a mountain; it was deflected by an angle ß, owing to the gravitational attraction of the mountain. The angle of deflection represents the ratio between the mass of the mountain and the mass of the whole Earth.

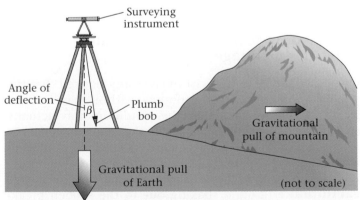

5.45 g/cm³ (more than 5 times the density of water), fairly close to modern estimates. Significantly, typical rocks (like granite and basalt) at the surface of the Earth have a density of only 2.2–2.5 g/cm³, so the average density of Earth exceeds that of its surface rocks. Certainly, the open voids that Jules Verne described could not exist!

Clues from Measuring Earth's Shape

Once they had determined that the density of the Earth's interior was greater than that of its surface rocks, nineteenth-century scientists asked, "Does this density increase *gradually* with depth, or does the Earth consist of a less dense shell surrounding a much denser core?" If the Earth's density increased only gradually with depth, most of its mass would lie far from the center, and the planet's spin would cause the Earth to flatten into a disk. Since this doesn't happen, scientists concluded that much of Earth's mass must be concentrated close to the center. The dense center came to be known as the *core*. Eventually, geologists determined that the density of this core approaches 13 g/cm³.

To further understand the nature of Earth's interior, geologists measured tides, the rise and fall of the Earth's surface in response to the gravitational attraction of the Moon and Sun. If the Earth were composed of a liquid surrounded by only a thin solid crust, then the surface of the land would rise and fall daily, like the surface of the sea. We don't observe such behavior, so the Earth's interior must be largely solid. Thus, the image perpetuated by the popular press that the outer skin of the Earth floats on a "sea" of molten rock is simply not true.

By the end of the nineteenth century, geologists had recognized that the Earth resembled a hard-boiled egg, in that it had three principal layers: a not-so-dense **crust** (like an eggshell; composed of rocks like granite, basalt, and gabbro), a denser solid **mantle** in between (the white; composed of a then-unknown material), and a very dense **core** (the yolk; also composed of a then-unknown material). Clearly, many questions remained. How thick are the layers? Are the boundaries between layers sharp or gradational? And what exactly are the layers composed of?

Clues from the Study of Earthquakes: Refining the Image

One day in 1889, a physicist in Germany noticed that the pendulum in his lab began to move without having been touched. He reasoned that the pendulum was actually standing still, because of its *inertia* (the tendency of an object at rest to remain at rest, and of an object in motion to remain in motion), and that the Earth was moving under it. A few days later, he read in a newspaper that a large **earthquake** (ground shaking due to the sudden break-

ing of rocks in the Earth) had taken place in Japan minutes before the movement of his pendulum began. The physicist deduced that vibrations due to the earthquake had traveled through the Earth from Japan and had juggled his laboratory in Germany. The energy in such vibrations moves in the form of waves, called either **seismic waves** or "earthquake waves," that resemble the shock waves you feel with your hands when you snap a stick (▶Fig. 2.13). The breaking of rock during an earthquake either produces a new fracture on which sliding occurs or causes sliding on a preexisting fracture. A fracture on which sliding occurs is called a **fault.**

Geologists immediately realized that the study of seismic waves traveling through the Earth might provide a tool for exploring the Earth's insides (much like doctors today use ultrasound to study a patient's insides). Specifically, laboratory measurements demonstrated that earthquake waves travel at different velocities (speeds) through different materials. Thus, by detecting depths at which velocities suddenly change, geoscientists pinpointed the boundaries between layers and even recognized subtler boundaries within layers. (We'll explain how in Interlude C, after we've had a chance to describe earthquakes in more detail.)

FIGURE 2.13 When the rock inside the Earth suddenly breaks and slips, forming a fracture called a fault, it generates shock waves that pass through the Earth and shake the surface (creating an earthquake), much as the sound waves from a stick snapping travel to you and make your eardrum vibrate.

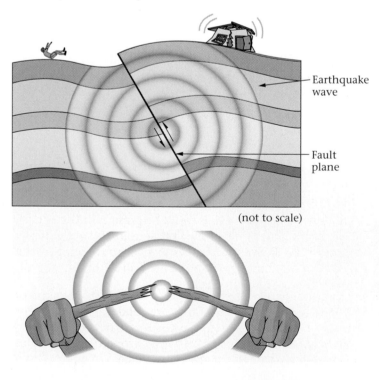

Earthquake wave

Fault plane

(not to scale)

Pressure and Temperature Inside the Earth

In order to keep underground tunnels from collapsing under the pressure created by the weight of overlying rock, mining engineers must design sturdy support structures. It is no surprise that deeper tunnels require stronger supports: the downward push from the weight of overlying rock increases with depth, simply because the mass of the overlying rock layer increases with depth. At the Earth's center, pressure probably reaches about 3,600,000 atm.

Temperature also increases with depth in the Earth. Even on a cool winter's day, miners who chisel away at gold veins exposed in tunnels 3.5 km below the surface swelter in temperatures of about 53°C (127°F). We refer to the rate of change in temperature with depth as the **geothermal gradient.** In the upper part of the crust, the geothermal gradient averages between 15° and 50°C per km. At greater depths, the rate decreases (to 10°C per km or less). Thus, 35 km below the surface of a continent, the temperature reaches "only" 400° to 700°C. No one has ever directly measured the temperature at the Earth's center, but recent calculations suggest it may reach 4,700°C, only about 800° less than the temperature at the surface of the Sun.

2.7 WHAT ARE THE LAYERS MADE OF?

We saw earlier that the material composing the Earth's insides must be much denser than familiar surface rocks like granite and basalt. To discover what this material was, geologists

- conducted laboratory experiments to determine what kinds of materials inside the Earth could serve as a source for magma;

- studied unusual chunks of rock that may have been carried up from the mantle in magma;

- conducted laboratory experiments to measure densities in samples of known rock types, so that they could compare these with observed densities in the Earth; and

- estimated which elements would be present in the Earth if the Earth had formed out of planetesimals similar in composition to **meteorites** (chunks of rock and/or metal alloy that fell from space and landed on Earth; Box 2.1).

As a result of this work, we now have a pretty clear sense of what the layers inside the Earth are made of, though this picture is constantly being adjusted when new findings become available. Let's now look at the properties of individual layers, starting with the Earth's surface.

The Crust

When you stand on the surface of the Earth, you are standing on the top of its outermost layer, the crust. The crust is our home and the source of all our resources. How thick is this all-important layer? Or, in other words, what is the depth to the crust-mantle boundary? An answer came from the studies of Andrija Mohorovičić, a researcher working in Zagreb, Croatia. In 1909, Mohorovičić discovered that the velocity of earthquake waves suddenly increased at a depth of about 50 km beneath the Earth's surface, and he suggested that this increase was caused by an abrupt change in the properties of rock (see Interlude C for further detail). Later studies showed that this change can be found most everywhere around our planet, though it actually occurs at different depths in different locations—it's deeper beneath continents than beneath oceans. Geologists now consider the change to be the crust-mantle boundary, and they refer to it as the **Moho** in Mohorovičić's honor. The relatively shallow depth of the Moho (7–70 km, depending on location) as compared to the radius of the Earth (6,371 km) emphasizes that the crust is very thin indeed. The crust is only about 0.1% to 1.0% of the Earth's radius, so if the Earth were the size of a balloon, the crust would be about the thickness of the balloon's skin.

Geologists distinguish between two fundamentally different types of crust—"oceanic crust," which underlies the sea floor, and "continental crust," which underlies continents (►Fig. 2.14a). The crust is not simply cooled mantle, like the skin on chocolate pudding, but rather consists of a variety of rocks that differ in composition (chemical makeup) from mantle rock.

Oceanic crust is only 7 to 10 km thick. At highway speeds (100 km per hour), you could drive a distance equal to the thickness of the oceanic crust in about five minutes. (It would take sixty-three hours, driving nonstop, to reach the Earth's center.) We have a good idea of what oceanic crust looks like in cross section, because geologists have succeeded in drilling down through its top few kilometers and have found places where slices of oceanic crust, known as "ophiolites," have been incorporated in mountains and therefore have been exposed on dry land. Studies of such examples show that oceanic crust consists of fairly uniform layers. At the top, we find a blanket of sediment, generally less than 1 km thick, composed of clay and tiny shells that have settled like snow. Beneath this blanket, the oceanic crust consists of a layer of basalt and, below that, a layer of gabbro.

Most continental crust is about 35 to 40 km thick—about four to five times the thickness of oceanic crust—but its thickness varies much more than does oceanic crust. Plate motions can cause continents to stretch like taffy in narrow bands called rifts, where crust can become only 25 km thick, or to squash and thicken in mountain belts, where crust can

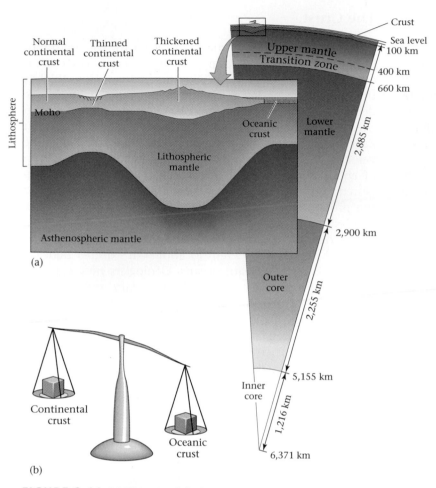

FIGURE 2.14 (a) This simplified cross section illustrates the differences between continental crust and oceanic crust. Note that the thickness of continental crust can vary greatly. (b) Oceanic crust is denser than continental crust.

become 70 km thick. In contrast to oceanic crust, continental crust contains a great variety of rock types, ranging from mafic to felsic in composition. On average, continental crust is less mafic than oceanic crust—it has a felsic (granite-like) to intermediate composition—so a block of average continental crust weighs less than a same-size block of oceanic crust (▶Fig. 2.14b).

Geologists have been able to calculate the overall chemical composition of the crust (▶Fig. 2.15). A glance at Figure 2.15 shows that regardless of whether you consider percentage by weight, percentage by volume, or percentage of atoms, *oxygen is by far the most abundant element in the crust!* This observation may surprise you, because most people picture oxygen as the colorless gas that we inhale when we breathe the atmosphere, not as a rock-forming chemical. But oxygen, when bonded to other elements, forms a great variety of minerals, and these minerals in turn form the bulk of the rock in the Earth's crust. Because oxygen atoms are relatively large in comparison with their mass, oxygen actually occupies about

93% of the crust's volume. If you compare the composition of the crust to that of the whole Earth (see Fig. 2.10), you'll notice that the composition of the crust differs markedly from that of the whole Earth. That's because the composition of the entire Earth takes into account the core and mantle, which (as we discuss next) do not have the same composition as the crust.

Finally, it is important to note that most rock in the crust contains pores (tiny open spaces), and in much of the upper several kilometers of the crust the pores are filled with liquid water. This subsurface water, or groundwater, provides the water that farmers pump out of wells for irrigation and that cities pump out for their water supplies. In places, the pores contain oil or gas.

The Mantle

The mantle of the Earth forms a 2,885-km-thick layer surrounding the core. In terms of volume, it is the largest part of the Earth. In contrast to the crust, the mantle consists entirely of an ultramafic rock called peridotite. This means that peridotite, though rare at the Earth's surface, is actually the most abundant rock in our planet! Overall, density in the mantle increases from about 3.5 g/cm^3 at the top to about 5.5 g/cm^3 at the base. On the basis of the occurrence of changes in the velocity of earthquake waves, geoscientists divide the mantle into three sublayers: the **upper mantle,** down to a depth of 400 km, the **transition zone,** from there down to a depth of 660 km, and the **lower mantle,** from there down to the core-mantle boundary.

Almost all of the mantle is solid rock. But even though it's solid, mantle rock below a depth of 100 to 150 km is so hot that it's soft enough to flow extremely slowly—at a rate of less than 15 centimeters a year. "Soft" here does not mean liquid, it simply means that over long periods of time mantle rock can change shape, like soft wax, without breaking. Note that we said *almost* all of the mantle is solid—in fact, up to a few percent of the mantle has melted in a layer that lies at depths of between 100 and 200 km beneath the ocean floor. This melt causes seismic waves to slow down, so geologists refer to this partly molten layer as the "low-velocity zone."

Though overall the temperature of the mantle increases with depth, it varies significantly with location even at the same depth. The warmer regions are less dense, while the cooler regions are denser. The blotchy pattern of warmer and cooler mantle indicates that the mantle convects like water in a simmering pot. Warm mantle gradually flows upward, while cooler, denser mantle sinks.

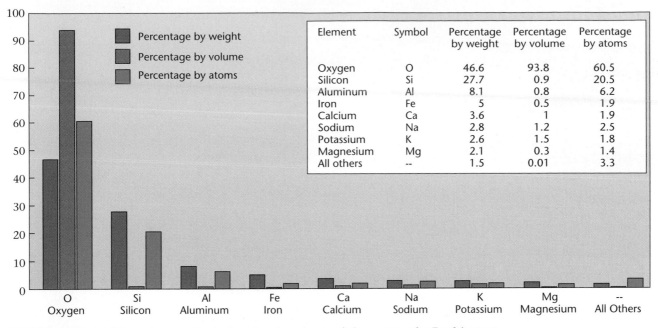

Element	Symbol	Percentage by weight	Percentage by volume	Percentage by atoms
Oxygen	O	46.6	93.8	60.5
Silicon	Si	27.7	0.9	20.5
Aluminum	Al	8.1	0.8	6.2
Iron	Fe	5	0.5	1.9
Calcium	Ca	3.6	1	1.9
Sodium	Na	2.8	1.2	2.5
Potassium	K	2.6	1.5	1.8
Magnesium	Mg	2.1	0.3	1.4
All others	--	1.5	0.01	3.3

FIGURE 2.15 A table and graph illustrating the abundance of elements in the Earth's crust.

The Core

Early calculations suggested that the core had the same density as gold, so for many years people held the fanciful hope that vast riches lay at the heart of our planet. Alas, geologists eventually concluded that the core consists of a far less glamorous material, iron alloy (iron mixed with lesser amounts of other elements). They arrived at this conclusion, in part, by comparing the properties of the core with the properties of metallic (iron) meteorites (see Box 2.1).

Studies of how earthquake waves bend as they pass through the Earth, along with the discovery that certain types of seismic waves cannot pass through the outer part of the core (see Interlude C), led geoscientists to divide the core into two parts, the **outer core** (between 2,900 and 5,155 km deep) and the **inner core** (from a depth of 5,155 km down to the Earth's center at 6,371 km). The outer core is a *liquid* iron alloy (composed of iron, nickel, and some other elements including silicon [Si], oxygen [O], and sulphur [S]) with a density of 10 to 12 g/cm^3. It can exist as a liquid because the temperature in the outer core is so high that even the great pressures squeezing the region cannot lock atoms into a solid framework. Because it is a liquid, the iron alloy of the outer core can flow (see art pp. 44–45); this flow generates Earth's magnetic field.

The inner core, with a radius of about 1,220 km and a density of 13 g/cm^3, is a *solid* iron-nickel alloy, which may reach a temperature of over 4,700°C. Even though it is hotter than the outer core, the inner core is a solid because it is

deeper and subjected to even greater pressure. The pressure keeps atoms from wandering freely, so they pack together tightly in very dense materials. The inner core probably grows through time, at the expense of the outer core, as the Earth slowly cools and the lower part of the outer core solidifies. Recent data suggests the inner core rotates slightly faster than the rest of the Earth because of the force applied to it by the Earth's magnetic field.

2.8 THE LITHOSPHERE AND ASTHENOSPHERE

So far, we have divided the insides of the Earth into layers (crust, mantle, and core) based on the velocity at which earthquake waves travel through the layers. The three major layers (crust, mantle, and core) differ compositionally from each other. An alternate way of thinking about Earth layers comes from studying the degree to which the material making up a layer can flow. In this context we distinguish between "rigid" materials, which can bend but cannot flow, and "plastic" materials, which are relatively soft and can flow.

Let's apply this concept to the outer portion of the Earth's shell. Geologists have determined that the outer 100 to 150 km of the Earth is relatively rigid; in other words, the Earth has an outer shell composed of rock that cannot flow easily. This outer layer is called the **lithosphere,** and it consists of the crust plus the uppermost part of the mantle. We refer to

The Earth, from Surface to Center

If we could remove all the air that hides much of the solid surface from view, we would see that both the land areas and the sea floor have plains and mountains.

If we could break open the Earth, we would see that its interior consists of a series of concentric layers, called (in order from the surface to the center) the crust, the mantle, and the core. The crust is a relatively thin skin (7–10 km beneath oceans, 25–70 km beneath the land surface). Oceanic crust consists of basalt (mafic rock), while the average continental crust is intermediate to silicic. The mantle, which overall has the composition of ultramafic rock, can be divided into three layers: upper mantle, transition zone, and lower mantle. The core can be divided into an outer core of liquid iron alloy and an inner core of solid iron alloy. Temperature increases progressively with depth, so at the Earth's center the temperature may approach that of the Sun's surface.

Within the mantle and outer core, there is swirling, convective flow. Flow within the outer core generates the Earth's magnetic field.

When discussing plate tectonics, it is convenient to call the outer part of the Earth, a relatively rigid shell composed of the crust and uppermost mantle, the lithosphere and the underlying warmer, more plastic portion of the mantle the asthenosphere. These are not shown in this painting.

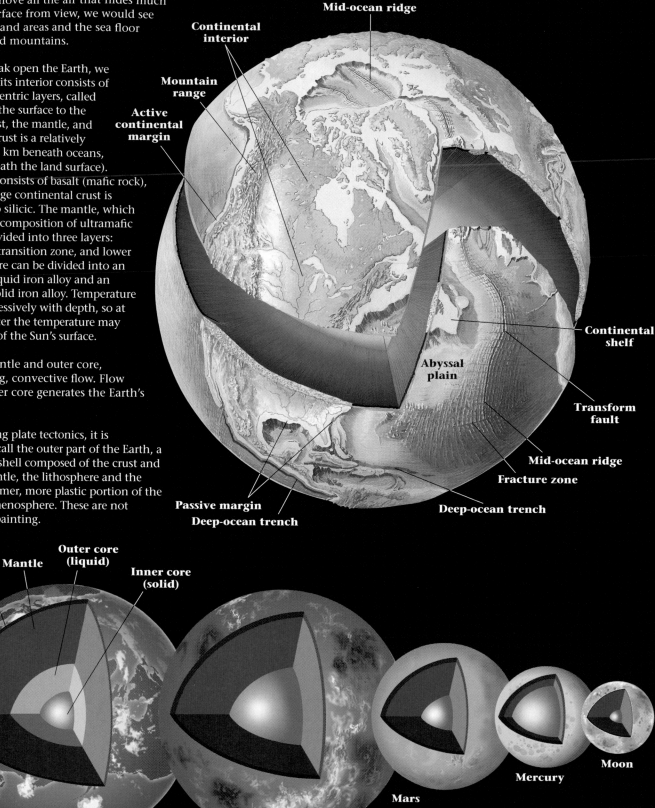

Mid-ocean ridge

Continental interior

Mountain range

Active continental margin

Continental shelf

Abyssal plain

Transform fault

Mid-ocean ridge

Fracture zone

Passive margin

Deep-ocean trench

Deep-ocean trench

Crust

Mantle

Outer core (liquid)

Inner core (solid)

Earth

Venus

Mars

Mercury

Moon

2,000 km

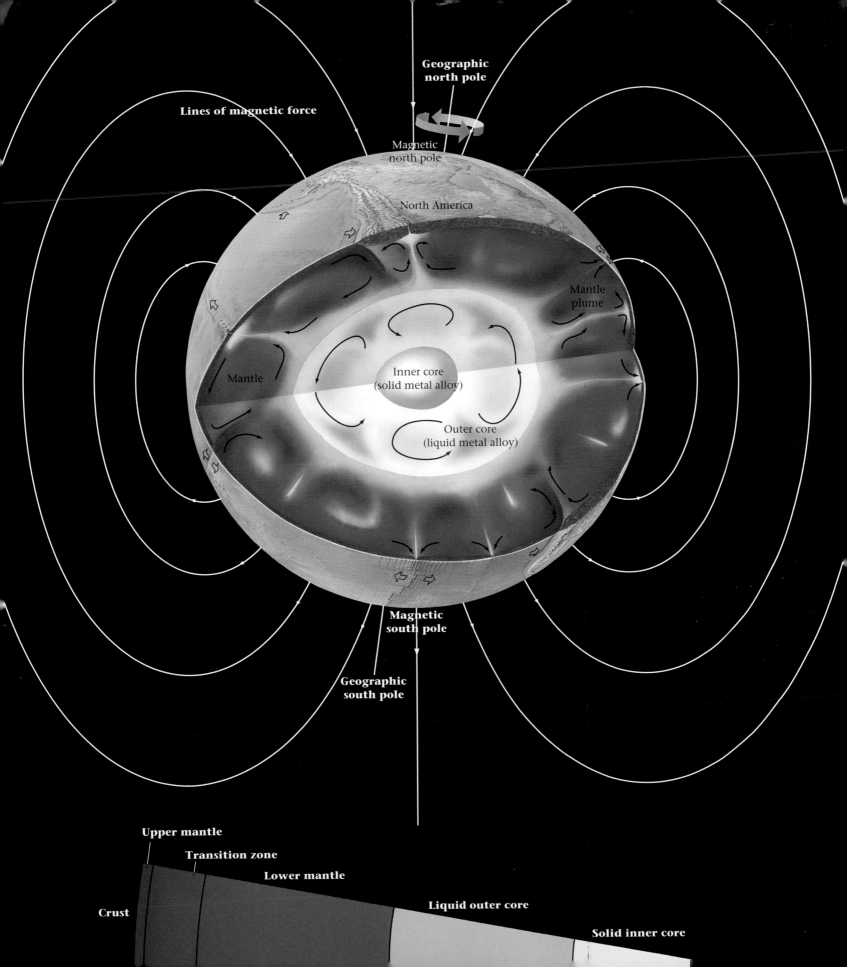

Geographic
north pole

Lines of magnetic force

Magnetic
north pole

North America

Mantle
plume

Mantle

Inner core
(solid metal alloy)

Outer core
(liquid metal alloy)

Magnetic
south pole

Geographic
south pole

Upper mantle

Transition zone

Lower mantle

Crust

Liquid outer core

Solid inner core

the portion of the mantle within the lithosphere as the *lithospheric mantle*. Note that the terms "lithosphere" and "crust" are not synonymous—the crust is just part of the lithosphere. The lithosphere lies on top of the **asthenosphere,** which is the portion of the mantle in which rock can flow. Notice that the asthenosphere is entirely in the mantle and lies below a depth of 100–150 km. We can't assign a specific depth to the base of the asthenosphere because all of the mantle below 150 km can flow, but for convenience, some geologists place the base of the asthenosphere at the upper mantle/transition zone boundary. One final point: even though the asthenosphere can flow, do not think of it as a liquid. It is not. Rather, the asthenosphere is largely solid. (A small amount of melt occurs in the low-velocity zone.) At its fastest, the asthenosphere flows at rates of 10–15 cm/year.

Oceanic lithosphere and continental lithosphere are somewhat different (▶Fig. 2.16). Oceanic lithosphere, topped by oceanic crust, generally has a thickness of about 100 km. In contrast, continental lithosphere, topped by continental crust, generally has a thickness of about 150 km.

The boundary between the lithosphere and asthenosphere occurs where the temperature is about 1,280°C, for at this temperature mantle rock becomes soft enough to flow. To see how temperature affects the ability of a material to flow, take a cube of candle wax and place it in the freezer. The wax becomes very rigid and can maintain its shape for long periods of time; in fact, if you were to drop the cold wax, it would shatter. But if you take another block of wax and place it in a warm (not hot) oven, it becomes soft, so that you can easily mold it into another shape. In fact, the force of gravity alone may cause the warm wax to flow and assume the shape of a pancake. Rock behaves somewhat similarly to the wax blocks. When rock is cool, it is quite rigid, but at high temperatures, rocks become soft and can flow, though much more slowly than wax. This ability to flow slowly can occur at a temperature much lower than is

---

BOX 2.1
THE REST OF THE STORY

Meteors and Meteorites

During the early days of the solar system, the Earth collided with and incorporated almost all of the planetesimals and smaller fragments of solid material lying in its path. Intense bombardment ceased about 4.0 Ga, but even today collisions with space objects continues, and over 1,000 tons of material (rock, metal, "dust," and "ice") fall to Earth, on average, every year. The vast majority of this material consists of fragments derived from comets and asteroids (see Box 1.2) sent careening into the path of the Earth after billiard ball–like collisions with one another out in space, or because of the gravitational pull of a passing planet. Some of the material, however, consists of chips of the Moon or Mars, ejected into space when large objects collided with these bodies.

Astronomers refer to any object from space that enters the Earth's atmosphere as a **meteoroid**. Meteoroids move at speeds of up to 75 km/s, so fast that when they reach an altitude of about 150 km, friction with the atmosphere causes them to begin to evaporate, leaving a streak of bright, glowing gas. The glowing streak, and atmospheric phenomenon, is a **meteor** (also known colloquially, though incorrectly, as a "falling star"). Most visible meteoroids completely evaporate by an altitude of about 30 km. But "dust"-sized ones may slow down sufficiently to float to Earth, and larger ones (fist-sized or bigger) can survive the heat of entry to reach the surface of the planet. Objects that strike the Earth are called meteorites. Most are asteroidal or planetary fragments, for the "icy" material of small cometary bodies is too fragile to survive the fall. In some cases, the meteoroids explode in brilliant fireballs; such particularly luminous objects are called bolides. In fact, the explosion of a cometary bolide over Siberia in 1908 flattened a region of forest with a force equivalent to that of 2,000 atomic bombs.

Scientists did not realize that meteors were the result of solid objects falling from space until 1803, when a spectacular meteor shower (the occurrence of a large number of meteors during a short time) lit the sky over Normandy, France, and over 3,000 meteorites were subsequently found on the ground. In the succeeding two centuries, many meteorites have been collected and studied in detail. Based on this work, researchers recognize three basic classes of meteorites: iron (made of iron-nickel alloy), stony (made of silicate rock), and stony iron (rock embedded in a matrix of metal). Of all known meteorites, about 93% are stony and 6% are iron. Based on their composition, researchers have concluded that some meteors (a special subcategory of stony meteorites called carbonaceous chondrites, because they contain carbon and small spherical nodules called chondrules) are asteroids derived from planetesimals that never underwent differentiation into a core and mantle. Others (other stony meteorites and all iron meteorites) are asteroids derived from planetesimals that differentiated into a metallic core and a rocky mantle early in solar system history but later shattered into fragments during collisions with other planetesimals. Most meteorites have yielded radiometric dates of 4.54 Ga, but carbonaceous chondrites are as old as 4.56 Ga, the oldest known material ever measured. Meteorites are important to geologists because they represent matter from the earliest days of the solar system. Rocks that formed on Earth, in contrast, melted or were otherwise modified directly or indirectly as a result of plate tectonics.

Although almost all meteors are small and have not caused notable damage on earth in historic time, a very few have smashed through houses, dented cars, and bruised people. During the longer term of Earth history, however, there have been some catastrophic collisions that left huge craters (Fig. 2.7a, b). As we will see later in the book, the largest collisions probably caused mass extinctions of life forms on our planet. It is likely that such collisions may happen again in the future—a large asteroid, for example, passed within 3 million miles of Earth in 2004.

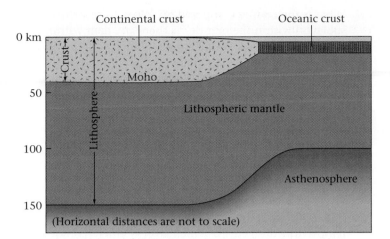

FIGURE 2.16 A cross section of the lithosphere, emphasizing the difference between continental and oceanic lithosphere.

necessary to cause rocks to melt. Rock of the lithosphere is cool enough to behave rigidly, whereas rock of the asthenosphere is warm enough to flow easily.

Now, with an understanding of Earth's overall architecture at hand, we can discuss geology's grand unifying theory—plate tectonics. The next two chapters introduce this key topic.

CHAPTER SUMMARY

• The Earth has a magnetic field, which shields it from solar wind. Closer to Earth, the field creates the Van Allen belts, which trap cosmic rays.

• A layer of gas surrounds the Earth. This atmosphere (78% nitrogen, 21% oxygen, 1% other) can be subdivided into distinct layers. All weather occurs in the troposphere, the layer we live in. Air pressure decreases with elevation, so 50% of the gas in the atmosphere resides below 5.5 km.

• The surface of the Earth can be divided into land (30%) and ocean (70%). Most of the land surface lies within 1 km of sea level.

• The Earth consists of organic chemicals, minerals, glasses, rocks, metals, melts, and volatiles. Most rocks on Earth contain silica (SiO_2) and thus are called silicate rocks. We distinguish between felsic, intermediate, mafic, and ultramafic rocks based on the proportion of silica.

• The Earth's interior can be divided into three compositionally distinct layers, named in sequence from the surface down: the crust, the mantle, and the core. The first recognition of this division came from studying the density and shape of the Earth.

• Pressure and temperature both increase with depth in the Earth. At the center, pressure is 3.6 million times greater than at the surface, and the temperature reaches over 4,700°C. The rate at which temperature increases as depth increases is the geothermal gradient.

• Studies of seismic waves have revealed the existence of sublayers in the core (outer core and inner core) and mantle (upper mantle, transition zone, and lower mantle).

• The crust is a thin skin that varies in thickness from 7–10 km (beneath the oceans) to 25–70 km (beneath the continents). Oceanic crust is mafic in composition, while average continental crust is felsic to intermediate. The mantle is composed of ultramafic rock. The core consists of iron alloy and consists of two parts—the outer core is liquid, and the inner core is solid. Flow in the outer core generates the magnetic field.

• The crust plus the upper part of the mantle constitute the lithosphere, a relatively rigid shell. The lithosphere lies over the asthenosphere, mantle that is capable of flowing.

KEY TERMS

alloy (p. 38)
asthenosphere (p. 46)
atmosphere (p. 34)
aurorae (p. 34)
basalt (p. 38)
core (p. 40)
crust (p. 40)
dipole (p. 33)
earthquake (p. 40)
fault (p. 40)
gabbro (p. 38)
geothermal gradient (p. 41)
glass (p. 38)
granite (p. 38)
groundwater (p. 36)
hydrosphere (p. 36)
inner core (p. 43)
lithosphere (p. 43)
lower mantle (p. 42)
magnetic field (p. 33)
magnetic field lines (p. 33)

mantle (p. 40)
melts (p. 38)
metals (p. 38)
meteor (p. 46)
meteorites (p. 41)
meteoroid (p. 46)
mineral (p. 38)
Moho (p. 41)
organic chemicals (p. 38)
outer core (p. 43)
periodotite (p. 38)
rocks (p. 38)
sediment (p. 38)
seismic waves (p. 40)
silicate minerals (p. 38)
solar wind (p. 33)
surface water (p. 36)
topography (p. 37)
transition zone (p. 42)
upper mantle (p. 42)
volatiles (p. 38)

REVIEW QUESTIONS

1. Why do astronomers consider the space between planets to be a vacuum, in comparison to the atmosphere near sea level?

2. What is the Earth's magnetic field? Draw a representation of the field on a piece of paper; your sketch should

illustrate the direction in which charged particles would flow if placed in the field.

3. How does the magnetic field interact with solar wind? Be sure to consider the magnetosphere, the Van Allen radiation belts, and the aurorae.

4. What is Earth's atmosphere composed of, and how does it differ from the atmospheres of Venus and Mars? Why would you die of suffocation if you were to eject from a fighter plane at an elevation of 12 km without taking an oxygen tank with you?

5. What is the proportion of land area to sea area on Earth? Based on studying the hypsometric curve, approximately what proportion of the Earth's surface lies at elevations above 2 km?

6. What are the two most abundant elements in the Earth? Describe the major categories of materials constituting the Earth.

7. What are silicate rocks? Give four examples of such rocks, and explain how they differ from one another in terms of their component minerals.

8. How did researchers first obtain a realistic estimate of Earth's average density? Based on this result, did they conclude that the inside of the Earth is denser than or less dense than rocks exposed at the surface? What observations led to the realization that the Earth is largely solid and that the Earth's mass is largely concentrated toward the center?

9. What are earthquake waves? Does the velocity at which an earthquake wave travels change or stay constant as the wave passes through the Earth? What are the principal layers of the Earth? What happens to earthquake waves when they reach the boundary between layers?

10. How do temperature and pressure change with increasing depth in the Earth? Be sure to explain the geothermal gradient.

11. What is the Moho, and how was it first recognized? Describe the differences between continental crust and oceanic crust. Approximately what percentage of the Earth's diameter is within the crust?

12. What is the mantle composed of? What are the three sublayers within the mantle? Is there any melt within the mantle?

13. What is the core composed of? How do the inner core and outer core differ from each other? We can't sample the core directly, but geologists have studied samples of material that are probably very similar in composition to the core. Where do these samples come from?

14. What is the difference among a meteoroid, a meteor, and a meteorite? Are all meteorites composed of the same material? Explain your answer.

15. What is the difference between lithosphere and asthenosphere? Be sure to consider material differences and temperature differences. Which layer is softer and flows easily? At what depth does the lithosphere/asthenosphere boundary occur? Is this above or below the Moho?

SUGGESTED READING

Bolt, B. A. 1982. *Inside the Earth*. San Francisco: Freeman.

Brown, G. C., and A. E. Mussett. 1993. *The Inaccessible Earth*. London: Chapman and Hall.

Ernst, W. G. 1990. *The Dynamic Planet*. New York: Columbia University Press.

Helffrich, G. R., and B. J. Wood. 2001. The Earth's mantle. *Nature* 412: 501–507.

Karato, S. I. 2003. *Rheology and Dynamics of Earth's Interior*. Princeton: Princeton University Press.

Stein, S., and M. Wysession. 2003. *An Introduction to Seismology, Earthquakes, and Earth Structure*. London: Blackwell.

Wysession, M. 1995. The inner workings of the Earth. *American Scientist* 83(2): 134–46.

CHAPTER 3

Drifting Continents and Spreading Seas

3.1 INTRODUCTION

In September 1930, fifteen explorers led by a German meteorologist, Alfred Wegener, set out across the endless snowfields of Greenland to resupply two weather observers stranded at a remote camp (▶Fig. 3.1). The observers were planning to spend the long polar night recording wind speeds and temperatures on Greenland's polar plateau. At the time, Wegener was well known, not only to researchers studying climate but also to geologists. Some fifteen years earlier, he had published a small book, *The Origin of the Continents and Oceans,* in which he had dared to challenge geologists' long-held assumption that the continents had remained fixed in position through geologic time (the time since the formation of the Earth). Wegener proposed, instead, that the present distribution of continents and ocean basins had evolved. According to Wegener, the continents had once fit together like pieces of a giant jigsaw puzzle, to make one vast supercontinent. He suggested that this supercontinent, which he named **Pangaea** (pronounced Pan-jee-ah; Greek for "all land"), later fragmented into separate continents that then drifted apart, moving slowly to their present positions (▶Fig. 3.2). This idea came to be known as the **continental drift** hypothesis.

A fossil leaf of Glossopteris *from an exposure in Antarctica. The presence of this fossil on many continents was one of the observations that led to the proposal of continental drift.*

FIGURE 3.1 Alfred Wegener, the German meteorologist who proposed a comprehensive model of continental drift and presented geologic evidence in support of the idea.

Wegener presented many observations in favor of the hypothesis, but he met with strong resistance. Drifting continents? Absurd! Or so proclaimed the leading geologists of the day. At a widely publicized 1926 geology conference in New York City, a phalanx of celebrated American professors scoffed: "What force could possibly be great enough to move the immense mass of a continent?" Wegener's writings didn't provide a good answer, so despite all the supporting observations he had provided, most of the meeting's participants rejected continental drift.

Now, four years later, Wegener faced his greatest challenge. As he headed into the interior of Greenland, the weather worsened and most of his party turned back. But

Wegener felt he could not abandon the isolated observers, and with two companions he trudged forward. On October 30, 1930, Wegener reached the observers and dropped off enough supplies to last the winter. Wegener and one companion set out on the return trip the next day, but they never made it home.

Had Wegener survived to old age, he would have seen his hypothesis become the foundation of a scientific revolution. Today, geologists accept Wegener's ideas and take it for granted that the map of the Earth constantly changes; continents indeed waltz around its surface, variously combining and breaking apart through geologic time. The revolution began in 1960, when Harry Hess, a Princeton University professor, proposed that continents drift apart because new ocean floor forms between them by a process that his contemporary Robert Dietz also described and labeled **sea-floor spreading,** and that continents move toward each other when the old ocean floor between them sinks back down into the Earth's interior, a process now called **subduction.** By 1968, geologists had developed a fairly complete model describing continental drift, sea-floor spreading, and subduction. In this model, Earth's lithosphere, its outer, relatively rigid shell, consists of about twenty distinct pieces, or **plates,** that slowly move relative to one another. Because we can confirm this model by many observations, it has gained the status of a theory (see Prelude), which we now call the theory of **plate tectonics,** from the Greek word *tekton,* which means "builder"; plate movements "build" regional geologic features. Geologists view plate tectonics as the grand unifying theory of geology, because it can successfully explain a great many geologic phenomena, as we will see.

In this chapter, we learn about the observations that led Wegener to propose his continental drift hypothesis. Then we look at paleomagnetism, the record of Earth's magnetic field in the past, which provides a key proof of continental drift. Finally, we learn how observations about the sea floor made by geologists during the mid-twentieth century led Harry Hess to propose the concept of sea-floor spreading. In Chapter 4, we will build on these concepts and describe the details of modern plate tectonics theory.

FIGURE 3.2 Wegener's image of Pangaea and its subsequent breakup and dispersal.

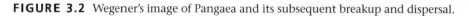

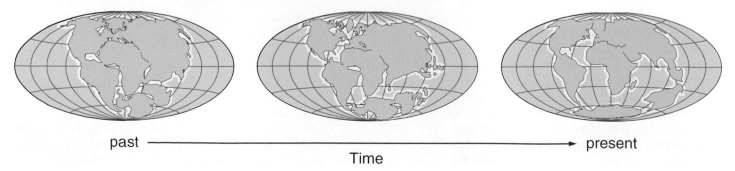

past ——————————————————————————→ present

Time

3.2 WEGENER'S EVIDENCE FOR CONTINENTAL DRIFT

Before Wegener, geologists viewed the continents and oceans as immobile—fixed in position throughout geologic time. According to Wegener, however, the positions of continents change through time. He suggested that a vast supercontinent, Pangaea, existed until the Mesozoic Era (the interval of geologic time, commonly known as the "age of dinosaurs," that lasted from 251 to 65 million years ago). During the Mesozoic, Pangaea broke apart to form the continents we see today; these continents then drifted away from each other. (Geologists now realize that supercontinents formed and dispersed at least a few times during Earth's history—the name Pangaea applies only to the most recent supercontinent.) Let's look at some of Wegener's arguments and see why he came to this conclusion.

The Fit of the Continents

Almost as soon as maps of the Atlantic coastlines became available in the 1500s, scholars noticed the fit of the continents. The northwestern coast of Africa could tuck in against the eastern coast of North America, and the bulge of eastern South America could nestle cozily into the indentation of southwestern Africa. Australia, Antarctica, and India could all connect to the southeast of Africa, while Greenland, Europe, and Asia could pack against the northeastern margin of North America. In fact, all the continents could be joined, with remarkably few overlaps or gaps, to create Pangaea. (Modern plate tectonics theory can now even explain the misfits.) Wegener concluded that the fit was too good to be coincidence.

Locations of Past Glaciations

Wegener was an Arctic meteorologist by training, it is no surprise that he had a strong interest in *glaciers,* the rivers or sheets of ice that slowly flow across the land surface. He realized that glaciers form mostly at high latitudes, and thus that by studying the past locations of glaciers, he might be able to determine the past locations of continents. We'll look at glaciers in detail in Chapter 22, but we need to know something about them here to understand Wegener's arguments.

When a glacier moves, it scrapes sediment (pebbles, boulders, sand, silt, and mud) off the ground and carries it along. The sediment freezes into the base of the glacier, so the glacier then becomes like a rasp and grinds exposed rock beneath it. In fact, rocks protruding from the base of the ice carve striations (scratches) into the underlying rock, and these striations indicate the direction in which the ice flowed. When the glacier eventually melts, the sediment collects on the ground and creates a distinctive layer of "till," a mixture of mud, sand, pebbles, and larger rocks. Later on, the till may be buried and preserved. Today, glaciers are found only in polar regions and in high mountains. But by studying the distribution and age of ancient till, geoscientists have determined that at several times during Earth's history, glaciers covered large areas of continents. We refer to these times as "ice ages." One of the major ice ages occurred about 260–280 million years ago, near the end of the Paleozoic Era (the interval of geologic time between 542 and 251 million years ago).

Why was the study of ancient glacial deposits important to Wegener? When he plotted the locations of Late Paleozoic till, he found that glaciers of this time interval occurred in southern South America, southern Africa, southern India, Antarctica, and southern Australia. These places are all now widely separated from one another and, with the exception of Antarctica, do not currently lie in cold polar regions (►Fig. 3.3a). To Wegener's amazement, all the late Paleozoic glaciated areas lie adjacent to each other on a map of Pangaea (►Fig. 3.3b). Furthermore, when he plotted the orientation of glacial striations, they all pointed roughly outward from a location in southeastern Africa, just as would be expected if an ice sheet comparable to the present-day Antarctic polar ice cap had developed in southeastern Africa and had spread outward from its origin. In other words, Wegener determined that the distribution of glaciations at the end of the Paleozoic Era could easily be explained if the continents had been united in Pangaea, with the southern part of Pangaea located over the South Pole, but could not be explained if the continents had always been in their present positions.

The Distribution of Equatorial Climatic Belts

If the southern part of Pangaea had straddled the South Pole at the end of the Paleozoic Era, then during this same time interval southern North America, southern Europe, and northwestern Africa would have straddled the equator and would have had tropical or subtropical climates. Wegener searched for evidence that this was so by studying sedimentary rocks that were formed at this time, for the material making up these rocks can reveal clues to the climate. Specifically, in the swamps and jungles of tropical regions, thick deposits of plant material accumulate, and when deeply buried, this material transforms into coal. Further, in the clear shallow seas of tropical regions, large reefs built from the shells of marine organisms develop offshore. Finally, subtropical regions, on either side of the tropical belt, contain deserts, an environment for sand-dune formation and the accumulation of salt from evaporating seawater or salt lakes. Wegener thought that the distribution of late

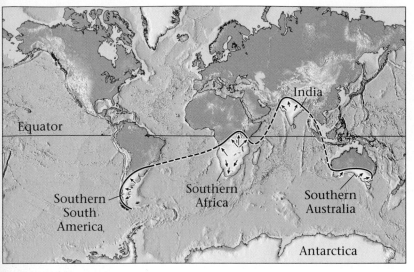

Equator

India

Southern
South
America

Southern
Africa

Southern
Australia

Antarctica

(a)

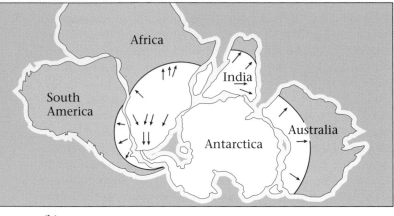

Africa

India

South
America

Australia

Antarctica

(b)

FIGURE 3.3 (a) The distribution of late Paleozoic glacial deposits on a map of the present-day Earth. The arrows indicate the orientation of striations. (b) The distribution of these glacial deposits on a map of the southern portion of Pangaea. Note that the glaciated areas fit together to define a polar ice cap.

Paleozoic coal, sand-dune deposits, and salt deposits could define climate belts on Pangaea.

Sure enough, in the belt of Pangaea that Wegener expected to be equatorial, late Paleozoic sedimentary rock layers included abundant coal and the relicts of reefs; and in the portions of Pangaea that Wegener predicted would be subtropical, late Paleozoic sedimentary rock layers included relicts of desert dunes and of salt (▶Fig. 3.4). On a present-day map of our planet, these deposits are scattered around the globe at a variety of latitudes—including high latitudes, where they cannot have formed. However, in Wegener's Pangaea, the deposits align in continuous bands that occupy appropriate latitudes.

The Distribution of Fossils

Today, different continents provide homes for different species. Kangaroos, for example, live only in Australia. Similarly, many kinds of plants grow only on one continent and not on others. Why? Because land-dwelling species of animals and plants cannot swim across vast oceans, and thus they evolve independently on different continents. During a period of Earth history when all continents were in contact, however, land animals and plants conceivably could have dispersed, so the same species might have appeared on many continents.

With this concept in mind, Wegener plotted locations of fossils of land-dwelling species that lived during the late Paleozoic and early Mesozoic eras (between about 300 and 210 million years ago) and found that they had indeed existed on several continents (▶Fig. 3.5). For example, an early Mesozoic land-dwelling reptile called *Cynognathus* lived in both southern South America and southern Africa. *Glossopteris,* a species of seed fern, flourished in regions that now constitute South America, Africa, India, Antarctica, and Australia (see the chapter opening photo). *Mesosaurus,*

FIGURE 3.4 Map of Pangaea, showing the distribution of coal deposits and reefs (indicating tropical environments), and sand-dune deposits and salt deposits (indicating subtropical environments). Note how deposits now on different continents align in distinct belts.

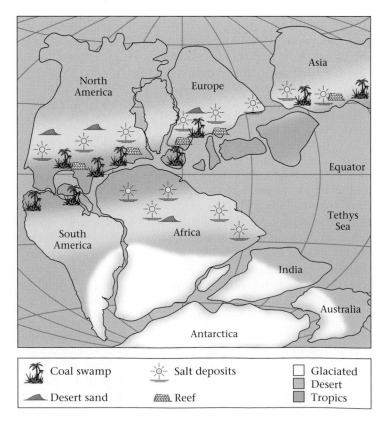

Asia

North
America

Europe

South
America

Africa

Equator

Tethys
Sea

India

Australia

Antarctica

🌴 Coal swamp	☀ Salt deposits	☐ Glaciated
🏔 Desert sand	🪸 Reef	☐ Desert
		☐ Tropics

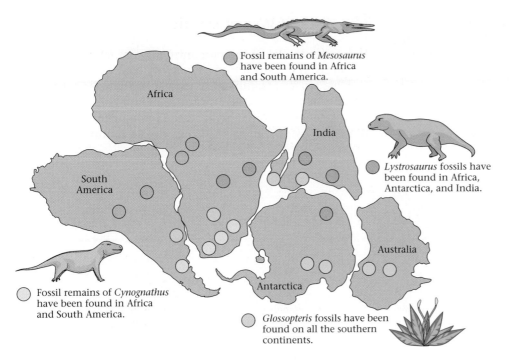

FIGURE 3.5 This map shows the distribution of terrestrial (land-based) fossil species. Note that creatures like *Lystrosaurus* could not have swum across the Atlantic to reach Africa. Sample locations are approximate.

FIGURE 3.6 (a) Distinctive areas of rock assemblages on South America link with those on Africa, as if they were once connected and later broke apart. (b) If the continents are returned to their positions in Pangaea by closing the Atlantic, mountain belts (shown in brown) of the Appalachians lie adjacent to similar-age mountain belts in Greenland, Great Britain, Scandinavia, and Africa. "Archean" is the older part of the Precambrian, and "Proterozoic" is the younger part.

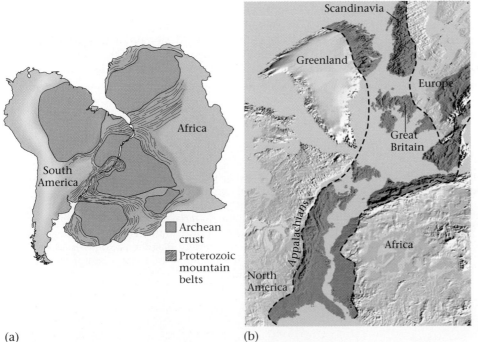

a freshwater reptile, inhabited portions of what is now South America and Africa. *Lystrosaurus,* another land-dwelling reptile, wandered through present-day Africa, India, and Antarctica. None of these species could have traversed a large ocean. Thus, Wegener argued, the distribution of these species required the continents to have been adjacent to one another in the late Paleozoic and early Mesozoic Eras.

Considering that paleontologists found fossils of species such as *Glossopteris* in Africa, South America, and India, Wegener suspected that they might also be found in Antarctica. The tragic efforts of Captain Robert Scott and his party of British explorers, who reached the South Pole in 1912, confirmed this proposal. On their return trip, the party died of starvation and cold, only 11 km from a food cache. When their bodies were found, their sled loads included *Glossopteris* fossils that they had hauled for hundreds of kilometers, in the process burning valuable calories that could possibly have kept them alive long enough to reach the cache. In 1969, paleontologists found fossils of *Lystrosaurus* in Antarctica, providing further confirmation that Wegener was right and that the continents had once been connected.

Matching Geologic Units

In the same way that an art historian can identify a Picasso painting and an architect a Victorian design, a geoscientist can identify a distinctive group of rocks. Wegener found that the same distinctive Precambrian (the interval of geologic time between Earth's formation and 542 million years ago) rock assemblages occurred on the eastern coast of South America and the western coast of Africa, regions now separated by an ocean (▶Fig. 3.6a). If the continents had been joined to create Pangaea in the past, then these matching rock groups would have been adjacent to one another, and thus could have composed continuous blocks. Wegener also noted that belts of rocks in the Appalachian Mountains of the United States and Canada closely resembled belts of rocks in mountains of southern Greenland, Great Britain, Scandinavia, and northwestern Africa (▶Fig. 3.6b), regions that would have lain adjacent to each other in Pangaea. Wegener thus demonstrated that not only did the coastlines of continents match—their component rocks did too.

Criticism of Wegener's Ideas

Wegener's model of a supercontinent that later broke apart explained the distribution of glaciers, coal, sand dunes, distinctive rock assemblages, and fossils we find today. Clearly, he had compiled a strong case for continental drift. But Wegener, as noted earlier, could not adequately explain how or why continents drifted. In his writings, Wegener

---

BOX 3.1
SCIENCE TOOLBOX

Some Fundamentals of Magnetism

If you hold a magnet over a pile of steel paper clips, it will lift the paper clips against the force of gravity. The magnet exerts an attractive force that pulls on the clips. A magnet can also create a repulsive force that pushes an object away. For example, when oriented appropriately, one magnet can levitate another. The push or pull exerted by a magnet is a **magnetic force**; this force creates an invisible **magnetic field** around the magnet. Magnetic forces can be created by a "permanent magnet," a special material that behaves magnetically for a long time all by itself. Magnetic forces can also be produced by an electric current passing through a wire. An electrical device that produces a magnetic field is an "electromagnet." The stronger the magnet, the greater its **magnetization.** When other magnets, special materials (such as iron), or electric charges enter a magnetic field, they feel a magnetic force. The strength of the pull that an object feels when placed in a magnet's field depends on the magnet's magnetization and on the distance of the object from the magnet.

Compass needles are simply magnetic needles that can pivot freely and that align with Earth's magnetic field. Recall from Chapter 2 that you can symbolically represent a magnetic field by a pattern of curving lines, known as "magnetic field lines." You can see the form of these lines by sprinkling iron filings on a sheet of paper placed over a bar magnet; each filing acts like a tiny magnetic compass needle and aligns itself with the magnetic field lines (▶Fig. 2.2).

All magnets have two **magnetic poles,** a north pole at one end and a south pole at the other. Opposite poles attract, but like poles repel. The imaginary line through the magnet that connects one pole to another represents the magnet's "dipole." Physicists specify the dipole by an arrow that points from the north to the south pole. The **polarity** of a magnet refers to the direction the arrow points; the dipoles of magnets with opposite polarity are represented by arrows with arrowheads at opposite ends. Because of the dipolar nature of magnetic fields, we can draw arrowheads on magnetic field lines oriented to form a continuous loop through the magnet.

An electron, which is a spinning, negatively charged particle that orbits the nucleus of an atom, behaves like a tiny electromagnet because its movement produces an electric current. Most of the magnetism is due to the electron's spin, but a little may come from its orbital motion (▶Fig. 3.7a). Each atom, therefore, can be pictured as a little dipole (▶Fig. 3.7b). But even though all materials consist of atoms, not all materials behave like strong, permanent magnets. In fact, most materials (wood, plastic, glass, gold, tin, etc.) are essentially nonmagnetic. That's because the atomic dipoles in the materials are randomly oriented, so overall the dipoles of the atoms cancel each other out (▶Fig. 3.7c). In a permanent magnet, however, all atomic dipoles lock into alignment with one another. When this happens, the magnetization of each atom adds to that of its neighbor, so the material as a whole becomes magnetic (▶Fig. 3.7d).

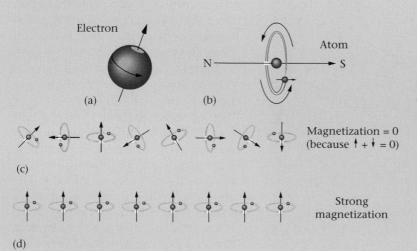

(a)   (b)

(c)   Magnetization = 0
(because ↑ + ↓ = 0)

(d)   Strong magnetization

FIGURE 3.7 (a) A spinning electron creates an electric current. (b) The magnetic dipole of an atom can be represented by an arrow that points from north to south. (c) In a nonmagnetic material, atoms tilt all different ways, so the dipoles cancel each other out, yielding a net magnetization of 0. (d) In a permanent magnet, the dipoles lock into alignment, so that they add to each other and produce a strong magnetization.

suggested that the force created by the rotation of the Earth could cause a supercontinent centered at a pole to break up into pieces that would move toward equatorial latitudes. He proposed that the continental crust (he didn't refer to the lithosphere, which includes the crust and the upper-most part of the mantle) moved by "plowing" through oceanic crust as a ship plows through water. But other geologists of the time found his explanation wholly unsatisfactory. Experiments showed that the relatively weak rock making up continents cannot plow through the relatively strong rock making up the ocean floor, and that force generated by Earth's spin is a million times too small to move a continent.

Wegener left on his final expedition to Greenland having failed to convince his peers, and he died in the icy wasteland never knowing that his ideas would smolder for decades before being reborn as the basis of the broader theory of plate tectonics. During these decades, a handful of iconoclasts continued to champion Wegener's notions. Among these was Arthur Holmes, a highly respected British geologist who argued that huge convection cells existed inside the Earth, slowly transporting hot rock from the deep interior up to the surface. Holmes suggested that continents might be split and the pieces dragged apart in response to convective flow in the mantle. But in general, geologists retreated to their subspecialties and remained indifferent to the possibility that a single bold idea could unify their work.

Fortunately, however, geologic research did not come to a halt. The middle decades of the twentieth century saw great discoveries about the nature of the Earth and its history; many of these discoveries relied on instruments and techniques that had not existed previously. Geologists produced maps showing the distribution of rock assemblages, refined the images of the Earth's interior, and described the nature of the ocean floor and its subsurface. But the key that ultimately proved continental drift, and opened the door to sea-floor spreading and then plate tectonics, came from the discovery of a phenomenon called paleomagnetism.

3.3 PALEOMAGNETISM AND APPARENT POLAR-WANDER PATHS

In 1853, an Italian physicist noticed that volcanic rock behaved like a very weak magnet, and proposed that it became magnetic when it solidified from melt. In the 1950s, instruments became available that could routinely measure such weak magnetization, so researchers in England began to study magnetization in *ancient* rocks. Their work showed that rocks preserve a record of Earth's past magnetic field. The record of ancient magnetism preserved in rock is called **paleomagnetism.** In order to understand paleomagnetism and how it enabled geologists to test the continental drift hypothesis, and later the sea-floor-spreading hypothesis, we must first look at the nature of Earth's magnetic field in a bit more depth than we did in Chapter 2. If you need to review background information on magnetism in general, please see Box 3.1.

Earth's Magnetic Field, Revisited

In Chapter 2, we learned that Earth has a magnetic field, which deflects the solar wind and traps cosmic rays. Why does this field exist? Geologists do not yet have a complete answer, but they have hypothesized that the field results from the circulation of liquid iron alloy, an electrical conductor, in the Earth's outer core—in other words, the outer core behaves like an electromagnet (▶Fig. 3.8a; see Box 3.2). For convenience, however, we can picture the planet as a giant bar magnet, with a north magnetic pole and a south magnetic pole (▶Fig. 3.8b). The "north-seeking end" of a compass points toward the north magnetic pole, while the "south-seeking end" points toward the south magnetic pole. We define the **dipole** of the Earth as an imaginary arrow that points from the north magnetic pole to the south magnetic pole, and passes through the planet's center.

Presently, Earth's dipole tilts at about 11° to the planet's "rotational axis" (the imaginary line through the center of the Earth around which Earth spins). Therefore, the "geographic poles" of the planet, the places where the rotational axis intersects the Earth's surface, do not coincide exactly with the magnetic poles. For example, the north magnetic pole currently lies in arctic Canada. As a consequence, the north-seeking end of a compass needle in New York points about 14° west of north. The angle between the direction that a compass needle points at a given location and the direction to "true" (geographic) north is called the **magnetic declination** (▶Fig. 3.9). Measurements over the past couple of centuries show that magnetic poles migrate very slowly through time, probably never straying more than about 15° of latitude from the geographic pole. In fact, the magnetic declination of a compass changes by 0.2° to 0.5° per year. Notably, when averaged over about 10,000 years, the magnetic poles are thought to coincide with the geographic poles.

▶Figure 3.10 illustrates the magnetic field lines in space around the Earth, as seen in cross section (without the warping caused by solar wind). Note that close to the Earth, the lines parallel Earth's surface at the equator; the lines tilt at an angle to the surface at mid-latitudes, and the lines are perpendicular to the surface at the magnetic poles. Thus, if we traveled to the equator and set up a magnetic needle such that it could pivot up and down freely, the needle would be horizontal. If we took the needle to mid-latitudes,

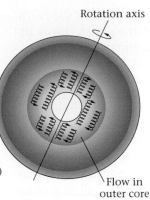

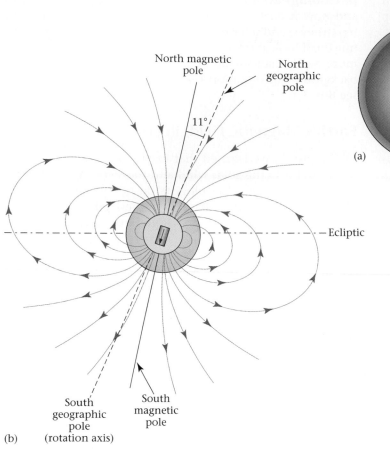

FIGURE 3.8 (a) The convective flow of liquid iron alloy in the Earth's outer core creates an electric current that in turn generates a magnetic field. Recent studies suggest the flow spirals up spring-like coils. (b) Earth's magnetism creates magnetic lines of force in space. We can picture Earth's magnetism by imagining that it contains a giant bar magnet. The dipole of this magnet points presently from the north magnetic pole to the south magnetic pole, and it pierces the Earth at the magnetic poles. Today, the magnetic poles do not coincide exactly with the Earth's geographic poles; they are 11° apart.

FIGURE 3.9 The projection of magnetic field lines in North America at present. Recall that lines of longitude run north-south, so in most places a compass needle will not parallel longitude. For example, a compass needle at New York would make an angle of about 14° to the west of true north. Note that along the circumference that passes through both magnetic north and geographic north, the magnetic declination = 0°. See Appendix B for magnetic declination maps for the United States and for the world.

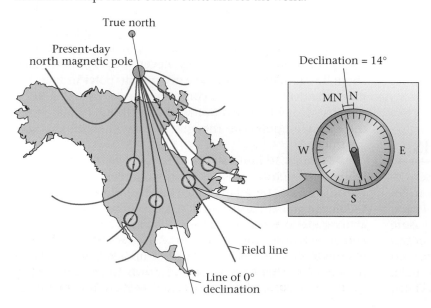

it would tilt at an angle to Earth's surface; and at a magnetic pole, the needle would point straight down. The needle's angle of tilt (which, as Fig. 3.10 shows, depends on latitude) is called the **magnetic inclination.** Note that a regular compass needle does not indicate inclination, because it cannot tilt—a compass needle aligns parallel to the "projection" of the magnetic field lines on the Earth's surface. (You can think of the "projection" as the shadow of a magnetic field line on Earth's surface.)

How Do Rocks Develop Paleomagnetism?

More than 1,500 years ago, Chinese sailors discovered that an elongated piece of lodestone suspended from a thread "magically" pivots until it points north, and thus that this rock could help guide their voyages. We now know that lodestone exhibits this behavior because it consists of **magnetite,** an iron-rich mineral that acts like a permanent magnet (see Box 3.1). Small crystals of magnetite or other magnetic

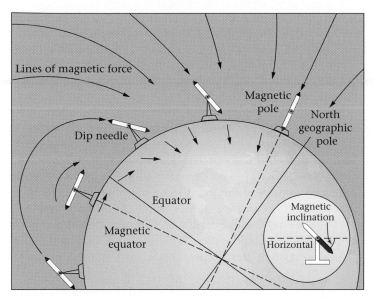

FIGURE 3.10 An illustration of magnetic inclination. A magnetic needle that is free to rotate around a horizontal axis aligns with magnetic field lines (here depicted in cross section). Because magnetic field lines curve in space, this needle is horizontal at the equator, tilts at an angle at mid-latitudes, and is vertical at the magnetic pole. Therefore, the angle of tilt depends on the latitude.

minerals occur in many rock types. Each crystal produces a tiny magnetic force. The sum of the magnetic forces produced by all the crystals makes the rock, as a whole, weakly magnetic.

To see how magnetic rocks preserve a record of Earth's past magnetic field, let's examine the development of magnetization in one type of rock, basalt. Basalt is dark-colored, magnetite-containing igneous rock that forms when lava, flowing out of a volcano, cools and solidifies. When lava first comes out of a volcano, it is very hot (up to about 1,200°C), and thermal energy makes its atoms wobble and tumble chaotically. Each atom acts like a mini-dipole, but the mini-dipoles of the wildly dancing atoms point in all different directions. When this happens, the magnetic force exerted by one atom cancels out the force of another with an oppositely oriented dipole, so the lava as a whole is not magnetic (▶Fig. 3.11a). However, as the temperature of the lava decreases to below the melting temperature (about 1,000°C), basalt rock starts to solidify. As the magnetite crystals in the basalt form and cool (i.e., as thermal energy decreases), their iron atoms slow down. The dipoles of all the atoms gradually become parallel with each other and with the Earth's magnetic field lines at the location where the basalt cools. Finally, at temperatures below 350°–550°C, well below the melting temperature, the dipoles lock into position, pointing in the direction of the magnetic pole, and the basalt becomes a permanent magnet (▶Fig. 3.11b).

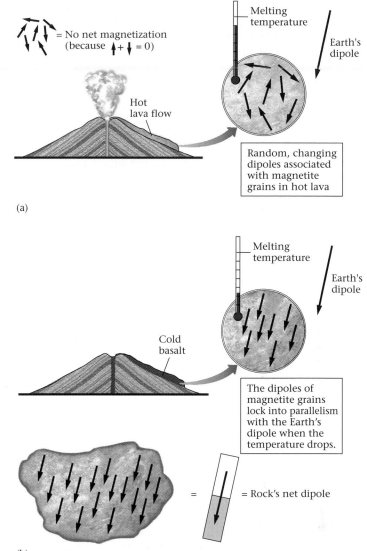

FIGURE 3.11 The formation of paleomagnetism. (a) At high temperatures (greater than 350°–550°C), thermal vibration makes atoms have random orientations; the dipoles thus cancel each other out and the sample has no overall magnetic dipole. (b) As the sample cools to below 350°–550°C, the atoms slow down and their dipoles lock into alignment with the Earth's field.

Since this alignment is permanent, the basalt provides a record of the orientation of the Earth's magnetic field lines, relative to the rock, at the time the rock cooled. This record is paleomagnetism.

Basalt is not the only rock to preserve a good record of paleomagnetism. Certain kinds of sedimentary rocks also can preserve a record. In some cases, the record forms when magnetic sedimentary grains align with the Earth's magnetic field as they settle to form a layer; this orientation can be preserved when the layer turns to rock (▶Fig. 3.12a). Paleomagnetism may also develop when magnetic minerals (magnetite or another iron mineral, hematite) grow in the spaces between grains after the sediment has accumulated. These minerals form from ions that had been dissolved in groundwater passing through the sediment (▶Fig. 3.12b).

Interpreting *Apparent* Polar-Wander Paths: Evidence for Continental Drift

When geologists measured paleomagnetism in samples of basalt that had formed millions of years ago, they were surprised to find that the imaginary "compass" representing this paleomagnetism did not point to the present-day magnetic poles of the Earth (▶Fig. 3.13). At first, they interpreted this observation to mean that the magnetic poles of Earth moved through time and thus had different locations in the past, a phenomenon that they called **polar wander,** and they introduced the term **paleopole** to refer to the supposed position of the Earth's magnetic pole at a time in the past. (See Box 3.3 to learn about locating paleopoles.) As we'll see, this initial interpretation of polar wander was quite wrong.

To explore the concept of polar wander, researchers decided to track the position of paleopoles through time. So they measured the magnetic field preserved in rocks of many different ages from about the same location. ▶Figure 3.15 shows how this would be done for a locality called X on an imaginary continent. Figure 3.15a shows that the paleomagnetic declination and inclination is different in different layers. The paleopoles can be calculated, as shown in Box 3.3. In Figure 3.15b, the paleopole for rock sample 1, which is 600 million years old, lies at position 1 on the map. In other

---

BOX 3.2
THE REST OF THE STORY

Generating Earth's Magnetic Field

Chinese scientists first studied Earth's magnetic field in 1040 B.C.E, yet almost a millennium later, Albert Einstein noted that the question of why Earth has a magnetic field remained one of the great physics questions of all times. Space exploration shows that planets, in fact, do not have to have magnetic fields. Neither Venus nor Mars presently has a significant field—so what is so special about the Earth that causes it to have a strong field? The path toward an answer became clear in 1926, when researchers proved that the Earth's outer core consists of liquid iron alloy. Flow of this liquid metal, presumably, can generate an electric current, which in turn can generate a magnetic field. In other words, the flow of iron alloy makes the Earth's outer core an electromagnet.

To better understand the generation of Earth's magnetic field, let's first consider how an electric power plant works. In a power plant, water or wind power spins a wire coil (an electrical conductor) around an iron bar (a permanent magnet). This apparatus is a **dynamo.** The motion of the wire in the bar's magnetic field generates an electric current in the wire, which in turn generates more magnetism. Applying this concept to the Earth, we can picture the flow of the outer core to play the role of a spinning wire coil. But what plays the role of the permanent magnet in the Earth? There can't be a permanent magnet in the core, because at the very high temperatures found in the core, thermal agitation causes atoms to vibrate and tumble so much that their atomic dipoles cannot lock into parallelism with each other—and without locked-in parallelism of atomic dipoles, permanent magnets can't exist (see Box 3.1). Thus, researchers suggest that the Earth is a "self-exiting dynamo." Somehow, in Earth's earlier history, flow in the outer core took place in the presence of a magnetic field. This flow generated an electric current. Once the current existed, it generated a magnetic field. Continued flow in the presence of this generated magnetic field produced more electric current, which in turn produced more magnetic field—once started, the system perpetuated.

Flow of iron alloy in the outer core must take place for a self-exiting dynamo to exist in the Earth. What causes this flow, and how does flow result in the geometry of the field that we measure today? This topic remains an area of active research, but recent work provides some possible answers. Calculations suggest that the inner core is growing in diameter, as the Earth cools, at a rate of 0.1 to 1 mm per year. Growth occurs as new crystals of solid iron form along the surface of the inner core. (An interesting observation is that at the present rate of growth, it appears that the inner core started forming only 1 to 2 billion years ago.) Solid iron crystals do not have room for lighter elements, such as silicon, sulfur, hydrogen, carbon, or oxygen, which had been contained in the liquid iron alloy of the outer core. Thus, these elements migrate into the base of the outer core as the inner core grows. The relatively high concentration of these lighter elements makes the base of the outer core less dense than the top. As a result, the base of the outer core is buoyant; like a block of styrofoam floating to the top of a pool, the outer core begins to rise, and this rise causes flow. We can consider the flow to be "convection." But unlike the familiar "thermal convection" that takes place in a pot of water on your stove, in which differences in density are caused by differences in temperature (warmer water is less dense and, thus, is buoyant), convection in the outer core is largely "chemical convection" caused by contrasts in composition.

Calculations suggest that convective motion in the outer core results in the flow of iron alloy in columnar spirals (resembling the coils of a spring), whose axes roughly parallel the spin axis of the Earth. That's because the spin of the Earth influences the geometry of convective flow in the mantle. Possibly for this reason, the magnetic dipole of the Earth roughly parallels the spin axis of the Earth. Because it is so hot, iron alloy in the outer core may flow at rates of up to 20 km per year.

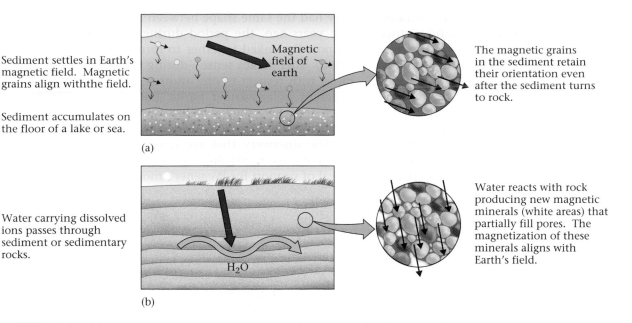

Sediment settles in Earth's magnetic field. Magnetic grains align withthe field.

Sediment accumulates on the floor of a lake or sea.

(a)

The magnetic grains in the sediment retain their orientation even after the sediment turns to rock.

Water carrying dissolved ions passes through sediment or sedimentary rocks.

H_2O

(b)

Water reacts with rock producing new magnetic minerals (white areas) that partially fill pores. The magnetization of these minerals aligns with Earth's field.

FIGURE 3.12 (a) Paleomagnetism can form during the settling of sediments. (b) Paleomagnetism can also form when iron-bearing minerals precipitate out of groundwater passing through sediment.

words, position 1 indicates the location of Earth's magnetic pole, *relative to locality X,* 600 million years ago. In sample 2, which is 500 million years old, the paleopole lies at position 2 on the map, and so on. When all the points are plotted, the resulting curving line shows the progressive change in the position of the Earth's magnetic pole, relative to locality X, *assuming* that the position of X on Earth has been fixed through time. This curve was called a polar-wander path. Note that the polar-wander path ends near the present North Pole, because recent rocks became magnetized when Earth's magnetic field was close to its position today.

In the late 1950s, geologists determined what they thought was the polar-wander path for Europe. When this path was first plotted, they did not accept the notion of continental drift, and assumed that the position of the continents was fixed. Thus, they *interpreted* the path to represent how the position of Earth's north magnetic pole migrated through time. Were they in for a surprise! When geologists then determined the polar-wander path for North America, they found that North America's path differed from Europe's (▶Fig. 3.16a). In fact, when paths were plotted for all continents, they all turned out to be different from one another (▶Fig. 3.16b). The hypothesis that the continents are fixed and the magnetic poles move simply *cannot* explain this observation. If the magnetic poles really moved while the continents stayed fixed, then all continents should have the same polar-wander paths.

Geologists suddenly realized that they were looking at polar-wander paths in the wrong way. *It's not the pole that*

FIGURE 3.13 A rock sample can maintain paleomagnetization for millions of years. In this example, the dipole representing the paleomagnetism in this rock sample, from a village on the equator, does not parallel the Earth's present field. Note that I (inclination) is not 0°, as it would be today for rock forming near the equator. (See Fig. 3.14a, b.)

Magnetic north

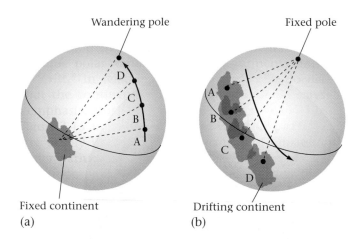

FIGURE 3.17 The two alternative explanations for an apparent polar-wander path. (a) In a "true polar-wander" model, the continent is fixed, so to explain polar-wander paths, the magnetic pole must move substantially. (In reality, the magnetic pole does move a little, but it never strays very far from the geographic pole.) (b) In a continental drift model, the magnetic pole is fixed near the geographic pole, and the continent drifts relative to the pole.

But how could continental drift take place? Until that question could be answered, many geologists were still uncomfortable with the concept. But the stage was set for the proposal of the sea-floor-spreading hypothesis, an idea that ultimately explains how continental drift occurs. The sea-floor-spreading concept culminated decades of sea-floor exploration, which we now review.

3.4 SETTING THE STAGE FOR THE DISCOVERY OF SEA-FLOOR SPREADING

New Images of Sea-Floor Bathymetry

Before World War II, we knew less about the shape of the ocean floor than we did about the shape of the Moon's surface. After all, we could at least see the surface of the Moon and could use a telescope to map its craters. But our knowledge of sea-floor **bathymetry** (the shape of the sea-floor surface) came only from scattered "soundings" of the sea floor. To sound the ocean depths, a surveyor let out a length of cable with a heavy weight attached. When the weight hit the sea floor, the length of the cable indicated the depth of the floor. Needless to say, it took many hours to make a single measurement, and not many could be made. Nevertheless, soundings carried out between 1872

and 1876 by the world's first oceanographic research vessel, the H.M.S. *Challenger,* did hint at the existence of submarine mountain ranges and deep troughs.

Military needs during World War II gave a boost to sea-floor exploration, for as submarine fleets grew, navies required detailed maps showing variations in the depth of the sea floor. The invention of "echo sounding" (sonar) permitted such maps to be made. Echo sounding works on the same principle that a bat uses to navigate and find insects. A sound pulse emitted from a ship travels down through the water, bounces off the sea floor, and returns up as an echo through the water to a receiver on the ship. Since sound waves travel at a known velocity, the time between the sound emission and the detection of the echo indicates the distance between the ship and the sea floor (velocity = distance/time, so distance = velocity × time). As the ship moves, echo sounding permits observers to obtain a continuous record of the depth of the sea floor; the resulting cross section showing depth plotted against location is called a "bathymetric profile" (▶Fig. 3.18a, b). By cruising back and forth across the ocean many times, investigators obtained a series of bathymetric profiles and from these constructed maps of the sea floor. Bathymetric maps revealed several important features of the ocean floor.

- *Mid-ocean ridges:* The floor beneath all major oceans includes two provinces: **abyssal plains,** the broad, relatively flat regions of the ocean that lie at a depth of about 4–5 km below sea level; and **mid-ocean ridges,** elongate submarine mountain ranges whose peaks lie only about 2–2.5 km below sea level (▶Figs. 3.19, 3.20a). Geologists call the crest of the mid-ocean ridge the "ridge axis." All mid-ocean ridges are roughly symmetrical—bathymetry on one side of the axis is nearly a mirror image of bathymetry on the other side. Some, like the Mid-Atlantic Ridge, include steep escarpments (cliffs) as well as a distinct axial trough, a narrow depression that runs along the ridge axis.

- *Deep-ocean trenches:* Along much of the perimeter of the Pacific Ocean, and in a few other localities as well, the ocean floor reaches astounding depths of 8–12 km—deep enough to swallow Mt. Everest. These deep areas define elongate troughs that are now referred to as **trenches.** Trenches border **volcanic arcs,** curving chains of active volcanoes.

- *Seamount chains:* Numerous volcanic islands poke up from the ocean floor: for example, the Hawaiian Islands lie in the middle of the Pacific. In addition to islands that rise above sea level, echo sounding has detected many **seamounts** (isolated submarine mountains), which were once volcanoes but no longer erupt. Oceanic islands and seamounts, typically occur in chains, but in contrast to the volcanic arcs that border deep-ocean trenches, only

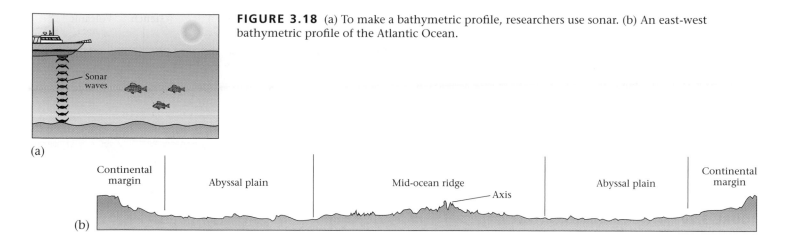

FIGURE 3.18 (a) To make a bathymetric profile, researchers use sonar. (b) An east-west bathymetric profile of the Atlantic Ocean.

one island at the end of a seamount chain is actively erupting today.

- *Fracture zones:* Surveys reveal that the ocean floor is diced up by narrow bands of vertical fractures. These **fracture zones** lie roughly at right angles to mid-ocean ridges, effectively segmenting the ridges into small pieces (►Fig. 3.20b).

New Observations on the Nature of Oceanic Crust

By the mid-twentieth century, geologists had discovered many important characteristics of the sea-floor crust. These discoveries led them to realize that oceanic crust is quite different from continental crust, and further that bathymetric features of the ocean floor provide clues to the origin of the crust. Specifically:

FIGURE 3.19 The mid-ocean ridges, fracture zones, and principal deep-ocean trenches of today's oceans.

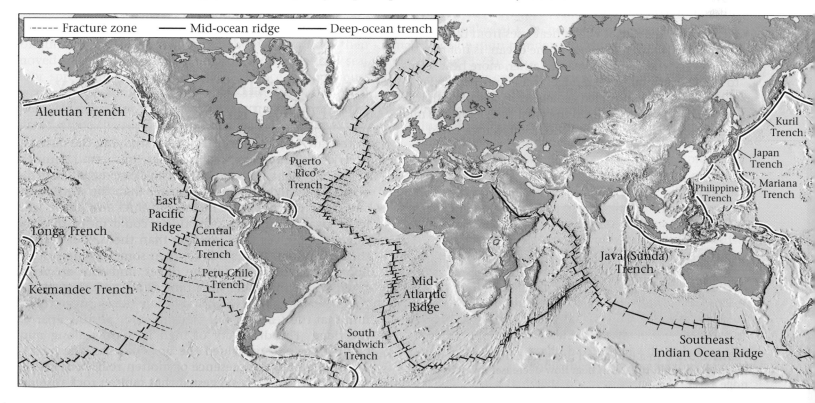

3.6 MARINE MAGNETIC ANOMALIES: EVIDENCE FOR SEA-FLOOR SPREADING

For a hypothesis to earn the status of being a theory, there must be proof. The proof of sea-floor spreading emerged from two discoveries. First, geologists found that the measured strength of Earth's magnetic field is not the same everywhere in the ocean basins; the variations are now called *marine magnetic anomalies*. Second, they found that Earth's dipole reverses direction every now and then; such sudden reversals of the Earth's polarity are now called *magnetic reversals*. To understand why geologists find the concept of sea-floor spreading so appealing, we first need to learn about anomalies and reversals.

Marine Magnetic Anomalies

Geologists can measure the strength of Earth's magnetic field with an instrument called a "magnetometer." At any given location on the surface of the Earth, the magnetic field that you measure includes two parts: one that is created by the main dipole of the Earth (which is caused in turn by the flow of liquid iron in the outer core), and another that is created by the magnetism of near-surface rock. A **magnetic anomaly** is the difference between the expected strength of the Earth's main field at a certain location and the actual measured strength of the magnetic field at that location. Places where the field strength is stronger than expected are "positive anomalies," and places where the field strength is weaker than expected are "negative anomalies."

On continents, the pattern of magnetic anomalies is very irregular. But magnetic anomalies on the sea floor yield a surprisingly different pattern. Geologists towed magnetometers back and forth across the ocean to map variations in magnetic field strength. As a ship cruised along its course, the magnetometer's gauge would first detect strong signals (a positive anomaly) and then weak signals (a negative anomaly). A graph of signal strength versus distance along the traverse, therefore, has a sawtooth shape (▶Fig. 3.24a). When data from many cruises was compiled on a map, these **marine magnetic anomalies** defined distinctive, alternating bands. And if we color positive anomalies dark and negative anomalies light, the pattern made by the anomalies resembles the stripes on a candy cane (▶Fig. 3.24b). The mystery of the marine magnetic anomaly pattern, however, remained unsolved until geologists recognized the existence of magnetic reversals.

FIGURE 3.24 (a) A ship sailing through the ocean dragging a magnetometer detects first a positive anomaly and then a negative one, then a positive one, then a negative one. (b) Magnetic anomalies on the sea floor off the northwestern coast of the United States. The dark bands are positive anomalies, the light bands negative anomalies. Note the distinctive stripes of alternating anomalies. A positive anomaly overlies the crest of the Juan de Fuca Ridge (a small mid-ocean ridge).

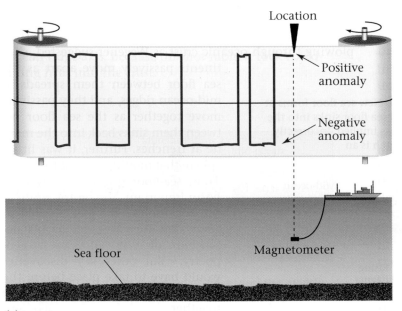

(a)

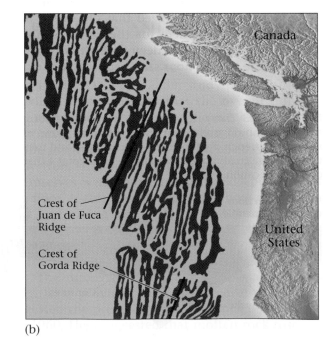

(b)

Magnetic Reversals

Soon after geologists began to study the phenomenon of paleomagnetism, they decided to see if the magnetism of rocks changed as time passed. To do this, they measured the paleomagnetism of many successive rock layers that represented a long period of time. To their surprise, they found that the *polarity* (which end of a magnet points north and which end points south; see Box 3.1) of the paleomagnetic field of some layers was the same as that of Earth's present magnetic field, while in other layers it was the opposite. Recall that Earth's magnetic field can be represented by an arrow, representing the dipole, that points from north to south; in some of the rock layers, the paleomagnetic dipole pointed south (these layers have **normal polarity**), but in others the dipole pointed north (these layers have **reversed polarity**) (▶Fig. 3.25).

At first, observations of reversed polarity were largely ignored, thought to be the result of lightning strikes or of chemical reactions between rock and water. But when repeated measurements from around the world revealed a systematic pattern of alternating normal and reversed polarity in rock layers, geologists realized that reversals were a global, not local, phenomenon. At various times during Earth history, the polarity of Earth's magnetic field has suddenly reversed! In other words, sometimes the Earth has normal polarity, as it does today, and sometimes it has reversed polarity (▶Fig. 3.26a, b). Times when the Earth's field flips from normal to reversed polarity, or vice versa, are called **magnetic reversals.** When the Earth has reversed polarity, the south magnetic pole lies near the north geo-

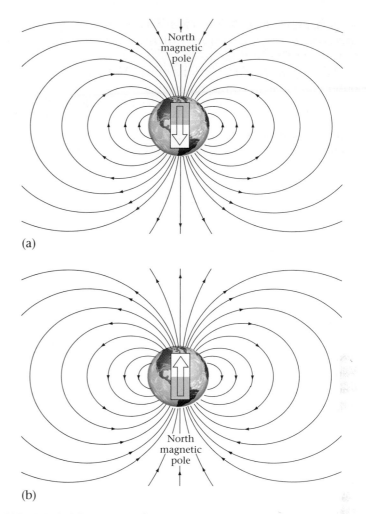

(a)

(b)

FIGURE 3.26 The magnetic field of the Earth has had reversed polarity at various times during Earth history. (a) If the dipole points from north to south, Earth has normal polarity. (b) If the dipole points from south to north, Earth has reversed polarity.

graphic pole, and the north magnetic pole lies near the south geographic pole. If you were to use a compass during periods when the Earth's magnetic field was reversed, the north-seeking end of the needle would point to the south geographic pole.

Note that magnetic reversals are not related to the gradual change in the inclination and declination of the paleomagnetic dipoles we find in successive layers of rock that results from continental drift (as represented by apparent polar-wander paths; see Fig. 3.15a). Also, magnetic reversals are not related to small changes in declination caused by the slight movement of the dipole with respect to Earth's rotational axis. Indeed, reversals do *not* occur by the slow migration of the dipole around the Earth and past the equator. Rather, geologists have found evidence indicating that reversals take place quickly, perhaps in as little as one thousand years.

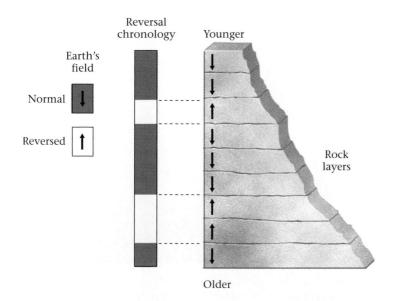

FIGURE 3.25 In a succession of rock layers on land, different flows exhibit different polarity (indicated here by whether the arrow points up or down). When these reversals are plotted on a time column, we have a magnetic-reversal chronology.

Though magnetic reversals have now been well documented, the mechanism by which they occur remains uncertain and continues to be a subject of research. They probably reflect changes in the configuration of flow in the outer core. Recent computer models of the Earth's magnetic field show that the field will flip spontaneously, without any external input, simply in response to the pattern of convection in the outer core. During the relatively short periods of time during which a reversal takes place, the magnetic field becomes disorganized and cannot be represented by a dipole (►Fig. 3.27a–c).

In the 1950s, about the same time geologists discovered polarity reversals, they developed a technique that permitted them to define the age of a rock in years, by measuring the rate of decay of radioactive elements in the rock. The technique is called "radiometric dating" (it will be discussed in detail in Chapter 12). Geologists applied the technique to determine the ages of rock layers in which they obtained their paleomagnetic measurements, and thus determined *when* the magnetic field of the Earth reversed. With this information, they constructed the history of magnetic reversals, now called the **magnetic-reversal chronology.** A diagram representing the Earth's magnetic-reversal chronology (►Fig. 3.28) shows that reversals do not occur regularly, so the lengths of different **polarity chrons,** the time intervals between reversals, are different. For example, we have had a normal-polarity chron for about the last 700,000 years. Before that, there was a reversed-polarity chron. Geologists named the youngest four polarity chrons (Brunhes, Matuyama, Gauss, and Gilbert) after scientists who had made important contributions to the study of rock magnetism. As more measurements became available, investigators realized that there were some short-duration reversals (less than 200,000 years long) within the chrons, and called these shorter reversals "polarity subchrons." Radiometric dating methods are not accurate enough to date reversals that are older than about 4.5 million years, because for rocks older than that, the uncertainty in a date exceeds the duration of a polarity subchron. (For example, a radiometric date of 5 million years may have an uncertainty of 300,000 years. We express this date as 5.0 ± 0.3 million years; that means the date is in the range of 4.7 to 5.3 million years.)

The Interpretation of Marine Anomalies

Armed with our knowledge of magnetic reversals, we can now understand the explanation for marine magnetic anomalies. A graduate student in England, Fred Vine, working with his adviser, Drummond Mathews, and a Canadian geologist, Lawrence Morley (working independently), discovered a solution to this riddle. Simply put, the three suggested that a positive anomaly occurs over areas of sea floor where basalt has normal polarity. In these areas, the mag-

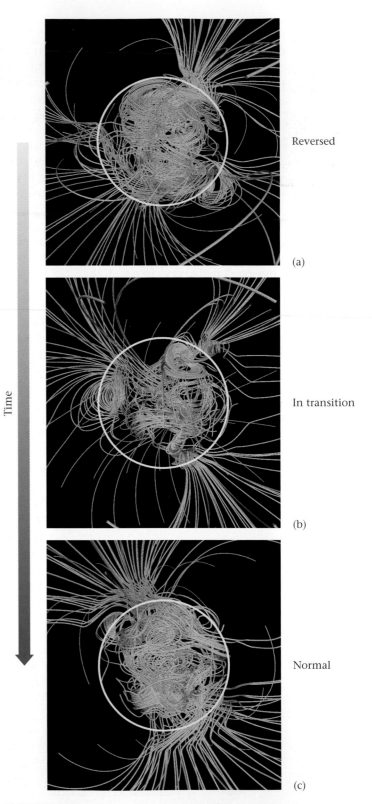

Time

Reversed

(a)

In transition

(b)

Normal

(c)

FIGURE 3.27 Images illustrating Earth's magnetic changes during a reversal, as calculated by a computer model. The colored lines represent magnetic field lines. The dipole points from yellow to blue. The white circle represents the outline of the Earth. (a) Reversed polarity. (b) Polarity during transition. (c) Normal polarity.

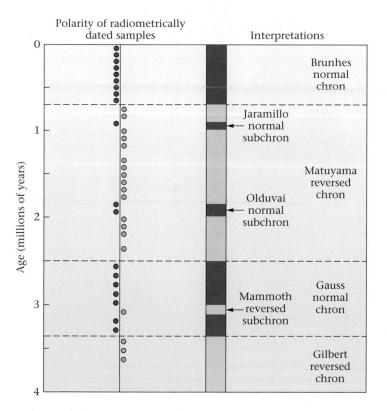

FIGURE 3.28 Radiometric dating of lava flows allows us to determine the age of magnetic reversals during the past 4 million years. Major intervals of a given polarity are referred to as polarity chrons, and are named after scientists who contributed to the understanding of Earth's magnetic field. Shorter-duration reversals are called subchrons.

netic force produced by the basalt *adds* to the force produced by Earth's dipole and creates a stronger magnetic signal than expected, as measured by the magnetometer. A negative anomaly occurs over regions of sea floor where basalt has reversed polarity. Here, the magnetic force of the basalt *subtracts* from the force produced by the dipole and results in a weaker magnetic signal (▶Fig. 3.29a).

Vine, Matthews, and Morley pointed out that marine magnetic anomalies could have formed as a consequence of sea-floor spreading—sea floor yielding positive anomalies developed at times when the Earth had normal polarity, while sea floor yielding negative anomalies formed when the Earth had reversed polarity. If this was so, then the seafloor-spreading hypothesis implies that the pattern of anomalies should be symmetric with respect to the axis of a mid-ocean ridge. With this idea in mind, geologists set sail to measure magnetic anomalies near mid-ocean ridges (▶Fig. 3.29b). By 1966, the story was complete. In the examples studied, the magnetic anomaly pattern on one side of a ridge was indeed a mirror image of the anomaly pattern on the other.

Let's look more closely at how marine magnetic anomalies are formed. Please refer to ▶Figure 3.30a. At time 1 (sometime in the past), a time of normal polarity, the dark stripe of sea floor forms. The tiny dipoles of magnetite grains in basalt making up this stripe align with the Earth's field. As it forms, the rock in this stripe migrates away from the ridge axis, half to the right and half to the left. Later, at time 2, the field has reversed, and the light-gray stripe forms with reversed polarity. As it forms, it too moves away from the axis, and still younger crust begins to develop along the axis. As the process continues over millions of years, many stripes form. A positive anomaly exists along the ridge axis today, because this represents sea floor that has developed during the most recent interval

FIGURE 3.29 (a) The explanation of marine anomalies. The sea floor beneath positive anomalies has the same polarity as Earth's field and therefore adds to it. The sea floor beneath negative anomalies has reversed polarity and thus subtracts from Earth's field. (b) The symmetry of the magnetic anomalies measured across the Mid-Atlantic Ridge south of Iceland. Note that individual anomalies are somewhat irregular, because the process of forming the sea floor, in detail, happens in discontinuous pulses along the length of the ridge.

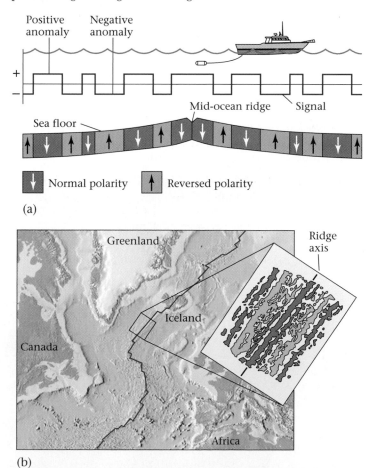

The Earth behaves like a giant magnet, and thus is surrounded by a magnetic field. The magnetism is due to the flow of liquid iron alloy in the outer core.

The age of oceanic crust varies with location. The youngest crust lies along a mid-ocean ridge, and the oldest along the coasts of continents. Here, the different color stripes correspond to different ages of oceanic crust. Red is youngest, purple is oldest.

The rock of oceanic crust preserves a record of the Earth's magnetic polarity at the time the crust formed. Eventually, a symmetric pattern of polarity stripes develops.

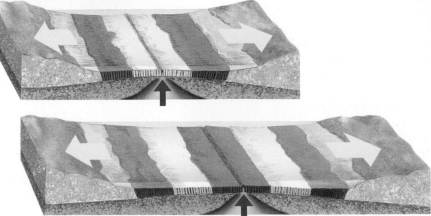

Marine magnetic anomalies are stripes representing alternating bands of oceanic crust that differ in the measured strength of the magnetic field above them. Stronger fields are measured over crust with normal polarity, while weaker fields are measured over crust with reversed polarity.

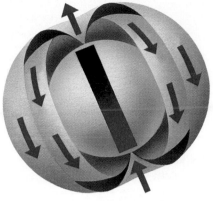

Normal polarity

Reversed polarity

Earth's magnetic field can be represented by a dipole that points from the north magnetic pole to the south. Every now and then, the magnetic polarity reverses.

Magnetic reversals are recorded in a succession of lava flows. Here, lavas with normal polarity are red, while lavas with reversed polarity are yellow.

Lava flows at a volcano.

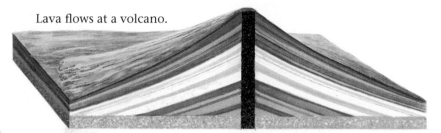

By dating successive lava flows, geologists can determine the timing and duration of magnetic reversals.

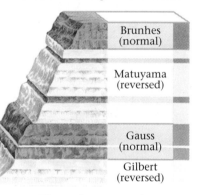

Brunhes (normal)

Matuyama (reversed)

Gauss (normal)

Gilbert (reversed)

The red stripes indicate rock with normal polarity, and the yellow stripes rock with reversed polarity.

Normal polarity

Reversed polarity

Mid-ocean ridge (normal polarity)

Paleomagnetism

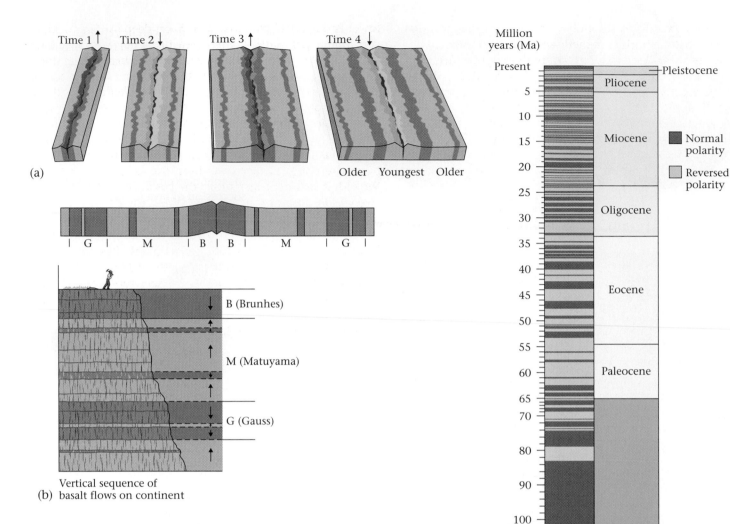

FIGURE 3.30 (a) The progressive development of stripes of alternating polarity in the ocean floor. Each time represents a successive stage of new sea floor forming at a mid-ocean ridge, while Earth's field undergoes magnetic reversals. (b) The observed stripes correlate with the polarity chrons and subchrons measured in lava flows on land. (c) The reversal chronology for the last 170 million years, based on marine magnetic anomalies.

of time, a chron of normal polarity. The magnetism of the rock along the ridge adds to the magnetism of the Earth's field.

Geologists realized that if the anomalies on the sea floor formed by sea-floor spreading as the Earth's magnetic field flipped between normal and reversed polarity, then the anomalies should correspond to the magnetic reversals that had been discovered and radiometrically dated in basalt layers on land. By relating the stripes on the sea floor to magnetic reversals found in dated basalt (▶Fig. 3.30b), geologists dated the sea floor back to an age of 4.5 million years, and they found that the relative widths of anomaly stripes on the sea floor exactly corresponded to the relative durations of polarity chrons in the magnetic-reversal chronology.

The relationship between anomaly-stripe width and polarity-chron duration provides the key for determining the rate (velocity) of sea-floor spreading, for it indicates that the rate of spreading has been constant for the last 4.5 million years. Remember that velocity = distance/time. In the North Atlantic Ocean, 4.5-million-year-old sea floor lies 45 km away from the ridge axis. Therefore, the velocity (v) at which the sea floor moves away from the ridge axis can be calculated as follows:

$$v = \frac{45 \text{ km}}{4{,}500{,}000 \text{ years}} = \frac{4{,}500{,}000 \text{ cm}}{4{,}500{,}000 \text{ years}} = 1 \text{ cm/y}$$

This means that the crust moves away from the Mid-Atlantic Ridge axis at a rate of 1 cm per year, or that a point on one side of the ridge moves away from a point on the other side by 2 cm per year. We call this number the **spreading rate.** In the Pacific Ocean, sea-floor spreading occurs at the East Pacific Rise (geographers named this a "rise" because it is not as rough and jagged as the Mid-Atlantic Ridge). The anomaly stripes bordering the East Pacific Rise are much wider, and 4.5-million-year-old sea floor lies about 225 km from the rise axis. This requires the sea floor to move away from the rise at a rate of about 5 cm per year, so the spreading rate for the East Pacific Rise is about 10 cm per year.

Geologists also realized that if they could assume that the rate of sea-floor spreading has remained fairly constant for a long time, then they could date the ages of magnetic-field reversals further back in Earth history, simply by measuring the distance of successive magnetic anomalies from the ridge axis (time = distance/velocity). Such analysis eventually defined the magnetic-reversal chronology back to about 170 million years ago (▶Fig. 3.30c).

The spectacular correspondence between the record of marine magnetic anomalies and the magnetic-reversal chronology can only be explained by the sea-floor-spreading hypothesis. Thus, the discovery and explanation of marine magnetic anomalies served as a proof of sea-floor spreading and allowed geologists to measure rates of spreading. At a rate of 5 cm per year, sea-floor spreading produces a 5,000-km-wide ocean in 100 million years.

3.7 DEEP-SEA DRILLING: FURTHER EVIDENCE

Soon after geologists around the world began to accept the idea of sea-floor spreading, an opportunity arose to really put the concept to the test. In the late 1960s, a drilling ship called the *Glomar Challenger* set out to sail around the ocean drilling holes into the sea floor. This amazing ship could lower enough drill pipe to drill in 5-km-deep water and could continue to drill until the hole reached a depth of about 1.7 km (1.1 miles) below the sea floor. Drillers brought up cores of rock or sediment that geoscientists then studied on board.

On one of its early cruises, the *Glomar Challenger* drilled a series of holes through sea-floor sediment to the basalt layer. These holes were spaced at progressively greater distances from the axis of the Mid-Atlantic Ridge. If the model of sea-floor spreading was correct, then the sediment layer should be progressively thicker away from the axis, and the age of the oldest sediment just above the basalt should be progressively older away from the axis. When the drilling and the analyses were complete, the predictions were confirmed.

So, by the early 1960s, it had become clear that Wegener had been right all along—continents do drift. But, though the case for drift had been greatly strengthened by the discovery of apparent polar-wander paths, it really took the proposal and proof of sea-floor spreading to make believers of most geologists. Very quickly, as we will see in the next chapter, these ideas became the basis of the theory of plate tectonics.

CHAPTER SUMMARY

• Alfred Wegener proposed that continents had once been joined together to form a single huge supercontinent (Pangaea) and had subsequently drifted apart. This idea is the continental drift hypothesis.

• Wegener drew from several different sources of data to support his hypothesis: (1) coastlines on opposite sides of the ocean match up; (2) the distribution of late Paleozoic glaciers can be explained if the glaciers made up a polar ice cap over the southern end of Pangaea; (3) the distribution of late Paleozoic equatorial climatic belts is compatible with the concept of Pangaea; (4) the distribution of fossil species suggests the existence of a supercontinent; (5) distinctive rock assemblages that are now on opposite sides of the ocean were adjacent on Pangaea.

• Despite all the observations that supported continental drift, most geologists did not initially accept the idea, because no one could explain *how* continents could move.

• Rocks retain a record of the Earth's magnetic field that existed at the time the rocks formed. This record is called paleomagnetism. By measuring paleomagnetism in successively older rocks, geologists found that the apparent position of the Earth's magnetic pole relative to the rocks changes through time. Successive positions of the pole define an apparent polar-wander path.

• Apparent polar-wander paths are different for different continents. This observation can be explained by continental drift: continents move with respect to one another, while the Earth's magnetic poles remain roughly fixed.

- The invention of echo sounding permitted explorers to make detailed maps of the sea floor. These maps revealed the existence of mid-ocean ridges, deep-ocean trenches, seamount chains, and fracture zones. Heat flow is generally greater near the axis of a mid-ocean ridge.

- Around 1960, Harry Hess proposed the hypothesis of sea-floor spreading. According to this hypothesis, new sea floor forms at mid-ocean ridges, above a band of upwelling mantle, then spreads symmetrically away from the ridge axis. As a consequence, an ocean can get progressively wider with time, and the continents on either side of the ocean basins drift apart. Eventually, the ocean floor sinks back into the mantle at deep-ocean trenches.

- Magnetometer surveys of the sea floor revealed marine magnetic anomalies. Positive anomalies, where the magnetic field strength is greater than expected, and negative anomalies, where the magnetic field strength is less than expected, are arranged in alternating stripes.

- During the 1950s, geologists documented that the Earth's magnetic field reverses polarity every now and then. The record of reversals, dated by radiometric techniques, is called the magnetic-reversal chronology.

- The proof of sea-floor spreading came from the interpretation of marine magnetic anomalies. Sea floor that forms when the Earth has normal polarity results in positive anomalies, and sea floor that forms when the Earth has reversed polarity results in negative anomalies. Anomalies are symmetric with respect to a mid-ocean ridge axis, and their widths are proportional to the duration of polarity chrons, observations that can only be explained by sea-floor spreading. Study of anomalies allows us to calculate the rate of spreading.

- Drilling of the sea floor confirmed its age and served as another proof of sea-floor spreading.

KEY TERMS

abyssal plains (p. 62)
apparent polar-wander path (p. 60)
bathymetry (p. 62)
continental drift (p. 49)
dipole (p. 55)
dynamo (p. 58)
fracture zones (p. 63)
heat flow (p. 64)
magnetic anomaly (p. 66)
magnetic declination (p. 55)
magnetic field (p. 54)
magnetic force (p. 54)
magnetic inclination (p. 56)
magnetic poles (p. 54)
magnetic-reversal chronology (p. 68)
magnetic reversals (p. 67)
magnetite (p. 56)
magnetization (p. 54)
marine magnetic anomalies (p. 66)
mid-ocean ridges (p. 62)
normal and reversed polarity (p. 67)
paleomagnetism (p. 55)
paleopole (p. 58)
Pangaea (p. 49)
plates (p. 50)
plate tectonics (p. 50)
polar wander (p. 58)
polarity (p. 54)
polarity chrons (p. 68)
sea-floor spreading (p. 50)
seamounts (p. 62)
spreading rate (p. 73)
subduction (p. 50)
trenches (p. 62)
volcanic arcs (p. 62)

REVIEW QUESTIONS

1. What was Wegener's continental drift hypothesis?

2. How does the fit of the coastlines around the Atlantic support continental drift?

3. Explain the distribution of glaciers as they occurred during the Paleozoic.

4. How does the evidence of equatorial climatic belts support continental drift?

5. Was it possible for a dinosaur to walk from New York to Paris when Pangaea existed? Explain your answer.

6. Why were geologists initially skeptical of Wegener's continental drift hypothesis?

7. Describe how the angle of inclination of the Earth's magnetic field varies with latitude. How could paleomagnetic inclination be used to determine the ancient latitude of a continent?

8. How does basalt develop paleomagnetism? Can paleomagnetism develop in sedimentary rock? If so, how?

9. How do apparent polar-wander paths show that the continents, rather than the poles, had moved?

10. Describe the basic characteristics of mid-ocean ridges, deep-ocean trenches, and seamount chains.

11. Describe the hypothesis of sea-floor spreading.

12. How did the observations of heat flow and seismicity support the hypothesis of sea-floor spreading?

13. How were the reversals of the Earth's magnetic field discovered? How did they corroborate the sea-floor-spreading hypothesis?

14. What is a marine magnetic anomaly? How is it detected?

15. Describe the pattern of marine magnetic anomalies across a mid-ocean ridge. How is this pattern explained?

16. How did geologists calculate rates of sea-floor spreading?

17. Did drilling into the sea floor contribute further proof of sea-floor spreading? If so, how?

SUGGESTED READING

Allégre, C. 1988. *The Behavior of the Earth: Continental and Seafloor Mobility.* Cambridge, Mass.: Harvard University Press.

Bird, J. M., ed. 1980. *Plate Tectonics.* Washington, D.C.: American Geophysical Union.

Butler, R. F. 1992. *Paleomagnetism: Magnetic Domains to Geologic Terranes.* Boston: Blackwell.

Campbell, W. H. 2001. *Earth Magnetism: A Guided Tour through Magnetic Fields.* New York: Harcourt/Academic Press.

Erikson, J. 1992. *Plate Tectonics: Unraveling the Mysteries of the Earth.* New York: Facts on File.

Glen, W. 1982. *The Road to Jaramillo: Critical Years of the Revolution in Earth Sciences.* Palo Alto, Calif.: Stanford University Press.

LeGrand, H. E. 1988. *Drifting Continents and Sifting Theories.* Cambridge: Cambridge University Press.

McFadden, P. L., and M. W. McElhinny. 2000. *Paleomagnetism: Continents and Oceans,* 2nd ed. San Diego: Academic Press.

McPhee, J. A. 1998. *Annals of the Former World.* New York: Farrar, Straus and Giroux.

Moores, E. M., ed. 1990. *Plate Tectonics: Readings from "Scientific American."* New York: Freeman.

Oreskes, N., ed. 2003. *Plate Tectonics: An Insider's History of the Modern Theory of the Earth.* Boulder: Westview Press.

Sullivan, W. 1991. *Continents in Motion: The New Earth Debate.* 2nd ed. New York: American Institute of Physics.

CHAPTER 4

The Way the Earth Works: Plate Tectonics

Astronauts in the space shuttle Endeavor could see the consequences of plate tectonics right outside their window. This photo from space shows the Sinai Peninsula, separated from Egypt to the west and the Arabian Peninsula to the east by rifts, narrow belts where the crust has stretched and broken apart. Note the green stripe in the middle—this is the Nile Valley, which flows into the southern Mediterranean Sea. The triangle of green is the Nile Delta.

4.1 INTRODUCTION

Thomas Kuhn, an influential historian of science working in the 1960s, argued that scientific thought evolves in fits and starts. According to Kuhn, scientists base their interpretation of the natural world on an established line of reasoning, a scientific *paradigm,* for many years. But once in a while, a revolutionary thinker or group of thinkers proposes a radically new point of view that invalidates the old paradigm. Almost immediately, the scientific community scraps old hypotheses and formulates others consistent with the new paradigm. Kuhn called such abrupt changes in thought a **scientific revolution.** Some new paradigms work so well that they become scientific law and will never be replaced.

In physics, Isaac Newton's mathematical description of moving objects established the paradigm that natural phenomena obey physical laws; older paradigms suggesting that natural phenomena followed the dictates of Greek philosophers had to be scrapped. In biology, Charles Darwin's proposal that species evolve by natural selection required biologists to rethink hypotheses based on the older paradigm that species never change. And in geology, a scientific revolution in the 1960s yielded the new paradigm that the outer layer of the Earth, the lithosphere, consists of separate pieces, or plates, that move with respect to each other. This idea, which we now call the "theory of plate tectonics," or simply **plate tectonics,** required geologists to cast aside hypotheses rooted in the paradigm of fixed continents, and thus led to a complete restructuring of how geologists think about Earth history. Compare this book with a geology textbook from the 1950s, and you will instantly see the difference.

Alfred Wegener planted the seed of plate tectonics theory with his proposal of continental drift in 1915, but between 1915 and 1960 this seed lay dormant while geoscientists focused on collecting new data about the Earth. Discoveries about the ocean floor and about apparent polar wander led to the germination of the seed in 1960, with Harry Hess's and Robert Dietz's proposal of sea-floor spreading. The roots took hold three years later when marine magnetic anomalies supplied strong evidence for sea-floor spreading. During the next five years, the study of geoscience turned into a feeding frenzy, as many investigators dropped what they'd been doing and turned their attention to testing the hypothesis of sea-floor spreading and describing its broader implications; by 1968, thanks primarily to the work of perhaps two dozen different investigators, the sea-floor-spreading hypothesis had bloomed into plate tectonics theory. Geologists clarified the concept of a plate, described the types of plate boundaries, calculated plate motions, related plate tectonics to earthquakes and volcanoes, showed how plate motions generate mountain belts and seamount chains, and defined the history of past plate motions. Between 1968 and 1970, after excited investigators had presented their new ideas to standing-room-only audiences at many conferences, the geoscience community, with few exceptions, embraced plate tectonics theory and has built on it ever since.

To begin our explanation of the key elements of plate tectonics theory, we first learn about lithosphere plates, the three types of plate boundaries, and the nature of geologic activity that occurs at each. We then look at hot spots and other special locations on plates. Finally, we see how continents break apart and how they collide, and we learn about what makes plates move. Because plate tectonics theory is geology's grand unifying theory, it is now an essential foundation for the discussion of all geology.

4.2 WHAT DO WE MEAN BY PLATE TECTONICS?

The Concept of a Lithosphere Plate

As we learned in Chapter 2, geoscientists divide the interior of the Earth into layers. If we want to distinguish layers according to the speed at which seismic waves pass through

them, we speak of the crust, upper mantle, lower mantle, outer core, and inner core, and we define the boundaries between these layers by abrupt changes in the speed of seismic waves. But if, instead, we want to distinguish layers according to whether they can flow relatively easily, we use the names "lithosphere" and "asthenosphere." Let's now clarify the definitions of these important terms.

The **lithosphere** consists of the crust *plus* the top (cooler) part of the upper mantle. It behaves rigidly and elastically, meaning that when a force pushes or pulls on it, overall it does not flow but rather bends and flexes, or breaks (▶Fig. 4.1a). The lithosphere floats on a relatively soft, or "plastic," layer called the **asthenosphere,** composed of warmer (>1,280°C) mantle that can flow (though very slowly) when acted on by force. Therefore, the asthenosphere can undergo convection, like water in a pot, but the lithosphere cannot.

Continental lithosphere and oceanic lithosphere differ markedly in their thickness. On average, continental lithosphere has a thickness of 150 km, whereas old oceanic lithosphere has a thickness of about 100 km; for reasons discussed later in this chapter, new oceanic lithosphere at a mid-ocean ridge is only 7 to 10 km thick. Recall that the crustal part of continental lithosphere ranges from 25 to 70 km thick and consists of relatively low density felsic and intermediate rock. In contrast, the crustal part of oceanic lithosphere is only 7 to 10 km thick and consists of relatively high density mafic rock.

The surface of continental lithosphere lies at a higher elevation than the surface of oceanic lithosphere. To picture why, imagine that we have two blocks of oak (a high-density wood), one 15 cm thick and one 10 cm thick. On top of the thicker block, we glue a 4-cm-thick layer of cork (a low-density wood), and on top of the thinner block, we glue a 1-cm-thick layer of pine (a medium-density wood). Now we place the two blocks in water (▶Fig. 4.1b). The total mass of the cork-covered block exceeds the total mass of the pine-covered block, so the base of the cork-covered block sinks deeper into the water. But because the cork-covered block is thicker and has a lower overall density, it floats higher. In our analogy, the cork-covered block represents continental lithosphere, with its thick crust of low-density rock, while the pine-covered block represents oceanic lithosphere, with its thin crust of dense rock. The oak represents the very high-density ultramafic rock (peridotite) constituting the mantle part of the lithosphere, thicker for the continent than for the ocean (▶Fig. 4.1c). Our analogy emphasizes that ocean floors are low and thus fill with water to form oceans, because continental lithosphere is more buoyant and floats higher than oceanic lithosphere (see Box 4.1 and Fig. 4.2).

The lithosphere forms the Earth's relatively rigid shell. But unlike the shell of a hen's egg, the lithosphere shell contains a number of major "breaks," which separate the

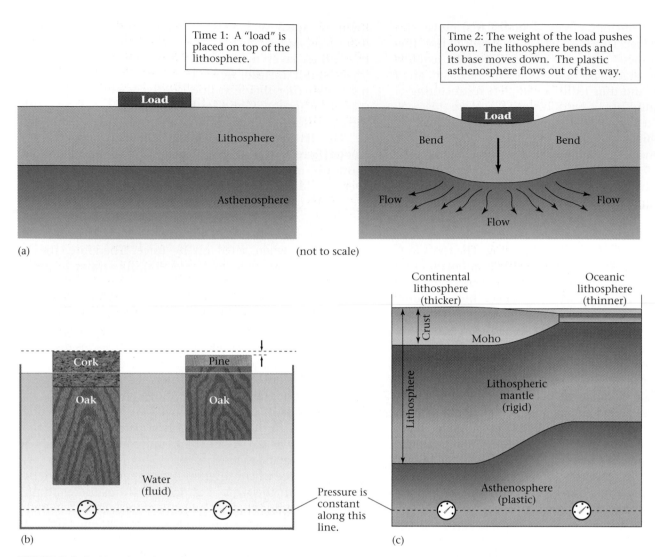

FIGURE 4.1 (a) Lithosphere bends when a load is placed on it, whereas asthenosphere flows. (b) We can picture continental lithosphere as a thick oak block (lithospheric mantle) overlaid by a layer of cork (continental crust), and oceanic lithosphere as a thinner block of oak overlaid by a layer of pine (oceanic crust). The pine layer is thinner than the cork layer. If both blocks float in a tub of water, the surface of the thick cork/oak block lies at a higher elevation than that of the pine/oak block. (c) Similarly, the ocean floor lies 4–5 km below the surface of the continents, on average, because lithosphere, like the wood blocks, floats on the asthenosphere.

lithosphere into distinct pieces.[1] We call the pieces "lithosphere plates," or simply **plates,** and we call the breaks **plate boundaries.** Geoscientists distinguish twelve major plates and several microplates. Some plates have fa-

miliar names (e.g., the North American Plate, the African Plate), while some do not (e.g., the Cocos Plate, the Juan de Fuca Plate).

As illustrated in ▶Figure 4.3, some plate boundaries follow "continental margins," the boundary between a continent and an ocean, while others do not. For this reason, we distinguish between **active margins,** which are plate boundaries, and **passive margins,** which are not plate boundaries. Along passive margins, continental crust is thinner than normal (▶Fig. 4.4). (As we discuss in Section 4.7, this thinning occurs during the initial formation of the ocean; during thinning, the upper part breaks into wedge-shaped slices.) Thick (10–15 km) ac-

---

[1]Note that the above definition equates the base of a plate with the base of the lithosphere. This definition works well for oceanic plates, because the asthenosphere directly beneath oceanic lithosphere is particularly weak. But some geologists now think that the upper 100–150 km of the asthenosphere beneath continents actually moves with the continental lithosphere. Thus, the base of continental plates—the base of the layer that moves during plate motion—may actually lie within the asthenosphere. To simplify our discussion we ignore this complication.

cumulations of sediment cover this thinned crust. The surface of this sediment layer is a broad, shallow (less than 500 m deep) region called the **continental shelf,** home to the major fisheries of the world. Note that some plates consist entirely of oceanic lithosphere or entirely of continental lithosphere, while some plates consist of both. For example, the Nazca Plate is made up entirely of ocean floor, while the North American Plate consists of North America plus the western half of the North Atlantic Ocean.

The Basic Premise of Plate Tectonics

We can now restate plate tectonics theory concisely as follows. The Earth's lithosphere is divided into plates that move relative to one another and relative to the underlying asthenosphere. Plate movement occurs at rates of about 1 to 15 cm per year. As a plate moves, its internal area remains largely rigid and intact, but rock along the plate's boundaries undergoes deformation (cracking, sliding, bending, stretching, and squashing) as the plate grinds or scrapes against its neighbors or pulls away from its neighbors. As plates move, so do the continents that form part of the plates, resulting in continental drift. Because of plate tectonics, the map of Earth's surface constantly changes.

Identifying Plate Boundaries

How do we recognize the location of a plate boundary? The answer becomes clear from looking at a map showing the locations of earthquakes (▶Fig. 4.5). Recall from Chapter 2 that earthquakes are vibrations caused by shock waves that are generated where rock breaks and suddenly shears (slides) along a fault (a fracture on which sliding occurs). The **hypocenter** (or focus) of the earthquake is the spot where the fault begins to slip, and the **epicenter** marks the point on the surface of the Earth directly above

---

BOX 4.1
SCIENCE TOOLBOX

Archimedes' Principle of Buoyancy

Archimedes (c. 287–212 B.C.E.), a Greek mathematician and inventor, left an amazing legacy of discoveries. He described the geometry of spheres, cylinders, and spirals, introduced the concept of a center of gravity, and was the first to understand buoyancy. **Buoyancy** refers to the upward force acting on an object immersed or floating in a fluid. According to legend, Archimedes recognized this concept suddenly, while bathing in a public bath, and was so inspired that he jumped out of the bath and ran home naked, shouting "Eureka!"

Archimedes realized that when you place a solid object in water, the object displaces a volume of water equal in mass to the object (▶Fig. 4.2a, b). An object denser than water, like a stone, sinks through the water, because even when completely submerged, the stone's mass exceeds the mass of the water displaced. When submerged, however, the stone weighs less than it does in air. (For this reason, a scuba diver can lift a heavy object underwater.) An object less dense than water, such as an iceberg, sinks only until the mass of the water displaced equals the total mass of the iceberg. This condition happens while part of the iceberg still protrudes up into the air. Put another way, an object placed in a fluid feels a "buoyancy force" that tends to push it up. If the object's weight is less than the buoyancy force, the object floats, but if its weight is greater than the buoyancy force, the object sinks.

FIGURE 4.2 Archimedes' principle of buoyancy. (a) Ice is less dense than water, so it is buoyant. When an iceberg floats, 80% of the ice lies underwater. (b) The ice sinks until the total mass of the water displaced equals the total mass of the whole iceberg. Since water is denser, the volume of the water displaced is less than the volume of the iceberg, so the iceberg protrudes above the water.

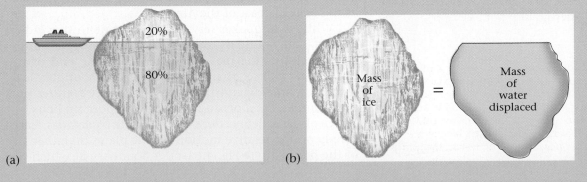

(a) (b)

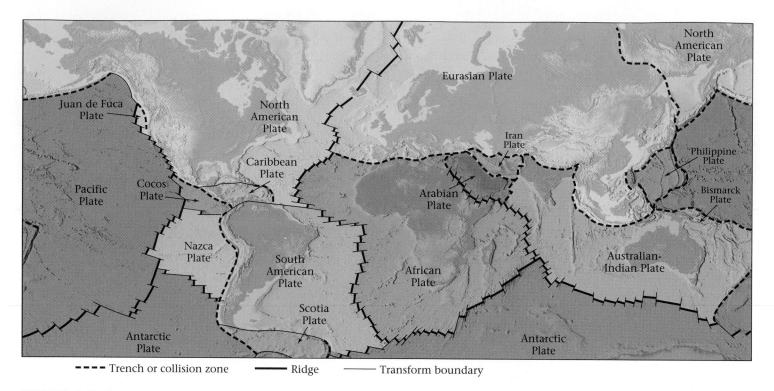

---- Trench or collision zone ——— Ridge ——— Transform boundary

FIGURE 4.3 The major plates making up the lithosphere. Note that some plates are all ocean floor, while some contain both continents and oceans. Thus, some plate boundaries lie along continental margins (coasts), while others do not. For example, the eastern border of South America is not a plate boundary, but the western edge is. For a more detailed map of plate boundaries, see the inside of this book's cover.

FIGURE 4.4 In this block diagram of a passive margin, note that the continental crust thins along the boundary (see Section 4.7, "Continental Rifting"). The sediment pile that accumulates over this thinned crust underlies the continental shelf.

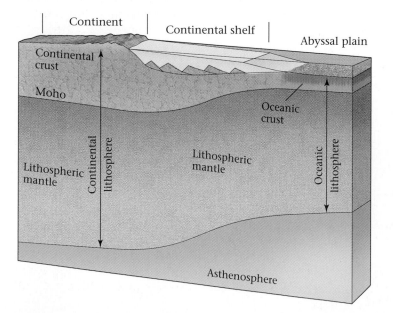

the focus. Earthquake epicenters do not speckle the Earth's surface randomly, like buckshot on a target. Rather, the majority occur in relatively narrow, distinct belts. These "earthquake belts" define the position of plate boundaries, because the fracturing and slipping that occur along plate boundaries as plates move generate earthquakes. (We will learn more about this process in Chapter 10.) **Plate interiors,** regions away from the plate boundaries, remain relatively earthquake free because they are stronger and do not accommodate much movement. Note that earthquakes occur frequently along active continental margins, such as the Pacific coast of the Americas, but not along passive continental margins, such as the eastern coast of the Americas.

While earthquakes serve as the most definitive indicator of a plate boundary, other prominent geologic features also develop along plate boundaries. By the end of this chapter, we will see that each type of plate boundary is associated with a diagnostic group of geologic features such as volcanoes, deep-ocean trenches, or mountain belts.

Geologists define *three types of plate boundaries,* based simply on the relative motions of the plates on either side of the boundary (▶Fig. 4.6a–c). A boundary at which two

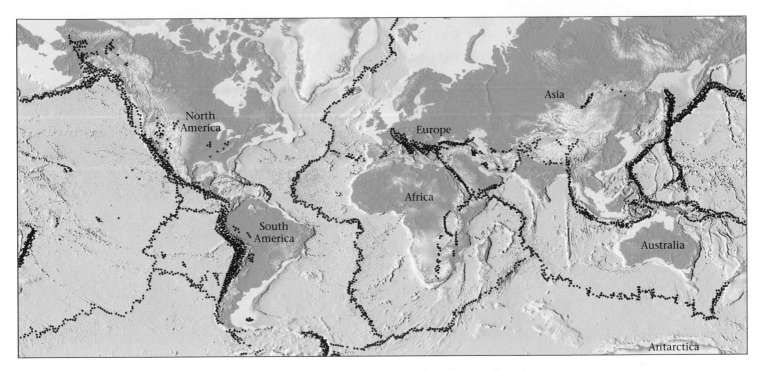

FIGURE 4.5 The locations of most earthquakes fall in distinct bands, or belts. These earthquake belts define the positions of the plate boundaries. Compare this map with the plate boundaries on Figure 4.3. For more detailed earthquake maps, see Appendix B.

plates move apart from each other is called a **divergent boundary.** A boundary at which two plates move toward each other so that one plate sinks beneath the other is called a **convergent boundary.** And a boundary at which one plate slips along the side of another plate is called a **transform boundary.** Each type looks and behaves differently from the others, as we will now see.

4.3 DIVERGENT PLATE BOUNDARIES AND SEA-FLOOR SPREADING

At divergent boundaries, or spreading boundaries, two oceanic plates move apart by the process of sea-floor spreading. Note that an open space does not develop between diverging plates. Rather, as the plates move apart, new oceanic lithosphere forms along the divergent boundary (▶Fig. 4.7). This process takes place at a submarine mountain range called a **mid-ocean ridge** (such as the Mid-Atlantic Ridge, the East Pacific Rise, and the Southeast Indian Ocean Ridge), which rises 2 km above the adjacent abyssal plains of the ocean. Thus, geologists also commonly call a divergent boundary a mid-ocean ridge, or simply a "ridge."

Characteristics of a Mid-Ocean Ridge

To better characterize a divergent boundary, let's look at one mid-ocean ridge, the Mid-Atlantic Ridge, in more detail (▶Figs. 4.8; 4.9). The Mid-Atlantic Ridge extends from the waters between northern Greenland and northern Scandinavia southward across the equator to the latitude opposite the southern tip of South America. For most of its length, the elevated area of the ridge is about 1,500 km wide.

Geologists have mapped segments of the Mid-Atlantic Ridge in detail, using sonar from ships and from research submarines. They have found that the formation of new sea floor takes place only across a remarkably narrow band—less than a few kilometers wide—along the axis (centerline) of the ridge. The axis lies at water depths of about 2–2.5 km. Along ridges, like the Mid-Atlantic, where sea-floor spreading occurs slowly, the axis lies in a narrow trough about 500 m deep and less than 10 km wide, bordered on either side by steep cliffs. Roughly speaking, the Mid-Atlantic Ridge is symmetrical—its eastern half looks like a mirror image of its western half. As illustrated by Figure 4.9, the ridge consists, along its length, of short segments (tens to hundreds of km long) linked by breaks called transform faults, which we will discuss later.

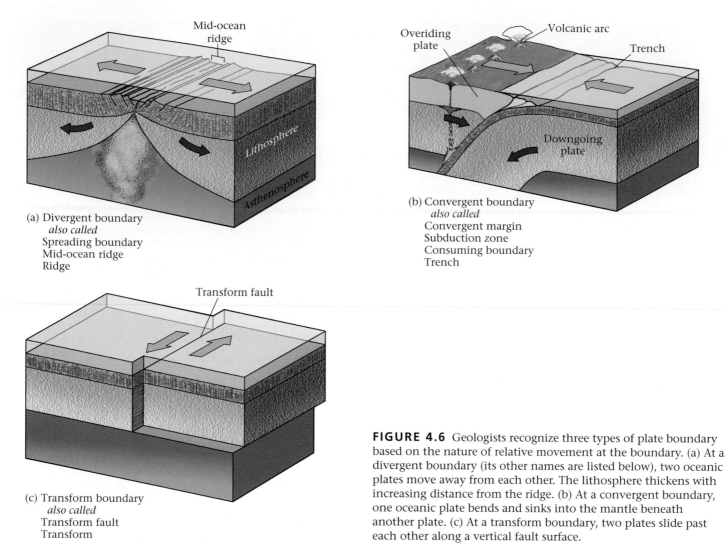

(a) Divergent boundary
   also called
   Spreading boundary
   Mid-ocean ridge
   Ridge

(b) Convergent boundary
   also called
   Convergent margin
   Subduction zone
   Consuming boundary
   Trench

(c) Transform boundary
   also called
   Transform fault
   Transform

FIGURE 4.6 Geologists recognize three types of plate boundary based on the nature of relative movement at the boundary. (a) At a divergent boundary (its other names are listed below), two oceanic plates move away from each other. The lithosphere thickens with increasing distance from the ridge. (b) At a convergent boundary, one oceanic plate bends and sinks into the mantle beneath another plate. (c) At a transform boundary, two plates slide past each other along a vertical fault surface.

Not all mid-ocean ridges look just like the Mid-Atlantic. For example, ridges at which spreading occurs rapidly, such as the East Pacific Rise, do not have the axial trough we see along the Mid-Atlantic Ridge. Also, the region of elevated sea floor of faster-spreading ridges is much wider.

The Formation of Oceanic Crust at a Mid-Ocean Ridge

As noted above, sea-floor spreading does not create an open space between diverging plates. Rather, as each increment of spreading occurs, new sea floor develops in the space. How does this happen?

As sea-floor spreading takes place, hot asthenosphere (the soft, flowable part of the mantle) rises beneath the ridge (Fig. 4.8). As this asthenosphere rises, it begins to melt, producing molten rock, or magma. Magma has a lower density than solid rock, so it behaves buoyantly and rises. It eventually accumulates in the crust below the ridge axis. The lower part of this region is a mush of crys-

tals, above which magma pools in a fairly small **magma chamber.** Some of the magma solidifies along the side of the chamber to make a coarse-grained, mafic igneous rock called gabbro. Some of the magma rises still higher to fill vertical cracks, where it solidifies and forms wall-like sheets, or **dikes,** of basalt. Finally, some magma rises all the way to the surface of the sea floor at the ridge axis and spills out of small submarine volcanoes. The resulting lava cools to form a layer of basalt blobs, called **pillow basalt,** on the sea floor. We can't easily see the submarine volcanoes because they occur at depths of more than 2 km beneath sea level, but they have been observed by the research submarine *Alvin* (see Box 4.2). *Alvin* has also detected chimneys spewing hot, mineralized water that rose through cracks in the sea floor, after being heated by magma below the surface. These chimneys are called **black smokers** because the water they emit looks like dark smoke (▶Fig. 4.10).

As soon as it forms, new oceanic crust moves away from the ridge axis, and as this happens, more magma rises from

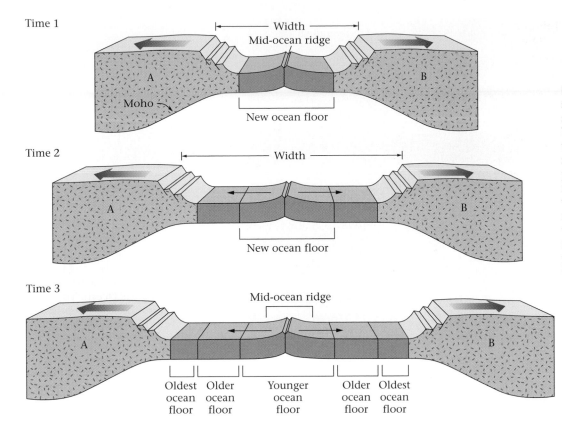

FIGURE 4.7 These sketches depict successive stages in sea-floor spreading along a divergent boundary (mid-ocean ridge); only the crust is shown. The top figure represents an early stage of the process, after the mid-ocean ridge formed but before the ocean grew very wide. With time, as seen in the next two figures, the ocean gets wider and continent A drifts way from continent B. Note that the youngest ocean crust lies closest to the ridge.

FIGURE 4.8 How new lithosphere forms at a mid-ocean ridge. Rising hot asthenosphere partly melts underneath the ridge axis. The molten rock, magma, rises to fill a magma chamber in the crust. Some of the magma solidifies along the side of the chamber, to make coarse-grained mafic rock called gabbro. Some magma rises still farther to fill cracks, solidifying into basalt that forms wall-like sheets of rock called dikes. Finally, some magma rises all the way to the surface of the sea floor at the ridge axis. This magma, now called lava, spills out and forms a layer of basalt. As sea-floor spreading continues, the oceanic crust breaks along faults. Also, as a plate moves away from a ridge axis and cools, the lithospheric mantle thickens.

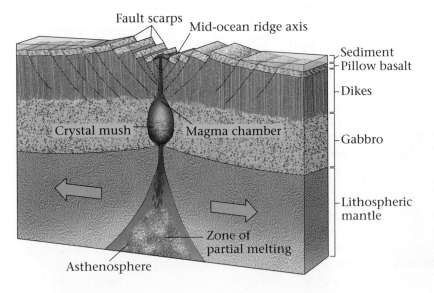

below, so still more crust forms. In other words, like a vast, continuously moving conveyor belt, magma from the mantle rises to the Earth's surface at the ridge, solidifies to form oceanic crust, and then moves laterally away from the ridge. Because all sea floor forms at mid-ocean ridges, the youngest sea floor occurs on either side of the ridge, and sea floor becomes progressively older away from the ridge (Fig. 4.7). In the Atlantic Ocean, the oldest sea floor lies adjacent to the passive continental margins on either side of the ocean (▶Fig. 4.11). The oldest ocean floor on our planet occurs in the western Pacific Ocean; this crust formed 200 million years ago.

The tension (stretching force) applied to newly formed solid crust as spreading takes place breaks this new crust, resulting in the formation of faults. Slip on the faults causes divergent-boundary earthquakes and creates numerous cliffs, or "scarps," that parallel the ridge axis.

The Formation of Lithospheric Mantle at a Mid-Ocean Ridge

So far, we've seen how oceanic crust forms at mid-ocean ridges. What about the formation of the mantle part of the oceanic lithosphere? Remember

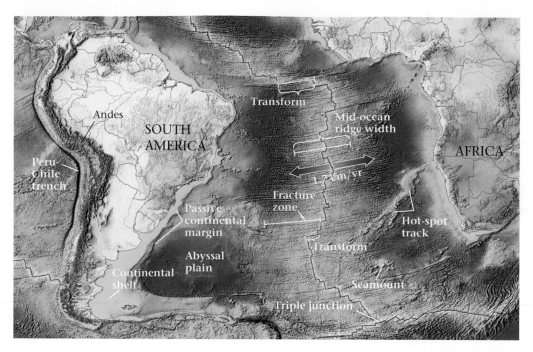

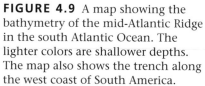

FIGURE 4.9 A map showing the bathymetry of the mid-Atlantic Ridge in the south Atlantic Ocean. The lighter colors are shallower depths. The map also shows the trench along the west coast of South America.

that this part consists of the cooler uppermost area of the mantle, in which temperatures are less than about 1,280°C. At the ridge axis, such temperatures occur almost at the base of the crust, because of the presence of rising hot asthenosphere and hot magma, so the lithospheric mantle be-

FIGURE 4.10 A column of superhot water gushing from a vent (known as a "black smoker") along the mid-ocean ridge. The water has been heated by magma (molten rock) just below the surface. The cloud of "smoke" actually consists of tiny mineral grains; the elements making these minerals had been dissolved in the hot water, but when the hot water mixes with cold water of the sea, they precipitate. Many exotic species of life, such as giant worms, live around these vents.

neath the ridge axis effectively doesn't exist. But as the newly formed oceanic crust moves away from the ridge axis, the crust and the uppermost mantle directly beneath it gradually cool as they lose heat to the ocean above. As soon as mantle rock cools below 1,280°C, it becomes, by definition, part of the lithosphere.

As oceanic lithosphere continues to move away from the ridge axis, it continues to cool, so the lithospheric mantle, and therefore the oceanic lithosphere as a whole, grows progressively thicker (▶Fig. 4.12a). This process doesn't change the thickness of the oceanic crust, for the crust formed entirely at the ridge axis. The rate at which cooling and thickening occur decreases with distance from the ridge axis. In fact, by the time the lithosphere is about 80 million years old it has just about reached its maximum thickness (▶Fig. 4.12b).

The Reason Mid-Ocean Ridges Are High

Why does the surface of the sea floor rise to form a mid-ocean ridge along divergent plate boundaries? The answer comes from considering the buoyancy of oceanic lithosphere (see Box 4.1). As sea floor ages, the asthenosphere below cools enough to become part of the lithosphere, and the lithospheric mantle thickens. Cooler rock is denser than warmer rock, so the process of cooling and thickening the lithosphere, like adding ballast to a ship, causes the lithosphere to sink deeper into the asthenosphere (▶Fig. 4.12c). Hot young lithosphere is less dense and floats higher; this high-floating lithosphere constitutes the mid-ocean ridge. Because lithosphere cools and thickens as it grows older, the depth of the sea floor depends on its age (Fig. 4.12b).

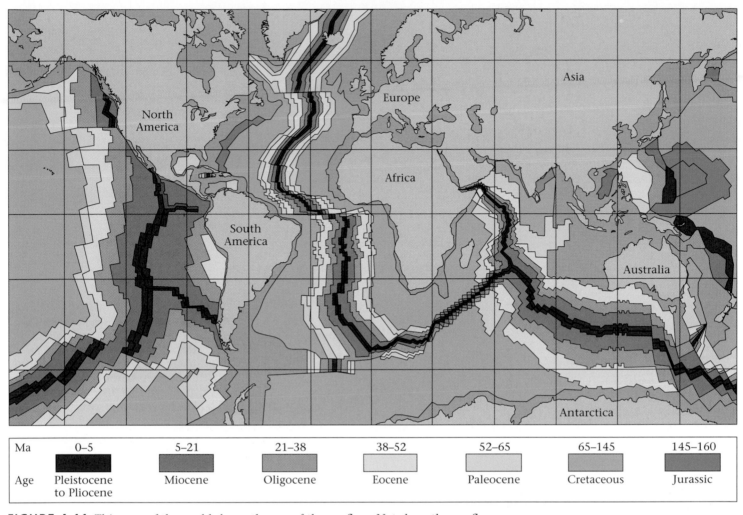

Ma	0–5	5–21	21–38	38–52	52–65	65–145	145–160
Age	Pleistocene to Pliocene	Miocene	Oligocene	Eocene	Paleocene	Cretaceous	Jurassic

FIGURE 4.11 This map of the world shows the age of the sea floor. Note how the sea floor grows older with increasing distance from the ridge axis. (Ma = million years ago)

4.4 CONVERGENT PLATE BOUNDARIES AND SUBDUCTION

At convergent plate boundaries, or convergent margins, two plates, at least one of which is oceanic, move toward each other. But rather than butting each other like angry rams, one oceanic plate bends and sinks down into the asthenosphere beneath the other plate. Geologists refer to the sinking process as **subduction,** so convergent boundaries are also known as subduction zones. Because subduction at a convergent boundary consumes old ocean lithosphere and thus closes (or "consumes") oceanic basins, geologists also refer to convergent boundaries as consuming boundaries, and because they are delineated by deep-ocean trenches, they are sometimes simply called **trenches** (Fig. 4.9). The amount of oceanic plate consumption worldwide, averaged over time, equals the amount of sea-floor spreading worldwide, so the surface area of the Earth remains constant through time.

Subduction occurs for a simple reason: oceanic lithosphere, once it has aged at least 10 million years, is denser than asthenosphere and thus can sink through the asthenosphere. When it lies flat on the surface of the asthenosphere, oceanic lithosphere doesn't sink because the resistance of the asthenosphere to flow is too great; however, once the end of the convergent plate bends down and slips into the mantle, it begins to sink like an anchor falling to the bottom of a lake (▶Fig. 4.13a, b). As the lithosphere sinks, asthenosphere flows out of the way, just as water flows out of the way of an anchor. Even though it is relatively soft and plastic the asthenosphere resists flow, so oceanic lithosphere can sink only very slowly, at a rate of less than 10–15 cm per year.

Note that the **downgoing plate** (or slab), the plate that has been subducted, *must* be composed of oceanic

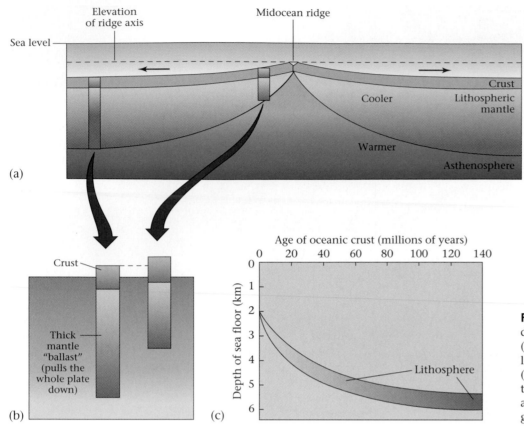

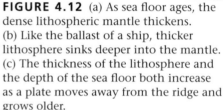

(a)

(b)

(c)

FIGURE 4.12 (a) As sea floor ages, the dense lithospheric mantle thickens. (b) Like the ballast of a ship, thicker lithosphere sinks deeper into the mantle. (c) The thickness of the lithosphere and the depth of the sea floor both increase as a plate moves away from the ridge and grows older.

FIGURE 4.13 (a) The concept of subduction. A plate bends, and one piece pushes over the other. Oceanic lithosphere is denser than the underlying asthenosphere, but when it lies flat on the surface of the asthenosphere, it can't sink because the resistance of the asthenosphere to flow is too great. However, once the end of the plate is pushed into the mantle, the lithosphere begins to sink. (b) The process of sinking is like an anchor pulling a floating anchor line down. As a consequence, the bend in the plate (or in the anchor line) progressively moves with time.

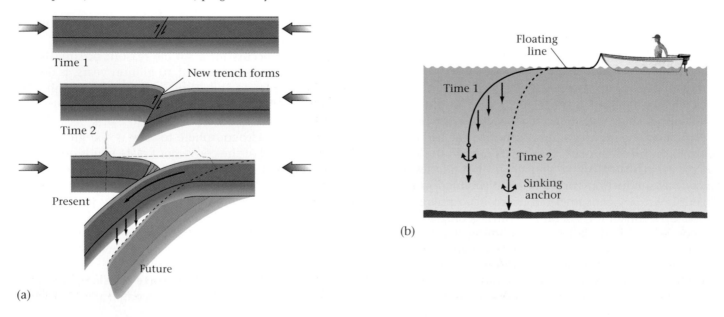

(a)

(b)

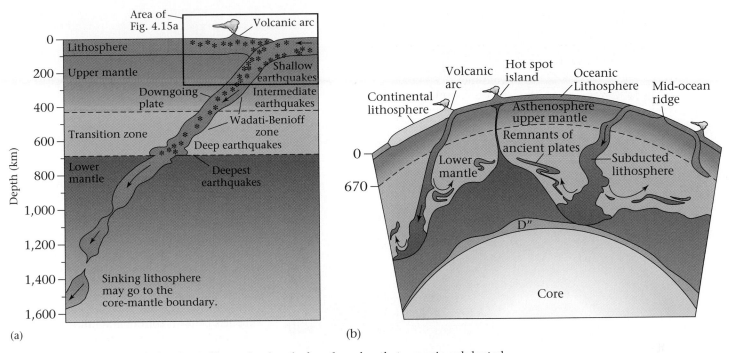

FIGURE 4.14 (a) The Wadati-Benioff zone is a band of earthquakes that occur in subducted oceanic lithosphere. The discovery of these earthquakes led to the proposal of subduction. (b) A model illustrating the ultimate fate of subducted lithosphere. In this model, the lower mantle contains two regions (shallower and deeper), which differ in density and possibly composition. D″ is the name for a hot region just above the core-mantle boundary.

lithosphere. The **overriding plate** (or slab), which does not sink, can consist of either oceanic or continental lithosphere. Continental crust cannot be subducted because it is too buoyant; the low-density rocks of continental crust act like a life preserver keeping the continent afloat. If continental crust moves into a convergent margin, subduction stops. Because of subduction, all ocean floor on the planet is less than about 200 million years old. Because continental crust cannot subduct, some continental crust has persisted at the surface of the Earth for over 3.8 billion years.

Earthquakes and the Fate of Subducted Plates

At convergent plate boundaries, the downgoing plate grinds along the base of the overriding plate, a process that generates large earthquakes. These earthquakes occur fairly close to the Earth's surface, so some of them trigger massive destruction in coastal cities. But earthquakes also happen in downgoing plates at greater depths, deep below the overriding plate. In fact, geologists have detected earthquakes within downgoing plates to a depth of 660 km; the belt of earthquakes in a downgoing plate is called a **Wadati-Benioff zone,** after its two discoverers (▶Fig. 4.14a).

At depths greater than 660 km, conditions leading to earthquakes evidently do not occur. Recent evidence, however, indicates that some downgoing plates do continue to sink below a depth of 660 km—they just do so without generating earthquakes. In fact, studies suggest that the lower mantle may be a graveyard for old subducted plates (▶Fig. 4.14b).

Geologic Features of a Convergent Boundary

To become familiar with the various geologic features that occur along a convergent plate boundary, let's look at an example, the boundary between the western coast of the South American Plate and the eastern edge of the Nazca Plate (a portion of the Pacific Ocean floor). A deep-ocean trench, the Peru-Chile Trench, delineates this boundary (Fig. 4.9). Such trenches form where the plate bends as it starts to sink into the asthenosphere.

In the Peru-Chile Trench, as the downgoing plate slides under the overriding plate, sediment (clay and plankton) that had settled on the surface of the downgoing plate, as well as sand that fell into the trench from the shores of South America, gets scraped up and incorporated in a wedge-shaped mass known as an **accretionary prism**

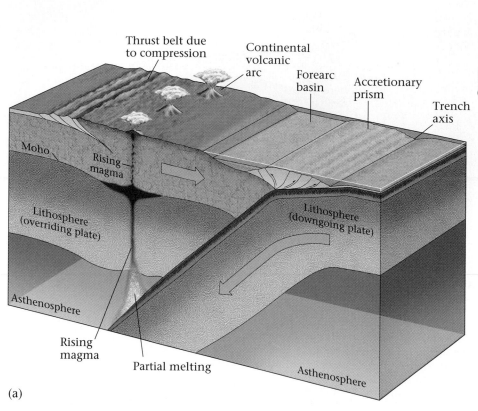

(a)

(b)

FIGURE 4.15 (a) This model shows the geometry of subduction along an active continental margin. The trench axis (lowest part of the trench) roughly defines the plate boundary. Numerous faults form in the accretionary prism, which is composed of material scraped off the sea floor. Behind the prism lies a basin (a forearc basin) of trapped sediment. A volcanic arc is created from magma that forms at or just above the surface of the downgoing plate. Here, the plate subducts beneath continental lithosphere, so the chain of volcanoes is called a continental arc. Faulting occurs on the backside of the arc. The Andes in South America and the Cascades in the United States are examples of such continental arcs. (b) The action of a bulldozer pushing snow or soil is similar to the development of an accretionary prism.

(►Fig. 4.15a). An accretionary prism forms in basically the same way as a pile of snow in front of a plow, and like the snow, the sediment tends to be squashed and contorted during the formation of the prism (►Fig. 4.15b).

A chain of volcanoes known as a **volcanic arc** develops behind the accretionary prism (see Box 4.2). As we will see in Chapter 6, the magma that feeds these volcanoes forms at or just above the surface of the downgoing plate when the plate reaches a depth of about 150 km below the Earth's surface. If the volcanic arc forms where an oceanic plate subducts beneath continental lithosphere, the resulting chain of volcanoes grows on the continent and forms a "continental volcanic arc." In some cases, the plates push together, causing mountains to rise (Fig. 4.15a). If, however, the volcanic arc forms where one oceanic plate subducts beneath another oceanic plate, the resulting volcanoes form a chain of islands known as a volcanic "island arc." A **marginal sea** (or back-arc basin), the small ocean basin between an island arc and the continent, forms either in cases where subduction happens to begin offshore, trapping ocean lithosphere behind the arc, or where stretching of the lithosphere behind the arc leads to the formation of a small spreading ridge between the arc and the continent (►Fig. 4.16).

4.5 TRANSFORM PLATE BOUNDARIES

We saw earlier that the spreading axis of a mid-ocean ridge consists of short segments. The ends of these segments are linked to each other by narrow belts of broken and irregular sea floor, known as **fracture zones** (see Fig. 4.9). Fracture zones lie roughly at right angles to the ridge segments and extend beyond the ends of the segments (►Fig. 4.17a). The geometric relationship of fracture zones to ridge segments, and evidence indicating that fracture zones are made up of broken-up crust, originally led geoscientists to conclude that fracture zones were faults. They then incorrectly assumed that sliding on faults in fracture zones broke an originally continuous ridge into segments and displaced the segments sideways (►Fig. 4.17b, c). This interpretation implies that one segment moves with respect to its neighbor, as shown by the arrows in Figure 4.17c. But soon after Harry Hess proposed his model of sea-floor spreading in 1960, a Canadian named J. Tuzo Wilson realized that if sea-floor spreading really occurred, then the notion that fracture zones offset an originally continuous ridge could not be correct.

In Wilson's alternative interpretation, the fracture zone formed *at the same time* as the ridge axis itself, and

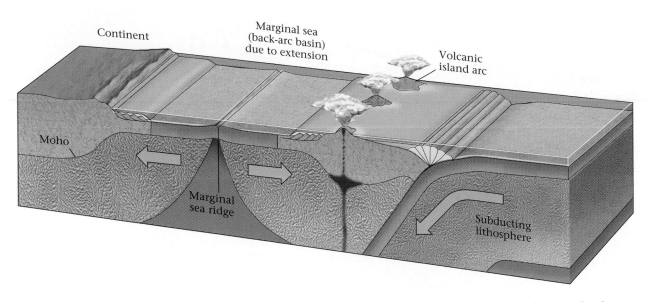

FIGURE 4.16 Subduction along an island arc. Here, the volcanoes build on the sea floor. Behind some island arcs, a marginal sea forms. This sea resembles a small ocean basin, with a spreading ridge that is created when the plate behind the arc moves away from the arc.

thus the ridge consisted of separate segments to start with. These segments were *linked* (not offset) by fracture zones. With this idea in mind, he drew a sketch map showing two ridge-axis segments linked by a fracture zone, and he drew arrows to indicate the direction that ocean crust was moving, relative to the ridge axis, as a result of sea-floor spreading (▶Fig. 4.17d). Look at Wilson's arrows. Clearly, the movement direction on the fracture zone must be opposite to the movement direction that geologists originally thought occurred on the structure. Further, in Wilson's model, slip occurs *only* along the segment of the fracture zone between the two ridge segments. Plate A moves with respect to plate B as sea-floor spreading occurs on the mid-ocean ridge. This movement results in slip along the segment of the fracture zone between points X and Y. But to the west of point X, the fracture zone continues merely as a boundary between two different parts of plate A. The portion of plate A at point 1, just to the north of the boundary (▶Fig. 4.17e), must be younger than the portion at point 2 just to the south, because point 1 lies closer to the ridge axis; but since points 1 and 2 move at the same speed, this segment of the fracture zone does not slip.

Wilson introduced the term **transform fault** for the actively slipping segment of a fracture zone between two ridge segments, and he pointed out that transform faults made a third type of plate boundary. Geologists now also call them transform boundaries, or simply "transforms."

At a transform boundary, one plate slides sideways past another, but no new plate forms and no old plate is consumed. Transform boundaries are therefore defined by a vertical fault on which slip parallels the Earth's surface (Fig. 4.17a).

Not all transforms link ridge segments. Some, such as the Alpine Fault of New Zealand, link trenches, while others link a trench to a ridge segment. Further, not all transform faults occur in oceanic lithosphere; a few cut across continental lithosphere. The San Andreas Fault, for example, which cuts across California, defines part of the plate boundary between the North American Plate and the Pacific Plate—the portion of California that lies to the west of the fault (including Los Angeles) is part of the Pacific Plate, while the portion that lies to the east of the fault is part of the North American Plate (▶Fig. 4.19a, b). On average, the Pacific Plate moves about 6 cm north, relative to North America, every year. If this motion continues, Los Angeles will become a suburb of Anchorage, Alaska, in about 100 million years (see Box 4.2).

The grinding of one plate past another along a transform fault generates frequent earthquakes, including some huge ones. Fortunately, most of these earthquakes occur out in the ocean basin, far from people. But large earthquakes along the transform faults that cut across continental crust can be very destructive. A huge earthquake on the San Andreas Fault, for example, in conjunction with the ensuing fire destroyed much of San Francisco in 1906.

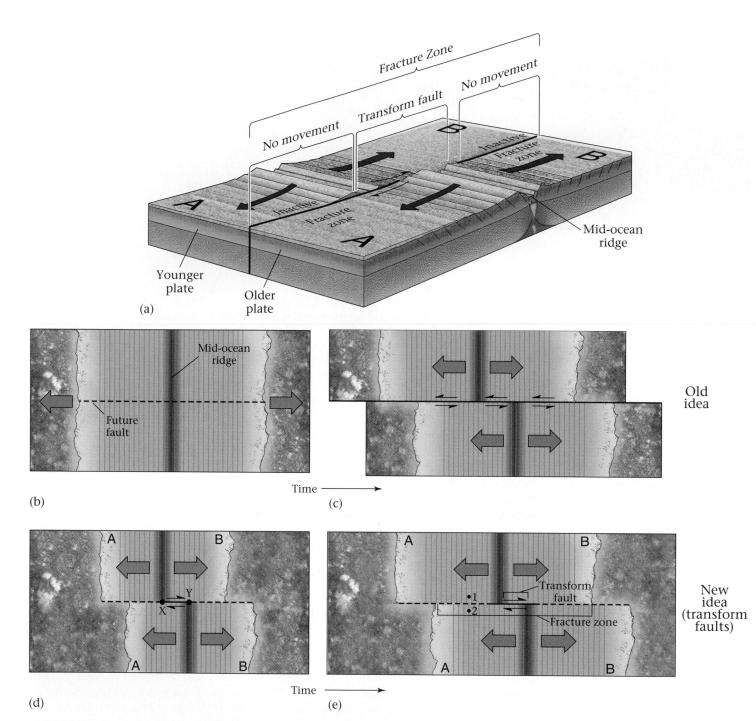

FIGURE 4.17 (a) The fracture zone beyond the ends of the transform fault does not slip, and thus is not a plate boundary. It does, however, mark the boundary between portions of the plate that are different in age. (b) In this *incorrect* interpretation of an oceanic fracture zone, the fault forms and cuts across an originally continuous ridge. (c) After slip on the fault, indicated by the arrows, the ridge consists of two segments. (d) In Wilson's *correct* interpretation, the ridge initiates at the same time as the transform fault, and thus was never continuous. Note that the way in which the fault slips (along the fracture zone between points X and Y) makes sense if sea-floor spreading takes place, but contrasts with the slip in (c). (e) Even though the ocean grows, the transform fault can stay the same length. Point 1 on plate A is younger than point 2 because it lies closer to the ridge axis.

So You Want to See a Plate Boundary?

BOX 4.2
THE HUMAN ANGLE

Plate boundaries are so important that geologists scrambled to study them when they were first recognized. But where do you go if you want to see one for yourself? First, it's important to realize that a plate boundary is not a single line on the surface of the Earth, but is actually a zone of faulting perhaps 40–200 km wide. Still, although you can't straddle a plate boundary the way you can a painted stripe in a road, there are in fact places where you can see a plate-boundary zone.

Let's start with a transform boundary. Most transform boundaries are on the sea floor, but a few cross continental crust. One of the most accessible of these is the San Andreas Fault in California. North of San Francisco, a visitor to Earthquake Park near Point Reyes will see the trace of a segment of the fault that ruptured during the great 1906 earthquake. A subtle depression on the ground marks the fault trace. This depression formed because rupturing weakened a band of rock and soil, which then collapsed. Small ponds and marshes also occur along the fault, marking places where a slight amount of stretching caused the surface of the ground to sink. In the San Francisco area, there are numerous faults related to the San Andreas. Some of these have moved slowly in recent decades, and have offset curbstones bordering streets. Farther south, in central California, the fault passes across an orchard, offsetting rows of trees, and beneath a winery, offsetting wine casks in the basement. Where the fault crosses a highway near the town of Palmdale, the layering of rock in road cuts (exposures) adjacent to the highway has been contorted wildly. Farther south still, movement on the fault has squeezed up ridges of broken rock (see Fig. 4.19b).

Convergent boundaries prove more elusive. You can get close to one by standing in the adjacent volcanic arc. If you visit Mt. Rainier or any of the other volcanoes in Washington or Oregon (▶Fig. 4.18a), for example, you are not far from the convergent boundary between the Juan de Fuca Plate and the North American Plate. The Olympic Peninsula, west of Seattle, exposes part of the accretionary prism. Likewise, when you hike in the Andes of South America, you are close to the convergent boundary between the Nazca Plate and the South American Plate. But to see the boundary, you would have to descend through 7 to 11 km of water to reach the deep darkness at the floor of the Peru-Chile Trench. And after all that effort, you would find that the boundary lies buried by sediment.

There are over 40,000 km of divergent plate boundaries on our planet, marked by mid-ocean ridges. But despite their height, almost all of the mid-ocean ridge system lies beneath nearly 2.5 km of water. Geologists in the research mini-submarine *Alvin* have explored the mid-ocean ridge. Because ocean crust has only just formed at the ridge axis, features of the plate boundary have not been buried by sediment, and they stand out in the glare of *Alvin*'s headlights. Peeking through the portholes of the submarine, geologists have seen molten rock oozing up through cracks in the crust, chimneys spewing out clouds of boiling water saturated with dissolved minerals, and steep cliffs created by movement on faults along the plate boundary.

Most of us will never have the chance to squeeze into *Alvin*'s cramped quarters for a claustrophobic drop to the sea floor (even though it sinks like a stone, *Alvin* takes almost two hours to reach bottom). But amazingly, there is one place in the world where we can stand on a divergent boundary: Iceland (▶Fig. 4.18b). This island straddles the Mid-Atlantic Ridge. A walking or driving tour of Iceland will take you across fault-bounded troughs and to fissures that frequently spew fountains of lava. When you stand in eastern Iceland, you are on the Eurasian Plate, but when you stand in western Iceland, you are on the North American Plate.

FIGURE 4.18 (a) The Cascades volcanic arc in the western United States. Here you can see a chain of volcanoes that are a consequence of subduction. (b) A portion of the Mid-Atlantic Ridge exposed in Iceland. Here, you can see scarps formed by faulting along the ridge. The gray rock is basalt formed from lava erupted along the ridge.

(a)

(b)

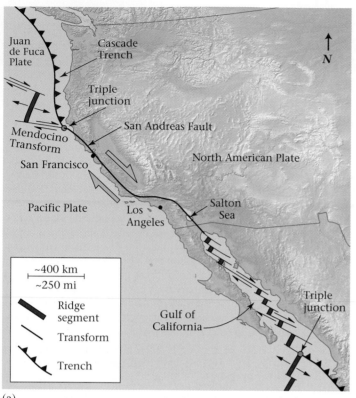

(a)

(b)

FIGURE 4.19 (a) The San Andreas Fault is a transform plate boundary between the Pacific Plate to the west and the North American Plate to the east. At its southeastern end, the San Andreas connects to spreading ridge segments in the Gulf of California. (b) The San Andreas Fault where it cuts across a dry landscape. The fault trace is the narrow valley running the length of the photo. The land has been pushed up slightly, along the fault; streams have cut small side canyons into this uplifted land.

4.6 SPECIAL LOCATIONS IN THE PLATE MOSAIC

Triple Junctions

So far, we've focused attention on boundaries—divergent (mid-ocean ridge), convergent (trench), and transform—between *two* plates. But there are several places where *three* plate boundaries intersect at a point. Geologists refer to these points as **triple junctions.** We name triple junctions after the types of boundaries that intersect. For example, the triple junction formed where the Southwest Indian Ocean Ridge intersects two arms of the Mid–Indian Ocean Ridge (this is the triple junction of the African, Antarctic, and Australian Plates) is a ridge-ridge-ridge triple junction (►Fig. 4.20a). The triple junction north of San Francisco is a trench-transform-transform triple junction (►Fig. 4.20b).

Hot Spots

Most subaerial (above sea level) volcanoes are situated in the volcanic arcs that border trenches. Small volcanoes also lie along mid-ocean ridges, but ocean water hides most of them. The volcanoes of volcanic arcs and mid-ocean ridges are "plate-boundary volcanoes," formed as a consequence of movement along the boundary. Not all volcanoes on Earth are plate-boundary volcanoes, however. For example, Hawaii, a huge active volcano, lies in the middle of the Pacific Plate, and Yellowstone National Park, site of a recent volcanic activity, lies in the northwestern corner of Wyoming, in the interior of the United States. Worldwide, geoscientists have identified about one hundred volcanoes that exist as isolated points and are not a consequence of movement at a plate boundary; these are called hot-spot volcanoes, or simply **hot spots** (►Fig. 4.21). Most hot spots are located in the interiors of plates, away from the boundaries, but a few lie on mid-ocean ridges.

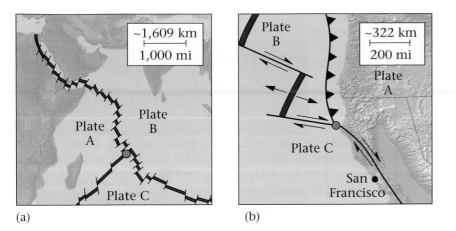

(a)   (b)

FIGURE 4.20 (a) A ridge-ridge-ridge triple junction (at the dot). (b) A trench-transform-transform triple junction (at the dot).

What causes hot-spot volcanoes? After examining hot spots in the interiors of ocean plates, J. Tuzo Wilson noted that an erupting hot-spot volcano formed an island at the end of a chain of now-dead (no-longer erupting) volcanic islands and seamounts. A volcano along a convergent plate boundary, in contrast, is one of many in a chain that are all active at about the same time. Wilson noted further that all the hot-spot volcanoes in the Pacific lie at the southeastern end of a north-west-southeast-lying chain of dead volcanoes (▶Fig. 4.22). With this image in mind, Wilson suggested in 1963 that a hot-

spot volcano develops over a heat source in the mantle that is fixed relative to the moving plate; an active volcano represents the location of the heat source. The chain of seamounts and inactive (dead) volcanic islands linked to the hot-spot volcano represent locations on the plate that were once over the source but have since moved off. Subsequently, the heat source came to be associated with a **mantle plume,** a column of very hot rock rising up through the mantle.

Let's look more closely at how Wilson's model works. Most mantle plumes are thought to originate just above the core-mantle boundary, where heat from the Earth's core warms the base of the mantle. (Recent evidence, however, suggests that some may originate in the upper mantle.) Because it expands as it heats, the hot rock above the core-mantle boundary becomes less dense than overlying mantle and begins to stream upward in a column-shaped mass—at the rate of a few centimeters per year. When the hot rock reaches the base of the lithosphere, it partly melts, for reasons discussed in Chapter 6. The magma formed by melting then rises through the lithosphere and erupts at a volcano on the Earth's surface (▶Fig. 4.23a). This location is the "hot spot." All the while, the plate on which the volcano grows continues to shift, so eventually the volcano moves off the plume and dies, or "goes extinct." Meanwhile, a new volcano grows over the plume. If the plume lasts for a long time in

FIGURE 4.21 The dots represent the locations of selected hot-spot volcanoes. The tails represent hot-spot tracks. The most recent volcano (dot) is at one end of this track. Some of these are extinct, indicating that the plume no longer exists. Some hot spots are fairly recent and do not have tracks. Dashed tracks are places where track was broken by sea-floor spreading.

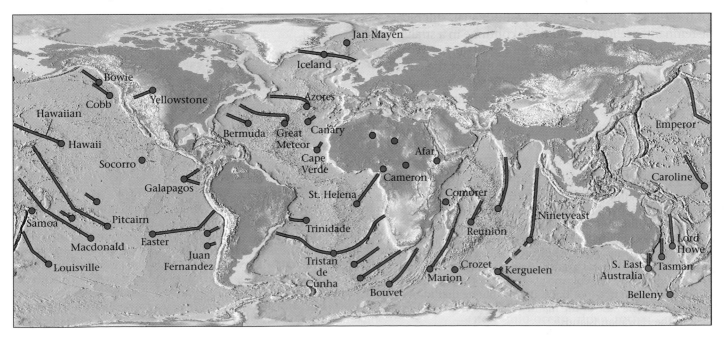

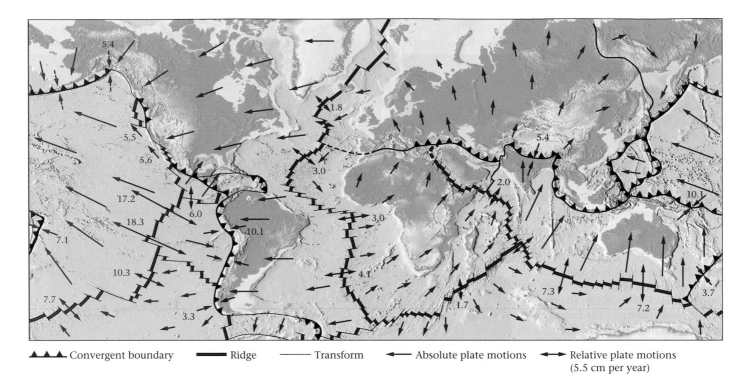

Convergent boundary ▲▲▲ — **Ridge** — **Transform** — **Absolute plate motions** ← — **Relative plate motions** ←→ (5.5 cm per year)

FIGURE 4.30 Relative plate velocities: the blue arrows show the rate and direction at which the plate on one side of the boundary is moving with respect to the plate on the other side. Outward-pointing arrows indicate spreading (divergent boundaries), inward-pointing arrows indicate subduction (convergent boundaries), and parallel arrows show transform motion. The length of an arrow represents the velocity. Absolute plate velocities: the red arrows show the velocity of the plates with respect to a fixed point in the mantle.

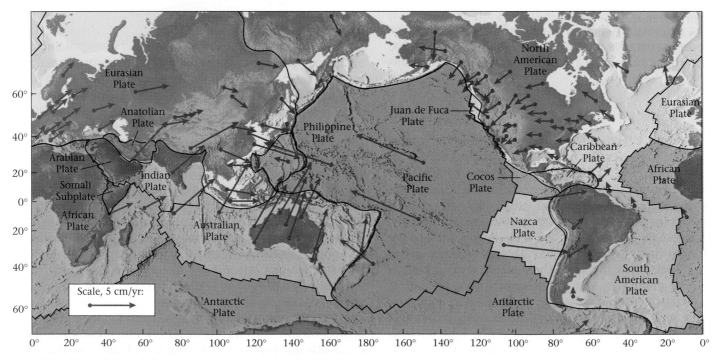

FIGURE 4.31 The Global Positioning System (GPS) is used to measure plate motions at many locations on Earth. The velocities shown here are determined for stations that continuously record GPS data.

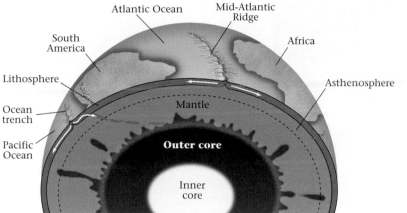

FIGURE 4.32 Plate tectonics involves the transfer of material from the mantle to the surface and back down again. The insides and surface of our dynamic planet are in constant motion.

their destinations, and geologists use an array of GPS receivers to monitor plate motions. If done carefully enough, we can detect displacements of millimeters per year (▶Fig. 4.31). In other words, we can now see the plates move!

4.10 THE DYNAMIC PLANET

Now, having completed our two-chapter introduction to plate tectonics, we can see more easily why plate tectonics holds the key to understanding most of geology. To start

with, plate tectonics explains the origin and distribution of earthquakes, major sea-floor features (mid-ocean ridges, deep-ocean trenches, seamount chains, and fracture zones), and volcanoes (▶Fig. 4.32). It also tells us why mountain belts form. Finally, plate tectonics explains the drift of continents and why the distribution of land changes with time, a change that significantly affects the evolution of life on Earth (▶Fig. 4.33). In coming chapters, we will explore these consequences and others in more detail.

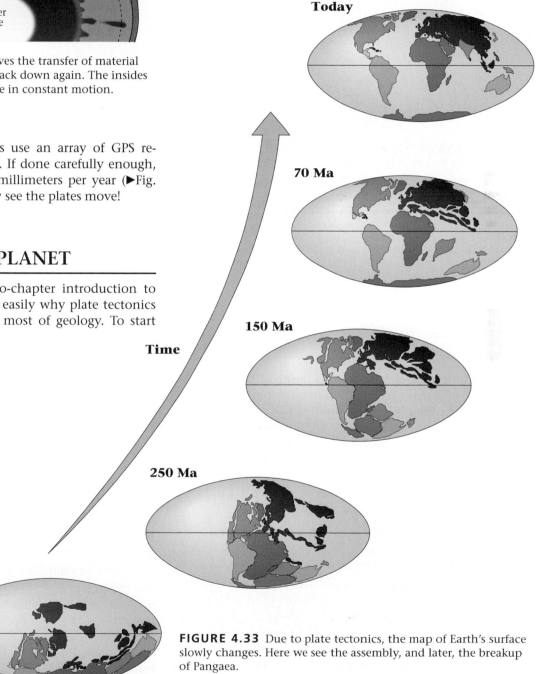

FIGURE 4.33 Due to plate tectonics, the map of Earth's surface slowly changes. Here we see the assembly, and later, the breakup of Pangaea.

THE VIEW FROM SPACE The Alps formed during the Cenozoic as a consequence of the collision between a "microplate" (including Italy) and the Eurasian plate. Such continental collisions build complex orogens. Because of the high elevation of its peaks, much of the Alps remains snow covered all year. Some Alpine valleys still contain glaciers. This image also shows the gentle curve of the Apennines on the Italian peninsula.

CHAPTER SUMMARY

• The lithosphere, the rigid outer layer of the Earth, is broken into discrete plates that move relative to one another. Plates consist of the crust and the uppermost (cooler) mantle. Lithosphere plates effectively float on the underlying soft asthenosphere. Continental drift and sea-floor spreading are manifestations of plate movement.

• Some continental margins are plate boundaries, but many are not. A single plate can consist of continental lithosphere, oceanic lithosphere, or both.

• Most plate interactions occur along plate boundaries; the interior of plates remain relatively rigid and intact. Earthquakes delineate the position of plate boundaries.

• There are three types of plate boundaries—divergent, convergent, and transform—distinguished from each other by the movement the plate on one side of the boundary makes relative to the plate on the other side.

• Divergent boundaries are marked by mid-ocean ridges. At divergent boundaries, sea-floor spreading takes place, a process that forms new oceanic lithosphere.

• Convergent boundaries, also called convergent margins or subduction zones, are marked by deep-ocean trenches and volcanic arcs. At convergent boundaries, oceanic lithosphere of the downgoing plate is subducted beneath an overriding plate. The overriding plate can consist of either continental or oceanic lithosphere. An accretionary prism forms out of sediment scraped off the downgoing plate as it subducts.

• Subducted lithosphere sinks back into the mantle. Its position can be tracked down to a depth of about 670 km by a belt of earthquakes known as the Wadati-Benioff zone.

• Transform boundaries, also called transform faults, are marked by large faults at which one plate slides past another. No new plate forms and no old plate is consumed at a transform boundary.

• Triple junctions are points where three plate boundaries intersect.

• Hot spots are places where a plume of hot mantle rock rises from just above the core-mantle boundary and causes anomalous volcanism at an isolated volcano. As a plate moves over the mantle plume, the volcano moves off the hot spot and dies, and a new volcano forms over the hot spot. As a result, hot spots spawn seamount/island chains.

• A large continent can split into two smaller ones by the process of rifting. During rifting, continental lithosphere stretches and thins. If it finally breaks apart, a new mid-ocean ridge forms and sea-floor spreading begins. Not all rifts go all the way to form a new mid-ocean ridge.

• Convergent plate boundaries cease to exist when a buoyant piece of crust (a continent or an island arc) moves into the subduction zone. When that happens, collision occurs. The collision between two continents yields large mountain ranges.

• Ridge-push force and slab-pull force drive plate motions. Plates move at rates of about 1–15 cm per year. Modern satellite measurements can detect these motions.

KEY TERMS

absolute plate velocity (p. 100)	marginal sea (p. 88)
accretionary prism (p. 87)	mid-ocean ridge (p. 81)
active margins (p. 78)	overriding plate (p. 87)
asthenosphere (p. 77)	passive margins (p. 78)
black smokers (p. 82)	pillow basalt (p. 82)
buoyancy (p. 79)	plate (p. 78)
collision (p. 97)	plate boundaries (p. 78)
continental rift (p. 95)	plate interiors (p. 80)
continental shelf (p. 79)	plate tectonics (p. 77)
convergent boundary (p. 81)	relative plate velocity (p. 100)
dikes (p. 82)	ridge-push force (p. 100)
divergent boundary (p. 81)	rifting (p. 95)
downgoing plate (p. 85)	scientific revolution (p. 76)
epicenter (p. 79)	slab-pull force (p. 100)
fracture zone (p. 88)	subduction (p. 85)
global positioning system (p. 101)	transform boundary (p. 81)
hot spot (p. 92)	transform fault (p. 89)
hypocenter (p. 79)	trenches (p. 85)
lithosphere (p. 77)	triple junctions (p. 92)
magma chamber (p. 82)	volcanic arc (p. 88)
mantle plume (p. 93)	Wadati-Benioff zone (p. 87)

REVIEW QUESTIONS

1. What is a scientific revolution? How is plate tectonics an example of a scientific revolution?

2. What are the characteristics of a lithosphere plate?

3. How does oceanic crust differ from continental crust in thickness, composition, and density?

4. Describe how Archimedes' principle of buoyancy can be applied to continental and oceanic lithosphere.

5. Contrast active and passive margins.

6. What are the basic premises of plate tectonics?

7. How do we identify a plate boundary?

8. Describe the three types of plate boundaries.

9. How does crust form along a mid-ocean ridge?

10. What happens to the mantle beneath the mid-ocean ridge?

11. Why are mid-ocean ridges high?

12. Why is subduction necessary on a nonexpanding Earth with spreading ridges?

13. What is a Wadati-Benioff zone, and how does it help to define the location of subducting plates?

14. Describe the major features of a convergent boundary.

15. Why are transform plate boundaries required on an Earth with spreading and subducting plate boundaries?

16. What are two examples of famous transform faults?

17. What is a triple junction?

18. Explain the processes that form a hot spot.

19. How is a hot-spot track produced, and how can hot-spot tracks be used to track the past motions of the overlying plate?

20. Describe the characteristics of a continental rift, and give examples of where this process is occurring today.

21. Describe the process of continental collision, and give examples of where this process has occurred.

22. Discuss the major forces that move lithosphere plates.

23. Explain the difference between relative plate velocity and absolute plate velocity.

24. Can we measure present-day plate motions directly?

SUGGESTED READING

Allegre, C. 1988. *The Behavior of the Earth: Continental and Seafloor Mobility.* Cambridge, Mass.: Harvard University Press.

Condie, K. C. 2001. *Mantle Plumes & Their Record in Earth History.* Cambridge: Cambridge University Press.

Condie, K. C. 1997. *Plate Tectonics and Crustal Evolution,* 4th ed. Boston: Butterworth–Heinemann.

Cox, A., and R. B. Hart. 1986. *Plate Tectonics: How It Works.* Palo Alto, Calif.: Blackwell.

Kearey, P., and F. J. Vine. 1996. *Global Tectonics,* 2nd ed. Cambridge, Mass.: Blackwell.

Moores, E. M., ed. 1990. *Shaping the Earth: Tectonics of Continents and Oceans.* Readings from *Scientific American.* New York: Freeman.

Moores, E. M., and R. J. Twiss. 1995. *Tectonics.* New York: Freeman.

Earth Materials

What is the Earth made of? There are four basic components: the solid Earth (the crust, mantle, and core), the biosphere (living organisms), the atmosphere (the envelope of gas surrounding the planet), and the hydrosphere (the liquid and solid water at or near the ground surface). In this part of the book, we focus on the materials that make up the crust and mantle of the solid Earth. We will find that these consist primarily of rock. Most rock, in turn, contains minerals, so minerals are, in effect, the building blocks of our planet.

We therefore begin in Chapter 5 by learning about minerals and how they grow. Then we see, in Interlude A, how geologists distinguish three categories of rock—igneous, sedimentary, and metamorphic—based on how the rocks form. In each of the next three chapters (6, 7, and 8), we look at one of these rock categories. Finally, Interlude B shows us how materials in the Earth System pass through a rock cycle, as atoms constituting one rock type may end up being incorporated into a succession of other rock types.

Patterns in Nature: Minerals

This photo is real, not a computer collage! We're seeing the world's largest known mineral crystals jutting from the walls of a cave near Chihuahua, Mexico. The crystals are of the mineral gypsum; they formed by precipitation from water solutions.

5.1 INTRODUCTION

Zabargad Island rises barren and brown above the Red Sea, about 70 km off the coast of southern Egypt. Nothing grows on Zabargad, except for scruffy grass and a few shrubs, so no one lives there now. But in ancient times many workers toiled on this 5-square-km patch of desert, gradually chipping their way into the side of its highest hill, seeking glassy green, pea-sized pieces of peridot, a prized gem. Carefully polished peridots were worn as jewelry by ancient Egyptians and may have been buried with them when they died. Eventually, some of the gems appeared in Europe, where jewelers set them into crowns and scepters (▶Fig. 5.1). These peridots now glitter behind glass cases in museums, millennia after first being pried free from the Earth, and perhaps 10 million years after first being formed by the bonding together of still more ancient atoms.

Peridot is one of about 4,000 minerals that have been identified on Earth, so far, and fascinate collectors and geologists alike. Fifty to one hundred new minerals are recognized every year. Each different mineral has a name. Some names come from Latin, Greek, German, or English words describing a certain characteristic (e.g., "albite" comes from the Latin word for

FIGURE 5.1 A royal crown containing a variety of valuable jewels. The large gemstone near the base of the crown is a green peridot.

"white," "orthoclase" comes from the German words meaning "splits at right angles," and olivine is olive-colored); some honor a person (sillimanite was named for Benjamin Silliman, a famous nineteenth-century mineralogist); some indicate the place where the mineral was first recognized (illite was first identified in rocks from Illinois); and some reflect a particular element in the mineral (chromite contains chromium). Several minerals have more than one name—for example, peridot is the gem-quality version of a common mineral named "olivine." Although the vast majority of mineral types are rare, forming only under special conditions, many are quite common and are found in a variety of rock types at Earth's surface.

Though ancient Greek philosophers pondered minerals and medieval alchemists puttered with minerals, true scientific study of minerals did not begin until 1556, when Georgius Agricola, a German physician, published *De Re Metallica,* in which he discussed mining and gave basic descriptions of minerals. (Agricola wrote the book in Latin. It's interesting to note that the book's first English translation was completed in 1912 by Herbert Hoover and his wife, Lou. At the time, Hoover was a successful geological engineer—he became president of the United States seventeen years later.) In 1669, more than a century after Agricola's work, Nicholas Steno, a Danish monk, discovered important geometric characteristics of minerals. Steno's work became the basis for systematic descriptions of minerals, a task that occupied many researchers during the next two centuries. These were the

first **mineralogists,** people who specialize in the study of minerals.

The study of minerals with an optical microscope began in 1828, but though such studies helped in mineral identification, they could not reveal the arrangement of atoms inside minerals. But in 1912, Max von Laue of Germany proposed that X-rays, electromagnetic radiation whose wavelength is comparable to the distance between atoms in a mineral, could be used to study the internal structure of minerals. A father-and-son team, W. H. and W. L. Bragg of England, published the first X-ray study of a mineral, work for which they shared the 1915 Nobel Prize in physics. In subsequent decades, researchers developed progressively more complex instruments to help study minerals. For example, beginning in the 1960s, mineralogists used electron microscopes to obtain actual images of the internal structure of minerals, and electron microprobes to analyze the chemical composition of grains that are almost too small to see. Both instruments emit beams of electrons, with wavelengths even smaller than that of X-rays.

Why study minerals? Without exaggeration, we can say that minerals are the building blocks of our planet. To a geologist, almost any study of Earth materials depends on an understanding of minerals, for minerals make up the rocks and sediments that make up the Earth and its landscapes. But minerals are also important from a practical standpoint. "Industrial minerals" serve as the raw materials for manufacturing chemicals, concrete, and wallboard. "Ore minerals" are the source of valuable metals like copper and gold and provide energy resources like uranium (▶Fig. 5.2a, b). Certain forms of minerals, gems, delight the eye as jewelry. Unfortunately, not all minerals are beneficial; some pose environmental hazards. No wonder **mineralogy,** the study of minerals, fascinates professionals and amateurs alike.

The word "mineral" has a broader meaning in everyday English than it does in geology. For example, when playing the game *Twenty Questions* you start by asking, "Is it animal, vegetable, or mineral?"—"mineral" referring here simply to an inanimate object. Nutritionists talk about the "vitamins and minerals" in various types of foods—to them, a mineral is simply a metallic compound. In geology, however, a mineral is a special kind of substance with certain distinctive characteristics. In this chapter, we begin with the geologic definition of a mineral, then look at how minerals form and the main characteristics that enable us to identify them. Finally, we note the basic scheme that geologists use to classify minerals. This chapter assumes that you understand the fundamental concepts of matter and energy, especially the nature of atoms, molecules, and chemical bonds. If you are rusty on these topics, please review Appendix A. Basic terms from chemistry are summarized in Box 5.1, for convenience.

(a)

(b)

FIGURE 5.2 (a) Museum specimen of malachite, a bright-green mineral containing copper (its formula is $Cu_2[CO_3][OH]_2$). Malachite is an ore mineral mined to produce copper, but because of its beauty, it is also used for jewelry. (b) Copper wire made by the processing of malachite and other ore minerals of copper.

5.2 WHAT IS A MINERAL?

To geologists, a **mineral** is a homogeneous, naturally occurring, solid substance with a definable chemical composition and an internal structure characterized by an orderly arrangement of atoms in a crystalline structure. Most minerals are inorganic. Let's pull apart this mouthful of a definition and examine what its components actually mean.

• *Homogeneous:* Homogeneous materials are the same through and through—they cannot be physically broken into simpler components. When you smash a mineral specimen with a hammer, you get many tiny fragments of the same mineral. (Note that rocks are aggregates [combinations] of many separate mineral grains, or

pieces. Fragments of a broken rock are not necessarily all the same mineral.)

• *Naturally occurring:* True minerals form by geological processes, not by the activity of a human. In recent decades, laboratory scientists have learned to manufacture materials that are essentially identical to naturally occurring minerals. For example, some companies routinely manufacture diamonds by squeezing carbon under very high pressure. We sometimes call such materials "synthetic minerals," to distinguish them from true minerals. Similar procedures can produce mineral-like substances that do not occur in nature.

• *Solid:* A solid is a form of matter that can maintain its shape indefinitely, and thus will not conform to the shape of its container. Liquids (like oil or water) and gases (like air) are not minerals.

• *Definable chemical composition:* This simply means that it is possible to write a chemical formula for a mineral (see Box 5.1). For example, the minerals diamond and graphite have the formula C because they consist entirely of carbon. Quartz has the formula SiO_2—it contains the elements silicon and oxygen in the proportion of one silicon atom for every two oxygen atoms. (Note that some minerals contain only one element, but most are compounds containing more than one element.) Some formulas are more complicated: for example, the formula for biotite is $K(Mg,Fe)_3(AlSi_3O_{10})(OH)_2$. While many minerals, like quartz, have only one composition, others have compositions that can vary slightly. In biotite, for example, iron (Fe) can substitute for magnesium (Mg)—which is why these elements are separated by a comma in the formula. Some samples of biotite contain more iron, while others contain more magnesium.

• *Orderly arrangement of atoms:* The atoms that make up a mineral are not distributed randomly and cannot move around easily. Rather, they are fixed in a specific pattern that repeats itself over a very large region, relative to the size of atoms. (To picture the contrast between a random arrangement and a fixed pattern, compare the distribution of people at a casual party with the distribution of people in a military regiment at attention. At the party, clusters of two or three people stand around chatting, and people or groups of people move around the room. But in the regiment at attention, everyone stands aligned in orderly rows and columns, and no one dares to move.) A material in which atoms are fixed in an orderly pattern is called a "crystalline solid." Mineralogists refer to the pattern itself (i.e., the imaginary framework representing the arrangement of atoms) as a **crystal lattice.** A lattice resembles the set of points constituting the intersections among bars making up scaffolding (▶Fig. 5.3a, b).

- *Generally inorganic:* Most, but not all, minerals are inorganic, by which we mean that most do not contain organic chemicals—there are only about thirty organic chemicals that are classified as minerals. To understand the above statement, we need to examine the meaning of "organic chemical." Organic chemicals are molecules composed of carbon bonded to hydrogen, along with varying amounts of oxygen, nitrogen, and other elements. The name "organic" emphasizes that many such chemicals occur in living organisms. But not all organic materials form in nature—plastic, for example, consists of organic chemicals. In this context, materials such as plastic, plant debris, sugar, fat, and protein are not minerals. It is important to note that the presence of carbon alone does not make a mineral organic. For example, diamond and graphite are minerals composed of pure carbon, and calcite is a mineral composed of carbon bonded to oxygen and calcium—neither are organic. We also emphasize that some

inorganic minerals are formed by biologic processes. For example, a clam shell consists mostly of the inorganic mineral calcite, constructed by the life activity of the clam.

With these definitions in mind, we can make an important distinction between a mineral and **glass.** Both minerals and glasses are solids, in that they can retain their shape indefinitely (see Appendix A). But a mineral is crystalline, while glass is not. Whereas atoms, ions, or molecules in a mineral are ordered into a crystal lattice, like soldiers standing in formation, those in a glass are arranged in a semichaotic way, like a crowd of people at a party, in small clusters or chains that are neither oriented in the same way nor spaced at regular intervals (▶Fig. 5.3c, d). Note that the chemical compound silica (SiO_2) forms the mineral quartz when arranged in a crystalline lattice, but forms common window glass when arranged in a semichaotic way.

Some Basic Definitions from Chemistry

BOX 5.1
SCIENCE TOOLBOX

To describe minerals, we need to use several terms from chemistry (for a more in-depth discussion, see Appendix A). To avoid confusion, terms are listed in an order that permits each successive term to utilize previous terms.

- **Element:** a pure substance that cannot be separated into other elements.

- **Atom:** the smallest piece of an element that retains the characteristics of the element. An atom consists of a nucleus surrounded by a cloud of orbiting electrons; the nucleus is made up of protons and neutrons (except in hydrogen, whose nucleus contains only one proton and no neutrons). Electrons have a negative charge, protons have a positive charge, and neutrons have a neutral charge. An atom that has the same number of electrons as protons is said to be "neutral," in that it does not have an overall electrical charge.

- **Atomic number:** the number of protons in an atom of an element.

- **Atomic weight:** the approximate number of protons plus neutrons in an atom of an element.

- **Ion:** an atom that is not neutral. An ion that has an excess negative charge (because it has more electrons than protons) is an **anion,** whereas an ion that has an excess positive charge (because it has more protons than electrons) is a **cation.** We indicate the charge with a superscript. For example, Cl^- (chlorine) has a single excess electron; Fe^{2+} is missing two electrons.

- **Chemical bond:** an attractive force that holds two or more atoms together. For example, *covalent bonds* form when atoms share electrons. *Ionic bonds* form when a cation and anion (ions with opposite charges) get close together and attract each other. In materials with **metallic bonds,** some of the electrons can move freely.

- **Molecule:** two or more atoms bonded together. The atoms may be of the same element or of different elements.

- **Compound:** a pure substance that can be subdivided into two or more elements. The smallest piece of a compound that retains the characteristics of the compound is a molecule.

- **Chemical:** a general name used for a pure substance (either an element or a compound).

- **Chemical formula:** a shorthand recipe that itemizes the various elements in a chemical and specifies their relative proportions. For example, the formula for water, H_2O, indicates that water consists of molecules in which two hydrogens bond to one oxygen.

- **Chemical reaction:** a process that involves the breaking or forming of chemical bonds. Chemical reactions can break molecules apart or create new molecules and/or isolated atoms.

- **Mixture:** a combination of two or more elements or compounds that can be separated without a chemical reaction. For example, a cereal composed of bran flakes and raisins is a mixture—you can separate the raisins from the flakes without destroying either.

- **Solution:** a type of material in which one chemical (the solute) dissolves (becomes completely incorporated) in another (the solvent). In solutions, a solute may separate into ions during the process. For example, when salt (NaCl) dissolves in water, it separates into sodium (Na^+) and chlorine (Cl^-) ions. In a solution, atoms or molecules of the solvent surround atoms, ions, or molecules of the solute.

- **Precipitate:** (noun) a compound that forms when ions in liquid solution join together to create a solid that settles out of the solution; (verb) the process of forming solid grains by separation and settling from a solution. For example, when saltwater evaporates, solid salt crystals precipitate and settle to the bottom of the remaining water.

properties of a mineral (e.g., the shape of its crystals, how hard it is, how it reacts chemically with other substances) depend both on the identity of the elements making up the mineral and on the way these elements are arranged and bonded in a crystal structure.

Because of its importance, we now look a little more closely at the nature of chemical bonding in minerals. Chemists recognize five different types of bonds (covalent, ionic, metallic, van der Waals', and hydrogen) based on the way in which atoms stick or link to each other. For example, in covalently bonded materials, atoms stick to each other by sharing electrons, whereas in ionically bonded materials, atoms either add electrons to become negative ions (anions) or lose electrons to become positive ions (cations)—the two kinds of ions stick to each other because opposite charges attract. (Appendix A illustrates these bonds and discusses

the other types as well.) Not all minerals have the same kind of bonding, and in some minerals more than one type of bonding occurs. The type of bonding, the ease with which bonds form or are broken, and the geometric arrangement of bonds play an important role in determining the characteristics of minerals. As your intuition might suggest, bonds are stronger in harder minerals (i.e., ones more resistant to breaking) and in minerals with higher melting temperatures (i.e., ones that transform from solid to liquid at a higher temperature). In some minerals, the nature and strength of bonding vary with direction in the mineral. If bonds form more easily in one direction than another, a crystal will grow faster in one direction than another. And if a mineral has weak bonds in one direction and strong bonds in another direction, it will break more easily in one direction than in the other.

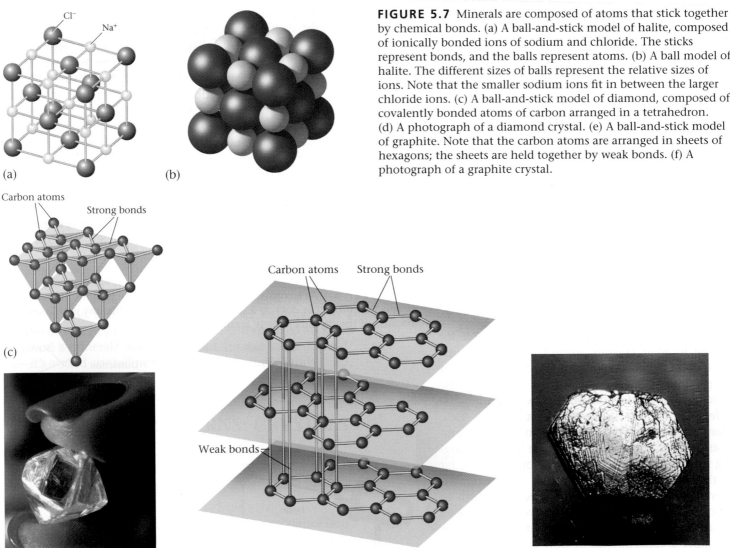

FIGURE 5.7 Minerals are composed of atoms that stick together by chemical bonds. (a) A ball-and-stick model of halite, composed of ionically bonded ions of sodium and chloride. The sticks represent bonds, and the balls represent atoms. (b) A ball model of halite. The different sizes of balls represent the relative sizes of ions. Note that the smaller sodium ions fit in between the larger chloride ions. (c) A ball-and-stick model of diamond, composed of covalently bonded atoms of carbon arranged in a tetrahedron. (d) A photograph of a diamond crystal. (e) A ball-and-stick model of graphite. Note that the carbon atoms are arranged in sheets of hexagons; the sheets are held together by weak bonds. (f) A photograph of a graphite crystal.

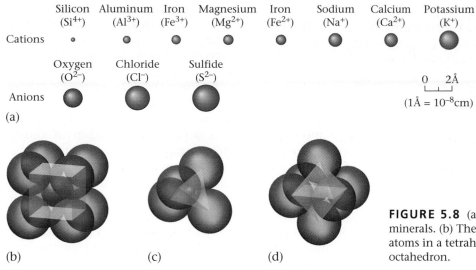

FIGURE 5.8 (a) Relative sizes of ions that are common in minerals. (b) The packing of atoms in a cube. (c) The packing of atoms in a tetrahedron. (d) The packing of atoms in an octahedron.

To illustrate crystal structures, we look at a few examples. Halite (rock salt) is an ionically bonded mineral in that it consists of oppositely charged ions (atoms that have gained or lost electrons) that stick together because opposite charges attract. In halite, the anions are chloride (Cl^-) and the cations are sodium (Na^+). In halite, six chloride ions surround each sodium ion, producing an overall arrangement of atoms that defines the shape of a cube (▶Fig. 5.7a, b). Diamond, by contrast, is a mineral made entirely of carbon. In diamond, each atom is covalently bonded to (i.e., shares electrons with) four neighbors. The atoms are arranged in the form a tetrahedron, so some naturally formed diamond crystals have the shape of a double tetrahedron (▶Fig. 5.7c, d). Covalent bonds are very strong, so diamond is very hard. Graphite is another mineral composed entirely of carbon, but it behaves very differently from diamond. In contrast to diamond, graphite is so soft that it can be used as the "lead" in a pencil; as you move a pencil across paper, tiny flakes of graphite peel off the pencil point and adhere to the paper. This behavior occurs because the carbon atoms in graphite are not arranged in tetrahedra, but rather occur in sheets (▶Fig. 5.7e, f). The sheets are bonded to each other by weak bonds (van der Waals' bonds) and thus can separate from each other easily. Note that two different minerals (e.g., diamond and graphite) that have the same composition but have different crystal structures are called **polymorphs.**

What determines how atoms pack together in a crystal? The size of an ion depends on the number of electrons orbiting the nucleus (▶Fig. 5.8a); so, since anions have extra electrons, they tend to be bigger than cations. Thus, cations nestle snugly in the spaces between anions in many crystal structures. As many anions will try to fit around a cation as there is room for. Depending on the identity of an ion, different

geometries of packing can occur (▶Fig. 5.8b–d). Note that in halite, as described above, each ion is a single atom. In many ionically bonded minerals, the ions building the minerals consist of more than one atom. For example, the mineral calcite ($CaCO_3$) consists of calcium (Ca^{2+}) cations and carbonate (CO_3^{2-}) anions—each carbonate anion, or "anionic group," consists of four atoms and thus is quite large.

The orderly arrangement of atoms inside a crystal, its crystal structure, provides one of nature's most spectacular examples of a pattern. The pattern on a sheet of wallpaper may be defined by the regular spacing of, say, clumps of flowers. Similarly, the pattern in a crystal is defined by the regular spacing of atoms (▶Fig. 5.9a–c). If the crystal contains more than one type of atom, the atoms alternate in a regular way. The orderly arrangement controls the outward shape, or morphology, of crystals. For example, if the atoms in a mineral are packed into the shape of a cube, a crystal of the mineral will have faces that intersect at 90° angles.

The pattern of atoms or ions in a mineral displays **symmetry,** meaning that the shape of one part of a mineral is a mirror image of the shape of another part. For example, if you were to cut a halite crystal in half and place one half against a mirror, the crystal would look whole again (▶Fig. 5.10a, b).

The Formation and Destruction of Minerals

New mineral crystals can form in five ways. First, they can form by the *solidification of a melt,* meaning the freezing of a liquid; for example, ice crystals, a type of mineral, are made by freezing water. Second, they can form by *precipitation from a solution,* meaning that atoms, molecules, or ions dissolved in water bond together and separate out of

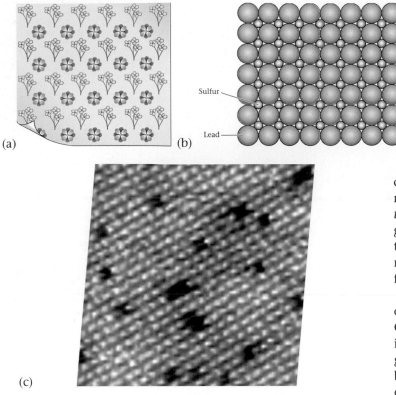

(a)

(b)

Sulfur

Lead

(c)

FIGURE 5.9 (a) Repetition of a flower motif in wallpaper illustrates a regular pattern. (b) On the face of a crystal of galena (a type of lead ore), lead and sulfur atoms pack together in a regular array. (c) An image of the surface of a galena crystal, taken with a scanning tunneling microscope.

the water—salt crystals, for example, develop when you evaporate saltwater. Third, they can form by *solid-state diffusion,* the movement of atoms or ions through a solid to arrange into a new crystal structure; this process takes place very slowly. (In Chapter 8, we'll discuss the importance of diffusion during the formation of minerals in metamorphic rocks.) Fourth, minerals can form at interfaces between the

FIGURE 5.10 (a) Crystals have symmetry: one half of a halite crystal is a mirror image of the other half. (b) Snowflakes, crystals of ice, are symmetrical hexagons.

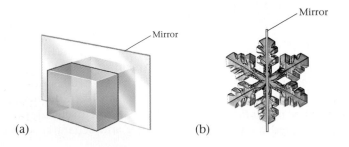

Mirror

Mirror

(a)

(b)

physical and biological components of the Earth system by a process called biomineralization. *Biomineralization* occurs when living organisms cause minerals to precipitate either within or on their bodies, or immediately adjacent to their bodies. For example, clams and other shelled organisms extract ions from water to produce mineral shells (a clam shell consists of two minerals: calcite and its polymorph, aragonite), and the metabolism of certain species of cyanobacteria produces chemicals that change the acidity of the water they live in and cause calcite crystals to precipitate. Fifth, minerals can form directly from a vapor. This process, called *fumerolic mineralization,* typically occurs around volcanic vents or around geysers, for at such locations volcanic gases or steam enter the atmosphere and cool abruptly, so certain elements cannot remain in gaseous form. Some of the bright yellow sulfur deposits found in volcanic regions form in this way.

The first step in forming a crystal is the chance formation of a "seed," or an extremely small crystal (▶Fig. 5.11a–c). Once the seed exists, other atoms in the surrounding material attach themselves to the face of the seed. As the crystal grows, crystal faces move outward from the center of the seed but maintain the same orientation. Thus, the youngest part of the crystal is always its outer edge (▶Fig. 5.12a, b).

In the case of crystals formed by the solidification of a melt, atoms begin to attach to the seed when the melt becomes so cool that thermal vibrations can no longer break apart the attraction between the seed and the atoms in the melt. Crystals formed by precipitation from a solution develop when the solution becomes saturated, meaning the number of dissolved atoms, ions, or molecules per unit volume of solution becomes so great that they can get close enough to one another to bond together. If a solution is not saturated, dissolved atoms, ions, or molecules are surrounded by solvent molecules and are shielded from the attractive forces of their neighbors. Sometimes crystals formed by precipitation from a solution grow from the walls of the solution's "container" (e.g., a crack or pore in a rock). This process can form a spectacular **geode,** a mineral-lined cavity in rock formed when water solutions pass through the rock (▶Fig. 5.13a).

As crystals grow, they develop their particular crystal shape, based on the geometry of their internal structure. The shape is defined by the relative dimensions of the crystal (needle-like, sheet-like, etc.) and the angles between crystal faces. If a mineral's growth is uninhibited so that it displays well-formed crystal faces, then it is a **euhedral crystal** (▶Fig. 5.13b). Typically, however, the growth of minerals is restricted in one or more directions, because existing crystals act as obstacles. In such cases, minerals grow to fill the space that is available, and their shape is controlled by the shape of the surroundings. Minerals without well-formed crystal faces are **anhedral grains** (▶Fig. 5.14a, b). In the case of rocks created by the solidification of melts, many crystals grow at about the

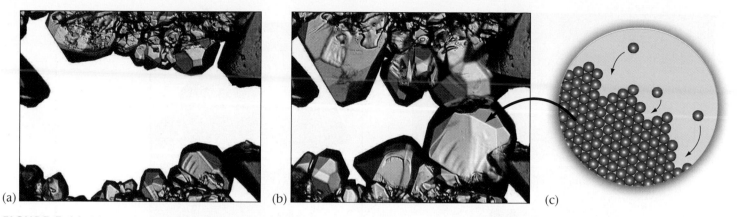

(a) (b) (c)

FIGURE 5.11 (a) New crystals nucleate (begin to form) in a water solution. They grow inward from the walls of the container. (b) At a later time, the crystals have grown larger. (c) On a crystal face, atoms in the solution are attracted to the surface and latch on.

same time, competing with each other for space. As a consequence, these minerals grow into one another, forming anhedral grains that interlock like pieces of a jigsaw puzzle.

A mineral can be destroyed by melting, dissolving, or some other chemical reaction. Melting involves heating a mineral to a temperature at which thermal vibration of the atoms or ions in the lattice can break the chemical bonds holding them to the lattice; the atoms or ions then separate, either individually or in small groups, to move around again freely. Different minerals have different melting temperatures. Dissolution occurs when you immerse a mineral in a solvent (like water). Atoms or ions then separate from the crystal face and are surrounded by solvent molecules. Some minerals, such as salt, dissolve easily, but most do not dissolve much at all. Chemical reactions can destroy a mineral when it comes in contact with reactive materials. For example, iron-bearing minerals react with air and water to form rust. The action of microbes in the environment can also destroy minerals. In effect, microbes can "eat" certain

minerals, using the energy stored in the chemical bonds that hold the atoms of the mineral together as their source of energy for metabolism.

5.4 HOW CAN YOU TELL ONE MINERAL FROM ANOTHER?

Amateur and professional geologists alike get a kick out of recognizing minerals. They'll be the show-offs in a museum who hover around the display case naming specimens without bothering to look at the labels. How do they do it? The trick lies in learning to recognize the basic **physical properties** (visual and material characteristics) that distinguish one mineral from another. Some physical properties, such as shape and color, can be seen from a distance—these are the properties that show-offs use to recognize specimens isolated behind the glass of a display case. Others, such as hardness and magnetization, can be determined only by handling the specimen or by performing an identification test on it. Identification tests include scratching the mineral against another object, placing it near a magnet, weighing it, tasting it, or placing a drop of acid on it. Let's examine some of the physical properties most commonly used in mineral identification. (Appendix B provides charts for identifying minerals, based on the physical properties of minerals.)

FIGURE 5.12 (a) Crystals grow outward from the central seed. (b) Crystals maintain their shape until they interfere with each other. When that happens, the crystal shapes can no longer be maintained.

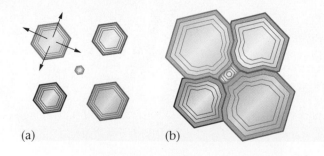

(a) (b)

- *Color:* **Color** results from the way a mineral interacts with light. Sunlight contains the whole spectrum of colors; each color has a different wavelength. A mineral absorbs certain wavelengths, so the color you see when looking at a specimen represents the wavelengths the mineral does not absorb. Certain minerals always have the same color (galena is always gray, for example), but many show a

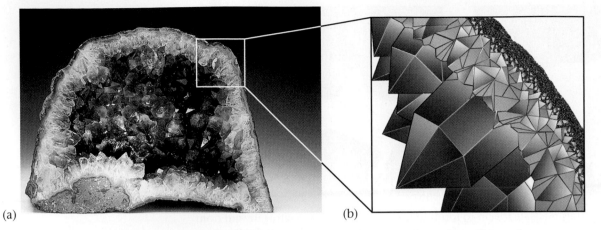

FIGURE 5.13 (a) A geode, in which euhedral crystals of purple quartz grow from the wall into the center. (b) An enlargement of a euhedral crystal, showing that the surfaces are crystal faces.

range of colors. For example, quartz can be clear, white, purple, gray (smoky), red (rose), or just about anything in between (▶Fig. 5.15). Purple quartz is known as amethyst (Fig. 5.13a). Color variations in a mineral reflect the presence of tiny amounts of impurities. For example, a trace of iron may produce a reddish or purple tint, and manganese may produce a pinkish tint.

• *Streak:* The **streak** of a mineral refers to the color of a powder produced by pulverizing the mineral. You can obtain a streak by scraping the mineral against an unglazed ceramic plate (▶Fig. 5.16). The color of a mineral powder tends to be less variable than the color of a whole crystal, and thus provides a fairly reliable clue to a mineral's identity. Calcite, for example, always yields a

white streak even though pieces of calcite may be white, pink, or clear.

• *Luster:* **Luster** refers to the way a mineral surface scatters light. Geoscientists describe luster simply by comparing the appearance of the mineral with the appearance of a familiar substance. For example, minerals that look like metal have "metallic luster," whereas those that do not have "nonmetallic luster"—the adjectives are self-explanatory (▶Fig. 5.17a, b). Types of nonmetallic luster may be further described, for example, as silky, glassy, satiny, resinous, pearly, or earthy.

• *Hardness:* **Hardness** is a measure of the relative ability of a mineral to resist scratching, and therefore represents the resistance of bonds in the crystal structure to being

FIGURE 5.14 A crystal growing in a confined space is anhedral, meaning its surface is not composed of crystal faces. (a) A crystal stops growing when it meets the surfaces of other grains and continues growing to fill in gaps. (b) The resulting mineral grain, if it were to be separated from other grains, would have an anhedral shape.

(a)

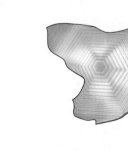

(b)

FIGURE 5.15 The range of colors of quartz, displayed by different crystals: milky, clear, and rose quartz.

FIGURE 5.16 A streak plate, showing the red streak of hematite.

• *Crystal habit:* The **crystal habit** of a mineral refers to the shape (morphology) of a single crystal with well-formed crystal faces, or to the character of an aggregate of many well-formed crystals that grew together as a group (▶Fig. 5.18). The habit depends on the internal arrangement of atoms in the crystal, for the arrangement, in turn, controls the geometry of crystal faces (e.g., triangular, square, rectangular, parallelogram) and the angular relationships among the faces. Mineralogists use a great many adjectives when describing habit. For example, a crystal may be compared to a geometric shape by using adjectives such as "cubic" or "prismatic." A description of habit generally includes adjectives that

broken. The atoms or ions in crystals of a hard mineral are more strongly bonded than those in a soft mineral. Hard minerals can scratch soft minerals, but soft minerals cannot scratch hard ones. Diamond is the hardest mineral known—it can scratch anything, which is why it is used to cut glass. In the early 1800s, a mineralogist named Friedrich Mohs listed some minerals in sequence of relative hardness; a mineral with a hardness of 5 can scratch all minerals with a hardness of 5 or less. This list, now called the **Mohs hardness scale,** helps in mineral identification. When you use the scale (Table 5.1), it might help to compare the hardness of a mineral with a common item like your fingernail, a penny, or a glass plate. Note that not all of the minerals in Table 5.1 are common or familiar. Also, it's important to realize that the numbers on the Mohs hardness scale do not specify the true relative differences in hardness of minerals. For example, on the Mohs scale, talc has a hardness of 1 and quartz has a hardness of 7. But this does not mean that quartz is 7 times harder than talc. Careful tests show that quartz is actually about 100 times harder than talc, as indicated by how difficult it is to make an indentation in the mineral (see Table 5.1).

• *Specific gravity:* **Specific gravity** represents the density of a mineral, as specified by the ratio between the weight of a volume of the mineral and the weight of an equal volume of water at 4°C. For example, one cubic centimeter of quartz has a weight of 2.65 grams, while one cubic centimeter of water has a weight of 1.00 gram. Thus, the specific gravity of quartz is 2.65. Divers use lead weights to help them sink down to great depths because lead is extremely heavy—it has a specific gravity of 11. In practice, you can develop a feel for specific gravity by hefting minerals in your hands. For example, a piece of galena (lead ore) "feels" heavier than a similar-sized piece of quartz.

FIGURE 5.17 (a) This specimen of pyrite looks like a piece of metal because of its shiny gleam; we call this metallic luster. (b) These specimens of feldspar have a nonmetallic luster. The white one on the left is plagioclase, and the pink one on the right is orthoclase ("potassium feldspar," or "K-spar").

(a)

(b)

BOX 5.2
THE REST OF THE STORY

Where Do Diamonds Come From?

As we saw earlier, diamond consists of the element carbon. Accumulations of carbon develop in a variety of ways: soot (pure carbon) results from burning plants at the surface of the Earth; coal (which consists mostly of carbon) forms from the remains of plants buried to depths of up to 15 km; and graphite develops from coal or other organic matter buried to still greater depths (15–70 km) in the crust during mountain building. Experiments demonstrate that the temperatures and pressures needed to form diamond are so extreme that, in nature, they generally occur only at depths of around 150 km below the Earth—that is, in the mantle. Under these conditions, the carbon atoms that were arranged in hexagonal sheets in graphite rearrange to form the much stronger and more compact structure of diamond. (Of note, engineers can duplicate these conditions in the laboratory; corporations manufacture several tons of diamonds a year.)

How does carbon get down into the mantle, where it transforms into diamond? Geologists speculate that the process of subduction provides the means. Carbon-containing rocks and sediments in oceanic lithosphere plates at the Earth's surface can be carried into the mantle at a convergent plate boundary. This carbon transforms into diamond, some of which becomes trapped in the lithospheric mantle beneath continents. But if diamonds form in the mantle, then how do they return to the surface? One possibility is that the process of rifting cracks the continental crust and causes a small part of the underlying lithospheric mantle to melt. Magma generated during this process rises to the surface, bringing the diamonds with it. Near the surface, the magma cools and solidifies to form a special kind of igneous rock called kimberlite (named for Kimberley, South Africa, where it was first found). Diamonds brought up with the magma are frozen into the kimberlite. Kimberlite magma contains a lot of dissolved gas and thus froths to the surface very rapidly. Kimberlite rock commonly oc- curs in carrot-shaped bodies 50–200 m across and at least 1 km deep that are called kimberlite pipes (▶Fig. 5.26a).

Controversial measurements suggest that many of the diamonds that sparkle on engagement rings today were created when subduction carried carbon into the mantle 3.2 billion years ago. The diamonds sat at depths of 150 km in the Earth until two rifting events, one of which took place in the late Precambrian and the other during the late Mesozoic, released them to the surface, like genies out of a bottle. The Mesozoic rifting event led to the breakup of Pangaea.

In places where diamonds occur in solid kimberlite, they can be obtained only by digging up the kimberlite and crushing it, to separate out the diamonds (▶Fig. 5.26b). But nature can also break diamonds free from the Earth. In places where kimberlite has been exposed at the ground surface for a long time, the rock chemically reacts with water and air (a process called weathering; see Chapter 7). These reactions cause most minerals in kimberlite to disintegrate, creating sediment that washes away in rivers. Diamonds are so strong that they remain as solid grains in river gravel. Thus, many diamonds have been obtained simply by separating them from recent or ancient river gravel.

Diamond-bearing kimberlite pipes are found in many places around the world, particularly where very old continental lithosphere exists. Southern and central Africa, Siberia, northwestern Canada, India, Brazil, Borneo, Australia, and the U.S. Rocky Mountains all have pipes. Rivers and glaciers, however, have transported diamond-bearing sediments great distances from their original sources. In fact, diamonds have even been found in farm fields of the midwestern United States. Not all natural diamonds are valuable: value depends on color and clarity. Diamonds that contain imperfections (cracks, or specks of other material), or are dark gray in color, won't be used for jewelry. These stones, called *industrial diamonds,* are used instead as abrasives, for diamond powder is so hard (10 on the Mohs hardness scale) that it can be used to grind away any other substance.

Gem-quality diamonds come in a range of sizes. Jewelers measure diamond size in carats, where one "carat" equals 200 milligrams (0.2 grams)—one ounce equals 142 carats. (Note that a carat measures gemstone weight, while a "karat" specifies the purity of gold. Pure gold is 24 karat, while 18-karat gold is an alloy containing 18 parts of gold and 6 parts of other metals.) The largest diamond ever found, a stone called the Cullinan Diamond, was discovered in South Africa in 1905. It weighed 3,106 carats (621 grams) before being cut into nine large gems (the largest weighing 516 carats) as well as many smaller ones. By comparison, the diamond on a typical engagement ring weighs less than 1 carat. Diamonds are rare, but not as rare as their price suggests. A worldwide consortium of diamond producers stockpile the stones so as not to flood the market and drive the price down.

FIGURE 5.26 (a) A mine in a kimberlite pipe. (b) A raw diamond still imbedded in kimberlite.

(a)

(b)

oysters when the oyster extracts calcium and carbonate ions from water and precipitates them around an impurity, like a sand grain, embedded in its body. Thus, they are a result of *biomineralization*. Most pearls used in jewelry today are "cultured" pearls, made by artificially introducing round sand grains into oysters in order to stimulate round pearl production. Amber is also formed by organic processes—it consists of fossilized tree sap. But because amber consists of organic compounds that are not arranged in a crystal structure, it does not meet the definition of a mineral.

Rare means hard to find, and some gemstones are indeed hard to find. Many diamond localities, for example, occur in isolated regions of Congo, South Africa, Brazil, Canada, Russia, India, and Borneo (see Box 5.2). In some cases, it is not the mineral itself but rather the "gem-quality" versions of the mineral that are rare. For example, garnets are found in many rocks in such abundance that people use them as industrial abrasives. But most garnets are quite small and contain inclusions (specks of other minerals and/or bubbles) or fractures, so they are not particularly beautiful. Gem-quality garnets—clean, clear, large unfractured crystals—are unusual. In some cases, gemstones are merely pretty and rare versions of more common minerals. For example, ruby is a special version of the common mineral corundum, and emerald is a special version of the common mineral beryl (▶Fig. 5.27). As for the beauty of a gemstone, this quality lies basically in its color and, in the case of transparent gems, its *fire*—the way the stone bends and internally reflects the light passing through it and disperses the light into a spectrum. "Fire" makes a diamond sparkle more than a similarly cut piece of glass.

FIGURE 5.27 Beryl crystals in rock.

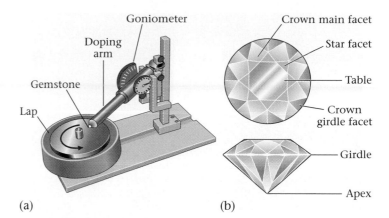

(a) (b)

FIGURE 5.28 The shiny faces on gems in jewelry are made by a faceting machine. (a) In this faceting machine, the gem is held against the face of the spinning lap. (b) Top and side views show the many facets of a brilliant-cut diamond, and names for different parts of the stone.

Gemstones form in many ways. Some solidify from a melt along with other minerals of igneous rock, some form by diffusion in a metamorphic rock, some precipitate out of a water solution in cracks, and some are a consequence of the chemical interaction of rock with water near the Earth's surface. Many gems come from pegmatites, particularly coarse-grained igneous rocks formed by the solidification of steamy melt.

Most gems when used in jewelry are "cut" stones. The smooth **facets** on a gem are ground and polished surfaces made with a faceting machine (▶Fig. 5.28a). Facets are not the natural crystal faces of the mineral, nor are they cleavage planes, though gem cutters sometimes make the facets parallel to cleavage directions and will try to break a large gemstone into smaller pieces by splitting it on a cleavage plane. A faceting machine consists of a doping arm, a device that holds a stone in a specific orientation, and a lap, a rotating disk covered with a wet paste of grinding powder and water. The gem cutter fixes a gemstone to the end of the doping arm and positions the arm so that it holds the stone against the moving lap. The movement of the lap grinds a facet. When the facet is complete, the gem cutter rotates the arm by a specific angle, lowers the stone, and grinds another facet. The geometry of the facets defines the cut of the stone. Different cuts have different names, like "brilliant," "French," "star," "pear," and "kite." Grinding facets is a lot of work—a typical engagement-ring diamond with a brilliant cut has fifty-seven facets (▶Fig. 5.28b)!

Some mineral specimens have special value simply because their geometry and color before cutting are beautiful. Prize specimens exhibit shapes and colors reminiscent of

FIGURE 5.29 A spectacular museum specimen of a mineral cluster. The arrangement of colors and shapes is like abstract art.

fine art and may sell for tens of thousands of dollars (▶Fig. 5.29). It's no wonder that mineral "hounds" risk their necks looking for a cluster of crystals protruding from the dripping roof of a collapsing mine or hidden in a crack near the smoking summit of a volcano.

CHAPTER SUMMARY

• Minerals are homogeneous, naturally occurring, solid substances with a definable chemical composition and an internal structure characterized by an orderly arrangement of atoms, ions, or molecules in a lattice. Most minerals are inorganic.

• In the crystalline lattice of minerals, atoms occur in a specific pattern—one of nature's finest examples of ordering.

• Minerals can form by the solidification of a melt, precipitation from a water solution, diffusion through a solid, the metabolism of organisms, and precipitation from a gas.

• There are close to 4,000 different known types of minerals, each with a name and distinctive physical properties (color, streak, luster, hardness, specific gravity, crystal form, crystal habit, and cleavage).

• The unique physical properties of a mineral reflect its chemical composition and crystal structure. By observing these physical properties, you can identify minerals.

• The most convenient way for classifying minerals is to group them according to their chemical composition. Mineral classes include: silicates, oxides, sulfides, sulfates, halides, carbonates, and native metals.

• The silicate minerals are the most common on Earth. The silicon-oxygen tetrahedron, a silicon atom surrounded by four oxygen atoms, is the fundamental building block of silicate minerals.

• There are several groups of silicate minerals, distinguished from one another by the ways in which the silicon-oxygen tetrahedra that constitute them are linked.

• Gems are minerals known for their beauty and rarity. The facets on cut stones used in jewelry are made by grinding and polishing the stone with a faceting machine.

KEY TERMS

anhedral grains (p. 116)
carbonates (p. 123)
cleavage (p. 120)
color (p. 117)
conchoidal fractures (p. 121)
crystal (p. 112)
crystal faces (p. 112)
crystal habit (p. 119)
crystal lattice (p. 110)
crystal structure (p. 113)
euhedral crystal (p. 116)
facets (p. 127)
gem (p. 125)
gemstone (p. 125)
geode (p. 116)
glass (p. 111)
halide (p. 123)
hardness (p. 118)

luster (p. 118)
mineral (p. 110)
mineral classes (p. 122)
mineralogists (p. 109)
mineralogy (p. 109)
Mohs hardness scale (p. 119)
native metals (p. 123)
oxides (p. 122)
physical properties (p. 117)
polymorphs (p. 115)
silicates (p. 122)
silicon-oxygen tetrahedron (p. 122)
specific gravity (p. 119)
streak (p. 118)
sulfates (p. 122)
sulfides (p. 122)
symmetry (p. 115)

REVIEW QUESTIONS

1. What is a mineral, as geologists understand the term? How is this definition different from the everyday usage of the word?

2. Why is glass not a mineral?

3. Salt is a mineral, but the plastic making an inexpensive pen is not. Why not?

4. Diamond and graphite have an identical chemical composition (pure carbon), yet they differ radically in physical properties. Explain in terms of their crystal structure.

5. In what way does the arrangement of atoms in a mineral define a pattern? How can X-rays be used to study these patterns?

6. Describe the several ways that mineral crystals can form.

7. Why do some minerals contain beautiful euhedral crystals, while others contain anhedral grains?

8. List and define the principal physical properties used to identify a mineral.

9. Give the chemical formulas of the following important minerals: quartz, halite, calcite.

10. What holds atoms together in a mineral?

11. Discuss the shape of crystals, including angular relations between crystal faces. What factors control crystal shape?

12. How can you determine the hardness of a mineral? What is the Mohs hardness scale?

13. How do you distinguish cleavage surfaces from crystal faces on a mineral? How does each type form?

14. What is the prime characteristic that geologists use to separate minerals into classes?

15. What is the principal anionic group in most familiar silicate minerals? On what basis are silicate minerals further divided into distinct groups?

16. What is the relationship between the way in which silicon-oxygen tetrahedra bond in micas and the characteristic cleavage of micas?

17. How do sulfate minerals differ from sulfides?

18. Why are some minerals considered gems? How do you make the facets on a gem?

SUGGESTED READING

Campbell, G. 2002. *Blood Diamonds: Tracing the Deadly Path of the World's Most Precious Stones.* Boulder: Westview Press.

Ciprianni, C., A. Borelli, and K. Lyman, eds. 1986. *Simon and Schuster's Guide to Gems and Precious Stones.* New York: Simon & Schuster.

Deer, W. A., J. Zussman, and R. A. Howie. 1996. *An Introduction to the Rock-Forming Minerals,* 2nd ed. Boston: Addison-Wesley.

Hall, C., J. J. Peters, and H. Taylor. 1994. *Gem Stones.* New York: DK Publishing.

Hart, M. 2002. *Diamond: A Journey to the Heart of an Obsession.* New York: Dutton/Plume.

Hibbard, J. J. 2001. *Mineralogy: A Geologist's Point of View.* New York: McGraw-Hill.

Klein, C., C. S. Hurlbut, and J. D. Dana. 2001. *The Manual of Mineral Sources,* 22nd ed. New York: John Wiley & Sons.

Kurlansky, M. 2003. *Salt: A World History.* New York: Penguin USA.

Matlins, A. L., and A. C. Bonanno. 2003. *Gem Identification Made Easy,* 2nd ed. Woodstock, Vt.: GemStone Press.

Nesse, W. D. 2000. *Introduction to Mineralogy.* New York: Oxford University Press.

Nesse, W. D. 2003. *Introduction to Optical Mineralogy,* 3rd ed. Oxford: Oxford University Press.

Perkins, D. 2001. *Mineralogy,* 2nd ed. Upper Saddle River, N.J.: Pearson Education.

Perkins, D., and K. R. Henke. 1999. *Minerals in Thin Sections.* Upper Saddle River, N. J.: Pearson Education.

Rock Groups

It took years of back-breaking labor for nineteenth-century workers to chisel and chip ledges and tunnels through the hard rock of the Sierra Nevadas, in their quest to run a rail line across this rugged range. In the process, the workers became very familiar with the nature of rock.

A.1 INTRODUCTION

During the 1849 gold rush in the Sierra Nevada Mountains of California, only a few lucky individuals actually became rich. The rest of the "forty-niners" either slunk home in debt or took up less glamorous jobs in new towns like San Francisco. These towns grew rapidly, and soon the American West Coast was demanding large quantities of manufactured goods from East Coast factories. Making the goods was no problem, but getting them to California meant either a stormy voyage around the southern tip of South America or a trek with stubborn mule teams through the deserts of Nevada or Utah. The time was ripe to build a railroad linking the East and West Coasts of North America, and, with much fanfare, the Central Pacific line decided to punch one right through the peaks of the Sierras. In 1863, while the Civil War raged elsewhere in the United States, the company transported six thousand Chinese laborers across the Pacific in the squalor of unventilated cargo holds and set them to work chipping ledges and blasting tunnels. Foremen measured progress in terms of feet per day—if they were lucky. Along the way, untold numbers of laborers died of frostbite, exhaustion, mistimed blasts, landslides, or avalanches.

Through their efforts, the railroad laborers certainly gained an intimate knowledge of how rock feels and behaves—it's solid, heavy, and hard! They also found that some rocks seemed to break easily into layers while others did not, and some rocks were dark while others were light. They realized, like anyone who looks closely at rock exposures, that rocks are not just gray, featureless masses, but rather come in a great variety of colors and textures. Why are there so many distinct types of rocks? The answer is simple: there are many different ways in which rocks can form, and many different materials out of which rocks can be made. Because of the relationship between rock type and the process of formation, *rocks provide a historical record of geologic events and give insight into interactions among components of the Earth System.* The next few chapters are devoted to a discussion of rocks and a description of how rocks form; this interlude serves as a general introduction. Here, we learn what the term "rock" means to geologists, what rocks are made of, and how to distinguish the three principal groups of rocks. We also look at how geologists study rocks. ESS

A.2 WHAT IS ROCK?

To geologists, **rock** is a coherent, naturally occurring solid, consisting of an aggregate of minerals or, less commonly, a mass of glass. Now let's take this definition apart.

- *Coherent:* A pile of unattached grains that can move around does not constitute a rock. A rock holds together, and it must be broken to separate into pieces. As a result of its coherence, rock can form cliffs or can be carved into sculptures.

- *Naturally occurring:* Geologists consider only naturally occurring materials to be rocks, so manufactured materials like concrete and brick do not qualify. (As a minor point, the term **stone** usually refers to rock used as a construction material.)

- *An aggregate of minerals or a mass of glass:* The vast majority of rocks consist of an aggregate (a collection) of many mineral grains, or crystals, stuck or grown together. (A **grain** is any fragment or piece of mineral, rock, or glass.

A *crystal* is a piece of a mineral that grew into its present shape. In casual discussion, geologists may use the word grain to include crystals.) Technically, a single mineral crystal is simply a "mineral specimen," not a rock, even if it is meters long. Some rocks contain only one kind of mineral, while others contain several different kinds. A few of the rock types that form at volcanoes (see Chapter 6) consist of glass, which may occur either as a homogeneous mass or as an accumulation of tiny shards (flakes).

What holds rock together? Grains in nonglassy rock stick together to form a coherent mass either because they are bonded by natural **cement,** mineral material that precipitates from water and fills the space between grains (▶Fig. A.1a–c), or because they interlock with one another like pieces in a jigsaw puzzle (▶Fig. A.2a–d). Rocks whose grains are stuck together by cement are called **clastic,** while rocks whose crystals interlock with one another are called **crystalline.** Glassy rocks hold together either because they originate as a continuous mass (i.e., they have no separate grains) or because glassy grains welded together while still hot.

(a)

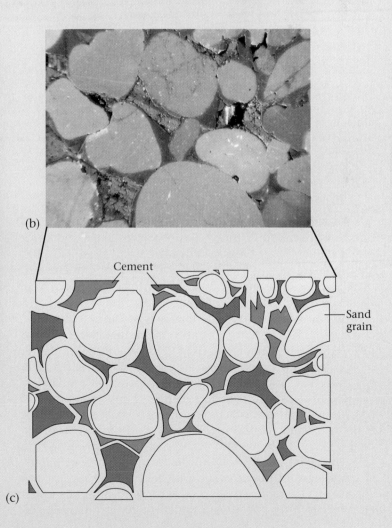

(b)

Cement

Sand grain

(c)

FIGURE A.1 (a) Hand specimens of sandstone. (b) A magnified image of the sandstone shows that it consists of round white sand grains, surrounded by cement. (c) This exploded image of the thin section emphasizes how the cement surrounds the sand grains. We depict the cement using two different colors because the cement probably formed at different times.

FIGURE A.2 (a) A hand specimen of granite, a rock formed when melt cools underground. (b) A photomicrograph (a photo taken through a microscope) shows the texture of granite is different from that of sandstone. In granite, the grains interlock with one another, like pieces of a jigsaw puzzle. (c) An artist's sketch emphasizes the irregular shapes of grains and how they interlock. (d) This exploded image highlights the grain shapes.

All rocks, in the most basic sense, are just masses of chemicals bonded in molecules of varying size and complexity. But not all rocks contain the same chemicals. For example, granite—a rock commonly used for gravestones, building facades, and kitchen counters—contains oxygen, silicon, aluminum, calcium, iron, magnesium, and potassium. Marble—a rock favored by fine sculptors—contains oxygen, carbon, and calcium. Note, as we pointed out in Chapter 2 (Fig. 2.15), that the elements oxygen and silicon are the most common elements in the Earth's crust; indeed, oxygen constitutes 93.8% of the volume of the crust. It is no surprise, therefore, that most of the rock in the crust as a whole consists of silicate minerals (minerals containing the silicon-oxygen tetrahedron). Very close to the Earth's surface, however, the activity of life plays a role in rock formation (see Chapter 7), so a significant proportion of bedrock exposed at the surface of the Earth consists of carbonate minerals (minerals containing the CO_3^- ion) extracted from water to form shells. Other minerals (e.g., oxides, sulfides, sulfates) are important as resources for metals and industrial materials, but they constitute only a small percentage of rocks in the crust.

A.3 ROCK OCCURRENCES

At the surface of the Earth, rock occurs either as broken chunks (pebbles, cobbles, or boulders; see Chapter 7) that have moved from their point of origin by falling down a slope or by being transported in ice, water, or wind; or as **bedrock,** which is still attached to the Earth's crust. Geologists refer to an exposure of bedrock as an **outcrop.** An outcrop may appear as a rounded knob out in a field, as a ledge along a cliff or ridge, on the face of a stream cut (where a river has cut down into bedrock), or along human-made road cuts and excavations (▶Fig. A.3a–d).

FIGURE A.3 (a) Outcrops (natural rock exposures) in a field along the coast of Scotland. (b) A stream cut, meaning an outcrop that forms when a stream's flow removes overlying soil and vegetation. Note that dense forest covers most of the adjacent hills, along this stream cut in Brazil, obscuring outcrops that may be exposed there. (c) Road cuts, like this one along a highway near Kingston, New York, are made by setting off dynamite placed at the bottom of drill holes. Note that the layers of rock exposed in this road cut are curved—such a bend is called a fold (see Chapter 11). (d) Mountain cliffs provide immense exposures of rock. These cliffs, in the Grand Teton Mountains of Wyoming, rise above a lowland in which bedrock has been covered by a layer of sediment, which hosts fields of sagebrush.

To people who live in cities or forests or on farmland, outcrops of bedrock may be unfamiliar, as they may be covered by vegetation, sand, mud, gravel, soil, water, asphalt, concrete, or buildings. Outcrops are particularly rare in regions like the midwestern United States, where, during the past million years, ice-age glaciers melted and left behind thick deposits of debris (see Chapter 22). These deposits completely buried preexisting valleys and hills, so today the bedrock surface lies as deep as 100 m below the ground. The depth of bedrock plays a key role in urban planning, because architects prefer to set the foundations of large buildings on bedrock rather than on loose sand or mud. Because of this preference, the skyscrapers of New York City rise in two clusters on the island of Manhattan, one at the south end and the other in the center, locations where bedrock lies close to the surface.

A.4 THE BASIS OF ROCK CLASSIFICATION

People have developed classification schemes for just about every group of materials on the planet—for insects, trees, airplanes, books, and so on. Why? Classification schemes help us organize information and remember significant details about materials or objects, and they help us recognize similarities and differences among them.

In the eighteenth century, geologists struggled to develop a sensible way to classify rocks, for they realized, as did miners from centuries past, that not all rocks are the same. One of the earliest classification schemes divided rocks into three groups—primary, secondary, and tertiary—on the basis of a perception (later proved incorrect) that the groups had formed in succession. In light of this concept, a German mineralogist named Abraham Werner proposed that a "universal ocean" containing dissolved and suspended minerals once had covered the Earth. According to Werner, the earliest rocks formed by precipitation from this solution; later, as sea level dropped, the action of rivers, waves, and wind wore down exposed rocks and produced debris that consolidated to form younger rocks. Werner was an influential teacher who developed a broad following of students and other geologists who came to be known as the Neptunists, after the Roman god of the sea.

At about the same time that Werner was developing his ideas, a Scottish gentleman farmer and doctor named James Hutton began exploring the outcrops of his native land. Hutton was a keen intellect who lived in Edinburgh, a hotbed of intellectual argument during the Age of Enlightenment where everything from political institutions to scientific paradigms became fodder for debate. He associated with prominent philosophers and scientists in Edinburgh and, like them, was open to new ideas. Hutton began to ponder the issue of how rocks formed; rather than force his perceptions to fit established dogma, he developed alternative ones based on his own observations. For example, he watched sand settle on a beach and realized that some rocks could have formed from accumulations of debris. He examined exposures in which bodies of granite and basalt appeared to have pushed into other rocks, heating the other rocks in the process, and concluded that some rocks could have formed by solidification from a melt. He also noticed that rocks adjacent to bodies of granite and basalt had somehow been altered, and he proposed that some rocks formed by change of preexisting rocks as a result of what he referred to as "subterranean heat." Hutton, like Werner, attracted followers—Hutton's group came to be known as the Plutonists after the Greek god of the underworld, because they favored the idea that the formation of certain rocks involved melts that had risen from deeper within the Earth.

In the last decades of the eighteenth century, as the armed rebellions that led to the formation of the United States and the Republic of France raged, a battle of ideas concerning the origin of rocks rattled the infant science of geology. This battle, pitting the Neptunists against the Plutonists, lasted for years. In the end the Plutonists won, for they demonstrated beyond a shadow of a doubt that rocks like granite and basalt must have been in molten form when emplaced. As a consequence, it became clear that different rocks formed in different ways and that as a starting point, rocks can best be classified on the basis of how they formed. This principle became the basis of the modern system of classification of rocks. For this contribution, among many others that we will describe later in the book, modern geologists revere Hutton as the "father of geology."

Hutton's scheme of rock classification is a "genetic classification," because it focuses on the genesis—the origin—of the rock. In modern terminology, geologists recognize three basic groups: (1) **igneous rocks,** which form by the freezing (solidification) of molten rock, or melt (▶Fig. A.4a); (2) **sedimentary rocks,** which form either by the cementing together of fragments (grains) broken off preexisting rocks or by the precipitation of mineral crystals out of water solutions at or near the Earth's surface (▶Fig. A.4b); and (3) **metamorphic rocks,** which form when preexisting rocks change into new rocks in response to a change in pressure and temperature conditions, and/or as a result of squashing, stretching, or shear (▶Fig. A.4c). Metamorphic change occurs in the "solid state," which means that it does not require melting. Each of the three groups contains many different individual rock types, distinguished from one another by physical characteristics such as

- *grain size:* The dimensions of individual grains (using the word here in a general sense to mean crystals, fragments of crystals, or fragments of preexisting rocks) in a rock may be measured in millimeters or centimeters. Some grains are so small that they can't be seen without a microscope, while others are as big as a fist or larger. Some grains are **equant,** meaning that they have the same dimensions in all directions, while some are **inequant,** meaning that the dimensions are not the same in all directions (▶Fig. A.5a–c). In some rocks, all the grains are the same size, while other rocks contain a variety of different-sized grains.

- *composition:* As we stated earlier, a rock is a mass of chemicals. These chemicals may be ordered into mineral grains or, less commonly, may be disordered and constitute glass. The term "rock composition" refers to the proportions of different chemicals making up the rock. The proportion of chemicals, in turn, affects the proportion of different minerals constituting the rock. As you will see, however, chemical composition does not completely control the minerals present in a rock. For example, two rocks with exactly the same chemical composition can consist of totally different assemblages of minerals, if each rock formed under different pressure and

(a)

(b)

FIGURE A.4 Examples of the major rock groups. (a) The volcano in the background erupted lava (molten rock) that flowed over the landscape and eventually froze, forming black basalt, an igneous rock. (b) Sand, deposited on a beach, eventually becomes buried to form layers of sandstone, a sedimentary rock, such as those exposed in the cliffs behind the beach. (c) When preexisting rocks become buried deeply during mountain building, the increase in temperature and pressure transforms them into metamorphic rocks. The lichen-covered outcrop in the foreground was once deeply buried beneath a mountain range, but was later exposed when glaciers and rivers stripped off the overlying rocks of the mountain.

(c)

temperature conditions. That's because the process of mineral formation is affected by environmental factors such as pressure and temperature.

- *texture:* This term refers to the arrangement of grains in a rock; that is, the way grains connect to one another and whether or not inequant grains are aligned parallel to one another. The concept of rock texture will become easier to grasp as we look at different examples of rocks in the following chapters.

- *layering:* Some rock bodies appear to contain distinct layering, defined either by bands of different compositions or textures, or by the alignment of inequant grains so that they trend parallel to one another. Different types of layering occur in different kinds of rocks. For example, the layering in sedimentary rocks is called **bedding,** while the layering in metamorphic rocks is called **metamorphic foliation** (▶Fig. A.6a, b).

Each individual rock type has a name. Names come from a variety of sources. Some come from the dominant component making up the rock, some from the region where the rock was first discovered or is particularly abundant, some from a root word of Latin origin, and some from a traditional name used by people in an area where the rock is found. All told, there are hundreds of different rock names, though in this book we introduce only about thirty.

A.5 STUDYING ROCK

Outcrop Observations

The study of rocks begins by examining a rock in an outcrop. If the outcrop is big enough, such an examination

CHAPTER 6

Up from the Inferno: Magma and Igneous Rocks

A river of molten rock (lava) weaves across a stark terrain of already solidified igneous rock. The volcano, the vent from which the lava has spilled out onto Earth's surface from the interior, can be seen in the distance.

6.1 INTRODUCTION

Every now and then, an incandescent liquid—molten rock, or melt—begins to fountain from a crater (pit) or crack on the big island of Hawaii, for Hawaii is a **volcano,** a vent at which melt from inside the Earth spews onto the planet's surface. Such an event is a "volcanic eruption." Some of the melt, which is called **lava** once it has reached the Earth's surface during an eruption, pools around the vent, while the rest flows down the mountainside as a viscous (syrupy) red-yellow stream called a lava flow. Near its source, the flow moves swiftly, cascading over escarpments at speeds of up to 60 km per hour (▶Fig. 6.1a). At the base of the mountain, the lava flow slows but advances nonetheless, engulfing any roads, houses, or vegetation in its path. At the edge of the flow, beleaguered plants incinerate in a burst of flames. As the flow cools, it slows down, and its surface darkens and crusts over, occasionally breaking to reveal the hot, sticky mass that remains within (▶Fig. 6.1b). Finally, it stops moving entirely, and within days or weeks the once red-hot melt has become a hard, black solid through and through (▶Fig. 6.1c). A new **igneous rock,** made by the *freezing* of a melt, has formed. Considering the fiery heat of the melt from which igneous rocks develop, the

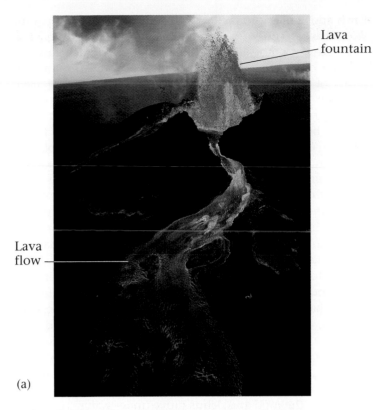

Lava
fountain

Lava
flow

(a)

(b)

(c)

FIGURE 6.1 (a) Lava fountains in this crater of a volcano on Hawaii, and a river of lava streams out of a gap in its side. As the lava moves rapidly away from the crater, it cools, and a black crust forms on the surface. (b) Farther down the mountain, the surface of the lava has completely crusted over with newborn rock, while the insides of the flow remain molten, allowing it to creep across the highway (in spite of the stop sign). Smoke comes from burning vegetation. (c) Eventually the flow cools through and through, and a new layer of basalt rock has formed. This rock is only a few weeks old.

name "igneous"—from the Latin *ignis,* meaning "fire"—makes sense.

It may seem strange to speak of freezing in the context of forming rock, when most people think of freezing as the transformation of liquid water to solid ice on, say, the surface of a lake when the temperature drops below 0°C (32°F). Nevertheless, the freezing of liquid melt to form solid igneous rock represents the same phenomenon, except that igneous rocks freeze at high temperatures—between 650°C and 1,100°C. To put such temperatures in perspective, remember that home ovens only attain a maximum temperature of 260°C (500°F).

Igneous rocks make up all of the oceanic crust and much of the continental crust. The oldest igneous rocks formed when the planet was still young, for after it coalesced out of planetesimals, the Earth was hot and part of it was probably molten. The heat of the early Earth came from five sources.

- Countless meteorites, pulled toward the Earth by gravity, slammed into the young planet. Their kinetic energy converted to heat. To picture this process, imagine repeatedly slamming two bricks together with your hands—the bricks will warm up because the slamming motion will cause the molecules in the brick to start jostling and this motion produces heat (see Appendix A). This process provided most of the heat in the early Earth.

- As the Earth grew, gravity compressed, or squeezed together, the matter making up the planet; and when materials undergo compression, they heat up.

- While the Earth was young and hot, heavy iron alloy sank toward the center of the planet; in this process potential energy transformed into heat (see Appendix A).

- The Earth contains radioactive elements which decay and generate heat (see Appendix A).

• The gravitational pull of the Moon and Sun caused tidal movements of the early Earth, for the planet was still fairly soft. Friction between grains of rock in the Earth, during these tidal movements, generated heat.

For a time, the surface of the young planet may even have turned into a sea of melt. But eventually, the Earth cooled sufficiently for its surface temperature to drop below the freezing point of rock. When this happened, a thin skin of igneous rock formed on its surface. Soon the entire mantle froze, becoming a thick shell of solid, though still very warm and plastic, igneous rock. Thus, igneous rocks were the first rocks to form on the Earth. But did they all originate early in Earth history? Clearly no. Igneous rocks continue to form today, for even though the crust and mantle are mostly solid, production of melt still takes place at special locations inside the Earth.

Although some igneous rocks solidify at the surface during volcanic eruptions like those on Hawaii, a vastly greater volume result from solidification of melt underground, out of sight. Geologists refer to melt that exists below the Earth's surface as **magma,** and melt that has erupted from a volcano at the surface of the Earth "lava." Rock made by the freezing of magma underground, after it has pushed its way ("intruded") into preexisting rock of the crust, is **intrusive igneous rock,** and rock that forms by the freezing of lava above ground, after it spills out ("extrudes") onto the surface of the

Earth and comes into contact with the atmosphere or ocean, is **extrusive igneous rock** (▶Fig. 6.2). Extrusive igneous rock includes both solid lava flows, formed when streams or mounds of lava solidify on the surface of the Earth, and deposits of **pyroclastic debris** (from the Latin word *pyro,* meaning "fire"). Some of the debris forms when clots of lava fly into the air in lava fountains and then freeze to form solid chunks before hitting the ground. And some debris forms when an explosion takes place during an eruption and blasts a fine spray of lava into the air and/or rips preexisting rock from the volcano and sends fragments into the sky. The fine spray of lava instantly freezes to form fine particles of glass called **ash.** Some of the ash billows up into the atmosphere, eventually drifting down from the sky as an **ash fall.** But some ash rushes down the side of the volcano in a scalding avalanche called an **ash flow** (see Chapter 9 for further detail).

Of note, a significant amount of the material making up a volcano actually consists of pyroclastic debris accumulations that move again *after* being initially deposited. Volcanic debris may slip down the mountain in a landslide; it may be carried away by streams and re-deposited as stream sediment; or it may mix with water to form a slurry of mud and debris, called a "volcanic" debris flow, that flows down the mountain like wet concrete. If the volcano is an island or seamount, the mud and debris move under water.

A great variety of igneous rocks exist on Earth. To understand why and how these rocks form, and why there are so many different kinds, we discuss why magma forms, why it rises, why it sometimes erupts as lava, and how it freezes in intrusive and extrusive environments. We then look at the scheme that geologists use to classify igneous rocks.

FIGURE 6.2 Extrusive igneous rocks, namely ash and lava, form above the Earth's surface, while intrusive rocks develop below. Melt that erupts from a volcano is lava, while underground melt is magma.

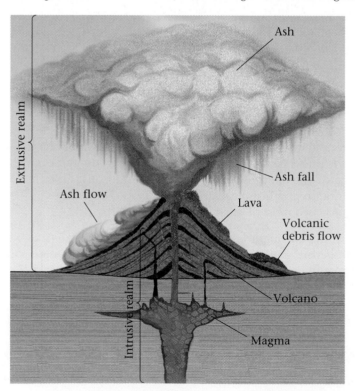

6.2 THE FORMATION OF MAGMA

As mentioned previously, the popular image that the solid crust of the Earth floats on a sea of molten rock is not correct. In reality, magma forms in the lithosphere and crust, but only in special places where preexisting solid rock melts. Following are the conditions that lead to melting.

Melting as a result of a decrease in pressure (decompression). The Earth is quite hot inside. At the continental Moho, 35 km below the surface, temperatures reach 500°–600°C; and at the base of the lithosphere, 100–150 km down, temperatures reach 1,280°C. The variation in temperature with depth can be expressed on a graph by a curving line, the **geotherm.** Beneath typical oceanic crust, temperatures comparable to those of lava (650°–1,100°C) generally occur in the upper mantle (▶Fig. 6.3). But even though the upper mantle is very hot, its rock stays solid because it is also under high pressure from the weight of overlying rock. Pressure at great depth, simplistically, prevents atoms from breaking free of mineral crystals.

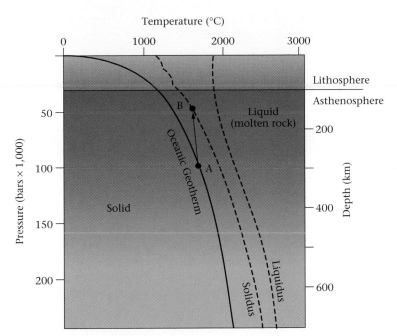

FIGURE 6.3 The graph plots the Earth's geotherm (solid line), which specifies the temperature at various depths below oceanic lithosphere, as well as the "liquidus" and "solidus" (dashed lines) for peridotite, the ultramafic rock that makes up the mantle. The "solidus" represents conditions of pressure and temperature at which a rock *begins to melt,* whereas the "liquidus" represents the conditions of pressure and temperature at which the *last solid disappears.* The region of the graph between the liquidus and solidus represents conditions under which there can be a mixture of solid and melt. Note that the geothermal gradient (the rate of change in temperature) decreases with greater depths; if it were constant, the geotherm would be a straight line. A rock that starts at pressure and temperature conditions indicated by point A, and then rises to point B, undergoes a significant decrease in pressure without much change in temperature. When it reaches the conditions indicated by point B, it begins to melt. This process is called decompression melting. Note that asthenosphere cools only slightly as it rises, because rock is such a good insulator.

Because pressure prevents melting, a decrease in pressure can permit melting. Specifically, if the pressure affecting hot mantle rock decreases while the temperature remains unchanged, a magma forms. This kind of melting, called "decompression melting" (see Box 6.1), occurs where hot mantle rock rises to shallower depths in the Earth, because pressure decreases toward the surface and rock is such a good insulator that it doesn't lose much heat as it rises (▶Fig. 6.4a).

Melting as a result of the addition of volatiles. Magma also forms at locations where chemicals called volatiles mix with hot mantle rock. Volatiles are elements or molecules, such as water (H_2O) and carbon dioxide (CO_2), that evaporate easily and can exist in gaseous forms at the Earth's surface. When volatiles mix with hot rock, they help break chemical bonds so

if you add volatiles to a solid, hot dry rock, the rock begins to melt (▶Fig. 6.4b). In effect, adding volatiles decreases a rock's melting temperature. (Melting due to addition of volatiles is sometimes called "flux melting.") Of the common volatiles, water plays the most important role in influencing melting.

We can represent the effect of volatiles by looking at the contrast between the melting curve (or "solidus"), which is the line defining the range of temperatures and pressures at which a rock melts, for wet basalt and that for dry basalt (▶Fig. 6.4c). Note that wet basalt (basalt containing volatiles) melts at much lower temperatures than dry (volatile-free) basalt. In fact, adding volatiles to rock to cause melting is somewhat like sprinkling salt on ice to make it melt.

Melting as a result of heat transfer from rising magma. When rock, or magma, from the mantle rises up into the crust, it brings heat with it. This heat flows into and raises the temperature of the surrounding crustal rock. In some cases, the rise in temperature may be sufficient to melt part of the crustal rock (Fig. 6.4a). Imagine injecting hot fudge into ice cream; the fudge transfers heat to the ice cream, raises its temperature, and causes it to melt. We call such melting "heat-transfer melting," because it results from the transfer of heat from a hotter material to a cooler one. Since mantle-derived magmas are very hot (over 1,100°C) and rocks of the crust melt at temperatures of about 650°–850°C, when mantle-derived magma intrudes into the crust, it can raise the temperature of the surrounding crust enough to melt it.

6.3 WHAT IS MAGMA MADE OF?

All magmas contain silicon and oxygen, which bond to form the silicon-oxygen tetrahedron. But magmas also contain varying proportions of other elements like aluminum (Al), calcium (Ca), sodium (Na), potassium (K), iron (Fe), and magnesium (Mg). Because magma is a liquid, its atoms do not lie in an orderly crystalline lattice but are grouped instead in clusters or short chains, relatively free to move with respect to one another.

"Dry" magmas contain no volatiles. "Wet" magmas, in contrast, include up to 15% dissolved volatiles such as water, carbon dioxide, nitrogen (N_2), hydrogen (H_2), and sulfur dioxide (SO_2). These volatiles come out of the Earth at volcanoes in the form of gas. Usually water constitutes about half of the gas erupting at a volcano. Thus, magma not only contains the elements that constitute solid minerals in rocks, but can also contain molecules that become water or air.

The Major Types of Magma

Imagine four pots of molten chocolate simmering on a stove. Each pot contains a different type of chocolate. One pot contains white chocolate, one milk chocolate, one semi-sweet

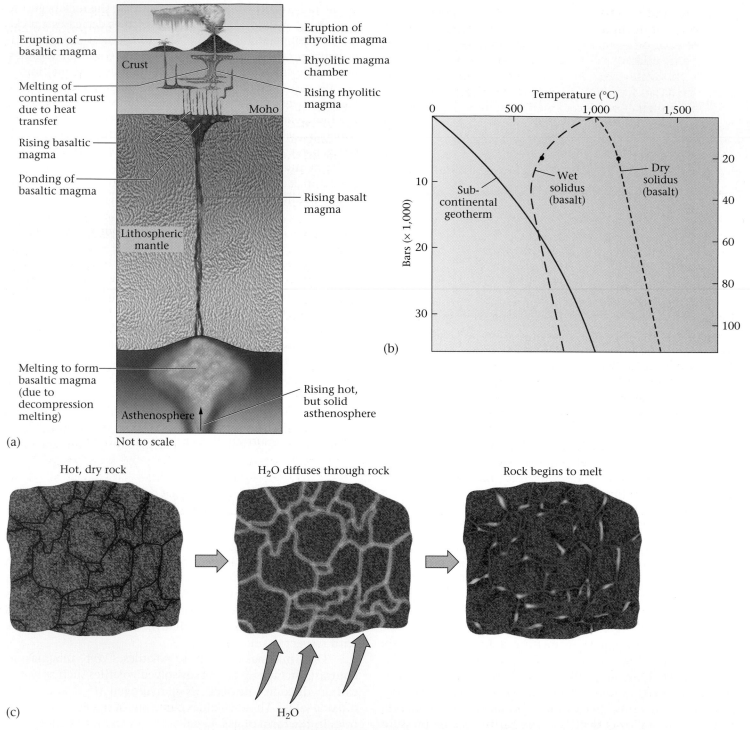

FIGURE 6.4 The three main causes of melting and magma formation in the Earth. (a) Decompression melting occurs when a hot rock rises to a shallow depth, where the pressure is lower. Melting as a result of heat transfer happens when hot magma rises into rock that has a lower melting temperature. For example, hot basaltic magma rising from the mantle can make the surrounding intermediate-composition crust melt. (b) The addition of volatiles decreases the melting temperature. For example, at a depth of 20 km, the melting temperature of wet basalt (basalt that contains volatiles) is about 500°C lower than the melting temperature of dry basalt. (c) Melting as a result of the addition of volatiles occurs when H_2O percolates into a solid hot rock. It's as if an "injection" of water triggers the melting.

chocolate, and one baker's chocolate. All the pots contain chocolate, but not all the chocolates are the same—they differ from each other in terms of proportions of sugar, cocoa butter, and milk. It's no surprise that different kinds of molten chocolate yield different kinds of solid chocolate; each type differs from the others in terms of taste and color.

Like molten chocolate, not all molten rock (magma) is the same. As we noted earlier, magmas are merely liquid mixtures of several chemicals, and magmas differ from one another in terms of the chemicals they contain. Geologists have found that the easiest way to describe the chemical makeup of magma is by specifying the proportions, in weight percent, of oxides. (By "weight percent" we mean the percentage of the magma's weight that consists of a given component, and by "oxide" we mean a combination of a cation with oxygen.) Common oxides include silica (SiO_2), iron oxide (FeO or Fe_2O_3), and magnesium oxide (MgO).

The chemicals present in a magma ultimately determine the identity of minerals that form when the magma cools. For convenience, many geologists use the same names for the compositional categories of magmas that they use for the compositional categories of igneous rocks, even though magmas, by definition, don't actually contain minerals. We can distinguish these categories from each other, most simply, in terms of the relative proportions of silica (SiO_2) they contain. The magma types are as follows. (1) **Felsic magma** contains about 66% to 76% silica. The name reflects the occurrence of felsic minerals (feldspar and quartz) in rocks formed from this magma. Because of its relatively high silica content, some geologists use the adjective "silicic" instead of "felsic" for this magma. (2) **Intermediate magma** contains about 52% to 66% silica. The name "intermediate" indicates that these magmas have a composition between that of felsic magma and mafic magma. (3) **Mafic magma** contains about 45% to 52% silica. Mafic magma produces rock containing abundant mafic minerals; that is, minerals with a relatively high proportion of MgO and FeO or Fe_2O_3. Recall that the "ma" in the word "mafic" stands for "magnesium," and the suffix "-fic" stands for "iron" (from the Latin *ferric*). (4) **Ultramafic magma** contains only 38% to 45% silica.

Magma properties depend on magma compositions. For example, the **viscosity** (resistance to flow) of magma reflects its silica content, for silica tends to polymerize, meaning it links up to form long chain-like molecules whose presence slows down the flow. Thus, felsic magmas have higher viscosity (i.e., are more sticky and flow less easily) than mafic magmas. The density of magma also reflects its composition, for SiO_2 is less dense than MgO or FeO. Thus, felsic magmas are less dense than mafic magmas. Finally, different magmas have different temperatures. Felsic magma can remain liquid at temperatures of only 650° to 800°C, whereas ultramafic magmas may reach temperatures of up to 1,300°C.

Why are there so many different kinds of magma? There are several factors.

Source rock composition. When you melt ice, you get water, and when you melt wax, you get liquid wax. There is no way to make water by melting wax. Clearly, the composition of a melt reflects the composition of the solid from which it was derived. Not all magmas form from the same source rock, so not all magmas have the same composition; magmas formed from crustal sources don't have the same composition as magmas formed from mantle sources, because the crust and mantle have different compositions to start with (see Chapter 2).

Understanding Decompression Melting

BOX 6.1
THE REST OF THE STORY

Look again at Figure 6.3. The horizontal axis on the graph represents temperature, in degrees Centigrade, and the left vertical axis represents pressure. Since pressure in the Earth results from the weight of overlying rock, we can calibrate the vertical axis either in units of depth (km below the surface) or in units of pressure, such as bars. (Note: 1 bar approximately equals 1 atm, where 1 atm, or atmosphere, is the air pressure at sea level.)

The solid line on this graph is the geotherm, which defines the temperature as a function of depth in the Earth. Notice that the rate of increase in temperature, a quantity called the **geothermal gradient** (expressed in degrees per km), decreases with increasing depth. The dashed lines represent the solidus and liquidus for mantle rock (peridotite). The solidus defines the conditions of pressure and temperature at which mantle rock begins to melt, as determined by laboratory measurements. Values to the left of the solidus indicate pressures and temperatures for which the rock stays entirely solid. The liquidus represents conditions at which all solid disappears and only melt remains.

To see what happens during decompression, imagine a mantle rock at a depth of 300 km (point A on the graph). According to the graph, the pressure ≈ 95,000 bars and the temperature ≈ 1,700°C at this depth. Now imagine that the rock moves closer to the Earth's surface without cooling, as may occur in a rising mantle plume, and reaches point B, where the pressure is only about 47,000 bars. It has undergone decompression. Notice that point B lies on the solidus, so the rock begins to melt—the thermal vibration of atoms in the rock, no longer countered by pressure, can cause the atoms to break free of crystals. Also notice that decompression melting takes place without adding new heat—in fact, because the rock expands as it rises, it actually cooled slightly.

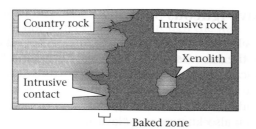

FIGURE 6.9 An intrusive contact, showing the baked zone, blocks of country rock, fingers of the intrusion protruding into the country rock, and a xenolith.

Intrusive Igneous Settings

Magma rises and intrudes into preexisting rock by slowly percolating upward between grains or by forcing open cracks. The magma that doesn't make it to the surface freezes solid underground in contact with preexisting rock and becomes intrusive igneous rock. Geologists commonly refer to the preexisting rock into which magma intrudes as country rock, or wall rock, and the boundary between wall rock and an intrusive igneous rock as an "intrusive contact" (▶Fig. 6.9). If the wall rock was cold to begin with, then heat from the intrusion "bakes" and alters it in a narrow band along an intrusive contact (see Chapter 8).

Geologists distinguish between different types of intrusions based on their shape. "Tabular intrusions," or sheet intrusions, are planar and are of roughly uniform thickness; they range in thickness from millimeters to tens of meters, and can be traced for meters to tens, or in a few cases hundreds, of kilometers. In places where tabular intrusions cut across rock that does not have layering, a nearly vertical, wall-like tabular intrusion is called a **dike**, whereas a nearly horizontal, tabletop-shaped tabular intrusion is a **sill.** In places where tabular intrusions enter rock that has layering (bedding or foliation), dikes are defined as intrusions that cut across layering, while sills are intrusions that are parallel to layering (▶Figs. 6.10a, b; 6.11a–g). Spectacular groups of dikes cut across the countrysides of interior Canada and western Britain, and a large sill, the Palisades Sill, makes up the cliff along the western bank of the Hudson River opposite New York City. Another sill forms the ledge on which Hadrian's Wall, which bisects Britain, was built. Some intrusions start to inject between layers, but then dome upwards, creating a blister-shaped intrusion known as a **laccolith** (▶Fig. 6.12a).

Plutons are irregular or blob-shaped intrusions that range in size from tens of meters across to tens of kilometers across (▶Figs. 6.12a–c; 6.13a, b). The intrusion of numerous plutons in a region creates a vast composite body that may be several hundred kilometers long and over 100 km wide; such immense masses of igneous rock are called **batholiths.** The rock making up the Sierra Nevada Mountains of California comprises a batholith created from plutons that intruded between 145 and 80 million years ago (▶Fig. 6.14a–d). Keep in mind that batholiths do not extend all the way down to the base of the crust. They are probably only a few kilometers to perhaps 10 km thick.

Where does the space for intrusions come from? Geologists continue to debate this issue, and over the past century they have suggested several models. Dikes form in regions where the crust is being stretched (e.g., in a rift). Thus, as the magma that makes a dike forces its

FIGURE 6.10 (a) Dikes and sills are vertical or horizontal bands, respectively, on the face of an outcrop. (b) If we were to strip away the surrounding rock, dikes would look like walls, and sills would look like tabletops.

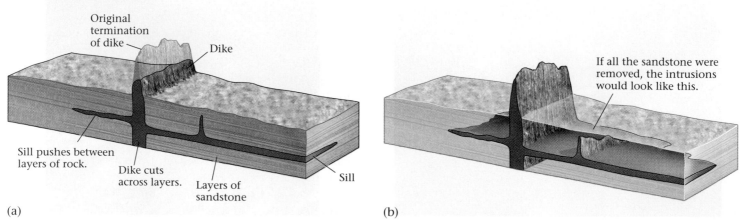

(a)

(b)

FIGURE 6.11 (a) A basalt dike looks like a black stripe painted on an outcrop of granite (here, in Arizona). But the dike actually intrudes, wall-like, into the outcrop. In this example, the dike happens to curve. (b) At this ancient volcano called Shiprock, in New Mexico, ash and lava flows have eroded away, leaving a "volcanic neck" (the solid igneous rock that cooled in a magma chamber within the volcano). Large dikes radiate outward from the center, like spokes of a wheel. The softer rocks that once surrounded the dikes have eroded away, leaving wall-like remnants of the dikes exposed. (c) Shiprock was once in the interior of a volcano or below a volcano. The dikes radiated from this neck and cut across preexisting rock layers. (d) These Precambrian dikes exposed in the Canadian Shield formed when the region underwent stretching over a billion years ago; at that time, numerous cracks in the crust filled with magma. (e) This dark sill, exposed on a cliff in Antarctica, is basalt; the white rock is sandstone. (f) This geologist's sketch shows the cliff face as viewed face on. (g) Map showing Cenozoic dikes in the United Kingdom and Ireland. Note that the dikes radiate from intrusive centers. When these dikes intruded, the region was undergoing stretching in the NE–SW direction.

(a)

(b)

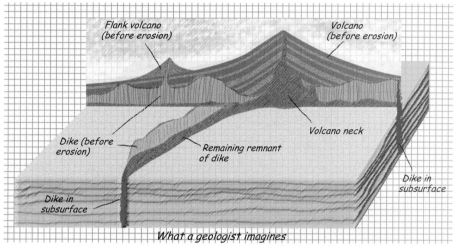

(c) *What a geologist imagines*

(d)

(e)

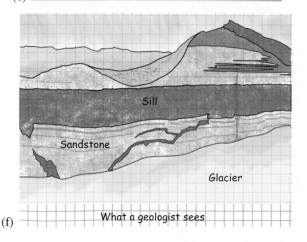

(f) *What a geologist sees*

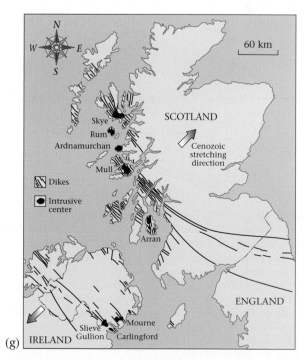

(g)

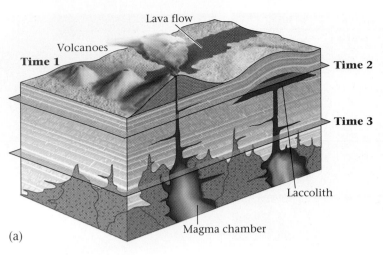

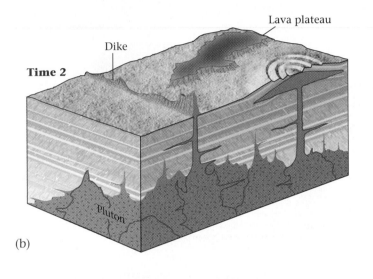

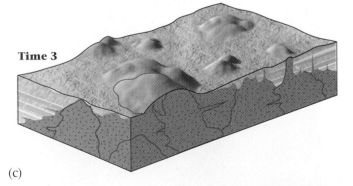

FIGURE 6.12 (a) While a volcano is active, a magma chamber exists underground; dikes, sills, and laccoliths intrude; and lava and ash erupt at the surface. Here, we see that the active magma chamber, the one currently erupting, is only the most recent of many in this area. Earlier ones are now solidified. Each mass is a pluton. A composite of many plutons is a batholith. (b) Later, the bulbous magma chamber freezes into a pluton. The soft parts of the volcano erode, leaving wall-like dikes and column-like volcanic necks. Hard lava flows create resistant plateaus. (c) With still more erosion, volcanic rocks and shallow intrusions are removed, and we see plutonic intrusive rocks.

way up through a crack (sometimes causing the crack to form in the first place), the crust stretches sideways (▶Fig. 6.15a). Intrusion of sills occurs near the surface of the Earth, so the pressure of the magma effectively pushes up the rock over the sill, leading to uplift of the Earth's surface (▶Fig. 6.15b).

How do plutons form? This question remains a topic of active research and much controversy (▶Fig. 6.15c). Some geologists propose that a pluton is a frozen "diapir", meaning a light-bulb-shaped blob of magma that pierced overlying rock and pushed it aside as it rose. Others suggest that pluton formation involves **stoping,** a process during which magma assimilates wall rock, and blocks of wall rock break off and sink into the magma (▶Fig. 6.15d). (If a stoped block does not melt entirely, but rather becomes surrounded by new igneous rock, it can be called a **xenolith,** after the Greek word *xeno,* meaning "foreign;" ▶Fig. 6.15e; see also Fig. 6.9.) More recently, geologists have proposed that plutons form by injection of several superimposed dikes or sills, which coalese to become a single larger intrusion. Finally, a few geologists speculate that plutons grow when chemical exchanges between magma and the wall rock transform the wall rock into granite.

If intrusive igneous rocks form beneath the Earth's surface, why can we see them exposed today? The answer comes from studying the dynamic activity of the Earth. Over long periods of geologic time, mountain building, driven by plate interactions, slowly uplifts huge masses of rock. Moving water, wind, and ice eventually strip away great thicknesses of overlying rock and expose the intrusive rock that had formed below.

6.6 TRANSFORMING MAGMA INTO ROCK

What makes magma freeze? In nature, two phenomena lead to the formation of solid igneous rock from a magma. Magma may freeze if the volatiles dissolved within it bubble out, for removal of volatiles (H_2O and CO_2) makes magma freeze at a higher temperature. But more commonly, magma freezes simply when it cools below its freezing temperature and crystals start to grow. For cooling to occur, magma must move to a cooler environment. Because temperatures decrease toward the Earth's surface, magma automatically enters a cooler environment when it rises. The cooler environment may be cool country rock if the magma intrudes underground, or it may be the atmosphere or ocean if the magma extrudes as lava at the Earth's surface.

The time it takes for a magma to cool depends on how fast it is able to transfer heat into its surroundings. To see

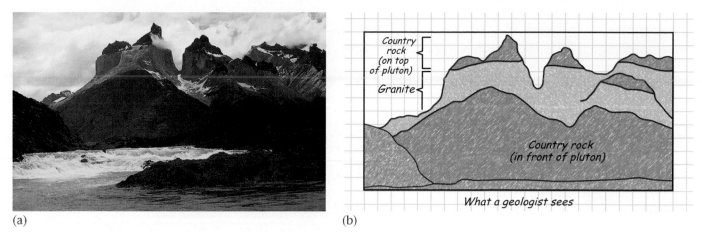

(a)

(b)

FIGURE 6.13 (a) Torres del Paines, a spectacular group of mountains in southern Chile. The light rock is a granite pluton, and the dark rock is the remains of the country rock (wall rock) into which the pluton intruded. A screen of country rock (in the lower half) hides the front of the pluton. (b) A geologist's sketch, labeling the two major rock units.

FIGURE 6.14 (a) The batholiths of western North America today. (b) The geography of western North America about 100 million years ago, showing the position of the subduction zone and volcanic arc responsible for the batholiths. Note that the west coast, south of Idaho, lay much farther east than it does today. (c) The Sierra Nevada Batholith as exposed today. The rounded, light-colored hills are all composed of granite-like intrusive igneous rock. (d) A geologist's sketch illustrates that today's land surface once lay several kilometers beneath a chain of volcanoes.

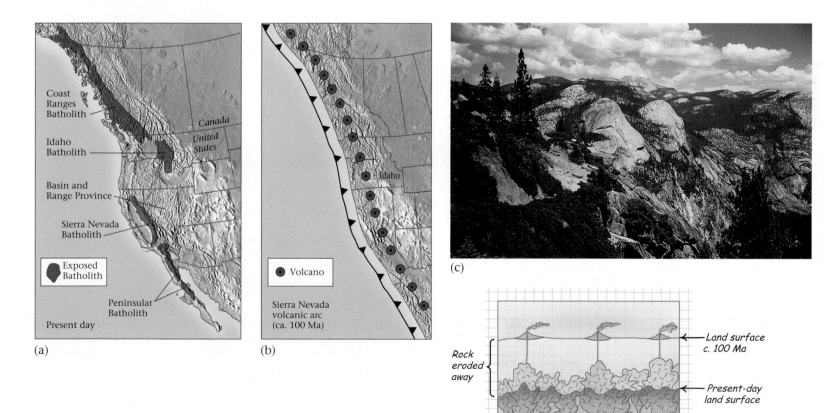

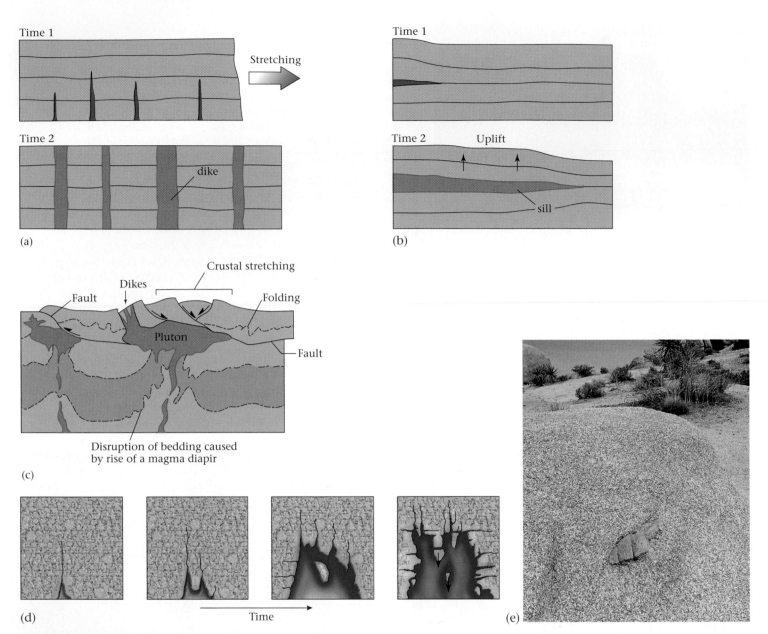

FIGURE 6.15 (a) Cross sections showing how the crust stretches sideways to accommodate dike intrusion. (b) Cross sections showing how intrusion of a sill may raise the surface of the Earth. (c) Ways in which crust accommodates emplacement of a pluton. (d) A magma stoping into country rock, gradually breaking off and digesting blocks as it moves. (e) Xenolith in a granite outcrop in the Mojave Desert.

why, think about the process of cooling coffee. If you pour hot coffee into a thermos bottle and seal it, the coffee stays hot for hours; because of insulation, the coffee in the thermos loses heat to the air outside only very slowly. Like the thermos bottle, rock acts as an insulator in that it transports heat away from a magma very slowly, so magma underground (in an intrusive environment) cools slowly. In contrast, if you spill coffee on a table, it cools quickly because it loses heat to the cold air. Similarly, lava that erupts at the ground surface, cools quickly because it is initially surrounded by air or water, which conduct heat away quickly.

Three factors control the cooling time of magma that intrudes below the surface.

- *The depth of intrusion:* Intrusions deep in the crust cool more slowly than shallow intrusions, because warm country rock surrounds deep intrusions whereas cold country rock surrounds shallow intrusions, and

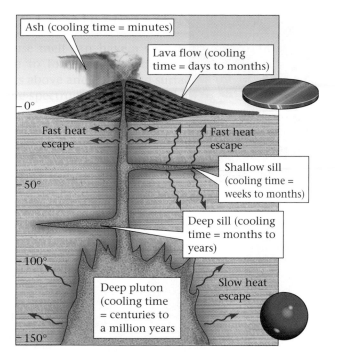

FIGURE 6.16 The cooling time of an intrusion decreases with greater depth (because country rock is hotter at greater depth). The cooling time also depends on the shape of the intrusion (a thin sheet cools faster than a sphere of the same volume) and on the size of the intrusion (a small intrusion cools faster than a large one). Thus, ash (formed when droplets of lava come in contact with air) cools fastest, followed by a thin sheet of lava, a shallow sill, a deep sill, and a deep pluton.

warmer country rock keeps the intrusion warmer for a longer time.

- *The shape and size of a magma body:* Heat escapes from an intrusion at the intrusion's surface, so the greater the surface area for a given volume of intrusion, the faster it cools. Thus, a pluton cools more slowly than a tabular intrusion with the same volume (because a tabular intrusion has a greater surface area across which heat can be lost), and, a large pluton cools more slowly than a small pluton of the same shape. Similarly, droplets of lava cool faster than a lava flow, and a thin flow of lava cools more quickly than a thick flow (▶Fig. 6.16).

- *The presence of circulating groundwater:* Water passing through magma absorbs and carries away heat, much like the coolant that flows around an automobile engine.

6.7 IGNEOUS ROCK TEXTURES

In Interlude A, we introduced the concept of a rock "texture." When talking about an igneous rock, a description of texture tells us whether the rock consists of glass, crystals, or

fragments. A description of texture may also indicate the size of the crystals or fragments. Geologists distinguish among three types of texture:

- *Glassy texture:* A rock made of a solid mass of glass, or of tiny crystals surrounded by glass, has a glassy texture. Rocks with this texture are **glassy igneous rocks** (▶Fig. 6.17a). They reflect light like glass does, and they tend to break along conchoidal fractures.

- *Interlocking texture:* Rocks that consist of mineral crystals that intergrow when the melt solidifies, and thus fit together like pieces of a jigsaw puzzle, have an interlocking texture (▶Fig. 6.17b). Rocks with an interlocking texture are called **crystalline igneous rocks.** The interlocking of crystals in these rocks occurs because once some grains have developed, they interfere with the growth of later-formed grains. The last grains to form end up filling irregular spaces between preexisting grains.

  Geologists distinguish subcategories of crystalline igneous rocks according to the size of the crystals. Coarse-grained ("phaneritic") rocks have crystals large enough to be identified with the naked eye. Typically, crystals in phaneritic rocks range in size from a couple millimeters across to several centimeters across. Fine-grained ("aphanitic") rocks have crystals too small to be identified with the naked eye. "Porphyritic rocks" have larger crystals surrounded by mass of fine crystals. In a porphyritic rock, the larger crystals are called "phenocrysts," while the mass of finer crystals is called "groundmass."

- *Fragmental texture:* Rocks with a fragmental texture consist of igneous fragments that are packed together, welded together, or cemented together after having solidified. Rocks with this texture are called "fragmental igneous rocks." The fragments can consist of glass, individual crystals, bits of crystalline rock, or a mixture of all of these.

Geologists study igneous textures carefully, because the texture provides a clue to the way in which the rock formed. For example, cooling time plays an important role in determining texture. Specifically, the presence of glass indicates that cooling happened so quickly that the atoms within a lava didn't have time to arrange into crystal lattices. Crystalline rocks form when a melt cools more slowly. In crystalline rocks, *grain size* depends on cooling time. A melt that cools rapidly, but not rapidly enough to make glass, forms fine-grained (aphanitic) rock, because many crystal seeds form but none has time to grow large. A melt that cools very slowly forms a coarse-grained (phaneritic) rock, because relatively few seeds form and each crystal has time to grow large.

Because of the relationship between cooling time and texture, lava flows, dikes, and sills tend to be composed of fine-grained igneous rock. In contrast, plutons tend to be composed of coarse-grained rock. Plutons that intrude into

The most famous active rift, the East African Rift, presently forming a 4,000-km-long gash in the crust of Africa, has produced numerous volcanoes, including Mt. Kilimanjaro. Recent rifting in North America has yielded the Basin and Range Province of Utah, Nevada, Arizona, and southeastern California. Though no currently erupting volcanoes exist in this region, the abundance of recent volcanic deposits suggests that igenous activity could occur again. Geoscientists are now monitoring the Mono Lakes volcanic area of California along the western edge of the Basin and Range Rift, because of the possibility that volcanoes in this area may erupt in the very near future.

In some cases, the bulbous head of a mantle plume underlies a rift. More partial melting can occur in a plume head

FIGURE 6.23 (a) The formation of pillow basalt. (b) This pillow basalt forms part of an ophiolite, a slice of sea floor that was pushed up onto the surface of a continent during mountain building. (c) A cross section through a single pillow shows the glassy rind, with a more crystalline center.

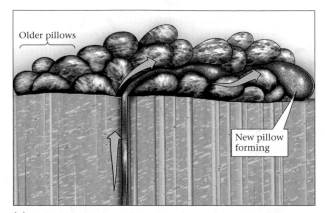

(a)

(b)

FIGURE 6.22 (a) Flood basalts underlie the Columbia River Plateau in Washington and Oregon, the dark area on this map. (b) Iguazu Falls, on the Brazil-Argentina border. The falls flow over the huge flood basalt sheet (the black rock) of the Paraná Plateau. Flood basalt underlies all of the region in view.

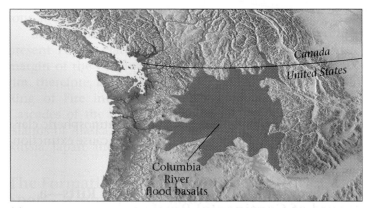

(a)

(b)

(c)

than in normal asthenosphere, because temperatures are higher in a plume head. Thus, an unusually large quantity of unusually hot magma forms where a rift overlies a plume head, so when volcanic eruptions begin in the rift, huge quantities of basaltic lava spew out of the ground, forming an LIP. The particularly hot basaltic lava that erupts at such localities has such low viscosity that it can flow tens to hundreds of kilometers across the landscape. Geoscientists refer to such flows as **flood basalts.** Flood basalts make up the bedrock of the Columbia River Plateau in Oregon and Washington (▶Fig. 6.22a), the Paraná Plateau in southeastern Brazil (▶Fig. 6.22b), the Karoo region of southern Africa, and the Deccan region of southwestern India.

The Formation of Igneous Rocks at Mid-Ocean Ridges—Hidden Plate Formation

Most igneous rocks at the Earth's surface form at mid-ocean ridges, that is, along divergent plate boundaries. Think about it—the entire oceanic crust, a 7- to 10-km-thick layer of basalt and gabbro that covers 70% of the Earth's surface, forms at mid-ocean ridges. And this entire volume gets subducted only to be replaced by new crust, over a period of about 200 million years.

Igneous magmas form at mid-ocean ridges for much the same reason they do at hot spots and rifts. As sea-floor spreading occurs and oceanic lithosphere plates drift away from the ridge, hot asthenosphere rises to fill the resulting space. As this asthenosphere rises, it undergoes decompression, which leads to partial melting and the generation of basaltic magma. As noted in Chapter 4, this magma rises into the crust and pools in a shallow magma chamber. Some cools slowly along the margins of the magma chamber to form massive gabbro, while some intrudes upward to fill vertical cracks that appear as newly formed crust splits apart (see Fig. 4.8). Magma that cools in the cracks forms basalt dikes, and magma that makes it to the sea floor and extrudes as lava forms basalt flows. The basalt flows of the sea floor don't look like those that erupt on land, because the seawater cools the lava so rapidly that it can't flow very far before solidifying into a pillow-shaped blob with a glassy rind. Eventually, the pressure of the lava inside a pillow breaks the glassy rind, and another pillow extrudes. Thus, sea-floor basalt is made up of a pile of pillows, known by geologists as **pillow basalt** (▶Fig. 6.23a–c).

In this chapter, we've focused on the diversity of igneous rocks, and why and where they form. We see that extrusive rocks develop at volcanoes. There's a lot more to say about volcanoes—eruptions have the potential to cause great harm. We will look at volcanic eruptions in detail in Chapter 9.

CHAPTER SUMMARY

- Magma is liquid rock (melt) under the Earth's surface. Lava is melt that has erupted from a volcano at the Earth's surface.

- Magma forms when hot rock in the Earth partially melts. This process only occurs under certain circumstances—where the pressure decreases (decompression), where volatiles (such as water or carbon dioxide) are added to hot rock, and where heat is transferred by magma rising from the mantle into the crust.

- Magma occurs in a range of compositions: felsic (silicic), intermediate, mafic, and ultramafic. The composition of magma is determined in part by the original composition of the rock from which the magma formed and in part by the way the magma evolves, by such processes as assimilation and fractional crystallization.

- During partial melting, only part of the source rock melts to create magma. Magma tends to be more silicic than the rock from which it was extracted.

- Magma rises from the depth because of its buoyancy and because the pressure caused by the weight of overlying rock squeezes magma upward.

- Magma viscosity (its resistance to flow) depends on its composition. Felsic magma is more viscous than mafic magma.

- Geologists distinguish between two types of igneous rocks. Extrusive igneous rocks form from lava that erupts out of a volcano and freezes in contact with air or the ocean. Intrusive igneous rocks develop from magma that freezes inside the Earth.

- Lava may solidify to form flows or domes, or it may explode into the air to form ash.

- Intrusive igneous rocks form when magma intrudes into preexisting rock (country rock) below Earth's surface. Blob-shaped intrusions are called plutons. Sheet-like intrusions that cut across layering in country rock are dikes, and sheet-like intrusions that form parallel to layering in country rock are sills. Huge intrusions, made up of many plutons, are known as batholiths.

- The rate at which intrusive magma cools depends on the depth at which it intrudes, the size and shape of the magma body, and whether circulating groundwater is present. The cooling time is reflected in the texture of an igneous rock.

- Crystalline (nonglassy) igneous rocks are classified according to texture and composition. Glassy igneous rocks are classified according to texture (a solid mass is obsidian, while ash that has cemented or welded together is a tuff).

- The origin of igneous rocks can readily be understood in the context of plate tectonics. Magma forms at continental or island volcanic arcs along convergent margins, mostly because of the addition of volatiles to the asthenosphere above the subducting slab. Igneous rocks form at hot spots owing to the decompression melting of a rising mantle plume. Igneous rocks form at rifts as a result of decompression melting of the asthenosphere below the thinning lithosphere. Igneous rocks form along mid-ocean ridges because of decompression melting of the rising asthenosphere.

KEY TERMS

ash (p. 140)
ash fall (p. 140)
ash flow (p. 140)
assimilation (p. 144)
batholith (p. 148)
Bowen's reaction series (p. 145)
crystalline igneous rocks (p. 153)
dike (p. 148)
extrusive igneous rock (p. 140)
felsic magma (p. 143)
flood basalts (p. 163)
fractional crystallization (p. 146)
geotherm (p. 140)
geothermal gradient (p. 143)
glassy igneous rocks (p. 153)
hot-spot track (p. 160)
hot-spot volcanoes (p. 160)
hyaloclasite (p. 156)
igneous rock (p. 138)
intermediate magma (p. 143)
intrusive igneous rock (p. 140)

laccolith (p. 148)
large igneous province (p. 160)
lava (p. 138)
mafic magma (p. 143)
magma (p. 140)
obsidian (p. 155)
partial melting (p. 144)
pegmatite (p. 154)
pillow basalt (p. 163)
plutons (p. 148)
pumice (p. 155)
pyroclastic debris (p. 140)
pyroclastic rocks (p. 155)
scoria (p. 155)
sill (p. 148)
stoping (p. 150)
superplumes (p. 161)
tachylite (p. 155)
tuff (p. 155)
ultramafic magma (p. 143)
vesicles (p. 155)
viscosity (p. 143)
volcanic breccia (p. 156)
volcano (p. 138)
xenolith (p. 150)

REVIEW QUESTIONS

1. How is the process of freezing magma similar to that of freezing water? How is it different?

2. What is the source of heat in the Earth? How did the first igneous rocks on the planet form?

3. Describe the three processes that are responsible for the formation of magmas.

4. Why are there so many different types of magmas?

5. Why do magmas rise from depth to the surface of the Earth?

6. What factors control the viscosity of a melt?

7. What factors control the cooling time of a magma within the crust?

8. How does grain size reflect the cooling time of a magma?

9. What does the mixture of grain sizes in a porphyritic igneous rock indicate about its cooling history?

10. Describe the way magmas are produced in subduction zones.

11. What processes in the mantle may be responsible for causing hot-spot volcanoes to form?

12. Describe how magmas are produced at continental rifts.

13. What is a large igneous province (LIP) and how might their formation have affected the Earth System?

14. Why does melting take place beneath the axis of a mid-ocean ridge?

SUGGESTED READING

Best, M. G., and E. H. Christiansen. 2001. *Igneous Petrology.* 2nd ed. Oxford, England: Blackwell Science.

Faure, G. 2000. *Origin of Igneous Rocks: The Isotopic Evidence.* New York: Springer-Verlag.

LeMaitre, R. E., ed. 2002. *Igneous Rocks: A Classification and Glossary of Terms,* 2nd ed. Cambridge: Cambridge University Press.

Leyrit, H., C. Montenat, and P. Bordet, eds. 2000. *Volcaniclastic Rocks, from Magmas to Sediments.* London, England: Taylor and Francis.

Mackenzie, W. S. 1982. *Atlas of Igneous Rocks and Their Textures.* New York: John Wiley & Sons.

Middlemost, E. A. K. 1997. *Magmas, Rocks and Planetary Development: A Survey of Magma/Igneous Rock Systems.* Boston: Addison/Wesley.

Philpotts, A. R. 2003. *Petrography of Igneous and Metamorphic Rocks.* Long Grove, IL: Waveland Press.

Thorpe, R., and G. Brown. 1985. *The Field Description of Igneous Rocks.* New York: John Wiley & Sons.

Winter, J. D. 2001. *Introduction to Igneous and Metamorphic Petrology.* Upper Saddle River, N.J.: Prentice-Hall.

Young, D. A. 2003. *Mind over Magma.* Princeton, N.J.: Princeton University Press.

A Surface Veneer: Sediments, Soils, and Sedimentary Rocks

7.1 INTRODUCTION

In the 1950s, the government of Egypt decided to build the Aswan High Dam to trap water of the Nile River in a huge reservoir. To identify a good site for the dam's foundation, geologists drilled holes into the ground to find the depth to bedrock. They discovered that the present-day Nile River flows on the surface of a 1.5-km-thick layer of loose debris (gravel, sand, and mud) that fills what was once a canyon as large as the Grand Canyon (▶Fig. 7.1). The carving and subsequent filling of this canyon baffled geologists, because today the river flows along a plain almost at sea level. How could the river have carved a canyon 1.5 km deep, and why did the canyon later fill with sediment?

The origin of the Nile "canyon" remained a mystery until the summer of 1970, when geologists drilled holes into the floor of the Mediterranean Sea to find out what lay beneath. They expected the sea floor to be covered with shells of plankton (tiny floating organisms) that had settled out of the water, or with clay that rivers had carried to the sea. To their surprise, however, they found that, in addition to clay and plankton shells, a 2 km-thick layer of halite and gypsum lies beneath the floor of the Mediterranean Sea. These minerals form when seawater dries up,

The colorful layers of sedimentary rock, exposed by erosion in the walls of Bryce Canyon, in Utah, were originally deposited in streams, lakes, and floodplains.

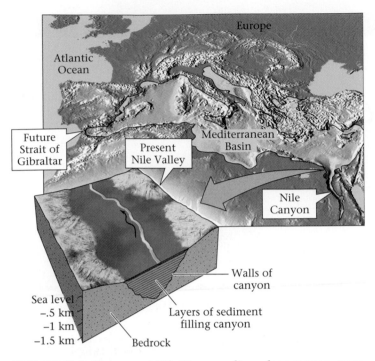

FIGURE 7.1 The present Nile River overlies a deep canyon, now filled with layers of sand and mud. The bottom of the canyon lies 1.5 km below sea level.

allowing the salt in seawater to precipitate (see Chapter 5). The researchers realized that to yield a layer that is 2 km thick, the entire Mediterranean would have had to dry up completely several times, with the sea refilling between each drying event. This discovery solved the mystery of the pre-Nile canyon. When the Mediterranean Sea dried up, the Nile

River was able to cut a canyon; and when the Sea refilled with water, this canyon flooded and filled with sand and gravel.

Why did the Mediterranean Sea dry up? Only 10% of the water in the Mediterranean enters the sea from rivers, and since the sea lies in a hot, dry region, ten times that amount of water evaporates from its surface each year. Thus, most of the water in the Mediterranean enters through the Strait of Gibraltar from the Atlantic Ocean—if this flow stops, the Mediterranean Sea evaporates. About 6 million years ago, the northward-drifting African Plate collided with the European Plate, forming a natural dam separating the Mediterranean from the Atlantic. When global sea level dropped, water stopped flowing in from the Atlantic and the Mediterranean evaporated. All the salt that had been dissolved in its water accumulated as a solid deposit of halite and gypsum on the floor of the resulting basin. When sea level rose, a gigantic flood rushed from the Atlantic into the Mediterranean, filling the basin again. This process repeated many times. About 5.5 million years ago, the Mediterranean rose to its present level and gravel, sand, and mud carried by the Nile River filled the Nile canyon to its present level.

Geologists refer to the kinds of deposits just described—sand, mud, gravel, halite and gypsum accumulates, shell fragments—as sediment. **Sediment,** in general, consists of loose fragments of rocks or minerals broken off of bedrock, mineral crystals that precipitate directly out of water, and shells (formed when organisms extract ions from water). Much of what we know about the history of the Earth (including the amazing story of the Mediterranean Sea) comes from studying sediments—not only

FIGURE 7.2 Near the bottom of the Grand Canyon, we can see the boundary between the sedimentary veneer, or cover (here, a succession of horizontal layers), and the older basement (here, the steep cliff of dark metamorphic rock that goes down to the river). The Colorado River flows along the floor of the canyon. A geologist's sketch emphasizes the contact, or boundary, between cover and basement.

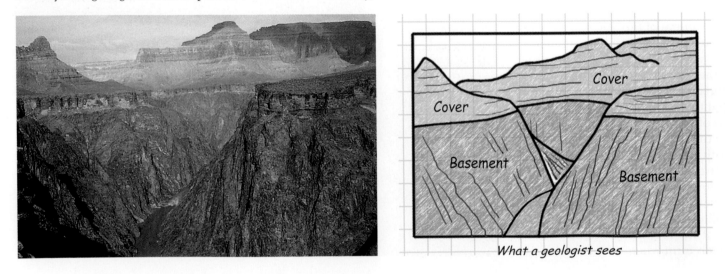

those that remain "unconsolidated" (loose and not connected), but also those that have been bound together into sedimentary rock. Formally defined, **sedimentary rock** is rock that forms at or near the surface of the Earth by the precipitation of minerals from water solutions, by the growth of skeletal material in organisms, or by the cementing together of shell fragments or loose grains derived from preexisting rock. Layers of sediment and sedimentary rock are like the pages of a book, recording tales of ancient events and ancient environments on the ever-changing face of the Earth.

Sediments and sedimentary rocks only occur in the upper part of the crust—in effect, they form a surface veneer, or **cover,** on older igneous and metamorphic rocks, which make up the **basement** of the crust (▶Fig. 7.2). This veneer ranges from nonexistent, in places where igneous and metamorphic rocks crop out at the Earth's surface, to several kilometers thick in regions called **sedimentary basins.** Though sediments and sedimentary rocks cover more than 80% of the Earth's surface, they actually constitute less than 1% of the Earth's mass. Nevertheless, they represent a uniquely important rock type, both because they contain a historic record and because they contain the bulk of the Earth's energy resources, (as we'll see in Chapter 14). Further, some sediments transform into soil, essential for life. Let's now look at how sediments, soils, and sedimentary rocks form, and what these materials can tell us about the Earth System.

7.2 WEATHERING: THE FORMATION OF SEDIMENT

The Mountains Crumble

If you ever have the chance to hike or drive through granitic mountains, like the Sierra Nevada of California or the Coast Mountains of Canada, you may notice that in some outcrops the granite surface looks hard and smooth and contains shining crystals of feldspar, biotite, and quartz, while in other outcrops the granite surface looks grainy and rough—feldspar crystals appear dull, biotite flakes have spots of rust, and the rock may peel apart like an onion. Why are these two types of outcrop different? The first type exposes fresh rock, rock whose mineral grains have kept their original composition and shape, while the second type exposes weathered rock, rock that has reacted with air and/or water at or near the Earth's surface and has thus been weakened (▶Fig. 7.3).

Weathering refers to the processes that break up and corrode solid rock, eventually transforming it into sediment. All mountains and other features on the Earth's surface sooner or later crumble away because of weather-

Weathered granite

Fresh granite

FIGURE 7.3 This outcrop shows the contrast between fresh and weathered granite. The rock below the notebook is fresh—the outcrop face is a fairly smooth fracture. The rock above the notebook is weathered—the outcrop face is crumbly, breaking into grains that have fallen and collected on the ledge.

ing. Geologists distinguish between two types of weathering: physical weathering and chemical weathering. Just as a plumber can unclog a drain by using physical force (with a plumber's snake) or by causing a chemical reaction (with a dose of liquid drain opener), nature can attack rocks in two ways.

Physical Weathering

Physical weathering, sometimes referred to instead as "mechanical weathering," breaks intact rock into unconnected grains or chunks, collectively called debris or **detritus.** Each size range of grains has a name (measurements are grain diameters):

- *boulders* more than 256 millimeters (mm)
- *cobbles*   between 64 mm and 256 mm
- *pebbles*   between 2 mm and 64 mm
- *sand*   between 1/16 mm and 2 mm
- *silt*   between 1/256 mm and 1/16 mm
- *mud*   less than 1/256 mm

For convenience, geologists refer to boulders, cobbles, and pebbles as "coarse-grained" sediment, sand as "medium-grained" sediment, and silt and mud as "fine-grained" sediment. Many different phenomena contribute to physical weathering.

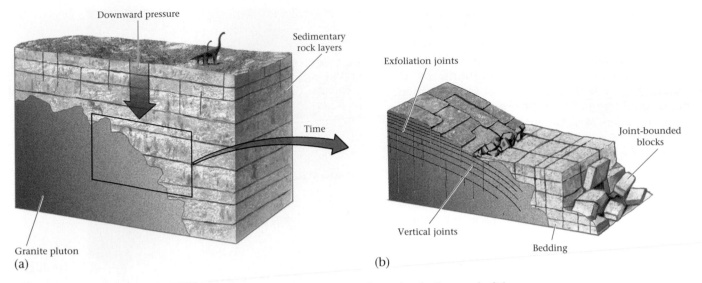

Downward pressure

Sedimentary rock layers

Time

Granite pluton

(a)

Exfoliation joints

Joint-bounded blocks

Vertical joints

Bedding

(b)

FIGURE 7.4 (a) The weight of overburden creates pressure on rocks at depth. Removal of the overburden by erosion allows once-deep rocks to be exposed at the Earth's surface. (b) The exposure of once-deep rocks causes them to crack. Different rock types crack in different ways. Here, the granite pluton develops exfoliation joints as well as vertical joints, while the sedimentary rock layers develop mostly vertical joints. Joint-bounded blocks break off the outcrop.

Jointing. Rocks buried deep in the Earth's crust endure enormous pressure because the "overburden" (overlying rock) weighs a lot and presses down on the buried rock. Rocks at depth are also warmer than rocks nearer the surface, because of Earth's geothermal gradient (see Chapter 6). Over long periods, moving water, air, and ice at the Earth's surface grind away and remove overburden, so rock formerly at depth rises closer to the Earth's surface. As a result, the pressure squeezing this rock decreases, and the rock becomes cooler. A change in pressure causes rock to change shape slightly, for the same reason that a rubber ball changes shape when you squeeze it and then let go. Similarly, a change in temperature causes rock to change shape for the same reason that a baked apple changes shape when it is removed from an oven and cools. But unlike a rubber ball or a soft apple, hard rock may break into pieces when it changes shape (▶Fig. 7.4a, b). Natural cracks that form in rocks due to removal of overburden or due to cooling (and for other reasons as well; see Chapter 11) are known as **joints.**

Almost all rock outcrops contain joints; some joints are fairly planar, some curving, and some irregular. The spacing between adjacent joints varies from less than a centimeter to tens of meters. Joints can break rock into large or small rectangular blocks, onion-like sheets, irregular chunks, or pillar-like columns. Typically, large granite plutons split into onion-like sheets along joints that lie parallel to the mountain face. This process is called **exfoliation** (▶Fig. 7.5a). Sedimentary rock layers tend to break into rectangular blocks (▶Fig. 7.5b). Regardless of their orientation, the formation of joints turns formerly intact bedrock into loose blocks.

Eventually, these blocks fall from the outcrop at which they formed. After a while, they may collect in an apron of **talus,** rock rubble at the base of a slope (▶Fig. 7.5c).

Frost wedging. Freezing water bursts pipes and shatters bottles, because water expands when it freezes and pushes the walls of the container apart. The same phenomenon happens in rock. When the water trapped in a joint freezes, it forces the joint open and may cause the joint to grow. Such **frost wedging** helps break blocks free from intact bedrock (▶Fig. 7.6a). Of course, frost wedging is most common where water periodically freezes and thaws, as occurs in temperate climates or at high altitudes in mountains.

Root wedging. Have you ever noticed how the roots of an old tree can break up a sidewalk? Even though the wood of roots doesn't seem very strong, as roots expand they apply pressure to their surroundings. Tree roots that grow into joints can push joints open in a process known as **root wedging** (▶Fig. 7.6b). Even the roots of small plants, fungi, and lichen get into the act, by splitting open small cracks and pores.

Salt wedging. In arid climates, dissolved salt in groundwater precipitates and grows as crystals in open pore spaces in rocks. This process, called **salt wedging,** pushes apart the surrounding grains and so weakens the rock that when exposed to wind and rain, the rock disintegrates into separate grains. The same phenomenon happens in coastal areas, where salt spray percolates into surface rock and then dries (▶Fig. 7.6c).

(a)

(b)

—— Joint

(c)

FIGURE 7.5 (a) Exfoliation joints in the Sierra Nevada. (b) Vertical joints in sedimentary rock (Brazil). (c) Talus has accumulated at the base of these cliffs near Mt. Snowdon in Wales.

Thermal expansion. When the heat of an intense forest fire bakes a rock, the outer layer of the rock expands. On cooling, the layer contracts. This change creates forces in the rock sufficient to make the outer part of the rock spall, or break off in sheet-like pieces.

Animal attack. Animal life also contributes to physical weathering: burrowing creatures, from earthworms to gophers, push open cracks and move rock fragments. And in the past century, humans have become perhaps the most energetic agent of physical weathering on the planet. When we excavate quarries, foundations, mines, or roadbeds by digging and blasting, we shatter and displace rock that might otherwise have remained intact for millions of years more.

Chemical Weathering

Up to now we've taken the "plumber's snake approach" to breaking up rock; now let's look at the "liquid drain opener approach." **Chemical weathering** refers to the chemical reactions that alter or destroy minerals when rock comes in contact with water solutions or air. Because many of these reactions proceed more quickly in warm, wet conditions,

chemical weathering takes place much faster in the tropics than it does in deserts, or near the poles. Chemical weathering in warm, wet climates can produce a layer of rotten rock, called **saprolite,** over 100 m thick. Common reactions involved in chemical weathering include the following.

Dissolution. Chemical weathering during which minerals dissolve into water is called "dissolution." Dissolution primarily affects salts and carbonate minerals, but even quartz dissolves slightly (▶Fig. 7.7a, b). Some minerals, like halite, can dissolve rapidly in pure rainwater. But some, like calcite, dissolve rapidly only when the water is acidic, meaning that it contains an excess of hydrogen ions (H^+). Acidic water reacts with calcite to form a solution and bubbles of CO_2 gas (see Chapter 5).

How does the water in rock near the surface of the Earth become acidic? As rainwater falls, it dissolves carbon dioxide gas in the atmosphere and as it sinks down through soil containing organic debris, it reacts with the debris—both processes yield carbonic acid. Because of the solubility of calcite, limestone and marble (two types of rock composed of calcite) dissolve, widening joints and leading to the formation of caverns (▶Fig. 7.7c; see Chapter 19).

an ancient arrowhead—these materials all have something in common. They all consist of rock formed primarily by the precipitation of minerals out of water solutions. We call such rocks chemical sedimentary rocks. They typically have a crystalline texture, partly formed during the original precipitation and partly the result of later recrystallization.

Evaporites: The Products of Saltwater Evaporation

In 1965, two daredevil drivers in jet-powered cars battled to be the first to break the land-speed record of 600 mph. On November 7, Art Arfons, in the "Green Monster," peaked at 576.127 mph, but eight days later Craig Breedlove, driving the "Spirit of America," reached 600.601 mph. Traveling at such speeds, a driver must maintain an absolutely straight course; any turn will catapult the vehicle out of control, because its tires simply can't grip the ground. Thus, high-speed trials take place on extremely long and flat racecourses. Not many places can provide such conditions —the Bonneville Salt Flats, near the Great Salt Lake of central Utah, do.

How did this vast salt plain come into existence? Streams bringing water from Utah's Wasatch Mountains into the Salt Lake basin, like all streams, carry trace amounts of dissolved ions, provided to the water by chemical weathering. Most lakes have an outlet, so the water in them constantly flushes out and the ion concentration stays low. But the Great Salt Lake has no such outlet, so water escapes from the lake only by evaporating. Evaporation removes just the water; dissolved ions stay behind, so over time, the lake water has become a concentrated solution of dissolved ions—in other words, very salty (▶Fig. 7.23a). In the past, when the region had a wetter climate, the Great Salt Lake was larger and covered the region of the Bonneville Salt Flats; this larger ancient lake was "Lake Bonneville." Along its shores, water dried up and salt precipitated. When the lake shrank to its present dimension, the vast extent of the Bonneville Salt Flats was left high and dry, and covered with salt (▶Fig. 7.23b). Such salt precipitation occurs wherever there is saturated saltwater—along desert lakes with no outlet (e.g., the Dead Sea) and along margins of restricted seas (e.g., the Persian Gulf). For thick deposits of salt to form, large volumes of water must evaporate (▶Fig. 7.23c, d). This may happen when plate tectonic movements temporarily cut off arms of the sea (as we saw in the case of the Mediterranean Sea) or during continental rifting, when seawater first begins to spill into the rift valley.

Because salt deposits form as a consequence of evaporation, geologists refer to them as **evaporites.** The specific type of salt constituting an evaporite depends on the amount of evaporation. When 80% of the water evaporates, gypsum forms; and when 90% of the water evaporates, halite precipitates. If seawater were to evaporate entirely, the resulting evaporite would consist of 80% halite, 13% gypsum, and the remainder of other salts and carbonates.

Travertine (Chemical Limestone)

Travertine is a rock composed of crystalline calcium carbonate (calcite and/or aragonite) formed by chemical precipitation out of groundwater that has seeped out at the ground surface (in hot- or cold-water springs) or on the walls of caves. What causes this precipitation? It happens, in part, when the groundwater "degases," meaning that some of the carbon dioxide that had been dissolved in the groundwater bubbles out of solution. Dissolved carbon dioxide makes water more acidic and better able to dissolve carbonate, so removal of carbon dioxide decreases the ability of the water to hold dissolved carbonate. Precipitation also occurs when water evaporates and leaves behind dissolved ions, thereby increasing the concentration of carbonate. Various kinds of microbes live in the environments in which travertine accumulates, so biologic activity may also contribute to the precipitation process.

Travertine produced at springs forms terraces and mounds that are meters or even hundreds of meters thick. Spectacular terraces of travertine grew at Mammoth Hot Springs in Yellowstone National Park (▶Fig. 7.24a), and amazing column-like mounds of travertine grew up from the floor of Mono Lake, California, where hot springs seeped into the cold water of the lake (▶Fig. 7.24b); the columns are now exposed because the water level of the lake has been lowered. Travertine also grows on the walls of caves where groundwater seeps out. In cave settings, travertine builds up beautiful and complex growth forms called speleothems (▶Fig. 7.24c; see Chapter 19).

Travertine has been quarried for millennia to make building stones and decorative stones. The rock's beauty comes in part because in thin slices it is translucent, and in part because it typically displays growth bands. Bands develop in response to changes in the composition of groundwater, or in the environment into which the water drains. Some travertines (a type called "tufa") contain abundant large pores, or "vugs."

Dolostone: Replacing Calcite with Dolomite

Dolostone differs from limestone in that it contains the mineral dolomite ($CaMg[CO_3]_2$). Most dolostone forms by a chemical reaction between solid calcite and magnesium-bearing groundwater. Much of the dolostone you may find in an outcrop actually originated as limestone but later recrystallized so that dolomite replaced the calcite. This recrystallization may take place beneath lagoons along a shore soon after the limestone formed, or a long time later, after the limestone has been buried deeply.

Replacement and Precipitated Chert

A tribe of Native Americans, the Onondaga, once inhabited the eastern part of New York State. In this region, outcrops of limestone contain layers of a black chert (▶Fig. 7.25). Because of the way it breaks, artisans could fashion sharp-edged tools (arrowheads and scrapers) from this chert, so the Onondaga collected it for their own toolmaking indus-try and for use in trade with other tribes. Unlike the deep-sea (biochemical) chert described earlier, the chert collected by the Onondaga formed when cryptocrystalline quartz gradually replaced calcite crystals within a body of lime-stone long after the limestone was deposited; geologists thus call it "replacement chert."

Chert comes in many colors (black, white, red, brown, green, gray), depending on the impurities it contains. Black chert, or flint, made the tools of the Onondaga. Red chert, or jasper, which like all chert takes on a nice polish, makes beautiful jewelry. Petrified wood is chert that's made when silica-rich sediment, such as ash from a volcanic eruption buries a forest. The silica dissolves in groundwater that then passes into the wood. Dissolved silica precipitates as cryp-tocrystalline quartz within wood, gradually replacing the wood's cellulose. The chert retains the shape of the wood and even its growth rings. Some chert, known as agate, pre-cipitates in concentric rings inside hollows in a rock and ends up with a striped appearance, caused by variations in the content of impurities while precipitation took place.

FIGURE 7.23 (a) In lakes with no outlet, the tiny amount of salt brought in by "fresh" water streams stays behind as the water evaporates. Along the margins of the lake, salts precipitate. If the whole lake evaporates, a flat surface of salt forms. (b) Recently deposited evaporites along the margin of a salt lake. (c) Salt precipitation can also occur along the margins of a restricted marine basin, if saltwater evaporates faster than it can be resupplied. The entire restricted sea may dry up if it is cut off from the ocean. (d) Thick layers of salt accumulate in a rift and later are buried deeply. The salt then recrystallizes. Here, thick salt layers are being mined.

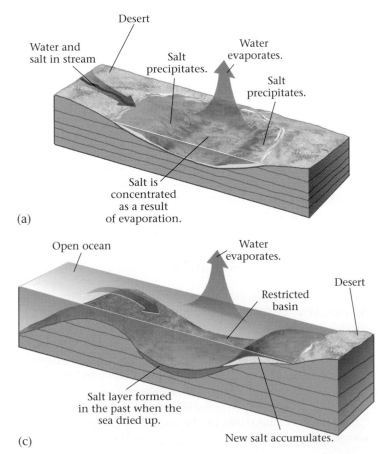

(a)

(c)

(b)

(d)

(a)　　　　　　　　　　　　(b)　　　　　　　　　　　　(c)

FIGURE 7.24 (a) A travertine buildup at Mammoth Hot Springs. Note the terraces. (b) Mounds of travertine forming at hot springs in Mono Lake, California. The material in these mounds is also called "tufa." (c) A travertine buildup on the wall of a cave in Puerto Rico.

FIGURE 7.25 Replacement chert occurring as nodules in a limestone. Chert forms the black band in these tilted layers.

7.8 SEDIMENTARY STRUCTURES

In the photo of a stark outcrop in Figure 7.10c, note the distinct lines across its face. In 3-D, we see that these lines are the traces of individual surfaces that separate the rock into sheets. In fact, sedimentary rocks in general contain distinctive layering. The layers themselves may have a characteristic internal arrangement of grains or distinctive markings on their surface. We use the term "sedimentary structure" for the layering of sedimentary rocks, surface features on layers formed during deposition, and the arrangement of grains within layers. Here, we examine some of the more important types.

Bedding and Stratification

Geologists have a jargon for discussing sedimentary layers. A single layer of sediment or sedimentary rock with a recognizable top and bottom is called a **bed;** the boundary between two beds is a "bedding plane;" several beds together constitute **strata;** and the overall arrangement of sediment into a sequence of beds is "bedding," or "stratification." From the word "strata," we derive other words, like "stratigrapher" (a geologist who specializes in studying strata) and "stratigraphy" (the study of the record of Earth history preserved in strata).

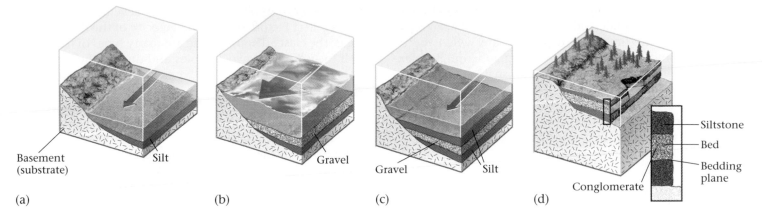

(a)   (b)   (c)   (d)

FIGURE 7.26 Bedding forms as a result of changes in the environment. (a) During a normal river flow, a layer of silt is deposited. (b) During a flood, turbulent water brings in a layer of gravel. (c) When the river returns to normal, another layer of silt is deposited. (d) Later, after lithification, uplift, and exposure, a geologist sees these layers as beds on an outcrop.

When you examine strata in a region with good exposure, the bedding generally stands out clearly. Beds appear as bands across a cliff face (Fig. 7.10c). Typically, a contrast in rock type distinguishes one bed from adjacent beds. For example, a sequence of strata may contain a bed of sandstone overlain by a bed of shale, overlain by a bed of limestone. Each bed has a definable thickness (from a couple of centimeters to tens of meters) and some contrast in composition, color, and/or grain size, which distinguishes it from its neighbors. But in many examples, adjacent beds all appear to have the same composition. In such cases, bedding may be defined by subtle changes in grain size, by surfaces that represent interruptions in deposition, or by cracks that have formed parallel to bed surfaces.

Why does bedding form? To find the answer, we need to think about how sediment is deposited. Changes in the source of sediment, climate, or water depth control the type of sediment deposited at a location at a given time. For example, on a normal day a slow-moving river may carry only silt, which collects on the river bed. During a flood, the river carries sand and pebbles, so a layer of sandy gravel forms over the silt layer. Then, when the flooding stops, more silt buries the gravel. If these sediments become lithified and exposed for you to see, they appear as alternating beds of siltstone and sandy conglomerate (▶Fig. 7.26a–d).

Bedding is not always well preserved. In some environments, burrowing organisms disrupt the layering. Worms, clams, and other creatures churn sediment and may leave behind burrows. This process is called "bioturbation."

During geologic time, long-term changes in a depositional environment can take place. Thus, a sequence of beds may differ markedly from sequences of beds above or below. If a sequence of strata is distinctive enough to be traced across a fairly large region, geologists call it a **stratigraphic formation,** or simply a "formation" (▶Fig. 7.27).

FIGURE 7.27 A particularly thick bed, a sequence of beds of the same composition, or a sequence of beds of alternating rock types can be called a stratigraphic formation, if the sequence is distinctive enough to be traced across the countryside. In this photo of the Grand Canyon, we can see five formations. Formations that consist primarily of one rock type may take the rock-type name (e.g., Kaibab Limestone), but formations containing more than one rock type may just be called "formation." The Supai Group is a group because it consists of several related formations, which are too thin to show here. Formations and groups are examples of stratigraphic units. Note that each formation consists of many beds, and that beds range greatly in thickness. The boundaries between units are called "contacts."

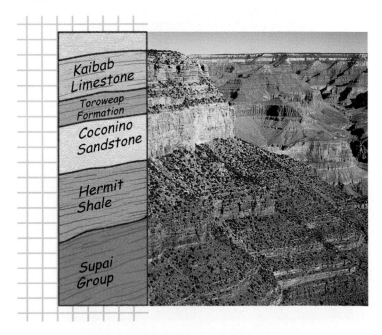

For example, a region may contain a succession of alternating sandstone and shale beds deposited by rivers, overlain by beds of marine limestone deposited later when the region was submerged by the sea. A stratigrapher might identify the sequence of sandstone and shale beds as one formation and the sequence of limestone beds as another. Formations are often named after the locality where they were first found and studied. For example, the Schoharie Formation was recognized and described from exposures near Schoharie, New York.

Ripples, Dunes, and Cross Bedding: A Consequence of Deposition in a Current

Many clastic sedimentary rocks accumulate in moving fluids (wind, rivers, or waves). The movement of the fluids creates fascinating sedimentary structures at the interface between the sediment and the fluid—these structures are called "bedforms." The bedforms that develop at a given location reflect factors such as the velocity of the flow and the size of the clasts. Though there are many types of bedforms, we'll focus on only two—ripples and dunes. The growth of both produces cross bedding, a special type of lamination within beds.

Ripples (or "ripple marks") are relatively small (generally no more than a few centimeters high), elongated ridges that form on a bed surface at right angles to the direction of current flow. If the current always flows in the same direction, the ripple marks are asymmetric, with a steeper slope on the downstream (lee) side (▶Fig. 7.28a). Along the shore, where water flows back and forth due to wave action, ripples tend to be symmetric. The crest (the high ridge) of a symmetric ripple is a sharp ridge, whereas the trough between adjacent ridges is a smooth, concave-up curve (▶Fig. 7.28b).

Dunes look like ripples, only they are much larger. For example, dunes on the bed of a stream may be tens of cen-

FIGURE 7.28 (a) A current that always flows in the same direction, as occurs in a stream, produces asymetric ripples. (b) A current that moves back and forth, as occurs on a wave-washed beach, produces symmetric ripples. (c) Modern ripples forming in the salt on a beach. (d) Ancient ripples preserved in a layer of quartzite that's over 1,500 million years old. This outcrop occurs in Wisconsin.

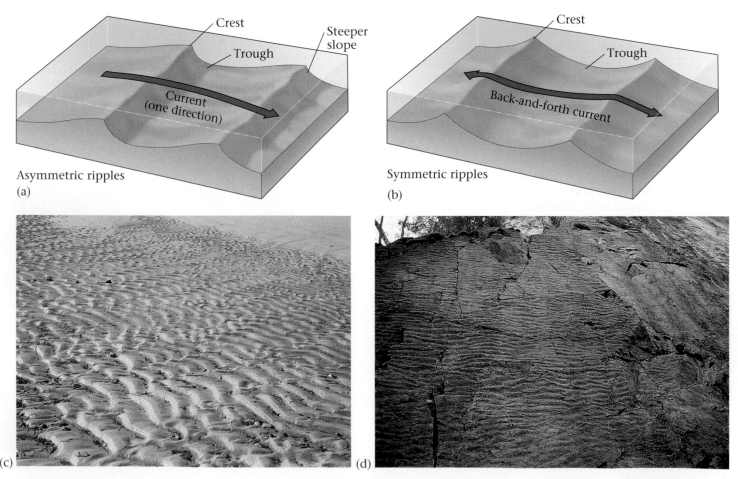

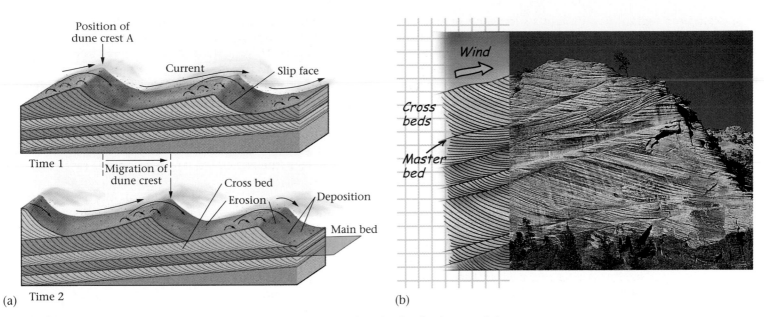

FIGURE 7.29 (a) Cross beds form as sand blows up the windward side of a dune and then accumulates on the slip face. At a later time, we see that dunes migrate, and eventually bury the layers below. (b) Successive layers, or master beds, of cross-bedded strata can be seen on this cliff face of sandstone in Zion National Park. We are looking at the remnants of ancient sand dunes. Cross beds indicate the wind direction during deposition.

timeters high, and wind-formed dunes formed in deserts may be tens to over 100 meters high. Small ripples often form on the surfaces of dunes.

If you slice into a ripple or dune and examine it in cross section, you will find distinct internal laminations that are inclined at an angle to the boundary of the main sedimentary layer. Such laminations are called **cross beds.** Cross bedding forms directly as a consequence of the evolution of ripples or dunes. To see how, imagine a current of air or water moving uniformly in one direction (▶Fig. 7.29a). The current erodes and picks up clasts from the upstream part of the bedform (because here the fluid moves quickly) and deposits them on the downstream or leeward part of the bedform, because here the fluid moves more slowly. The face of the downstream side of the bedform is called the "slip face," for the accumulation of sediment causes this face to become so steep that gravity causes sediment to slip downward. The upper part of the slip face becomes steeper than its base, so the slip face becomes curved, with a concave-up shape. The process repeats as more sediment builds up on the leeward side of the bedform and then slips down, so with time the leeward side of the bedform builds in the downstream direction. The curving surface of the slip face establishes the shape of the cross beds. During slippage events, heavier clasts (of denser minerals or of larger grains) remain stranded along the slip face—these mark the visible boundaries between successive

cross beds. Erosion by the current usually clips off the top of the cross bed, so only the bottom half or two-thirds will be preserved and buried.

With time, a new cross-bedded layer builds out over a preexisting one. The boundary between two successive layers is called the "main bedding," and the internal curving surfaces within the layer constitute the "cross bedding" (▶Fig. 7.29b). Note that the shape of the cross bed indicates both the direction in which the current was flowing during deposition and the direction in a stratigraphic sequence in which beds are younger.

Turbidity Currents and Graded Beds

Sediment deposited on a submarine slope might not stay in place forever. For example, an earthquake or storm might disturb this sediment and cause it to slip downslope. If the sediment is loose enough, it mixes with water to create a murky, turbulent cloud. This cloud is denser than clear water, and thus flows downslope like an underwater avalanche (▶Fig. 7.30a). We call this moving submarine suspension of sediment a "turbidity current." Turbidity currents can be powerful enough to snap undersea phone cables and displace shipwrecks.

Eventually, in deeper water where the slope becomes gentler, or the turbidity current spreads out, the turbidity current slows. When this happens, the sediment that it has

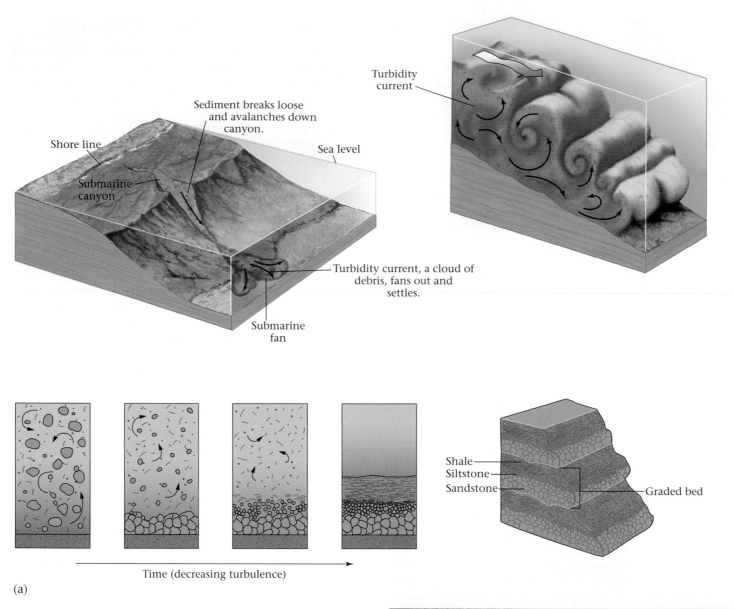

(a)

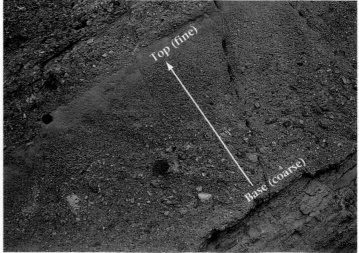

(b)

FIGURE 7.30 (a) An earthquake or storm triggers an underwater avalanche (turbidity current), which mixes sediment of different sizes together. When the current slows, the larger grains settle faster, gradually creating a graded bed. (b) In this example of a graded bed, pebbles lie at the bottom of the bed, and silt at the top. The bed was tilted after deposition.

carried starts to settle out. Larger grains sink faster through a fluid than do finer grains, so the coarsest sediment settles out first. Progressively finer grains accumulate on top, with the finest sediment (clay) settling out last. This process forms a **graded bed**—that is, a layer of sediment in which grain size varies from coarse at the bottom to fine at the top (▶Fig. 7.30b).

Typically, turbidity currents flow down submarine canyons—in fact, their flow contributes to scouring and deepening the canyon. The graded beds thus form an apron, called a "submarine fan," at the mouth of the canyon (see Fig. 7.30a). Successive turbidites deposit successive graded beds, creating a sequence of strata called a **turbidite.**

Bed-Surface Markings

A number of features appear on the surface of a bed as a consequence of events that happen during deposition or soon after, while the sediment layer remains soft. These "bed-surface markings" include the following.

- *Mud cracks:* If a mud layer dries up after deposition, it cracks into roughly hexagonal plates that typically curl up at their edges. We refer to the openings between the plates as **mud cracks.** Later, these fill with sediment and can be preserved (▶Fig. 7.31).

- *Scour marks:* As currents flow over a sediment surface, they may scour out small troughs called **scour marks** parallel to the current flow. These indentations can be buried and preserved.

- *Fossils:* **Fossils** are relicts of past life. Some fossils are shell imprints or footprints on a bedding surface (see Interlude D).

FIGURE 7.31 Mud cracks in dried red mud, from Utah, as viewed from the top. Note how the edges of the mud cracks curl up.

The Value of Studying Sedimentary Structures

Sedimentary structures are not just a curiosity, but rather are key clues that help geologists understand the environment in which clastic sedimentary beds were deposited. For example, the presence of ripple marks and cross bedding indicates that layers were deposited in a current. The presence of mud cracks indicates that the sediment layer was exposed to the air on occasion. Graded beds indicate deposition by turbidity currents. And fossil types can tell us whether sediment was deposited along a river or in the sea, for different species of organisms live in different environments. In the next section of this chapter, we examine these environments in greater detail.

7.9 SEDIMENTARY ENVIRONMENTS

Geologists refer to the conditions in which sediment was deposited as the **sedimentary environment** (or "depositional environment"). Examples include beach environments, glacial environments, and river environments. To identify these environments, geologists, like detectives, look for such clues as grain size, composition, sorting, and roundness of clasts, which can tell us how far the sediment has traveled from its source and whether it was deposited from the wind, from a fast-moving current, or from a stagnant body of water. Clues like fossil content and sedimentary structures can tell us whether the sediments were deposited subaerially, just off the coast, or in the deep sea.

Now let's look at some examples of different sedimentary environments and the sediments deposited in them by imagining that we are taking a journey from the mountains to the sea, examining sediments as we go. We begin with "terrestrial sedimentary environments," those formed on dry land, and end with "marine sedimentary environments," those formed along coasts and under the waters of the ocean. (See art on pp. 200–201.) Of note, sediments deposited in terrestrial environments may oxidize (rust) when undergoing lithification in oxygen-bearing water, or if in contact with air. If this happens, the sedimentary beds develop a reddish color, and can be called **redbeds.** The red comes from a film of iron oxide (hematite) that forms on grain surfaces.

Terrestrial (Nonmarine) Sedimentary Environments

Glacial environments. We begin high in the mountains, where it's so cold that more snow collects in the winter than melts away, so glaciers—rivers of ice—develop and slowly flow downslope. Because ice is a solid, it can move sediment of any size. So as a glacier moves down a valley in the mountains, it carries along *all* the sediment that falls on its surface from adjacent cliffs or gets plucked from the ground at its

(a)   (b)   (c)

FIGURE 7.32 (a) Glacial till deposited at the end of a melting glacier in New Zealand. (b) Coarse boulders deposited by a flooding mountain stream in California. (c) An alluvial fan in California. Note the road for scale.

base. At the end of the glacier, where the ice finally melts away, it drops its load and makes a pile of "glacial till" (►Fig. 7.32a). Till is unsorted and unstratified—it contains clasts ranging from clay size to boulder size all mixed together, with large clasts distributed through a matrix of silt and clay. Thus, in a sequence of strata, a layer of diamictite would be the record of an ancient episode of glaciation.

Mountain stream environments. As we walk down beyond the end of the glacier, we enter a realm where turbulent streams rush downslope in mountain valleys. This fast-moving water has the power to carry large clasts; in fact, during floods, boulders and cobbles tumble down the stream bed. Between floods, when water flow slows, the largest clasts settle out to form gravel and boulder beds, while the stream carries finer sediments like sand and mud away (►Fig. 7.32b). Sedimentary deposits of a mountain stream would, therefore, include coarse conglomerate.

Alluvial fan environments. Our journey now takes us to the mountain front, where the fast-moving stream empties onto a plain. In arid regions, where there is insufficient water for the stream to flow continuously, the stream deposits its load of sediment right at the mountain front, creating a large, wedge-shaped apron called an "alluvial fan" (►Fig. 7.32c). Deposition takes place here because when the stream pours from a canyon mouth and spreads out over a broader region, friction with the ground causes the water to slow down, and slow-moving water does not have the power to move coarse sediment. Because we are still so close to the mountains—the source of the sediment—the sand still contains feldspar grains, for these have not yet broken up and have not yet weathered into clay. Alluvial-fan sediments, when later buried and transformed into sedimentary rock, become arkose and conglomerate.

Sand-dune environments. In deserts, relatively few plants grow, so the ground lies exposed to the wind. The strongest winds can transport sand. As a result, large "sand dunes," of well-sorted sand accumulate (Fig. 7.19g, h). Thus, thick layers of well-sorted sandstone, in which we see large (meters-high) cross beds, are relics of desert sand-dune environments (Fig. 7.29b).

Lake environments. From the dry regions, we continue our journey into a temperate realm, where water remains at the surface throughout the year. Some of this water collects in lakes, in which relatively quiet water is unable to move coarse sediment; any coarse sediment brought into the lake by a stream settles out along the shore. Only fine clay makes it out into the center of the lake, where it eventually settles to form mud on the lake bed. Thus, lake sediments (also called "lacustrine sediments") typically consist of finely laminated shale (►Fig. 7.33a).

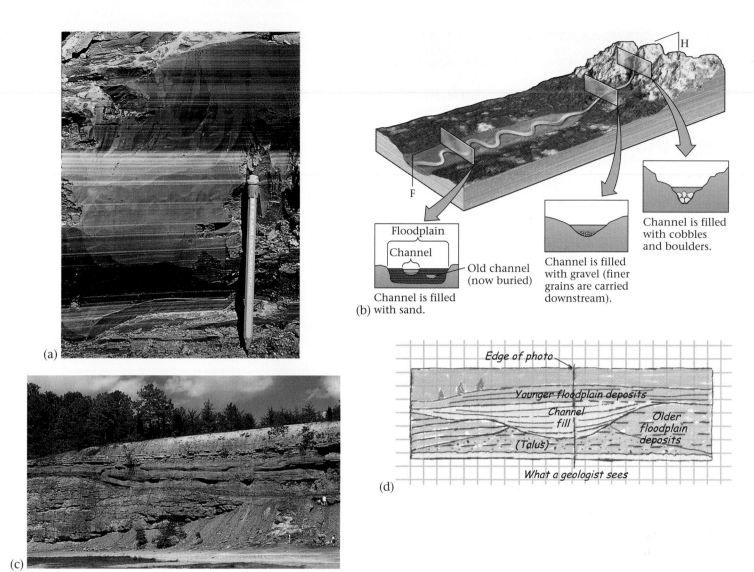

FIGURE 7.33 (a) Finely laminated lake-bed shales in Grenoble, France. (b) The character of river sediment varies with distance from the source. In the steep channel, the turbulent river can carry boulders and cobbles. As the river slows, it can only carry sand and gravel. And as the river winds across the floodplain, it carries sand, silt, and mud. The coarser sediment is deposited in the river channel, the finer sediment on the floodplain. (c) This exposure shows the lens-like shape of an ancient gravel-filled river channel in cross section. (d) A geologist's sketch emphasizes the channel shape.

River environments. The lake drains into a stream that carries water onward toward the sea. As we follow the stream, it merges with tributaries to become a large river, winding back and forth across a plain. Rivers transport sand, silt, and mud. The coarser sediments tumble along the bed in the river's channel, while the finer sediments drift along, suspended in the water (►Fig. 7.33b). This fine sediment settles out along the banks of the river, or on the "floodplain," the flat region on either side of the river that is covered with water only during floods. Since river sediment is deposited in a current, the sediment surface develops ripple marks, and the sediment layers internally have small cross beds.

On the floodplain, mud layers dry out between floods, leading to the formation of mud cracks.

Because the river has transported sediment for a great distance, the minerals making up its sediment have undergone chemical weathering. As a result, very little feldspar remains—most of it has changed into clay. Thus, river sediments (also called "fluvial sediments," from the Latin word for "river") lithify to form sandstone, siltstone, and shale. Typically, channels of coarser sediment (sandstone) are surrounded by layers of fine-grained floodplain deposits; in cross section, the channel has a lens-like shape (►Fig. 7.33c, d).

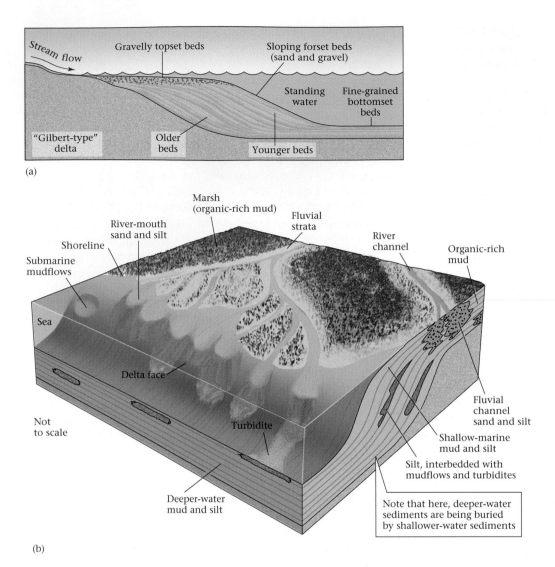

FIGURE 7.34 Sedimentary deposits of deltas. (a) A simple "Gilbert-type" delta formed where a small stream carrying gravel, sand, and silt enters a standing water body. The delta contains topset, foreset, and bottomset beds. (b) A larger river delta is very complex and doesn't fit the simple Gilbert-type model. The great variety of local depositional environments in a delta setting are labeled. Note that as time passes, the delta builds out seaward.

Marine Sedimentary Environments

Marine Delta deposits. After following the river downstream for a long distance, we reach its mouth where it empties into the sea. Here, the river builds a "delta" of sediment out into the sea. Deltas were so named because the map shape of some deltas (e.g., the Nile Delta of Egypt) resembles the Greek letter *delta* (Δ), as we will discuss further in Chapter 17. River water stops flowing when it enters the sea, so sediment settles out.

In 1885, an American geologist named G. K. Gilbert studied small deltas that formed where mountain streams (carrying gravel, sand, and silt) emptied into lakes. He showed that the deltas contained three components (▶Fig. 7.34a): nearly horizontal "topset beds" composed of gravel, sloping "forset beds" of gravel and sand (deposited on the sloping face of the delta), and nearly horizontal silty "bottomset beds," formed at depth on the floor of the water body. Gilbert's model makes intuitive sense but does not adequately describe the complexity of large river deltas. In large river deltas there are many different sedimentary environments, ranging from fluvial and marsh environments to deeper-water marine environments (▶Fig. 7.34b). In addition, storms may cause masses of sediment to slip down the seaward-sloping face of the delta, creating mudflows (slurries of mud), or turbidity currents. Finally, sea-level changes may cause the position of the different environments to move with time. Nevertheless, deposits of a delta can be identified in the stratigraphic record, as thick sequences in which deeper-water (offshore) sediments of a given age grade progressively into fluvial sediments in a shoreward direction.

Coastal beach sands. Now we leave the delta and wander along the coast. Oceanic currents transport sand along the coastline. The sand washes back and forth in the surf, so it becomes well sorted (waves winnow out mud and silt) and well rounded, and because of the back-and-forth movement of ocean water over the sand, the sand surface may become rippled. Thus, if you find well-sorted, medium-grained sandstone, perhaps with ripple marks, you may be looking at the remnants of a beach environment.

Shallow-marine clastic deposits. From the beach, we proceed offshore. Wave action transports coarser sediment shoreward, so in deeper water, where wave energy does not stir the sea floor, finer sediment accumulates. Also, finer sediment gets washed out to sea by the waves. As the water here may be only meters to a few tens of meters deep, geologists refer to this depositional setting as a shallow-marine environment. Clastic sediments that accumulate in this environment tend to be fine-grained, well-sorted, well-rounded silt, and they are inhabited by a great variety of organisms like mollusks and worms. Thus, if you see smooth beds of siltstone and mudstone containing marine fossils, you may be looking at shallow marine clastic deposits.

Shallow-water carbonate environments. In shallow-marine settings far from the mouth of a river, where relatively little clastic sediment (sand and mud) enters the water, the warm, clear, and nutrient-rich water hosts an abundant number of organisms. Their shells, which consist of carbonate minerals make up most of the sediment that accumulates, so we call such environments carbonate environments. The margins of tropical islands, away from the clastic debris of land, provide ideal carbonate environments (▶Fig. 7.35a). In carbonate environments, the nature of sediment depends on the water depth. Beaches collect sand composed of shell fragments, lagoons (quiet water) are sites where lime mud accumulates, and reefs consist of coral and coral debris. Farther offshore of a reef, we can find a sloping apron of reef fragments (▶Fig. 7.35b). Shallow-water carbonate environments transform into sequences of fossiliferous limestone and micrite.

Deep-marine deposits. We conclude our journey by sailing offshore. Along the transition between coastal regions and the deep ocean, turbidity currents deposit turbidites (Fig. 7.30). Farther offshore, in the deep-ocean realm, only fine clay and plankton provide a source for sediment. The clay eventually settles out onto the deep sea floor, forming deposits of finely laminated mudstones, and plankton shells settle to form chalk (from calcite shells; ▶Fig.

7.36a, b) or chert (from siliceous shells). Thus, deposits of mudstone, chalk, or bedded chert indicate a deep-marine origin.

You Can Be a Sedimentary Detective!

Now you should be able to look at most sedimentary rocks in quarries, road cuts, and cliffs and take a pretty good guess as to what ancient environments they represent. From now on, when you see fossiliferous limestone in a quarry, it's not just a limestone, it's the record of a tropical reef. And a sandstone cliff—think ancient dune or beach. Coarse conglomerates should scream "alluvial fan!" or "mountain stream!" and a shale should bring to mind a floodplain, a lake bed, or the floor of the deep sea. Every sequence of strata has a story to tell.

7.10 SEDIMENTARY BASINS

The sedimentary veneer on the Earth's surface varies greatly in thickness. If you stand in central Siberia or south-central Canada, you will find yourself on igneous and metamorphic "basement" rocks that are over a billion years old—there are no sedimentary rocks anywhere in sight. Yet if you stand along the southern coast of Texas, you would have to drill through over 10 km of sedimentary beds before reaching igneous and metamorphic basement. Thick accumulations of sediment form only in regions where the surface of the Earth's lithosphere sinks, as sediment collects. Geologists use the term **subsidence** to refer to the sinking of lithosphere, and the term *"sedimentary basin"* for the sediment-filled depression. In what geologic settings do sedimentary basins form? An understanding of plate tectonics theory provides some answers.

Types of Sedimentary Basins in the Context of Plate Tectonics Theory

Geologists distinguish among different kinds of sedimentary basins on the basis of the region of a lithosphere plate in which they formed. Let's consider a few examples.

* *Rift basins:* These form in continental rifts, regions where the lithosphere has been stretched. During the early stages of rifting, the surface of the Earth subsides simply because crust becomes thinner as it stretches. (To picture this process, imagine pulling on either end of a block of clay with your hands—as the clay stretches, the central region of the block thins and becomes lower than the ends.) As the rift grows, slip on faults drops blocks of crust down, creating low areas bordered by narrow mountain ridges. Alluvial fan deposits form along the

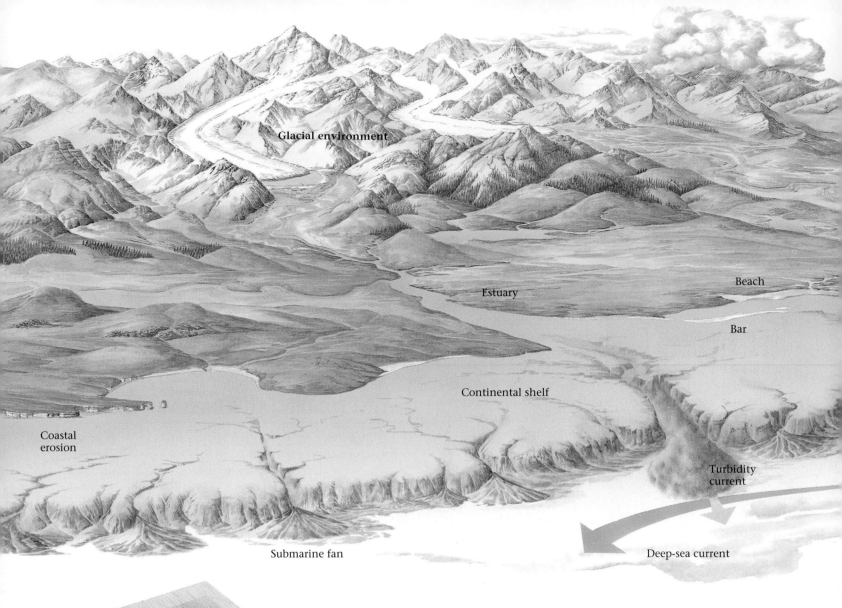

Glacial environment

Estuary

Beach

Bar

Continental shelf

Coastal
erosion

Turbidity
current

Submarine fan

Deep-sea current

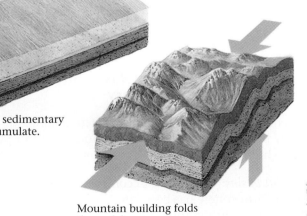

Layers of sedimentary
rock accumulate.

Mountain building folds
the rock layers.

Forming an unconformity

The mountains are eroded; the
folded layers are submerged.

New sedimentary layers
accumulate.

The Formation of Sedimentary Rocks

Categories of sedimentary rocks include clastic sedimentary rocks, chemical sedimentary rocks (formed from the precipitation of minerals out of water), and biochemical sedimentary rocks (formed from the shells of organisms). Clastic sedimentary rocks develop when grains (clasts) break off preexisting rock by weathering and erosion and are transported to a new location by wind, water, or ice; the grains are deposited to create sediment layers, which are then cemented together. We distinguish among types of clastic sedimentary rocks based on grain size.

The character of a sedimentary rock depends on the composition of the sediment and on the environment in which it accumulated. For example, glaciers carry sediment of all sizes, so

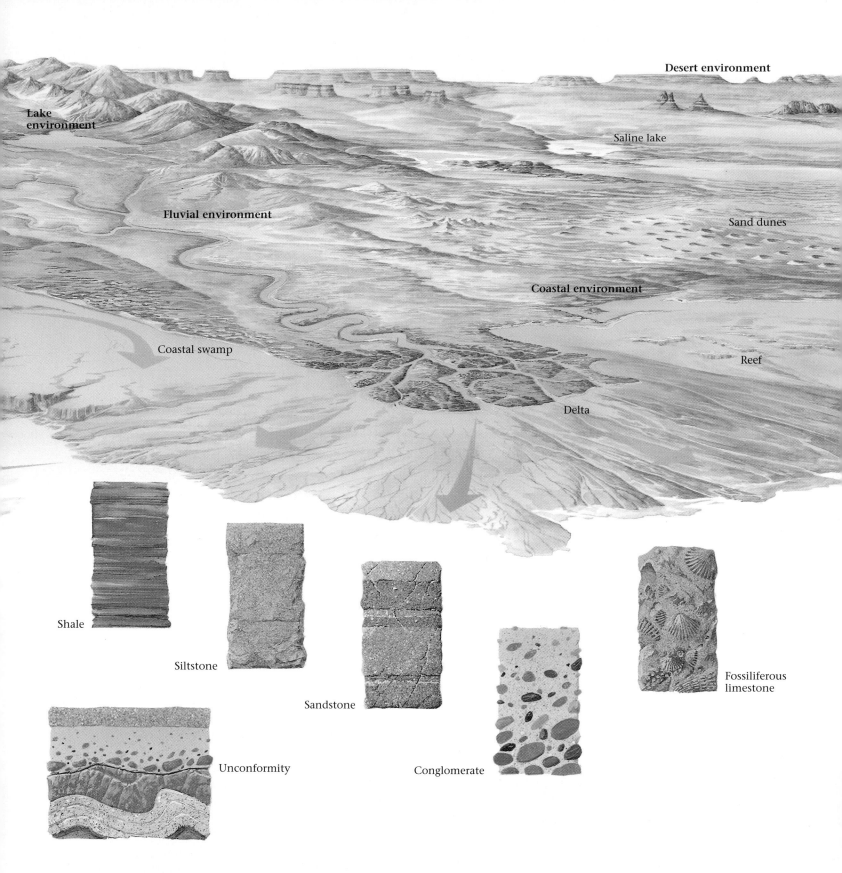

Lake environment

Desert environment

Saline lake

Fluvial environment

Sand dunes

Coastal environment

Coastal swamp

Reef

Delta

Shale

Siltstone

Sandstone

Unconformity

Conglomerate

Fossiliferous limestone

they leave deposits of poorly sorted (different-sized) till; streams deposit coarser grains in their channels and finer ones on floodplains; a river slows down at its mouth and deposits an immense pile of silt in a delta. Fossiliferous limestone develops on coral reefs. In desert environments, sand accumulates into dunes and evaporates precipitate in saline lakes. Offshore, submarine canyons channel avalanches of sediment, or turbidity currents, out to the deep-sea floor.

Sedimentary rocks tell the history of the Earth. For example, the layering, or bedding, of sedimentary rocks is initially horizontal. So where we see layers bent or folded, we can conclude that the layers were deformed during mountain building. Where horizontal layers overlie folded layers, we have an unconformity: for a time, sediment was not deposited, and/or older rocks were eroded away.

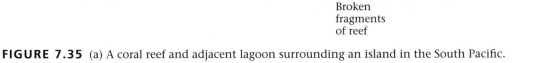

FIGURE 7.35 (a) A coral reef and adjacent lagoon surrounding an island in the South Pacific. (b) The different carbonate environments associated with a reef.

base of the mountains, and salt flats or lakes develop in the low areas between the mountains.

Thinning is not the only reason that rifted lithosphere subsides. During rifting, warm asthenosphere rises beneath the rift and heats up the thin lithosphere. When rifting ceases, the rifted lithosphere then cools, thickens, and becomes denser. This "heavier" lithosphere sinks down, causing more subsidence, just as the deck of tanker ship drops to a lower elevation when the ship is filled with ballast. Subsidence due to cooling of the lithosphere is called "thermal subsidence."

- *Passive-margin basins:* These form along the edges of continents that are *not* plate boundaries. They are underlain by stretched lithosphere, the remnants of a rift whose evolution ultimately led to the formation of a mid-ocean ridge (see Chapter 4). Passive-margin basins form because thermal subsidence of stretched lithosphere continues long after rifting ceases and sea-floor spreading begins. They fill with sediment carried to the sea by rivers and with carbonate rocks formed in coastal reefs. Sediment in a passive-margin basin can reach an astounding thickness of 15 to 20 km.

- *Intracontinental basins:* These develop in the interiors of continents, initially because of thermal subsidence over an unsuccessful rift. They may continue to subside for discrete episodes of time, even hundreds of millions of years after they first formed, for reasons that are not well understood. Illinois and Michigan are each underlaid with

FIGURE 7.36 (a) These plankton shells, which make up deep-marine sediment, are so small that they could pass through the eye of a needle. (b) The chalk cliffs of Dover, England. These were originally deposited on the sea floor and later uplifted.

As indicated by the above description, different assemblages of sedimentary rocks form in different sedimentary basins. So geologists may be able to determine the nature of the basin in which ancient sedimentary deposits accumulated by looking at the character of the deposits.

Transgression and Regression

Sea-level changes control the succession of sediments that we see in a sedimentary basin. At times during Earth history, sea level has risen by as much as a couple hundred meters, creating shallow seas that covered the interiors of continents; there have also been times when sea level has fallen by a couple hundred meters, exposing even the continental shelves to air. Sea-level changes may be due to a number of factors, including climate changes, which control the amount of ice stored in polar ice caps and changes in the volume of ocean basins.

When sea level rises, the coast migrates inland—we call this process **transgression.** As the coast migrates, the sandy beach migrates with it, and the site of the former beach gets buried by deeper-water sediment. Thus, as transgression occurs, an extensive layer of beach sand eventually forms. This layer may look like a blanket of sand that was deposited all at once, but in fact the sand deposited at one location differs in age from the sand deposited at another location. When sea level falls, the coast migrates seaward— we call this process **regression** (▶Fig. 7.37). Typically, the record of a regression will not be well preserved, because as sea level drops, areas that had been sites of deposition become exposed to erosion. A succession of strata deposited during a cycle of transgression and regression is called a "depositional sequence."

an intracontinental basin (the Illinois basin and the Michigan basin, respectively) in which up to 7 km of sediment has accumulated. Most of this sediment is fluvial, deltaic, or shallow marine. At times, extensive swamps formed along the shoreline in these basins. The plant matter of these swamps was buried to form coal.

• *Foreland basins:* These form on the continent side of a mountain belt because as the mountain belt grows, large slices of rock are pushed up and out onto the surface of the continent. Such movement takes place by slip along large faults. The weight of these slices pushes down on the surface of the lithosphere, creating a wedge-shaped depression adjacent to the mountain range that fills with sediment eroded from the range. Fluvial and deltaic strata accumulate in foreland basins.

7.11 DIAGENESIS

Earlier in this chapter we discussed the process of lithification, by which sediment hardens into rock. Lithification is an aspect of a broader phenomenon called diagenesis. Geologists use the term **diagenesis** for all the physical, chemical, and biological processes that transform sediment into sedimentary rock and that alter characteristics of sedimentary rock once the rock has formed.

In the depositional environment, diagenesis includes bioturbation, growth of minerals in pore spaces and around grains, and replacement of existing crystals with new crystals. In buried sediment, diagenesis involves the compaction of sediment and growth of cement that leads to complete lithification. Of note, pressure due to the weight of overburden may cause a process called pressure solution, during which the faces of grains dissolve where they are squeezed against neighboring grains. In fully lithified sedimentary

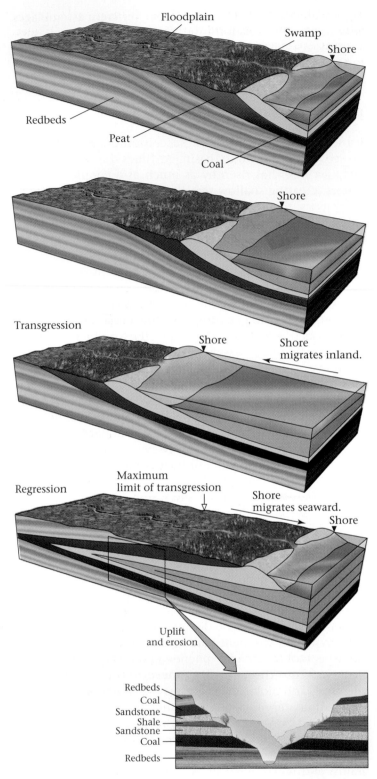

FIGURE 7.37 The concept of transgression and regression. As sea level rises and the shore migrates inland, coastal sedimentary environments overlap terrestrial environments. Eventually, deeper-water environments overlap shallower ones. Thus, a regionally extensive layer does not all form at the same time. During regression, sea level falls and the shore moves seaward.

THE VIEW FROM SPACE A great variety of depositional environments occur on Earth. In these environments. sediments accumulate and may be buried deeply enough to transform into sedimentary rock. The type of sediment, and the form of a deposit is an indication of the depositional environment. Here, we see an alluvial fan in the Taklimakan Desert of China, formed when an intermittant stream spreads out over a flat basin and divides into many small channels within and along which sediment accumulates.

rocks, diagenesis continues both as a result of chemical reactions between the rock and groundwater passing through the rock, and as a result of increases in temperature and pressure. These reactions may dissolve existing cement and/or form new cement, and may grow new minerals. Though diagenesis may alter the texture and mineral composition of sedimentary rock, it usually does not destroy all sedimentary structures in the rock.

As temperature and pressure increase still deeper in the subsurface, the changes that take place in rocks become more profound. At sufficiently high temperature and pressure, a whole new assemblage of minerals forms, and/or mineral grains become aligned parallel to one another. Geologists consider such changes to be examples of *metamorphism*. The transition between diagenesis and metamorphism in sedimentary rocks is gradational. Most geologists consider changes that take place in rocks at temperatures of below about 150°C to be clearly "diagenetic reactions," and those that occur in rocks at temperatures above about 300°C to be clearly "metamorphic reactions." In the temperature range between 150°C and 300°C, whether diagenesis or metamorphism takes place depends on rock type. In the next chapter, we enter the true realm of metamorphism.

CHAPTER SUMMARY

• Sediment consists of detritus (mineral grains and rock fragments derived from preexisting rock), mineral crystals that precipitate directly out of water, and shells (formed when organisms extract ions from water).

• Rocks at the surface of the Earth undergo physical and chemical weathering. During physical weathering, intact rock breaks into pieces. During chemical weathering, rocks react with water and air to produce new minerals like clay, and ions in solution.

• The covering of loose rock fragments, sand, gravel, and soil at the Earth's surface is regolith. Soil differs from other types of regolith in that it has been changed by the activities of organisms, by downward-percolating rainwater, and by the mixing in of organic matter. Distinct horizons can be identified in soil. The type of soil that forms depends on factors such as climate and source material.

• Geologists recognize four major classes of sedimentary rocks. Clastic (detrital) rocks form from cemented-together detritus (mineral grains and rock fragments) that were first produced by weathering, then were transported, deposited, and lithified. Biochemical rocks develop from the shells of organisms. Organic rocks consist of plant debris, or of altered plankton remains. Chemical rocks, such as evaporites, precipitate directly from water.

• Sedimentary structures include features such as bedding, cross bedding, graded bedding, ripple marks, and dunes. Their presence provides clues to depositional settings.

• Glaciers, mountain streams and fronts, sand dunes, lakes, rivers, deltas, beaches, shallow seas, and deep seas each accumulate a different assemblage of sedimentary strata. Thus, by studying sedimentary rocks, we can reconstruct the characteristics of past environments.

• Thick piles of sedimentary rocks accumulate in sedimentary basins, regions where the lithosphere sinks, creating a depression at the Earth's surface.

• Sea level changes with time. Transgressions occur when sea level rises and the coastline migrates inland. Regressions occur when sea level falls and the coastline migrates seaward.

• Diagenesis involves processes leading to lithification and processes that alter sedimentary rock once it has formed.

KEY TERMS

arkose (p. 185)
basement (p. 167)
bed (p. 190)
biochemical sedimentary
  rocks (p. 180)
breccia (p. 182)
caliche (calcrete) (p. 178)
cement (p. 182)
cementation (p. 182)
chemical sedimentary rocks
  (p. 180)
chemical weathering
  (p. 169)
chert (p. 187)
clastic (detrital) sedimentary
  rocks (p. 180)
clasts (p. 181)
coal (p. 187)
compaction (p. 181)
conglomerate (p. 183)
cover (p. 167)
cross beds (p. 193)
deposition (p. 181)
detritus (p. 167)
diagenesis (p. 203)
diamictite (p. 185)
dolostone (p. 188)
dunes (p. 192)
erosion (p. 181)
evaporites (p. 188)
exfoliation (p. 168)
fossils (p. 195)
frost wedging (p. 168)
graded bed (p. 195)
horizons (p. 175)
joints (p. 168)
laterite soil (p. 178)
limestone (p. 186)
lithification (p. 181)
mud cracks (p. 195)

mudstone (p. 185)
organic sedimentary rocks
  (p. 180)
pedalfer soil (p. 178)
pedocal soil (p. 178)
physical (mechanical)
  weathering (p. 167)
redbeds (p. 195)
regolith (p. 174)
regression (p. 203)
ripples (ripple marks) (p. 192)
root wedging (p. 168)
salt wedging (p. 168)
sandstone (p. 180)
saprolite (p. 169)
scour marks (p. 195)
sediment (p. 166)
sedimentary basin (p. 167)
sedimentary environment
  (p. 195)
sedimentary rock (p. 167)
shale (p. 185)
siltstone (p. 185)
soil (p. 174)
soil erosion (p. 179)
soil profile (p. 175)
sorting (p. 182)
strata (p. 190)
stratigraphic formation
  (p. 191)
subsidence (p. 199)
talus (p. 168)
transgression (p. 203)
travertine (p. 188)
turbidite (p. 195)
wacke (p. 185)
weathering (p. 167)
zone of accumulation
  (p. 175)
zone of leaching (p. 175)

REVIEW QUESTIONS

1. Explain the circumstances that allowed the Mediterranean Sea to dry up.

2. How does physical weathering differ from chemical weathering?

3. Describe the processes that produce joints in rocks.

4. Feldspars are among the most common minerals in igneous rocks, but they are relatively rare in sediments. Why are they more susceptible to weathering and what common sedimentary minerals are produced from weathered feldspar?

5. What types of minerals tend to weather more quickly?

6. Describe the different horizons in a typical soil profile.

7. What factors determine the nature of soils in different regions?

8. Describe how a clastic sedimentary rock is formed from its unweathered parent rock.

9. Clastic and chemical sedimentary rocks are both made of material that has been transported. How are they different?

10. Explain how biochemical sedimentary rocks form.

11. Describe how grain size and shape, sorting, sphericity, and angularity change as sediments move downstream.

12. Describe the two different kinds of chert. How are they similar? How are they different?

13. What kinds of conditions are required for the formation of evaporites?

14. What minerals precipitate out of seawater first? next? last? What does this suggest when geologists find huge volumes of pure gypsum in the Earth's crust?

15. How is dolostone different from limestone, and how does it form?

16. Describe how cross beds form. How can you read the current direction from cross beds?

17. Describe how a turbidity current forms and moves. How does it produce graded bedding?

18. Compare the deposits of an alluvial fan with those of a typical river environment and with those of a deep-marine deposit.

19. Why don't sediments accumulate everywhere? What types of tectonic conditions are required to create basins?

20. What happens during diagenesis and how does diagenesis differ from metamorphism?

SUGGESTED READING

Birkeland, P. W. 1999. *Soils and Geomorphology,* 3rd ed. New York: Oxford University Press.

Blatt, H., W. B. N. Berry, and S. Brande. 1991. *Principles of Stratigraphic Analysis.* Boston: Blackwell Scientific.

Boggs, S., Jr. 1998. *Petrology of Sedimentary Rocks.* Englewood Cliffs, N.J.: Prentice-Hall.

Boggs, S., Jr. 2000. *Principles of Sedimentology and Stratigraphy,* 3rd ed. Upper Saddle River, N.J.: Pearson Education.

Brady, N. C., and R. R. Weil. 2001. *The Nature and Properties of Soils,* 13th ed. Upper Saddle River, N.J.: Prentice-Hall.

Collinson, J. D., and D. B. Thompson. 1989. *Sedimentary Structures.* London: Allen and Unwin.

Einsele, G. 2000. *Sedimentary Basins: Evolution, Facies, and Sediment Budget.* New York: Springer-Verlag.

Harpstead, M. I., et al. 2001. *Soil Science Simplified,* 4th ed. Ames: Iowa State University Press.

Julien, P. Y. 1995. *Erosion and Sedimentation.* New York: Cambridge University Press.

Leeder, M. R. 1999. *Sedimentology and Sedimentary Basins: From Turbulence to Tectonics.* Oxford, England: Blackwell.

Martini, I. P., and W. Chesworth, eds. 1992. *Weathering, Soils, and Paleosols.* New York: Elsevier.

Ollier, C., and C. Pain. 1996. *Regolith, Soils, and Land Forms.* New York: Freeman.

Prothero, D. R., and F. L. Schwab. 1998. *Sedimentary Geology.* New York: Freeman.

Reading, H. G., ed. 1996. *Sedimentary Environments: Processes, Facies, and Stratigraphy.* Oxford, England: Blackwell.

Robinson, D. A., and R. B. G. Williams, eds. 1994. *Rock Weathering and Landform Evolution.* New York: Wiley.

Tucker, M. E. 2001. *Sedimentary Petrology,* 3rd ed. Oxford, England: Blackwell.

Metamorphism: A Process of Change

Nothing in the world lasts, save eternal change.
— HONORAT DE BUEIL (1589–1650)

8.1 INTRODUCTION

Cool winds sweep across Scotland for much of the year. In this blustery climate, vegetation has a hard time taking hold, so the landscape provides countless outcrops of barren rock. During the latter half of the eighteenth century, a gentleman farmer and doctor named James Hutton became fascinated with the Earth and examined these outcrops, hoping to learn how rock formed. Hutton found that many features in the outcrops resembled the products of present-day sediment deposition or of volcanic activity, and soon he came to an understanding of how sedimentary and igneous rock forms. But Hutton also found rock that contained minerals and textures quite different from those in sedimentary and igneous samples. He described this puzzling rock as "a mass of matter which had evidently formed originally in the ordinary manner . . . but which is now extremely distorted in its structure . . . and variously changed in its composition."

The rock that so puzzled Hutton is now known as metamorphic rock, from the Greek words *meta*, meaning "beyond" or "change," and *morphe*, meaning "form." In modern terms, a **metamorphic rock** is a rock that forms from a preexisting rock, or **protolith,** when the protolith undergoes

An outcrop of Precambrian metamorphic rock, exposed in the Wasatch Mountains in Utah. The layering, or "foliation," formed during metamorphism and then later was bent into an S-shape.

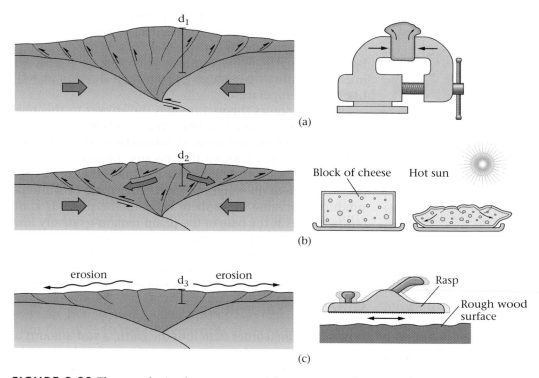

FIGURE 8.28 Three geologic phenomena, together, can contribute to exhumation in a collisional mountain belt. Because of exhumation, the vertical distance between a point at depth in the belt decreases with time (i.e., $d_3 < d_2 < d_1$). (a) Collision squeezes rock in the mountain belt upward, like dough pressed in a vise. (b) Eventually, the crust beneath the mountain range becomes warm and weak, so the mountain belt collapses, like a block of soft cheese placed in the hot sun. This process thins the upper crust. (c) Throughout the history of the mountain belt, erosion grinds rock off the surface and removes it, much like a giant rasp.

from impact, and when astronauts sampled the Moon, they discovered that the regolith covering the lunar surface contains the products of shock metamorphism.

Metamorphism in the Mantle

Discussions of metamorphic rocks rarely mention the mantle, because mantle rocks rarely crop out on Earth. But if you think about the process of convection in the mantle, it becomes clear that rocks in the mantle must have undergone metamorphic change several times during Earth history. Specifically, as peridotite (ultramafic rock) from shallower depths in the asthenosphere slowly cools, it becomes denser and sinks. As it sinks, pressure acting on it increases until, at a depth of about 440 km, minerals in the peridotite undergo a phase change and transform into different minerals that are stable at higher pressure. The process happens again when the rock sinks below 660 km, and it occurs at other depths as well. Later in Earth history, when the rock reaches great depth in the mantle, it heats up and becomes relatively buoyant. Therefore, it slowly rises back toward the base of

the lithosphere and, in the process, undergoes phase changes again, only this time the phase changes produce minerals that are stable at lower pressures.

8.7 BRINGING METAMORPHIC ROCKS BACK TO THE GROUND SURFACE

When you stand on an outcrop of metamorphic rock, you are standing on material that once lay many kilometers beneath the surface of the Earth. In areas of regional metamorphism, high-grade rocks rose, generally from a greater depth than did low-grade rocks—some high-grade rocks were once tens of kilometers below the surface. How does metamorphic rock return to the Earth's surface? Geologists refer to the overall process by which deeply buried rocks end up back at the surface as **exhumation.**

Exhumation results from several processes in the Earth System that happen simultaneously. Let's look at the specific processes that contribute to bringing high-grade metamorphic rocks from below a collisional mountain range back to the surface. First, as two continents progressively push together, the rock caught between them squeezes upward, or is "uplifted," much like a ball of dough pressed in a vise (▶Fig. 8.28a); the upward movement takes place by slip on faults and by plastic-like flow of rock. Second, as the mountain range grows, the crust at depth beneath it warms up and gets weak. Eventually, the range starts to collapse under its own weight, much like a block of soft cheese placed in the hot sun, a process called "extensional collapse" (▶Fig. 8.28b; see Chapter 11). As a result of this collapse, the upper crust spreads out laterally. This movement stretches the upper part of crust in the horizontal direction and causes it to become thinner in the vertical direction. As the upper part of the crust becomes thinner, the deeper crust ends up closer to the surface. Third, erosion takes place at the surface (▶Fig 8.28c); weathering, landslides, river flow, and glacial flow together play the role of a giant rasp, stripping away rock at the surface and exposing rock that was once below the surface.

8.8 WHERE DO YOU FIND METAMORPHIC ROCKS?

If you want to study metamorphic rocks, you can start by taking a hike in a collisional or convergent mountain range (e.g., Fig. 8.21). During the formation of such ranges, rocks undergo both contact metamorphism and regional metamorphism; because of exhumation, these rocks are exposed as towering cliffs of gneiss and schist. The process of exhumation can occur relatively quickly, geologically speaking—some metamorphic rocks now visible in active mountains formed only a few million years ago. Where ancient mountain ranges once existed, we still find belts of metamorphic rocks cropping out at the ground surface even though the high peaks of the range have long since eroded away.

Huge expanses of metamorphic rock crop out in continental shields. A **shield** represents the older portion of a continent, where extensive areas of Precambrian rock crop out at the ground surface, because overlying younger rock has eroded away (▶Fig. 8.29a). The shield of North America

FIGURE 8.29 (a) The distribution of shield areas (exposed Precambrian metamorphic and igneous rock) on the Earth. *(continued)*

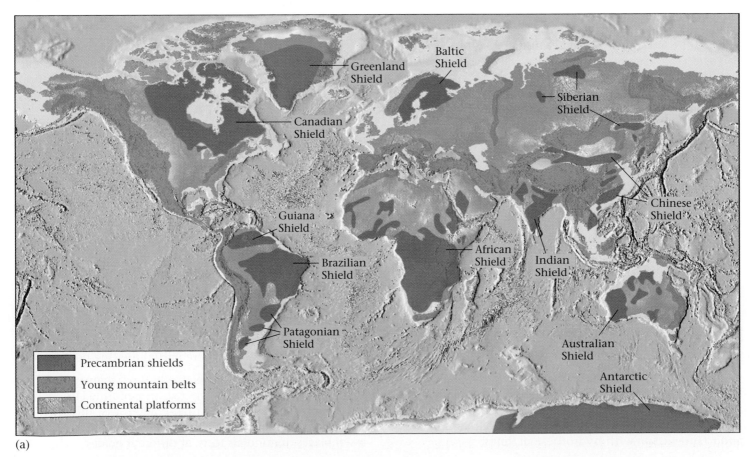

(a)

(b)

(c)

FIGURE 8.29 *(cont'd.)* (b) The Canadian Shield as viewed from the air. (c) The walls of the Black Canyon of the Gunnison River in Colorado display high-grade metamorphic rocks. These rocks underlie the Rocky Mountains of Colorado. The stripes are pegmatite dikes that intruded the dark rock.

is called the Canadian Shield because it encompasses about half the land area of Canada (▶Fig. 8.29b). Large shields also occur in South America, northern Europe, Africa, India, and Siberia. Rocks in shields were metamorphosed during a succession of Precambrian mountain-building events that were responsible for building the continent in the first place—the oldest rocks on Earth occur in shields.

In the United States and much of Europe, a veneer of Paleozoic and Mesozoic sedimentary rocks covers most of the Precambrian metamorphic rocks, so you can see these metamorphic rocks only where they were uplifted and exposed by erosion in younger mountain belts, or where rivers have cut down deeply enough to expose basement (▶Fig. 8.29c; see Chapter 7). For example, at the base of the Grand Canyon, erosion by the Colorado River exposes dark cliffs of the 1.8-billion-year-old Vishnu Schist (Fig. 7.2).

CHAPTER SUMMARY

• Metamorphism refers to changes in a rock that result in the formation of a metamorphic mineral assemblage and/or a metamorphic foliation, in response to change in temperature and/or pressure, to the application of differential stress, and to interaction with hydrothermal fluids.

• Metamorphism involves recrystallization, metamorphic reactions (neocrystallization), phase changes, pressure solution, and/or plastic deformation. If hot-water solutions bring in or remove elements, we say that metasomatism has occurred.

• Metamorphic foliation can be defined either by preferred mineral orientation (aligned inequant crystals) or by compositional banding. Preferred mineral orientation develops where differential stress causes the squashing and shearing of a rock, so that its inequant grains align parallel with each other.

• Geologists separate metamorphic rocks into two classes, foliated rocks and nonfoliated rocks, depending on whether the rocks contain foliation.

• The class of foliated rocks includes slate, metaconglomerate, phyllite, schist, amphibolite, and gneiss. The class of nonfoliated rocks includes hornfels, quartzite, marble, and amphibolite, though the latter three can have a foliation. Migmatite, a mixture of igneous and metamorphic rock, forms under conditions where partial melting begins.

• Rocks formed under relatively low temperatures are known as low-grade rocks, whereas those formed under high temperatures are known as high-grade rocks. Intermediate-grade rocks develop between these two extremes. Different assemblages of minerals form at different grades.

- Geologists track the distribution of different grades of rock by looking for index minerals. Isograds indicate the location at which index minerals first appear. A metamorphic zone is the region between two isograds.

- A metamorphic facies is a group of metamorphic mineral assemblages that develop under a specified range of temperature and pressure conditions. The assemblage in a given rock depends on the composition of the protolith, as well as on the metamorphic conditions.

- Thermal metamorphism (also called contact metamorphism) occurs in an aureole surrounding an igneous intrusion. Dynamically metamorphosed rocks form along faults, where rocks are only sheared, under metamorphic conditions. Dynamothermal metamorphism (also called regional metamorphism) results when rocks are buried deeply during mountain building.

- Metamorphism occurs because of plate interactions: the process of mountain building in either convergent or collisional zones causes dynamothermal metamorphism; shearing along plate boundaries causes dynamic metamorphism; and igneous plutons in rifts cause thermal metamorphism. The circulation of hot water causes hydrothermal metamorphism of oceanic crust at mid-ocean ridges. Unusual metamorphic rocks called blueschists form at the base of accretionary prisms. Metamorphism also results from the shock of meteorite impact and from convection in the mantle.

- We find extensive areas of metamorphic rocks in mountain ranges. Vast regions of continents known as shields expose ancient (Precambrian) metamorphic rocks.

KEY TERMS

amphibolite (p. 217)
burial metamorphism (p. 228)
compositional layering (p. 216)
contact metamorphism (p. 224)
differential stress (p. 211)
dynamic metamorphism (p. 228)
dynamothermal (regional) metamorphism (p. 228)
exhumation (p. 230)
flattened-clast conglomerate (p. 215)
foliation (p. 213)
gneiss (p. 216)
hydrothermal metamorphism (p. 229)
index minerals (p. 221)
isograd (p. 221)
marble (p. 218)
metamorphic aureole (p. 224)

metamorphic facies (p. 222)
metamorphic grade (p. 219)
metamorphic mineral (p. 208)
metamorphic rock (p. 207)
metamorphic texture (p. 208)
metamorphic zones (p. 221)
metamorphism (p. 208)
metasomatism (p. 213)
migmatite (p. 217)
mylonite (p. 228)
nonfoliated metamorphic rock (p. 217)
phyllite (p. 215)
preferred mineral orientation (p. 211)
prograde metamorphism (p. 219)
protolith (p. 207)
quartzite (p. 218)

retrograde metamorphism (p. 220)
schist (p. 215)
schistosity (p. 215)
shield (p. 231)
shock metamorphism (p. 229)

slate (p. 214)
slaty cleavage (p. 214)
supercritical fluid (p. 212)
thermal (contact) metamorphism (p. 224)
vein (p. 213)

REVIEW QUESTIONS

1. How are metamorphic rocks different from igneous and sedimentary rocks?

2. What two features characterize most metamorphic rocks?

3. What phenomena cause metamorphism?

4. What is metamorphic foliation, and how does it form?

5. How is a slate different from a phyllite? How does a phyllite differ from a schist? How does a schist different from a gneiss?

6. Why are hornfels nonfoliated?

7. What is a metamorphic grade, and how can it be determined? How does grade differ from "facies"?

8. How does prograde metamorphism differ from retrograde metamorphism?

9. Describe the geologic settings where thermal, dynamic, and dynamothermal metamorphism take place.

10. Why does metamorphism happen at the site of meteor impacts, along mid-ocean ridges, and deep in the mantle?

11. How does plate tectonics explain the peculiar combination of low-temperature but high-pressure minerals found in a blueschist?

12. Where would you go if you wanted to find exposed metamorphic rocks, and how did such rocks return to the surface of the Earth after being at depth in the crust?

The Rock Cycle

The three rock types of the Earth System. (a) Igneous rock, here formed by the cooling of lava at a volcano. (b) Sedimentary rock, here eroding to form sediment. (c) Metamorphic rock, here exposed in a mountain belt. Over time, materials composing one rock type may be incorporated in another.

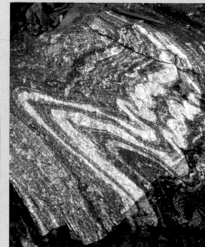

(a) (b) (c)

B.1 INTRODUCTION

"Stable as a rock." This familiar expression implies that a rock is permanent, unchanging over time—it isn't. In the time frame of Earth history, a span of over 4.5 billion years, atoms making up one rock type may be rearranged or moved elsewhere, eventually becoming part of another rock type. Later, the atoms may move again to form a third rock type, and so on. Geologists call the progressive transformation of Earth materials from one rock type to another the **rock cycle** (▶Fig. B.1), one of many examples of cycles acting in or on the Earth. (James Hutton, the eighteenth-century Scottish geologist, was the first person to visualize and describe the rock cycle.) We focus on the rock cycle here because it illustrates the relationships among the three rock types described in the previous three chapters.

A **cycle** (from the Greek word for "circle" or "wheel") is a series of interrelated events or steps that occur in suc-

cession and can be repeated, perhaps indefinitely. During **temporal cycles,** such as the phases of the Moon or the seasons of the year, events happen according to a time-table, but the materials involved do not necessarily change. The rock cycle, in contrast, is an example of a geologic **mass-transfer cycle,** one that involves the transfer or movement of materials (mass) to different parts of the Earth System. (The hydrologic cycle, which we will learn about in Interlude E, is another mass-transfer cycle.)

There are many paths around or through the rock cycle. For example, igneous rock may weather and erode to produce sediment, which lithifies to form sedimentary rock. The new sedimentary rock may become buried and form metamorphic rock, which then could partially melt to create magma. This magma later solidifies to form new igneous rock. We can symbolize this path as igneous → sedimentary → metamorphic → igneous. But alternatively, the metamorphic rock could be uplifted and eroded to

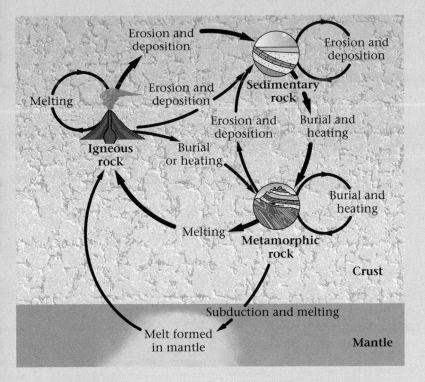

FIGURE B.1 The stages of the rock cycle, showing various alternative pathways.

form new sediment and then new sedimentary rock without melting, taking a shortcut path through the cycle that we can symbolize as igneous → sedimentary → metamorphic → sedimentary. Likewise, the igneous rock could be metamorphosed directly, without first turning to sediment. This metamorphic rock could again be turned into sedimentary rock, defining another shortcut path: igneous → metamorphic → sedimentary. To get a clearer sense of how the rock cycle works, we'll look at one example.

B.2 THE ROCK CYCLE IN THE CONTEXT OF THE THEORY OF PLATE TECTONICS

Material can enter the rock cycle when basaltic magma rises from the mantle. Suppose the magma erupts and forms basalt (an igneous rock) at a continental hot-spot volcano (▶Fig. B.2a). Interaction with wind, rain, and vegetation gradually weathers the basalt, physically breaking it into smaller fragments and chemically altering it to create clay. As water washes over the newly formed clay, it carries the clay away and transports it downstream—if you've ever seen a brown-colored river, you've seen clay en route to a site of deposition. Eventually the river reaches the sea, where the water slows down and the clay settles out.

Let's imagine, for this example, that the clay settles out along the margin of continent X and forms a deposit of mud. Gradually, through time, the mud becomes progressively buried and the clay flakes pack tightly together, resulting in a new sedimentary rock, shale. The shale resides 6 km below the continental shelf for millions of years, until the adjacent oceanic plate subducts and a neighboring continent, Y, collides with X. The shale gets buried *very* deeply when the edge of the encroaching continent pushes over it. As the mountains grow, the shale that had once been 6 km below the surface now ends up 20 km below the surface, and under the pressure and temperature conditions present at this depth, it metamorphoses into schist (▶Fig. B.2b).

The story's not over. Once mountain building stops, erosion grinds away the mountain range, and exhumation brings some of the schist to the ground surface. This schist erodes to form sediment, which is carried off and deposited elsewhere to form new sedimentary rock—this material takes a shortcut through the rock cycle. But other schist remains preserved below the surface. Eventually, continental rifting takes place at the site of the former mountain range, and the crust containing the schist begins to split apart. When this happens, some of the schist partially melts and a new felsic magma forms. This felsic magma rises to the surface of the crust and freezes to create rhyolite, a new igneous rock (▶Fig. B.2c). In terms of the rock cycle, we're back at the beginning, having once again made igneous rock (▶Fig. B.2d).

Note that atoms, as they pass through the rock cycle, do not always stay within the same mineral. In our example, a silicon atom in a pyroxene crystal of the basalt may become part of a clay crystal in the shale, part of a muscovite crystal in the schist, and part of a feldspar crystal in the rhyolite.

B.3 RATES OF MOVEMENT THROUGH THE ROCK CYCLE

We saw that not all atoms pass through the rock cycle in the same way. Similarly, not all atoms pass through the rock cycle at the same rate, and for that reason we find rocks of many different ages at the surface of the Earth. Some rocks remain in one form for less than a few million years, while others stay unchanged for most of Earth history. Rocks exposed on Precambrian shields have remained unchanged for billions of years—the Canadian Shield of North America includes rocks as old as 3.9 billion years. In contrast, a rock with an Appalachian Mountain address has passed through stages of the rock cycle many times in the past few billion years, because the eastern margin of North

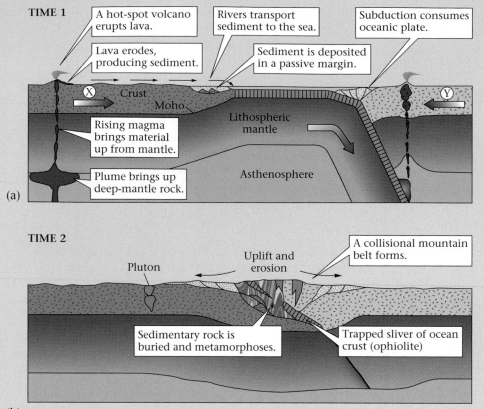

TIME 1

A hot-spot volcano erupts lava.

Rivers transport sediment to the sea.

Subduction consumes oceanic plate.

Lava erodes, producing sediment.

Sediment is deposited in a passive margin.

X   Crust
        Moho
                    Lithospheric mantle

Rising magma brings material up from mantle.

Asthenosphere

Plume brings up deep-mantle rock.

Y

(a)

TIME 2

Pluton

Uplift and erosion

A collisional mountain belt forms.

Sedimentary rock is buried and metamorphoses.

Trapped sliver of ocean crust (ophiolite)

(b) ▨ Metamorphic rock   ☐ Sediment eroded from mountains

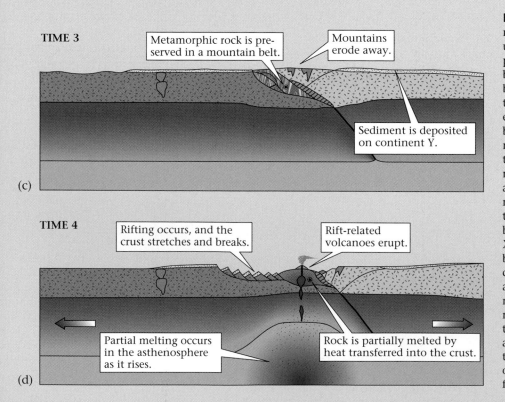

TIME 3

Metamorphic rock is pre-served in a mountain belt.

Mountains erode away.

Sediment is deposited on continent Y.

(c)

TIME 4

Rifting occurs, and the crust stretches and breaks.

Rift-related volcanoes erupt.

Partial melting occurs in the asthenosphere as it rises.

Rock is partially melted by heat transferred into the crust.

(d)

FIGURE B.2 (a) At the beginning of the rock cycle (time 1), atoms, originally making up peridotite in the mantle, rise in a mantle plume. The peridotite partially melts at the base of the lithosphere, and the atoms become part of a basaltic magma that rises through the lithosphere of continent X and erupts at a volcano. At this time, the atoms become part of a lava flow; that is, an igneous rock. Weathering breaks the lava down, and the resulting clay is transported to a passive-margin basin. After the clay is buried, the atoms become part of a shale—a sedimentary rock. Note that the ocean floor to the east of the passive-margin basin is being consumed beneath continent Y. (b) At time 2, continents X and Y collide, and the shale is buried deeply beneath the resulting mountain range (at the dot). Now the atoms become part of a schist— a metamorphic rock. (c) At time 3, the mountain range erodes away and the schist rises but does not reach the surface. (d) At time 4, rifting begins to split the continents apart, and igneous activity occurs again. At this time, the atoms of the schist become part of a new melt, which eventually freezes to form a rhyolite, another igneous rock.

America has been subjected to multiple events of basin formation, mountain building, and rifting since the shield to the west developed.

Studies of the past two decades suggest that most of the rock now making up the Earth's continental crust contains atoms that were extracted from the mantle over 2.5 billion years ago. Yet we see rocks of many different ages in the continents today. That is because geologic processes recycle these atoms again and again, similar to the way people recycle the metal of old cars to make new ones. And just as the number of late-model cars on the road today exceeds the number of vintage cars, younger rocks are more common than ancient rocks. At the surface of continents, sedimentary rocks created during the last several hundred million years are the most widespread type, whereas rocks recording the early history of the Earth are quite rare. But even though most continental crustal rocks are recycled, some new ones continue to be freshly extracted from the mantle each year, adding to the continent at volcanic arcs or hot spots.

Do the atoms in continental rocks ever get a chance to start the rock cycle all over, by returning to the mantle? Yes. Some sediment that erodes off a continent ends up in deep-ocean trenches, and some of this is dragged back into the mantle by subduction. In fact, recent research suggests that metamorphic and igneous rocks at the base of the continental crust may be removed and carried back down into the mantle at subduction zones.

Our tour of the rock cycle has focused on continental rocks. What about the oceans? Oceanic crust consists of igneous rock (basalt and gabbro) overlaid with sediment. Because a layer of water blankets the crust, erosion does not affect it, so oceanic crustal rock does not follow the path into the sedimentary loop of the rock cycle. But sooner or later, oceanic crust subducts. When this happens, the rock of the crust first undergoes metamorphism, for as it sinks, it is subjected to progressively higher temperatures and pressures. And eventually, a little of the rock may melt and become new magma, which then rises at a volcanic arc.

B.4 WHAT DRIVES THE ROCK CYCLE IN THE EARTH SYSTEM?

The rock cycle occurs because the Earth is a dynamic planet. The planet's internal heat and gravitational field drive plate movements. Plate interactions cause the uplift of mountain ranges, a process that exposes rock to weathering, erosion, and sediment production. Plate interactions also generate the geologic settings in which metamorphism occurs, where rock melts to provide magma, and where sedimentary basins develop.

At the surface of the Earth, the gases released by volcanism collect to form the ocean and atmosphere. Heat (from the Sun) and gravity drive convection in the atmosphere and oceans, leading to wind, rain, ice, and currents—the agents of weathering and erosion. Weathering and erosion grind away at the surface of the Earth and send material into the sedimentary loop of the cycle. In sum, external energy (solar heat), internal energy (Earth's internal heat), and gravity all play a role in driving the rock cycle, by keeping the mantle, crust, atmosphere, and oceans in constant motion.

Tectonic Activity of a Dynamic Planet

*E*arlier in this book, we learned that the map of the Earth constantly, but ever so slowly, changes in response to plate movements and interactions. We now turn our attention to the dramatic consequences of such tectonic activity in the Earth System: volcanoes (Chapter 9), earthquakes (Chapter 10), and mountains (Chapter 11).

Why does molten rock rise like a fountain out of the ground, or explode into the sky at a volcano? Why does the ground shake and heave, in some cases so violently that whole cities topple, during an earthquake? How can the energy released by an earthquake tell us about the insides of the Earth, thousands of kilometers below the surface? What processes cause the land surface to rise several kilometers above sea level to form mountain belts? How do rocks bend, squash, stretch, and break in response to forces caused by plate interactions? Read on, and you will not only be able to answer these questions, but you will also see how the answers help people deal with some of the most deadly natural hazards that threaten society.

The Wrath of Vulcan: Volcanic Eruptions

During a volcanic eruption, molten rock rises from inside the Earth and flows out as lava or blasts into the air as ash. This 1989–1990 eruption of Redoubt Volcano, Alaska, a consequence of the subduction of the Pacific Plate beneath Alaska, produced clouds of ash. A jumbo jet flew through the ash and lost power in all four engines. Fortunately, after losing about 2.5 km (8,000 feet) of elevation, the engine restarted and the plane was able to land. Volcanoes are hazards, and volcanoes are drama! In the words of Hans Cloos, who witnessed the eruption of Mt. Vesuvius (Italy) in 1944, "Glowing waves rise and flow, burning all life on their way, and freeze into black, crusty rock which adds to the height of the mountain and builds the land, thereby adding another day to the geologic past . . . I became a geologist forever, by seeing with my own eyes: the Earth is alive!"

9.1 INTRODUCTION

Every few hundred years, one of the hills on Vulcano, an island in the Mediterranean Sea off the western coast of Italy, rumbles and spews out molten rock, glassy cinders, and dense "smoke" (actually a mixture of various gases, fine ash, and very tiny liquid droplets). Ancient Romans thought that such eruptions happened when Vulcan, the god of fire, fueled his forges beneath the island to manufacture weapons for the other gods. Geologic study suggests, instead, that eruptions take place when hot magma, formed by melting inside the Earth, rises through the crust and emerges at the surface. No one believes the myth anymore, but the island's name evolved into the English word **volcano,** which geologists use to designate either an erupting vent through which molten rock reaches the Earth's surface, or a mountain built from the products of eruption.

On the main peninsula of Italy, not far from Vulcano, another volcano, Mt. Vesuvius, towers over the nearby Bay of Naples. Two thousand years ago, Pompeii was a prosperous Roman resort and trading town of 20,000 inhabitants, sprawled at the foot of Vesuvius (▶Fig. 9.1a, b). Then one morning in 79 C.E., earthquakes signaled the mountain's awakening. At 1:00 P.M., a dark mottled cloud boiled up above

(a)

(b)

What a geologist imagines

(c)

FIGURE 9.1 (a) Pompeii, once buried by 6 m of volcanic debris from Mt. Vesuvius, was excavated by archaeologists in the late nineteenth century. Vesuvius rises in the distance. (b) What a geologist imagines: When Mt. Vesuvius erupted in 79 C.E., it was probably much larger, as depicted in this sketch. The pellets are hot volcanic bombs and lapilli. (c) A plaster cast of an unfortunate inhabitant of Pompeii, found buried by ash in the corner of a room, where the person crouched for protection. The flesh rotted away, leaving only an open hole that could be filled by plaster.

Mt. Vesuvius's summit to a height of 27 km. As lightning sparked in its crown, the cloud drifted over Pompeii, turning day into night. Blocks and pellets of rock fell like hail, while fine ash and choking fumes enveloped the town. Frantic people rushed to escape, but for many it was too late. As the growing weight of volcanic debris began to crush buildings, an avalanche of ash swept over Pompeii, and by the next day the town had vanished beneath a 6-m-thick gray-black blanket. The ruins of Pompeii were protected so well by their covering that when archaeologists excavated the town 1,800 years later, they found an amazingly complete record of Roman daily life. During their work, archaeologists discovered open spaces in the debris. Out of curiosity, they filled the spaces with plaster, and realized that the spaces were fossil casts of Pompeii's unfortunate inhabitants, their bodies twisted in agony or huddled in despair (▶Fig. 9.1c).

Clearly, volcanoes are unpredictable and dangerous. Volcanic activity can build a towering, snow-crested mountain or can blast one apart. It can provide the fertile soil that enables a civilization to thrive, or it can snuff out a civilization in a matter of minutes. Because of the diversity of volcanic activity and its consequences, this chapter sets out ambitious goals. We first review the products of volcanic eruptions and the basic characteristics of volcanoes. Then we look again at the different kinds of volcanic eruptions on Earth. Volcanoes are *not* randomly distributed around the globe—their positions reflect the locations of plate boundaries, rifts, and hot spots. Finally, we examine the hazards posed by volcanoes, efforts by geoscientists to predict eruptions and help minimize the damage they cause, the possible influence of eruptions on climate and civilization.

volcano. The process is similar to the way the rapidly expanding gas accompanying the explosion of gunpowder in a cartridge shoots a bullet out of a gun.

In some cases, an explosive eruption blasts the volcano apart and leaves behind a large caldera. Such explosions, awesome in their power and catastrophic in their consequences, eject cubic kilometers of igneous particles upward at initial speeds of up to 90 m per second. Convection in the cloud can carry ash up through the entire troposphere and into the stratosphere. The resulting plume of debris resembles the mushroom cloud above a nuclear explosion. Coarse-grained ash and lapilli settle from the cloud close to the volcano, while finer ash settles farther away.

Some explosive eruptions take place when water gains access to the magma chamber and suddenly transforms into steam—the steam pressure blasts the volcano apart and energetically expels debris. Geologists refer to pyroclastic eruptions involving the reaction of water with magma as **phreatomagmatic eruptions** (▶Fig. 9.13b).

Notably, the type of volcano (shield, cinder cone, or composite) depends on its eruptive style. Volcanoes that have only effusive eruptions become shield volcanoes, those that generate small pyroclastic eruptions yield cinder cones, and those that alternate between effusive and large pyroclastic eruptions become composite volcanoes. Large explosions yield calderas and blanket the surrounding countryside with sheets of ignimbrite.

Why are there such contrasts in eruptive style, and therefore in volcano shape? Eruptive style depends on the viscosity and gas pressure of the magma in the volcano. These characteristics, in turn, depend on the composition and temperature of the magma and on the environment (subaerial or submarine) in which the eruption occurs. Let's look at these controls in detail.

- *The effect of viscosity on eruptive style:* Low-viscosity (basaltic) lava flows out of a volcano easily, while high-viscosity (andesitic and rhyolitic) lava can clog up a volcano's plumbing and lead to a buildup of pressure. Thus, basaltic eruptions are typically effusive and produce shield volcanoes, whereas rhyolitic eruptions are explosive.

- *The effect of gas pressure on eruptive style:* The injection of magma into the magma chamber and conduit generates an outward "push" or pressure inside the volcano. The presence of gas within the magma increases this pressure, because gas expands greatly as it rises toward the Earth's surface. In runny (basaltic) magma, gas bubbles can rise to the surface of the magma and pop, causing the lava to fountain into the sky; a small cinder cone of bombs and cinders results. In a viscous (andesitic or rhyolitic) magma, however, the gas bubbles cannot escape and thus move with the magma toward the Earth's surface. As pressure on

the magma from overlying rock decreases, the gas bubbles expand and create a tremendous outward pressure. Eventually, the gas pressure shatters the partially solidified magma and sends a large cloud of pyroclastic debris into the sky or down the flank of the volcano, causing a pyroclastic eruption. Rhyolitic and andesitic magmas contain more gas, and thus eruptions of these magmas are more explosive than are eruptions of basaltic magmas.

- *The effect of the environment on eruptive style:* The ability of a lava to flow depends on where it erupts. Lava flowing on dry land cools more slowly than lava erupting beneath water. Thus, a basaltic lava that could flow easily down the flank of a subaerial volcano will pile up in a mound of pillows around the vent of a submarine volcano.

Traditionally, geologists have classified volcanoes according to their eruptive style, each style named after a well-known example (Hawaiian, Vulcanian, etc.) as described in books focused on volcanoes (see art on pp. 252–53). Below, we focus on relating eruptive styles to the geologic setting in which the volcano forms, in the context of plate tectonics theory (▶Fig. 9.16).

9.5 HOT-SPOT ERUPTIONS

Hot-spot volcanoes develop above finger-like plumes of hot mantle that rise from near the core-mantle boundary (see Chapters 4 and 6). When the plume, composed of hot, plastic, but still solid rock, reaches the base of the lithosphere, it partially melts, generating quantities of basaltic magma that rise and fuel a volcano at the Earth's surface above. Let's look at two well-known examples of hot-spot volcanism.

Oceanic Hot-Spot Volcanoes (Hawaii)

When a mantle plume rises beneath oceanic lithosphere, basaltic magma erupts at the surface of the sea floor and forms a submarine volcano. At first, such submarine eruptions yield an irregular mound of pillow lava. With time, the volcano grows up above the sea surface and becomes an island. But when the volcano emerges from the sea, the basalt lava that erupts no longer freezes so quickly, and thus flows as a thin sheet over a great distance. Thousands of thin basalt flows pile up, layer upon layer, to build a broad, dome-shaped shield volcano with gentle slopes (Fig. 9.11a, b). Note that such shield volcanoes develop their distinctive shape because the low-viscosity, hot basaltic lava that constitutes them spreads out like pancake syrup and cannot build up into a steep cone. As the volcano grows, portions of it can't resist the pull of gravity and slip seaward, creating large slumps. Thus, in cross section, hot-spot volcanoes are quite complex (▶Fig. 9.17).

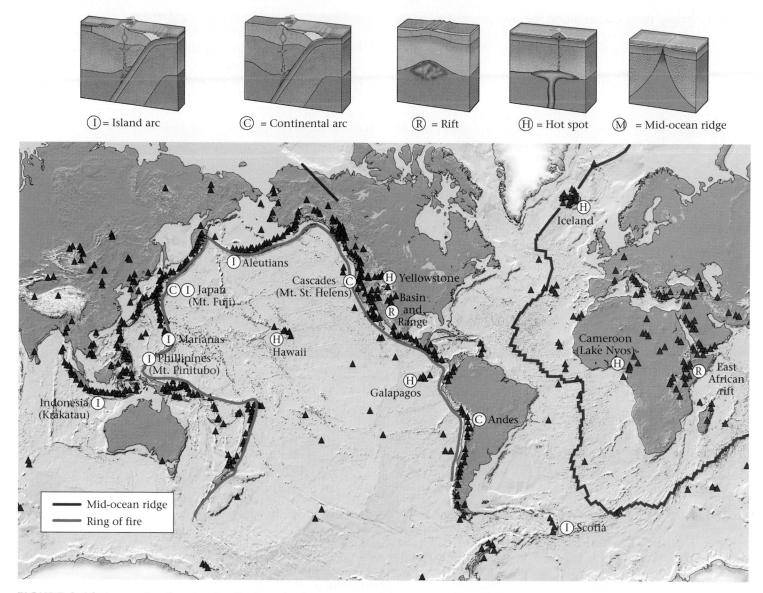

Ⓘ = Island arc   Ⓒ = Continental arc   Ⓡ = Rift   Ⓗ = Hot spot   Ⓜ = Mid-ocean ridge

FIGURE 9.16 A map showing the distribution of volcanoes around the world, and the basic geologic settings in which volcanoes form, in the context of plate-tectonics theory.

FIGURE 9.17 The inside of an oceanic hot-spot volcano is a mound of pillow basalt built on the surface of the oceanic crust. When the mound emerges above sea level, a shield volcano forms on top. Volcanic debris accumulates along the margin of the volcano. The weak material occasionally slumps seaward on sliding surfaces (indicated with arrows).

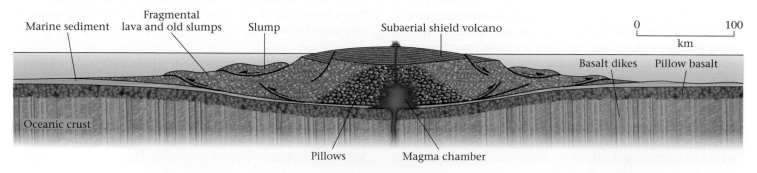

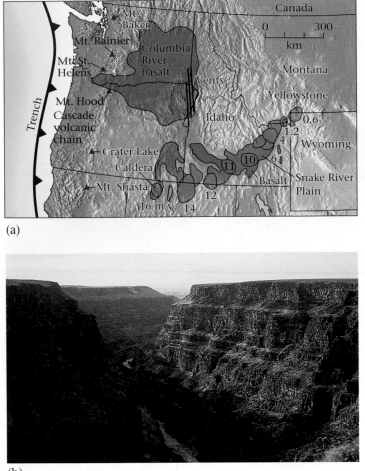

(a)

(c)

(b)

FIGURE 9.18 (a) Volcanic rocks from hot spots formed in two places in the western United States. First, the Columbia River basalt plateau erupted about 17 million years ago (m.y. = million years), when the plume rose beneath a rift. Then, either a new plume or a different part of the same plume rose beneath northern Nevada. As North America drifted to the west-southwest, eruptions yielded the calderas along the Snake River Plain; the numbers indicate the age of the calderas (millions of years). The hot spot now lies underneath Yellowstone National Park. (b) Numerous flows of basalt piled one on top of the other in the Snake River Plain. The Snake River has cut a canyon through these basalts. (c) The "yellow stone" of Yellowstone National Park consists of silicic tuffs. The Yellowstone River has been able to cut a deep canyon through these tuffs, because they are relatively soft.

The big island of Hawaii, one of the largest oceanic hot-spot volcanoes on Earth today, currently consists of four shield volcanoes, each built around a different vent. The island now towers over 9 km above the adjacent ocean floor (about 4.2 km above sea level), the greatest relief from base to top of any mountain on Earth; by comparison, Mt. Everest rises 8.85 km above the plains of India. Calderas up to 3 km wide have formed at the summit, and basaltic lava has extruded from both conduits and fissures. During some eruptions, the lava fountains into the air, or fills deep "lava lakes" in craters (Fig. 9.13a). The lakes gradually drain to feed streams of lava that cas-

cade down the flanks of the volcano. Lava tubes within flows carry lava all the way to the sea, where the glowing molten rock drips into the water and instantly disappears in a cloud of steam.

Continental Hot-Spot Volcanoes (Yellowstone National Park)

North America has also been drifting westward above a mantle plume for millions of years (▶Fig. 9.18a, b). Today, the plume (whose source remains a subject of debate) lies beneath Yellowstone National Park, yielding fascinating landforms, rock deposits, and geysers. Eruptions at the Yellowstone hot spot differ from those on Hawaii in an important way: unlike Hawaii, the Yellowstone hot spot erupts both basaltic lava and rhyolitic pyroclastic debris. This happens because basaltic magma rising from the top of the mantle plume must pass through thirty or so kilometers of continental crust before it reaches the surface. Rhyolitic magma forms when rising basaltic magma heats up and partially melts the continental crust through which it rises. Rhyolitic eruptions produce thick tuffs that now crop out as yellow and red rocks in the canyon of the Yellowstone River (▶Fig. 9.18c).

About 627,000 years ago, an immense pyroclastic flow, as well as a cloud of ash, blasted out of the Yellowstone region. Close to the eruption, ignimbrites up to tens of meters thick formed, and ash from the giant cloud sifted down over the United States as far east as the Mississippi River. The eruption, 1,000 times more powerful than that of Mt. St. Helens, left a huge caldera, almost 100 km across, which now dominates the landscape of Yellowstone. Magma remains beneath Yellowstone today, causing geyser activity. The Yellowstone caldera is only one of at least a dozen calderas that underlie the Snake River Plain (Fig. 9.18a).

Flood-Basalt Eruptions

When a plume first rises, it has a bulbous head in which there is a huge amount of partially molten rock. If the crust above the plume stretches and rifts, voluminous amounts of lava erupt along fissures. A particularly large amount of magma is available because of the size of the plume head and because the very hot asthenosphere of the plume undergoes a greater amount of partial melting than does the cooler asthenosphere that normally under-

lies rifts. The low-viscosity lava spreads out in sheets over vast areas. Geologists refer to these sheets as **flood basalt.** Over time, eruption of basalts builds a broad plateau (▶Fig. 9.19a, b). Geologists refer to broad areas covered by flood basalt as large igneous provinces (LIPs; see Chapter 6). Once the plume head has drained, the volume of eruption decreases, and normal hot-spot eruptions take place.

About 15 million years ago, rifting above a plume created the region that now constitutes the Columbia River Plateau of Washington and Oregon (Fig. 9.18a). Eruptions yielded sheets of basalt up to 30 m thick that flowed as far as 550 km from the source. Gradually, layer upon layer erupted, creating a pile of basalt up to 500 m thick over a region of 220,000 square km. Even larger flood-basalt provinces formed elsewhere in the world, notably the Deccan Plateau of India (▶Fig. 9.20), the Paraná Basin of Brazil, and the Karroo Plateau of South Africa.

9.6 ERUPTIONS ALONG MID-OCEAN RIDGES

Most eruptions of lava occur along mid-ocean ridges, plate boundaries at which new sea-floor crust forms. In fact, products of mid-ocean ridge volcanism cover 70% of our planet's surface. We don't generally see this volcanic activity, however, because the ocean hides most of it beneath a blanket of water. Mid-ocean ridge volcanoes, which develop along fissures parallel to the ridge axis, are not all continuously active. Each one turns on and off in a time scale measured in tens to hundreds of years. They erupt basalt, which,

FIGURE 9.19 (a) When the lithosphere cracks and rifts above the bulbous head of a plume, huge amounts of magma rise and erupt through fissures, producing sheets of basalt that pile up to form a plateau. (b) Later, the bulbous plume head no longer exists, leaving only a narrower plume stalk. The lithosphere has moved relative to the plume, so a track of hot-spot volcanoes begin to form.

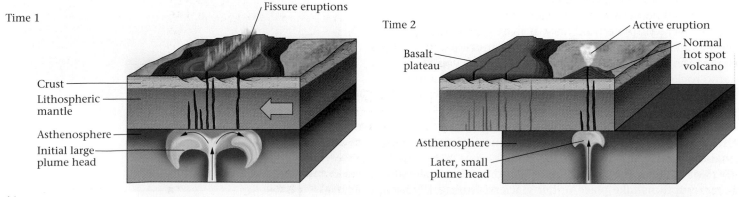

FIGURE 9.20 The flood basalts of western India, known as the "Deccan traps," are exposed in a canyon near the village of Ajanta. Between about 100 B.C.E. and 700 C.E., Buddhists carved a series of monasteries and meeting halls into the solid basalt. These are decorated by huge statues, carved in place, as well as spectacular frescoes, painted on cow-dung plaster.

because it's underwater, forms pillow-lava mounds. Water that heats up as it circulates through the crust near the magma chamber bursts out of hydrothermal (hot-water) vents (see Chapter 4).

Iceland—a Hot Spot on a Ridge

Iceland is one of the few places on Earth where mid-ocean ridge volcanism built a mound of basalt that protrudes above the sea. The island formed where a mantle plume lies beneath the Mid-Atlantic Ridge—the presence of this plume means that far more magma erupted here than beneath other places along mid-ocean ridges. Because Iceland straddles a divergent plate boundary, it is being stretched apart, with faults forming as a consequence. Indeed, the central part of the island is a narrow rift, in which the youngest volcanic rocks of the island have erupted (▶Fig. 9.21a, b); this rift *is* the trace of the Mid-Atlantic Ridge. Faulting cracks the crust and so provides a conduit to a magma chamber. Thus, eruptions on Iceland tend to be fissure eruptions, yielding either curtains of lava that are many kilometers long or linear chains of small cinder cones (Fig. 9.9b).

Not all volcanic activity on Iceland occurs subaerially. Some eruptions take place under glaciers. During 1996, for

example, an eruption at the base of a 600-m-thick glacier melted the ice and produced a column of steam that rose several kilometers into the air. Meltwater accumulated under the ice for six days, until it burst through the edge of the glacier and became a flood that lasted two days and destroyed roads, bridges, and telephone lines. Some of Iceland's volcanic activity occurs under the sea. Continuing eruptions off the coast yielded the island of Surtsey, whose birth was first signaled by huge quantities of steam bubbling up from the ocean. Eventually, steam pressure explosively

FIGURE 9.21 (a) Iceland consists of volcanic rocks that erupted from a hot spot along the Mid-Atlantic Ridge. Because the island straddles a divergent boundary, it gradually stretches, leading to the formation of faults. The central part of the island is an irregular northeast-trending rift, where we find the youngest rocks of the island. (b) The surface of Iceland has dropped down along the faults that bound the central rift. This low-altitude air photo shows an escarpment formed where slip occurred on a fault.

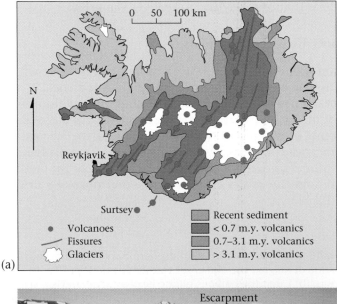

(a)

(b)

ejected ash as much as 5 km into the atmosphere. Surtsey finally emerged from the sea on November 14, 1963, building up a cone of ash and lapilli that rose almost 200 m above sea level in just three months (Fig. 9.13b). Waves could easily have eroded the cinder cone away, but the island has survived because lava erupted from the vent and flowed over the cinders, effectively encasing them in an armor-like blanket of solid rock.

9.7 ERUPTIONS ALONG CONVERGENT BOUNDARIES

Most of the subaerial volcanoes on Earth lie along convergent plate boundaries (subduction zones). The volcanoes form when volatile compounds like water and carbon dioxide released from the subducting plate rise into the overlying hot mantle, causing melting and producing magma which then rises through the lithosphere and erupts. Some of these volcanoes start out as submarine volcanoes and later grow into volcanic island arcs (such as the Aleutians of Alaska), while others grow on continental crust, building continental volcanic arcs [such as the Cascade mountain chain of Washington and Oregon (Fig. 9.18a)]. Typically, individual volcanoes in volcanic arcs lie about 50–100 km apart. Subduction zones border over 60% of the Pacific Ocean, creating a 20,000-km-long chain of volcanoes known as the **"Ring of Fire."** See Chapters 4 and 6 for illustrations.

Many different kinds of magma form at volcanic arcs. As a result, these volcanoes sometimes have effusive eruptions and sometimes pyroclastic eruptions—and occasionally they explode. Such eruptions yield composite volcanoes, like the elegant symmetrical cone of Mt. Fuji (Fig. 9.11e), and the blasted-apart hulk of Mt. St. Helens (Fig. 9.14c).

Volcanoes of island arcs initially erupt underwater. Thus, their foundation consists of volcanic material that froze in contact with water, or of volcanic debris that was deposited underwater. The layers that make up the foundation include pillow basalts, hyaloclastites, and submarine debris flows.

9.8 ERUPTIONS IN CONTINENTAL RIFTS

The rifting of continental crust yields a wide array of different types of volcanoes, because (as in the case of continental hot spots) the magma that feeds these volcanoes comes both from the partial melting of the mantle and from the partial melting of the crust. Thus, rifts host basaltic fissure eruptions, in which curtains of lava fountain up or linear chains of cinder cones develop. But, they also host explosive rhyolitic volcanoes, and in some places even stratovolcanoes.

Rift volcanoes are active today in the East African Rift (Fig. 9.10d). During the past 25 million years, rift volcanoes were active in the Basin and Range Province of Nevada, Utah, and Arizona. About 1 billion years ago, a narrow but deep rift formed in the middle of the United States and filled with over 15 km of basalt; this Mid-Continent Rift runs from the tip of Lake Superior to central Kansas.

9.9 VOLCANOES IN THE LANDSCAPE

Why do volcanoes look the way they do? First of all, the shape of a volcano depends on whether it has been erupting recently or ceased erupting long ago. For erupting volcanoes, the shape (shield, stratovolcano, or cinder cone) depends primarily on the eruptive style, because at an erupting volcano, the process of construction happens faster than the process of erosion. For example, in southern Mexico, the volcano Paricutín began to spatter out of a cornfield on February 20, 1943. Its eruption continued for nine years, and by the end, 2 cubic km of tephra had piled up into a cone almost half a kilometer high. But once a volcano stops erupting, erosion attacks. The rate at which a volcano is destroyed depends on whether it's composed of pyroclastic debris or lava. Cinder cones and ash piles can wash away quickly. For example, in the summer of 1831, a cinder cone grew 60 m above the surface of the Mediterranean Sea. As soon as the island appeared, Italy, Britain, and Spain laid claim to it, and shortly the island had at least seven different names. But the volcano stopped erupting, and within six months it was gone, fortunately before a battle for its ownership had begun. In contrast, composite or shield volcanoes, which have been armor-plated by lava flows, can withstand the attack of water and ice for quite some time.

In the end, however, erosion wins out, and you can tell an old volcano that has not erupted for a long time from a volcano that has erupted recently by the extent to which river or glacial valleys have been carved into its flanks. In some cases, the softer exterior of a volcano completely erodes away, leaving behind the plug of harder frozen magma that once lay just beneath the volcano, as well as the network of dikes that radiate from this plug (▶Fig. 9.22a–c). You can see good examples of such landforms at Shiprock, New Mexico (Fig. 6.11b), and at Devil's Tower, Wyoming (▶Fig. 9.22d).

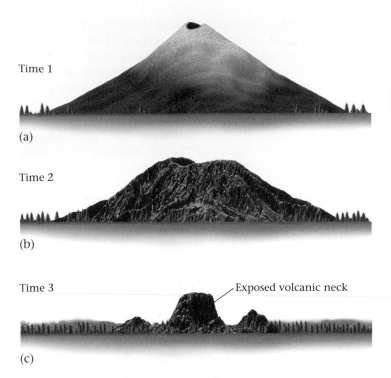

Time 1

(a)

Time 2

(b)

Time 3 — Exposed volcanic neck

(c)

(d)

FIGURE 9.22 (a) The shape of an active volcano is defined by the surface of the most recent lava flow or ash fall. Little erosion affects the surface. (b) An inactive volcano that has been around long enough for the surface to be modified by erosion. In humid climates, these volcanoes have gullies carved into their flanks and may be partially covered with forest. (c) A long-dead (extinct) volcano has been so deeply eroded that only the neck of the volcano may remain. (d) Devil's Tower, Wyoming, rises 260 m above the surrounding land surface. It formed when a mass of magma cooled beneath a volcano, about 40 million years ago. Huge columnar joints, 2.5 m wide at the base, developed when the magma cooled. Subsequently, erosion stripped away overlying softer tuff and flows and exposed the mass. In Native American legend, the ribbed surface of Devil's Tower represents the claw marks of a giant bear, trying to reach a woman who sought refuge on the Tower's summit.

9.10 BEWARE: VOLCANOES ARE HAZARDS!

Like earthquakes, volcanoes are natural hazards that have the potential to cause great destruction to humanity, in both the short term and the long term. According to one estimate, volcanic eruptions in the last two thousand years have caused about a quarter-million deaths—much fewer than those caused by earthquakes, but nevertheless a sizable number. Considering the rapid expansion of cities, far more people live in dangerous proximity to volcanoes today than ever before, so if anything, the hazard posed by volcanoes has gotten worse—imagine if a Krakatau-like explosion were to occur next to a major city today. Let's now look at the different kinds of threats posed by volcanic eruptions.

Threat of lava flows. When you think of an eruption, perhaps the first threat that comes to mind is the lava that flows from a volcano. Indeed, on many occasions lava has overwhelmed towns. Basaltic lava from effusive eruptions is the greatest threat, because it can flow quickly and spread over a broad area. In Hawaii, recent lava flows have buried roads, housing developments, and cars (Fig. 9.3c). In one place, basalt almost completely submerged a parked (and empty) school bus (▶Fig. 9.23a). Usually people have time to get out of the way of such flows, but not necessarily with their possessions. All they can do is watch helplessly from a distance as an advancing flow engulfs their home (▶Fig. 9.23b). Before the lava even touches it, the building may burst into flames from the intense heat. Similarly, forests, orchards, and sugarcane fields are burned and then buried by rock, their verdure replaced by blackness. The most disastrous lava flow in recent time came from the eruption of Mt. Nyiragongo, a 3.7-km-high volcano in the Congo in 2002 (▶Fig. 9.23c). Lava flows traveled almost 50 km and flooded the streets of Goma, encasing the streets with a 2-m-thick layer of basalt. The flows destroyed almost half the city and turned 300,000 people into refugees.

Threat of ash and lapilli. During a pyroclastic eruption, large quantities of ash erupt into the air, later to fall back to Earth. Close to a volcano, pumice and lapilli tumble out of

(a)

(b)

(c)

(d)

(e)

(f)

FIGURE 9.23 (a) This empty school bus was engulfed by a basalt flow in Hawaii. (b) When lava at over 1,000°C comes close to a house, the house erupts in flame. (c) Residents of Goma, in west-central Africa, walking over lava-filled streets after a 2002 eruption of a nearby volcano. (d) A blizzard of ash falling from the eruption of Mt. Pinatubo, in the Philippines, blankets a nearby town in ghostly white. (e) A pyroclastic flow rushes toward fleeing firefighters in Japan, during the eruption of Mt. Unzen. (f) A devastating lahar buried the town of Armero, Colombia.

the sky, smashing through or crushing roofs of nearby buildings (for this reason, Japanese citizens living near volcanoes keep hard hats handy), and can accumulate into a blanket up to several meters thick. Winds can carry fine ash over a broad region. In the Philippines, for example, a typhoon spread heavy air-fall ash from the 1991 eruption of Mt. Pinatubo so that it covered a 4,000-square-km area (▶Fig. 9.23d). Because of heavy rains, the ash became soggy and heavy, and it was particularly damaging to roofs. Ash buries crops, may spread toxic chemicals that poison the soil, and insidiously infiltrates machinery causing moving parts to wear out.

Fine ash from an eruption can also present a hazard to airplanes. Ash clouds rise so fast that they may be at airplane heights (11 km) long before the volcanic eruption has been reported, especially if the eruption occurs in a remote locality; and at high elevations, the ash cloud may be too dilute for a pilot to see. Like a sandblaster, the sharp, angular ash abrades turbine blades, greatly reducing engine efficiency, and the ash, along with sulfuric acid formed from the volcanic gas, scores windows and damages the fuselage. Also, when heated inside a jet engine, the ash melts, creating a liquid that sprays around the turbine and freezes; the resulting glassy coating restricts the air flow and causes the engine to flame out.

For example, in 1982, a British Airways 747 flew through the ash cloud over a volcano on Java. Corrosion turned the windshield opaque, and ingested ash caused all four engines to fail. For thirteen minutes, the plane glided earthward, dropping from 11.5 km (37,000 feet) to 3.7 km (12,000 feet) above the black ocean below. As passengers assumed a brace position for ditching at sea, the pilots tried repeatedly to restart the engines. Suddenly, in the oxygen-rich air of lower elevations, the engines roared back to life. The plane swooped back into the sky and headed for an emergency landing in Jakarta, where, without functioning instruments and with an opaque windshield, the pilot brought his 263 passengers and crew back to the ground safely. To land, he had to squint out an open side window, with only his toes touching the controls. In 1989, the same fate befell a KLM 747 en route to Anchorage. The plane encountered ash from the Redoubt Volcano (see Chapter Opener), lost power in all four engines and all instruments, and sank about 2.6 km (8,000 feet) before the pilot could restart the engines and bring the plane in for a landing. During the month after the 1991 eruption of Mt. Pinatubo, fourteen jets flew through the resulting ash cloud, and of these, nine had to make emergency landings because of engine failure.

Threat of pyroclastic flows. Pyroclastic flows race down the flanks of a volcano at speeds of 100–300 km per hour (▶Fig. 9.23e). The largest can travel tens to hundreds of kilometers. The volume of ash contained in such glowing avalanches is not necessarily great—St. Pierre on Martinique was covered

only by a thin layer of dust after the pyroclastic flow from Mt. Pelée had passed (see Chapter 6)—but the cloud can be so hot and poisonous that it means instant death to anyone caught in its path, and because it moves so fast, the force of its impact can flatten buildings and forests (Fig. 9.6a–c).

Threat of the blast. Most exploding volcanoes direct their fury upward. But some, like Mt. St. Helens, explode sideways. The forcefully ejected gas and ash, like the blast of a bomb, flattens everything in its path. In the case of Mt. St. Helens, the region around the volcano had been a beautiful pine forest; but after the eruption, the once-towering trees were stripped of bark and needles and lay scattered over the hill slopes like matchsticks (Fig. 9.15e).

Threat of landslides. Eruptions commonly trigger large landslides along the volcano's flanks. The debris, composed of ash and solidified lava that erupted earlier, can move quite fast (250 km per hour) and far. During the eruption of Mt. St. Helens, 8 billion tons of debris took off down the mountainside, careered over a 360-m-high ridge, and tumbled down a river valley, until the last of it finally came to rest over 20 km from the volcano.

Threat of lahars. When volcanic ash and other debris mix with water, the result is a slurry that resembles freshly mixed concrete. This slurry, known as a *lahar,* can flow downslope at speeds of over 50 km per hour. Because lahars are denser and more viscous than water, they pack more force than flowing water and can literally carry away everything in their path. The lahars of Mt. St. Helens traveled more than 40 km from the volcano, following existing drainages. When they had passed, they left a gray and barren wake of mud, boulders, broken bridges, and crumpled houses, as if a giant knife had scraped across the landscape. Widespread lahars also swept down the flanks of Mt. Pinatubo in 1991, the water provided by typhoonal and monsoonal rains.

Lahars may develop in regions where snow and ice cover an erupting volcano, for the eruption melts the snow and ice, thereby creating an instant supply of water. Perhaps the most destructive lahar of recent times accompanied the eruption of snow-crested Nevado del Ruiz in Colombia on the night of November 13, 1985. The lahar surged down a valley of the Rio Lagunillas like a 40-m-high wave, hitting the sleeping town of Armero, 60 km from the volcano. Other pulses of lahar followed. When they had passed, 90% of the buildings in the town were gone, replaced by a 5-m-thick layer of mud (▶Fig. 9.23f), which now entombs the bodies of 25,000 people.

Threat of earthquakes. Earthquakes accompany almost all major volcanic eruptions, for the movement of magma breaks rocks underground. Such earthquakes may trigger landslides on the volcano's flanks, and can cause buildings to collapse and dams to rupture, even before the eruption itself begins.

Threat of tsunamis (giant waves). Where explosive eruptions occur in the sea, the blast and the underwater collapse of a caldera generate huge sea waves, tens of meters (in rare cases, over 100 meters) high. Most of the 36,000 deaths attributed to the 1883 eruption of Krakatau were due not to ash or lava, but rather to tsunamis that slammed into nearby coastal towns (see Box 9.1).

Threat of gas. We have already seen that volcanoes erupt not only solid material, but also large quantities of gases like water, carbon dioxide, sulfur dioxide, and hydrogen sulfide. Usually the gas eruption accompanies the lava and ash eruption, with the gas contributing only a minor part of the calamity. But occasionally gas erupts alone and snuffs out life in its path without causing any other damage. Such an event occurred in 1986 near Lake Nyos in Cameroon, western Africa.

Lake Nyos is a small but deep lake filling the crater of an active hot-spot volcano in Cameroon. Though only 1 km across, the lake reaches a depth of over 200 m. Because of its depth, the cool bottom water of the lake does not mix with warm surface water, and for many years the bottom water remains separate from the surface water. During this time, carbon dioxide gas slowly bubbles out of cracks in the floor of the crater and dissolves in the cool bottom water. Apparently, by August 21, 1986, the bottom water had become supersaturated in carbon dioxide. On that day, perhaps triggered by a landslide or wind, the lake "burped" and, like an exploding seltzer bottle, expelled a forceful froth of CO_2 bubbles (together constituting 1 cubic km of gas). Because it is denser than air, this invisible gas flowed down the flank of the volcano and spread out over the countryside for about 23 km, before dispersing. While not toxic, carbon dioxide cannot provide oxygen for metabolism or oxidation (for this reason, it is the principal component of dry fire extinguishers). When the gas cloud engulfed the village of Nyos, it quietly put out the cooking fires and suffocated the sleeping inhabitants, most of whom died where they lay. The next morning, the landscape looked exactly as it had the day before, except for the lifeless bodies of 1,742 people and about 6,000 head of cattle (▶Fig. 9.24). Recently, engineers have been testing methods to de-gas the lake gradually, to avoid a similar disaster in the future.

Which threats are most dangerous? It's hard to compile statistics on how fatalities occur as a consequence of volcanic eruption. But a recent study has done just that and has produced some surprising results. Lava flows, though dramatic, actually cause only a small percentage of the fatalities, because the flows generally move slowly enough that

FIGURE 9.24 Cattle near Lake Nyos, Cameroon, fell where they stood, victims of a cloud of carbon dioxide.

people can get out of their way. The greatest number of fatalities (almost 30%) result from pyroclastic flows, because these can strike so fast that people cannot escape. Other leading causes of death are mudflows (about 15%), tsunamis (about 20%), and indirect causes (almost 25%). The last item in the list recognizes the fact that when eruptions blanket and kill crops, disrupt transportation, and destroy communities, they cause starvation and illness that can lead to death. All other effects of volcanoes (earthquakes, floods, gas, ash falls, lava) together account for about 10% of fatalities.

9.11 PROTECTION FROM VULCAN'S WRATH

Active, Dormant, and Extinct Volcanoes

In the geologic record, volcanoes come and go. For example, while a particular convergent plate boundary exists, a volcanic arc exists; but if subduction ceases, the volcanoes in the arc die and erode away. Even when alive, individual volcanoes erupt only intermittently. In fact, the average time between successive eruptions (the **recurrence interval**) ranges from a few years to a few centuries, and in some cases millennia.

Geologists refer to volcanoes that are erupting, have erupted recently, and are likely to erupt soon as **active volcanoes,** and distinguish them from **dormant volcanoes,** which have not erupted for hundreds to thousands of years but do have the potential to erupt again in the future. Volcanoes that were active in the past but have shut off entirely and will never erupt in the future are called **extinct volcanoes.** As examples, geologists consider Hawaii's Kilauea

who was an accomplished scientist as well as a statesman, couldn't resist seeking an explanation for this phenomenon, and soon learned that in June of 1783, a huge volcanic eruption had taken place in Iceland. He wondered if the "smoke" from the eruption had prevented sunlight from reaching the Earth, thus causing the cooler temperatures. Franklin reported this idea at a meeting, and by doing so, may well have been the first scientist ever to suggest a link between eruptions and climate.

Franklin's idea seemed to be confirmed in 1815, when Mt. Tambora in Indonesia exploded. Tambora's explosion ejected over 100 cubic km of ash and pumice into the air (compared with 1 cubic km for Mt. St. Helens). Ten thousand people were killed by the eruption and associated tsunami. Another 82,000 died of starvation. The sky became so hazy that stars dimmed by a full magnitude. Temperatures dipped so low in the Northern Hemisphere that 1816 became known as "the year without a summer." The unusual weather of that year left a permanent impact on Western culture. Memories of fabulous sunsets and the hazy glow of the sky may have inspired the luminous and atmospheric quality that made the landscape paintings of the English artist Joseph M. W. Turner so famous (►Fig. 9.27). Two English writers also documented the phenomenon. Lord Byron's 1816 poem "Darkness" contains the gloomy lines "The bright Sun was extinguish'd, and the stars / Did wander darkling in the eternal space . . . Morn came and went—and came, and brought no day"; and two years later, Mary Shelley, trapped in her house by bad weather, wrote *Frankenstein*, with its numerous scenes of gloom and doom.

Geoscientists have witnessed other examples of eruption-triggered coolness more recently. In the months following the 1883 eruption of Krakatau and the 1991 eruption of Pinatubo, global temperatures noticeably dipped.

Classical literature provides more evidence of the volcanic impact on climate. For example, Plutarch wrote around 100 C.E., "Among events of divine ordering there was . . . after Caesar's murder . . . the obscuration of the Sun's rays. For during all the year its orb rose pale and without radiance . . . and the fruits, imperfect and half ripe, withered away." Similar conditions appear to have occurred in China the same year, as described in records from the Han dynasty, and may be a consequence of volcanic eruption.

To study the effect of volcanic activity on climate even further in the past, geologists have studied ice from the glaciers of Greenland and Antarctica. Glacial ice has layers, each of which represents the snow that fell in a single year. Some layers contain concentrations of sulfuric acid, formed when SO_2 from volcanic gas dissolves in the water from which snow forms. These layers indicate years in which major eruptions occurred. Years in which ice contains acid correspond to years during which the thinness of tree rings elsewhere in the world indicates a cool growing season.

How can a volcanic eruption create these cooling effects? When a large explosive eruption takes place, fine ash and aerosols enter the stratosphere. It takes only about two weeks for the ash and aerosols to circle the planet. They stay suspended in the stratosphere for many months to years, because they are above the weather and do not get washed away by rainfall. The haze they produce causes cooler average temperatures, because it absorbs incoming visible solar radiation during the day but does not absorb the infrared radiation that rises from the Earth's surface at night. A Krakatau-scale eruption can lead to a drop in global average temperature of about 0.3° to 1°C, and, according to some calculations, a series of large eruptions over a short period of time could cause a global average temperature drop of 6°C.

Notably, the observed effect of volcanic eruptions on the climate provides a model with which to predict the consequences of a nuclear war. Researchers have speculated that so much dust and gas would be blown into the sky in the mushroom clouds of nuclear explosions that a "nuclear winter" would ensue.

FIGURE 9.27 The glowing sunset depicted in this 1840 painting by the English artist Joseph M. W. Turner was typical in the years following the 1815 eruption of Mt. Tambora in Indonesia.

9.13 VOLCANOES AND CIVILIZATION

Not all volcanic activity is bad. Over time, volcanic activity has played a major role in making the Earth a habitable planet. Eruptions and underlying igneous intrusions have produced the rock making up the Earth's crust, and gases emitted by volcanoes provided the raw materials from which the atmosphere and oceans formed. The black smokers surrounding vents along mid-ocean ridges may have served as a birthplace for life, and volcanic islands in the oceans have hosted populations whose evolution adds to the diversity of life on the planet. Volcanic activity continues to bring nutrients (potassium, sulfur, calcium, and phosphorous) from Earth's interior to the surface, and provides fertile soils that nurture plant growth. And in more recent time, people have exploited the mineral and energy resources generated by volcanic eruptions.

Volcanoes and people have lived in close association since the first human-like ancestors walked the Earth 3 million years ago. In fact, one of the earliest relics of human ancestors consists of footprints fossilized in a volcanic ash layer in East Africa. Since volcanic ash contains abundant nutrients that make crops prosper, people tend to populate volcanic regions. It's amazing how soon after an eruption a volcanic soil in a humid climate sprouts a cloak of green. Only twenty years after the eruption of Mt. St. Helens, new plants covered much of the affected area.

But, as we have seen, volcanic eruptions also pose a hazard. Eruptions may even lead to the demise of civilizations. The history of the Minoan people, who inhabited several islands in the eastern Mediterranean during the Bronze Age, illustrates this possibility. Beginning around 3000 B.C.E., the Minoans built elaborate cities and prospered. Then their civilization waned and disappeared (▶Fig. 9.28a). Geologists have discovered that the disappearance of the Minoans came within 150 years of a series of explosive eruptions of the Santorini volcano in 1645 B.C.E. Remnants of the volcano now constitute Thera, one of the islands of Greece. After a huge eruption, the center of the volcano collapsed into the sea, leaving only a steep-walled caldera (▶Fig. 9.28b). Archaeologists speculate that pyroclastic debris from the eruptions periodically darkened the sky, burying Minoan settlements and destroying crops. In addition, related earthquakes crumbled homes, and large tsunamis generated by the eruptions washed away Minoan seaports. Perhaps the Minoans took these calamities as a sign of the gods' displeasure, became demoralized, and left the region. Or, perhaps trade was disrupted, and bad times led to political unrest. Eventually the Mycenaeans moved in, bringing the culture that evolved into that of classical Greece. The Minoans, though, were not completely forgotten. Plato, in his *Dialogues,* refers to a lost city, home of an advanced civilization that bore many similarities to that of the Minoans. According to Plato, this city, which he named "Atlantis," disappeared beneath the waves of the sea. Perhaps this legend evolved from the true history of the Minoans, as modified by Egyptian scholars who passed it on to Plato.

Numerous cultures living along the Pacific Ring of Fire have evolved religious practices that are based on volcanic activity—no surprise, considering the awesome might of a volcanic eruption in comparison with the power of hu-

(a)

(b)

FIGURE 9.28 (a) Archaeologists have uncovered Minoan cities in the eastern Mediterranean, remnants of a culture that disappeared before the rise of classical Greece. (b) The cataclysmic eruption of the Santorini volcano in about 1645 B.C.E. may have contributed to the demise of the Minoan culture. All that is left of Santorini is a huge caldera, whose rim still lies above sea level, forming the island of Thera.

mans. In some cultures, this reverence took the form of sacrifice in hopes of preventing an eruption that could destroy villages and bury food supplies. In traditional Hawaiian culture, Pelé, goddess of the volcano, created all the major landforms of the Hawaiian Islands. She gouged out the craters that top the volcano mountains, her fits and moods bring about the eruptions, and her tears are the smooth, glassy lapilli ejected from the lava fountains.

9.14 VOLCANOES ON OTHER PLANETS

We conclude this chapter by looking beyond the Earth, for our planet is not the only one in the solar system to have hosted volcanic eruptions. We can see the effects of volcanic activity on our nearest neighbor, the Moon, just by looking up on a clear night. The broad darker areas of the Moon, the **mare** (after the Latin word for "sea"), consist of flood basalts that erupted over 3 billion years ago (▶Fig. 9.29a). They cover 17% of the lunar surface. Geologists propose that the flood basalts formed when huge meteors collided with the Moon, blasting out giant craters. The craters decreased the pressure in the Moon's mantle by so much that the mantle partially melted, generating hot and fluid basaltic magma that rose to the surface and filled the craters.

On Venus, about 22,000 volcanic edifices have been identified. Some of these even have caldera structures at their crests. Though no volcanoes currently erupt on Mars, the planet's surface displays a record of a spectacular volcanic past. The largest known mountain in the solar system, Olympus Mons (▶Fig. 9.30a), is an extinct shield volcano on Mars. The base of Olympus Mons is 600 km across, and its peak rises 25 km above the surrounding plains, making it three times as high as Mt. Everest. A caldera 65 km in diameter developed on its summit.

Active volcanism currently occurs on Io, one of the many moons of Jupiter. Cameras in the *Galileo* spacecraft have recorded huge volcanoes on Io in the act of spraying plumes of sulfur gas into space (▶Fig. 9.30b) and have tracked immense, moving lava flows. Different colors of erupted material make the surface of this moon resemble a pizza. What causes the heat that produces all the melt? Researchers have proposed that the volcanic activity is due to tidal power: Io moves in an elliptical orbit around the

FIGURE 9.29 The Mare of the Moon, the broad dark areas, are composed of flood basalts.

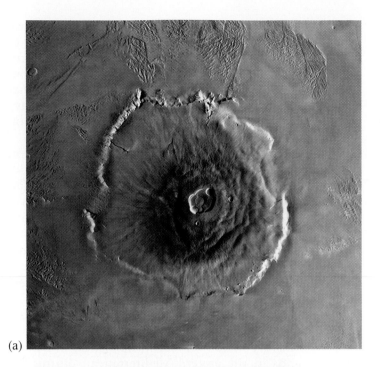

(a)

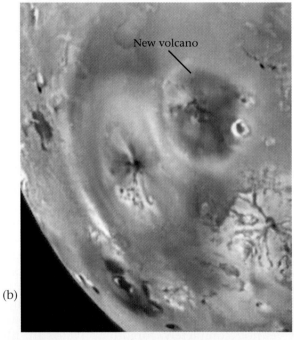

New volcano

(b)

FIGURE 9.30 (a) Satellites orbiting Mars have provided this digital image of Olympus Mons, an immense shield volcano. Notice the caldera at the summit. (b) A satellite image caught a volcano on Io, one of the moons of Jupiter, in the act of erupting. The bluish bubble is a cloud of erupting gas.

huge mass of Jupiter and near Jupiter's other, larger moons. The gravitational pull exerted by these objects alternately stretches and then squeezes Io, creating sufficient friction to keep Io's mantle hot.

THE VIEW FROM SPACE Subduction of the Pacific Ocean floor beneath Mexico causes partial melting in the Earth's mantle. The melting yields magma which rises into the crust of Mexico. Some magma makes it to the ground surface and erupts, forming a mound of ash and lava called a volcano. This example, Colima Volcano, has two eruptive centers. The northern one is older. A radial drainage network has formed on the flanks of the volcanoes.

CHAPTER SUMMARY

• Volcanoes are vents at which molten rock (lava), pyroclastic debris (ash, pumice, and fragments of volcanic rock), gas, and aerosols erupt at the Earth's surface. A hill or mountain created from the products of an eruption is also called a volcano.

• The characteristics of a lava flow depend on its viscosity, which in turn depends on its temperature and composition. Rhyolitic lavas tend to be more viscous than basaltic lavas.

• Basaltic lavas can flow great distances. Pahoehoe flows have smooth, ropy surfaces, while a'a' flows have rough, rubbly surfaces. Andesitic and rhyolitic lava flows tend to pile into mounds at the vent.

• Pyroclastic debris includes powder-sized ash, marble-sized lapilli, and apple- to refrigerator-sized blocks. Some falls from the air, whereas some forms glowing avalanches that rush down the side of the volcano.

• Eruptions may occur at a volcano's summit or from fissures on its flanks. The summit of an erupting volcano may collapse to form a bowl-shaped depression called a caldera.

• A volcano's shape depends on the type of eruption. Shield volcanoes are broad, gentle domes. Cinder cones are steep-sided, symmetrical hills composed of tephra. Composite volcanoes can become quite large, and consist of alternating layers of pyroclastic debris and lava.

• The type of eruption depends on several factors, including the lava's viscosity and gas content. Effusive eruptions produce only flows of lava, while explosive eruptions produce clouds and flows of pyroclastic debris.

• Different kinds of volcanoes form in different plate-tectonic settings.

• Volcanic eruptions pose many hazards: lava flows overrun roads and towns, ash falls blanket the landscape, pyroclastic flows incinerate towns and fields, landslides and lahars bury the land surface, earthquakes topple structures and rupture dams, tsunamis wash away coastal towns, and invisible gases suffocate nearby people and animals.

• Eruptions can be predicted through changes in heat flow, changes in shape of the volcano, earthquake activity, and the emission of gas and steam.

• We can minimize the consequences of an eruption by avoiding construction in danger zones and by drawing up evacuation plans. In a few cases, it may be possible to divert flows.

• Volcanic gases and ash, erupted into the stratosphere, may keep the Earth from receiving solar radiation and thus may affect climate. Eruptions bring nutrients from inside the Earth to the surface, and eruptive products may evolve into fertile soils.

• Immense mare of flood basalts cover portions of the Moon. The largest known volcano in the solar system, Olympus Mons, towers over the surface of Mars. Satellites have documented eruptions on Io, a moon of Jupiter.

KEY TERMS

a'a' (p. 244)
active volcanoes (p. 267)
aerosols (p. 250)
blocks (p. 248)
bombs (p. 248)
caldera (p. 251)
cinder cone (p. 254)
columnar jointing (p. 246)
crater (p. 251)
dormant volcanoes (p. 267)
effusive eruption (p. 255)
explosive (pyroclastic) eruption (p. 255)
extinct volcanoes (p. 267)
fissure (p. 250)

flood basalt (p. 261)
ignimbrite (p. 248)
lahar (p. 248)
lapilli (p. 248)
lava dome (p. 246)
lava flow (p. 244)
lava tube (p. 244)
magma chamber (p. 250)
mare (p. 272)
pahoehoe (p. 244)
phreatomagmatic eruption (p. 258)
pyroclastic debris (p. 246)
pyroclastic flow (p. 248)
recurrence interval (p. 267)

Ring of Fire (p. 263)
shield volcano (p. 251)
stratovolcano (composite
    volcano) (p. 254)
tephra (p. 248)
tuff (p. 248)

vesicles (p. 250)
volcanic ash (p. 246)
volcanic danger assessment
    map (p. 268)
volcano (p. 242)

REVIEW QUESTIONS

1. Describe the three different kinds of material that can erupt from a volcano.

2. Describe different types of lava flows.

3. Describe the differences between a pyroclastic flow and a lahar.

4. How is a crater different from a caldera?

5. Describe the differences among shield volcanoes, stratovolcanoes, and cinder cones. How are these differences explained by the composition of their lavas and other factors?

6. Why do some volcanic eruptions consist mostly of lava flows, while others are explosive and have no flow?

7. Explain how viscosity, gas pressure, and the environment affect the eruptive style of a volcano.

8. Describe the activity in the mantle that leads to hot-spot eruptions.

9. How do continental rift eruptions form flood basalts?

10. Contrast an island volcanic arc with a continental volcanic arc.

11. How have volcanoes affected civilization?

12. Identify some of the major volcanic hazards, and explain how they develop.

13. How do scientists predict volcanic eruptions?

14. Explain how steps can be taken to protect people from the effects of eruptions.

15. Describe the nature of volcanism on the other planets and moons in the solar system.

SUGGESTED READING

Cattermole, P. 1996. *Planetary Volcanism,* 2nd ed. Chichester, England: John Wiley & Sons.

Chester, D. 1993. *Volcanoes and Society.* London: Edward Arnold.

De Boer, J. Z., and D. T. Sanders. 2001. *Volcanoes in Human History: The Far-Reaching Effects of Major Eruptions.* Princeton: Princeton University Press.

Decker, R. W., and B. B. Decker. 1997. *Volcanoes.* New York: W. H. Freeman.

Fisher, R. V., G. Heiken, and J. B. Hulen. 1997. *Volcanoes: Crucibles of Change.* Princeton: Princeton University Press.

Francis, P. 1993. *Volcanoes: A Planetary Perspective.* Oxford, England: Clarendon Press.

Grove, N. 1992. Crucibles of creation. *National Geographic* 182: 5–41.

Krakaner, J. 1996. Geologists worry about dangers of living "under the volcano." *Smithsonian* (July): 33–125.

McGuire, B., C. Kilburn, and J. Murray. 1995. *Monitoring Active Volcanoes:* London: UCL Press.

Pinna, M. 2002. Etna ignites. *National Geographic* (February): 68–87.

Scarth, A. 1994. *Volcanoes.* London: UCL Press.

Schminck, H. U. 2004. *Volcanism.* Berlin: Springer.

Sigurdsson, H., et al., eds. 2000. *Encyclopedia of Volcanoes.* San Diego: Academic Press.

Stone, R., et al., contributors. 2003. Volcanology special section. *Science* 299 (March 28): 2015–30.

CHAPTER 10

A Violent Pulse: Earthquakes

10.1 INTRODUCTION

As the morning of January 17, 1994, approached, residents of Northridge, a suburb near Los Angeles, slept peacefully in anticipation of the Martin Luther King Day holiday. But beneath the quiet landscape, a disaster was in the making. For many years, the imperceptibly slow movement of the Pacific Plate relative to the North American Plate had been bending the rocks making up the California crust. But like a stick that you flex with your hands, rock can bend only so far before it snaps (▶Fig. 10.1a, b). Under California, the "snap" happened at 4:31 A.M., 10 km down; it sent pulses of energy racing through the crust at an average speed of 11,000 km (7,000 miles) per hour, ten times the speed of sound.

When the first energy pulse, or "shock," reached the Earth's surface, it pushed the ground up with a violent jolt. Sleepers bounced off their beds, homes slipped off their foundations, and freeway bridges disconnected from their supports. As more and more shocks arrived, the ground bucked up and down and side to side, causing walls to sway and cave in, roofs to fall, and rail lines to buckle (▶Fig. 10.2). Early risers brewing coffee in their kitchens tumbled to the floor, under attack by dishes and cans catapulting out of cupboards. Trains

An oblique air view, generated by a computer, of the San Francisco region. The red lines are faults on which earthquakes have occurred and will in the future. The complex topography of this region results from fault movement.

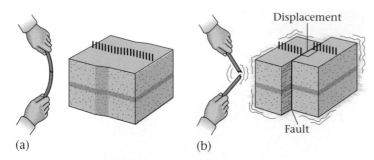

(a)   (b)

FIGURE 10.1 Most earthquakes happen when rock in the ground first bends slightly and then suddenly snaps and breaks, like a stick you flex in your hands. (a) Before an earthquake, the crust bends (the amount of bending is greatly exaggerated here). (b) When the crust breaks, sliding suddenly occurs on a fault, generating vibrations.

FIGURE 10.2 In the 1994 Northridge, California, earthquake, this building facade tore free of its supports and collapsed.

careered off their tracks, and steep hill slopes bordering the coast gave way, dumping heaps of rock, mud, and broken houses onto the beach below. Ruptured gas lines fed fires that had been ignited by water heaters and sparking wires in the rubble. Then, forty seconds after it started, the motion stopped, and the shouts and sirens of rescuers replaced the crash and clatter of breaking masonry and glass. A major **earthquake**—an episode of ground shaking—had occurred.

Earthquakes have affected the Earth since the formation of its solid crust almost 4 billion years ago. Most are a consequence of lithosphere-plate movement; they punctuate each step in the growth of mountains, the drift of continents, and the opening and closing of ocean basins. And, perhaps of more relevance to us, earthquakes have afflicted human civilization since the construction of the first village and they have directly caused the deaths of over 3.5 million people during the past two millennia (see Table 10.1). Ground shaking, giant waves, landslides, and fires during earthquakes turn cities to rubble. The destruction caused by some earthquakes may even have changed the course of civilization.

What does an earthquake feel like? When you're in one, time seems to stand still, so even though most earthquakes take less than a minute, they seem much longer. Because of the lurching, bouncing, and swaying of the ground and buildings, people become disoriented, panicked, and even seasick. Some people recall hearing a dull rumbling or a series of dull thumps, as well as crashing and clanging. Earthquakes may even shake dust into the air, creating a fine, fog-like mist.

Earthquakes are a fact of life on planet Earth: almost 1 million detectable earthquakes happen every year. Fortunately, most cause no damage or casualties, because they are too small or they occur in unpopulated areas. But a few hundred earthquakes per year rattle the ground sufficiently to damage buildings and injure their occupants, and every five to twenty years, on average, a great earthquake triggers

a horrific calamity. What geologic phenomena generate earthquakes? Why do earthquakes take place where they do? How do they cause damage? Can we predict when earthquakes will happen, or even prevent them from happening? These questions have puzzled **seismologists** (from *seismos,* Greek for "shock" or "earthquake"), geoscientists who study earthquakes, for decades. In this chapter, we seek some of the answers, answers that can help those living in earthquake-prone regions to cope.

10.2 FAULTS AND THE GENERATION OF EARTHQUAKES

Ancient cultures offered a variety of explanations for **seismicity** (earthquake activity), most of which involved the action or mood of a giant animal or god. For example, in Japanese folklore, a giant catfish, Namazu, is said to have lived in the mud below the surface of the ground (►Fig. 10.3). If the gods did not restrain him, he would thrash about and shake the ground. In Indian folklore, earthquakes happened when one of eight elephants holding up the Earth shook its head, and in Siberia, earthquakes were thought to happen when a dog hauling the Earth in a sled stopped to scratch. Native American cultures of the West Coast thought earthquakes were caused by arguments among the turtles holding up the Earth. Scientific study suggests that seismicity can occur for several reasons, including

- the sudden formation of a new **fault** (a fracture on which sliding occurs),

TABLE 10.1 Some Notable Earthquakes

Year	Location	Number of Deaths
2003	Bam, Iran	41,000
2001	Bhuj, India	20,000
1999	Calaraca/Armenia, Colombia	2,000
1999	Izmit, Turkey	17,000
1995	Kobe, Japan	5,500
1994	Northridge, California	51
1990	Western Iran	50,000
1989	Loma Prieta, California	65
1988	Spitak, Armenia	24,000
1985	Mexico City	9,500
1983	Turkey	1,300
1978	Iran	15,000
1976	T'ang-shan, China	255,000
1976	Caldiran, Turkey	8,000
1976	Guatemala	23,000
1972	Nicaragua	12,000
1971	Los Angeles	50
1970	Peru	66,000
1968	Iran	12,000
1964	Anchorage, Alaska	131
1963	Skopje, Yugoslavia	1,000
1962	Iran	12,000
1960	Agadir, Morocco	12,000
1960	Southern Chile	6,000
1948	Turkmenistan, USSR	110,000
1939	Erzincan, Turkey	40,000
1939	Chillan, Chile	30,000
1935	Quetta, Pakistan	60,000
1932	Gansu, China	70,000
1927	Tsinghai, China	200,000
1923	Tokyo, Japan	143,000
1920	Gansu, China	180,000
1915	Avezzano, Italy	30,000
1908	Messina, Italy	160,000
1906	San Francisco	500
1898	Japan	22,000
1886	Charleston, South Carolina	60
1866	Peru and Ecuador	25,000
1811–12	New Madrid, Missouri (3 events)	few
1783	Calabria, Italy	50,000
1755	Lisbon, Portugal	70,000
1556	Shen-shu, China	830,000

FIGURE 10.3 Painting depicting the Japanese legend of Namazu, the giant catfish whose thrashings were thought to cause earthquakes.

- sudden slip on an existing fault,
- a sudden change in the arrangement of atoms in the minerals of rock,
- movement of magma in a volcano,
- the explosion of a volcano,
- giant landslides,
- a meteorite impact, or
- underground nuclear-bomb tests.

As we learned in Chapter 2, the place in the Earth where rock ruptures and slips, or the place where an explosion occurs, is the **hypocenter** (or **focus**) of the earthquake. Energy radiates from the hypocenter. The point on the surface of the Earth that lies directly above the hypocenter is the **epicenter** (▶Fig. 10.4).

The formation and movement of faults cause the vast majority of destructive earthquakes, so typically the hypocenter of an earthquake lies on a fault plane (the surface of the fault). Thus, we'll begin our investigation with a look at how faults develop and why their movement generates earthquakes.

Faults in the Crust

Faults are fractures on which slip or sliding occurs (see Chapter 11). They can be pictured simplistically as planes that cut through the crust. Some faults are vertical, but most slope at an angle. Nineteenth-century miners who encountered faults in mine tunnels referred to the rock mass above a sloping fault plane as the "hanging wall," because it

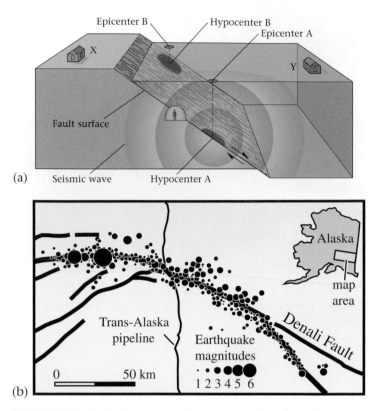

FIGURE 10.4 (a) The energy of an earthquake radiates from the hypocenter (focus), the place underground where rock has suddenly broken. The point on the ground surface (i.e., on a map) directly above the hypocenter is the epicenter. During a single earthquake, only part of a fault may slip. House X is on the footwall, and house Y is on the hanging wall. The miner excavating a tunnel along the fault has the hanging wall over his head and the footwall under his feet. (b) A simplified map of the Denali National Park region, Alaska, shows the trace of the Denali fault and epicenters of earthquake events along the fault during a period in late 2002. The diameter of a circle represents the size (magnitude) of the shock it represents.

hung over their heads, and the rock mass below the fault plane as the footwall, because it lay beneath their feet (Fig. 10.4). The miners described the direction in which rock masses slipped on a fault by specifying the direction that the hanging wall moved in relation to the footwall, and we still use these terms today. When the hanging wall slips down the slope of the fault, it's a "normal fault," and when the hanging wall slips up the slope, it's a "reverse fault" if steep and a "thrust fault" if shallowly sloping (▶Fig. 10.5a–d). "Strike-slip faults" have near-vertical planes on which slip occurs parallel to an imaginary horizontal line, called a strike line, on the fault plane—no up or down motion takes place here (▶Fig. 10.5e). In Chapter 11, we will introduce other types of faults.

Normal faults form in response to stretching or extension of the crust, reverse or thrust faults develop in response

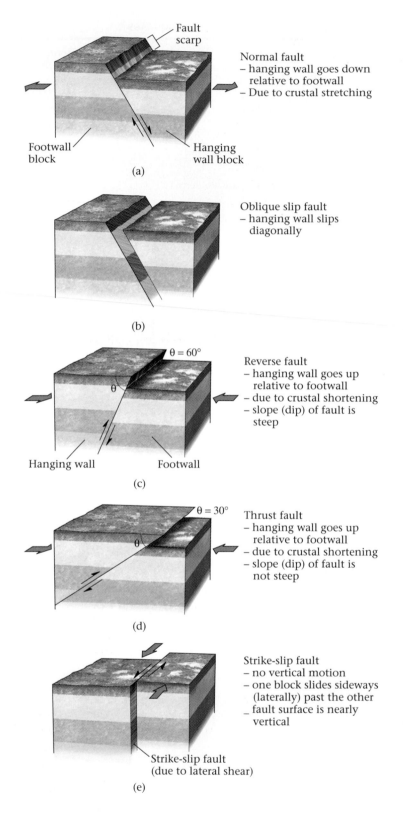

FIGURE 10.5 The basic types of faults. Note that faults are distinguished from each other by the nature of the slip. (a) Normal fault, (b) Oblique-slip fault, (c) Reverse fault, (d) Thrust fault, (e) Strike-slip fault.

(a)

Displacement

Fault trace

What a geologist sees

(b)

(c)

FIGURE 10.6 (a) This wooden fence was built across the San Andreas Fault. During the 1906 San Francisco earthquake, slip on the fault broke and offset the fence; the displacement of the fence indicates that the fault is strike-slip, as we see no evidence of up or down motion. The rancher quickly connected the two ends of the fence so no cattle could escape. (b) The amount the fence was offset indicates the displacement on the fault. (c) A photo taken looking down from a helicopter, showing the trace of the strike-slip fault that ruptured the ground surface during the Hector Mine earthquake in the southern California desert in 1999. Note the cracks in the ground and the small ridges and depressions, and how the fault offsets the dirt road.

to squeezing (compression) and shortening of the crust, and strike-slip faults form where one block of crust slides past another laterally. By measuring the distance between the two ends of a distinctive sedimentary bed or igneous dike that's been offset by a fault, geologists define the **displacement,** the amount of slip, on the fault (▶Fig. 10.6a, b).

Faults are found almost everywhere—but don't panic! Not all of them are likely to be the source of earthquakes. Faults that have moved recently or are likely to move in the near future are called "active faults" (and if they generate earthquakes, news media sometimes refer to them as "earth-

FIGURE 10.7 A hidden (or "blind") fault does not intersect the ground surface. Rather, it dies out at the fault tip. In this example, a fold (curving beds) has developed in response to slip on the fault.

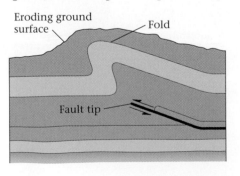

Eroding ground surface

Fold

Fault tip

quake faults"); faults that last moved in the distant past and probably won't move again in the near future (but are still recognizably faults because of the displacement across them) are called "inactive faults." Some faults have been inactive for billions of years.

The intersection between a fault and the ground surface is a line we call the **fault trace,** or "fault line" (Fig. 10.6b). In places where an active normal or reverse fault intersects the ground, one side of the fault moves vertically with respect to the other side, creating a small step called a **fault scarp** (Fig. 10.5a). Active strike-slip faults tend to form narrow bands of low ridges and narrow depressions, because they break up the ground when they move (▶Fig. 10.6c). Not all active faults, though, intersect the ground surface. Those that don't are called "blind faults" or "hidden faults" (▶Fig. 10.7). Many fault traces that we see on maps represent inactive faults. The portion of the fault that we now see may once have been far below the surface of the Earth, becoming visible only because of erosion.

Formation of Faults, Friction, and Stick-Slip

As we learned in Chapter 8, "stress" is the push, pull, or shear that a material feels when subjected to a force. In a drawing, we can represent stress by arrows showing the direction in which the stress acts. Objects can change shape

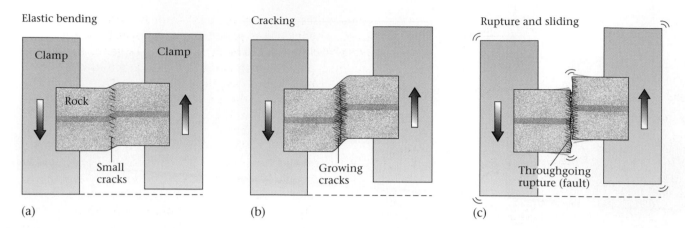

(a) Elastic bending

(b) Cracking

(c) Rupture and sliding

Clamp   Clamp

Rock

Small cracks

Growing cracks

Throughgoing rupture (fault)

FIGURE 10.8 The stages in the development of a fault can be illustrated by breaking a rock block gripped at each end by a clamp. (a) As one clamp moves relative to the other (indicated by arrows), the rock feels a stress and begins to bend elastically and develop strain. (b) If bending continues, the rock begins to crack, and then the cracks grow and start to connect. (c) When the cracks connect sufficiently to make a throughgoing fracture, the rock ruptures into two pieces. The instant one piece slides past another, the rupture becomes a fault.

in response to the application of a stress; this change in shape is referred to as a "strain." (We will learn more about stress and strain in Chapter 11.)

What does stress have to do with fault formation? Stress causes faulting. To see why, imagine that you grip each side of a brick-shaped rock with a clamp (▶Fig. 10.8a–c). Now, suppose you apply a stress to the rock by moving one side upward and the other side downward. As soon as the movement begins, the rock begins to change shape (a line traced across the middle of the brick bends into an S-like curve), but it doesn't break, and if you were to remove the stress at this stage, the rock would return to its original shape, just as a stretched rubber band returns to its original shape when you let go. A change in the shape of an object that disappears when stress is removed is called an "elastic strain". If you apply a larger stress, so that the sides of the rock shift further, the rock starts to crack. First, a series of small cracks form, but as movement continues, the cracks grow and connect to one another to create a fracture that cuts across the entire block of rock. The instant this thoroughgoing fracture forms, the rock on one side of the fracture slides past the rock on the other side, and the fracture becomes a fault. And as soon as the fault forms, the once-bent line across the middle separates into two segments that no longer align with each other, and the stress in the rock decreases (i.e., there is a "stress drop"). Also, the elastic strain disappears.

If you slide a book across a tabletop, it eventually slows down and stops because of friction. Similarly, once a fault forms and rock starts to slip, it doesn't slip forever because of friction. Friction, the resistance to sliding on a surface, regulates movement. Friction occurs because,

in reality, no surface can be perfectly smooth—rather, all surfaces contain little bumps or depressions, called "asperities". As one surface moves against another, the bumps on one surface snag on the bumps of the opposing surface, acting like little anchors that slow down and eventually stop the movement (▶Fig. 10.9a, b).

Once friction stops movement on a fault, the stress begins to increase again. Eventually, the magnitude of stress becomes so great that friction can no longer prevent movement, and the instant this happens, the fault slips once again and the stress drops once again. Notably, the stress re-

FIGURE 10.9 At a microscopic scale, real surfaces have protrusions and indentations, also known as asperities. (a) Before movement, the protrusions in rock lock together, causing friction that prevents sliding. One block is "anchored" to the other. (b) Like a boat whose anchor cable snaps, when the protrusions break off, sliding can take place.

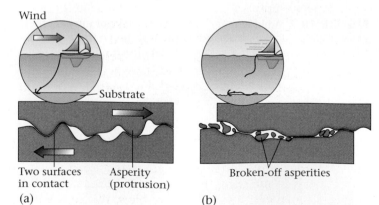

Wind

Substrate

Two surfaces in contact

Asperity (protrusion)

(a)

Broken-off asperities

(b)

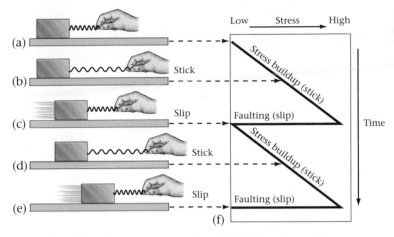

FIGURE 10.10 The concept of stick-slip behavior can be illustrated by means of a model consisting of a heavy block with a spring attached. (a) The block starts at rest. (b) We pull on the spring and the spring stretches, but friction keeps the block in place. Stretching the spring builds up stress. (c) Suddenly, the block slides along the tabletop (analogous to the slip on a fault), and the spring relaxes (analogous to a drop in stress). (d) We pull on the spring again, and the stress builds up once more. (e) The block slides again, and the stress drops. (f) The graph shows how stress gradually builds up and then suddenly drops during stick-slip behavior.

quired to overcome friction and reactivate an existing fault tends to be less than the stress that's needed to fracture intact rock and form a new fault; in effect, once a fault has formed, it's like a permanent scar that is weaker than the surrounding crust. Thus, existing faults may reactivate many times.

In sum, between faulting events, stress builds up. In some cases, the stress causes intact rock to rupture and a new fault to form. In other cases, stress overcomes friction on an existing fault and the fault slips again. In either case, after a faulting event, stress drops and elastic strain stored in rock decreases—the rock "rebounds" so layers near the fault are no longer bent. Friction stops the movement and the fault locks, until stress builds up enough again to cause slip. This overall image of how earthquakes occur is now known as **elastic-rebound theory,** and the start-stop movement on a fault as **stick-slip behavior** (▶Fig. 10.10a–f). When a fault slips, the whole fault does not move at once; rather, the slipped area starts at a certain point and then grows outward (▶Fig. 10.11a).

How Faulting Generates Earthquakes

How does the formation of a fault generate an earthquake? The moment a fault slips, the rock around the fault feels a sudden push or pull. Like a hammer blow, this movement sends pulses of energy (shocks) into the surrounding rock. As the energy pulses pass, the rock moves back and forth like a vibrating bell. The series of shocks generated by the sudden slip on the fault and the subsequent vibration create the shaking we feel as an earthquake. The bigger the amount of slip and the greater the amount of rock that moves, the greater the size of the vibrations and, therefore, the larger the earthquake.

A major earthquake may be preceded by smaller ones, called **foreshocks,** which possibly represent the development of the smaller cracks that will eventually link up to form a major rupture. Smaller earthquakes that follow a major earthquake, called **aftershocks,** may occur for days to several weeks. The largest aftershock tends to be ten times smaller than the main shock; most are much smaller. Aftershocks happen because the movement of rock during the main earthquake creates new stresses, which may be large enough to reactivate small portions of the main fault or to activate small, nearby faults.

The Amount of Slip on Faults, and Ground Distortion due to Earthquakes

How much of a fault slips during an earthquake? The answer depends on the size of the earthquake: the larger the earthquake, the larger the slipped area. For example, the major earthquake that hit San Francisco, California, in 1906 ruptured a 430-km-long (measured parallel to the Earth's surface) by 15-km-high (measured perpendicular to the Earth's surface) segment of the San Andreas Fault. The fault slip (the amount of sliding along the fault) died out at the ends of the segment, but toward the middle it reached 7 m, in a strike-slip sense. Slip on a thrust fault caused the 1964 "Good Friday" earthquake in southern Alaska: at depth in the Earth, slip reached a maximum of 12 m, and at the Earth's surface, the faulting uplifted the ground over a 500,000-square-km area by as much as 2 m. Smaller earthquakes, like the one that hit Northridge in 1994, resulted in about 0.5 m slip on a break that was about 5 km long and 5 km wide. The smallest-felt earthquakes (which rattle the dishes but not much more) represent displacements measured in millimeters to centimeters.

Because of the displacement that takes place on faults during an earthquake, the ground surface in the vicinity of the epicenter may undergo a change in shape. For example, slip on a thrust fault may cause a region to wrap upwards, even if the fault plane itself does not cut the ground surface. Recently, researchers have developed a new technique, called "Interferometric Synthetic Aperture Radar" (abbreviated InSAR), to help detect subtle ground-surface distortion associated with earthquakes. To create an InSAR image, a satellite uses radar to make a precise map of ground elevation of a region at two different times (weeks to years apart). A computer compares the two images and

Faulting in the Crust

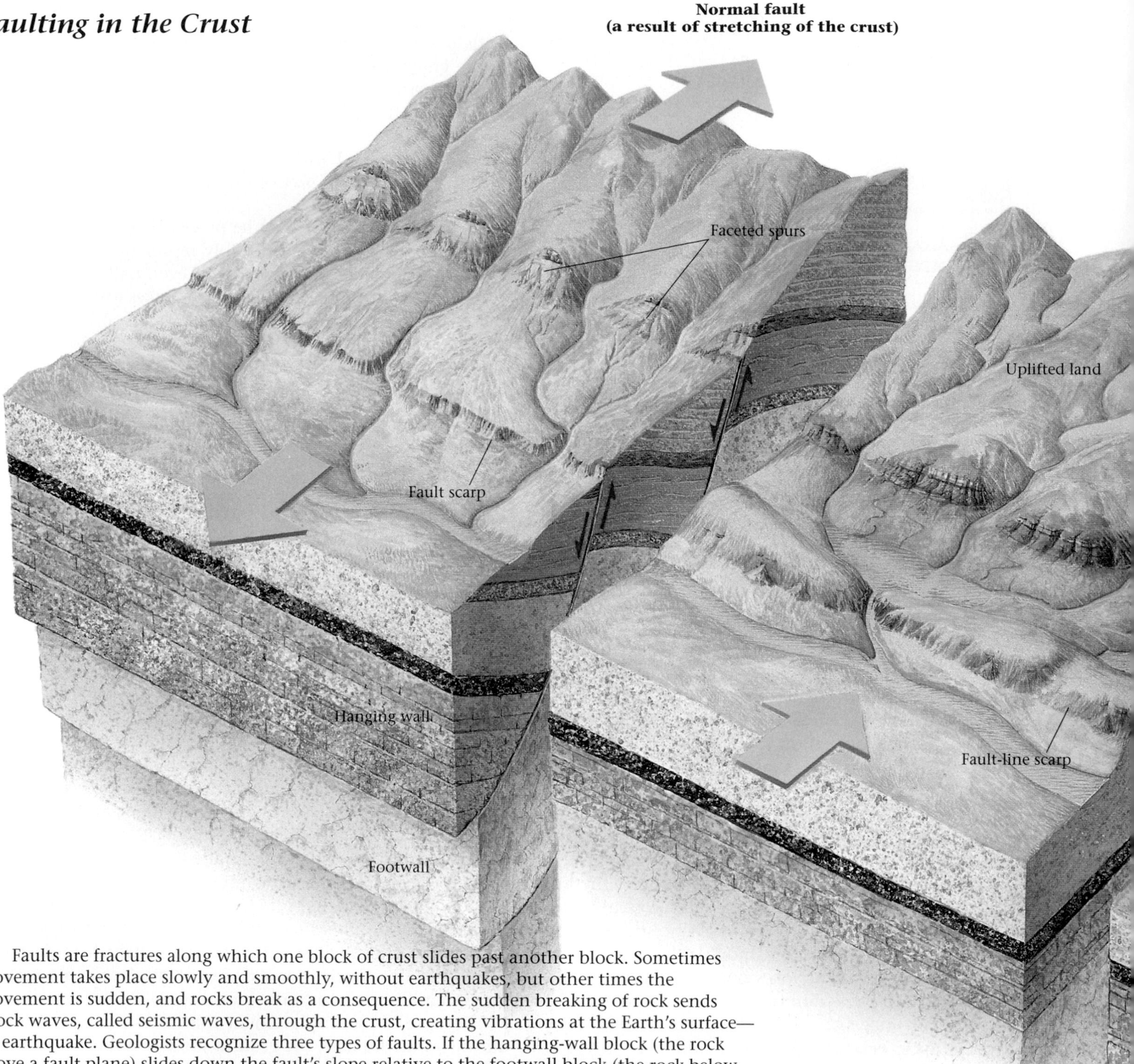

Normal fault
(a result of stretching of the crust)

Faceted spurs

Uplifted land

Fault scarp

Hanging wall

Fault-line scarp

Footwall

Faults are fractures along which one block of crust slides past another block. Sometimes movement takes place slowly and smoothly, without earthquakes, but other times the movement is sudden, and rocks break as a consequence. The sudden breaking of rock sends shock waves, called seismic waves, through the crust, creating vibrations at the Earth's surface—an earthquake. Geologists recognize three types of faults. If the hanging-wall block (the rock above a fault plane) slides down the fault's slope relative to the footwall block (the rock below the fault plane), the fault is a normal fault. (Normal faults form where the crust is being stretched apart, as in a continental rift.) If the hanging-wall block is being pushed up the slope of the fault relative to the footwall block, then the fault is a reverse fault. (Reverse faults develop where the crust is being compressed or squashed, as in a collisional mountain belt.) If one block of rock slides past another and there is no up or down motion, the fault is a strike-slip fault. Strike-slip fault planes tend to be nearly vertical.

If a fault displaces the ground surface, it creates a ledge called a fault scarp. Sometimes we can identify the trace (or line) of a fault on the land surface because the rock of the hanging wall has a different resistance to erosion than the rock of the footwall; a ledge formed along this line due to erosion is a fault-line scarp. Where fault scarps cut a system of rivers and valleys, the ridges are truncated to make triangular facets. Strike-slip faults may offset ridges, streams, and orchards sideways. If there is a slight extension along the fault, the land surface sinks, and a sag pond develops.

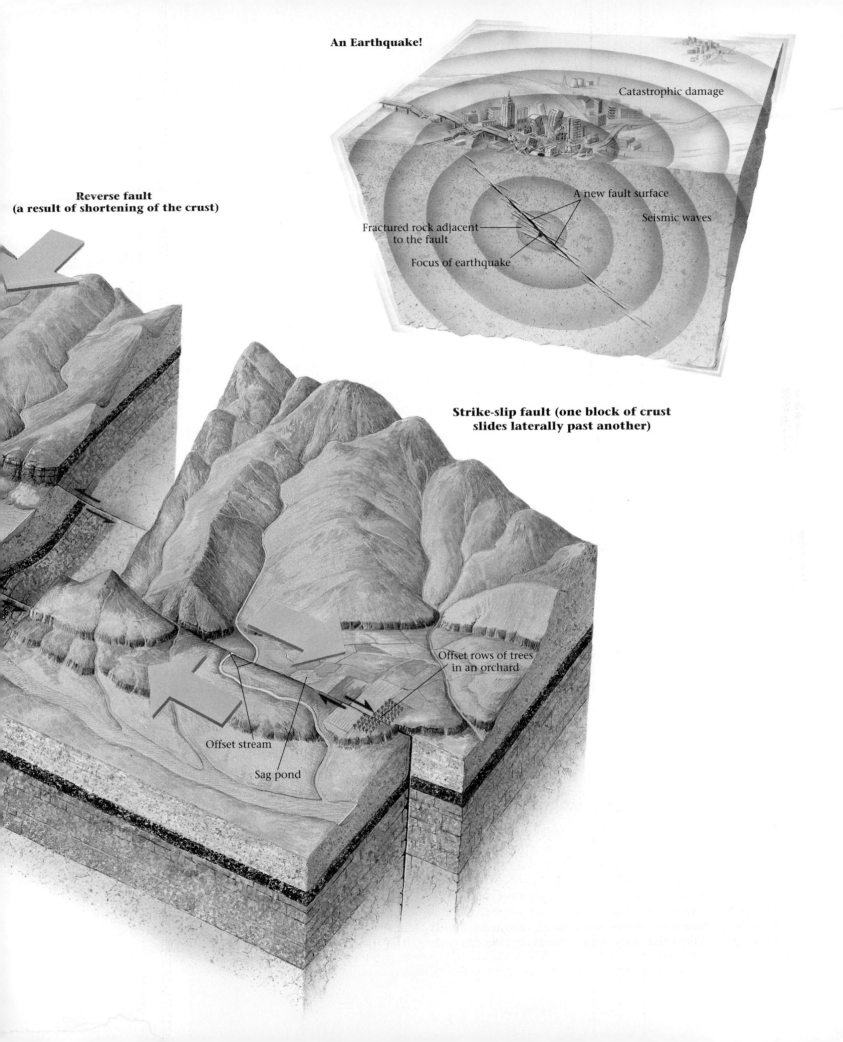

An Earthquake!

Catastrophic damage

A new fault surface

Seismic waves

Fractured rock adjacent to the fault

Focus of earthquake

**Reverse fault
(a result of shortening of the crust)**

**Strike-slip fault (one block of crust
slides laterally past another)**

Offset rows of trees
in an orchard

Offset stream

Sag pond

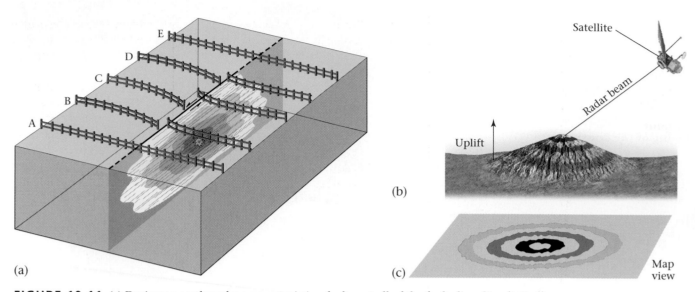

FIGURE 10.11 (a) During an earthquake on a preexisting fault, not all of the fault slips. Simplistically, slip starts at the hypocenter, and then the slipped area grows outward; on a large fault this growth takes tens of seconds. In this example, the slipped area on a strike-slip fault intersects the ground surface; fences beyond the end of this intersection have not been offset, whereas fences near the epicenter have been offset the most. (b) A satellite can use radar to map uplift of the ground related to faulting. (c) An InSAR map of the hill. Dark bands can be thought of as contour lines.

detects differences in elevation as small as the wavelength of radar energy. A printout of the result portrays these differences as color bands which indicate the change in elevation between the time the first image was taken and the time the second image was taken (▶Fig. 10.11b, c). Each band represents a certain amount of change in elevation.

While the cumulative movement on a fault during a human life span may not amount to much, over geologic time the cumulative movement becomes significant. For example, if earthquakes occurring on a reverse fault cause 1 cm of uplift over ten years, on average, the fault's movement will yield 1 km of uplift after 1 million years. Thus, earthquakes mark the incremental movements that create mountains. To take another example, movement on the San Andreas Fault in California averages around 6 cm per year. As a result, Los Angeles, which is to the west of the fault, will move northward by 6,000 km in 100 million years.

Can Faults Slip without Earthquakes?

When a material subjected to stress cracks and fractures, we say that it has undergone "brittle deformation." For example, a glass plate shattering on the floor is a type of brittle deformation. Similarly, faulting that generates earthquakes represents brittle deformation. Generally, rock must be fairly cool and must be stressed fairly quickly to behave in a brittle way. If a rock is warm or weak, or stress builds slowly, it can bend and flow without breaking. We call such

behavior "ductile deformation." If the glass plate were heated to a high temperature, it could be bent in a ductile manner, like chewing gum. Ductile deformation does not cause earthquakes.

Because the temperature of the Earth increases with depth, most brittle deformation and, therefore, earthquake-generating faulting in continental crust occurs in the upper 15–20 km of the crust. At greater depths, shear and movement can take place, but they are accomplished by means of ductile deformation. In oceanic plates, earthquake faulting happens even in plates that have been subducted.

In some cases, movement on faults in the upper 15–20 km of the crust takes place slowly and steadily, without generating earthquakes. When movement on a fault happens without generating earthquakes, we call the movement **fault creep.** Seismologists do not completely understand fault creep, but speculate that it occurs in particularly weak rock, which can change shape without breaking or can slip smoothly without creating shock waves.

10.3 SEISMIC WAVES

How does the energy emitted at the hypocenter of an earthquake travel to the surface or even pass through the entire Earth? Like other kinds of energy, earthquake energy travels in the form of waves. We call these waves **seismic** waves (or "earthquake waves"). You feel them if you touch one end of a brick and tap the other end with a hammer—the energy of the hammer blows travels to your fingertip in the form of waves.

Seismologists distinguish among different types of seismic waves based on where and how the waves move. **Body waves** pass through the interior of the Earth (i.e., within the body of the Earth), while **surface waves** travel along the Earth's surface. Waves in which particles of material move back and forth parallel to the direction in which the wave itself moves are called **compressional waves.** As a compressional wave passes, the material first compresses (or squeezes) together, then dilates (or expands). To see this kind of motion in action, push on the end of a spring and watch as the little pulse of compression moves along the length of the spring. Waves in which particles of material move back and forth perpendicular to the direction in which the wave itself moves are called **shear waves.** To see shear-wave motion, jerk the end of a rope up and down and watch how the up-and-down motion travels along the rope. With these concepts in mind, we can define four basic types of seismic waves (►Fig. 10.12a–f):

- **P-waves** (P stands for "primary") are compressional body waves.

- **S-waves** (S stands for "secondary") are shear body waves.

- **R-waves** (R stands for Rayleigh, the name of a physicist) are surface waves that cause the ground to ripple up and down.

- **L-waves** (L stands for Love, the name of a seismologist) are surface waves that cause the ground to ripple back and forth, creating a snake-like movement.

P-waves travel the fastest and thus arrive first. S-waves travel more slowly, at about 60% of the speed of P-waves, so they arrive later. Surface waves (R- and L-waves) are the slowest of all.

Friction absorbs energy as waves pass through a material, and waves bounce off layers and obstacles in the Earth, so the amount of energy carried by seismic waves decreases the farther they travel. People near the epicenter of a large earthquake may be thrown off their feet, but those 100 km away barely feel it. Similarly, an earthquake caused by slip on a fault deep in the crust causes less damage than one caused by slip on a fault near the surface.

10.4 MEASURING AND LOCATING EARTHQUAKES

Most news reports about earthquakes provide information on the "size" and "location" of an earthquake. What does this information mean, and how do we obtain it? What's the difference between a great earthquake and a minor one? How do seismologists locate an epicenter? Understanding how a seismograph works and how to read the information it provides will allow you to answer these questions.

Seismographs and the Record of an Earthquake

When, in 1889, a German physicist realized that a pendulum in his lab had moved in reaction to a deadly earthquake that had occurred in Japan, his observation confirmed speculations that earthquake energy can pass through the planet. On reading of this discovery, other researchers saw a

---

Moonquakes

BOX 10.1
THE REST OF THE STORY

When the *Apollo* astronauts landed on the Moon in the 1960s and 1970s, they left behind seismic instruments that could measure moonquakes, shaking events on the Moon. The seismic instruments found that moonquakes happen far less often than earthquakes (only about 3,000 a year) and are very small. Geologists were not surprised, because plate movement does not occur on the Moon, so there's no volcanism, rifting, subduction, or collisions to generate the forces that cause earthquakes. The Moon has a diameter of about 1,738 km. Of this, the outer 1,000 km constitutes the lithosphere, which we know because instruments have detected moonquakes with hypocenters as deep as 1,000 km. Thus, rigid lithosphere accounts for almost

60% of the Moon's diameter (as opposed to only about 1.7% of Earth's), and this ultra-thick lithosphere is simply too strong to break into plates in response to the flow in the Moon's thin asthenosphere.

But if plates don't move on the Moon, then what causes moonquakes? Undoubtedly some are caused by the impacts of meteorites, which hit the Moon's surface like a large hammer. Most moonquakes, though, seem to occur when the Moon reaches its closest distance to the Earth, as it travels along its elliptical orbit, suggesting that gravitational attraction between the Earth and the Moon creates tidal-like motions that crack the Moon's lithosphere or make the faults in the Moon slip.

(a)

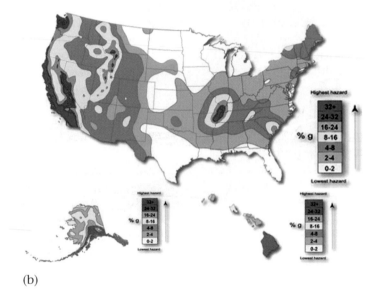

(b)

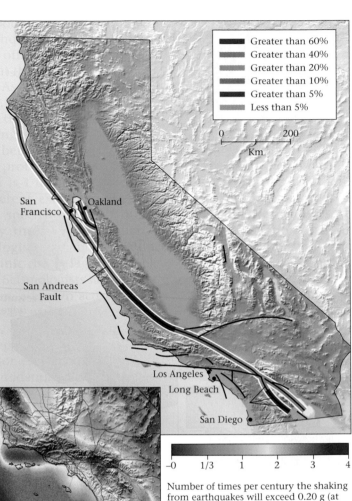

(c)

Number of times per century the shaking from earthquakes will exceed 0.20 g (at this level, there is significant damage to older buildings).

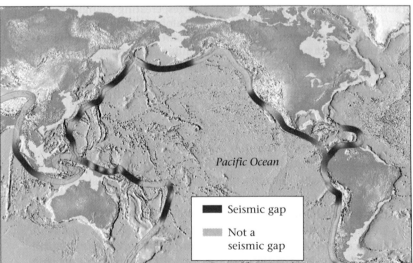

(d)

FIGURE 10.38 (a) A Global Seismic Hazard Map. The regions with darker (redder) colors have a greater probability of experiencing a large earthquake. (b) Map of seismic hazard in the United States. Darker (redder) colored areas are regions facing the greatest hazard. (c) This map shows the probability (in percentage) of a strong-to-great earthquake occurring along segments of the San Andreas Fault during the next thirty years. When an earthquake takes place here, only part of the fault slips. Not all segments of the fault have the same likelihood of producing a large earthquake, because some segments have slipped more recently than others; seismologists assume that stress is lower across segments that have slipped more recently. (d) Earthquakes may be more likely to occur in the seismic gaps around the Pacific in the near future.

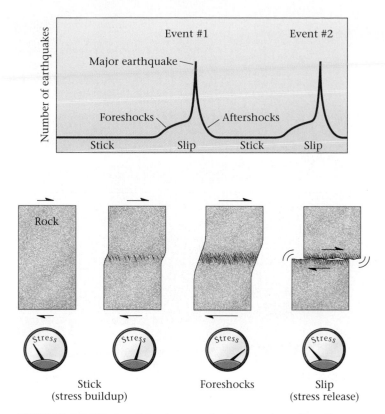

FIGURE 10.39 A sudden increase in the number of small earthquakes along a fault segment may be a possible precursor to a large earthquake along that segment; that is, the small earthquakes may be foreshocks. The pattern may represent stick-slip behavior.

earthquakes. Use of such "stress-triggering models" has provided substantial insight into seismic activity along the North Anatolian Fault, in Turkey (▶Fig. 10.40a). This fault is a large strike-slip fault along which Turkey slips westward. (Turkey is essentially being squeezed like a watermelon seed out of the way of Arabia, which is moving northward and is colliding with Asia.) Earthquakes happen again and again along the fault—in fact, ruins of several ancient cities lie along the trace of the fault. Since 1939, 11 major earthquakes have occurred along the fault—each ruptured a different portion of the fault (▶Fig. 10.40b). Overall, there has been a westward progression of faulting. By modeling the stress changes resulting from each earthquake, geologists have determined places where stress accumulates. The model predicted high stresses at the western end of the fault, and it predicted that the next big earthquake would happen here. It did—in August 1999, a magnitude (M_w) 7.4 earthquake devastated Izmit, a city just south of Istanbul, killing more than 17,000 people and leaving more than 600,000 homeless. The segment of the North Anatolian Fault that ruptured during this terrible event exactly fills a seismic gap.

Other techniques that have been explored but have *not* yet been proven to be precursors of earthquakes include the following: changes in the water level in wells; appearance of gases, such as radon or helium, in wells; changes in the electrical conductivity of rock underground; and unusual animal behavior (e.g., dogs howling). Believers in these proposed clues suggest that they all reflect the occurrence of cracking in the crust prior to an earthquake. But most investigators remain skeptical.

As long as short-term predictions remain questionable, emergency service planners must ask: "What if a prediction is wrong?" Should schools and offices be shut because of a prediction? Should millions be spent to evacuate people? Should a city be deserted, allowing for the possibility of looters? Should the public be notified, or should only officials be notified, creating a potential for rumor? If the prediction proves wrong, can seismologists be sued? No one really knows the answers to these questions.

10.8 EARTHQUAKE ENGINEERING AND ZONING

If we can't avoid earthquakes, can we prepare for them? A glance at Table 10.1 illustrates that the loss of human life from earthquakes varies widely. The loss depends on a number of factors, most notably the proximity of an epicenter to a population center, the depth of the hypocenter, the style of construction in the epicentral region, whether or not the earthquake occurred in a region of steep slopes or along the coast, whether building foundations are on solid bedrock or on weak substrate, whether the earthquake happened when people were outside or inside, and whether the government was able to provide emergency services promptly.

For example, the 1988 earthquake in Armenia was not much bigger than a 1971 earthquake in southern California, but it caused almost five hundred times as many deaths (24,000 versus 50). The difference in death toll reflects differences in the style and quality of construction, and the characteristics of the substrate. The unreinforced concrete-slab buildings and masonry houses of Armenia collapsed, whereas the structures in California had, by and large, been erected according to building codes that take into account stresses caused by earthquakes. Most flexed and twisted but did not fall down and crush people. The terrible 1976 earthquake in T'ang-shan, China, was such a calamity because the ground beneath the epicenter had been weakened by coal mining and collapsed, and because buildings were so poorly constructed. Mexico City's 1985 earthquake, as we have seen, proved disastrous because the city lies over a sedimentary basin whose composition and bowl-like shape focused seismic energy, and caused buildings of a certain height to

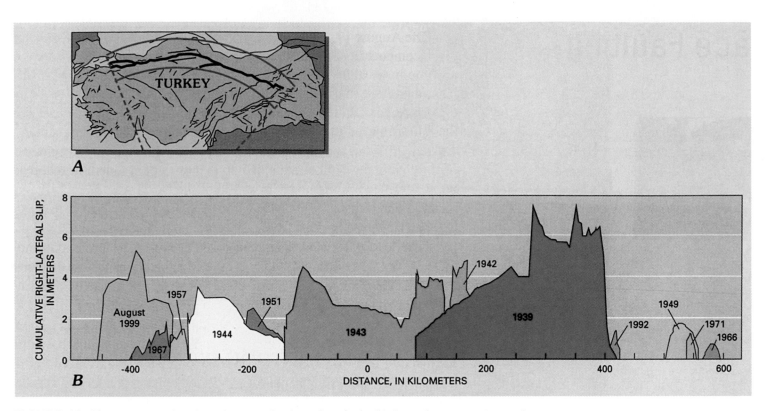

FIGURE 10.40 (a) A map of Turkey, showing the Anatolian fault. (b) A graph representing regions that slipped during various earthquakes. The horizontal axis represents location along the fault, and the vertical axis represents the amount of slip.

resonate. During the 1989 Loma Prieta, California, quake, portions of Route 880 in Oakland that were built on a weak substrate collapsed, while portions built on bedrock or gravel remained standing.

The differences in the destructiveness of earthquakes demonstrate that we can mitigate or diminish their conse-

FIGURE 10.41 How to prevent damage and injury during an earthquake. (a) By wrapping a bridge's support columns in cable (preventing buckling of the columns) and bolting the span to the columns (preventing the span from separating from the columns), the bridge will not collapse as easily. (b) Buildings will be stronger if they are wider at the base and if cross beams are added inside. (c) Placing buildings on rollers (or shock absorbers) will lessen the severity of the vibrations.

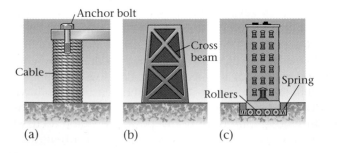

quences by taking sensible precautions. Clearly, **earthquake engineering,** the designing of buildings that can withstand shaking, and **earthquake zoning,** the determination of where land is stable and where it might collapse, can help save lives and property.

In regions prone to large earthquakes, buildings should be constructed so they are able to withstand vibrations without collapsing (▶Fig. 10.41a–c). They should be somewhat flexible so that ground motions can't crack them, and supports should be strong enough to maintain loads far in excess of the loads caused by the static (nonmoving) weight of the building. Bridge support columns should also be constructed with earthquakes in mind. By wrapping steel cables around the columns, they become many times stronger. Bolting the bridge spans to the top of a column prevents the spans from bouncing off. Concrete-block buildings, unreinforced concrete, and unreinforced brick buildings crack and tumble under conditions where wood-frame, steel-girder, or reinforced concrete buildings remain standing. Traditional heavy, brittle tile roofs shatter and bury the inhabitants inside, while sheet-metal or asphalt shingle roofs do not. Loose decorative stone and huge open-span roofs do not fare well when vibrated, and should be avoided.

Similarly, developers should avoid construction on land underlain by weak, sedimentary mud that could liquefy.

FIGURE 10.42 If an earthquake strikes, take cover under a sturdy table near a wall.

They should not build on top of, on, or at the base of steep escarpments (which could fail and produce landslides), and they should avoid locating large population centers downstream of dams (which could fail, causing a flood). And they should avoid constructing vulnerable buildings directly over active faults, whose movement could crack and destroy the structure. Cities in seismic zones need to draw up emergency plans to deal with disaster. Communication centers should be located in safe localities, and strategies need to be implemented for providing supplies under circumstances where roads may be impassable.

Finally, communities and individuals should learn to protect themselves during an earthquake. In your home, keep emergency supplies, bolt bookshelves to walls, strap the water heater in place, install locking latches on cabinets, know how to shut off the gas and electricity, know how to find the exit, have a fire extinguisher handy, and know where to go to find family members. Schools and offices should have earthquake-preparedness drills (►Fig. 10.42). If an earthquake strikes, stay outdoors and away from buildings if you can. If you're inside, stand near a wall or in a doorway near the center of the building, or crouch under a heavy table. And if you're on the road, stay away from bridges. As long as plates continue to move, earthquakes will continue to shake. But we can learn to live with them.

CHAPTER SUMMARY

• Earthquakes are episodes of ground shaking, caused when earthquake waves reach the ground surface. Earthquake activity is called seismicity.

• Most earthquakes happen when rock breaks during faulting. A fault is a fracture on which sliding occurs. The place where rock breaks and earthquake energy is released is called the hypocenter (focus), and the point on the ground directly above the hypocenter is the epicenter.

• We can distinguish among normal, reverse, thrust, and strike-slip faults, based on the relative motion of rock across the fault. The amount of movement is called the displacement.

• Active faults are faults on which movement is currently taking place. Inactive faults ceased being active long ago, but can still be recognized because of the displacement across them. Displacement on active faults that intersect the ground surface may yield a fault scarp. The intersection of the fault with the ground is the fault trace.

• According to elastic-rebound theory, during fault formation, rock elastically strains, then cracks. Eventually, cracks link to form a throughgoing rupture on which sliding occurs. When this happens, the elastically strained rock breaks and vibrates, and this generates an earthquake.

• Faults exhibit stick-slip behavior, in that they move in sudden increments.

• Earthquakes in the continental crust can only happen in the brittle, upper part of the crust. At depth, where rocks become ductile, earthquakes don't occur.

• Earthquake energy travels in the form of seismic waves. Body waves, which pass through the interior of the Earth, include P-waves (compressional waves) and S-waves (shear waves). Surface waves, which pass along the surface of the Earth, include R-waves (Rayleigh waves) and L-waves (Love waves).

• We can detect earthquake waves by using a seismograph.

• Seismograms demonstrate that different earthquake waves arrive at different times, because they travel at different velocities. Using the difference between P-wave and S-wave arrival times, seismologists can pinpoint the epicenter location.

• The Mercalli intensity scale is based on documenting the damage caused by an earthquake. Magnitude scales, such as the Richter scale, are based on measuring the amount of ground-motion, as indicated by traces of waves on a seismogram. The seismic-moment magnitude scale takes into account the amount of slip, the length and depth of the rupture, and the strength of the ruptured rock.

• A magnitude 8 earthquake yields about ten times as much ground motion as a magnitude 7 earthquake, and releases about thirty-two times as much energy.

• Most earthquakes occur in seismic belts, or zones, of which the majority lie along plate boundaries. Intraplate earthquakes happen in the interior of plates. Different kinds of earthquakes happen at different kinds of plate boundaries. Shallow-focus earthquakes associated with normal faults occur at divergent plate boundaries and in rifts. Earthquakes associated with thrust and reverse faulting occur at convergent and collisional boundaries. At convergent plate boundaries, we also observe intermediate- and deep-focus earthquakes, which define the Wadati-Benioff zone. Shallow-focus strike-slip earthquakes occur along transform boundaries.

• Earthquake damage results from ground shaking (which can topple buildings), landslides (set loose by vibration), sediment liquefication (the transformation of compacted clay into a muddy slurry), fire, and tsunamis (giant waves).

• Seismologists predict that earthquakes are more likely in seismic zones than elsewhere, and can determine the recurrence interval (the average time between successive events) for great earthquakes. But it may never be possible to pinpoint the exact time and place at which an earthquake will take place.

• Earthquake hazards can be reduced with better construction practices and zoning, and by knowing what to do during an earthquake.

KEY TERMS

aftershock (p. 281)
arrival time (p. 287)
body waves (p. 285)
compressional waves (p. 285)
displacement (p. 279)
earthquake (p. 276)
earthquake engineering (p. 314)
earthquake zoning (p. 314)
elastic-rebound theory (p. 281)
epicenter (p. 277)
fault (p. 276)
fault creep (p. 284)
fault scarp (p. 279)
fault trace (p. 279)
focus (p. 277)
foreshock (p. 281)
hypocenter (focus) (p. 277)
induced seismicity (p. 300)
intraplate earthquakes (p. 299)
L-waves (Love waves) (p. 285)
liquefaction (p. 304)
long-term predictions (p. 309)
magnitude (p. 291)
Mercalli intensity scale (p. 289)

moment magnitude (p. 292)
P-waves (p. 285)
R-waves (Rayleigh waves) (p. 285)
recurrence interval (p. 310)
resonance (p. 302)
Richter scale (p. 292)
S-waves (p. 285)
seiche (p. 302)
seismic belts (zones) (p. 294)
seismic gaps (p. 310)
seismicity (p. 276)
seismic-hazard assessment (p. 309)
seismic waves (p. 285)
seismogram (p. 287)
seismograph (seismometer) (p. 287)
seismologist (p. 276)
shear waves (p. 285)
short-term predictions (p. 309)
stick-slip behavior (p. 281)
surface waves (p. 285)
travel-time curve (p. 289)
tsunami (p. 307)
Wadati-Benioff zone (p. 294)

REVIEW QUESTIONS

1. Compare normal, reverse, and strike-slip faults.
2. Describe elastic rebound theory and the concept of stick-slip behavior.
3. Compare brittle and ductile deformation.
4. Describe the motions of the four types of seismic waves. Which are body waves, and which are surface waves?
5. Explain how the vertical and horizontal components of an earthquake are detected on a seismograph.
6. Explain the contrasts among the different scales used to describe the size of an earthquake.
7. How does seismicity on mid-ocean ridges compare with seismicity at convergent or transform boundaries? Do all earthquakes occur at plate boundaries?
8. What is the Wadati-Benioff zone, and why was it important in understanding plate tectonics?
9. Describe the types of damage caused by earthquakes.
10. What is a tsunami, and why does it form? What is a seiche?
11. Explain how liquefaction occurs in an earthquake, and how it can cause damage.
12. How are long-term and short-term earthquake predictions made? What is the basis for determining recurrence interval, and what does a recurrence interval mean?
13. Why is it difficult to make accurate short-term predictions? What clues might suggest an earthquake may happen fairly soon?
14. What types of structure are most prone to collapse in an earthquake? What types are most resistant to collapse?
15. What should you do when you feel an earthquake starting?

SUGGESTED READING

Bolt, B. A. 1999. *Earthquakes,* 4th ed. New York: Freeman.

Bryant, E. 2001. *Tsunami: The Underrated Hazard.* Cambridge, England: Cambridge University Press.

Fradkin, P. L. 1999. *Magnitude 8.* Berkeley: University of California Press.

Geschwind, C. H. 2001. *California Earthquakes: Science, Risk & the Politics of Hazard Mitigation.* Baltimore: Johns Hopkins University Press.

Hough, S. E. 2002. *Earthshaking Science: What We Know (and Don't Know) about Earthquakes.* Princeton: Princeton University Press.

Ritchie, D., and A. E. Gates. 2001. *Encyclopedia of Earthquakes and Volcanoes.* New York: Facts on File.

Shearer, P. 1999. *Introduction to Seismology.* Cambridge, England: Cambridge University Press.

Stein, S., and M. Wysession. 2002. *An Introduction to Seismology, Earthquakes and Earth Structure.* Boston: Blackwell Science.

U.S. Geological Survey. 2000. *Implications for Earthquake Risk Reduction in the United States from the Kocaeli, Turkey, Earthquake of August 17, 1999.* Circular 1193. Washington, D.C.: U.S. Government Printing Office.

Yeats, R. S. 2001. *Living with Earthquakes in California: A Survivor's Guide.* Corvallis: Oregon State University Press.

The azure waters and palm-fringed islands of the Indian Ocean's east coast hide one of the most complicated and seismically-active plate boundaries on Earth. Along the Sunda Trench, north of central Sumatra (Indonesia), the Indian Plate subducts obliquely beneath the Burma Plate at a rate of 61 mm/year (▶Fig. 10.43a). As a result of this oblique motion, huge thrust faults have developed between the trench and volcanic arc, and a system of small rifts and transform faults evolved in the backarc region. Stress, generated by subduction, triggered massive slip along one of the thrust faults at 7:59 a.m. on December 26, 2004. The hypocenter occurred at a depth of 10–30 km beneath the Earth's surface. During this event, a huge area of the fault, measuring 1100 km long (parallel to the trench) by 100 km-wide, slipped, causing the hanging wall to lurch westward by as much as 15 m. The resulting shock waves generated a great earthquake (moment magnitude = 9.0), the fourth largest of the past century anywhere on Earth.

In terms of destruction, a great earthquake can be bad enough near the epicenter. But because the December 26th Indonesia earthquake occurred beneath the sea floor, it ultimately led to devastation over a much broader region. Slip of the hanging-wall caused the ocean floor to rise by tens of centimeters to a few meters, and this rise, in turn, shoved the overlying water upwards, producing tsunami! These waves raced outwards from the epicenter (▶Fig. 10.43b), reaching speeds of about 800 km/h (500 mph) in the open ocean. They struck Sumatra soon after the earthquake, and reached Sri Lanka and the coast of India 2 to 3 hours later (▶Fig. 10.43c). Eventually, waves traversed the entire Indian Ocean and came ashore along the coast of Africa almost 6 hours after the earthquake.

As noted earlier in this chapter, the sea surface rises by only a few centimeters to a few tens of centimeters as a tsunami passes in the open ocean, and thus would be unnoticeable to boats. But unlike familiar wind-driven waves, the width of the sea surface uplifted as a tsunami passes can be kilometers wide or more, so the wave affects an immense volume of water. When the front of a tsunami reaches the shallows near shore, it begins to slow down, to perhaps 20 – 30 km/h. As this happens, the back part of the wave catches up, and water builds into an immense, broad breaker. The largest waves of December 26, 2004 reached heights of 10 to 15 m (▶Fig. 10.43d). Storm-driven waves may be comparably high, but they are fairly narrow (as measured perpendicular to the wave face) and involve much less water, so when they crash on a beach, the wave's energy disperses before water can traverse the beach. A tsunami, however, is different—because it is much wider, the wave front is more like a wall at the edge of a water plateau, and when the front washes over the shore, more water just keeps on coming, like a tide that doesn't stop rising.

Along the flat coastal plains bordering parts of the Indian Ocean, the waves of December 26th swirled inland for hundreds of meters, carrying boats, debris, cars, and people. They crushed and submerged buildings, or tore them right off their foundations. When the inland-directed movement finally stopped, the water surged back to sea, in some cases transporting debris and people hundreds of meters to kilometers offshore. In many localities, multiple waves hit. Satellite images taken before and after the event show how the tsunami breached sandbars, flooded fields, and transformed coastal communities into jumbles of flotsam and jetsam. The total death toll will never be known for sure, but sadly, will likely exceed 150,000 people. Victims included both local inhabitants and foreign tourists; the latter had come to enjoy the winter holiday break. Tragically, many victims died from simple cuts that became infected.

As long as people build along the coast, tsunamis will cause devastation. But the death toll of that December 26th disaster would not have been nearly as large if there had been an effective tsunami-warning network in the Indian Ocean region. Officials established such a network in the Pacific Ocean region in 1946. The Pacific Tsunami Warning Center monitors seismographs and tidal gauges to detect tsunami and give warnings, sometimes hours before the waves strike. The Center did issue a tsunami warning for the Indian Ocean within 90 minutes of the 9.0 earthquake. But unfortunately, no one in the countries bordering the Indian Ocean had been designated to receive this information, and even when contacts were made, officials didn't know what to do, as no evacuation plans had been formulated. As relief efforts raced to avert disease and starvation caused by destruction of clean water supplies and transportation infrastructure, geologists and officials set to the task of developing a tsunami warning system for the Indian Ocean, so that when the next tsunami strikes, fewer lives may be lost.

Please see Appendix B for further images and diagrams regarding this event.

FIGURE 10.43 (a) The geometry of plate interactions near the epicenter. Location is shown by the box in part (c). Note the oblique motion of India relative to Burma (indicated by the arrow), and that the epicenter occurred near a triple junction. (b) A computer model showing the tsunami about 2 hours after the earthquake. The yellow areas have the greatest wave height. (c) A map of the Indian Ocean, showing the position of the wave front as time passes. (d) A cross-section sketch showing the building of a tsunami at 2 times. The inset shows the wave front in relation to a person.

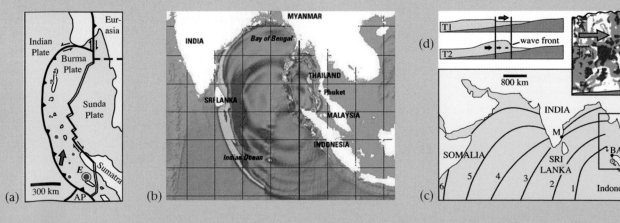

Seeing Inside the Earth

C.1 INTRODUCTION

We live on the Earth's skin, and can see light years into space just by looking up. But when we look down—that's another story! We can't use our eyes to look through rock, because it is opaque. So how do we learn about what's inside this planet? Tunneling and drilling aren't much help because they do little more than prick the surface—the deepest mine (a gold mine in South Africa that reaches a depth of 3.6 km below the surface) represents less than 0.06% of the Earth's radius, and the deepest drill hole (drilled to a depth of 12.3 km in Precambrian rock of the Kola Peninsula in northwest Russia) represents less than 0.19% of the Earth's radius. Fortunately, as discussed in Chapter 2, nineteenth-century geologists realized that measurements of the Earth's mass and shape provide indirect clues to the mystery of what's inside and, from these clues, determined that the Earth is not homogeneous but rather consists of three concentric layers: a crust (of low density), a mantle (of intermediate density), and a core (of high density). Further study showed that the crust beneath continents differs from the crust beneath oceans. Continental crust consists of a variety of felsic, intermediate, and mafic rock, while oceanic crust consists almost entirely of mafic-composition rock (▶Fig. C.1). However, the determination of the depths of the boundaries between Earth's layers and the division of the layers into sublayers with distinct properties could not be done until the twentieth century, when studies of seismic waves became available. By studying the speed and direction in which seismic waves travel through the Earth, seismologists can effectively "see" details of the planet's internal layers.

In this interlude, we look at the behavior of seismic waves as they pass through our planet, and we learn how this behavior characterizes Earth's interior. We begin by reviewing a few key points about seismic waves, then move on to the phenomena of wave reflection and refraction. Finally, we witness the discoveries of the different layer boundaries in the Earth. This interlude, incorporating the information about earthquakes and seismic waves provided in Chapter 10, completes the journey to the center of the Earth that we began in Chapter 2.

FIGURE C.1 The nineteenth-century three-layer image of the Earth, showing the crust, mantle, and core. The inset shows the contrast between continental crust (silicic, intermediate, and mafic rock) and oceanic crust (mafic rock).

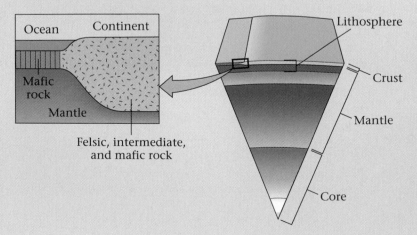

C.2 THE MOVEMENT OF SEISMIC WAVES THROUGH THE EARTH

Recall that a sudden rupture of intact rock or the frictional slip of rock on a fault produces seismic waves. These waves move outward from the point of rupture, the earthquake hypocenter, in all directions at once. A single earthquake produces many kinds of waves, distinguished from one another by where and how they move. Notably, surface waves (R-waves and L-waves) propagate along the planet's surface, while body waves (P-waves and S-waves) pass through the interior. P-waves are compressional, and resemble the waves generated when you push a spring back and forth in a direction parallel to the length of the spring. S-waves are shear, and resemble the waves

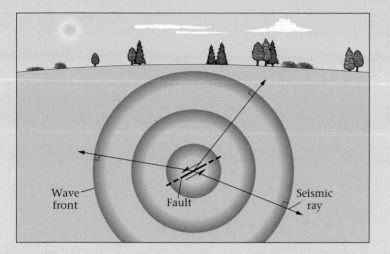

FIGURE C.2 An earthquake sends out waves in all directions. Seismic rays are perpendicular to the wave fronts.

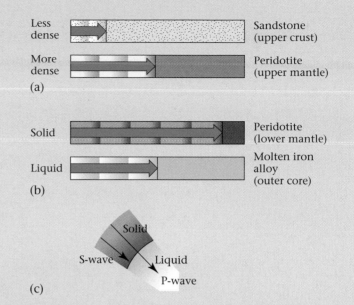

FIGURE C.3 (a) Seismic waves travel at different velocities in different rock types. For example, they travel faster in peridotite than in sandstone. (b) Seismic waves travel faster in solid peridotite than in a liquid like molten iron alloy. (c) Both P-waves and S-waves can travel through a solid, but only P-waves can travel through a liquid.

generated when you wiggle a rope back and forth perpendicular to the length of the rope (see Fig. 10.12).

The boundary between the rock through which a wave has passed and the rock through which it has not yet passed is called a **wave front.** A wave front expands outward from the earthquake focus like a growing bubble. We can represent a succession of waves in a drawing by a series of concentric wave fronts. The changing position of an imaginary point on a wave front as the front moves through rock is called a **seismic ray.** Note that seismic rays are perpendicular to wave fronts, so that each point on the wave front follows a slightly different ray (▶Fig. C.2). The time it takes for a wave to travel from the focus to a seismograph station along a given ray is the **travel time** along that ray.

The ability of a seismic wave to travel through a certain material and the velocity at which it travels depend on the character of the material. Factors such as density (mass per unit volume), rigidity (how stiff or resistant to twisting a material is), and compressibility (how much a material's volume changes in response to squashing) all affect seismic-wave movement. Studies of seismic waves reveal the following:

- Seismic waves travel at different velocities in different rock types (▶Fig. C.3a). For example, P-waves travel at 8 km per second in peridotite (an ultramafic rock), but at only 3.5 km per second in low-density sandstone. Therefore, waves accelerate or slow down if they pass from one rock into another. P-waves in rock travel about ten to twenty-five times faster than sound waves in air. But even at this rate, they take about twenty minutes to pass entirely through the Earth along a diameter.

- In general, seismic waves travel faster in a solid than in a liquid. For example, they travel more slowly in magma than in solid rock, and more slowly in molten iron alloy than in solid peridotite (▶Fig. C.3b).

- Both P-waves and S-waves can travel through a solid, but only P-waves can travel through a liquid (▶Fig. C.3c). To see why, picture what happens if you push down on the water surface in a pool—you send a pulse of compression (a P-wave) to the bottom of the pool. Now move (shear) your hand sideways through the water. The water in front of your hand simply slides or flows past the water deeper down—your shearing motion has no effect on the water at the bottom of the pool (▶Fig. C.4a, b).

C.3 THE REFLECTION AND REFRACTION OF WAVE ENERGY

Shine a flashlight into a container of water so that the light ray hits the boundary (or interface) between water and air at an angle. Some of the ray bounces off the water surface and heads back up into the air, while some enters the water (▶Fig. C.5a). The light ray that enters the water bends at the air-water boundary, so that the angle between the ray and the boundary in the air is different from the angle between the ray and the boundary in the water. Physicists refer to the light ray that bounces off the air-water boundary and heads back into the air as the "reflected ray," and the ray that bends at the boundary as the "refracted ray." The phenomenon of bouncing off is **reflection,** and the phenomenon of bending is **refraction.** Wave reflection and refraction take place at the interface between two materials, if the wave travels at different velocities in the two materials.

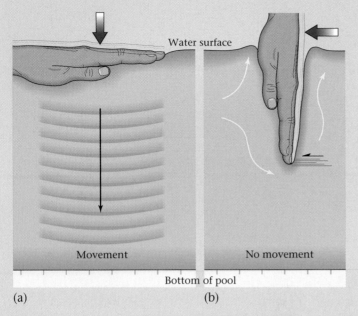

FIGURE C.4 (a) Pushing down on a liquid creates a compressive pulse (P-wave) that can travel through a liquid. (b) Shearing your hand through water does not generate a shear wave; the moving water simply flows past the water deeper down.

The amount and direction of refraction at a boundary depend on the contrast in wave velocity across the boundary, and on the angle at which a wave hits the interface. As a rule, if waves enter a material through which they will travel more slowly, the rays representing the waves bend away from the interface, while if the waves enter a material through which they will travel faster, the rays bend toward the interface. For example, the light ray in ▶Figure C.5b bends down when hitting the air-water boundary, because light travels more slowly in water. This relation makes sense if you picture a car driving from a paved surface diagonally onto a sandy beach—the wheel that rolls onto the sand first slows down relative to the

wheel still on the pavement, causing the car to turn. If the ray were to pass from a material in which it travels slowly into one in which it travels more rapidly, it would bend up (▶Fig. C.5c).

Like light, seismic energy travels in the form of waves, so seismic waves, like rays of light in water, reflect and/or refract when reaching the interface between two rock layers if the waves travel at different velocities in the two layers. For example, imagine a layer of sandstone overlying a layer of peridotite. Seismic velocities in sandstone are faster than in peridotite, so as seismic waves reach the boundary, some reflect while some refract.

C.4 DISCOVERING THE CRUST-MANTLE BOUNDARY

The concept that seismic waves refract at boundaries between different layers led to the first documentation of the core-mantle boundary. In 1909, Andrija Mohorovičić, a Croatian seismologist, noted that P-waves arriving at seismograph stations less than 200 km from the epicenter traveled at an average speed of 6 km per second, while P-waves arriving at seismographs more than 200 km from the epicenter traveled at an average speed of 8 km per second. To explain this observation, he suggested that P-waves reaching nearby seismographs followed a shallow path through the crust, in which they traveled more slowly, while P-waves reaching distant seismographs followed a deeper path through the mantle, in which they traveled more rapidly.

To understand Mohorovičić's proposal, examine ▶Figure C.6a, which shows P-waves, depicted as rays, generated by an earthquake in the crust. Ray A, the shallower wave, travels through the crust directly to a seismograph. Ray B, the deeper wave, heads downward, refracts at the crust-mantle boundary, curves through the mantle, refracts again at the boundary, and then proceeds through

FIGURE C.5 (a) A ray of light, when it reaches the boundary between water and air, partly reflects and partly refracts. The refracted ray bends down as it enters the water. (b) A ray that enters a slower medium bends away from the boundary (like light reaching water from air). (c) A ray that enters a faster medium bends toward the boundary.

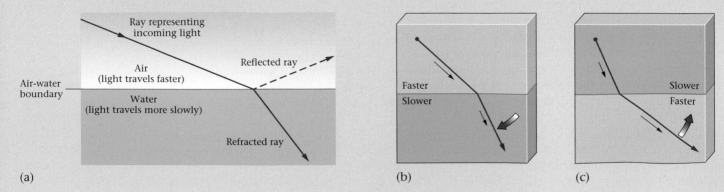

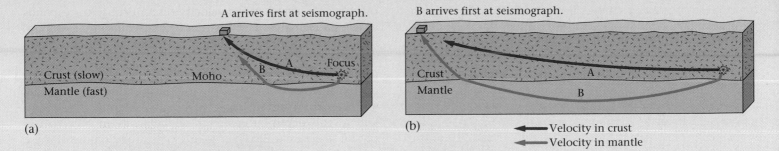

(a)

(b)

⬅— Velocity in crust
⬅— Velocity in mantle

FIGURE C.6 (a) At a nearby seismograph station, seismic waves traveling through the crust reach the seismograph first. Seismic rays refract at the Moho, the crust-mantle boundary. (b) At a distant station, seismic waves traveling through the mantle reach the seismograph first, which means that seismic waves travel at a faster velocity in the mantle than in the crust. The Moho lies at a depth of 35–40 km beneath continental interiors.

the crust up to the seismograph. At stations less than 200 km from the epicenter, ray A arrives first, because it has a shorter distance to travel. But at stations more than 200 km from the epicenter (▶Fig. C.6b), ray B arrives first, even though it has farther to go, because it travels faster for much of its length. Calculations based on this observation requires the crust-mantle boundary beneath continents to be at a depth of about 35–40 km. As we learned in Chapter 2, this boundary is now called the **Moho,** in honor of Mohorovičić.

C.5 DEFINING THE STRUCTURE OF THE MANTLE

After studying materials erupted from volcanoes, geologists concluded that the entire mantle has roughly the chemical composition of the ultramafic igneous rock called peridotite. If the density, rigidity, and compressibility of peridotite were exactly the same at all depths, seismic velocities would be the same everywhere in the mantle, and seismic rays would be straight lines. But by studying travel times, seismologists have determined that seismic waves travel at different velocities at different depths. Let's now look at variations in seismic velocity that depend on mantle depth, and consider how these variations affect the shape of seismic rays.

Between about 100 and 200 km deep in the mantle beneath oceanic lithosphere, seismic velocities are slower than in the overlying lithospheric mantle (▶Fig. C.7). In this **low-velocity zone,** the prevailing temperature and pressure conditions cause peridotite to partially melt, by up to 2%. The melt, a liquid, coats solid grains and fills voids between grains. Because seismic waves travel more slowly through liquids than through solids, the coatings of melt slow seismic waves down. In the context of plate tectonics theory, the low-velocity zone is the weak layer on which oceanic lithosphere plates move.

Below the low-velocity zone, the mantle does not contain melt. Geologists do not find a well-developed low-velocity zone beneath continents.

Below about 200 km, seismic-wave velocities everywhere in the mantle increase with depth. Seismologists interpret this increase to mean that mantle peridotite becomes progressively less compressible, more rigid, and denser with depth. This proposal makes sense, considering that the weight of overlying rock increases with depth, and as pressure increases, the atoms making up rock squeeze together more tightly and

FIGURE C.7 The velocity of P-waves in the mantle changes with depth. Note the low-velocity zone between 100 and 200 km, and the sudden jumps in velocity defining the transition zone between 410 and 660 km.

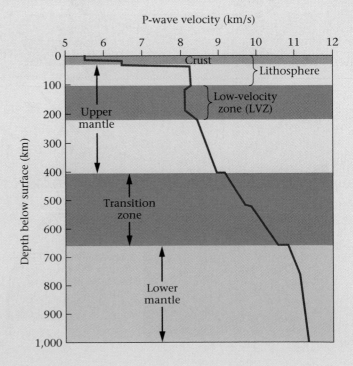

(a)

(b)

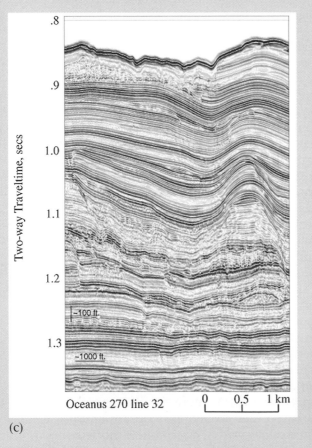
(c)

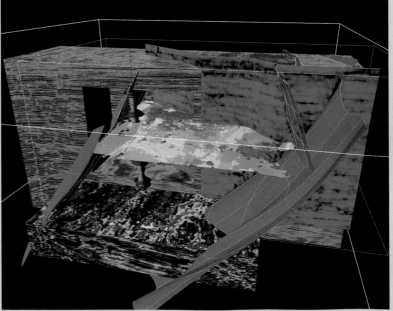
(d)

(e)

FIGURE C.13 (a) Trucks thumping on the ground to generate the signal needed for making a seismic-reflection profile. (b) Analyzing data with a computer. (c) A seismic-reflection profile. The colored stripes are layers of strata. (d) A ship collecting seismic data at sea. (e) This image shows layers of subsurface strata in 3-D. Computers can expose different cross-section and map-view slices of the image. From such data, important features like faults (indicated by colored surfaces) can be located.

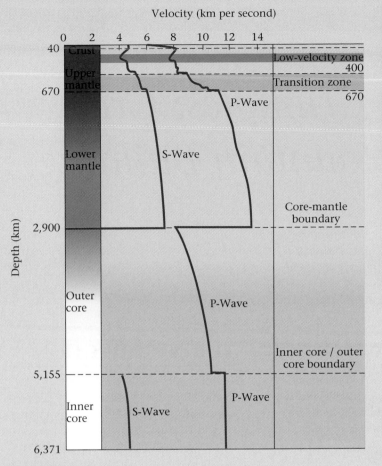

Velocity (km per second)

FIGURE C.14 The velocity-versus-depth profile for the Earth.

FIGURE C.15 A laboratory apparatus for studying the characteristics of minerals under very high pressures and temperatures. The green laser beam is heating up a microscopic sample being squeezed between two diamonds hidden at the center of the metal cylinder.

KEY TERMS

core-mantle boundary (p. 322)
inner core (p. 323)
low-velocity zone (p. 321)
lower mantle (p. 322)
Moho (p. 321)
outer core (p. 323)
P-wave shadow zone (p. 322)
reflection (p. 319)
refraction (p. 319)
S-wave shadow zone (p. 323)
seismic ray (p. 319)

seismic-reflection profile (p. 324)
seismic tomography (p. 324)
seismic-velocity discontinuities (p. 322)
transition zone (p. 322)
travel time (p. 319)
upper mantle (p. 322)
velocity-versus-depth curve (p. 325)
wave front (p. 319)

FIGURE C.16 The modern view of a complex and dynamic Earth interior. Note the convecting cells, the mantle plumes, and the subducted-plate graveyards.

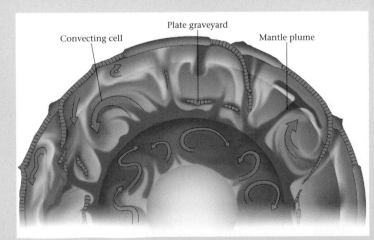

Convecting cell   Plate graveyard   Mantle plume

Crags, Cracks, and Crumples: Crustal Deformation and Mountain Building

Innumerable peaks, black and sharp, rose grandly into the dark blue sky, their bases set in solid white, their sides streaked and splashed with snow, like ocean rocks with foam. . . . [Mountains] are nature's poems carved on tables of stone. . . . How quickly these old monuments excite and hold the imagination!
—JOHN MUIR, from *WILDERNESS ESSAYS*

Some of the world's most beautiful scenery can be found in mountainous regions. Here, gazing on Mt. Cook in the southern Alps of New Zealand, we see evidence of the many processes that contribute to the development of mountain scenery. Compression between two plates uplifted rock to an elevation of over 3.7 km above sea level. Landslides, along with erosion by glaciers and rivers, create jagged peaks towering above debris choked valleys.

11.1 INTRODUCTION

Geographers call the peak of Mt. Everest "the top of the world," for this mountain, which lies in the Himalayas of south Asia, rises higher than any other on Earth (▶Fig. 11.1). The cluster of flags on Mt. Everest's summit flap at 8.85 km (29,029 feet) above sea level—almost at the cruising height of modern jets. No one can live very long at the top, for the air there is too thin to breathe. In fact, even after spending weeks acclimating to high-altitude conditions at a base camp a couple of kilometers below the summit, most climbers need to use bottled oxygen during their summit attempt. In 1953, the British explorer Sir Edmund Hillary and Tenzing Norgay, a Nepalese guide, were the first to reach the summit. By 1999, about 750 more people had also succeeded—but 150 died trying. So many climbs end in death because success depends not just on the skill of the climber, but also on the path of the jet stream, a 200-km-per-hour current of air that flows at high elevations (see Chapter 20). If the jet stream crosses the summit, it engulfs climbers in heat-robbing winds that can freeze a person's face, hands, and feet even if they're swaddled in high-tech clothing.

FIGURE 11.1 Mt. Everest (the large peak in the center) and the surrounding Himalayas, as viewed from the space shuttle *Atlantis*.

Mountains draw nonclimbers as well, for everyone loves a vista of snow-crested peaks (see chapter-opening photo). Their stark cliffs, clear air, meadows, forests, streams, and glaciers provide a refuge from the mundane. For millennia, mountain beauty has inspired the work of artists and poets, and in some cultures mountains served as a home to the gods. Geologists feel a special fascination with mountains, for they provide one of the most obvious indications of dynamic activity on Earth. To make a mountain, Earth forces lift cubic kilometers of rock skyward against the pull of gravity. This uplift then provides the fodder for erosion, which, over time, grinds away at a mountain to make sediment, and in the process sculpts jagged topography.

The process of forming a mountain not only uplifts the surface of the crust, but also causes rocks to undergo **deformation,** a process by which rocks squash, stretch, bend, or break in response to squeezing, stretching, or shearing. Deformation produces **geologic structures,** including **joints** (cracks), **faults** (fractures on which one body of rock slides past another), **folds** (bends or wrinkles), and **foliation** (layering resulting from the alignment of mineral grains or the creation of compositional bands). Mountain building may also involve metamorphism and melting. In this chapter, we learn about the phenomena that happen during mountain building—deformation, igneous activity, sedimentation, metamorphism, uplift, and erosion—and discover why they occur, in the context of plate tectonics theory.

11.2 MOUNTAIN BELTS AND THE CONCEPT OF OROGENY

With the exception of the large volcanoes formed over hot spots, mountains do not occur in isolation, but rather as part of linear ranges variously called **mountain belts,** oro-

genic belts, or **orogens** (from the Greek words *oros*, meaning "mountain," and *genesis*, meaning "formation"). Geographers define about a dozen major mountain belts and numerous smaller ones worldwide (▶Fig. 11.2). Some large orogens contain smaller ranges within.

A mountain-building event, or **orogeny,** has a limited lifetime. The process begins, lasts for tens of millions of years, and then ceases. After an orogeny ceases, erosion may eventually bevel the land surface almost back to sea level and can do so in as little as 50 million years. Thus, the mountain ranges we see today are comparatively young; most of Earth's present mountainous topography didn't exist before the Cretaceous Period. But even long after erosion has eliminated its peaks, a belt of "deformed" (contorted or broken) and metamorphosed rocks remains to define the location of an ancient orogen.

Why do orogens form? Scientific attempts to answer this question date back to the birth of geology, but explanations of the origin and distribution of mountains became available only with the discovery of plate tectonics theory: orogens develop because of subduction at convergent plate boundaries, rifting, continental collisions, and, locally, because of motion on transform faults.

11.3 ROCK DEFORMATION IN THE EARTH'S CRUST

Deformation and Strain

As noted above, orogeny causes deformation (bending, breaking, squashing, stretching, or shearing), which in turn yields geologic structures. To get a visual sense of deformation, let's compare a road cut along a highway in the central Great Plains of North America, a region that has not undergone orogeny, with a cliff in the Alps.

The road cut, which lies at an elevation of only about 100 m above sea level, exposes nearly horizontal beds of sandstone and shale—these beds have the same orientation that they had when first deposited (▶Fig. 11.3). Notably, sand grains in sandstone beds of this outcrop have a nearly spherical shape (the same shape they had when deposited), and clay flakes in the shale lie roughly parallel to the bedding, because of compaction. Rock of this outcrop is *undeformed*, meaning that it contains no geologic structures other than a few joints.

In the Alpine cliff, exposed at an elevation of 3 km, rocks look very different. Here, we find layers of quartzite and slate (the metamorphic equivalent of sandstone and shale) in contorted beds whose shapes resemble the wrinkles in a rug that has been pushed across the floor. These wrinkles are *folds*. Quartz grains in the quartzite are not spheres, but resemble flattened eggs, and the clay flakes in slate are aligned parallel

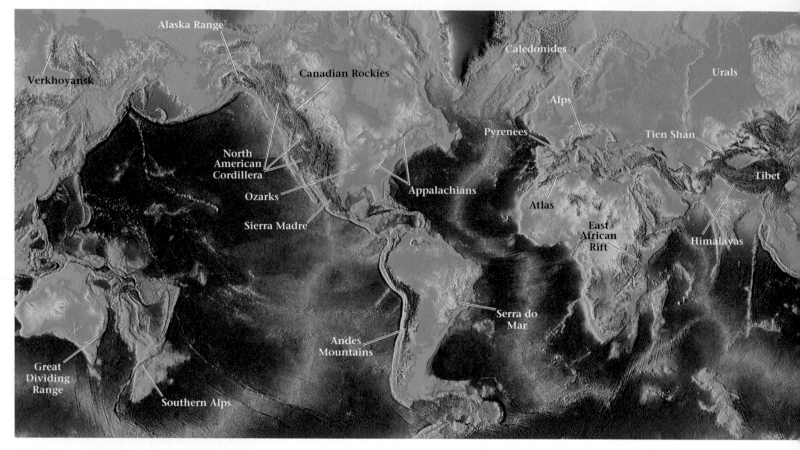

FIGURE 11.2 Digital map of world topography, showing the locations of major mountain ranges.

FIGURE 11.3 (a) This road cut exposes flat-lying beds of Paleozoic shale and sandstone along a highway, occurring in the interior region of North America. The region has *not* been involved in orogeny subsequent to the deposition of the beds. A few vertical joints cut the beds. Inset: An enlargement showing that the undeformed sandstone has spherical grains. (b) In this diagram of an Alpine mountain cliff, note the folded layers of quartzite and slate and the fault. Inset: Grains of sand in the quartzite have become flattened and are aligned parallel to one another. The slate has slaty cleavage.

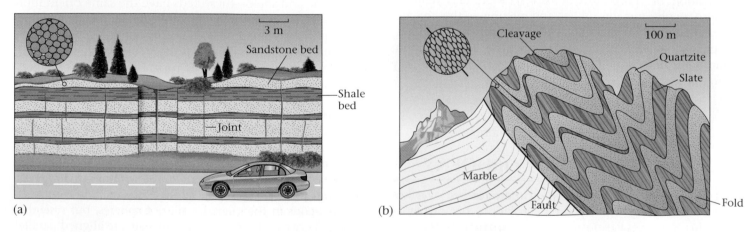

to one another and tilt at a steep angle to the bedding. In fact, the rock splits on planes called *slaty cleavage* that parallel the flattened sand grains and clay flakes, and thus cut across the bedding at a steep angle. (As you recall from Chapter 6, slaty cleavage is a type of *foliation*, or metamorphic layering.) Finally, if we try tracing the quartzite and slate layers along the outcrop face, we find that they abruptly terminate at a sloping surface marked by shattered rock. This surface is a *fault*. In this example, thick beds of marble lie below this surface, so the fault juxtaposes two different rock units—the quartzite and slate must have moved from where they formed to get to their present location.

Clearly, the beds in the Alpine cliff have been deformed, and as a result the cliff exposes a variety of geologic structures. Beds no longer have the same shape and position that they had when first formed, and the shape and orientation of grains has changed. In sum, deformation includes one or more of the following (►Fig. 11.4a–d): (1) a change in location ("translation"), (2) a change in orientation ("rotation"), (3) a change in shape ("distortion"). Deformation can be fairly obvious when observed in an outcrop (►Fig. 11.5a–c).

Geologists refer to the change in shape that deformation causes as **strain.** We distinguish among different kinds of strain according to how the rock changes shape. If a layer of rock becomes longer, it has undergone "stretching," but if the layer becomes shorter, it has undergone "shortening" (►Fig. 11.6a–c). If a change in shape involves the movement of one part of a rock body past another so that angles between features in the rock change, the result is called "shear strain" (►Figs. 11.6d; 11.7).

Kinds of Deformation: Brittle and Ductile Behavior

We saw in Chapter 10 that rocks can temporarily change shape when subjected to force (push, pull, or shear), developing an *elastic strain*, and then change back when the force that caused the strain is removed. But rocks can also develop a *permanent strain*, in two fundamentally different ways. During **brittle deformation,** a material breaks into two or more pieces, like a plate shattering on the floor, while during **ductile deformation,** a material changes shape without breaking, like a ball of dough squeezed beneath a book (►Fig. 11.8a–d). Joints and faults are brittle structures, while folds and foliations are ductile structures. What actually happens in a rock during the different kinds of deformation? Recall that rocks are solids in which chemical bonds, like little springs, link atoms together. During elastic deformation, bonds stretch and bend, but do not break. During brittle deformation, many bonds break at once so that rocks can no longer hold together, while during ductile deformation, some bonds break but new ones quickly form, so that rocks do not separate into pieces as they change shape.

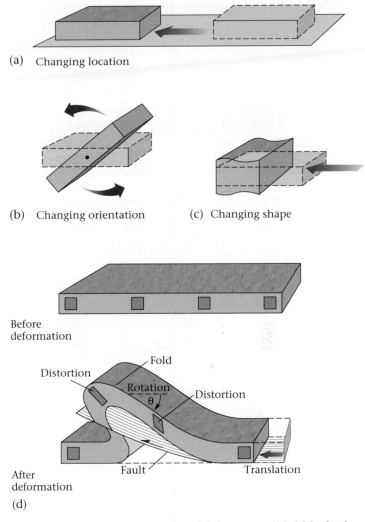

(a) Changing location

(b) Changing orientation   (c) Changing shape

Before deformation

After deformation

(d)

FIGURE 11.4 The components of deformation. (a) A block of rock changes location when it moves from one place to another. (b) It changes orientation when it tilts or rotates around an axis. (c) It changes shape when its dimensions change, or once planar surfaces become curved. (d) Folds and faults represent deformation, because they involve changes in location (e.g., sliding has occurred on a fault), orientation (a layer has tilted to form a fold), and shape (the squares in the undeformed layer have become rectangles or parallelograms in the deformed layer).

Why do rocks inside the Earth sometimes deform brittlely and sometimes ductilely? The behavior of a rock depends on:

• *Temperature:* Warm rocks tend to deform ductilely, while cold rocks tend to deform brittlely. To see this contrast, try an experiment with a candle. Chill a candle in a freezer, then press its middle against the edge of a table; the candle will brittlely snap in two. But if you first warm the candle in an oven, it will ductilely bend without breaking when pressed against the table.

(a)

(b)

(c)

FIGURE 11.5 (a) Undeformed, flat-lying beds of sediment in Badlands National Monument, South Dakota. (b) Tilted beds of strata in Arizona. The tilting is a manifestation of deformation. (c) Folded layers of quartzite and schist in Australia. The folding is also a manifestation of deformation.

- *Pressure:* Under great pressures deep in the Earth, rock behaves more ductilely than it does under low pressures near the surface. Pressure effectively prevents rock from separating into fragments.

- *Deformation rate:* A sudden change in shape causes brittle deformation, while a slow change in shape causes ductile deformation. For example, if you hit a thin marble bench with a hammer, it shatters, but if you leave it alone for a century, it gradually sags without breaking.

- *Composition:* Some rock types are softer than others; for example, halite (rock salt) can deform ductilely under conditions in which granite behaves brittlely.

Considering that pressure and temperature both increase with depth in the Earth, geologists find that in typical continental crust, rocks behave brittlely above about 10–15 km, while they behave ductilely below; we call this

FIGURE 11.6 Different kinds of strain. (a) An unstrained cube and an unstrained fossil shell (brachiopod). (b) Horizontal stretching changes the cube into a brick whose long dimension parallels the direction of stretching, and it makes the brachiopod longer. (c) Horizontal shortening changes the cube into a brick whose long dimension lies perpendicular to the shortening direction, and it makes the brachiopod taller. (d) Shear strain tilts the cube over and transforms it into a parallelogram, and it changes the angular relationships in the brachiopod.

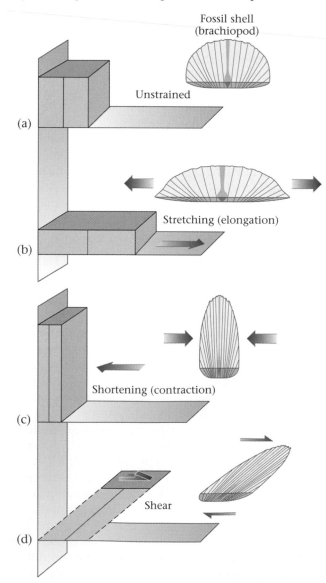

Fossil shell
(brachiopod)

Unstrained

(a)

Stretching (elongation)

(b)

Shortening (contraction)

(c)

Shear

(d)

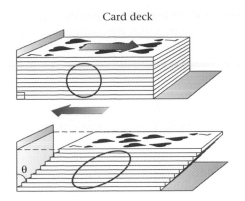

Card deck

FIGURE 11.7 You can simulate shear strain by moving a deck of cards so that each card slides a little with respect to the one below. Note how a circle drawn on the side of the deck changes shape to become an ellipse, and that the angle between the bottom of the deck and the back side of the deck has changed from a right angle into an acute angle.

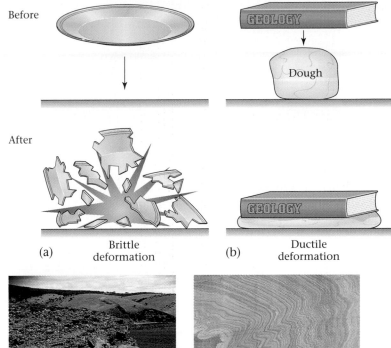

Before

After

(a) Brittle deformation

(b) Ductile deformation

(c)

(d)

FIGURE 11.8 (a) Brittle deformation occurs when you drop a plate and it shatters. (b) Ductile deformation takes place when you squash a soft ball of dough beneath a book and the dough flattens into a pancake without breaking. (c) Cracks (joints) in an outcrop result from brittle deformation. (d) Folds, like these in the marble of a quarry wall, form without breaking a rock, and thus represent ductile deformation.

depth the "brittle-ductile transition." Earthquakes in continental crust happen only above this depth because these earthquakes involve brittle breaking.

In some cases, both brittle and ductile structures occur in the same outcrop. For example, in our Alpine cliff (Fig. 11.3b), you can see both faulting (a brittle structure) and folding (a ductile structure). Such an occurrence may seem like a paradox at first. But the juxtaposition of styles happens simply because of changes in the deformation rate during orogeny. Slow deformation yielded the folds, while a pulse of rapid deformation caused the fault to form suddenly.

Force, Stress, and the Causes of Deformation

Up to this point, we've focused on picturing the *consequences* of deformation. Describing the *causes* of deformation is a bit more challenging, in the context of an introductory geology book. In captions for displays about mountains, museums and national parks typically dispense with the issue by using the phrase "The mountains were caused by forces deep within the Earth." But what does this mean? Isaac Newton defined force by using the equation: force = mass × acceleration. According to this equation, if you apply a force to an object, the object speeds up, slows down, or changes direction. Applying this concept to geology, we see that phenomena such as plate interactions (e.g., continent-continent collisions) apply forces to rock and thus cause rock to change location, orientation, or shape. In other words, the application of forces in the Earth indeed causes deformation.

However, geologists use the word "stress" instead of "force" when talking about the cause of deformation. We define the stress acting on a plane as the force applied *per unit area* of the plane. Written as an equation this becomes:

stress = force/area. The need to distinguish between stress and force arises because the actual consequences of applying a force depend not just on the amount of force but also on the area over which the force acts. A simple pair of experiments shows why. Experiment 1: Stand on a single, empty aluminum can (▶Fig. 11.9a). All of your weight—a force—focuses entirely on the can, and the can crushes. Experiment 2: Place a board over 100 cans, and stand on the board (▶Fig. 11.9b). In this case, your weight is distributed across 100 cans, so the force acting on any one can is not

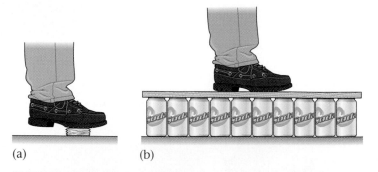

(a)   (b)

FIGURE 11.9 (a) When you stand on a single can, you apply enough force to the can to crush it, for the can feels a large stress. (b) When you stand on a board resting on 100 cans, you apply the same force to the board, but now it is spread out over 100 cans. Therefore, each can feels only a small stress and does not crumple.

enough to crush it. In both experiments, the force caused by the weight of your body was the same, but in experiment 1 the force was applied over a small area so the single can felt a *large* stress, whereas in experiment 2 the same force was applied over a large area so only a *small* stress developed. How does this concept apply to geology? During mountain building, the force of one plate interacting with another is distributed across the area of contact between the two plates, so the deformation resulting at any specific location actually depends on the stress (force/area) developed at that location, not on the total force involved in the plate interaction.

Different kinds of stress occur in rock bodies. As we learned in Chapter 10, **compression** develops when a rock is squeezed, **tension** occurs when a rock is pulled apart, and **shear stress** develops when one side of a rock body moves sideways past the other side (▶Fig. 11.10a–d).

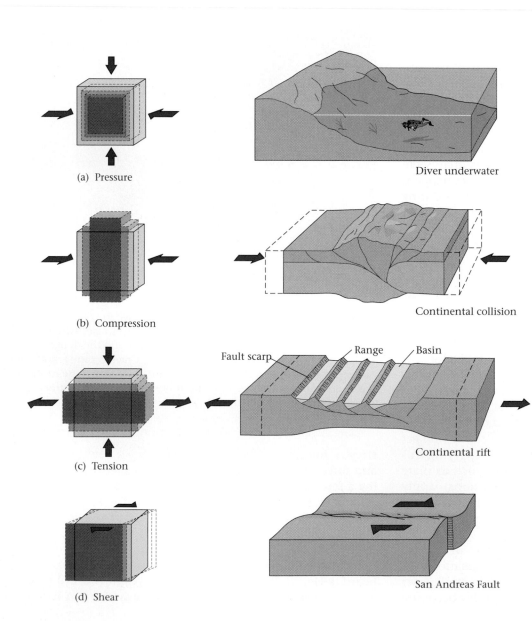

FIGURE 11.10 We represent the direction and magnitude of stress acting on each face of an object by arrows; the lengths of the arrows represent the magnitude of the stress. (a) Pressure occurs when an object feels the same stress on all sides. A diver feels pressure when submerged. (b) Compression takes place when an object is squeezed. Compression occurs during growth of a collisional mountain range. (c) Tension is created when the opposite ends of an object are pulled in opposite directions. Tension occurs during growth of a continental rift. (d) Shear stress occurs when one surface of an object slides relative to the other surface (we depict the shear direction with half arrows). Shear stress parallel to the Earth's surface causes slip on the San Andreas Fault.

Describing the Orientation of Structures

BOX 11.1
THE REST OF THE STORY

When discussing geologic structures, it's important to be able to communicate information about their orientation. For example, does a fault exposed in an outcrop at the edge of town continue beneath the nuclear power plant 3 km to the north, or does it go beneath the hospital 2 km to the east? If we knew the fault's orientation, we might be able to answer this question. To describe the orientation of a geologic structure, geologists picture the structure as a simple geometric shape, then specify the angles that the shape makes with respect to a horizontal plane (a flat surface parallel to sea level), a vertical plane (a flat surface perpendicular to sea level), and the north direction (a line of longitude).

Let's start by observing *planar* structures like faults, beds, and joints. We call these structures planar because they resemble a geometric plane. A planar structure's orientation can be specified by its strike and dip. The strike is the angle between an imaginary horizontal line (the strike line) on the plane and the direction to true north (▶Fig. 11.11a, b). We measure the strike with a magnetic compass (▶Fig. 11.11d). The **dip** is the angle of the plane's slope (more precisely, the angle between a horizontal plane and the dip line, an imaginary line parallel to the steepest slope on the plane, as measured in a vertical plane perpendicular to the strike). We measure the dip angle with a clinometer, a type of protractor that measures slope angles. A horizontal plane has a dip of 0°, and a vertical plane has a dip of 90°. We represent strike and dip on a geologic map using the symbol shown in Figure 11.11b.

A linear structure resembles a line rather than a plane; examples of linear structures include scratches or grooves on a rock surface. Geologists specify the orientation of linear structures by giving their plunge and bearing (▶Fig. 11.11c). The **plunge** is the angle between a line and horizontal, as measured with a clinometer, in the vertical plane that contains the line. A horizontal line has a plunge of 0°, and a vertical line has a plunge of 90°. The **bearing** is the compass heading of the line (more precisely, the angle between the projection of the line on the horizontal plane and the direction to true north).

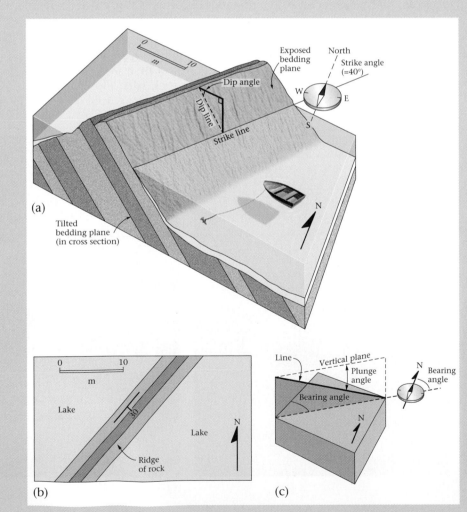

FIGURE 11.11 (a) We use strike and dip to measure the orientation of planar structures like these tilted beds. The strike is the compass angle between the strike line (an imaginary horizontal line on the plane) and true north. The dip is the angle between the strike line and the dip line (an imaginary line parallel to the steepest slope on the plane) as measured in a vertical plane. Note that the strike line and the dip line are perpendicular to each other. (b) On a map, the line segment represents the strike direction, while the tick on the segment represents the dip direction. The number indicates the dip angle as measured in degrees. (c) To specify the orientation of a line, we use plunge and bearing. The plunge is the angle between the line and horizontal as measured in a vertical plane, whereas the bearing is the compass orientation of the line. (d) A geologist measuring the strike of a moderately dipping bed, next to a riverbed. Note the water line is a strike line.

Pressure refers to a special stress condition in which the same push acts on all sides of an object. Note that "stress" and "strain" have different meanings to geologists (though we tend to use them interchangeably in everyday English): stress refers to the amount of force per unit area of a rock, while strain refers to the change in shape of a rock. Thus, stress *causes* strain. Specifically, compression causes shortening, tension leads to stretching, and shear stress creates shear strain. Pressure can cause an object to become smaller (i.e., pull in on all sides equally), but will not cause it to change shape. With our knowledge of stress and strain, we can now look at the nature and origin of various classes of geologic structures.

11.4 JOINTS: NATURAL CRACKS IN ROCKS

If you look at the photographs of rock outcrops in this book, you'll notice thin black lines that cross the rock faces (▶Fig. 11.8c). These lines represent traces of natural cracks, along which the rock broke and separated into two pieces during brittle deformation. Geologists refer to such natural cracks as *"joints."* Rock bodies do *not* slide past each other on joints. (Since joints are roughly planar structures, we define their orientation by their strike and dip; see Box 11.1.)

Joints develop in response to tensional stress in brittle rock: a rock splits open because it has been pulled slightly apart. They may form for a variety of geologic reasons. For example, some joints form when a rock cools and contracts, because contraction makes one part of a rock pull away from the adjacent part. Others develop when rock layers formerly at depth feel a decrease in pressure as overlying rock erodes away, and thus change shape slightly. Still others form when rock layers bend.

Rock bodies may contain two categories of joints. "Systematic joints" are long planar cracks that occur fairly regularly through a rock body, while "nonsystematic joints" are short cracks that occur in a range of orientations and are randomly spaced. A group of systematic joints constitutes a "joint set," a spectacular example of which can be seen in Arches National Park, in Utah (▶Fig. 11.12a). Thick sandstone beds in the park cracked and developed joints. Erosion has created narrow gullies along the joints. In sedimentary rocks, systematic joints typically are vertical planes (▶Fig. 11.12b).

If groundwater seeps through joints for a long period of time, minerals like quartz or calcite can precipitate out of the groundwater and fill the joint. Such mineral-filled joints are called **veins** and look like white stripes cutting across a body of rock (▶Fig. 11.12c). Some veins contain small quantities of valuable metals, like gold.

Geotechnical engineers, people who study the geologic setting of construction sites, pay close attention to jointing when recommending where to put roads, dams, and buildings. Water flows much more easily through joints than it does through solid rock, so it would be a bad investment to situate a water reservoir over rock with closely spaced

FIGURE 11.12 (a) This bedding plane in sandstone of Arches National Park, Utah, contains many systematic joints. (b) These vertical joints exposed on a cliff face near Ithaca, New York, run from the surface down into bedrock. (c) The veins in this outcrop, composed of milky white quartz, fill fractures in gray shale.

(a)

(b)

(c)

joints—the water would leak down into the joints. Also, building a road on a steep cliff composed of jointed rock could be risky, for joint-bounded blocks separate easily from bedrock, and the cliff might collapse.

11.5 FAULTS: FRACTURES ON WHICH SLIDING HAS OCCURRED

After the San Francisco earthquake of 1906, geologists found a rupture that ripped across the landscape near the city. Where this rupture crossed orchards, it offset rows of trees, and where it crossed a fence, it broke the fence in two; the western side of the fence moved northward by about 2 m (Fig. 10.6a). The rupture represented the trace of the San Andreas Fault (▶Fig. 11.13a, b). As we have seen, a "fault" is a fracture on which sliding occurs. Slip events, or "faulting," generate earthquakes. Faults, like joints, are planar structures, so we represent their orientation by strike and dip.

Faults riddle the Earth's crust. Some are currently active (sliding has been occurring on them in recent geologic time), but most are inactive (sliding on them ceased millions of years ago). Some faults, like the San Andreas, intersect the ground surface and thus displace the ground when they move. Others accommodate the sliding of rocks in the

FIGURE 11.13 (a) An oblique air photo showing the San Andreas Fault displacing a creek flowing from the Tremblor Range (background) into the Carizzo Plain, California. (b) What a geologist sees in the previous photo. (c) A road cut in the Rocky Mountains of Colorado, showing a fault offsetting strata in cross section. Note that the fault is actually a band of broken rock about 50 cm wide. (d) What a geologist sees looking at the Rocky Mountain road cut.

(a)

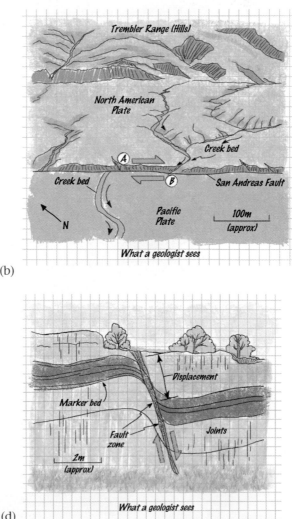

(b)

(c)

(d)

crust at depth, and remain invisible at the surface unless they are later exposed by erosion (Fig. 11.13c, d).

Geologists study faults not only because the movement on some faults causes earthquakes, but also because they juxtapose bodies of rock that did not originally lie adjacent to each other and thus complicate the arrangement of rocks at the Earth's surface. (For example, in our Alpine cliff, Fig. 11.3b, movement on a fault placed quartzite and slate beds against marble beds.) We must understand these rearrangements in order to predict where resources lie underground.

Fault Classification

Geologists have developed terminology to classify faults and describe movement on them. (We introduced this terminology in Chapter 10, and add to it here.) The fault plane can be vertical, horizontal, or at some angle in between, and we can describe its orientation by a strike and dip measurement. In the case of nonvertical faults (those that slope at an angle), we can define the **hanging-wall block** as the rock above the fault plane, and the **footwall block** as the rock below the fault plane (▶Fig. 11.14a). If you stand in a tunnel along a fault plane, the hanging-wall block looms over your head, and the footwall block lies under your feet. We distinguish several types of faults.

- *Dip-slip vs. strike-slip vs. oblique-slip faults:* On **dip-slip faults,** sliding occurs up or down the slope of the fault (therefore, up or down the dip); on **strike-slip faults,** one block slides past another (therefore, parallel to the strike line); and on **oblique-slip faults,** sliding occurs diagonally on the fault plane (▶Fig. 11.14b–d).

- *Types of dip-slip faults:* We subdivide dip-slip faults into two kinds, depending on which way the hanging-wall block moves relative to the footwall block. On

FIGURE 11.14 (a) A hanging-wall block and footwall block, relative to a sloping fault surface. The weathered fault scarp is an exposure of the fault at the ground surface. (b) Three types of dip-slip faults, on which sliding parallels the dip line. (c) Two types of strike-slip faults, on which sliding parallels the strike line. (d) Two examples of oblique-slip faults, on which sliding takes place diagonally along the surface.

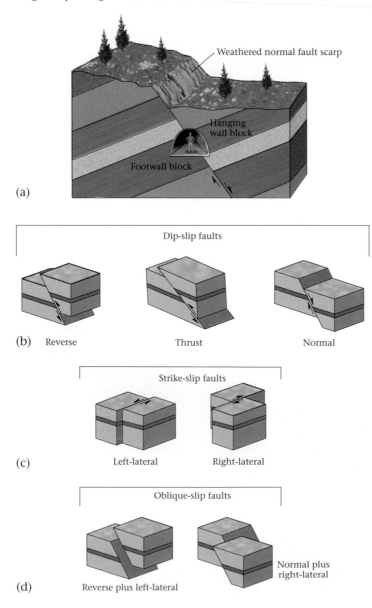

(a)

Dip-slip faults

(b) Reverse — Thrust — Normal

Strike-slip faults

(c) Left-lateral — Right-lateral

Oblique-slip faults

(d) Reverse plus left-lateral — Normal plus right-lateral

FIGURE 11.15 This large thrust fault (the Lewis thrust) puts older rock (Precambrian) over younger rock (Mesozoic). Erosion has removed much of the hanging-wall block, but a small remnant still lies to the east of the mountains. On the geologic map of the region, the triangular barbs point to the hanging-wall block. The hanging wall has moved about 100 km relative to the footwall.

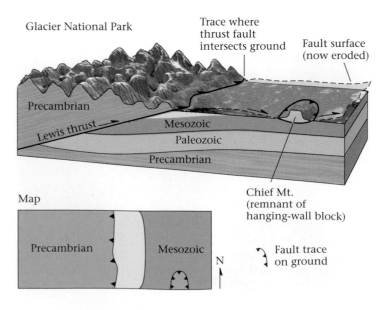

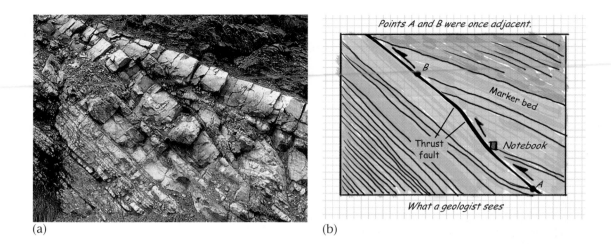

FIGURE 11.16 (a) A thrust fault, on which a distinct layer has been offset. (b) A geologist's sketch emphasizes the offset. Point B was originally adjacent to point A. (c) A fault scarp formed after an earthquake in Nevada. (d) This fault breccia along a fault consists of broken-up rock. (e) Slip lineations on a fault surface.

thrust faults and **reverse faults,** the hanging-wall block moves up the slope of the fault. Thrust faults differ from reverse faults only in terms of the fault-plane's slope (or dip)—thrust faults have a slope (or dip) of less than about 35°, while reverse faults have a slope of greater than 35° (▶Fig. 11.15). On **normal faults,** the hanging-wall block moves down the slope of the fault. "Normal" and "reverse" are relics of nineteenth-century miners' jargon. Normal faults were simply more common in the mines where faults were first recognized. But globally, normal faults aren't any more common or typical than reverse faults.

• *Types of strike-slip faults:* Geologists distinguish between two types of strike-slip faults, based on the relative movement of one side of the fault with respect to the other. If you stand facing the fault, you can say that it is a "left-lateral strike-slip fault" if the block on the far side slipped to your left, and that it is a right-lateral strike-slip fault if the block on the far side slipped to your right. Note that

strike-slip faults commonly have a vertical dip, so we generally cannot define the hanging-wall or footwall block on such faults.

Recognizing Faults

How do you recognize a fault when you see one? The most obvious criterion is the appearance of **displacement,** or offset, meaning the amount of movement across a fault plane (Fig. 11.13c; ▶Fig. 11.16a, b). Displacement disrupts the layers in rocks, so that layers on one side of a fault are not continuous with layers on the other side. In our Alpine cliff example (Fig. 11.3b), we can spot the fault as the plane where quartzite and slate beds are juxtaposed against marble beds. Typically, thrust or reverse faults cutting sedimentary beds place older beds on younger ones, while normal faults place younger beds on older. In some cases, layers of rock cut by a fault undergo folding during, or just before slip; the resulting folds are informally called "drag folds" (Fig. 11.4d).

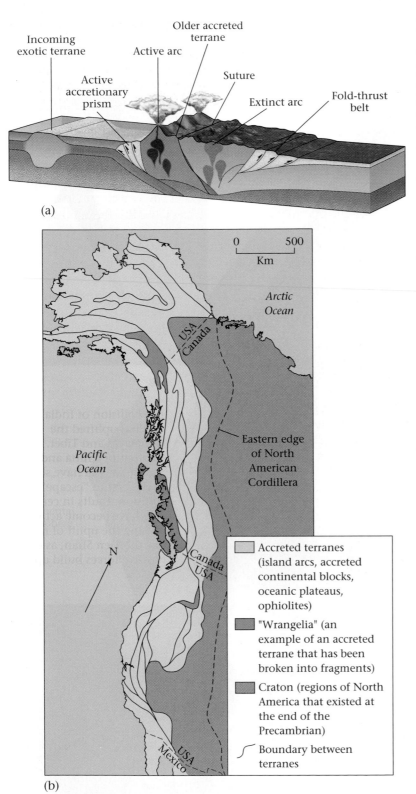

(a)

(b)

FIGURE 11.33 (a) In a convergent-margin orogen, volcanic arcs form, and there may be compression. Where this occurs, a large mountain range develops. Exotic terranes may collide with the convergent margin and accrete to the orogen. (b) Much of the western portion of the North American Cordillera consists of accreted terranes.

Mountains Related to Continental Collision

Once the oceanic lithosphere between two continents completely subducts, the continents themselves collide with one another. Continental collision results in the creation of large mountain ranges such as the present-day Himalayas or the Alps (▶Fig. 11.34) and the Paleozoic Appalachian Mountains. The final stage in the growth of the Appalachians happened when Africa and North America collided.

During collision, intense compression generates fold-thrust belts on the margins of the orogen. In the interior of the orogen, where one continent overrides the edge of the other, high-grade metamorphism occurs, accompanied by formation of flow folds and tectonic foliation. During this process, the crust below the orogen thickens to as much as twice its normal thickness. Gradually, rocks squeeze upward in the hanging walls of large thrust faults and later become exposed by exhumation. Finally, as noted earlier, rock at depth in the orogen heats up and becomes so weak that the mountain belt may collapse and spread out sideways. The broad Tibet Plateau may have formed in part when crust, thickened during the collision of India with Asia, spread to the northeast (see art on pp. 352–53).

Mountains Related to Continental Rifting

Continental rifts are places where continents are splitting in two. When rifts first form, there is generally significant uplift, and this uplift contributes to creating mountainous topography (▶Fig. 11.35). Uplift occurs, in part, because as the lithosphere thins, hot asthenosphere rises, making the remaining lithosphere less dense. Because the lithosphere is less dense, it becomes more buoyant and thus rises to reestablish isostatic equilibrium.

FIGURE 11.34 In a collisional orogen, two continents collide. The compression that results from the collision shortens and thickens the continental crust so that a large mountain range develops. Fold-thrust belts form along the margins of the orogen.

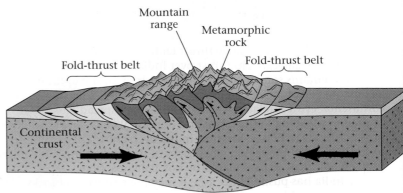

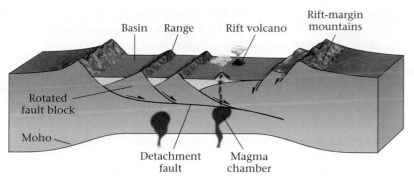

FIGURE 11.35 When the crust stretches in a continental rift, rift-related mountains form, as do normal faults. Displacement on the faults leads to the tilting of crustal blocks and the formation of half-grabens. The half-grabens fill with sediment eroded from the adjacent mountains. Later, the exposed ends on the tilted blocks create long, narrow ranges, and the half-grabens become flat basins. In the United States, the region containing such a structure is called the Basin and Range Province (located in Nevada, Utah, and Arizona).

As heating in a rift takes place, stretching causes normal faulting in the brittle crust above, creating a normal-fault system. Movement on the normal faults drops down blocks of crust, creating deep basins separated by narrow, elongate mountain ranges that contain tilted rocks. These ranges are sometimes called **fault-block mountains.** In addition, the rising asthenosphere beneath the rift partially melts, generating magmas that rise to form volcanoes within the rift. Today, the East African Rift clearly shows the configuration of rift-related mountains and volcanoes. And in North America, rifting yielded the broad Basin and Range Province of Utah, Nevada, and Arizona. If you drive across the province from east to west, you'll pass over two dozen fault-block mountain ranges, separated from each other by sediment-filled basins.

11.11 CRATONS AND THE DEFORMATION WITHIN THEM

A **craton** consists of crust that has not been affected by orogeny for at least 1 billion years. Because orogeny happened so long ago in cratons, their crust has become quite cool, and therefore relatively strong and stable. We can divide cratons into two provinces: **shields,** in which Precambrian metamorphic and igneous rocks crop out at the ground surface, and the **cratonic platform,** where a relatively thin layer of Phanerozoic sediment covers the Precambrian rocks (▶Fig. 11.36; see Fig. 13.9).

In shield areas, we find intensively deformed metamorphic rocks—abundant examples of shear zones, flow folds, and tectonic foliation. That's because the crust making the cratons was deformed during a succession of orogenies in the precambrian. Recent studies of the Canadian Shield, which occupies much of the eastern two-thirds of Canada, for example, reveal the traces of Himalaya-like collision zones, Andean-like convergent boundaries, and East African–like rifts, all formed more than 1 billion years ago (some over 3 billion years ago). These orogens are so old that erosion has worn away the original topography, in the process exposing deep crustal rocks at the Earth's surface.

In the cratonic platform, we can't see the Precambrian rocks and structures, except where they are exposed by deep erosion. Younger strata do display deformation features, but in contrast to the deformation of orogens, cratonic-platform deformation is less intense. The cratonic platform of the U.S. Midwest region includes two classes of structures: regional basins and domes, and local zones of folds and faults.

Regional basins and regional domes are broad areas that gradually sank or rose, respectively (▶Fig. 11.37a, b). They are illustrated by a slice of the upper crust running across Missouri and Illinois. In Missouri, strata arch across a broad dome, the Ozark Dome, whose diameter is 300 km. Individual sedimentary layers thin toward the top of the dome, because less sediment accumulated on the dome than in adjacent basins. Erosion during more recent geologic history has produced the characteristic bull's-eye

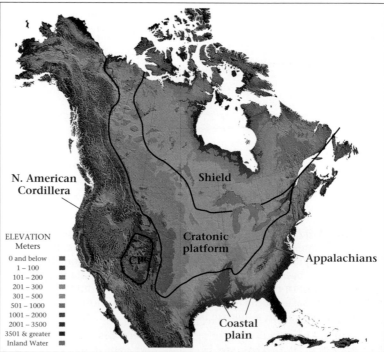

FIGURE 11.36 Digital elevation map of North America showing the platform and shield areas. CP=Colorado Plateau.

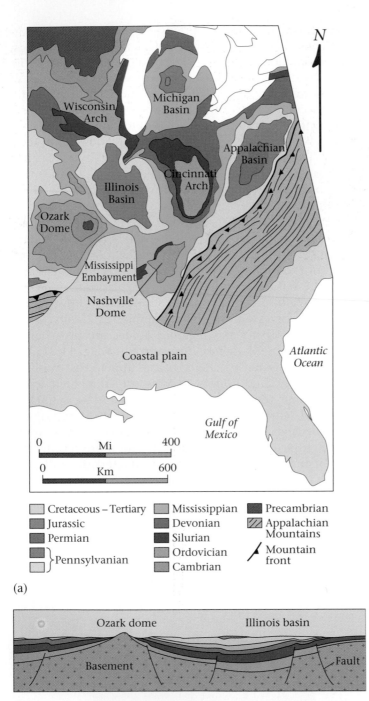

(a)

(b)

FIGURE 11.37 (a) Geologic map of the mid-continent region of the United States, showing the basins and domes and the faults that cut the region. (b) This cross section illustrates the geometry of the regional basins and domes. Note how layers of strata thin toward the crest of the Ozark Dome and thicken toward the center of the Illinois Basin. Some of the faults originated as normal faults and later moved again as reverse faults.

pattern of a dome, with the oldest rocks (Precambrian granite) exposed near the center. In the Illinois Basin, strata appear to warp downward into a huge bowl that is also about 300 km across. Strata get thicker toward the center, indicating that the floor of the basin sank so there was more room for sediment to accumulate. The Illinois Basin also has a bull's-eye shape, but here the youngest strata are in the center. Geologists refer to the broad vertical movements that generate huge, but gentle, mid-continent domes and basins as **epeirogeny.**

Folds and faults are hard to find in the cratonic platform, because most do not cut the ground surface. But subsurface studies indicate that faults do occur at depth. Monoclines, step-shaped folds, develop over these faults; the folds formed as a block of basement pushes up. Most of these zones were likely active when major orogenies happened along the continental margin. This relation suggests that the orogenies created enough stress in the craton to cause faults to move, but not enough to generate large mountains or to create foliation.

11.12 LIFE STORY OF A MOUNTAIN RANGE: A CASE STUDY

Perhaps the easiest way to bring together all the information in this chapter is to look at the life story of one particular mountain range—let's take the Appalachian Mountains of North America as our example. (We've simplified the story a bit, for ease of reading.) Geologists have constructed the range's life story by studying its structures, by determining the ages of igneous and metamorphic rocks, and by searching for strata formed from sediment eroded from the range.

About 1 billion years ago, the Appalachian region was involved in a massive collision with another continent (►Fig. 11.38). This event, called the Grenville orogeny, yielded a belt of deformed and metamorphosed rocks that underlie the eastern fifth of the continent. For a while, after the Grenville event, the Appalachian region lay in the middle of a supercontinent. But this supercontinent rifted apart around 600 million years ago. Eventually, new ocean formed to the east, and the former rifted margin of eastern North America cooled, sank, and evolved into a passive-margin sedimentary basin. From 600 to about 420 million years ago, this basin filled with a thick sequence of sediment.

Between around 420 and 370 million years ago, two collisions took place between North America and exotic terranes. During the first convergent event, called the Taconic orogeny, a crustal block and volcanic arc collided with eastern North America, and during the second

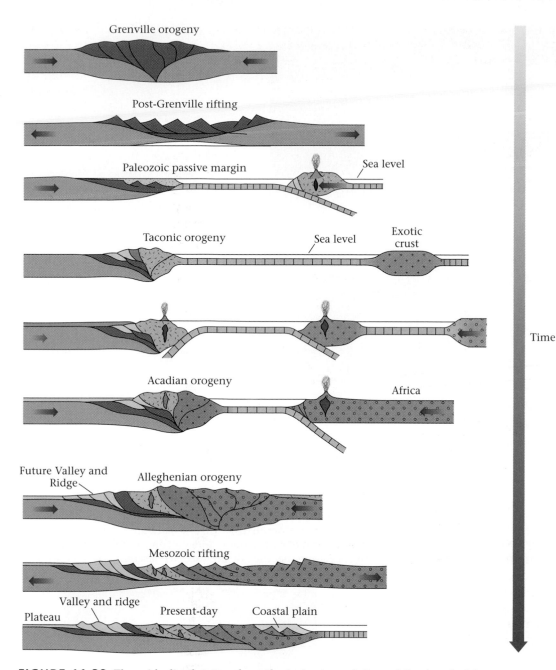

FIGURE 11.38 These idealized stages show the tectonic evolution of the Appalachian Mountains. Note that while mountains do form during rifting events, geologists traditionally assign names only to the collisional or convergent events.

convergent event, the Acadian orogeny, continental crustal slivers accreted to the continent. The accretion of these terranes deformed the sediment that had accumulated in the passive-margin basin, and made the continent grow eastward. Significant strike-slip displacement occurred during these events; thus, slivers of crust were transported along the margin of the continent. Then, 270 million years ago, Africa collided with North America. This event, the Alleghenian orogeny, yielded a huge mountain range resembling the present-day Himalayas and created a wide fold-thrust belt along the mountains' western margin. Eroded folds of this belt make up the topography of the present Valley and Ridge Province in Pennsylvania (▶Fig. 11.39).

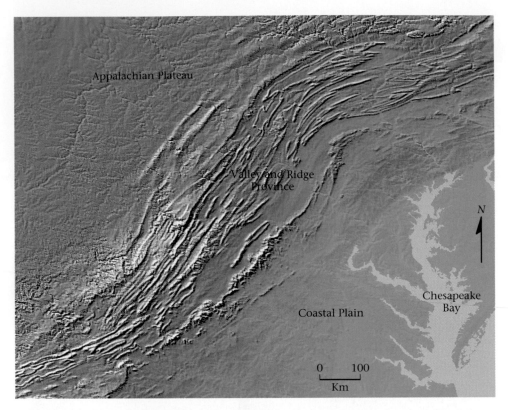

FIGURE 11.39 Relief map of the Valley and Ridge Province, Pennsylvania to Virginia. The ridges, which outline the shapes of plunging folds in the fold-thrust belt, are composed of resistant sandstone beds.

When the Alleghenian orogeny ceased, the Appalachian region once again lay in the interior of a supercontinent (Pangaea), where it remained until about 180 million years ago. At that time, rifting split the region open again, creating the Atlantic Ocean. As you can see from this example, major ranges like the Appalachians incorporate the products of multiple orogenies and reflect the opening and closing of ocean basins (a sequence of events called the "Wilson cycle" after J. Tuzo Wilson).

11.13 MEASURING MOUNTAIN BUILDING IN PROGRESS

Mountains are not just "old monuments," as John Muir mused. The rumblings of earthquakes and the eruptions of volcanoes attest to present-day movements in mountains. Geologists can measure the rates of these movements through field studies and satellite technology. For example, geologists can determine where coastal areas have been rising relative to the sea level by locating ancient beaches that

FIGURE 11.40 Convergence between the Nazca Plate and South America creates the broad Andean orogen along the western coast. Here, the blue arrow indicates the rate of motion of Nazca relative to the interior of South America. The black arrows indicate the rate of movement of places in the Andes relative to the interior of South America. Note that the rate decreases to the east.

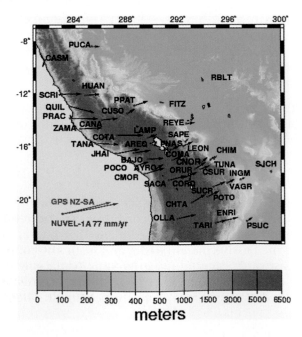

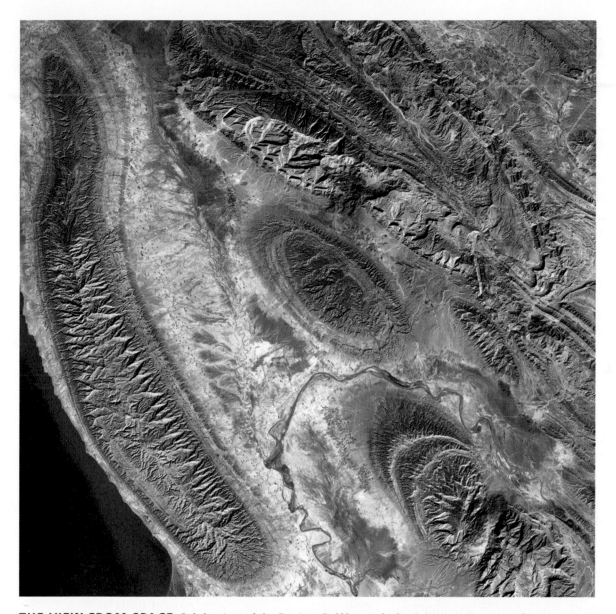

THE VIEW FROM SPACE Subduction of the Persian Gulf beneath the Asian continent, in the vicinity of Konari, Iran, produced doubly plunging folds. Dipping strata outline the ellipsoidal shape of the folds. The large one, near the coast, is about 10 km wide. Some of the folds are cored by salt. This is a tectonically active landscape, so deformation is warping the land surface.

now lie high above the water. And they can tell where the land surface has risen relative to a river by identifying places where a river has recently carved a new valley down into sediments that it had previously deposited. In addition, geologists now can use the satellite **global positioning system (GPS)** to measure rates of uplift and horizontal shortening in orogens. With present technology, we can "see" the Andes shorten horizontally at a rate of a couple of centimeters per year, and we can "watch" as mountains along this convergent boundary rise by a couple of millimeters per year (▶Fig. 11.40).

CHAPTER SUMMARY

• Mountains occur in linear ranges called mountain belts, orogenic belts, or orogens. An orogen forms during an orogeny, or mountain-building event. Orogenies, which last for millions of years, are a consequence of continental collision, subduction at a convergent plate boundary, or rifting.

• Mountain building causes rocks to bend, break, squash, stretch, and shear. Because of such deformation, rocks change their location, orientation, and shape.

- During brittle deformation, rocks crack and break into two or more pieces. During ductile deformation, rocks change shape without breaking.

- Rocks undergo three kinds of stress: compression, tension, and shear.

- Strain refers to the way rocks change shape when subjected to a stress. Compression causes shortening, tension causes stretching, and shear stress leads to shear strain.

- Deformation results in the development of geologic structures. Brittle structures include joints and faults, while ductile structures include folds and foliation.

- Structures can be visualized as geometric lines or planes. We can define the orientation of a plane by giving its strike and dip and the orientation of a line by giving its plunge and bearing.

- Joints are natural cracks in rock, formed in response to tension under brittle conditions. Some joints develop when rock cools and contracts; others form when erosion decreases the pressure on rocks buried at depth.

- Veins develop when minerals precipitate out of water passing through joints.

- Faults are fractures on which there has been shearing. In the case of nonvertical faults, the rock above the fault plane is the hanging-wall block, and the rock below the fault plane is the footwall block.

- On normal faults, the hanging-wall block slides down the surface; on reverse faults, the hanging-wall block slides up the surface; on strike-slip faults, rock on one side of the fault slides horizontally past the other; and on oblique-slip faults, rock slides diagonally across the surface.

- Faults can be recognized by the presence of broken rock (breccia) or fine powder (gouge). Scratches or grooves on fault surfaces are called slip lineations.

- Folds are curved layers of rock. Anticlines are arch-like folds, synclines are trough-like, monoclines resemble the shape of a carpet draped over a stair step, basins are shaped like a bowl, and domes are shaped like an overturned bowl.

- Tectonic foliation forms when grains are flattened or rotated so that they align parallel with one another, or when new platy grains grow parallel to one another.

- The process of orogeny may yield new igneous, metamorphic, and sedimentary rocks.

- Large mountain ranges are underlaid with relatively buoyant roots. The height of such mountains is controlled by isostasy.

- Once uplifted, mountains are sculpted by the erosive forces of glaciers and rivers. Also, when the crust thickens during mountain building, the lower part eventually becomes warm and weak and begins to flow, leading to orogenic collapse.

- Mountain belts formed by convergent margin tectonism may incorporate accreted terranes.

- Continental collision, which resulted in the Alps, Himalayas, and Appalachians, generates metamorphic rocks and tectonic foliation. Fold-thrust belts form on the continental edge of collisional and convergent-margin orogens.

- Tilted blocks of crust in rifts become narrow, elongate mountain ranges, called fault-block mountains.

- Cratons are the old, relatively stable parts of continental crust. They include shields, where Precambrian rocks are exposed at the surface, and platforms, where Precambrian rocks are buried by a thin layer of sedimentary rock. Broad regional domes and basins form in platform areas because of epeirogeny.

- With modern satellite technology, it is now possible to measure the slow movements of mountains.

KEY TERMS

accretionary orogens (p. 351)
anticline (p. 341)
basin (p. 342)
bearing (p. 335)
brittle deformation (p. 331)
compression (p. 334)
craton (p. 355)
cratonic platform (p. 355)
crustal root (p. 348)
cuesta (p. 350)
deformation (p. 329)
detachment fault (p. 340)
dip (p. 335)
dip-slip fault (p. 338)
displacement (p. 339)
dome (p. 342)
ductile deformation (p. 331)
epeirogeny (p. 356)
exotic terranes (p. 351)
fault (p. 329)
fault-block mountains (p. 355)
fault scarp (p. 340)
fold (p. 329)
fold-thrust belt (p. 351)
foliation (p. 329)
footwall block (p. 338)
geologic structures (p. 329)
global positioning system (GPS) (p. 359)
hanging-wall block (p. 338)

hinge (p. 340)
hogbacks (p. 350)
isostasy (isostatic equilibrium) (p. 349)
joint (p. 329)
limbs (of fold) (p. 341)
monocline (p. 341)
mountain belts (p. 329)
normal fault (p. 339)
oblique-slip fault (p. 338)
orogenic belts (orogens) (p. 329)
orogenic collapse (p. 351)
orogeny (p. 329)
plunge (p. 335)
pressure (p. 336)
reverse fault (p. 339)
shear stress (p. 334)
shear zone (p. 340)
shield (p. 355)
slickensides (p. 340)
slip lineations (p. 340)
strain (stretching, shortening) (p. 331)
strike-slip fault (p. 338)
syncline (p. 341)
tectonic foliation (p. 346)
tension (p. 334)
thrust fault (p. 339)
uplift (p. 348)
veins (p. 336)

REVIEW QUESTIONS

1. What are the changes that rocks undergo during formation of an orogenic belt like the Alps?

2. What is the difference between brittle and ductile deformation?

3. What factors influence whether a rock will behave in brittle or ductile fashion?

4. How are stress and strain different?

5. How is a fault different from a joint?

6. Compare the motion of normal, reverse, and strike-slip faults.

7. How do you recognize faults in the field?

8. Describe the differences among an anticline, a syncline, and a monocline.

9. Discuss the relationship between foliation and deformation.

10. Explain how certain kinds of igneous, sedimentary, and metamorphic rocks are formed during orogeny.

11. Describe the principle of isostasy.

12. What happens to the isostatic equilibrium of a mountain range as it is eroded away?

13. What happens to a mountain range when its uplift rate slows down?

14. Discuss the processes by which mountain belts are formed in convergent margins, in continental collisions, and in continental rifts.

15. How are the structures of a craton different from those of an orogenic belt?

SUGGESTED READING

Condie, K. 1997. *Plate Tectonics and Crustal Evolution*. 4th ed. Woburn, Mass.: Butterworth-Heinemann.

Davis, G. H., and S. J. Reynolds. 1996. *Structural Geology of Rocks and Regions*. 2nd ed. New York: Wiley.

Hancock, P. L., ed. 1994. *Continental Deformation*. Oxford, England: Pergamon Press.

Hatcher, R. D., Jr. 1990. *Structural Geology: Principles, Concepts, and Problems*. Upper Saddle River, N.J.: Prentice-Hall.

McPhee, J. 1998. *Annals of the Former World*. New York: Farrar, Straus and Giroux.

Moores, E. M., and R. J. Twiss. 1995. *Tectonics*. New York: Freeman.

Park, R. G. 1988. *Geologic Structures and Moving Plates*. New York: Chapman and Hall.

Van der Pluijm, B. A., and S. Marshak, 2004. *Earth Structure: An Introduction to Structural Geology and Tectonics*, 2nd ed. New York: W. W. Norton.

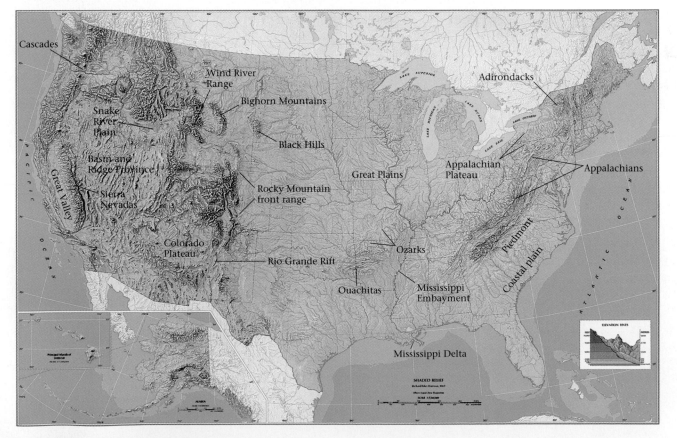

REFERENCE MAP Topography of the United States.

History before History

*P*erhaps the most important contribution that the science of geology has made to our understanding of Earth is the demonstration that our planet existed long, long before humans took their first footsteps. In Part IV, we peer back into this history. First we look at fossils, remnants of ancient life that allow us to correlate life's evolution with that of Earth. Then, in Chapter 12, we learn how geologists gaze into "deep time"—geologic time, or the time since Earth formed—first by determining the relative ages of geologic features (whether one feature is older or younger than another), and then by learning how to calculate numerical ages (ages in years) by measuring the ratios of radioactive elements to their daughter products in minerals. With the background provided in Chapter 12, we're ready for Chapter 13's brief synopsis of Earth history—from the birth of the planet to the present. We see how plate tectonics has redistributed continents and built mountains, how sea level has risen and fallen, and climate has changed.

Every rock exposure has a tale to tell about Earth history. On the coast of France at Etretat, a site that inspired several famous paintings of the Impressionist school in the nineteenth century, the layers of chalk record a time before the cliffs rose and the entire region in view was submerged by the sea.

Memories of Past Life: Fossils and Evolution

D.1 THE DISCOVERY OF FOSSILS

Rocks throughout the world contain shapes that closely resemble shells, bones, stems, or leaves (▶Fig. D.1). How do these shapes form? People initially assumed that they were the handiwork of supernatural beings, placed in rock as pranks. Then, about 450 B.C.E., the Greek historian Herodotus suggested that imprints of shells in rocks were the remains of sea creatures that lived when the ocean covered what is now dry land. Beginning around 300 B.C.E., however, students of Aristotle abandoned this hypothesis and argued that lifelike shapes grew within rocks and had nothing to do with life; this concept of inorganic growth became dogma throughout Medieval times.

At the dawn of the Renaissance, lifelike shapes, along with any other items dug up from the Earth (such as mineral and rock specimens), came to be known as "fossils" (from the Latin word *fossilis,* which means "dug up"), and their origin once again became an issue for debate. Leonardo da Vinci favored Herodotus's idea, but Georgius Agricola (born George

Bauer), in his 1546 *De natura fossilium* (considered the first geology textbook), maintained the medieval idea that the shapes grew in place. Some authors suggested another alternative, that fossil shells formed when seeds dropped by clouds washed down cracks and grew inside rocks.

In the early 1600s, the term **fossil** assumed its present definition: the remnant or trace of an ancient living organism that has been preserved in rock or sediment. Fossils include preserved tracks (like footprints and worm burrows), bones that have been transformed into rock, insects encased in amber, and the imprints of shells. The modern era of fossil study began in 1669, when Nicholaus Steno suggested that fossil-containing rocks originated as loose sediment incorporating the remains of organisms; when the sediment hardened into rock, the remains became fossils. Robert Hooke (1635–1703) described fossils in detail (▶Fig. D.2a) and was the first to examine them with a microscope. Hooke and his contemporaries also realized that most fossils represent extinct species, meaning species that have died out entirely and have vanished forever from the Earth. During the next two centuries, geologists (mostly amateurs who studied rocks as an avocation) described thousands of fossils and established museum collections (▶Fig. D.2b).

By 1800, geologists had found that different fossils occur in different layers of strata within a sequence of sedimentary rock. A certain fossil species present in the lower layers is absent in the higher layers. The level at which the species disappears represents the time it went extinct. Extinct species never reappear higher in the sequence. As we will see in Chapter 12, this concept, now called the principle of fossil succession, means that extinction lasts forever.

The nineteenth century saw **paleontology,** the study of fossils, ripen into a science. (Note that paleontology differs from *archaeology:* the former focuses on the remains of ancient organisms, while the latter focuses on the remains, or artifacts, of human culture.) William Smith, an engineer supervising canal construction in England during the 1830s, showed that because of fossil succession, the group of fossils found in a particular layer of strata can be used as a basis for determining the age of that layer relative to other layers. Thus, fossils became an indispensable tool for study-

FIGURE D.1 This bedding surface in limestone contains fossils of organisms that lived about 420 million years ago. These particular species no longer exist on Earth.

(a)

(b)

FIGURE D.2 (a) Robert Hooke published these sketches of fossil ammonites (an organism with a chambered shell) in 1703. They are among the first such sketches to be published. (b) A drawer of labeled fossils in a museum. Paleontologists from around the world study such collections to help identify unknown specimens.

ing geologic time and the evolution of life. As this subject is the focus of the next two chapters, we learn here what fossils are and how they develop.

D.2 FOSSILIZATION

What Kinds of Rocks Contain Fossils?

Most fossils are found in sediments or sedimentary rocks, for they form when organisms die and become buried by sediment, or when organisms travel over or through sediment and leave their mark. Rocks formed from sediments deposited under anoxic (oxygen-free) conditions in quiet water (such as lake beds or lagoons) preserve particularly fine specimens. Rocks made from sediments deposited in high-energy environments, on the other hand—where strong currents tumble shells and bones and break them up—contain at best only small fragments of fossils mixed with other clastic grains.

Fossils can survive low grades of metamorphism, but not the recrystallization and new mineral growth, and in some

cases shearing, that occur during intermediate- and high-grade metamorphism. Similarly, fossils generally do not occur in igneous rocks that crystallize directly from melt, for organisms can't live in molten rock, and if engulfed by molten rock, they will be incinerated. Occasionally, however, lava flows preserve the shapes of tree trunks, because lava may surround a tree and freeze before the tree completely burns up—the resulting hole in the lava is, strictly speaking, a fossil. Also, fossils can occur in deposits formed from air-fall ash, for the ash settles just like sediment and can bury an organism or a footprint. In fact, ash preserved the footprints of 3.6-million-year-old human ancestors in ash deposits now exposed in Olduvai Gorge, in the East African Rift (▶Fig. D.3).

Forming a Fossil

Paleontologists refer to the process of forming a fossil as **fossilization.** To see how a typical fossil develops in sedimentary rock, let's follow the fate of an old dinosaur as it searches for food along a riverbank (▶Fig. D.4). On a scalding summer day, the hungry dinosaur, plodding through the muddy ground, succumbs to the heat and collapses dead into the

FIGURE D.3 The famous fossil footprints at a site called Laetoli, in Olduvai Gorge, Tanzania. They were left when *Australopithecus* walked—on two feet—over ash that had recently been erupted by a nearby volcano, and then had been dampened by rain. A second ash eruption buried, and thereby preserved, the footprints.

mud. Scavengers over the coming days strip the skeleton of meat and scatter the bones among the dinosaur footprints. But before the bones have time to weather away, the river floods and buries the bones, along with the footprints, under a layer of silt. More silt from succeeding floods buries the bones and prints still deeper in a chemically stable environment, below the depth that can be reworked by currents or disrupted by burrowing organisms. Later, sea level rises and a thick sequence of marine sediment buries the fluvial sediment. The weight of overlying sediment squashes and flattens the bones somewhat.

Eventually, the sediment containing the bones turns to rock (siltstone). The footprints remain outlined by the contact between the siltstone and mud, while the bones reside within the siltstone. Minerals precipitating from groundwater passing through the siltstone gradually replace some of the chemicals constituting the bones, until the bones themselves have become rock-like. The buried bones and footprints are now fossils. One hundred million years later, uplift and erosion exposes the dinosaur's grave. Part of a fossil bone protrudes from a rock outcrop. A lucky paleontologist observes the fragment and starts excavating, gradually uncovering enough of the bones to permit reconstruction of the beast's skeleton. Further digging uncovers footprints. The dinosaur rises again, but this time in a museum. In recent years, bidding wars have made some fossil finds extremely valuable. For example, a skeleton of a *Tyrannosaurus rex,* a 67-million-year-old dinosaur found in South Dakota, sold at auction for $7.6 million in 1997. The specimen, named Sue after its discoverer, now stands in the Field Museum of Chicago.

Similar tales can be told for fossil seashells buried by sediment settling in the sea, for insects trapped in hardened tree sap (amber), and for mammoths drowned in the muck of a tar pit. In all cases, fossilization involves the burial and preservation of an organism or the trace (like a footprint) of an organism. Once buried, the organism may be altered to varying degrees by pressure from overlying rock and chemical interaction with groundwater.

The Many Different Kinds of Fossils

Perhaps when you think of a fossil, you picture either a dinosaur bone or the imprint of a seashell in rock. But there are many types of fossils: **body fossils** are whole bodies or pieces of bodies; **trace fossils** are features left by an organism as it passed by; and **chemical fossils** are chemicals first formed by organisms and now preserved in rock. Let's look at examples of these categories.

- *Frozen or dried body fossils:* In a few environments, whole bodies of organisms may be preserved. Most of these fossils are fairly young, by geologic standards—their ages can be measured in thousands, not millions, of years. Examples include woolly mammoths that became incorporated in the permafrost (permanently frozen ground) of Siberia and have stayed frozen since their death. (▶Fig. D.5a). In desert climates, organisms become desiccated (dried out) and can last for a long time.

- *Body fossils preserved in amber or tar:* Insects landing on the bark of trees may become trapped in the sticky sap or resin the trees produce. This golden syrup envelops the insects and over time hardens into **amber,** the semiprecious "stone" used for jewelry. Amber can preserve insects, as well as other delicate organic material such as feathers, for 40 million years or more (▶Fig. D.5b).

   Tar similarly serves as a preservative. In isolated regions where oil has seeped to the surface, the more

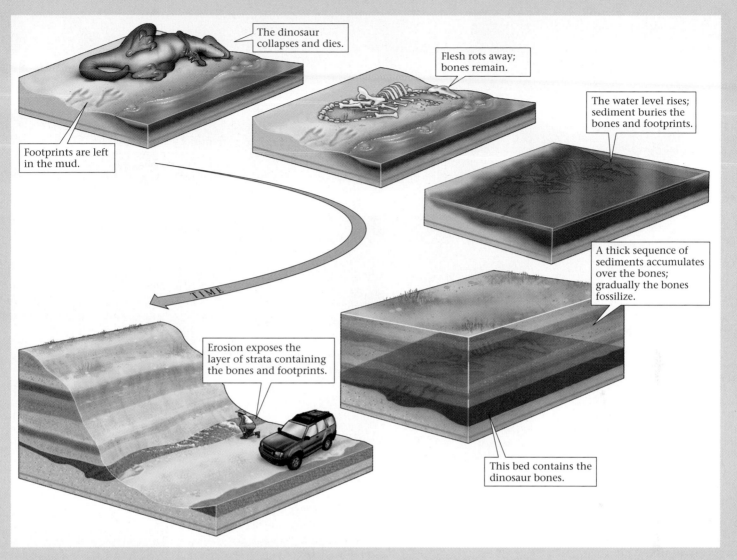

The dinosaur collapses and dies.

Footprints are left in the mud.

Flesh rots away; bones remain.

The water level rises; sediment buries the bones and footprints.

A thick sequence of sediments accumulates over the bones; gradually the bones fossilize.

Erosion exposes the layer of strata containing the bones and footprints.

This bed contains the dinosaur bones.

TIME

FIGURE D.4 How a dinosaur eventually becomes a fossil.

volatile components of the oil evaporate away and bacteria degrade what remains, leaving behind a sticky residue. At one such locality, the La Brea Tar Pits in Los Angeles, tar accumulated in a swampy area. While grazing or drinking at the swamp, or while attacking at the swamp, animals became mired in the tar and sank into it. Their bones have been remarkably well preserved for over 40,000 years.

• *Preserved or replaced bones, teeth, and shells:* Bones (the internal skeletons of vertebrate animals) and shells (the external skeletons of invertebrate animals) consist of durable minerals, which may survive in rock. Some bone or shell minerals are not stable, and they recrystallize (►Fig. D.5c). But even when this happens, the shape of the bone or shell remains in the rock.

• *Permineralized organisms:* **Permineralization** refers to the process by which minerals precipitate in porous

material, like wood or bone, from groundwater solutions that have seeped into the pores. **Petrified wood,** for example, forms by permineralization of wood, so that cell interiors are replaced with silica, causing the wood to become chert. In fact, the word "petrified" literally means "turned to stone." The more resistant cellulose of the wood transforms into an organic film that remains after permineralization, so that the fine detail of the wood's cell structure can be seen in a petrified log (►Fig. D.5d). The colorful bands in a petrified log come from impurities such as iron or carbon. Petrified wood typically forms when a volcanic eruption rapidly buries a forest in siliceous (silica-containing) ash.

• *Molds and casts of bodies:* As sediment compacts around a shell, it conforms to the shape of the shell or body. If the shell or body later disappears, because of weathering and

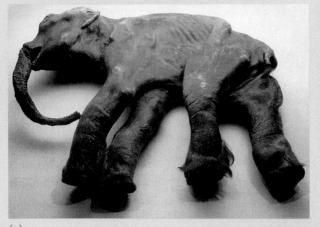

(a)

(b)

(c)

(d)

(e)

(f)

FIGURE D.5 (a) A frozen mammoth, found in the permafrost (permanently frozen ground) of Siberia about 100,000 years after it died. It still had flesh and fur. (b) A piece of amber containing two fossil insects. (c) Fossil dinosaur bones exposed on a tilted bed of sandstone in Dinosaur National Monument, Utah. (d) Petrified wood from the Petrified National Forest, Arizona. Petrified wood is much harder than the surrounding tuff and thus remains after the tuff has eroded away. (e) Molds and casts of organisms. (f) The carbonized impression of fern fronds. (g) Dinosaur footprints in mudstone. (h) Worm burrows on a block of siltstone. (Lens cap for scale.)

(g)

(h)

dissolution, a cavity called a **mold** remains (▶Fig. D.5e). (Sculptors use the same term to refer to the receptacle into which they pour bronze or plaster.) A mold preserves the delicate shape of the organism's surface; it looks like an indentation on a rock bed. The sediment that had filled the mold also preserves the organism's shape; this **cast** protrudes from the surface of the adjacent bed.

- *Carbonized impressions of bodies:* Impressions are simply flattened molds created when soft or semisoft organisms (leaves, insects, shell-less invertebrates, sponges, feathers, jellyfish) get pressed between layers of sediment. Chemical reactions eventually remove the organic chemicals that composed the organism, leaving only a thin film of carbon on the surface of the impression (▶Fig. D.5f).

- *Trace fossils:* These include footprints, feeding traces, burrows, and dung ("coprolites") that organisms leave behind in sediment (▶Fig. D.5g, h).

- *Chemical fossils:* Organisms consist of complex organic chemicals. When buried with sediment and subjected to diagenesis, some of these chemicals are destroyed, but some either remain intact or break down to form different, but still distinctive, chemicals. A distinctive chemical derived from an organism and preserved in rock is called a *chemical fossil.* (Such chemicals may also be called molecular fossils or biomarkers.)

Paleontologists also find it useful to distinguish among different fossils on the basis of their size. **Macrofossils** are fossils large enough to be seen with the naked eye. But some rocks and sediments also contain abundant **microfossils,** which can be seen only with a microscope or even an electron microscope (▶Fig. D.6). Microfossils include remnants of plankton, algae, bacteria, and pollen. Some deep-sea sediments consist almost entirely of microfossils derived from plankton. Pollen proves to be a particularly valuable microfossil for studying ancient climates (paleoclimates), as pollen can be used to identify the types of plants that live in a certain area.

Fossil Preservation

Not all living organisms become fossils when they die. In fact, only a small percentage do, for it takes special circumstances—namely, one or more of the following four—to create a fossil.

- *Death in an anoxic (oxygen-poor) environment:* A dead squirrel by the side of the road won't become a fossil. As time passes, birds, dogs, or other scavengers may come along and eat the carcass. And if that doesn't happen, maggots, bacteria, and fungi will infest the carcass and gradually digest it. Flesh that has not been eaten or does not rot reacts with oxygen in the atmosphere and is transformed into carbon dioxide gas. The remaining

FIGURE D.6 Fossil plankton from deep-marine sediment. Because of their small size (a fraction of a millimeter), these specimens are considered to be microfossils.

skeleton weathers in air and turns to dust. Thus, before the dead squirrel can become incorporated in sediment, it has vanished. In order for fossilization to occur, a carcass must settle into an oxygen-poor environment, where oxidation reactions happen slowly, where scavenging organisms aren't as abundant, and where bacterial metabolism takes place very slowly. In such environments the organism won't rot away before it has a chance to be buried and preserved.

- *Rapid burial:* If an organism dies in a depositional environment where sediment accumulates rapidly, it may be buried before it has time to rot, oxidize, be eaten, be completely broken up, or be consumed by burrowing organisms. For example, if a storm suddenly buries an oyster bed with a thick layer of silt, the oysters die and become part of the sedimentary rock derived from the sediment.

- *The presence of hard parts:* Organisms without durable shells or skeletons, collectively called "hard parts," commonly won't be fossilized, for soft flesh decays long before hard parts do under most depositional conditions. For this reason, paleontologists know much more about the fossil record of bivalves (a class of organisms, including clams and oysters, with strong shells) than they do about the fossil record of jellyfish (which have no shells) or spiders (which have very fragile shells).

- Lack of diagenesis or metamorphism: Processes such as diagenesis or metamorphism may destroy fossils, either by dissolving them away, or by causing recrystallization or neocrystallization sufficient to obscure the fossil's shape.

Paleontologists, by carefully studying modern organisms, have been able to provide rough estimates of the **preservation potential** of organisms, meaning the likelihood that an organism will be buried and eventually transformed into a fossil. For example, in a typical modern-day shallow-marine environment, such as the mud-and-sand sea floor close to a beach, about 30% of the organisms have sturdy shells and thus a high preservation potential, 40% have fragile shells and a low preservation potential, and the remaining 30% have no hard parts at all and are not likely to be fossilized except in special circumstances. Of the 30% with sturdy shells, though, few happen to die in a depositional setting where they actually *can* become fossilized.

Extraordinary Fossils: A Special Window to the Past

Though only hard parts survive in most fossilization environments, paleontologists have discovered a few special locations where rock contains relics of soft parts as well; such fossils are known as **extraordinary fossils.** We've already seen how extraordinary fossils such as insects and even feathers can be preserved in amber, and how complete skeletons have been found in tar pits. Another environment in which extraordinary fossils are found is the anoxic floor of lakes or lagoons or the deep ocean. Here, organic-rich mud accumulates, oxidation cannot occur, and flesh does not rot before burial. Carcasses of animals that settle into the mud gradually become fossils, but because they were buried before the destruction of their soft parts, fossil impressions of their soft parts surround the fossils of their bones.

A small quarry near Messel, in western Germany, for example, has revealed extraordinary fossils of 49-million-year-old mammals, birds, fish, and amphibians that died in a shallow-water lake (▶Fig. D.7a). Bird fossils from the quarry include the delicate imprints of feathers, bat fossils come complete with impressions of ears and wing flaps, and mammal fossils have an aura of carbonized fur. In southern Germany, exposures of the Solenhofen Limestone, an approximately 150-million-year-old rock derived from lime mud deposited in a stagnant lagoon, contain extraordinary fossils of six hundred species, including *Archaeopteryx*, one of the earliest birds (▶Fig. D.7b). And exposures of the Burgess Shale in the Canadian Rockies of British Columbia have yielded a plentitude of fossils showing what shell-less invertebrates that inhabited the deep-sea floor about 510 million years ago looked like (▶Fig. D.7c). The Burgess Shale fauna (animal life) is so strange—for example, it includes organisms with circular jaws—that it has been hard to determine how these organisms are related to present-day ones. Indeed, many of the forms of life represented by Burgess Shale fossils went extinct *without* leaving any descendants that evolved into modern species.

In a few cases, extraordinary fossils include actual tissue, a discovery that has led to a research race to find the oldest pre-

(a)

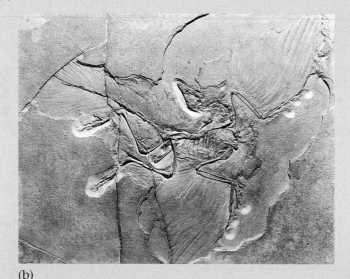

(b)

(c)

FIGURE D.7 Extraordinary fossils: (a) a mammal from Messel, Germany; (b) *Archaeopteryx* from the Solenhofen Limestone; (c) reproduction of a soft-bodied creature from the Burgess Shale, in Canada.

served DNA. (DNA, short for deoxyribonucleic acid, is the complex molecule, shaped like a double helix, that contains the code that guides the growth and development of an organism. Individual components of this code are called genes, and the study of genes and how they transmit information is called genetics.) Paleontologists have isolated small segments of DNA, from amber-encased insects, that is over 40 million years old. The amounts are not enough, however, to clone extinct species, as suggested in the popular movie *Jurassic Park*.

D.3 CLASSIFYING LIFE

The classification of fossils follows the same principles used for classifying living organisms, so before we learn how paleontologists classify fossils, we must first examine how biologists classify life forms. The principles of classification were first proposed in the eighteenth century by Carolus Linnaeus, a Swedish biologist. The study of how to classify organisms is now referred to as **taxonomy.** Linnaeus's scheme has a hierarchy of divisions. First, all life is divided into kingdoms, and each kingdom consists of one or more phyla. A phylum, in turn, consists of several classes; a class, of several orders; an order, of several families; a family, of several genera; and a genus, of one or more species (▶Fig. D.8). Kingdoms, then, are the broadest category and species the narrowest. Where needed, biologists and paleontologists separate out subsets of individual categories—thus, there are subphyla and subclasses. Biologists now recognize six kingdoms (▶Fig. D.9):

- *Archaea:* micro-organisms typically found in extreme environments such as hot springs and in very acidic or very alkaline water;

- *Eubacteria:* "true bacteria," including the kind that cause infections;

- *Protista:* various unicellular and simple multicellular organisms; these include algae (such as diatoms) and forams, two of the major plankton types in the oceans;

- *Fungi:* mushrooms and yeast, etc.;

- *Plantae:* trees, grasses, and ferns, etc.; and

- *Animalia:* sponges, corals, snails, dinosaurs, ants, and people, etc.

More recently, biologists have realized that Archaea are as different from Eubacteria as both are from the other kingdoms. Specifically, the last four kingdoms share enough common characteristics, at the cellular level, to be lumped together as a group called **Eukarya.** Thus, biologists now divide life into three "domains": Archaea, Eubacteria, and Eukarya. The cells of Eukaryotic organisms have distinct nuclei and internal membranes, whereas those of Eubacteria and Archaea do not. (The latter two domains have "prokaryotic" cells.)

Figure D.8 shows the subdivisions below the level of kingdom that apply to humans. Humans, together with monkeys, apes, and lemurs, are primates because they have opposable thumbs and similar facial characteristics. Primates, together with dogs, cats, horses, cows, elephants, whales, and many other animals, are mammals because they nurse their young with milk and most of them have fur or hair. Mammals, along with reptiles, birds, and fish, are chordates because they have an internal skeleton with a spinal canal. And chordates, along with clams, insects, lobsters, and worms, are animals because they share a similar cell structure. Note that genus and species names are always given in Latin and are italicized.

FIGURE D.8 The taxonomic subdivisions. The right-hand column shows how the names apply to human beings.

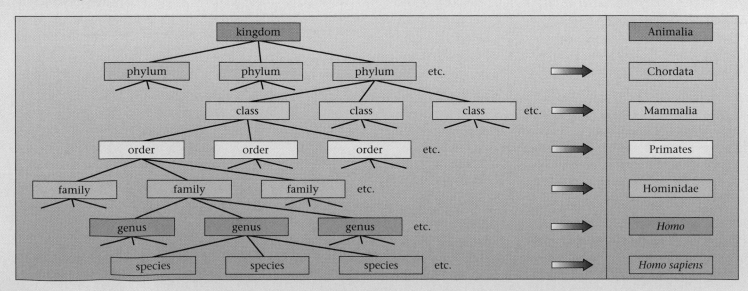

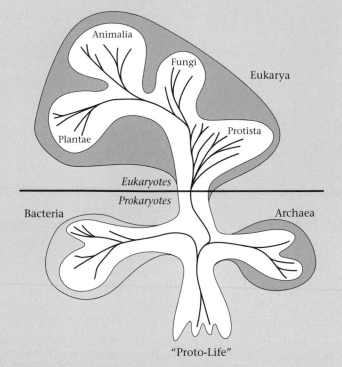

FIGURE D.9 The basic kingdoms of life on Earth. Archaea and Eubacteria are both prokaryotes, because among other characteristics they don't have nuclei, while all other life forms are eukaryotes, because they consist of cells with a nucleus.

D.4 CLASSIFYING FOSSILS

As you'll recall from Chapter 5, mineralogists recognize thousands of different types of minerals. A given mineral has a specific crystal structure and a chemical composition that varies within a definable range. In contrast, paleontologists traditionally distinguish different fossil species from one another according to **morphology** (the form or shape) alone. A fossil clam is a fossil clam because it looks like one. In some cases, the characteristics that distinguish one fossil species from another may be pretty subtle (e.g., the number of ridges on the surface of a shell, or the relative length of different leg bones), but even beginners can distinguish the major groups of fossils from one another on sight. Though presently morphology provides the primary basis for fossil identification, paleontologists may someday be able to determine relationships among organisms by specifying the percentages of shared protein sequences constituting the fossil's DNA.

When it came to naming fossils, paleontologists began by comparing fossil forms with known organisms. They looked at the nature of the skeleton (was it internal or external?), the symmetry of the organism (was it bilaterally symmetric like a mammal, or did it have five-fold symmetry like a starfish?), the design of the shell (in the case of invertebrates), and the design of the jaws or feet (in the case of vertebrates). For example, a fossil organism with a spiral shell

that does not contain internal chambers is classified a member of the class Gastropoda (the snails) within the phylum Mollusca (▶Fig. D.10).

Not all fossil species resemble living families of organisms. In such cases, the comparisons must take place at the level of orders or even higher. For example, a group of extinct organisms called trilobites have no close living relatives. But they were clearly segmented invertebrate animals, and as such they resemble arthropods like insects and crustaceans. Thus, they are considered to be an extinct member of the phylum Arthropoda.

Sometimes there's nothing magical about classifying fossils; major characteristics at the level of class or higher are fairly easy to distinguish just from the way they look. For instance, skeletons of birds (class Aves) are hard to mistake for skeletons of mammals (class Mammalia), and snail shells (class Gastropoda) are hard to mistake for clam shells (class Pelecypoda). But classification can also be difficult, especially if specimens are incomplete, and in many cases classification can be controversial.

▶Figure D.11 shows examples of some of the major classes and subphyla of invertebrate fossils. With this chart, you should be able to identify many of the fossils you'll find in a common bed of limestone. Particularly common invertebrate fossils include the following.

- *Trilobites:* These have a segmented shell that is divided lengthwise into three parts. They are a type of arthropod.

- *Gastropods* (snails): Most fossil specimens of gastropods have a shell that does not contain internal chambers. Slugs are gastropods without shells.

- *Bivalves* (clams and oysters): These have a shell that can be divided into two similar halves. The plane of symmetry is parallel to the plane of the shell.

FIGURE D.10 Examples of the diversity of gastropods (snails). Note that although all these shells have a spiral shape, they differ in detail from each other. Some gastropods have no shells at all.

- *Brachiopods* (lamp shells): The top and bottom parts of these shells have different shapes and the plane of symmetry is perpendicular to the plane of the shell. Shells typically have ridges radiating out from the hinge.

- *Bryozoans:* These are colonial animals. Their fossils resemble a screen-like grid of cells. Each cell is the shell of a single animal.

- *Crinoids* (sea lilies): These organisms look like a flower but actually are animals. Their shells have a stalk consisting of numerous circular plates stacked on top of each other.

- *Graptolites:* These look like tiny carbon saw blades in a rock. They are remnants of colonial animals that floated in the sea.

- *Cephalopods:* These include ammonites, with a spiral shell, and nautiloids, with a straight shell. Their shells contain internal chambers and have ridged surfaces. These organisms were squid-like.

- *Corals:* These include colonial organisms that form distinctive mounds or columns. Paleozoic examples are solitary.

FIGURE D.11 Common types of invertebrate fossils.

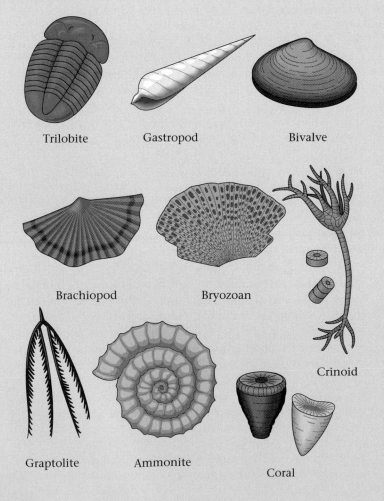

Trilobite   Gastropod   Bivalve

Brachiopod   Bryozoan

Graptolite   Ammonite   Crinoid

Coral

D.5 THE FOSSIL RECORD

A Brief History of Life

As we will see in Chapter 12, the fossil record defines the long-term evolution of life on planet Earth. Archaea and bacteria fossils are found in rocks as old as about 3.7 billion years, indicating that these organisms are the earliest forms of life on Earth. Researchers once thought that archaea somehow developed by chemical reactions in concentrated "soups" that formed when seawater trapped in shallow pools evaporated. Growing evidence suggests, however, that they may instead have developed in warm groundwater beneath the Earth's surface or at hydrothermal vents on the sea floor.

For the first billion years or so of life history, archaea and bacteria were the only type of life on Earth. Then, about 2.5 Ga (billion years ago), organisms of the protist kingdom first appeared. Early multicellular organisms, shell-less invertebrates of the animal kingdom, and fungi of the plant kingdom came into existence perhaps 1.0–1.5 Ga. Within each kingdom, life radiated (divided) into different phyla, and within each phylum, life radiated into different classes. The great variety of shelly invertebrate classes appeared a little over 540 million years ago, and many organisms that persist today appeared at different times since then: first came invertebrates, then fish, land plants, amphibians, reptiles, and finally mammals and birds. Paleontologists and geomicrobiologists have been working hard to understand the phylogeny (the evolutionary relationships among organisms), using both the morphology of organisms and, more recently, the study of genetic material (e.g., DNA). Ideas about which groups radiated from which ancestors is shown in a chart called the "tree of life," or, more formally, the **phylogenetic tree**. Study of the DNA of different organisms is beginning to enable researchers to understand the relationship between molecular processes and evolutionary change.

Is the Fossil Record Complete?

By some estimates, more than 250,000 species of fossils have been collected and identified to date, by thousands of geologists working on all continents during the past two centuries. These fossils define the framework of life evolution on planet Earth. But the record is not complete—every intermediate step in the evolution of every organism cannot be accounted for by known fossils. Considering that there may be as many as 5 million species living on Earth today (not counting bacteria), over the billions of years that life has existed there may have been 5 billion to 50 billion species. Clearly, known fossils represent at most a tiny percentage of these species. Why is the record so incomplete?

First, despite all the fossil-collecting efforts of the past two centuries, paleontologists have not even come close to sampling every cubic centimeter of sedimentary rock exposed on Earth. Just as biologists have not yet identified every living species of insect, paleontologists have not yet identified every species of fossil. New species and even genera of fossils continue to be discovered every year.

Second, not all organisms are represented in the rock record, because not all organisms have a high preservation potential. As noted earlier, fossilization occurs only under special conditions, and thus only a minuscule fraction of the organisms that have lived on Earth become fossilized. There may be few, if any, fossils of a vast number of extinct species, so we have no way of knowing that they ever existed.

Finally, as we will learn in Chapter 12, the sequence of sedimentary strata that exists on Earth does not account for every minute of time since the formation of our planet. Sediments only accumulate in environments where conditions are appropriate for deposition and not for erosion—sediments do not accumulate, for example, on the dry great plains or on mountain peaks, but do accumulate in the sea and in the floodplains and deltas of rivers. Because Earth's climate changes through time and because the sea level rises and falls, certain locations on continents are sometimes sites of deposition and sometimes aren't, and on occasion are sites of erosion. Therefore, the sequence of strata records only part of geologic time.

In sum, a rock sequence provides an incomplete record of Earth history, organisms have a low probability of being preserved, and paleontologists have found only a small percentage of the fossils preserved in rock. So the incompleteness of the fossil record should come as no surprise.

D.6 EVOLUTION AND EXTINCTION

Darwin's Grand Idea

As a young man in England in the early nineteenth century, Charles Darwin had been unable to settle on a career but had developed a strong interest in natural history. As a consequence, he jumped at the opportunity to serve as a naturalist aboard the H.M.S. *Beagle* on an around-the-world surveying cruise. During the five years of the cruise, from 1831 to 1836, Darwin made detailed observations of plants, animals, and geology in the field and amassed an immense specimen collection from South America, Australia, and Africa. Just before departing on the voyage, a friend gave him a copy of Charles Lyell's 1830 textbook *Principles of Geology*, which argued in favor of James Hutton's proposal that the Earth had a long history and that geologic time extended much farther in the past than

did human civilization. A visit to the Galápagos Islands, off the coast of Peru, was a turning point in Darwin's thinking. The naturalist was most impressed with the variability of Galápagos finches, and he marveled not only at the fact that different varieties of the bird occurred on different islands, but at how each variety had adapted to utilize a particular food supply. With Lyell's writings in mind, Darwin developed a hypothesis that the finches had begun as a single species but had branched into several different species when isolated on different islands. This proposal implied that a species could change, or undergo **evolution,** throughout long periods of time and that new species could appear.

On his return to England, Darwin discussed his hypothesis with fellow biologists and geologists, and over succeeding years he gathered supporting evidence. When Alfred Russell Wallace, a naturalist working in Indonesia, wrote to Darwin (in 1858) outlining almost identical thoughts about evolution, Darwin realized that he needed to publish. Darwin and Wallace jointly presented the concept of evolution at a scientific meeting, so as to share credit, and in 1859 Darwin published the revolutionary book *On the Origin of Species by Means of Natural Selection*, in which he outlined evolution and proposed how it occurred.

The crux of Darwin's argument is simply this: because populations of organisms cannot grow exponentially forever, they must be limited by competition for scarce resources in the environment. In nature, only organisms capable of survival can pass on their characteristics to the next generation. In each new generation, some individuals have characteristics that make them more fit, whereas some have characteristics that make them less fit. The fitter organisms are more likely to survive and produce offspring. Thus, beneficial characteristics that they possess get passed on to the next generation. Darwin called this process **natural selection,** because it occurs on its own in nature. (He noted that the same process occurs when farmers artificially select animals or plants for breeding to develop new kinds of domestic livestock or crops.) According to Darwin, when natural selection takes place over long periods of time—geologic time—natural selection eventually produces new organisms that differ so significantly from their distant ancestors that new organisms can be considered to constitute a new species. If environmental conditions change, or if competitors enter the environment, species that do not evolve and become better adapted to survive and compete eventually die off.

Darwin's view of evolution was not just a philosophical speculation—it was a scientific proposal that made predictions that could be tested by observations. Darwin himself provided many observations that support evolution. And in succeeding decades paleontologists, biologists, and anthropologists have made countless observations at countless sites around the world that support evolution. Thus, the idea has gained the status of a theory—it is an idea that has been successfully supported by many observations, that can be used to make testable predictions, and that so far has not been definitively disproved by any observation or experiment. Thus, we now refer to Darwin's idea as the **theory of evolution.**

In the century and a half since Darwin published his work, the science of genetics has developed and has provided insight into *how* evolution works. Progress began in the late nineteenth century when an Austrian monk, Gregor Mendel, studied peas in the garden of his monastery and showed that genetic mutations led to new traits that could be passed on to offspring. Traits that make an organism less likely to survive are not passed on, either because the organism dies before it has offspring or because the offspring themselves cannot survive, but traits that make an organism better suited to survival are passed on to succeeding generations. With the discovery of DNA in 1953, biologists began to understand the molecular nature of genes and mutations, and thus of evolution. And with the genome projects of the twenty-first century, which define the detailed architecture of DNA molecules for a given species, it is now possible to pinpoint the exact arrangement of genes responsible for specific traits.

The theory of evolution provides a conceptual framework in which to understand paleontology. By studying fossils in sequences of strata, paleontologists are able to observe progressive changes in species through time, and can determine when some species go extinct and other species appear. But because of the incompleteness of the fossil record, questions remain as to the rates at which evolution takes place during the course of geologic time, and new proposals of specific phylogenies often are met with skepticism. Traditionally, it was assumed that evolution happened at a constant, slow rate—this concept is called **gradualism.** More recently, however, researchers have suggested that evolution takes place in fits and starts: evolution occurs very slowly for quite a while (the species are in equilibrium), and then during a relatively short period it takes place very rapidly. This concept is called **punctuated equilibrium.** Factors that could cause sudden pulses of evolution include (1) a sudden mass extinction event, during which many organisms disappear, leaving ecological niches open for new species to colonize; (2) a sudden change in the Earth's climate that puts stress on organisms—organisms that evolve to survive the new stress survive, while others go extinct; (3) the sudden formation of new environments, as may happen when rifting splits apart a continent and generates a new ocean with new coastlines; and (4) the isolation of a breeding population.

Extinction: When Species Vanish

Extinction occurs when the last members of a species die, so there are no parents to pass on their genetic traits to offspring. Some species become extinct as they evolve into new species, whereas others just vanish, leaving no hereditary offspring. These days, we take for granted that species become extinct, because a great number have, unfortunately, vanished from the Earth during human history. Before the 1770s, however, few geologists thought that extinction occurred; they thought that fossils that didn't resemble known species must have living relatives somewhere on Earth. Considering that large parts of the Earth remained unexplored, this idea wasn't so far-fetched. But by the end of the eighteenth century, it became clear that numerous fossil organisms did not have modern-day counterparts. The bones of mastodons and woolly mammoths, for example, were too different from those of elephants to be of the same species, and the animals were too big to hide.

What causes extinction? Initially, paleontologists assumed that Noah's flood (recounted in the Old Testament) was the cause. But as the fossil record began to accumulate, and the principle of fossil succession was documented, they realized that not all extinction could have happened in a single event. Some authors proposed that extinction was the consequence of a series of global catastrophic floods, but continued geologic research demonstrated that such floods have not taken place.

Twentieth-century studies concluded that many different phenomena can contribute to extinction. Some extinctions may happen suddenly, when all members of a species die off in a short time, whereas others may occur over longer periods, when the replacement rate of a population simply becomes lower than the mortality rate. By examining the number of species on Earth through time (i.e., by studying variations in the diversity of life, or biodiversity), paleontologists have found that there is a varying rate of extinction. Generally, the rate is fairly slow, but on occasion a **mass extinction event** occurs, during which a large number of species worldwide disappear. At least five major mass extinction events have happened during the past half billion years (►Fig. D.12). These events define the boundaries between some of the major intervals into which geologists divide time. For example, a major extinction event marks the end of the Cretaceous Period, 65 million years ago. During this event, dinosaur species (with the exception of their modified descendants, the birds) vanished, along with many marine invertebrate species. Some researchers have suggested that extinction events are periodic, but this idea remains controversial.

Following are some of the geologic factors that may cause extinction.

- *Global climate change:* At times, the Earth's mean temperature has been significantly colder than today's, while at other times it has been much hotter. These shifts affect ocean currents and sea level, and may trigger ice ages or droughts. Because of a change in climate, an individual species may lose its habitat, and if it cannot

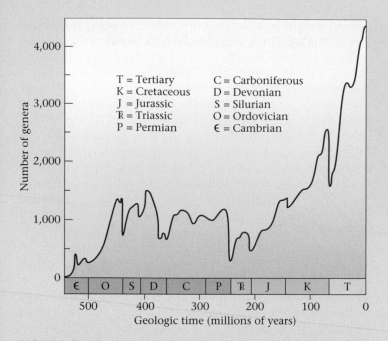

FIGURE D.12 This graph illustrates the variation in diversity of life with time. Steep dips in the curve mean that the number of species on Earth suddenly decreased substantially. The largest dips represent major mass extinctions.

adapt to the new habitat or migrate to stay with its old one, the species will disappear.

- *Tectonic activity:* Tectonic activity causes mountain building, the gradual vertical movement of the crust over broad regions, changes in sea-floor spreading rates, and changes in the amount of volcanism. These phenomena can make sea level rise or fall, or can bring about changes in elevation, thus modifying the distribution and area of habitats. Species that cannot adapt die off.

- *Asteroid or comet impact:* Many geologists have concluded that impacts of large meteorites or asteroids with the Earth have been catastrophic for life (▶Fig. D.13). An impact would send so much dust and debris into the

FIGURE D.13 Painting of an asteroid impact in Yucatán that possibly eliminated most if not all dinosaur species.

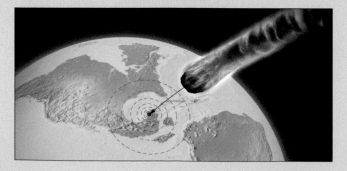

atmosphere that it would blot out the Sun and plunge the Earth into darkness and cold. Such a change, though relatively short-lived, could interrupt the food chain.

- *The appearance of a predator or competitor:* Some extinctions may happen simply because a new predator appears on the scene. Some researchers suggest that this phenomenon explains the mass extinction that occurred during the past 20,000 years, when a vast number of large mammal species vanished from North America. The timing of these extinctions appears to coincide with the appearance of the first humans (fierce predators) on the continent. If a more efficient competitor appears, the competitor steals an ecological niche from the weaker species, whose members can't obtain enough food and thus die out.

KEY TERMS

amber (p. 366)
body fossils (p. 366)
cast (p. 369)
chemical fossils (p. 366)
Eukarya (p. 371)
evolution (p. 374)
extinction (p. 375)
extraordinary fossils (p. 370)
fossil (p. 364)
fossilization (p. 365)
gradualism (p. 375)
macrofossils (p. 369)
mass extinction event (p. 375)
microfossils (p. 369)

mold (p. 369)
morphology (p. 372)
natural selection (p. 374)
paleontology (p. 364)
permineralization (p. 367)
petrified wood (p. 367)
phylogenetic tree (p. 373)
preservation potential (p. 370)
punctuated equilibrium (p. 375)
taxonomy (p. 371)
theory of evolution (p. 375)
trace fossils (p. 366)

SUGGESTED READING

Benton, M. J. 2003. *When Life Nearly Died: The Greatest Mass Extinction of All Time.* London: Thames & Hudson.

Clark, J. A., and J. O. Farlow. 2002. *Gaining Ground: The Origin and Early Evolution of Tetrapods.* Bloomington: Indiana University Press.

Cutle, A. 2003. *The Seashell on the Mountaintop: A Story of Science, Sainthood, and the Humble Genius Who Discovered a New History of the Earth.* New York: E. P. Dutton.

Knoll, A. H. 2003. *Life on a Young Planet: The First Three Billion Years of Evolution on Earth.* Princeton: Princeton University Press.

Prothero, D. R. 1998. *Bringing Fossils to Life: An Introduction to Paleontology.* Boston: WCB/McGraw-Hill.

Deep Time: How Old Is Old?

If the Eiffel Tower were now representing the world's age, the skin of paint on the pinnacle-knob at its summit would represent man's share of that age; and anybody would perceive that that skin was what the tower was built for. I reckon they would, I dunno.
—MARK TWAIN (1835–1910)

12.1 INTRODUCTION

In May of 1869, a one-armed Civil War veteran named John Wesley Powell set out with a team of nine geologists and scouts to explore the previously unmapped expanse of the Grand Canyon, the greatest gorge on Earth. Though Powell and his companions battled fearsome rapids and the pangs of starvation, most managed to emerge from the mouth of the canyon three months later (▶Fig. 12.1). During their voyage, seemingly insurmountable walls of rock both imprisoned and amazed the explorers, and led them to pose important questions about the Earth and its history, questions that even casual tourists to the canyon ponder today: Did the Colorado River sculpt this marvel, and if so, how long did it take? When did the rocks making up the walls of the canyon form? Was there a time *before* the colorful layers accumulated? These questions pertain to **geologic time,** the span of time since Earth's formation. Powell realized that, like the pages in a book, rock layers of the Grand Canyon contain a record of Earth's history.

In this chapter, we first learn how geologists developed the concept of geologic time and thus a frame of reference for describing the ages of rocks, fossils, structures, and landscapes. Then we look at the tools

How do geologists determine the age of rocks? This cliff face in Missouri shows two rock units. The sedimentary layers on the left were deposited on the rhyolite to the right. From this relationship, we can determine the relative age of the two units—the rhyolite is older. But to determine the age of the granite in years, its numerical age, we must date it radiometrically.

FIGURE 12.1 Woodcut illustration of the "noonday rest in Marble Canyon," from J. W. Powell, *The Exploration of the Colorado River and Its Canyons* (1895). The explorers have just entered a quiet stretch of the river: "We pass many side canyons today that are dark, gloomy passages back into the heart of the rocks."

geologists use to determine the age of the Earth and its features. With the concept of geologic time in hand, a hike down a trail into the Grand Canyon becomes a trip into the distant past, into what authors call *deep time*. The geological discovery that our planet's history extends billions of years into the past changed humanity's perception of the Universe as profoundly as did the astronomical discovery that the limit of space extends billions of light years beyond the edge of our solar system.

12.2 TIME: A HUMAN OBSESSION

When you plan your daily schedule, you have to know not only where you need to be, but when you need to be there. Because time assumes such significance in human consciousness today, we have developed elaborate tools to measure it and formal scales to record it. We use a "second" as the basic unit of time measurement. What exactly is a second? From 1900 to 1968, we defined the second as 1/31,556,925.9747 of the year 1900, but now we define it as the duration of time that it takes for a cesium atom to

change back and forth between two energy states 9,192,631,770 times. This change is measured with a device called an atomic clock. An atomic clock has an accuracy of about one second per million years. We sum sixty seconds into one minute, sixty minutes into one hour, and twenty-four hours into one "day," about the time it takes for Earth to spin once on its axis.

In the preindustrial era, each locality kept its own time, setting noon as the moment when the Sun reached the highest point in the sky. But with the advent of train travel and telegraphs, people needed to calibrate schedules from place to place. So in 1883, countries around the globe agreed to divide the world into 15°-wide bands of longitude called "time zones"—in each time zone, all clocks keep the same "standard time." The times in each zone are set in relation to Greenwich Mean Time (GMT), the time at the astronomical observatory in Greenwich, England. Today, the world standard for time is determined by a group of about 200 atomic clocks that together define Coordinated Universal Time (CUT). CUT is the basis for the global positioning system (GPS), used for precise navigation.

12.3 THE CONCEPT OF GEOLOGIC TIME

The Birth of Geologic Time

Most people develop a sense for the duration of time by remembering when events took place in terms of human lifetimes. So if an average generation spans twenty years, then recorded history began about two to four hundred generations ago. Many cultures viewed geologic time to be about the same duration as historical time. For example, in 1654, the Irish archbishop James Ussher added up the generations of patriarchs described in the Judeo-Christian Old Testament and concluded that the Earth came into being on October 23, 4004 B.C.E. Scholars at the time assumed that the Earth formed essentially just as we see it.

Beginning in the Renaissance, however, scientists studying geological features began to speculate that geological time might far exceed historical time. The discovery of seashell-like shapes in rocks triggered this revolution in thinking. Although most people dismissed these shapes, which came to be known as fossils (see Interlude D), as a coincidence or a supernatural trick, some concluded that they were the relics of ancient sea creatures trapped in the rocks. The idea did not gain acceptance until the work of Nicolaus Steno (1638–1686).

Steno, born Niels Stenson in Denmark, served as court physician to the Grand Duke of Tuscany in Florence, Italy. During walks in the nearby Apennine Mountains, Steno fre-

FIGURE 12.2 Fossilized shark teeth. Before Steno explained the origin of these fossils, they were known as dragons' tongues.

quently came upon fossils and wondered how they ended up in solid rock hundreds of meters above sea level. One day, local fishermen gave Steno a shark's head, which he dissected out of curiosity. Inside its mouth, he found rows of distinctive triangular teeth, identical to so-called tongue stones, fossils thought to be the petrified tongues of dragons (►Fig. 12.2). Steno became convinced that these fossils, and by implication others as well, did not form supernaturally but were the relics of ancient life, and introduced this idea in his 1669 book *Forerunner to a Dissertation on a Solid Naturally Occurring within a Solid.* The title may seem peculiar until you think about the puzzle Steno was trying to solve: How did a solid object (a fossil) get into another solid object (a sedimentary rock)? Steno concluded that the rock must have once been loose sand that incorporated the teeth and shells while soft and only later hardened into rock. His discovery implied that geologic features developed not all at once, but rather by a series of events that took a long time, longer than human history.

The next major step in the development of the idea that geologic time exceeded historical time came from the work of a Scottish doctor and gentleman farmer named James Hutton (1726–1797) based on relationships that he observed in the rocky crags of his native land. Hutton lived during the Age of Enlightenment, when philosophers like Voltaire, Kant, Hume, and Locke encouraged people to cast aside the constraints of dogma and think for themselves, and when the discovery of physical laws by Sir Isaac Newton made people look to natural, not supernatural, processes to explain the Universe. As Hutton wandered around Scotland, he came up with an idea that provides the foundation of geology and led to his title: "father of geology." This idea, the **principle of uniformitarianism,** states simply that physical processes we observe today also operated in the past and were responsible for the formation of the geologic features we see in outcrops; more concisely, *the present is the key to the past.* If this principle is correct, Hutton reasoned, the Earth must be much older than human history allowed, for observed geologic processes work very slowly. In his 1785 book *Theory of the Earth with Proofs and Illustrations,* Hutton mused on the issue of geologic time, and suggested that "there is no vestige of a beginning, no prospect of an end."

Relative versus Numerical Age

In the early nineteenth century, geologists struggled to develop ways to divide and describe geologic time. Like historians, geologists want to establish both the sequence of events that created an array of geologic features (such as rocks, structures, and landscapes) and the exact dates on which the events happened. We specify the age of one feature with respect to another as its **relative age** and the age of a feature given in years as the **numerical age** (or, in older literature, the "absolute age") (►Fig. 12.3a, b). Geologists developed ways of defining relative age long before they did so for numerical age, so we will look at relative-age determination first.

FIGURE 12.3 The difference between relative and numerical age. (a) The relative ages of selected wars in the last 100 or so years. (b) The numerical ages of these same wars. Clearly, this chart provides more information, for it displays the duration of events and indicates the amount of time between events.

(a)   (b)

12.4 PHYSICAL PRINCIPLES FOR DEFINING RELATIVE AGE

Nineteenth-century geologists such as Charles Lyell recast the ideas of Steno and Hutton into formal, usable geological principles. These principles, defined below, continue to provide the basic framework within which geologists read the record of Earth history.

- **The principle of uniformitarianism:** As noted earlier, physical processes we observe operating today also operated in the past, at roughly comparable rates (▶Fig. 12.4a, b). Geologists emphasize that physical processes do not occur at exactly the same rate through time; factors such as climate can change, and occasional catastrophic events, such as a large meteor impact, affect the Earth.

- **The principle of superposition:** In a sequence of sedimentary rock layers, each layer must be younger than the one below, for a layer of sediment cannot accumulate unless there is already a substrate on which it can collect. Thus, the layer at the bottom of a sequence is the oldest, and the layer at the top is the youngest, unless the layers have been overturned during mountain building (▶Fig. 12.4c).

- **The principle of original horizontality:** Sediments on Earth settle out of a fluid in a gravitational field. Typically, the surfaces on which sediments accumulate (such as a floodplain or the bed of a lake or sea) are fairly horizontal. If sediments were deposited on a steep slope, they would likely slide downslope before lithification, and so would not be preserved as sedimentary layers. Therefore, layers of sediment when originally deposited are fairly horizontal. With this principle in mind, we know that when we see folds and tilted beds, we are seeing the consequences of deformation that postdates deposition (▶Fig. 12.4d, e).

- **The principle of original continuity:** Sediments generally accumulate in continuous sheets. If today you find a sedimentary layer cut by a canyon, then you can assume that the layer once spanned the canyon but was later eroded by the river that formed the canyon (▶Fig. 12.4f, g).

- **The principle of cross-cutting relations:** If one geologic feature cuts across another, the feature that has been cut is older. For example, if an igneous dike cuts across a sequence of sedimentary beds, the beds must be older than the dike (▶Fig. 12.4h). If a fault cuts across and displaces layers of sedimentary rock, then the fault must be younger than the layers. But if a layer of sediment buries a fault, the sediment must be younger than the fault.

- **The principle of inclusions:** If an igneous intrusion contains fragments of another rock, the fragments must be older than the intrusion. If a layer of sediment deposited on an igneous layer includes pebbles of the igneous rock, then the sedimentary layer must be younger. The fragments (xenoliths) in an igneous body and the pebbles in the sedimentary layer are inclusions, or pieces of one material incorporated in another. The rock containing the inclusion must be younger than the inclusion (▶Fig. 12.4i).

- **The principle of baked contacts:** An igneous intrusion "bakes" (metamorphoses) surrounding rocks. The rock that has been baked must be older than the intrusion (▶Fig. 12.4j).

Now let's use these principles to determine the relative ages of features shown in ▶Figure 12.5a. In so doing, we develop a **geologic history** of the region, defining the relative ages of events that took place there.

First, note what kinds of rocks and structures the figure contains. Most of the rocks constitute a sequence of folded sedimentary beds, but we also see a granite pluton, a basalt dike, and a fault. Let's start our analysis by looking at the sedimentary sequence. By the principle of superposition, we know that the oldest layer, the limestone labeled 1, occurs at the bottom and that progressively younger beds lie above the limestone. We can confirm that layer 1 predates layer 2 by applying the principle of inclusions, because layer 2 contains pebbles (inclusions) of layer 1. As the same holds true for the other layers, the sedimentary beds from oldest to youngest are 1, 2, 3, 4, 5, 6, 7. (There are other rocks below layer 1, but we do not see them in our cross section.) And considering the principle of original horizontality, we conclude that the layers were folded sometime after deposition.

Now let's look at the relationships between the igneous rocks and the sedimentary rocks. The granite pluton cuts across the folded sedimentary rocks, so the intrusion of the pluton occurred after the deposition of the sedimentary beds and after they were folded. The layer of igneous rock that parallels the sedimentary beds could be either a sill that intruded between the sedimentary layers or a flow that spread out over the sandstone and solidified before the shale was deposited. From the principle of inclusions, we deduce that the layer is a sill, because it contains xenoliths of both the underlying sandstone and the overlying shale. Since the sill is folded, it intruded before folding took place. By applying the principle of baked contacts, we also can tell that the sill intruded before the pluton did, because the baked zones (metamorphic aureole) surrounding the pluton affected the sill. The dike cuts across both the pluton and the sill, as well as all the sedimentary layers, and thus formed later.

(a)

(b)

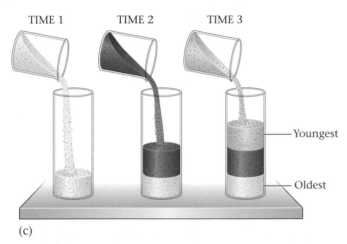

TIME 1   TIME 2   TIME 3

Youngest

Oldest

(c)

(d)

(e)

FIGURE 12.4 Geologic principles: (a, b) Uniformitarianism: Geologists assume that the physical processes that formed the mud cracks in a tidal flat of (a) also formed the mud cracks preserved on the surface of a Paleozoic siltstone bed shown in (b). Note that the mud cracks in (b) were inside solid rock and are visible now only because the overlying bed has been removed by erosion. (c) Superposition: If you fill a glass cylinder with different colors of sand, the oldest sand (white) must be on the bottom. (d, e) Original horizontality: When sediment is originally deposited, like this silt on the tidal flat near Mont St. Michel, France, it forms a flat layer. If layers do not become deformed, they can remain horizontal for hundreds of millions of years, as illustrated by these beds of Paleozoic sandstone in Wisconsin. *(continued)*

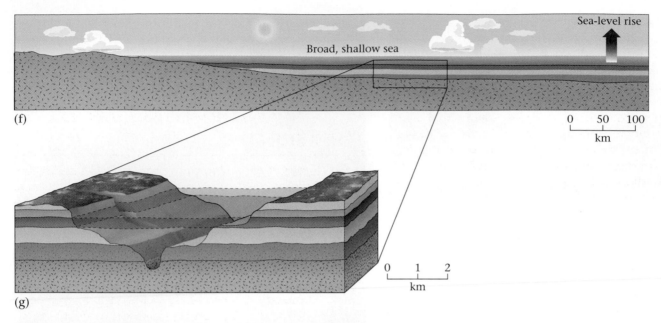

(f)

0 50 100
km

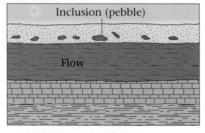

(g)

0 1 2
km

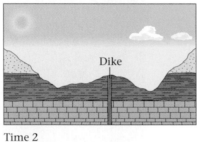

Dike

Time 1
(h)

Time 2

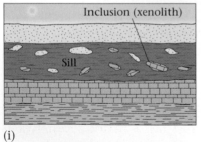

Inclusion (xenolith)

Sill

(i)

Inclusion (pebble)

Flow

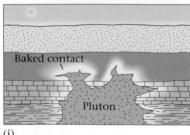

Baked contact

Pluton

(j)

FIGURE 12.4 *(cont'd.)* (f, g) Original continuity: The sediments on the floor of this broad, shallow sea (f), which formed when the sea level rose and flooded the continent, accumulated in nearly horizontal layers, that are continuous over a broad region. We assume that the layers of sediment were originally continuous across the canyon before the canyon developed (g). (h) Cross-cutting relations: The sedimentary beds existed first; then they were cut by igneous magma rising to a volcano, so today we see a dike cutting across the bedding. Since the dike does the cutting, it must be younger. (i) Inclusions: In the left-hand example, a sill intrudes between a limestone layer and a sandstone. The sill incorporates fragments (inclusions) of limestone and sandstone, so it must be younger. In the right-hand example, the igneous rock is a lava flow that existed before the sandstone was deposited; the sandstone contains pebbles (inclusions) of lava. (j) Baked contacts: The intrusion of the pluton creates a metamorphic aureole (baked contact) in the surrounding rock, so the pluton must be younger. Note that the pluton also crosscuts bedding, confirming this interpretation.

Finally, let's consider the fault and the land surface. Because the fault offsets the granite pluton and the sedimentary beds, by the principle of cross-cutting relations the fault must be younger than those rocks. But the fault itself has been cut by the dike, and so must be older than the dike. The present land surface erodes all rock units and the fault, and thus must be younger. We can now propose the following geologic history for this region (▶Fig. 12.5b): (a) deposition of the sedimentary sequence, in order from layers 1 to 7; (b) intrusion of the sill; (c) folding of the sedimentary layers and the sill; (d) intrusion of the granite pluton; (e) faulting; (f) intrusion of the dike; (g) formation of the land surface.

12.5 ADDING FOSSILS TO THE STORY: FOSSIL SUCCESSION

As England entered the Industrial Revolution in the late eighteenth and early nineteenth centuries, new factories demanded coal to fire their steam engines. The government decided to build a network of canals to transport coal and iron, and hired an engineer named William Smith (1769–1839) to survey the excavations. These excavations provided fresh exposures of bedrock, which previously had been covered by vegetation. Smith learned to recognize distinctive layers of sedimentary rock and to identify the **fossil assemblage** (the group of fossil species) that they contained (▶Fig. 12.6). He also realized that a particular fossil species can be found only in a limited interval of strata, and not above or below this interval. Thus, once a fossil species disappears at a horizon in a sequence of strata, it never reappears higher in the sequence. In other words, extinction is forever. Smith's observation has been repeated at millions of locations around the world, and has been codified as the **principle of fossil succession.**

To see how this principle works, examine ▶Figure 12.7a, which depicts a sequence of strata. Bed 1 at the base contains fossil species A, bed 2 contains fossil species A and B, bed 3 contains B and C, bed 4 contains C, and so on. From these data, we can define the **range** of specific fossils in the sequence, meaning the interval in the sequence in which the fossils occur. Note that the sequence contains a definable succession of fossils (A, B, C, D, E, F), that the range in which a particular species occurs may overlap with the range of other species, and that once a species vanishes, it does not reappear higher in the sequence.

Because of the principle of fossil succession, we can define the relative ages of strata by looking at fossils. For example, if we find a bed containing fossil A, we can say that the bed is older than a bed containing, say, fossil F (▶Fig. 12.7b). Geologists have now identified and determined the relative ages of millions of fossil species.

12.6 UNCONFORMITIES: GAPS IN THE RECORD

James Hutton often strolled along the coast of Scotland because the shore cliffs provided good exposures of rock, stripped of soil and shrubbery. He was particularly puzzled by an outcrop along the shore at Siccar Point. One sequence of rock exposed there consisted of alternating beds of gray sandstone and shale, while another consisted of red sandstone and conglomerate (▶Fig. 12.8a, b). The beds of gray sandstone and shale were nearly vertical, while the beds of red sandstone and conglomerate were roughly horizontal. Further, the horizontal layers seemed to lie across the truncated ends of the vertical layers, like a handkerchief lying across a row of books. We can imagine that as Hutton examined this odd geometric relationship, the tide came in and deposited a new layer of sand on top of the rocky shore. With the principle of uniformitarianism in mind, Hutton suddenly realized the significance of what he saw. Clearly, the gray sandstone/shale sequence had been deposited, then tilted, and then truncated by erosion before the red sandstone/conglomerate beds were deposited. Just as rock outcrops were being covered by sand along the shoreline as Hutton watched, at some time in the past the gray sandstone/shale sequence had formed the rocky shore, and the red sand and conglomerate had been deposited above.

Hutton deduced that the surface between the gray and red rock sequences represented a long interval of time during which new strata were not deposited at Siccar Point and older strata may have been eroded away. We now call such a surface, representing a period of nondeposition and possibly erosion, an **unconformity.** The interval of time between deposition of the youngest rock below an unconformity and deposition of the oldest rock above is called a **hiatus.** Essentially, an unconformity forms wherever the land surface does not receive and accumulate sediment. Geologists recognize three kinds of unconformity:

- *Angular unconformity:* Rocks below an "angular unconformity" were tilted or folded before the unconformity developed (▶Fig. 12.9a). Thus, an angular unconformity cuts across the underlying layers; the layers below have a different orientation from the layers above. (We can see an angular unconformity in the outcrop at Siccar Point.) Angular unconformities form where rocks were either folded, or tilted by faulting, before being uplifted and eroded.

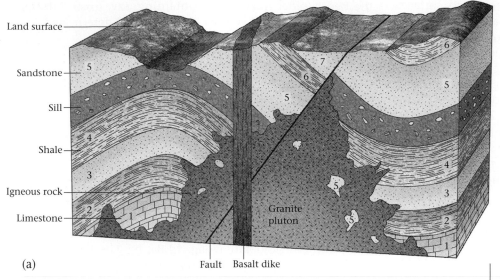

Land surface

Sandstone

Sill

Shale

Igneous rock

Limestone

(a)

Fault Basalt dike

Granite pluton

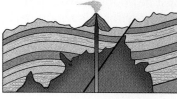

Today

Erosion to form present land surface

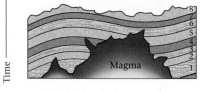

Intrusion of dike

Faulting

Intrusion of granite

Magma

Time

Folding, uplift, erosion

Deposition of strata

(b) Oldest

FIGURE 12.5 (a) Geologic principles allow us to interpret the sequence of events leading to the development of the features shown here. Beds 1–7 were deposited first. Intrusion of the sill came next, followed by folding, intrusion of the granite, faulting, intrusion of the dike, and erosion to yield the present land surface. (b) Sequence of geologic events leading to the geology shown in (a).

1 cm

FIGURE 12.6 A bedding surface containing many fossils.

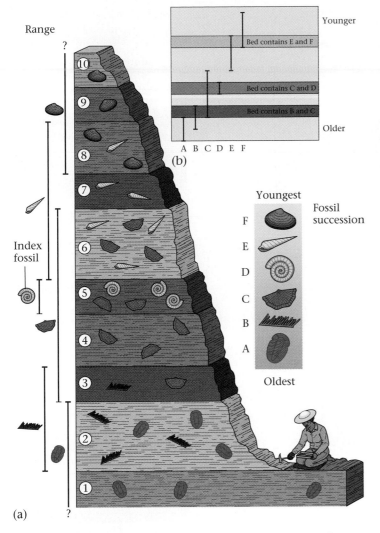

(a)

(a)

FIGURE 12.7 (a) The principle of fossil succession. Note that each species has only a limited range in a succession of strata, and ranges of different fossils may overlap. Widespread fossils with a short range are index fossils. (b) Overlapping fossil ranges can be used to limit the relative age of a given bed and to determine the relative ages of beds. For example, a bed containing fossils E and F must be younger than a bed containing B and C. Note that a bed containing C alone could be older than or younger than a bed containing D alone.

(b)

FIGURE 12.8 (a) The Siccar Point unconformity in Scotland, on the coast about 60 km east of Edinburgh. (b) A geologic interpretation of the unconformity.

• *Nonconformity:* A "nonconformity" is a type of unconformity at which sedimentary rocks overlie intrusive igneous rocks and/or metamorphic rocks (►Fig. 12.9b). The igneous or metamorphic rocks must have cooled, been uplifted, and been exposed by erosion to form the substrate on which new sedimentary rocks were deposited. At a nonconformity, you typically find pebbles of the igneous or

metamorphic rock in the lowermost bed of the sedimentary sequence.

• *Disconformity:* Imagine that a sequence of sedimentary beds has been deposited beneath a shallow sea. Then sea level drops, and the recently deposited beds are exposed for some time. During this time, no new sediment accumulates, and some of the preexisting sediment gets eroded away. Later, sea level rises, and a new sequence of

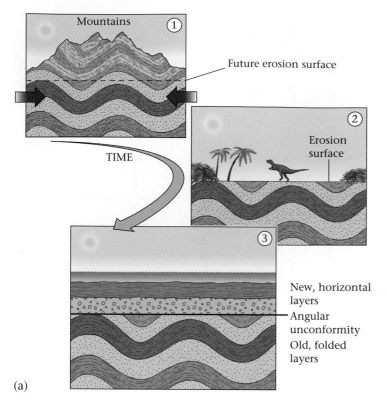

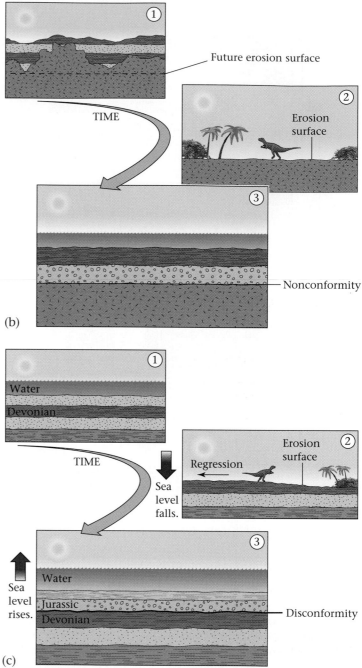

FIGURE 12.9 (a) The stages during the development of an angular unconformity: (1) mountains form and layers are folded; (2) erosion removes the mountains, creating an erosion surface; (3) sea level rises and new horizontal layers of sediment are deposited. (b) The stages during the development of a nonconformity: (1) a pluton intrudes sedimentary rocks; (2) erosion removes all the sedimentary layers and cuts down into the crystalline rock, making an erosion surface; (3) sea level rises and new sedimentary layers are deposited above the erosion surface. (c) The stages during the development of a disconformity: (1) layers of sediment are deposited; (2) sea level drops and an erosion surface forms; (3) sea level rises and new sedimentary layers accumulate. Note that regardless of the details, an unconformity represents a surface of erosion and/or a period of nondeposition.

sediment accumulates over the old. The boundary between the two sequences is a **disconformity** (▶Fig. 12.9c). Even though the beds above and below the disconformity are parallel, the contact between them represents an interruption in deposition. Disconformities may be hard to recognize unless you notice evidence of erosion (such as stream channels) or soil formation at the disconformity surface, or can identify a distinct gap in the succession of fossils.

The succession of strata at a particular location provides a record of Earth history there. But because of unconformities, the record preserved in the rock layers is incomplete (▶Fig. 12.10). It's as if geologic history is being chronicled by a tape recorder that turns on only intermittently—when it's on (times of deposition), the rock record accumulates, but when it's off (times of nondeposition and possibly erosion), an unconformity develops. Because of unconformities, no single location on Earth contains a complete record of Earth history.

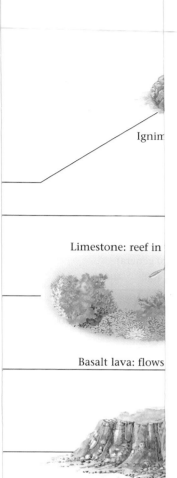

Ignim

Limestone: reef in

Basalt lava: flows

Conglomera
eroded from

-- Unconformity

Gneiss: metamor
beneath a moun

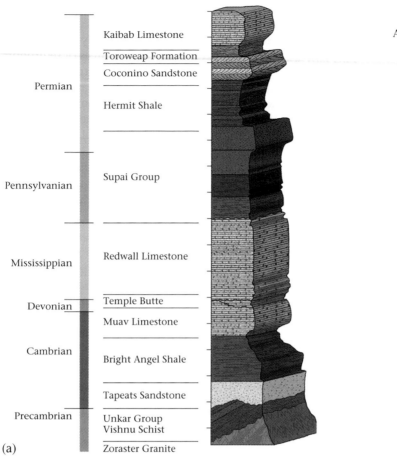

Permian
- Kaibab Limestone
- Toroweap Formation
- Coconino Sandstone
- Hermit Shale

Pennsylvanian
- Supai Group

Mississippian
- Redwall Limestone

Devonian
- Temple Butte
- Muav Limestone

Cambrian
- Bright Angel Shale
- Tapeats Sandstone

Precambrian
- Unkar Group
- Vishnu Schist
- Zoraster Granite

(a)

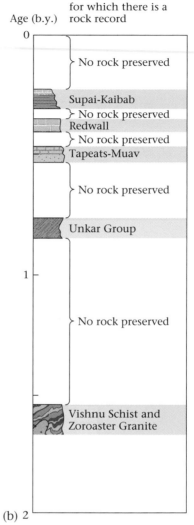

Age (b.y.) | The actual interval for which there is a rock record

0

No rock preserved

Supai-Kaibab
No rock preserved
Redwall
No rock preserved
Tapeats-Muav

No rock preserved

Unkar Group

1

No rock preserved

Vishnu Schist and
Zoroaster Granite

(b) 2

FIGURE 12.10 (a) The sequence of strata in the Grand Canyon (shown beneath the arrow in Fig. 12.11a) can be represented on a stratigraphic column. The vertical scale gives relative thicknesses. The right-hand edge of the column represents resistance to erosion (e.g., Coconino Sandstone is more resistant than Hermit Shale). (b) Because of unconformities, the stack of strata exposed in the Grand Canyon represents only bits and pieces of geologic history. If the strata are projected on a numerical time scale, you can see that large intervals of time are not accounted for.

12.7 STRATIGRAPHIC FORMATIONS AND THEIR CORRELATION

Geologists summarize information about the sequence of strata at a location by drawing a **stratigraphic column.** Typically, we draw columns to scale, so that the relative thicknesses of layers portrayed on the column represent the thicknesses of layers in the outcrop. Then, for ease of reference, geologists divide the sequence of strata represented on a column into **stratigraphic formations** ("formations" for short), recognizable intervals of a specific rock type or group of rock types deposited during a specific time interval, that can be traced over a fairly broad region. The boundary surface between two formations is a type of geologic **contact.**

Let's see how the concept of a stratigraphic formation applies to the Grand Canyon. The walls of the canyon look striped, because they expose a variety of rock types that differ in color and in resistance to erosion. Geologists identify major contrasts and use them as a basis to divide the strata into formations, each of which may consist of many beds (►Fig. 12.11a, b). Note that some formations include a single rock type, while others include interlayered beds of two or more rock types. Also, note that not all formations have the same thickness. Typically, geologists name a formation after a locality where it was first identified. If the formation consists of only one rock type, we may incorporate that rock type in the name (e.g., Kaibab Limestone), but if the formation contains more than one rock type, we use the word "formation" in the name (e.g., Toroweap Formation; note that both words are capitalized). Several related formations in a succession may be lumped together as a "group."

Fault scarp:
a consequence
of recent faulting

Basalt dike:
a result of
igneous activity

Trilobite

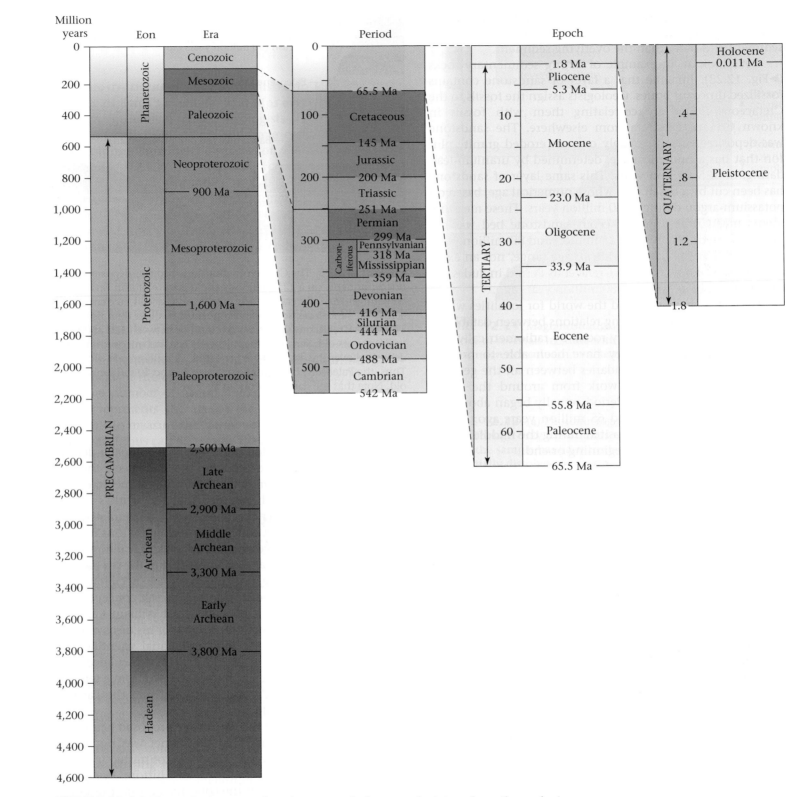

FIGURE 12.24 The geologic time scale assigns numerical ages to the intervals on the geologic column. Note that we have to change to a larger scale to portray the ages of intervals higher in the column. This time scale utilizes 2004 numbers favored by the International Commission on Stratigraphy. For further information, see the *GeoWhen Database* on the web, prepared by Robert A. Rohde.

route to a solution came in 1896, when the physicist Henri Becquerel announced the discovery of radioactivity. Geologists immediately realized that the Earth's interior was producing heat from the decay of radioactive material. This realization uncovered the flaw in Kelvin's argument: Kelvin had assumed that no heat was added to the planet after it first formed. Because radioactivity constantly generates new heat in the Earth, the planet has cooled down much more slowly than Kelvin had calculated. In 1904, the British physicist Ernest Rutherford presented this discovery to an audience that included Kelvin, as he later recounted:

> I came into the room, which was half dark, and presently spotted Lord Kelvin in the audience and realized that I was in trouble at the last part of the speech dealing with the age of the Earth, where my views conflicted with his. To my relief, Kelvin fell fast asleep, but as I came to the important point, I saw the old bird sit up, open an eye and cock a baleful glance at me!

The discovery of radioactivity not only invalidated Kelvin's estimate of the Earth's age, it also led to the development of radiometric dating.

Since the 1950s, geologists have scoured the planet to identify its oldest rocks. Samples from several localities (Wyoming, Canada, Greenland, and China) have yielded dates as old as 3.96 billion years. And sandstones found in Australia contain clastic grains of zircon that yielded dates of 4.1–4.2 billion years, indicating that rock as old as 4.2 billion years did once exist (▶Fig. 12.25a). Models of the Earth's formation assume that all objects in the solar system developed at roughly the same time from the same nebula. Radiometric dating of meteors and Moon rocks have yielded ages as old as 4.57 billion years (▶Fig. 12.25b); geologists take this to be the approximate age of the Earth, leaving more than enough time for the rocks and life forms of the Earth to have formed and evolved.

We don't find 4.57 billion-year-old (Ga) rocks in the crust because during the first half-billion years of Earth history, rocks in the crust remained too hot for the radiometric clock to start (their temperature stayed above the blocking temperature), and/or crust that has formed before about 4.0 Ga was destroyed by an intense meteorite bombardment at about 4.0 Ga. Geologists have named the time interval between the birth of the Earth and the formation of the oldest dated rock the Hadean Eon (see Fig. 12.24).

12.12 PICTURING GEOLOGIC TIME

The mind grows giddy gazing so far back into the abyss of time.
—JOHN PLAYFAIR (1747–1819), British geologist
who popularized the works of Hutton

The number 4.57 billion is so staggeringly large that we can't begin to comprehend it. If you lined up this many pennies in a row, they would make an 87,400-km-long

FIGURE 12.25 (a) One of the oldest dated rocks on Earth, this 3.96-billion-year-old gneiss comes from the Northwest Territories of Canada. (b) A Moon rock.

(a)

(b)

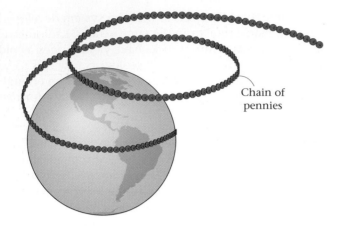

Chain of pennies

FIGURE 12.26 We can use the analogy of distance to represent the duration of geologic time.

line that would wrap around the Earth's equator more than twice (▶Fig. 12.26). Notably, at the scale of our penny chain, human history is only about 100 city blocks long.

Another way to grasp the immensity of geologic time is to make a scale model, which we do by equating the entire 4.57 billion years to a single calendar year. On this scale, the oldest rocks preserved on Earth date from early February, and the first bacteria appear in the ocean on February 21. The first shelly invertebrates appear on October 25, and the first amphibians crawl out onto land on November 20. On December 7, the continents coalesce into the supercontinent of Pangaea. The first mammals and birds appear about December 15, along with the dinosaurs, and the Age of Dinosaurs ends on December 25th. The last week of December represents the last 65 million years of Earth history, covering the entire Age of Mammals. The first human-like ancestor appears on December 31 at 3 P.M., and our species, *Homo sapiens*, shows up an hour before New Year's Eve. The last ice age ends a minute before midnight, and all of recorded human history takes place in the last 30 seconds. Put another way, human history occupies the last 0.000001% of Earth history.

CHAPTER SUMMARY

- The concept of geologic time, the span of time since the Earth's formation, developed when early geologists suggested that the Earth must be very old if geologic features formed by same natural processes we see today.

- Relative age specifies whether one geologic feature is older or younger than another; numerical age provides the age of a geologic feature in years.

- Using such principles as uniformitarianism, superposition, original horizontality, original continuity, cross-cutting relations, inclusions, and baked contacts, we can construct the geologic history of a region.

- The principle of fossil succession states that the assemblage of fossils in a sequence of strata changes from base to top. Once a fossil species becomes extinct, it never reappears.

- Strata are not deposited continuously at a location. An interval of nondeposition and/or erosion is called an unconformity. Geologists recognize three kinds: angular unconformity, nonconformity, and disconformity.

- A stratigraphic column shows the succession of formations in a region. A given succession of strata that can be traced over a fairly broad region is called a stratigraphic formation. The boundary surface between two formations is a contact. The process of determining the relationship between strata at one location and strata at another is called correlation. A geologic map shows the distribution of formations.

- A composite stratigraphic column that represents the entirety of geologic time is called the geologic column. The geologic column's largest subdivisions, each of which represents a specific interval of time, are eons. Eons are further subdivided into eras, eras into periods, and periods into epochs.

- The numerical age of rocks can be determined by radiometric dating. This is because radioactive elements decay at a constant rate. During radioactive decay, parent isotopes transform into daughter isotopes. The decay rate for a given element is known as its half-life, the time it takes for half of the parent isotopes to decay. The ratio of parent to daughter isotopes in a mineral grain indicates the age.

- The radiometric date of a mineral specifies the time at which the mineral cooled below a certain temperature. We can use radiometric dating to determine when an igneous rock solidified and when a metamorphic rock cooled from high temperatures. To date sedimentary strata, we must examine cross-cutting relations with dated igneous or metamorphic rock.

- Other methods for dating materials include counting rings in trees, layers in shells and glaciers, and fission tracks in mineral grains. We can also study the sequence of magnetic reversals in strata.

- From the radiometric dating of meteors and Moon rocks, geologists conclude that the Earth formed about 4.57 billion years ago. Our species, *Homo sapiens,* has been around for only 0.000001% of geologic time.

KEY TERMS

baked contacts (p. 380)
blocking temperature (p. 398)
Cambrian explosion (p. 391)
carbon-14 (^{14}C) dating (p. 398)
contact (p. 387)
correlation (p. 388)
cross-cutting relations (p. 380)
daughter isotope (p. 396)
disconformity (p. 386)
eon (p. 390)
epoch (p. 390)
era (p. 390)
fission track (p. 400)
formation (stratigraphic formation) (p. 387)
fossil assemblage (p. 383)
fossil succession (p. 383)
geologic column (p. 389)
geologic history (p. 380)
geologic map (p. 389)
geologic time (p. 377)
geologic time scale (p. 401)
growth rings (p. 399)

half-life (p. 396)
hiatus (p. 383)
inclusions (p. 380)
magnetostratigraphy (p. 400)
numerical age (absolute age) (p. 379)
original continuity (p. 380)
original horizonality (p. 380)
parent isotope (p. 396)
period (p. 390)
Precambrian (p. 390)
radioactive decay (p. 396)
radiometric dating (goochronology) (p. 391)
range (p. 383)
relative age (p. 379)
rhythmic layering (p. 399)
stratigraphic column (p. 387)
superposition (p. 380)
unconformity (p. 383)
uniformitarianism (p. 379, 380)

REVIEW QUESTIONS

1. Compare numerical age and relative age.

2. Describe the principles that allow us to determine the relative ages of geologic events.

3. How does the principle of fossil succession allow us to determine the relative ages of strata?

4. How does an unconformity develop?

5. Describe the differences among the three kinds of unconformities.

6. Describe two different methods of correlating rock units. How was correlation used to develop the geologic column? What is a stratigraphic formation?

7. What does the process of radioactive decay entail?

8. How do geologists obtain a radiometric date? What are some of the pitfalls in obtaining a reliable one?

9. Why can't we date sedimentary rocks directly?

10. Why is carbon-14 dating useful in archaeology, but useless for dating dinosaur fossils?

11. How are growth rings and ice cores useful in determining the ages of geologic events?

12. How are the reversals of the Earth's magnetic field useful in dating strata?

13. Why did early scientists think the Earth was less than 100 million years old?

14. How did the discovery of radioactivity invalidate Kelvin's assumptions about the Earth's age and also provide a method for obtaining its true age?

15. What is the age of the oldest rocks on Earth? What is the age of the oldest rocks known? Why is there a difference?

SUGGESTED READING

Berggren, W. A., D. V. Kent, M.-P. Aubry, and J. Hardenbol, eds. 1995. *Geochronology, Time Scales, and Global Stratigraphic Correlation.* SEPM Special Publication 54.

Berry, W. B. N. 1987. *Growth of a Prehistoric Time Scale.* 2nd ed. Palo Alto, Calif.: Blackwell Scientific Publications.

Burchfield, J. D. 1975. *Lord Kelvin and the Age of the Earth.* New York: Science History Publications.

Dalrymple, G. B. 1991. *The Age of the Earth.* Palo Alto, Calif.: Stanford University Press.

Dott, R. H., Jr., and D. R. Prothero. 1994. *Evolution of the Earth.* 5th ed. New York: McGraw-Hill.

Faul, H., and C. Faul. 1983. *It Began with a Stone: A History of Geology from the Stone Age to Plate Tectonics.* New York: Wiley.

Faure, G. 1986. *Principles of Isotope Geology.* 2nd ed. New York: Wiley.

Gould, S. J. 1987. *Time's Arrow, Time's Cycle.* Cambridge, Mass.: Harvard University Press.

Gradstein, F. J., and A. Smith. 2005. *A Geologic Time Scale 2004.* Cambridge, UK: Cambridge University Press.

Prothero, D. R. 1990. *Interpreting the Stratigraphic Record.* New York: Freeman.

A Biography of Earth

I weigh my words well when I assert that the man who should know the true history of the bit of chalk which every carpenter carries about in his breeches pocket, though ignorant of all other history, is likely, if he will think his knowledge out to its ultimate results, to have a truer and therefore a better conception of this wonderful universe and of man's relation to it than the most learned student who [has] deep-read the records of humanity [but is] ignorant of those of nature.
—THOMAS HENRY HUXLEY, from ON A PIECE OF CHALK (1868)

A geologist at work in the field, here on the shore of Kangaroo Island, Australia. Such field observations have allowed geologists to work out the history of Earth.

13.1 INTRODUCTION

In 1868, a well-known British scientist, Thomas Henry Huxley, presented a public lecture on geology to an audience in Norwich, England. Seeking a way to convey his fascination with Earth history to people with no previous geological knowledge, he focused his audience's attention on the piece of chalk he had been writing with (see epigraph above). And what a tale the chalk has to tell! Chalk, a type of limestone, consists of microscopic marine algae shells and shrimp feces. The specific chalk that Huxley held came from beds deposited in Cretaceous time (the name "Cretaceous," in fact, derives from the Latin word for "chalk") and now exposed along the White Cliffs of Dover (▶Fig. 13.1). Geologists in Huxley's day knew of similar chalk beds in outcrops throughout much of Europe, and had discovered that the chalk contains not only plankton shells but also fossils of bizarre swimming reptiles, fish, and invertebrates—species absent in the seas of today. Clearly, when the chalk was deposited, warm seas holding unfamiliar creatures covered much of what is dry land today.

Clues in his humble piece of chalk allowed Huxley to demonstrate to his audience that the geography and inhabitants of the Earth in the past differed markedly from those today, and

FIGURE 13.1 Horizontal chalk beds, created from layers of deep-sea sediment, exposed along the coast of England.

thus that *the Earth has a history*. In the many decades since Huxley's lecture, field and laboratory studies worldwide, have allowed geologists to develop an overall image of Earth's history. With every new geologic discovery, this image, once a blur, comes progressively more into focus. We now see a complex, evolving "Earth System" in which physical and biological components interact pervasively in ways that have transformed a formerly barren, crater-pocked surface into one of countless environments supporting a diversity of life. Fossil finds of the past two decades demonstrate that life began when the Earth was still quite young; life has affected surface and near-surface processes ever since. And, as we pointed out near the beginning of this book, the map of our planet constantly changes as continents rift, drift, and collide, and as ocean basins open and close.

We will discuss the nature of overall global change further in Chapter 23. In this chapter, we offer a concise geological biography of the planet, from its birth 4.57 billion years ago to the present. After learning the methods geologists use to interpret the past, we see how continents came into existence and have waltzed across the globe ever since. We also learn when mountain-building events took place and how Earth's climate and sea level have changed through time—in fact, the climate alternates from being relatively warm (greenhouse conditions) to being relatively cool (icehouse conditions). And while these physical changes have taken place, life has evolved. To simplify the discussion, the following abbreviations are used: Ga (for "billion years ago"), Ma ("million years ago"), and Ka ("thousand years ago").

13.2 METHODS FOR STUDYING THE PAST

When historians outline human history, they describe daily life, wars, economics, governments, leaders, inventions, and explorations. When geologists outline Earth history, we describe the distribution of depositional environments, mountain-building events (orogenies), past climates, life evolution, the changing positions of continents, the past configuration of plate boundaries, and changes in the composition of the atmosphere and oceans. Historians collect data by reading written accounts, examining relics and monuments, and, for more recent events, listening to recordings or watching videos. Geologists collect data by examining rocks, geologic structures, and fossils and, for more recent events, by studying sediments, ice cores, and tree rings.

Figuring out Earth's past hasn't been an easy task for geologists, because the available record isn't complete—the Earth materials that hold the record of the past don't form continuously through time, and erosion may destroy some materials. Also, it is no surprise that the record of early Earth history is vaguer than the record for more recent history, for older rocks are more likely to have been eroded away, and because the uncertainty attached to age measurement may become larger as the rocks get older. Nevertheless, we can find enough of the record to at least outline major geologic events of the past. Following are a few examples of how we use observational data to study Earth history.

- *Identifying ancient orogens:* We identify present-day orogens (mountain belts) by looking for regions of high, rugged topography. However, since it takes as little as 50 million years to erode the peaks of a mountain range entirely away, we cannot identify orogens of the past simply by studying topog-raphy; rather, we must look for the rock record they leave behind. Orogeny causes igneous activity, deformation (folds, faults, and foliation), and metamorphism. Thus, we can recognize an ancient orogen by looking for a belt of crust containing these features (▶Fig. 13.2), and we can date rocks formed during orogeny by using radiometric dating.

    Orogeny also leads to the development of unconformities, for uplift exposes rocks to erosion, and "foreland sedimentary basins" (basins located on the continent adjacent to the mountain front), for the weight of the mountain belt pushes crust down, creating a depression that traps sediment. This sediment provides a record of the erosion of the mountains.

- *Recognizing the growth of continents:* Not all continental crust formed at the same time. In order to determine how a continent grew, geologists find the ages of different regions of the crust by using modern radiometric dating techniques. They can figure out not only when rocks

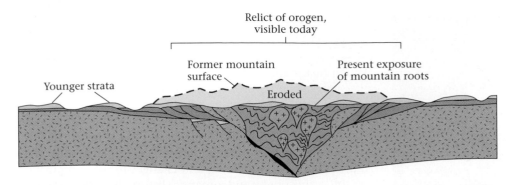

FIGURE 13.2 Even after the topography of a mountain range has been eroded away, the deformed and metamorphosed rocks and plutons that formed within it can still be recognized. These complex rocks contrast with surrounding regions and define the position of the orogen.

originally formed from magmas rising out of the mantle, but also when the rocks were metamorphosed during a subsequent orogeny. The identities of the rock types making up the crust indicate the tectonic environment in which the crust formed.

- *Recognizing past depositional environments:* The environment at a particular location changes through time. To learn about these changes, we study successions of sedimentary rocks, for the environment controls both the type of sediment deposited at a location and the type of organisms that live there.

- *Recognizing past changes in relative sea level:* We can determine whether sea level has gone up or down by looking for changes in the depositional environment. For example, where a marine limestone overlies an alluvial-fan conglomerate, the relative sea level rose at that site.

- *Recognizing positions of continents in the past:* To help us find out where a continent lay in the past, we have three sources of information. First, the study of paleomagnetism can tell us the latitude of a continent in the past (see Chapter 3; ▶Fig. 13.3a, b). Second, we can study marine magnetic anomalies to reconstruct the change over time in the width of an ocean basin between continents (▶Fig. 13.3c). Third, we can compare rocks and/or fossils from different continents to see if there are correlations that could indicate that the continents were adjacent.

- *Recognizing past climates:* We can gain insight into past climates by looking at fossils and rock types that formed at given latitudes. For example, if organisms requiring semitropical conditions lived near the poles during a given time period, then the atmosphere overall must have been warmer. Geologists have also learned how to use the ratios of isotopes for certain elements in fossil shells (e.g., $^{18}O/^{16}O$) as a measure of past temperatures.

- *Recognizing life evolution:* Progressive changes in the assemblage of fossils in a sequence of strata represent changes in the assemblage of organisms inhabiting Earth through time, and thus indicate the occurrence of evolution.

13.3 THE HADEAN EON: HELL ON EARTH?

A viscid pitch boiled in the fosse below
and coated all the bank with gluey mire.
I saw the pitch; but I saw nothing in it
except the enormous bubbles of its boiling
which swelled and sank, like breathing, through all the pit.
—DANTE, *THE INFERNO*, CANTO XXI

James Hutton, the eighteenth-century Scottish geologist who was the first to provide convincing evidence that the Earth was vastly older than human civilization, could not measure Earth's age directly, and indeed speculated that there may be "no vestige of a beginning." But radiometric dating studies of recent decades have shown that it is possible, in fact, to assign a numerical age to Earth's formation. Specifically, dates obtained for a class of *meteorites* thought to be representative of the planetesimal cloud out of which the Earth and other planets formed consistently yield an age of 4.57 billion years. Geologists currently take this age to be Earth's birth date; the Earth formed within 30 million years of the time when the Sun's nuclear furnace first fired up. But a clear record of Earth history, as recorded in continental crustal rocks, does not begin until about 3.80 Ga—crustal rocks older than 3.80 billion years are exceedingly rare. Geologists refer to the mysterious time interval between the birth of Earth and 3.80 Ga as the

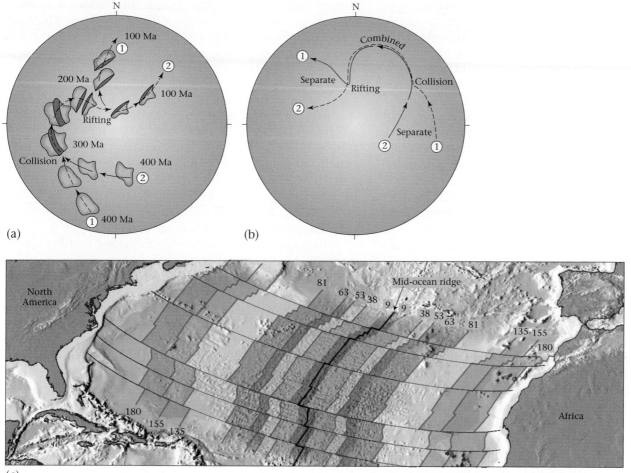

FIGURE 13.3 Geologists use apparent polar-wander paths as clues to the past positions of continents. (a) This hypothetical example shows the movement of two continents (1 and 2). At 400 Ma (million years ago), the continents are separate. They collide at 300 Ma and move together as a supercontinent until 200 Ma. Then they rift apart and drift away from each other. (b) The apparent polar-wander paths for the two continents. Note that the paths coincide when the continents move together as a supercontinent. (c) Geologists use marine magnetic anomalies to define relative motions of continents for the past 200 million years. This map shows the sea floor at various times in the past (in Ma) corresponding with specific marine magnetic anomalies. If you remove the strip of sea floor between the coast and the 81-Ma anomaly (or between the coast and the 63-Ma anomaly, etc.), you will see the relative positions of the continents at 81 Ma (or 63 Ma, etc.).

"Hadean Eon" (from the Greek *Hades*, the god of the underworld). A number of clues provide the basic framework of Hadean history.

The Hadean Eon began with the formation of the Earth by the accretion of planetesimals (see Chapter 1). As the planet grew, collection and compression of matter into a dense ball generated substantial heat, and each time another meteorite collided with the Earth, its kinetic energy added more heat. Radioactive decay produced still more heat within the newborn planet. Also, during Earth's first 150 million years of existence, many short-lived radioactive isotopes still existed, so physicists estimate that radioactive decay produced 5 times as much heat then as it does today.

Eventually, the Earth became hot enough to partially melt, and when this happened, by about 4.5 Ga, it underwent **internal differentiation**—gravity pulled molten iron down to the center of the Earth, where it accumulated to form the core. A mantle, composed of ultramafic rock, remained as a thick shell surrounding the core (see Chapter 2). Internal differentiation, which may have occurred relatively quickly, generated sufficient heat to make the Earth even hotter. Researchers suggest that soon after—or perhaps during—differentiation, a Mars-sized proto-planet collided with the Earth. The energy of this collision blasted away a significant fraction of Earth's mantle, which mixed with the shattered fragments of the colliding planet's mantle to form a ring of silicate-rock debris orbiting the Earth. Heat generated by the collision substantially melted Earth's remaining mantle, making it so weak that the iron core of the colliding body sank through it and merged with the Earth's core. Meanwhile, the ring of debris surrounding the Earth coalesced to form the Moon, which, when first formed, was less than 20,000 km away. (By comparison, the Moon is 384,000 km from Earth today.)

In the wake of differentiation and Moon formation, the Earth was so hot that its surface was likely an ocean of seething magma, supplied by intense eruption of melts rising from the mantle—it resembled Dante's Inferno (▶Fig. 13.4). Here and there, rafts of solid rock formed temporarily on the surface of the magma ocean, but these eventually sank and remelted. This stage lasted at least until about 4.4 Ga. After that time, because of the decrease in radioactive heat generation, Earth *might* have become cool enough for solid rocks to form at its surface. The evidence for this statement comes from western Australia, where geologists have found 4.4 Ga grains of a durable mineral called zircon in sandstone beds. The zircons must have originally formed in igneous rock, but were eroded and later deposited in sedimentary beds. Existence of these grains suggests that at least some solid igneous rocks had formed by 4.4 Ga.

During the hot early stages of the Hadean Eon, rapid "outgassing" of the Earth's mantle took place. This means that volatile elements or compounds originally incorporated in mantle minerals were released and erupted at the Earth's surface, along with lava. The gases accumulated to constitute an unbreathable atmosphere of water (H_2O), methane (CH_4), ammonia (NH_3), hydrogen (H_2), nitrogen (N_2), carbon dioxide (CO_2), sulfur dioxide (SO_2), and other gases. Some researchers have speculated that gases from

comets colliding with Earth during the Hadean may have contributed components of the early atmosphere. The early atmosphere was probably much denser—perhaps 250 times denser—than our present atmosphere.

If the Hadean Earth's surface was sufficiently cool for an extensive solid crustal rock to form beginning at 4.4 Ga, then the first oceans may have accumulated soon thereafter, when water in the atmosphere condensed and fell as rain. Evidence favoring early ocean formation at this time is subtle—it comes from studying the ratios of oxygen isotopes in zircon grains. What the surface of the Earth looked like between 4.4 and 3.8 Ga remains a subject of debate. Very few rocks remain to provide insight—the oldest *whole rock* yet found is a gneiss (high-grade metamorphic rock) that yielded an age of 4.03 Ga.

Studies of cratering on the Moon suggest that the rate of meteorite bombardment of Earth may have increased drastically between about 4.0 and 3.9 Ga. This late "bombardment" would have obliterated any crust or ocean that had formed up to that time. After this bombardment ceased, new crust formed, some of which has survived ever since. Geologists have found marine sediments in rock sequences as old as 3.85 billion years, clearly indicating that liquid-water oceans existed at least by this time, and oceans have existed ever since. Thus, if you could see the Earth as the Hadean Eon drew to a close, you would see barren, crater-pocked lands poking above oceans; both land and sea may have been partially covered by ice, and both lay beneath a CO_2- and SO_2-rich atmosphere.

FIGURE 13.4 A painting of the early Hadean Earth. Note the magma ocean, with small crusts of solid basalt floating about. If the sky had been clear, streaks of meteors would have filled it, and the Moon would have appeared much larger than it does today, because it was closer. Probably, however, an observer on Earth could not have seen the sky, because the atmosphere contained so much water vapor and volcanic ash.

13.4 THE ARCHEAN EON: THE BIRTH OF THE CONTINENTS AND THE APPEARANCE OF LIFE

The date that marks the boundary between the **Archean Eon** (from the Greek word for "ancient") and the Hadean has been placed at about 3.8 Ga. Effectively, it marks the time at which intense meteorite bombardment of Earth had ceased and the record of crustal rocks, including rocks that originated as marine sediments, had started to become progressively more complete.

Geologists still argue about whether plate tectonics operated in the early part of the Archean Eon in the same form as today. Some researchers picture an early Archean Earth with rapidly moving small plates, numerous volcanic island arcs, and abundant hot-spot volcanoes. Others propose that early Archean lithosphere was too warm and buoyant to subduct, and that plate tectonics could not have operated until the later part of the Archean; these authors argue that plume-related volcanism was the main source of new crust until the late Archean. Regardless of which model ultimately proves better, it is clear that the Archean was a time of signif-

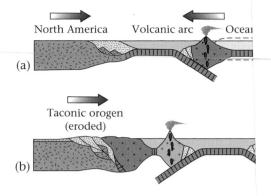

FIGURE 13.17 (a) A volcanic island arc col[lided with the]
eastern margin of North America to cause the [Taconic orogeny.]
Some geologists have suggested that South A[merica lay beyond the]
volcanic arc, so that at the end of the Taconic [orogeny,]
North America bordered western South Amer[ica. But]
after the collision, the high land of the Tacon[ic eroded]
away, and other blocks, such as the Avalon m[icrocontinent,]
approached from the east. Their eventual coll[ision with North]
America caused the Acadian orogeny.

The first animals to appear in the [Cambrian seas]
had simple tube- or cone-shaped shells. [Soon there-]
after, the shells became more complex (▶[Fig. 13.19). Tiny]
fossils called conodonts, which resembl[e teeth, also appear]
in Cambrian strata. Their presence sugg[ests that animals]
with jaws had appeared at this time, h[aving perhaps]
evolved as a means of protection against [predators. By the]
end of the Early Cambrian, trilobites we[re crawling on the sea]
floor. Trilobites shared the environmen[t with]
brachiopods, and echinoderms (see Inte[rlude B).]

FIGURE 13.18 A museum diorama illustrate[s the kinds of]
organisms living during the early Paleozoic ma[rine realm.]
Here, we see creatures like trilobites and nautil[oids.]

icant change in the map of the Earth. During that time the
volume of continental crust increased significantly.

How did the continental crust come to be? According to
a compromise model, relatively buoyant (felsic and intermediate) crustal rocks formed both at subduction zones and at
hot-spot volcanoes. Frequent collisions sutured volcanic arcs
and hot-spot volcanoes together, creating progressively larger
blocks called **protocontinents** (▶Fig. 13.5a, b). Some of
these protocontinents developed rifts that filled with basalt.
As the Earth gradually cooled, protocontinents became
cooler and stronger, and by 2.7 Ga the first **cratons,** long-
lived blocks of durable continental crust, had developed. By
the end of the Archean Eon, about 80% of continental crust
had formed (▶Fig. 13.6). Volcanic activity during this time
continued to supply gas to the atmosphere.

Archean cratons contain five principal rock types: *gneiss*
(relicts of Archean metamorphism in collisional zones),
greenstone (metamorphosed relicts of ocean crust trapped between colliding blocks; or of basalts that had filled early continental rifts, or of flood basalts produced at hot spots),
granite (formed from magmas generated by the partial melting of the crust in continental volcanic arcs or above hot
spots), *graywacke* (a mixture of sand and clay eroded from the
volcanic areas and dumped into the ocean), and *chert*
(formed by the precipitation of silica in the deep sea).

FIGURE 13.5 (a) In the Archean Eon, rocks that would
eventually make up the continental crust began to form. At early
subduction zones, volcanic island arcs, composed of relatively
buoyant rocks, developed, and large shield volcanoes formed over
hot spots. Larger crustal blocks underwent rifting, creating rift
basins that filled with volcanic rocks, and the erosion of crustal
blocks deposited graywacke on the sea floor. (b) Successive
collisions brought all the buoyant fragments together to form a
protocontinent. Melting at the base of the protocontinent may
have caused the formation of plutons.

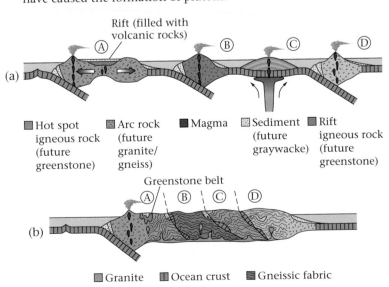

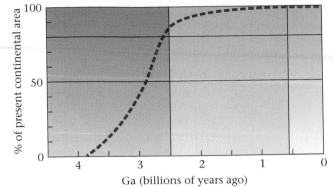

FIGURE 13.6 As time progressed, the area of the Earth covered
by continental crust increased, though not all geologists agree
about the rate of increase. This model shows growth beginning
about 3.9 Ga and continuing rapidly until the end of the Archean
Eon. The rate slowed substantially in the Proterozoic Eon.

Archean shallow-water sediments are rare, either because
continents were so small that depositional environments in
which such sediments could accumulate didn't exist, or because any that were present have since eroded away.

Once land areas had formed, rivers flowed over their
stark, unvegetated surfaces. Geologists reached this conclusion because sedimentary beds from this time contain clastic grains that were clearly rounded by transport in liquid
water. Salts weathered out of rock and transported to the sea
in rivers made the oceans salty.

Clearly, the Archean Eon saw many firsts in Earth history—the first continents, and probably the first life.
Geologists use three sources of evidence to identify early life:

- *Chemical (molecular) fossils, or biomarkers:* These are
  durable chemicals that are produced only by the
  metabolism of living organisms.

- *Isotopic signatures:* Carbon can occur as ^{12}C, ^{13}C, or ^{14}C.
  Organisms preferentially incorporate ^{12}C instead of ^{13}C,
  by a slight amount. Thus, by analyzing the ratio of ^{12}C to
  ^{13}C in carbon-rich sediment geologists can determine if
  the sediment once contained the bodies of organisms.

- *Fossil forms:* Given appropriate depositional conditions,
  shapes representing bacteria or archaea cells can be
  preserved in rock. However, identification of such fossil
  forms can be controversial—similar shapes can result
  from inorganic crystal growth.

The search for the earliest evidence of life continues to make
headlines in the popular media. It is a story of high hopes and
intense frustrations. For example, isotopic signatures of life
were discovered in 3.8-billion-year-old metamorphosed sedimentary rocks of western Greenland, suggesting that life
started right at the beginning of the Archean Eon. But this
proposal has since been disputed by researchers who argue

rifted away from the future North Am
margin basin formed along North Amer
At the beginning of the Paleozoic Era
Pannotia broke up, yielding smaller c
Laurentia (composed of North Ameri
Gondwana (South America, Africa, Ar
Australia), Baltica (Europe), and Siberia (
continents drifted apart, Laurentia, Baltic
at low latitudes, but Gondwana drifted to
and for a brief interval in the Late Ordovi
it became ice-covered.

Following the breakup of Pannoti
sive-margin basins formed around the g
on the eastern coast of Laurentia (w
North America). In addition, sea leve
areas of continents were flooded with
epicontinental seas. By the end of th
the only dry land in Laurentia was an c
on Hudson's Bay, so most of what is no
lay beneath water (▶Fig. 13.16). In r
depths in epicontinental seas reached
creating a well-lit environment in wh
Deposition in the seas yielded a layer c
ment that you can see today near the
Canyon. While a relatively thin layer of
Laurentia's interior (the region that wou
tinental platform), a much thicker layer
passive-margin basins that fringed the c
however, did not stay high for the entire
regressions and transgressions occurred,

FIGURE 13.15 The distribution of contine
Period (about 510 Ma). Note that Gondwana
and Laurentia straddled the equator. The mar
Earth might have looked as viewed from the S
Florida had not yet connected to North Amer

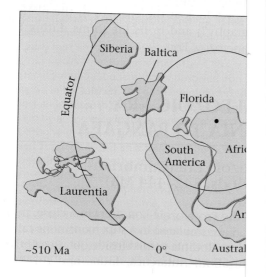

without interruption until the Eocene Epoch, yielding, as we have seen, the Laramide orogen. Then, because of the re-arrangement of plates off the western shore of North America, beginning about 40 Ma, a transform boundary replaced the convergent boundary in the western part of the continent (▶Fig. 13.32a–c). When this happened, volcanism and compression ceased in western North America, the San Andreas fault system formed along the coast of the United States, and the Queen Charlotte fault system developed off the coast of Canada. Along the San Andreas and Queen Charlotte Faults today, the Pacific Plate moves northward with respect to North America at a rate of about 6 cm per year. The Queen Charlotte Fault links to the Aleutian subduction zone along southern Alaska, where the Pacific Ocean floor undergoes subduction. In the western United States, convergent-boundary tectonics continues only in Washington, Oregon, and northern California where subduction of the Juan de Fuca Plate produces the Cascade volcanic chain.

As convergent tectonics ceased in the western United States south of the Cascades, the region began to undergo rifting (stretching) in roughly an east-west direction. The result was the formation of the **Basin and Range Province,** a broad continental rift that has caused the region to stretch to twice its original width (▶Fig. 13.33). The Basin and Range gained its name from its topography—the province contains long, narrow mountain ranges separated from each other by flat, sediment-filled basins. This geometry reflects the normal faulting resulting from stretching: crust of the region was broken up by faults, and movement on the faults created elongate depressions. The Basin and Range Province terminates just north of the Snake River Plain, the track of the hot spot that now lies beneath Yellowstone National Park.

Recall that in the Cretaceous Period, the world experienced greenhouse conditions and sea level rose so that extensive areas of continents were submerged. During the Cenozoic Era, however, the global climate rapidly shifted to icehouse conditions, and by the early Oligocene Epoch, Antarctic glaciers reappeared for the first time since the Triassic. The climate continued to grow colder through the Late Miocene Epoch, leading to the formation of grasslands in temperate climates. About 2.5 Ma, the Isthmus of Panama formed, separating the Atlantic completely from the Pacific, changing the configuration of oceanic currents, and allowing the Arctic Ocean to freeze over.

In the overall cold climate of the last 2 million years, the Quaternary Period, continental glaciers have expanded and retreated across northern continents at least twenty times, resulting in the **Pleistocene ice age** (▶Fig. 13.34). Each time the edge of the glacier advanced, sea level fell so much that the continental shelf became exposed to air, and a land bridge formed across the Bering Strait, west of Alaska, providing migration routes for animals and people from Asia into North America; a partial land bridge also formed

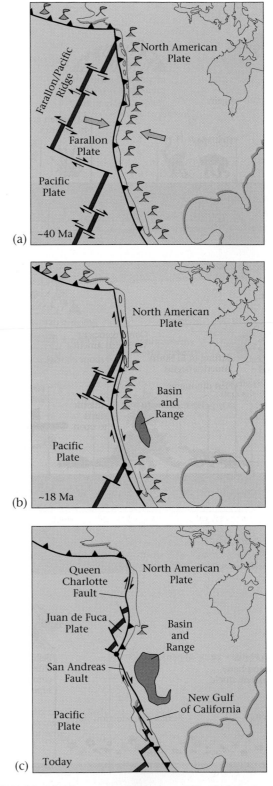

FIGURE 13.32 The western margin of North America changed from a convergent-plate boundary into a transform-plate boundary when the Farallon-Pacific Ridge was subducted. (a, b) The Farallon Plate was moving toward North America, while the Pacific Plate was moving parallel to the western margin of North America. (c) The Basin and Range Province opened as the San Andreas Fault developed. Subduction along the West Coast today only occurs where the Juan de Fuca Plate, a remnant of the Farallon Plate, continues to subduct.

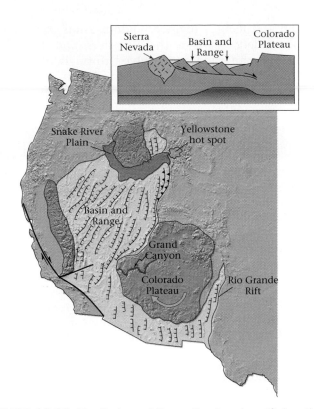

FIGURE 13.33 The Basin and Range Province is a rift (inset). The northern part has opened more than the southern part, causing the Sierran arc to swing westward and rotate. The Rio Grande Rift is a small rift that links to the Basin and Range. The Colorado Plateau is a block of craton bounded by the Rio Grande Rift to the east and the Basin and Range to the west.

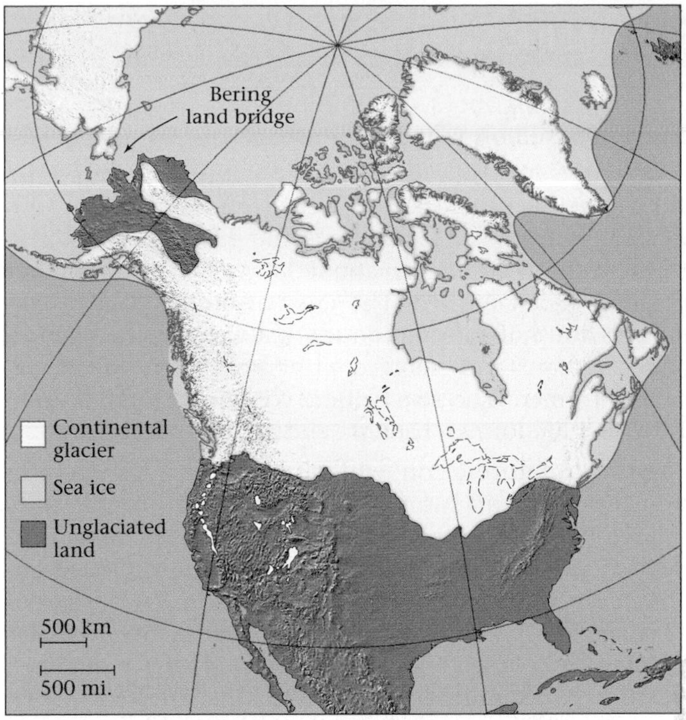

FIGURE 13.34 North America during the maximum advance of the Pleistocene ice sheet. Because sea level was so low (water was stored in the ice sheet on land), a land bridge formed across the Bering Strait, allowing people and animals to migrate from Asia to America.

from southeast Asia to Australia, making human migration to Australia easier. Erosion and deposition by the glaciers created much of the landscape we see today in northern temperate regions. About 11,000 years ago, the climate warmed, and we entered the interglacial time interval we are still experiencing today (see Chapter 22). This most recent interval if time is the Holocene.

Life evolution. When the skies finally cleared in the wake of the K-T boundary catastrophe, plant life recovered, and soon forests of both angiosperms and gymnosperms reappeared. A new group of plants, the grasses, sprang up and began to dominate the plains in temperate and subtropical climates by the middle of the Cenozoic Era. The dinosaurs, however, were gone for good, for no examples of such great beasts survived; their descendants are the birds of today. Mammals rapidly diversified into a variety of forms to take their place. In fact, most of the modern groups of mammals that exist today originated at the beginning of the Cenozoic Era, giving this time the nickname "Age of Mammals." During the latter part of the era, remarkably huge mammals appeared (such as mammoths, giant beavers, giant bears, and giant sloths), but these went extinct during the past 10,000 years, perhaps because of hunting by humans.

It was during the Cenozoic that our own ancestors first appeared. Ape-like primates diversified in the Miocene Epoch (about 20 Ma), and the first human-like primate appeared at about 4 Ma, followed by the first members of the human genus, *Homo,* at about 2.4 Ma. (Note that people and dinosaurs did *not* inhabit Earth at the same time!) Perhaps the evolution of *Homo,* with its significantly larger brain, accompanied climate changes that led to the spread of grasslands, allowing primates to leave the trees—life on the ground provides a longer time for infant development and the growth of a large brain. Fossil evidence, primarily from Africa, indicates that *Homo erectus,* capable of making stone axes, appeared about 1.6 Ma, and the line leading to *Homo sapiens* (our species) diverged from *Homo neanderthalensis* (Neanderthal man) about 500,000 years ago. According to the fossil record, modern people appeared about 150,000 years ago. Thus, much of human evolution took place during the radically shifting climatic conditions of the Pleistocene Epoch.

CHAPTER SUMMARY

• Earth formed about 4.57 billion years ago. For part of the first 600 million years, the Hadean Eon, the planet was so hot that its surface was a magma ocean. We have hardly any rock record of this time interval, but we can gain insight into it by studying the Moon and meteorites.

• The Archean Eon began about 3.8 Ga, when permanent continental crust formed. The crust assembled out of volcanic arcs and hot-spot volcanoes that were too buoyant to subduct. Oceans and stable continental blocks called cratons also formed. The atmosphere contained little oxygen, but the first life forms—bacteria and archaea—appeared.

• In the Proterozoic Eon, which began at 2.5 Ga, Archean cratons collided and sutured together along orogenic belts and large Proterozoic cratons. Photosynthesis by organisms added oxygen to the atmosphere, and iron precipitated out of the ocean to form banded-iron formation. By the end of the Proterozoic, complex but shell-less marine invertebrates populated the planet. Most continental crust accumulated to form a supercontinent called Rodinia at about 1 Ga. Rodinia broke apart and reorganized to form another supercontinent at the end of the Proterozoic Eon.

• At the beginning of the Paleozoic Era, Pannotia broke apart, yielding several smaller continents. Sea level rose and fell a number of times, creating sequences of strata in continental interiors. Continents began to collide and coalesce again, leading to orogenies and, by the end of the era, another supercontinent, Pangaea. Early Paleozoic evolution produced many invertebrates with shells, and jawless fish. Land plants and insects appeared in middle Paleozoic. And by the end of the eon, there were land reptiles and gymnosperm trees.

• In the Mesozoic Era, Pangaea broke apart and the Atlantic Ocean formed. Convergent-boundary tectonics dominated along the western margin of North America, causing orogenic events like the Sevier and Laramide orogenies. Dinosaurs appeared in Late Triassic time and became the dominant land animal through the Mesozoic Era. During the Cretaceous Period, sea level was very high, and the continents flooded. Angiosperms (flowering plants) appeared at this time, along with modern fish. A huge mass extinction event (the K-T boundary event), which wiped out the dinosaurs, occurred at the end of the Cretaceous Period, probably because of the impact of a large bolide with the Earth.

• In the Cenozoic Era, continental fragments of Pangaea began to collide again. The collision of Africa and India with Asia and Europe formed the Alpine-Himalayan orogen. Convergent tectonics has persisted along the margin of South America, creating the Andes, but ceased in North America when the San Andreas Fault formed. Rifting in the western United States during the Cenozoic Era produced the Basin and Range Province. Various kinds of mammals filled niches left vacant by the dinosaurs, and the human genus, *Homo*, appeared and evolved throughout the radically shifting climate of the Pleistocene Epoch.

KEY TERMS

Acadian orogeny (p. 419)
accretionary orogens (p. 414)
Alleghenian orogeny (p. 421)
Alpine-Himalayan chain (p. 429)
Ancestral Rockies (p. 424)
angiosperms (p. 428)
Antler orogeny (p. 420)
banded-iron formation (BIF) (p. 416)
basement uplifts (p. 427)
Basin and Range Province (p. 432)
Caledonian orogeny (p. 419)
Cambrian explosion (p. 418)
Continental platform (p. 414)
craton (p. 411)
cratonic (continental) platform (p. 414)
Ediacaran fauna (p. 416)
epicontinental seas (p. 418)

greenhouse conditions (p. 419)
Grenville orogen (p. 414)
Hercynian orogen (p. 424)
internal differentiation (p. 409)
Laramide orogeny (p. 427)
Nevadan orogeny (p. 424)
Pangaea (p. 421)
Pannotia (p. 415)
Pleistocene ice age (p. 432)
protocontinents (p. 411)
Rodinia (p. 414)
Sevier orogeny (p. 427)
shield (p. 413)
Sierran arc (p. 426)
snowball Earth (p. 417)
Sonoma orogeny (p. 424)
stratigraphic sequence (p. 420)
stromatolites (p. 412)
superplumes (p. 427)
Taconic orogeny (p. 418)

REVIEW QUESTIONS

1. List some methods by which geologists study the past.

2. Why are there no rocks on Earth that yield radiometric dates older than 4 billion years?

3. Describe the condition of the crust, atmosphere, and oceans during the Hadean Eon.

4. Describe the five principal rock types found in Archean protocontinents. Under what kind of environmental conditions did each type form?

5. What are stromatolites? How do they form?

6. How did the atmosphere and tectonic conditions change during the Proterozoic Eon?

7. How do banded-iron formations tell us about atmospheric oxygen levels?

8. What evidence do we have that the Earth nearly froze over twice during the Proterozoic Eon?

9. How did the Cambrian explosion of life change the nature of the living world?

10. How did the Taconic and Acadian orogenies affect the eastern coast of North America?

11. How did the Alleghenian and Ancestral Rocky orogenies affect North America?

12. What are the major classes of organisms to appear in the Paleozoic?

13. Compare the typical sedimentary deposits of the early Paleozoic greenhouse Earth with those of the late Paleozoic/early Mesozoic icehouse Earth.

14. Describe the plate tectonic conditions that led to the formation of the Sierran arc and the Sevier thrust belt.

15. How did the plate tectonic conditions of the Laramide orogeny differ from more typical subduction zones?

16. What life forms appeared during the Mesozoic?

17. What caused the flooding of the continents during the Cretaceous Period?

18. What could have caused the K-T extinctions?

19. What continents formed as a result of the breakup of Pangaea?

20. What are the causes of the uplift of the Himalayas and the Alps?

21. What events led to the end of the Mesozoic greenhouse and the development of glaciers on the Arctic and Antarctic during the Cenozoic Era?

SUGGESTED READING

Cloud, P. 1988. *Oasis in Space: Earth History from the Beginning.* New York: Norton.

Condie, K. C. 1989. *Plate Tectonics and Crustal Evolution.* New York: Pergamon.

Condie, K. C., ed. 1994. *Archean Crustal Evolution.* New York: Elsevier.

Condie, K. C., ed. 1994. *Proterozoic Crustal Evolution.* New York: Elsevier.

Dott, R. H., Jr., and D. R. Prothero. 1994. *Evolution of the Earth.* 5th ed. New York: McGraw-Hill.

Gould, S. J., et al. 1993. *The Book of Life.* New York: Norton.

Gradstein, F., Ogg, J., and Smith, A. 2004. *A Geologic Time Scale 2004.* Cambridge (UK): Cambridge University Press.

Hartmann, J., and R. Miller. 1991. *The History of the Earth: An Illustrated Chronicle of an Evolving Planet.* New York: Workman.

Hoffman, P. 1988. United Plates of America: The birth of a craton. *Annual Review of Earth and Planetary Sciences* 16: 543–604.

Knoll, A. H. 2003. *Life on a Young Planet.* Princeton: Princeton University Press.

Nisbet, E. G. 1987. *The Young Earth: An Introduction to Archean Geology.* Boston: Allen and Unwin.

Rodgers, J. J. W. 1994. *A History of the Earth.* Cambridge: Cambridge University Press.

Stanley, S. M. 1999. *Earth System History.* New York: Freeman.

Windley, B. F. 1995. *The Evolving Continents.* 3rd ed. New York: Wiley.

Earth Resources

Many of the materials we use in our daily lives come from geologic materials—the Earth itself is, essentially, a natural resource. In Chapter 14, we look at the energy resources that come from the Earth. These include fossil fuels, such as oil and coal, as well as nuclear fuel and moving water. Chapter 15 focuses on nonenergy resources, particularly the mineral deposits from which we obtain metals.

By the end of Part V, we'll realize that many natural resources are not renewable and thus must be conserved if we are to avoid shortages. Also, we'll see how our use of resources has had and continues to have an impact on the environment, and on transnational politics.

CHAPTER 14

Squeezing Power from a Stone: Energy Resources

This inferno resulted when Kuwaiti oil wells were set ablaze at the end of the Gulf War. The flames speak of the energy the oil holds.

14.1 INTRODUCTION

In the extreme chill of a midwinter night in northwestern Canada, a pan of water freezes almost instantly. But the low temperature doesn't stop a wolf from stalking its prey—the wolf's legs move through the snow, its heart pumps rapidly, and its body radiates heat. These life processes require energy. **Energy,** simply defined, means the capacity to do work, to cause something to happen, or to cause change in a system (see Appendix A). The wolf's energy comes from the metabolism of special chemicals like sugar, protein, and carbohydrates in its body. These chemicals, in turn, came from the food the animal eats—in order to survive, a wolf must catch and eat mice and rabbits, so to a wolf, these animals are "energy resources." In a general sense, we use the term **resource** for any item that can be employed for a useful purpose, and, more specifically, **energy resource** for something that can be used, for example, to produce heat, power muscles, produce electricity, or move automobiles. If the energy resource consists of matter that stores energy in a usable form, such as wood or oil, we can also call it a **fuel.**

The earliest humans needed about the same quantity of fuel per capita as a wolf, and thus could maintain themselves by hunting and gathering. But when people discovered how fire could be used for cooking and heating, their need for energy resources began to exceed that of other animals, for they now needed fuel to feed their fires. Before the dawn of civilization, wood and dried dung provided adequate fuel. But as people began to congregate in towns, they also required energy for agriculture and transportation, and new resources such as animal power, wind, and flowing water came into use. By the end of the seventeenth century, energy-resource needs began to outpace the supplies available at the Earth's surface, for not only did the population continue to grow, but new industries such as iron smelting became commonplace. In fact, to feed the smelting industry, woodcutters mowed down most European forests, so that when the Industrial Revolution began in the eighteenth century, workers had to mine supplies of underground coal to feed the new steam engines that were driving factories.

Since the Industrial Revolution, society's hunger for energy has increased almost unabated (▶Fig. 14.1). In the United States today, for example, the average person uses more than 110 times the amount of energy needed by a prehistoric hunter-gatherer. Most energy for human con-

sumption in the industrial world now comes from oil and natural gas. To a lesser extent, we also continue to use coal, wind, and flowing water, and in the last half century we've added nuclear energy, geothermal energy, and solar energy to the list. As the twenty-first century dawns, we have started to investigate and work toward developing new ways of obtaining energy resources, such as: the extraction of natural gas from coal beds and from gas hydrates (strange, ice-like substances on the sea floor); the gasification of coal (turning coal into various burnable gases); the production of hydrogen fuel cells; and the production of alcohol from plant crops.

Why is a chapter in a geology book devoted to such energy resources? Simply because these resources originate in geologic materials or processes: oil, gas, coal, and the fuel for atomic power plants come from rocks, geothermal energy is a product of Earth's internal heat, and the movement of wind and water involves cycles in the Earth System. To understand the source and limitations of energy resources and to find new resources, we must understand their geology. That's why the multibillion-dollar-a-year energy industry employs tens of thousands of geologists—these are the people who find new energy-resource supplies.

In this chapter, we begin by looking at various types of energy resources on Earth. Then we focus on **fossil fuels** (oil, gas, and coal), combustible materials derived from organisms that lived in the past, in detail and other types of energy sources more briefly. The chapter concludes by outlining the dilemmas we will face as energy resources begin to run out, and as the products of energy consumption enter our environment.

FIGURE 14.1 The graph demonstrates how energy needs have increased in the past 150 years, and how different energy resources have been used to fill those needs. Oil and natural gas together now account for more than half the world's energy usage.

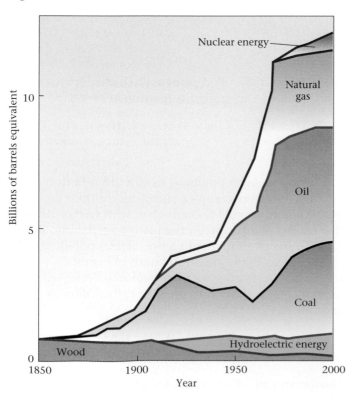

14.2 SOURCES OF ENERGY IN THE EARTH SYSTEM

When you get down to basics, there are only five fundamental sources of energy on the Earth: (1) energy generated by nuclear fusion in the Sun and transported to Earth via electromagnetic radiation; (2) energy generated by the pull of gravity; (3) energy generated by nuclear fission reactions; (4) energy that has been stored in the interior of the Earth since the planet's beginning; and (5) energy stored in the chemical bonds of compounds. Let's look at the different ways these forms of energy become resources we can use (▶Fig. 14.2).

• *Energy directly from the Sun:* Solar energy, resulting from nuclear fusion reactions in the Sun, bathes the Earth's surface. It may be converted directly into electricity, using solar-energy panels, or it may be used to heat water or warm a house. (No one has yet figured out how to produce controlled nuclear fusion on Earth, but we *can*

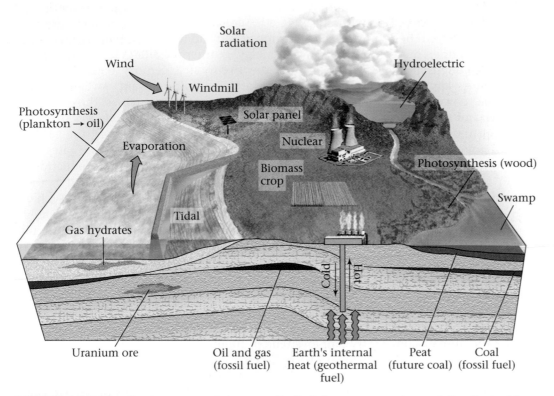

FIGURE 14.2 The diverse sources of energy on Earth. Solar energy can be used directly, to drive wind or to cause water to evaporate and make rain and ultimately running water. It also provides energy for photosynthesis, which produces wood and plankton; these ultimately become fossil fuels (coal and oil, respectively). Radioactive material from the Earth powers nuclear reactors, and the Earth's internal heat creates geothermal resources. Gravity plays a role in producing wind and water power.

produce uncontrolled fusion—this is the energy of a hydrogen bomb.)

- *Energy directly from gravity:* The gravitational attraction of the Moon, and to a lesser extent the Sun, causes ocean tides, the daily up-and-down movement of the sea surface. The flow of water in and out of channels during tidal changes can drive turbines.

- *Energy involving both solar energy and gravity:* Solar radiation heats the air, which becomes buoyant and rises. As this happens, gravity causes cooler air to sink. The resulting air movement, wind, powers sails and windmills. Solar energy also evaporates water, which enters the atmosphere. When the water condenses, it rains; the water accumulates in streams that flow downhill in response to gravity. This moving water powers waterwheels and turbines.

- *Energy via photosynthesis:* Green plants absorb some of the solar energy that reaches the Earth's surface. Their green color comes from a pigment called chlorophyll. With the aid of chlorophyll, plants produce sugar through a

chemical reaction called **photosynthesis.** In chemist's shorthand, we can write this reaction as

$$6CO_2 + 12H_2O + \text{light} \rightarrow 6O_2 + C_6H_{12}O_6 + 6H_2O.$$
carbon   water   oxygen   sugar   water
dioxide

Plants use the sugar produced by photosynthesis to manufacture more complex chemicals, or they metabolize it to provide themselves with energy. Animals that eat the plants convert the plant material into flesh.

Burning plant matter in a fire releases potential energy stored in the chemical bonds of organic chemicals. During burning, the molecules react with oxygen and break apart to produce carbon dioxide, water, and carbon (soot):

$$\text{plant} + O_2 \xrightarrow{\text{burning}} CO_2 + H_2O + C \text{ (soot)}$$
$$+ \text{ other gases} + \text{heat energy.}$$

The flames you see in fire consist of glowing gases released and heated by this reaction.

People have burned wood to produce energy for centuries. More recently, plant material (biomass) from crops such as corn and sugar cane has been used to produce ethanol, a flammable alcohol.

- *Energy from chemical reactions:* A number of inorganic chemicals can burn to produce light and energy. The energy results from exothermic (heat-producing) chemical reactions. A dynamite explosion is an extreme example of such energy production. Recently, researchers have been studying electrochemical devices, such as hydrogen fuel cells, that produce electricity directly from chemical reactions.

- *Energy from fossil fuels:* Oil, gas, and coal come from organisms that lived long ago, and thus store solar energy that reached the Earth long ago. We refer to these substances as "fossil fuels," to emphasize that they were derived from ancient organisms and have been preserved in rocks for geologic time. Burning fossil fuels produces energy in the same way that burning plants does.

- *Energy from nuclear fission:* Atoms of radioactive elements can split into smaller pieces, a process called nuclear fission (see Appendix A). During fission, a tiny amount of mass is transformed into a large amount of energy, called nuclear energy. This type of energy runs nuclear power plants and nuclear submarines. Remember that fission is essentially the opposite of fusion.

- *Energy from Earth's internal heat:* Some of Earth's internal energy dates from the birth of the planet, while some is produced by radioactive decay in minerals. Some of this internal energy heats water underground. The resulting hot water, when transformed to steam, provides **geothermal energy** that can drive turbines.

14.3 OIL AND GAS

What Are Oil and Gas?

For reasons of economics and convenience, industrialized societies today rely primarily on oil (petroleum) and natural gas for their energy needs. Oil and natural gas consist of **hydrocarbons,** chain-like or ring-like molecules made of carbon and hydrogen atoms. For example, bottled gas (propane) has the chemical formula C_3H_9. Chemists consider hydrocarbons to be a type of **organic chemical,** so named because similar chemicals make up living organisms.

Some hydrocarbons are gaseous and invisible, some resemble a watery liquid, some appear syrupy, and some are solid (▶Fig. 14.3). The "viscosity" (ability to flow) and the "volatility" (ability to evaporate) of a hydrocarbon product depend on the size of its molecules. Hydrocarbon products composed of short chains of molecules tend to be less viscous

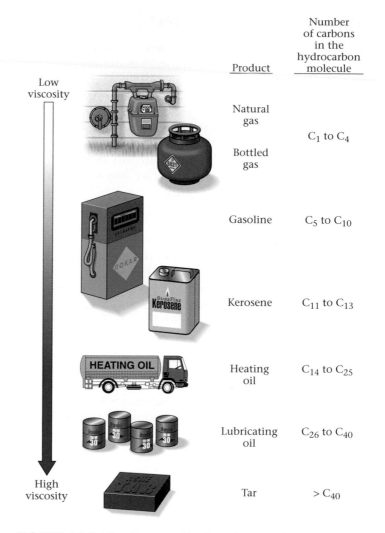

Product	Number of carbons in the hydrocarbon molecule
Natural gas	
	C_1 to C_4
Bottled gas	
Gasoline	C_5 to C_{10}
Kerosene	C_{11} to C_{13}
Heating oil	C_{14} to C_{25}
Lubricating oil	C_{26} to C_{40}
Tar	> C_{40}

FIGURE 14.3 The diversity of hydrocarbon products we use: natural gas piped to houses for heating and cooking, bottled gas (propane), gasoline for cars, kerosene for heating or illumination, diesel fuel for trucks, lubricating oil for motors, and solid tar, which melts when heated and can be used to make asphalt. Note that these products are listed in order of increasing viscosity.

(they can flow more easily) and more volatile (they evaporate more easily) than products composed of long chains, simply because the long chains tend to tangle up with one another. Thus, short-chain molecules occur in gaseous form at room temperature, moderate-length-chain molecules occur in liquid form, and long-chain molecules occur in solid form (**tar**).

Why can we use hydrocarbons as fuel? Simply because hydrocarbons, like wood, burn—they react with oxygen to form carbon dioxide, water, and heat. For example, we can describe the burning of gasoline by the reaction

$$2C_8H_{18} + 25O_2 \rightarrow 16CO_2 + 18H_2O + \text{heat energy.}$$

During such reactions, the potential energy stored in the chemical bonds of the hydrocarbon molecules converts into usable heat energy.

Where Do Oil and Gas Form?

Many people entertain the false notion that hydrocarbons come from buried trees or the carcasses of dinosaurs. In fact, the primary sources of the organic chemicals in oil and gas are dead algae and plankton bodies. Plankton, as we have seen, are the tiny plants and animals (typically around 0.5 mm in diameter) that float in sea or lake water. When algae and plankton die, they settle to the bottom of a lake or sea. Because their cells are so tiny, they can only be deposited in quiet-water environments in which clay also settles, so typically the cells mix with clay to create an organic-rich, muddy ooze. For this ooze to be preserved, it must be deposited in oxygen-poor water; otherwise, the organic chemicals in the ooze would react with oxygen or be eaten by bacteria, and thus would decompose quickly and disappear. But in some quiet-water environments (in oceans, lagoons, or lakes), dead algae and plankton can get buried by still more sediment before being destroyed. Eventually, the muddy organic ooze lithifies and becomes black organic shale (in contrast to regular shale, which consists only of clay), which contains the raw materials from which hydrocarbons eventually form. Thus, we refer to organic shale as a **source rock.**

If organic shale is buried deeply enough (2 to 4 km), it becomes warmer, since temperature increases with depth in the Earth. Chemical reactions slowly transform the organic mate-

rial in the shale into waxy molecules called **kerogen** (▶Fig. 14.4). Shale containing kerogen is called **oil shale.** If the oil shale warms to temperatures of greater than about 90°C, the kerogen molecules break down to form oil and natural gas molecules. At temperatures over about 160°C, any remaining oil breaks down to form natural gas, and at temperatures over 250°C, organic matter transforms into graphite. Thus, oil itself forms only in a relatively narrow range of temperatures, called the **oil window** (▶Fig. 14.5). For regions with a geothermal gradient of 25°C/km, the oil window lies at depths of about 3.5 to 6.5 km whereas gas can exist down to 9 km. If the geothermal gradient is low (e.g. 15°C/km), oil exists only below about 11 km. Thus, hydrocarbon reserves can only exist in the topmost 15–25% of the crust.

14.4 HYDROCARBON SYSTEMS: THE MAKING OF A RESERVE

Oil and gas do not occur in all rocks at all locations. That's why the desire to control "oil fields," regions that contain significant amounts of accessible oil underground, has played a role in sparking bitter wars. The known supply of oil and gas held underground is a **hydrocarbon reserve;** if the reserve

FIGURE 14.4 Plankton, algae, and clay settle out of water and become progressively buried and compacted, gradually being transformed into black organic shale. When heated for a long time, the organic matter in black shale is transformed into oil shale, which contains kerogen. Eventually, the kerogen transforms into oil and gas. The oil may then start to seep upward out of the shale. The red arrow indicates pressure, which increases as more sediment accumulates above.

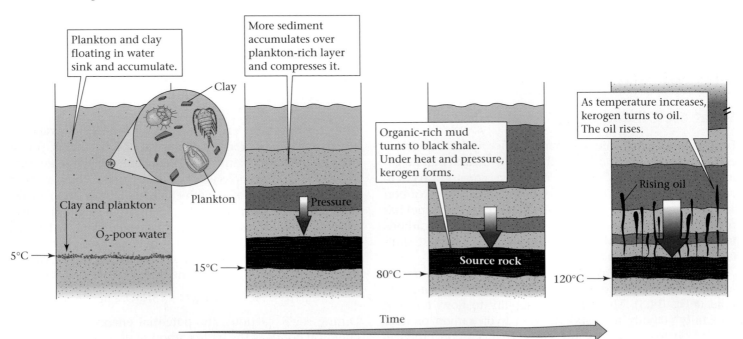

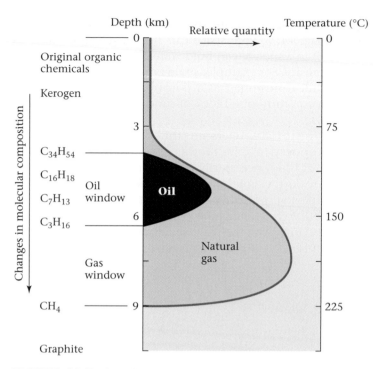

FIGURE 14.5 The oil window is the range of temperature conditions (i.e., depth) at which hydrocarbons form. In regions with a geothermal gradient of 25°C per km, oil occurs only at depths of less than about 6.5 km, the gradient portrayed on this graph, Note that gas can be found at greater depths. The length of hydrocarbon chains decreases with increasing depth, because at higher temperatures longer chains break to form smaller ones.

consists dominantly of oil, it is an "oil reserve." Reserves are not randomly distributed around the Earth (▶Fig. 14.6). For example, countries bordering the Persian Gulf contain the world's largest reserves. In this section, we learn that the development of a reserve requires the existence of four geologic features: a source rock, a reservoir rock, a migratory pathway, and a trap. A particular association of all of these components—along with the processes of hydrocarbon generation, migration, and accumulation that ultimately produce a reserve from a given source—is now called a **hydrocarbon system.**

Source Rocks and Hydrocarbon Generation

We saw above that the chemicals that become oil and gas start out in algae and plankton bodies. Their cells accumulate along with clay to form an organic ooze, which when lithified becomes black organic shale. Geologists refer to organic-rich shale as a *source rock* because it serves as the source for the organic chemicals that ultimately become oil and gas. If black shale resides in the oil window, the organic material within transforms into kerogen, and then oil and gas. This process is **hydrocarbon generation.**

Reservoir Rocks and Hydrocarbon Migration

Wells drilled into source rocks do not yield much oil because kerogen can't flow easily from the rock into the well;

FIGURE 14.6 The distribution of oil reserves around the world. The largest fields occur in the region surrounding the Persian Gulf.

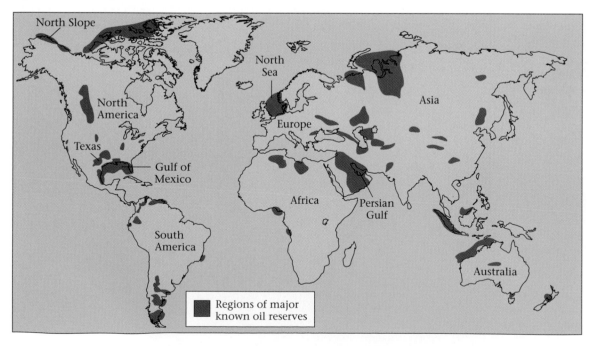

any organic matter in an oil shale remains trapped among the grains and can't move easily. So oil companies instead drill into **reservoir rocks,** rocks that contain (or could contain) an abundant amount of *easily accessible* oil and gas, meaning hydrocarbons that can be extracted out of the ground.

To be a reservoir rock, a body of rock must have space in which the oil or gas can reside, and must provide channels through which the oil or gas can move. The space can be in the form of open spaces, or **pores,** between clastic grains (which exist because the grains didn't fit together tightly and because cement didn't fill all the spaces grains during cementation) or in the form of cracks and fractures that developed after the rock formed. In some cases, groundwater passing through rock dissolves minerals and creates new space. **Porosity** refers to the amount of open space in a rock (▶Fig. 14.7). Pore space can hold oil or gas, much as the holes in a sponge can hold water. Not all rocks have the same porosity. For example, shale typically has a porosity of 10%, while sandstone has a porosity of 35%—that means that about a third of a block of sandstone actually consists of open space.

Permeability refers to the degree to which pore spaces connect to one another. Even if a rock has high porosity, it is not necessarily permeable (Fig. 14.7). In a permeable rock, the holes and cracks (pores) are linked, so a fluid is able to flow slowly through the rock, following a tortuous pathway. Keeping the concepts of porosity and permeability in mind, we can see that a poorly cemented sandstone makes a good reservoir rock, because it is both

FIGURE 14.7 The porosity and permeability of a sedimentary rock depends on the character of the rock. For example, poorly cemented sandstone can have high porosity and permeability, whereas well cemented sandstone does not. Shale tends to be impermeable and have low porosity, and the porosity in limestone is commonly due to the presence of cracks. Rocks with high porosity and permeability make the best reservoir rocks. Such rocks not only can hold a lot of oil, but the oil can flow relatively easily and thus can be pumped out efficiently.

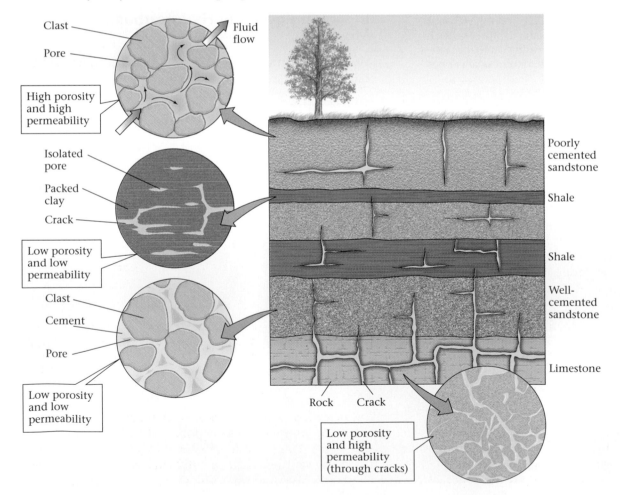

porous and permeable. A highly fractured rock can be porous and permeable, even if there is no pore space between individual grains. The greater the porosity, the greater the capacity of a reservoir rock to hold oil; and the greater the permeability, the easier it is for the oil to be extracted.

In an oil well, which is simply a hole drilled into the ground to where it penetrates reservoir rock, oil flows from the permeable reservoir rock into the well and then up to the ground surface. If the oil in the rock is under natural pressure, it may move by itself, but usually producers must set up a pump to literally suck the oil up and out of the hole.

To fill the pores of a reservoir rock, oil and gas must first "migrate" (move) from the source rock into a reservoir rock, which it will do over millions of years of geologic time (▶Fig. 14.8). Why do hydrocarbons migrate? Oil and gas are less dense than water, so they try to rise toward the Earth's surface to get above groundwater, just as salad oil rises above the vinegar in a bottle of salad dressing. Natural gas, being less dense, ends up floating above oil. In other words, buoyancy drives oil and gas upward. Typically, a hydrocarbon system must have a good **migration pathway,** such as a permeable set of fractures, in order for large volumes of hydrocarbons to move.

Traps and Seals

The existence of a reservoir rock alone does not create a reserve, because if hydrocarbons can flow easily into a reservoir rock, they can also flow out. If oil or gas escapes from the reservoir rock and ultimately reaches the Earth's surface, where it leaks away at an **oil seep,** there will be none left underground to pump. Thus, for an oil reserve to exist, oil and gas must be *trapped* underground in the reservoir rock, by means of a geologic configuration called a **trap.** A field contains one or more traps.

There are two components to an oil or gas trap. First, a **seal rock,** a relatively impermeable rock such as shale, salt, or unfractured limestone, must lie above the reservoir rock and stop the hydrocarbons from rising further. Second, the seal and reservoir rock bodies must be arranged in a geometry that collects the hydrocarbons in a restricted area. Geologists recognize several types of hydrocarbons trap geometries, four of which are described in Box 14.1.

Note that when we talk about trapping hydrocarbons underground, we are talking about a temporary process in the context of geologic time. Oil and gas may be trapped for millions of years, but eventually they will manage to pass

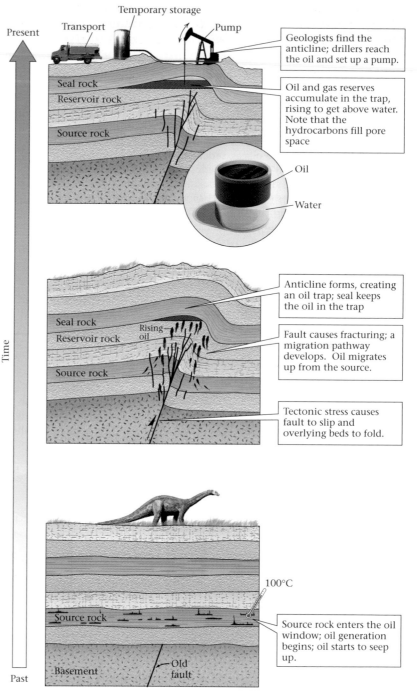

FIGURE 14.8 Initially, oil resides in the source rock. Oil gradually migrates out of the source rock and rises into the overlying water-saturated reservoir rock. Oil rises because it is buoyant relative to water—it tries to float on water. The oil is trapped beneath a seal rock. If gas exists, it floats to the top of the oil.

through a seal rock, because no rock is absolutely impermeable. Also, in some cases, microbes eat hydrocarbons in the subsurface. Thus, innumerable oil fields that existed in the past have vanished, and the oil fields we find today, if left alone, will disappear millions of years in the future.

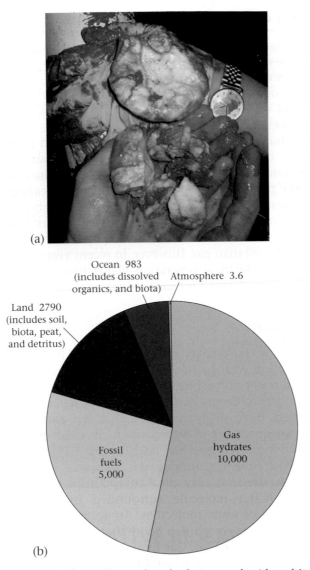

(a)

Ocean 983
(includes dissolved
organics, and biota)   Atmosphere 3.6

Land 2790
(includes soil,
biota, peat,
and detritus)

Fossil
fuels
5,000

Gas
hydrates
10,000

(b)

FIGURE 14.13 (a) Photo of gas hydrate samples (the white, icy material) brought up from the sea floor. (b) Graph showing the proportion of organic matter in different materials. Note that gas hydrate may contain the most organic carbon. (Numbers are $\times 10^5$ tons of carbon.)

14.7 COAL: ENERGY FROM THE SWAMPS OF THE PAST

Coal, a black, brittle, sedimentary rock that burns, consists of elemental carbon mixed with minor amounts of organic chemicals, quartz, and clay (▶Fig. 14.14). Note that coal and oil do not have the same composition or origin. In contrast to oil, coal forms from plant material (wood, stems, leaves) that once grew in **coal swamps,** regions that resembled the wetlands and rain forests of modern tropical to semitropical coastal areas (▶Fig. 14.15a). Like oil and gas,

FIGURE 14.14 Chunks of coal. Since 1,000 C.E., coal has been a major source of energy.

coal is a fossil fuel because it stores solar energy that reached Earth long ago. The United States burns about 1 billion tons of coal per year, mostly at electrical-generating stations, yielding about 23% of the country's energy supply.

Significant coal deposits could not form until vascular land plants appeared in the late Silurian Period, about 420 million years ago. The most extensive deposits of coal in the world occur in Carboniferous-age strata (deposited between 286 and 354 million years ago; ▶Fig 14.15b)—in fact, geologists coined the name "Carboniferous" because strata representing this interval of the geologic column contain so much coal. The abundance of Carboniferous coal reflects (1) the past position of the continents (during the Carboniferous Period, North America, Europe, and northern Asia straddled the equator, and thus had warm climates in which vegetation flourished) and (2) the height of sea level (at this time, shallow seas bordered by coal swamps covered vast parts of continental interiors). Because of their antiquity, Carboniferous coal deposits contain fossils of long-extinct species such as giant tree ferns, primitive conifers, and giant horsetails (▶Fig. 14.15c). Extensive coal deposits can also be found in strata of Cretaceous age (about 65 to 144 million years ago).

The Formation of Coal

How do the remains of plants transform into coal? The vegetation of an ancient swamp must fall and be buried in an *oxygen-poor* environment, such as stagnant water, so that it can be incorporated in a sedimentary sequence without first reacting with oxygen or being eaten. Compaction and partial decay of the vegetation transforms it into **peat.** Peat, which contains about 50% carbon, itself serves as a fuel in many parts of the world, where deposits formed from moss and grasses in bogs during the last several thousand years; it

(a)

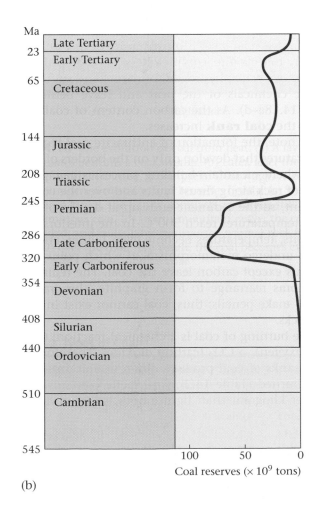

(b)

(c)

FIGURE 14.15 (a) This museum diorama depicts a Carboniferous coal swamp. (b) The graph shows the distribution of coal reserves in rocks of different ages. Notice that most reserves occur in late Paleozoic (Carboniferous) strata, a time when the continents were locked together to form Pangaea, a supercontinent that straddled the equator. (c) Fossils of fern leaves from a Carboniferous coal bed.

Furthermore, this energy supply can be destroyed, for if people pump groundwater out of the ground faster than it can be replenished, the hot-water supply diminishes.

Hydroelectric and Wind Power

As water flows downslope, its potential energy converts into kinetic energy. In a modern **hydroelectric power plant,** the water flow drives turbines, which in turn drive generators that produce electricity. In order to increase the rate and volume of water flow, engineers build dams to create a reservoir that retains water and raises it to a higher elevation—the water flows through pipes down to turbines at the foot of the dam (▶Fig. 14.27a).

At first glance, hydroelectric power seems ideal, because it produces no smoke or radioactive waste, and because reservoirs can also be used for flood control, irrigation, and recreation. But unfortunately, reservoirs may also bring unwanted changes to a region's landscape and ecology. Damming a fast-moving river may flood a spectacular canyon, eliminate exciting rapids, and destroy a river's ecosystem. Further, the reservoir traps sediment, so floodplains downstream lose their sediment and nutrient supply, and reservoirs have not all been remaining full of water, because planners underestimated water supplies.

Not all hydroelectric power plants utilize river water. In a few places, engineers have employed the potential energy stored in ocean water at high tide (▶Fig. 14.27b, c). To do this, they build a floodgate dam across an inlet. Water flows into the inlet when the tide rises, only to be trapped when the gate is closed. After the tide has dropped outside the floodgate, the water retained by the floodgate flows back to sea via a pipe that carries it through a power-generating turbine.

Wind power has been used for millennia as a way to provide power for mills and water pumps. In recent decades, numerous wind farms have been established around the world to generate electricity (▶Fig. 14.28a). The electricity is clean, but wind production has a serious drawback: it requires construction of large, somewhat noisy towers. Not everyone wants a giant wind farm in their "backyard"—because of the visual effect, because of the noise, and because the towers pose a hazard for birds.

Solar Power

The Sun drenches the Earth with energy in quantities that dwarf the amounts stored in fossil fuels. Were it possible to harness this energy directly, humanity would have a reliable and totally clean solution for powering modern technology. But using solar energy is not quite so simple, for the energy is diffuse—on the sunniest days, each square meter of the Earth's surface receives about 1,000 watts of energy. How can this energy be concentrated sufficiently to produce heat

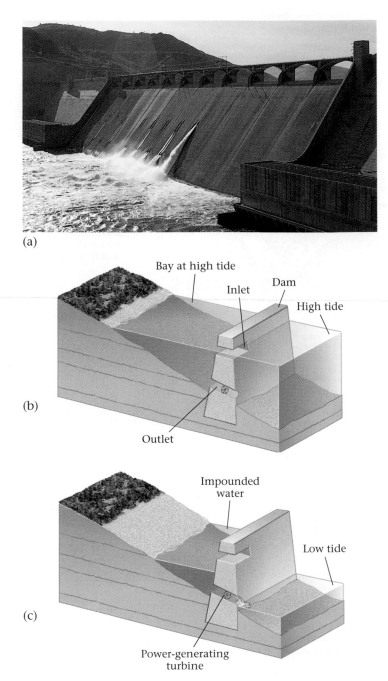

FIGURE 14.27 (a) A hydroelectric dam. This example is the Grand Coulee Dam, located on the Columbia River in central Washington State; it is the third-largest producer of hydroelectric power in the world. (b, c) Oceanic tides can be used to generate electricity. At high tide, water fills the volume behind the dam. At low tide, water flows out via a turbine.

and electricity? At present, energy consumers have two options for the direct use of solar energy.

Solar collectors constitute the first option. A **solar collector** is a device that collects energy to produce heat. One class of solar collectors includes mirrors or lenses that focus

FIGURE 14.28 (a) A wind farm in southwestern England. The towers are about 50 m high. (b) An example of a photovoltaic cell; (c) Diagram illustrating the way in which a photovoltaic cell produces electricity. (d) Diagram illustrating the way in which a hydrogen fuel cell works.

light striking a broad area into a smaller area. At a small scale, such devices can be used for cooking. Another class of solar collectors consists of a black surface placed beneath a glass plate. The black surface absorbs light that has passed through the glass plate and heats up. The glass does not let the heat escape. When a consumer runs water between the glass and the black surface, the water heats up; the hot water then can be stored in an insulated tank.

Photovoltaic cells constitute the second option. The use of **photovoltaic cells** (solar cells) allows light energy to convert directly into electricity (▶Fig. 14.28b). Most photovoltaic (PV) cells consist of two wafers of silicon pressed together. Silicon is a semiconductor, meaning that it can only conduct electricity when doped with impurities; "doping" means that manufacturers intentionally add atoms other than silicon to the wafer. One wafer is doped with atoms of arsenic, which have extra electrons and can serve as electron donors. The other wafer is doped with boron, an element that lacks electrons and can serve as an electron acceptor. Electrons would like to flow from the donor wafer to the acceptor wafer, but their way is blocked at the contact surface between the two wafers by neutral atoms. When light strikes the cell, the electrons gain enough energy to cross the boundary. If a wire loop connects the back side of one wafer to the other, an electrical current flows when

light strikes the cell (▶Fig. 14.28c). Although PV cells sound like a great way to produce electricity, they are fairly inefficient—a typical cell only converts about 10% to 15% of the energy that it receives into electricity. Thus, until the technology improves significantly, huge areas of land would have to be covered by sheets of PV cells if the cells were to provide a significant proportion of global energy supplies.

Biomass Production

Until the nineteenth century, the burning of biomass—wood, charcoal, or other plant and animal materials—provided most of the energy needed to drive civilization. But, as noted earlier, wood cannot regenerate quickly enough to provide a sufficient energy source for modern needs. In recent years, farmers have begun to produce rapidly growing crops specifically for the purpose of creating biomass for fuel production.

The most commonly used biomass fuel, *ethanol* (a type of alcohol), is produced from the fermentation of corn or sugar cane. Ethanol can substitute for gasoline in car engines. The process of producing ethanol includes the following steps. (1) Producers grind grain into a fine powder, mix it with water, and cook it to produce a mash of starch. (2) They add an enzyme to the mash, which converts the mash into sugar. (3) They mix with yeast and allow it to ferment. Fermentation

FIGURE 15.15 Stained rock is an indicator of ore. The stain comes when ore reacts with air and water. The photo encompasses about 1 m of section.

that the bedrock shatters into appropriate-sized blocks for handling. When the dust settles, large front-end loaders dump the ore into giant ore trucks, which can carry as much as 200 tons of ore in a single load (for comparison, a loaded cement mixer weighs about 70 tons). The tires on these trucks are so huge that a tall person comes up only to the base of the hub. The trucks transport waste rock (rock that doesn't contain ore) to a tailings pile and the ore to a crusher, a giant set of moving steel jaws that smash rock into small fragments. Workers then separate ore minerals from other minerals and send the ore-mineral concentrate to a processing plant, where the ore undergoes smelting or treatment with acidic solutions to separate metal atoms from other atoms. Eventually, workers melt the metal and then pour it into molds to make ingots (brick-shaped blocks) for transport to a manufacturing facility.

If the ore deposit lies more than about a hundred meters below the Earth's surface, miners must make an **underground mine.** To do so, they first dig a tunnel into the side of a mountain. (The entrance to the tunnel is called an adit.) Then the miners sink a vertical shaft in which they install an elevator. At the level in the crust where the ore body appears, they build a maze of tunnels

into the ore by drilling holes into the rock and then blasting. The rock removed must be carried back to the surface. Rock columns between the tunnels hold up the ceiling of the mine. The deepest mine on the planet currently reaches a depth of 3.5 km, where temperatures exceed 55°C, making mining there a very uncomfortable occupation. Miners face the danger from mine collapse and rock falls. Some miners have been killed or injured by "rock bursts," sudden explosions of rock off the ceiling or walls of a tunnel. These explosions happen because the rock surrounding the adit is under such great pressure that it sometimes spontaneously fractures.

15.5 NONMETALLIC MINERAL RESOURCES

So far this chapter has focused on resources that contain metal. But society uses many other geological materials, commonly known as industrial minerals, as well. From the ground we get the stone used to make roadbeds and buildings, the chemicals for fertilizers, the gypsum in drywall, the salt filling salt shakers, and the sand used to make glass—the list is endless. This section looks at a few of these geological materials and explains where they come from.

Dimension Stone

The Parthenon, a colossal stone temple rimmed by forty-six carved columns, has stood atop a hill overlooking the city of Athens for almost 2,500 years. No wonder—"stone," an architect's word for rock, outlasts nearly all other construction materials. We use stone to make facades, roofs, curbs and steps, and countertops and floors. It is used for its visual appeal as well as its durability. The names that architects give to various types of stone may differ from the formal names used by geologists. For example, architects refer to any polished carbonate rock as marble, whether or not it has been metamorphosed. Likewise, they refer to any rocks containing feldspar and quartz as granite, regardless of whether the rock has an igneous or a metamorphic texture.

To obtain intact slabs and blocks of rock (granite or marble)—known as **dimension stone** in the trade—for architectural purposes, workers must carefully cut rock out of the walls of quarries (▶Fig. 15.17a). Note that a "quarry" provides stone, while a "mine" supplies ore. To cut stone slabs, quarry operators split rock blocks from bedrock by hammering a series of wedges into the rock, or cut it off bedrock by using a wireline saw, thermal lance, or a water jet. A wireline saw consists of a loop of braided wire moving between two pulleys. In some cases, as the wire moves along the rock surface, the quarry operator spills abrasive (sand or

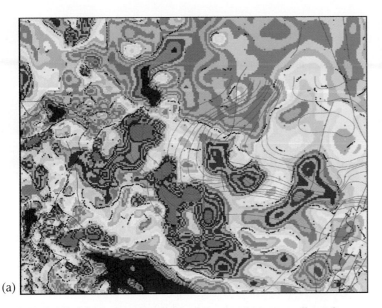

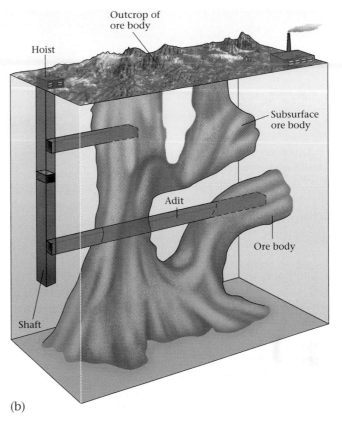

FIGURE 15.16 (a) On this magnetic-anomaly map, yellowish areas are regions with normal intensity, reddish areas are positive anomalies (greater-than-magnetic intensity), and purplish areas are negative anomalies (less-than-expected magnetic intensity). In some cases, ore-bearing rocks occur where there are positive anomalies. (b) The three-dimensional shape of an ore body underground, and the workings that miners dig to access the ore body. Note that shafts are vertical and tunnels are horizontal.

garnet grains) and water onto the wire. The movement of the wire drags the abrasive along the rock and grinds a slice into it. Alternatively, the quarry operator may use a diamond-coated wire, cooled with pure water. A thermal lance looks like a long blow torch—a flame of burning diesel fuel stoked by high-pressure air pulverizes rock, and thereby cuts a slot. More recently, quarry operators have begun to use an abrasive water jet, which squirts out water and abrasives at very high pressure, to cut rock.

Crushed Stone and Concrete

Crushed stone forms the substrate of highways and railroads and serves as the raw material for manufacturing cement, concrete, and asphalt. In crushed-stone quarries (▶Fig. 15.17b), operators use high explosives to break up bedrock into rubble that they then transport by truck to a jaw crusher, which reduces the rubble into usable-size fractions.

Much modern construction utilizes mortar and concrete (see Box 15.1), human-made rock-like materials formed when a slurry composed of sand and/or gravel mixed with cement and water is allowed to harden. The hardening takes place when a complex assemblage of minerals grows by chemical reactions in the slurry; these minerals bind together the grains of sand or gravel in mortar or concrete. (Note that the word **mortar** refers to the sub-

stance that holds bricks or stone blocks together, whereas **concrete** refers to the substance that workers shape into roads or walls by spreading it out into a layer or by pouring it into a form.) The **cement** in mortar or concrete starts out as a powder composed of lime (CaO), quartz (SiO_2), aluminum oxide (Al_2O_3), and iron oxide (Fe_2O_3). Typically, lime accounts for 66% of cement, silica for 25%, and the remaining chemicals for about 9%.

It appears that the ancient Romans were the first to use cement—they made it from a mixture of volcanic glass and limestone. During recent centuries, most cement has been produced by heating specific types of limestone (which happened to contain calcite, clay, and quartz in the correct proportions) in a kiln up to a temperature of about 1,450°C; the heating releases CO_2 gas and produces "clinker," chunks consisting of lime and other oxide compounds. Manufacturers crush the clinker into cement powder and pack it in bags for transport. But natural limestone with the exact composition of cement is fairly rare, so most cement used today is **Portland cement,** made by intentionally mixing limestone, sandstone, and shale in just the right proportions to provide the right chemical makeup. Isaac Johnson, an English engineer, came up with the recipe for Portland cement in 1844, and he named it after the town of Portland, England, because he thought it resembled rock exposed there.

(a)

(b)

FIGURE 15.17 (a) An active quarrying operation, showing large blocks of cut stone; (b) a crushed-limestone quarry.

Nonmetallic Minerals in Your Home

We use an astounding variety of nonmetallic geologic resources (see Table 15.2) without ever realizing where they come from. Consider the materials in a house or apartment. The concrete foundation consists of cement, made from limestone mixed with sand or gravel. The bricks in the exterior walls originated as clay, formed from the chemical weathering of silicate rocks and perhaps dug from the floodplain of a stream. To make *bricks*, workers mold wet clay into blocks, which they then bake. Baking drives out water and causes metamorphic reactions that recrystallize the clay. Clay also serves as the raw material from which pottery, porcelain, and other ceramic materials are made.

The glass used to glaze windows consists largely of silica, formed by first melting and then freezing pure quartz

---

BOX 15.1
THE HUMAN ANGLE

The Sidewalks of New York

Untold tons of concrete have gone into the construction of New York City. In fact, with the exception of a few city parks, most of the walking space in the city consists of concrete. And concrete skyscrapers tower above the concrete plain. Where does all this concrete come from?

Much of the sand used in New York concrete was deposited during the last ice age. As vast glaciers moved southward over 14,000 years ago, they ground away the igneous and metamorphic rocks that constituted central and eastern Canada. These ancient rocks contained abundant quartz, and since quartz lasts a long time (it does not undergo chemical weathering easily), the sediment transported by the glaciers retained a large amount of quartz. Glaciers deposited this sediment in huge piles called moraines (see Chapter 22). As the glaciers melted, fast-moving rivers of meltwater washed the sediment, sorting sand from mud and pebbles. The sand was deposited in bars in the meltwater rivers, and these relict bars now provide thick lenses of sand that can be economically excavated.

What about the cement? Cement contains a mixture of lime, derived from limestone, and other elements (such as silica) derived from shale and sandstone. The bedrock of New York, though, consists largely of schist and gneiss, not sedimentary rocks. Fortunately, a source of rocks appropriate for making cement lies up the Hudson River. A rock unit called the Rosendale Formation, which naturally contains exactly the right mixture of lime and silica needed to make a durable cement, crops out in low ridges just to the west of the river. Beginning in the late 1820s, workers began quarrying the Rosendale Formation for cement, creating a network of underground caverns. Quarry operators followed the Rosendale beds closely, making horizontal mine tunnels where the beds were horizontal, tilted mine tunnels where the beds tilted, and vertical mine tunnels where the beds were vertical. They then dumped the excavated rock into nearby kilns and roasted it to produce lime mixed with other oxides. The resulting powder was packed into barrels, loaded onto barges, and shipped downriver to New York. As demand for cement increased, operators eventually dug open-pit quarries in which they excavated other limestone and shale units, mixing them together in the correct proportion to make Portland cement.

The rocks making up the Rosendale Formation consist of cemented-together shell fragments and small, reef-like colonies of organisms. In other words, the lime in the concrete of New York sidewalks was originally extracted from seawater by living organisms—brachiopods, crinoids, and bryozoans—over 350 million years ago.

TABLE 15.2 Common Nonmetallic Resources

Limestone	Sedimentary rock made of calcite; used for gravel or cement.
Crushed stone	Any variety of coherent rock (limestone, quartzite, granite, gneiss).
Siltstone	Beds of sedimentary rock, used to make flagstone.
Granite	Coarse igneous rock, used for dimension stone.
Marble	Metamorphosed limestone; used for dimension stone.
Slate	Metamorphosed shale; used for roofing shingles.
Gypsum	A sulfate salt precipitated from saltwater; used for wallboard.
Phosphate	From the mineral apatite; used for fertilizer.
Pumice	Frothy volcanic rock; used to decorate gardens and paths.
Clay	Very fine mica-like mineral in sediment; used to make bricks or pottery.
Sand	From sandstone, beaches, or riverbeds; quartz sand is used for construction and for making glass.
Salt	From the mineral halite, formed by evaporating saltwater; used for food, melting ice on roads.
Sulfur	Occurs either as native sulfur, typically above salt domes, or in sulfide minerals; used for fertilizer and chemicals.

sions. The copper in the house's electrical wiring most likely comes from the ores in a porphyry copper deposit; the iron in the nails probably comes from banded-iron formation; and the plastic used in everything from countertops to light fixtures comes from oil formed from the bodies of plankton and algae that died millions of years ago. As you can see, geologic processes acting over millions to billions of years have been at work to produce the materials making your home.

Chemicals employed for agricultural purposes also come from the ground. For example, potash (K_2CO_3) comes from the minerals in evaporite deposits. Phosphate (PO_4^{-3}) comes from the mineral apatite, formed by diagenesis of mud on the ocean floor. Truly, without the geologic resources of the Earth, modern society would grind to a halt.

sand from a beach deposit or a sandstone formation. Quartz may also be used in the construction of photovoltaic cells for solar panels. Gypsum board (drywall), used to construct interior walls, comes from a slurry of water and the mineral gypsum sandwiched between sheets of paper. Gypsum ($CaSO_4 \cdot 2H_2O$) occurs in evaporite strata precipitated from seawater or saline lake water. Evaporites provide other useful minerals as well, such as halite.

The asbestos that was once used to make roof shingles is derived from serpentine, a rock created by the reaction of olivine with water. The olivine may come from oceanic lithosphere, thrust onto continental crust during continental colli-

15.6 GLOBAL MINERAL NEEDS

How Long Will Resources Last?

The average citizen of an industrialized country uses 25 kilograms (kg) of aluminum, 10 kg of copper, and 550 kg of iron and steel in a year's time (▶Fig. 15.18; Table 15.3). If you combine these figures with the quantities of energy resources and nonmetallic geologic resources a person uses, you get a total of about 15,000 kg (15 metric tons) of resources used per capita each year. Thus, the population of the United States consumes about 4 billion metric tons

FIGURE 15.18 We consume vast quantities of mineral resources in a year, as the diagram indicates. The numbers indicate the weight of the material used per person per year.

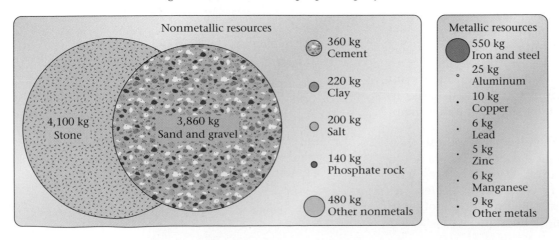

TABLE 15.3 Yearly Per Capita Usage of Geologic Materials in the U.S.A.

4,100 kg	Stone
3,860 kg	Sand and gravel
3,050 kg	Petroleum
2,650 kg	Coal
1,900 kg	Natural gas
550 kg	Iron and steel
360 kg	Cement
220 kg	Clay
200 kg	Salt
140 kg	Phosphate
25 kg	Aluminum
10 kg	Copper
6 kg	Lead
5 kg	Zinc

1 kg = 2.205 pounds

of geologic material per year, and to create this supply, workers must mine, quarry, or pump 18 billion metric tons. By comparison, the Mississippi River transports 190 million metric tons of sediment per year into the Gulf of Mexico.

Mineral resources, like oil and coal, are nonrenewable resources. Once mined, an ore deposit or a limestone hill disappears forever. Natural geologic processes do not happen fast enough to replace the deposits as quickly as we use them. Geologists have calculated **reserves** (measured quantities of a commodity) for various mineral deposits just as they have for oil. Based on current definitions of reserves (which depend on today's prices) and rates of consumption, supplies of some metals may run out in only decades to centuries (see Table 15.4). But these estimates may change as supplies become depleted and prices rise (making previously uneconomical deposits worth mining). And supplies could increase if geologists discover new reserves or if new ways of mining become available (e.g., providing access to nodules on the sea floor or to deeper parts of the crust). Further, increased efforts at conservation and recycling can cause a dramatic decrease in rates of consumption, and thereby stretch the lifetime of existing reserves.

Ore deposits do not occur everywhere, because their formation requires special geological conditions. As a result, some countries possess vast supplies, while others have none. In fact, no single country owns all the mineral resources it needs, so nations must trade with one another to maintain supplies, and global politics inevitably affects prices. Many wars have their roots in competition for mineral reserves, and it is no surprise that the outcomes of some wars have hinged on who controls these reserves.

The United States worries in particular about supplies of so-called **strategic metals,** which include manganese, platinum, chromium, and cobalt—metals alloyed with iron to make the special-purpose steels needed in the aerospace industry. At present, the country must import 100% of the manganese, 95% of the cobalt, 73% of the chromium, and 92% of the platinum it consumes. Principal reserves of these metals lie in the crust of countries that have not always practiced open trade with the United States. As a defense precaution, the United States stockpiles these metals in case supplies are cut off.

Mining and the Environment ESS

Mining leaves a big footprint in the Earth System. Some of the gaping holes that open-pit mining creates in the landscape have become so big that astronauts can see them from space. Both open-pit and underground mining yield immense quantities of waste rock, which miners dump in "tailings piles," some of which grow into artificial hills 200 meters high and many kilometers long. Lacking soil, tailings piles tend to remain unvegetated for a long time. Mining also exposes ore-bearing rock to the atmosphere, and since many ore minerals are sulfides, they react with rainwater to produce **acid mine runoff,** which can severely damage vegetation downstream (▶Fig. 15.19a). Ore processing tends to release noxious chemicals that can mix with rain and spread over the countryside, damaging life. Before the installation of modern environmental controls,

TABLE 15.4 Lifetimes (in years) of Currently Known Ore Resources

Metal	World Resources	U.S. Resources
Iron	120	40
Aluminum	330	2
Copper	65	40
Lead	20	40
Zinc	30	25
Gold	30	20
Platinum	45	1
Nickel	75	less than 1
Cobalt	50	less than 1
Manganese	70	0
Chromium	75	0

smoke from ore-processing plants caused severe air pollution; plumes of smoke from the old smelters in Sudbury, Ontario, for example, created a wasteland for many kilometers downwind (▶Fig. 15.19b). In recent years, there have been efforts to reclaim mining spoils, and new technologies have been developed to extract metals in ways that are less deleterious to the environment and that treat waste more efficiently. Clearly, a mine has the potential to become a scar on the landscape, the size of which depends on the efforts of miners to minimize damage.

CHAPTER SUMMARY

• Industrial societies use many types of minerals, all of which must be extracted from the upper crust. We distinguish two general categories: metallic resources and nonmetallic resources.

• Metals are materials in which atoms are held together by metallic bonds. They are malleable and make good conductors.

• Metals come from ore. An ore is a rock containing native metals or ore minerals (sulfide, oxide, or carbonate minerals with a high proportion of metal) in sufficient quantities to be worth mining. An ore deposit is an accumulation of ore.

• Magmatic deposits form when sulfide ore minerals settle to the floor of a magma chamber. In hydrothermal deposits, ore minerals precipitate from hot-water solutions. Secondary-enrichment deposits form when groundwater carries metals away from a preexisting deposit. Mississippi Valley–type deposits precipitate from groundwater that has passed long distances through the crust. Sedimentary deposits precipitate out of the ocean. Residual mineral deposits in soil are the result of severe leaching in tropical climates. Placer deposits develop when heavy metal grains accumulate in sediment along a stream.

• Many ore deposits are associated with igneous activity in subduction zones, along mid-ocean ridges, along continental rifts, or at hot spots.

• Nonmetallic resources include dimension stone for decorative purposes, crushed stone for cement and asphalt production, clay for brick making, sand for glass production,

FIGURE 15.19 (a) The orange color in this acid mine runoff is from dissolved iron in the water. (b) This vegetation-free zone near Sudbury, Ontario, developed in response to acidic smelter smoke. A large tailings pile can be seen in the distance. Once the tall smoke stack came into use, the smoke blew further away, and the land in the foreground became revegetated.

(a)

(b)

and many others. A large proportion of materials in your home have a geological ancestry.

• Mineral resources are nonrenewable. Many are now or may soon become in short supply.

KEY TERMS

acid mine runoff (p. 486)
alloy (p. 473)
banded-iron formation (BIF) (p. 477)
base metals (p. 473)
cement (p. 483)
concrete (p. 483)
dimension stone (p. 482)
disseminated deposit (p. 476)
grade (p. 475)
hydrothermal deposit (p. 476)
magmatic deposit (p. 475)
manganese nodules (p. 477)
massive-sulfide deposit (p. 475)
metallic bonds (p. 472)
metals (p. 472)
mineral resources (p. 472)
Mississippi Valley–type (MVT) ores (p. 477)

mortar (p. 483)
native metals (p. 472)
open-pit mine (p. 479)
ore (p. 474)
ore deposit (p. 475)
ore minerals (economic minerals) (p. 474)
placer deposit (p. 478)
Portland cement (p. 483)
precious metals (p. 473)
reserves (p. 486)
residual mineral deposit (p. 477)
secondary-enrichment deposit (p. 477)
smelting (p. 473)
strategic metals (p. 486)
underground mine (p. 482)
vein deposit (p. 476)

REVIEW QUESTIONS

1. Describe how people have used copper, bronze, and iron throughout history.

2. Why don't we use an average granite as a source for useful metals?

3. What kinds of concentrations of a metal are required for it to be economically minable?

4. Describe various kinds of economic mineral deposits.

5. What procedures are used to locate and mine mineral resources today?

6. How is stone cut from a quarry?

7. What are the ingredients of cement? How is Portland cement made?

8. How many kilograms of mineral resources does the average person in an industrialized country use in a year?

9. Compare the estimated lifetimes of ore supplies (worldwide and in the United States) of iron, aluminum, copper, gold, and chromium.

10. What are some environmental hazards of large-scale mining?

THE VIEW FROM ABOVE A large limestone quarry cut into a forested landscape. The material gouged out of the ground at this location has been transferred elsewhere to make roadbeds and buildings. Use of mineral resources does make a mark on the landscape.

SUGGESTED READING

Brands, H. W. 2002. *The Age of Gold: The California Gold Rush and the New American Dream.* New York: Doubleday.

Carr, D. D., and N. Herz, eds. 1988. *Concise Encyclopedia of Mineral Resources.* Cambridge, Mass.: MIT Press.

Craddock, P., and J. Lang, eds. 2003. *Mining and Metal Production through the Ages.* London: British Museum Publications.

Craig, J. R., D. J. Vaughan, and B. J. Skinner. 1989. *Resources of the Earth.* Englewood Cliffs, N.J.: Prentice-Hall.

Dorr, A. 1987. *Minerals: Foundations of Society.* Alexandria, Va.: American Geological Institute.

Evans, A. M. 1992. *Ore Geology and Industrial Minerals: An Introduction.* 3rd ed. Malden, Mass.: Blackwell Science.

Evans, A. M. 1997. *An Introduction to Economic Geology and Its Environmental Impact.* Malden, Mass.: Blackwell Science.

Guilbert, J. M., and C. F. Park, Jr. 1986. *The Geology of Ore Deposits.* New York: Freeman.

Kesler, S. E. 1994. *Mineral Resources, Economics, and the Environment.* New York: Macmillan.

Manning, D. A. C. 1995. *Introduction to Industrial Minerals.* London: Chapman & Hall.

National Research Council. 1990. *Competitiveness of the U.S. Minerals and Metals Industry.* Washington, D.C.: National Academy Press.

Sawkins, F. J. 1984. *Metal Deposits in Relation to Plate Tectonics.* New York: Springer-Verlag.

Stone, I. 1956. *Men to Match My Mountains.* New York: Doubleday; reprinted 1982 by Berkley Books. See pp. 128–51.

PART VI

Processes and Problems at the Earth's Surface

*I*n the last part of this book, we focus on Earth's surface and near-surface realms. This portion of the Earth System, which encompasses the boundaries among the lithosphere, hydrosphere, and atmosphere, displays great variability, for the dynamic interplay between internal processes (driven by Earth's internal heat) and external processes (driven by the warmth of the Sun), under the influence of Earth's gravitational field, has resulted in a diverse array of landscapes. In Chapters 16 through 22, we examine five of these landscapes, plus groundwater and the atmosphere, and finally, in Chapter 23, we see how forces at work in the Earth System cause the planet to constantly change.

Ever-Changing Landscapes and the Hydrologic Cycle

Talk of mysteries! Think of our life in nature—daily to be shown matter, to come in contact with it—rocks, trees, wind on our cheeks! the solid earth! the actual world! the common sense! Contact! Contact! Who are we? Where are we?
—HENRY DAVID THOREAU (1817–1862)

This view of a windy inlet along the coast of Norway illustrates various reservoirs for water in the Earth System. Most of our planet's water resides in the sea, but some forms the clouds of the sky, some exists as snow or ice, some flows in streams across the land, and some hides underground. Tectonic forces uplift mountains, but moving water, ice, and air erodes the land as time passes.

E.1 INTRODUCTION

The Earth's surface is at once a place of endless variety and intricate detail. Observe the height of its mountains, the expanse of its seas, the desolation of its deserts, and you may be inspired, frightened, or calmed. It's no wonder that artists and writers across the ages have sought inspiration from the **landscape**—the character and shape of the land surface in a region—for landscapes display the diversity of human emotion (▶Fig. E.1a–e). Geologists, like artists and writers, savor the impression of a dramatic landscape, but on seeing one, they can't help but ask, "How did it come to be, and how will it change in the future?"

The subject of landscape development and evolution, and the **landforms** (individual shapes such as mesas, valleys, cliffs, and dunes) that constitute it, dominates many of the chapters in Part VI. Geologists who study landscape development are known as geomorphologists. This interlude, a general introduction to the topic, explains the driving forces behind landscape development, identifies factors that control which landscape develops in a given locality, and describes the hydrologic cycle, the pathway water molecules follow as they move from ocean to air to land and back to ocean. We focus on the hydrologic cycle here because so many processes on or near Earth's surface involve water in its various forms.

E.2 SHAPING THE EARTH'S SURFACE

If the Earth's surface were totally flat, the great diversity of landscapes that embellish our vistas would not exist. But

(a)

(c)

(b)

(d)

(e)

FIGURE E.1 A great variety of landscapes on Earth. (a) The desert ranges of the Mojave Desert, in southeastern California. (b) The peaks of the Grand Tetons, in Wyoming. (c) A rock and sand seascape along the coast of Brazil. (d) Buttes at Monument Valley, Arizona. (e) Steep cliffs, in Australia's Blue Mountains.

the surface isn't flat, because a variety of geologic processes cause portions of the surface to move up or down relative to adjacent regions. We refer to the upward movement of the land surface as **uplift,** and the sinking or downward movement of the land surface as **subsidence.** Both uplift and subsidence occur for a variety of reasons (Table E.1).

When uplift or subsidence takes place, the elevation difference does not remain the same forever, because other components of the Earth System kick into action—material at higher elevations becomes unstable and susceptible to *downslope movement* (the tumbling or sliding of rock and sediment from higher elevations to lower ones); moving water, ice, and air cause *erosion* (the grinding away and removal of the Earth's surface); and where moving fluids slow down, *deposition* of sediment takes place. Downslope movement, erosion, and deposition redistribute rock and sediment, ultimately stripping it in from higher areas and collecting it in low areas. As a consequence, a great variety of both **erosional landforms** (those carved by erosion) and **depositional landforms** (those built from an accumulation of sediment) can form.

TABLE E.1 Causes of Uplift and Subsidence

Causes of Uplift

- *Thickening of the crust due to deformation.* At convergent and collisional boundaries, compression causes the crust to shorten horizontally (by development of folds, faults, and foliations) and thicken in the vertical direction. Because of isostasy (see Chapter 11), lithosphere with thickened crust floats relatively higher on the asthenosphere, with the result that the surface of the crust in mountain belts rises.

- *Heating of the lithosphere.* Heating decreases the thickness and density of the lithosphere, so to maintain isostatic equilibrium, it floats higher. Intrusion or extrusion of igneous rocks thickens the crust or builds volcanoes on top of the surface; these phenomena, therefore, cause uplift.

- *Rebound due to unloading.* Removal of a heavy load (such as a glacier or mountain) causes the Earth's surface to rise, somewhat the way the surface of a trampoline rises when you step off of it.

Causes of Subsidence

- *Thinning of the crust due to deformation.* In rifts, where the crust undergoes horizontal stretching, the axis of the rift drops down by slip on normal faults.

- *Cooling of the lithosphere.* Cooling thickens the lithosphere and makes it denser, so to maintain isostatic equilibrium, the lithosphere sinks down and its surface lies at a lower elevation.

- *Sinking due to loading.* Where a heavy load (such as a glacier or volcano) forms on the Earth's surface, the lithosphere warps downward, somewhat the way the surface of a trampoline warps down when you stand on it.

Notably, because lithosphere floats on asthenosphere (see the discussion of isostasy in Chapter 11), the removal of 1 km of rock of the top of a mountain range causes the crust to rise by about 1/3 km, just as the removal of heavy containers from the deck of a cargo ship in a body of water makes the ship rise. Because the crust rises as erosion takes place, more than 5 km of rock must erode from a 5-km-high mountain range to return the land surface to sea level. Similarly, in a depositional setting, a 1-km-thick layer of sediment depresses the crust by about 1/3 km.

The energy that drives landscape evolution comes from three sources: **internal energy,** the heat within the Earth, which drives the plate motions and mantle plumes that cause displacement of the crust's surface; **external energy,** energy coming to the Earth from the Sun, which causes the atmosphere and ocean to flow; and **gravitational energy,** which pulls rock down slopes at the surface and causes convective flow. Landscape evolution, in fact, reflects a "battle" between (1) tectonic processes such as collision, convergence, and rifting, which create **relief** (▶Fig. E.3a) (an elevation difference between two locations) in an area, and (2) processes such as downslope movement, erosion, and deposition, which destroy relief by removing material from high areas and depositing it low ones. If, in a particular region, the rate of uplift exceeds the rate of erosion, the land surface rises; if the rate of subsidence exceeds

the rate of deposition, the land surface sinks. Without uplift and subsidence, Earth's surface would long ago have been beveled to a flat plain, and without erosion and deposition, high and low areas would have lasted for the entirety of Earth history.

How rapidly do uplift, subsidence, erosion, and deposition take place? The Earth's surface can rise or sink by as much as 3 m during a single major earthquake; but averaged over time, the rates of uplift and subsidence range between 0.01 and 10 mm per year (▶Fig. E.2a). Similarly, erosion can carve out several meters of *substrate,* the material just below the ground surface, during a single flood, storm, or landslide (▶Fig. E.2b). And deposition during a single event can create a layer of debris tens of meters thick in a matter of minutes to days. But, averaged over time, erosional and depositional rates also vary between 0.10 and 10 mm per year. Although these rates seem small, a change in surface elevation of just 0.5 mm (the thickness of a fingernail) per year can yield a net change of 5 km in 10 million years. Uplift can build a mountain range, and erosion can whittle one down to near sea level—it just takes time!

E.3 TOOLS OF THE TRADE: TOPOGRAPHIC MAPS AND PROFILES

We can distinguish one landform from another by its shape—for example, as you will see in succeeding chapters, a river-carved valley simply does not look like a glacially carved valley. Landform shapes are manifested by variations in elevation within a region. Geologists use the term **topography** to refer to such variations.

How can we convey information about topography—a three-dimensional feature—on a two-dimensional sheet of paper? Geologists do this by means of a **topographic map,** which uses contour lines to represent the variations in elevation (▶Fig. E.3a–d). A **contour line** is an imaginary line along which all points have the same elevation. For example, if you walk along the 200-m contour line on a hill slope, you stay at exactly the same elevation. As another example, the shoreline on a flat, calm body of water is a contour line. In other words, you can picture the contour line as the intersection between the land surface and an imaginary

(a)

(b)

FIGURE E.2 (a) Uplifted beach terraces along a coast appear where the coast is rising relative to sea level. Wave erosion eats into the land, creating a terrace. Eventually, the terrace rises out of the reach of the waves, which then cut a new one. (b) So much erosion can take place during a single hurricane that the foundations of houses built along the beach become undermined.

horizontal plane. Contour lines form a closed loop around a hill, and they form a V shape that points upstream where they cross a river valley.

The elevation difference between two adjacent contour lines on a topographical map is called the **contour interval.** For a given topographic map, the contour interval is constant, so the spacing between contour lines represents the steepness of a slope. Specifically, closely spaced contour lines represent a steep slope, whereas widely spaced contour lines represent a gentle slope. Think of it this way: if contour lines were painted on the ground surface and you were walking along a straight line on the ground surface, you would cross many contour lines while covering only a

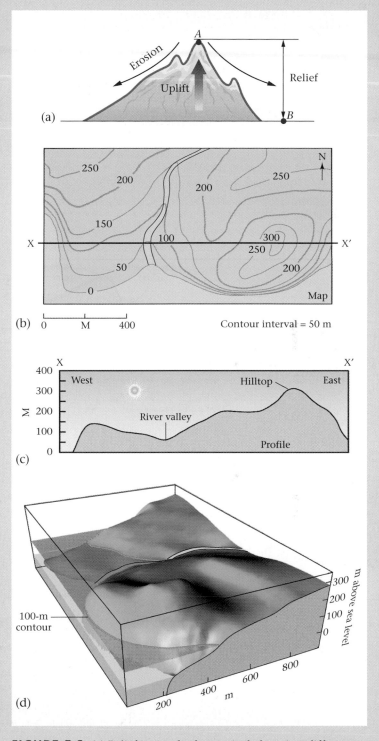

FIGURE E.3 (a) Relief is simply the vertical elevation difference between two points (A and B) at the surface of the Earth. (b) A topographic map depicts the 3-D shape of the land surface on a 2-D map through the use of contour lines. The difference in elevation between two adjacent lines is the contour interval. (c) A topographic profile (along section line X-X') shows the shape of the land surface as seen in a vertical slice. (d) You can picture a contour line as the intersection of a horizontal plane with the land surface. This block diagram shows the area of (a) in 3-D.

example, agriculture greatly increases the rate of erosion, because for much of the year farm fields have no vegetation cover.

E.5 THE HYDROLOGIC CYCLE

As is evident from the discussion above, water in its various forms (liquid, gas, and solid) plays a major role in erosion and deposition on Earth's surface. Our planet's water is found in certain distinct reservoirs (containers), namely: the oceans, glacier ice (today, mostly in Antarctica and Greenland), groundwater, lakes, soil moisture, living organisms, the atmosphere, and rivers (see Table E.2). Together, these constitute the "hydrosphere". Water constantly flows from reservoir to reservoir, and this never-ending passage is called the **hydrologic cycle** (see art on pp. 500–501). Without the hydrologic cycle, the erosive forces of running water (rivers and streams) and flowing ice (glaciers) would not exist.

The *average* length of time that water stays in a particular reservoir during the hydrologic cycle is called the **residence time.** Water in different reservoirs has different residence times. For example, a typical molecule of water remains in the oceans for 4,000 years or less, in lakes and ponds for 10 years or less, in rivers for 2 weeks or less, and in the atmosphere for 10 days or less. Groundwater residence times are highly variable and depend on how deep the groundwater flows. Water can stay underground for anywhere from 2 weeks to 10,000 years before it inevitably moves on to another reservoir.

To get a clearer sense of how the hydrologic cycle operates, let's follow the fate of seawater that has just reached the surface of the ocean. Solar radiation heats the water, and the increased thermal energy of the vibrating water molecules allows them to evaporate (break free from the liquid) and drift upward in a gaseous state to become part of the atmosphere. About 417,000 cubic km (102,000 cubic miles), or about 30% of the total ocean volume, evaporates every year. Atmospheric water vapor moves with the wind to higher elevations, where it cools, undergoes condensation (the molecules link together to form a liquid), and rains or snows. About 76% of this water precipitates (falls out of the air) directly back into the ocean. The remainder precipitates onto land; most of this becomes trapped temporarily in the soil, or in plants and animals, and soon returns directly to the atmosphere by what is called evapotranspiration: the sum of evaporation from bodies of water, evaporation from the ground surface, and transpiration (release as a metabolic byproduct) from plants and animals. Rainwater that did not become trapped in the soil or in living organisms either enters lakes or rivers and ultimately flows back to the sea as surface water, becomes trapped in glaciers, or sinks deeper into the ground

TABLE E.2 Major Water Reservoirs of the Earth

Reservoir	% of Total Water
Oceans	95.0
Glaciers and ice sheets	2.97
Groundwater	1.05
Lakes and rivers	0.009
Atmosphere	0.001
Living organisms and soil	0.0001

to become groundwater. Groundwater also flows and ultimately returns to the Earth's surface reservoirs. In sum, during the hydrologic cycle, water moves among the ocean, the atmosphere, reservoirs on or below the land, and living organisms.

E.6 LANDSCAPES OF OTHER PLANETS

The dynamic, ever-changing landscape on Earth contrasts markedly with those of other terrestrial planets. Each of the terrestrial planets and moons has its own unique surface landscape features, reflecting the interplay between the object's particular tectonic and erosional processes. Let's look at a few examples: the Moon, Mars, and Venus.

Our Moon has a static, pockmarked landscape generated exclusively by meteorite impacts and volcanic activity. Because no plate tectonics occurs on the Moon, no new mountains form, and because no atmosphere or ocean exists, there is no hydrologic cycle and no erosion from rivers, glaciers, or winds. Therefore, the lunar surface has remained largely unchanged for over billions of years. The landscape can be divided into the Lunar Highlands, the heavily cratered light-colored regions of the moon exposing rocks over 4.0 billions of years old, and the Mare, vast plains of flood basalt possibly formed in response to impacts over 3.8 billion years ago that were so huge that they caused melting in the Moon's mantle and extrusion of flood basalts (▶Fig. E.7a, b).

Landscapes on Mars differ from those of the Moon because Mars *does* have an atmosphere (though much less dense than that of Earth) whose winds generate huge dust storms, some of which obscure nearly the entire surface of the planet for months at a time. The landscapes of Mars also differ from the moon's because Mars probably once had surface water (Box E.1 and ▶Fig. E.9a, b). Thus, the Martian surface appears to have four kinds of materials: volcanic flows

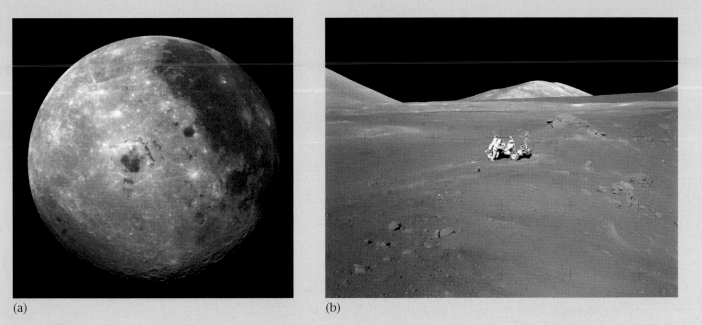

(a)   (b)

FIGURE E.7 (a) This lunar landscape is virtually the same landscape that would have been visible 2 billion years ago, because the Moon's surface is static. (b) A close-up view of the lunar landscape, with the lunar rover and an astronaut for scale.

and deposits (primarily of basalt), debris from impacts, wind-blown sediment, and water-laid sediment. There is even evidence that soil-forming processes affected surface materials. Martian winds not only deposit sediment, they slowly erode impact craters, and polish surface rocks.

Landscapes on Mars also differ from those on Earth, because Mars does not have plate tectonics. So, unlike Earth, Mars has no mountain belts or volcanic arcs. In fact, most landscape features on Mars, with the exception of wild-related ones, are over three billion-years old. Notably, long ago, a huge mantle plume formed. This plume caused the uplift of a 9-km-high bulge (the Tharsis Ridge) that covers an area comparable to that of North America (▶Fig. E.8). Thermal activity also led to the eruption of gargantuan hot-spot volcanoes, such as the 22-km-high Olympus Mons. Mars also boasts the largest known canyon, the Valles Marineris, a gash over 3,000 km long and

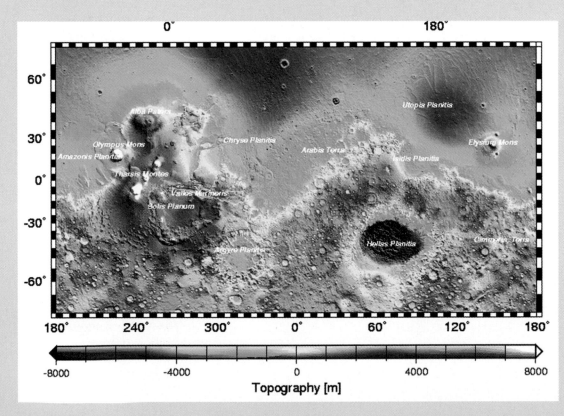

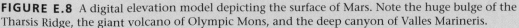

FIGURE E.8 A digital elevation model depicting the surface of Mars. Note the huge bulge of the Tharsis Ridge, the giant volcano of Olympic Mons, and the deep canyon of Valles Marineris.

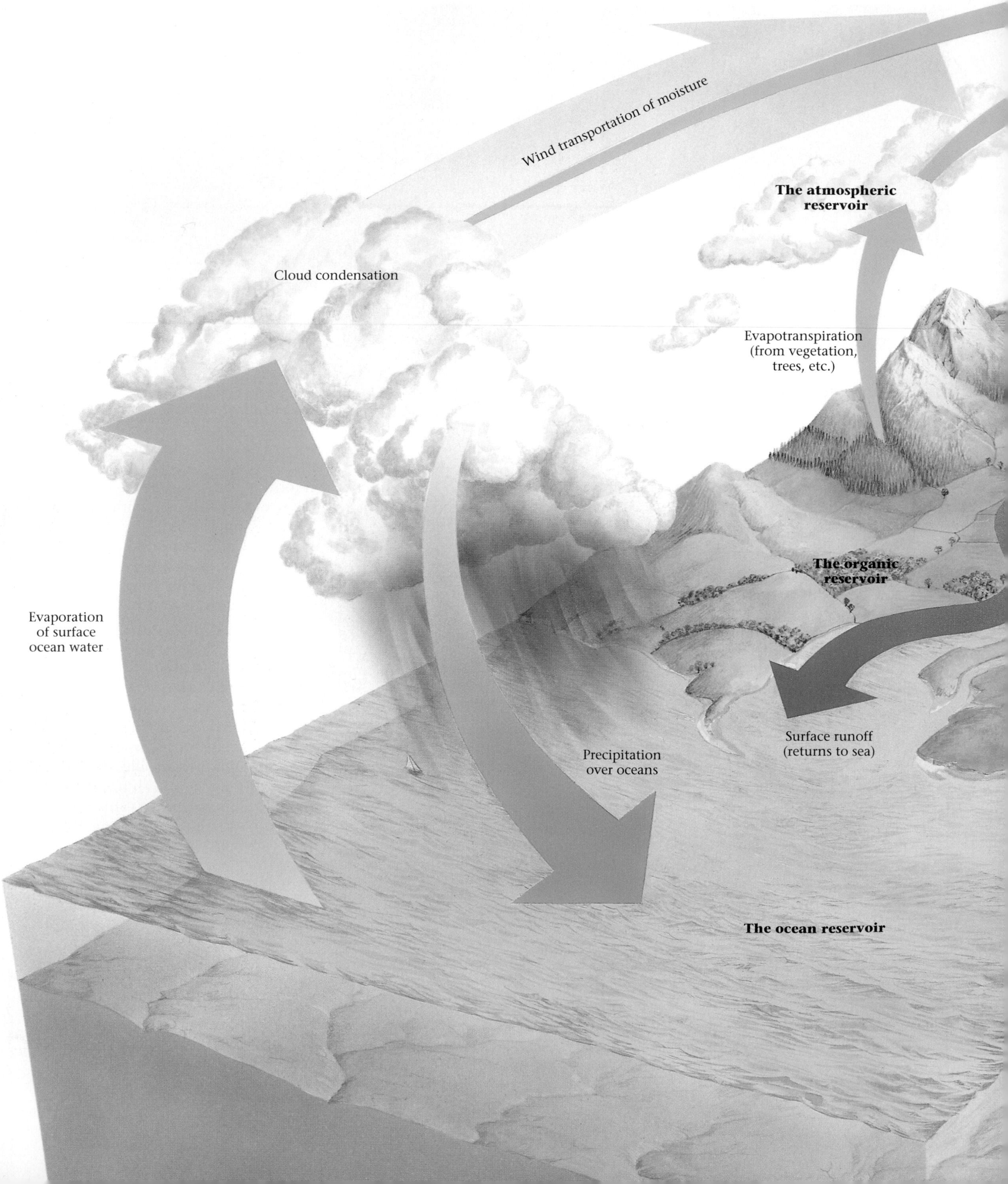

Wind transportation of moisture

The atmospheric
reservoir

Cloud condensation

Evapotranspiration
(from vegetation,
trees, etc.)

The organic
reservoir

Evaporation
of surface
ocean water

Surface runoff
(returns to sea)

Precipitation
over oceans

The ocean reservoir

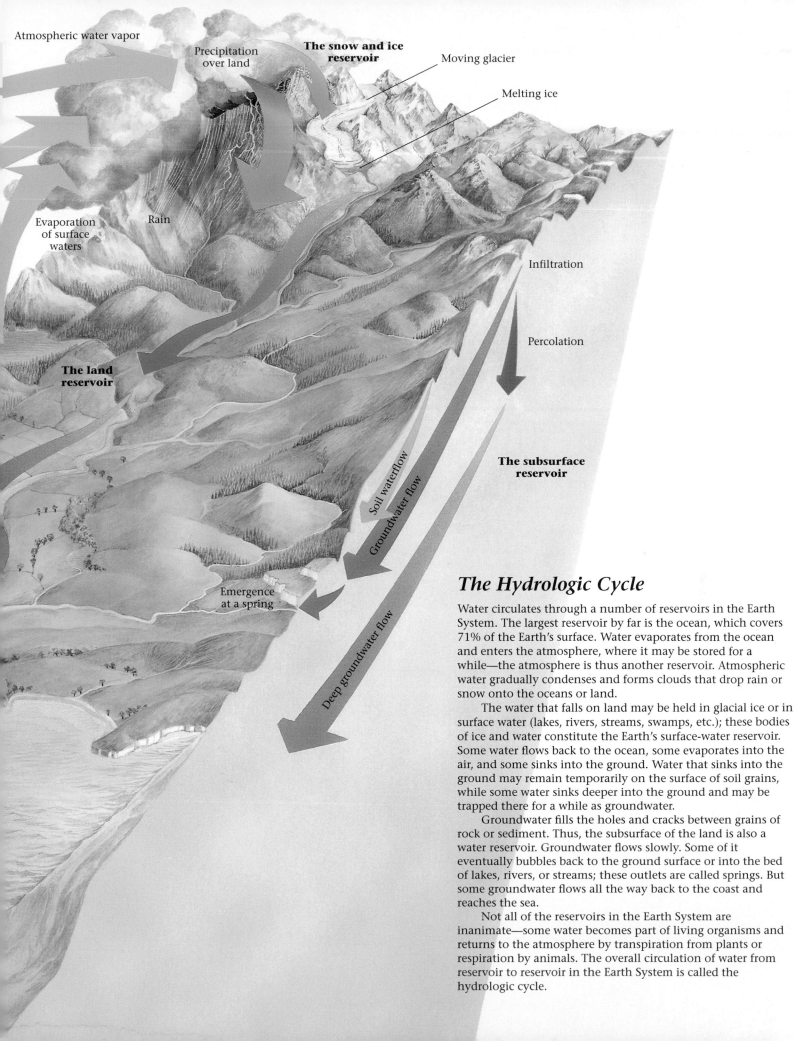

Atmospheric water vapor

Precipitation over land

The snow and ice reservoir

Moving glacier

Melting ice

Evaporation of surface waters

Rain

Infiltration

Percolation

The land reservoir

Soil waterflow

Groundwater flow

The subsurface reservoir

Emergence at a spring

Deep groundwater flow

The Hydrologic Cycle

Water circulates through a number of reservoirs in the Earth System. The largest reservoir by far is the ocean, which covers 71% of the Earth's surface. Water evaporates from the ocean and enters the atmosphere, where it may be stored for a while—the atmosphere is thus another reservoir. Atmospheric water gradually condenses and forms clouds that drop rain or snow onto the oceans or land.

The water that falls on land may be held in glacial ice or in surface water (lakes, rivers, streams, swamps, etc.); these bodies of ice and water constitute the Earth's surface-water reservoir. Some water flows back to the ocean, some evaporates into the air, and some sinks into the ground. Water that sinks into the ground may remain temporarily on the surface of soil grains, while some water sinks deeper into the ground and may be trapped there for a while as groundwater.

Groundwater fills the holes and cracks between grains of rock or sediment. Thus, the subsurface of the land is also a water reservoir. Groundwater flows slowly. Some of it eventually bubbles back to the ground surface or into the bed of lakes, rivers, or streams; these outlets are called springs. But some groundwater flows all the way back to the coast and reaches the sea.

Not all of the reservoirs in the Earth System are inanimate—some water becomes part of living organisms and returns to the atmosphere by transpiration from plants or respiration by animals. The overall circulation of water from reservoir to reservoir in the Earth System is called the hydrologic cycle.

BOX E.1
GEOLOGIC CASE STUDY

Water on Mars?

In 1877, an Italian astronomer named Giovanni Schiaparelli studied the surface of Mars with a telescope and announced that long, straight *canali* criss-crossed the planet's surface. *Canali* should have been translated into the English word "channel," but perhaps because of the recent construction of the Suez Canal, newspapers of the day translated the word into the English "canal," with the implication that the features were constructed by intelligent beings. An eminent American astronomer, Percival Lowell, began to study the "canals" and suggested that they had been built to carry water from polar ice caps to Martian deserts.

Late-twentieth-century satellite mapping of Mars showed that the "canals" do not exist—they were simply optical illusions. There are no lakes, oceans, rainstorms, or flowing rivers on the surface of Mars today. The atmosphere of Mars has such low density, and thus exerts so little pressure on the planet's surface, that any liquid water released at the surface in recent time would quickly evaporate. Thus, there is no hydrologic cycle on Mars, the way there is on Earth. But three crucial questions remain: Does liquid water ever form, even for short periods of time, on the Martian surface today? Was there ever a significant amount of running water or standing water on Mars in the past? If there once was significant water on the planet, where is the water now? The question of the presence of water lies at the heart of the even more basic question: Is there, or was there, life on Mars, for even the simplest life as we know it requires water?

Many planetary geologists believe that the case for liquid water on Mars is quite strong. Much of the evidence comes from comparing landforms on the planet's surface with landforms of known origin on Earth. High-resolution images of Mars reveal a number of landforms that look like they formed in response to the action of water. Examples include networks of channels resembling river networks on Earth (▶Fig. E.9a, b) scour features, deep gullies, and streamlined deposits of sediment. Dark streaks on the walls of craters and canyons look like the products of short-term floods emanating from springs in the crater walls. Some researchers speculate that the northern third of the planet was once a vast ocean.

Studies by the *Odyssey* satellite in 2003, and by Mars rovers (*Spirit* and *Opportunity*) that landed on the planet in 2004, have added intriguing new data to the debate. *Odyssey* detected hints that hydrogen, an element in water, exists beneath the surface of the planet over broad regions, and the Mars rovers have documented the existence of hematite and gypsum minerals that form in the presence of water. The rovers have also found sedimentary deposits that appear to have been deposited in water. Researchers speculate that Mars was much wetter in its past, perhaps billions of years ago, when it had active volcanic activity and a denser atmosphere. But since the atmosphere became less dense, the water evaporated and now lies hidden underground or trapped in polar ice caps. Sudden local ruptures of the ice layer may produce short-term releases of water on the surface of Mars.

FIGURE E.9 Water-related landscape features on Mars, as photographed by satellites. (a) A streamlined island in the middle of what looks like a broad stream channel. Water flow could have carved this island. (b) A river-like network of channels. Small tributary channels join a large trunk channel.

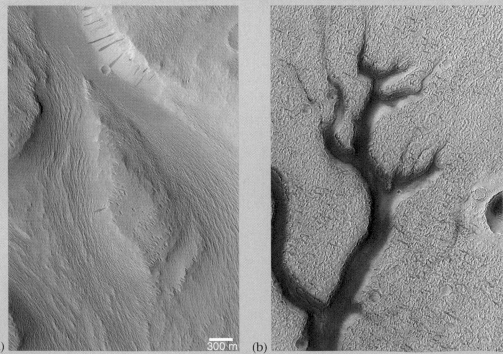

(a) 300 m (b)

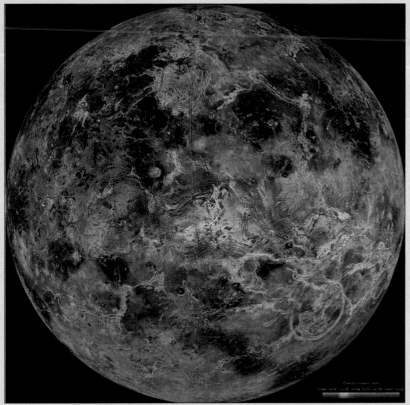

FIGURE E.10 A radar map of the surface of Venus as it would appear if the atmosphere were removed. All these features are invisible to Earth-bound observers because of cloud cover. The colors represent elevation. The light tan is high, and the dark blue is low.

8 km deep—no comparable feature exists on Earth. Because Mars has no vegetation and no longer has rain, its surface does not weather and erode like that of Earth, so it still bears the scars of impact by swarms of meteors earlier in the history of the solar system, scars that have long since disappeared on Earth. Mars does have a hydrologic cycle, of sorts, in that it has ice caps that grow and recede on an annual basis.

Venus is closer to the size of the Earth, and may still have operating mantle plumes. Virtually the entire surface of Venus was resurfaced by volcanic eruptions about 300 to 1,600 Ma, making the planet's surface much younger than that of the Moon and Mars. Further, Venus has a dense atmosphere that protects it from impacts by smaller objects. Because there has been relatively little cratering since the resurfacing event, volcanic and tectonic features dominate the landscape of Venus (▶Fig. E.10). Satellites have used radar to reveal a variety of volcanic constructions (e.g., shield volcanoes, lava flows, calderas). Rifting on Venus produced faults, some of which occur in association with volcanic features. Liquid water cannot survive the scalding temperatures of Venus's surface, so no hydrologic cycle operates there and no life exists. Because of the density of the atmosphere, winds are too slow to cause much erosion or deposition, thus leaving volcanic landforms virtually unchanged.

After this brief side trip to other planets, let's now return to Earth. Our planet has the greatest diversity of landscapes in the solar system.

KEY TERMS

agents of erosion (p. 496)	hydrologic cycle (p. 498)
contour interval (p. 495)	internal energy (p. 494)
contour line (p. 494)	landforms (p. 492)
depositional landforms (p. 493)	landscape (p. 492)
digital elevation map (DEM) (p. 496)	relief (p. 494)
	residence time (p. 498)
erosional landforms (p. 493)	subsidence (p. 493)
external energy (p. 494)	topographical map (p. 494)
gravitational energy (p. 494)	topography (p. 494)
	uplift (p. 493)

THE VIEW FROM SPACE: Satellites can provide images of the great variety of landscape features that decorate the Earth's surface. Textures, colors, and shapes all tell us about geologic processes happening at interfaces between the land, air, and sea. This image shows rugged hills, alluvial fans, sand dunes, and salt lakes in the Bogda Mountains and adjacent Turpan depression of China.

CHAPTER 16

Unsafe Ground: Landslides and Other Mass Movements

What goes up must come down. As a consequence, rock and regolith forming the substrate of hill slopes occasionally give way and slide downslope. The results can be disastrous, as happened when the La Conchita, California, landslide buried nearby homes.

16.1 INTRODUCTION

It was Sunday, May 31, 1970, a market day, and thousands of people had crammed into the Andean town of Yungay, Peru, to shop. Suddenly they felt the jolt of an earthquake, strong enough to topple some masonry houses. But worse was to come. This earthquake also broke an 800-m-wide ice slab off the end of a glacier at the top of Nevado Huascarán, a nearby 6.6-km-high mountain peak. Gravity instantly pulled the ice slab down the mountain's steep slopes. As it tumbled down over 3.7 km, the ice disintegrated into a chaotic avalanche of chunks traveling at speeds of over 300 km per hour. Near the base of the mountain, most of the avalanche channeled into a valley and thickened into a moving sheet as high as a ten-story building that ripped up rocks and soil along the way. Friction transformed the ice into water, which when mixed with rock and dust created 50 million cubic meters of mud, a slurry viscous enough to carry boulders larger than houses. This mass, sometimes floating on a compressed air cushion that allowed it to pass without disturbing the grass below, traveled over 14.5 km in less than four minutes.

At the mouth of the valley, most of the mass overran the village of Ranrahica before coming to rest and

(a)

(b)

FIGURE 16.1 (a) Before the May 1970 earthquake, the town of Yungay, Peru, perched on a hill within view of the ice-covered mountain Nevado Huascarán. (b) Three months after the earthquake, the town lay buried beneath debris. A landslide scar remains visible on the mountain.

creating a dam that blocked the Santa River. But part of it shot up the sides of the valley and became airborne for several seconds, flying over the ridge bordering Yungay. As the town's inhabitants and visitors stumbled out of earthquake-damaged buildings, they heard a deafening roar and looked up to see the churning mud cloud bursting above the nearby ridge. The town was completely buried under several meters of mud and rock. When the dust had settled, only the top of the church and a few palm trees remained visible to show where Yungay once lay (▶Fig. 16.1a, b); 18,000 people are forever entombed beneath the mass. Today, the site is a grassy meadow with a hummocky (irregular and lumpy) surface, spotted with crosses left by mourning relatives.

Could the Yungay tragedy have been prevented? Perhaps. A few years earlier, climbers had recognized the insta-

bility of the glacial ice on Nevado Huascarán, and Peruvian newspapers had published a warning, but alas, no one took notice. In the aftermath of the event, geologists discovered that Yungay had been built on ancient layers of debris, from past avalanches. The government has since prevented new towns from rising in the danger zone.

People often assume that the earth beneath them is *terra firma,* a solid foundation on which they can build their lives. But the catastrophe at Yungay says otherwise—much of the Earth's surface is unstable ground, land capable of moving downslope in a matter of seconds to weeks. Geologists refer to the gravitationally caused downslope transport of rock, regolith (soil, sediment, and debris), snow, and ice as **mass movement,** or mass wasting. Like earthquakes, volcanic eruptions, storms, and floods, mass movements are a type of **natural hazard,** meaning a natural feature of the environment that can cause damage to living organisms and to buildings. Unfortunately, mass movement becomes more of a threat every year, because as the world's population grows, cities expand into areas of unsafe ground. In fact, by some estimates, mass movements may, on average, be *the most costly* natural hazard. But mass movement also plays a critical role in the rock cycle, for it's the first step in the transportation of sediment. And it plays a critical role in the evolution of landscapes: it's the most rapid means of modifying the shapes of slopes.

In this chapter, we look at the types, causes, and consequences of mass movement, and precautions society can take to protect people and property from its dangers. You might want to consider this information when selecting a site for your home or when voting on land-use propositions for your community.

16.2 TYPES OF MASS MOVEMENT

Though in everyday language people commonly refer to all mass-movement events as "landslides," geologists and civil engineers tend to distinguish different types of mass movement based on four factors: the type of material involved (rock, regolith, or snow and ice), the velocity of the movement (fast, intermediate, or slow), the character of the moving mass (chaotic cloud, slurry, or coherent body), and the environment in which the movement takes place (subaerial or submarine). Below, we look at mass movements that occur on land roughly in order from slow to very fast.

Creep, Solifluction, and Rock Glaciers

In temperate climates, the upper few centimeters of ground freeze during the winter, only to thaw again the following spring. Because water increases in volume by about 9.2%

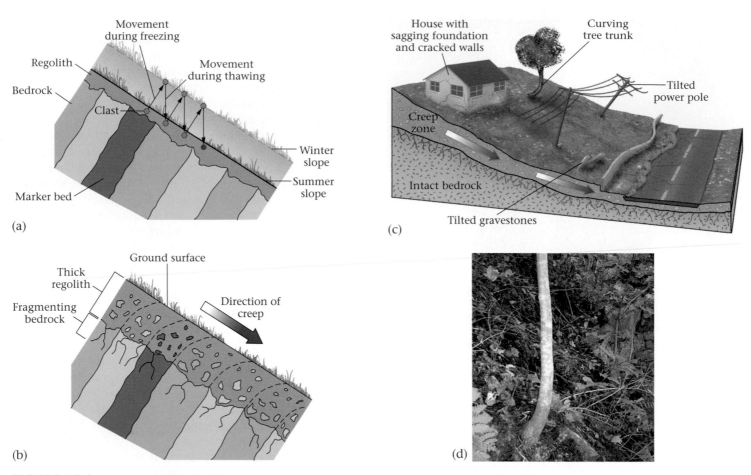

FIGURE 16.2 (a) Creep on hill slopes accompanies the annual freeze-thaw cycle. The clast originally came from the marker bed. It rose perpendicular to the slope when the slope froze and dropped down when the slope thawed. After three years, it has migrated downslope to the position shown. (b) As rock layers weather and break up, the resulting debris creeps downslope. (c) Soil creep causes walls to bend and crack, building foundations to sink, trees to bend, and power poles and gravestones to tilt. (d) Trees that grow in creeping soil gradually develop pronounced curves.

when it freezes, the water-saturated soil and underlying fractured rock expand outward, and particles in the regolith move out perpendicular to the slope during the winter. During the spring thaw, water becomes liquid again, and gravity makes the particles sink vertically and thus migrate downslope slightly. This gradual downslope movement of regolith is called **creep.** You can't see creep by staring at a hill slope because it occurs too slowly, but over a period of years creep causes trees, fences, gravestones, walls, and foundations built on a hillside to tilt downslope. Notably, trees that continue to grow after they have been tilted display a pronounced curvature at their base (▶Fig. 16.2a–d).

In Arctic or high-elevation regions, regolith freezes solid to great depth during the winter. In the brief summer thaw, only the uppermost 1–3 m of the ground thaws. Since meltwater cannot sink into permanently frozen ground, or **permafrost,** the melted layer becomes soggy and weak and flows slowly downslope in overlapping

sheets. Geologists refer to this kind of creep, characteristic of tundra regions (cold, treeless regions), as **solifluction** (▶Fig. 16.3a). Another type of slow mass movement in cold regions takes place in **rock glaciers,** which consist of a mixture of rock fragments and ice, with the rock fragments making up the major proportion (▶Fig. 16.3b). Rock glaciers develop where the volume of debris falling into a valley equals or exceeds the volume of glacial ice forming from snow. They move downslope with the ice.

Slumping

Near Pacific Palisades, along the coast of southern California, Highway 1 runs between the beach and a 120-m-high cliff. Between March 31 and April 3, 1958, a 1-km-long section of the highway disappeared beneath a mass of regolith that had moved down the adjacent cliff. When the movement stopped, the face of the cliff lay 200 m farther inland than it

FIGURE 16.3 (a) Solifluction on a hill slope in the tundra. (b) A rock glacier in Alaska.

(a)

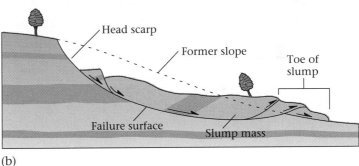

(b)

FIGURE 16.4 (a) A slump threatening a highway in California. (b) Note the curving failure surface in this slump. The dashed line indicates the slope's shape before slumping.

previous elevation. Farther downslope, at the toe, or end, of the slump, the ground elevation rises as the slump rides up and over the preexisting land surface. The toe may break into a series of slices that form curving ridges at the ground surface. Slumps come in all sizes, from only a few meters across to tens of kilometers across. Slumps move at speeds of millimeters per day to tens of meters per minute. They typically break up as they move, and structures (such as houses, patios, and swimming pools) built on them crack and fall apart.

Mudflows and Debris Flows

Rio de Janeiro, Brazil, originally occupied only the flatlands bordering beautiful crescent beaches between towering mountains. But in recent decades, the population has grown so much that the city has expanded up the steep sides of the mountains, and in many places densely populated communities of makeshift shacks cover the slopes. These communities, which have no storm drains, were built on the thick regolith that resulted from long-term weathering of bedrock in Brazil's tropical climate. In 1988,

had before. It took weeks for bulldozers to uncover the road. During such **slumping,** a mass of regolith detaches from its substrate along a spoon-shaped sliding surface and slips semicoherently downslope (▶Fig. 16.4a). We call the moving mass a slump, and the surface on which it slips a **failure surface** (▶Fig. 16.4b). On average, the sea cliffs of southern California retreat (move inland) by up to a few meters a year because of slumping.

The distinct, curving step at the upslope edge of a slump, where the regolith detached, is called a **head scarp.** Immediately below the head scarp, the land surface sinks below its

FIGURE 16.12 Perfectly intact rock is rare at the surface of the Earth. Most outcrops, like this one on the western coast of Ireland, are highly jointed.

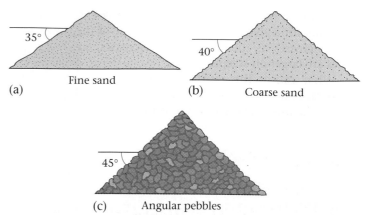

FIGURE 16.14 The angle of repose is the steepest slope that a pile of unconsolidated sediment can have and remain stable. Angles of repose depend on the size and shape of grains. (a) Fine, well-rounded sand has a small angle of repose. (b) Coarse, angular sand has a larger angle. (c) Large, irregularly shaped pebbles have a large angle of repose.

value of between 30° and 37°. The angle depends partly on the shape and size of grains, which determine the amount of friction across boundaries. For example, larger angles of repose (up to 45°) tend to form on slopes composed of large, irregularly shaped grains, for these grains interlock with one another (▶Fig. 16.14a–c).

In many locations, the resistance force is less than might be expected because a weak surface exists at some depth below ground level. This weak surface separates unstable rock and debris above from the substrate below. If downslope movement begins on the weak surface, we say that "failure" has occurred and that the weak surface has become a "failure surface." Geologists recognize several different kinds of weak surfaces that are prone to become failure surfaces (▶Fig. 16.15a–c). These include: wet clay layers; wet, unconsolidated sand layers; surface-parallel joints (also known as exfoliation joints); weak bedding planes (shale beds and evaporite beds are particularly weak); and metamorphic foliation planes.

Failure surfaces that dip parallel to the slope are particularly likely to fail because the downslope force is parallel to the surface. As an example, consider the 1959 landslide that occurred in Madison Canyon, in southwestern Montana. On August 17 of that year, shock waves from a strong earthquake jarred the region. The southern wall of the canyon is underlaid with metamorphic rock with a strong foliation that provided a plane of weakness, and when the ground vibrated, rock detached along a foliation plane and tumbled downslope. Unfortunately, twenty-eight campers lay sleeping on the valley floor. They were awakened by the hurricane-like winds blasting in front of the moving mass, but seconds later were buried under 45 m of rubble.

Fingers on the Trigger: Factors Causing Slope Failure

What triggers an individual mass-wasting event? In other words, what causes the balance of forces to change so that the downslope force exceeds the resistance force, and a slope suddenly fails? Here, we look at various phenomena—natural and human-made—that trigger slope failure.

FIGURE 16.13 (a) Gravity, represented by the black arrow, pulls a block toward the center of the Earth. The gravitational force has two components, the downslope force parallel to the slope and the normal force perpendicular to the slope. On gentle slopes, the normal force is larger than the downslope force. The resistance force, caused by friction and represented by an arrow pointing upslope, is larger in this example than the downslope force. (b) If the slope angle increases, the normal force becomes smaller than the downslope force. If the downslope force then becomes greater than the resistance force, then the block starts to move.

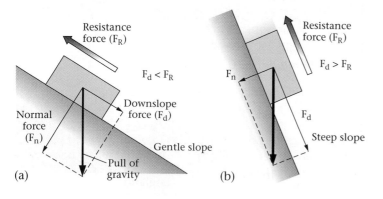

(a)

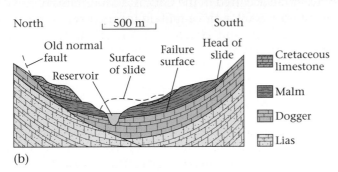

(b)

FIGURE 16.6 (a) Vaiont Dam and the debris now behind it. The exposed failure surface is visible in the distance. (b) A cross-section parallel to the face of Vaiont Dam before the landslide. Mount Toc is to the right. Note the failure surface at the base of the weak Malm Shale. The landslide completely filled the reservoir. Its surface after movement is indicated by the dashed line. The names Malm, Dogger, and Lias refer to epochs in the Jurassic Period.

When the flood had passed, nothing of Longarone and its 1,500 inhabitants remained. Though the dam itself still stands, it holds back only debris and has never provided any electricity.

Geologists refer to such a sudden movement of rock and debris down a nonvertical slope as a **landslide.** If the mass consists only of rock, it may also be called a **rock slide** (the case in the Vaiont Dam disaster), and if it consists mostly of regolith, it may also be called a **debris slide.** Once a landslide has taken place, it leaves a landslide scar on the slope and forms a debris pile at the base of the slope.

Slides happen when bedrock and/or regolith detaches from a slope and shoots downhill on a *failure surface* (slide surface) roughly parallel to the slope surface. Thus, landslides generally occur where a weak layer of rock or sediment

at depth below the ground parallels the land surface. (At the Vaiont Dam, the plane of weakness that would become the failure surface was a weak shale bed.) Slides may move at speeds of up to 300 km per hour; they are particularly fast when a cushion of air gets trapped beneath, so there is virtually no friction between the slide and its substrate, and the mass moves like a hovercraft. Rock and debris slides sometimes have enough momentum to climb the opposite side of the valley into which they fell.

Landslides, like slumps, come at a variety of scales. Most are small, involving blocks up to a few meters across. Some, like the Vaiont slide, are large enough to cause a catastrophe. But even the Vaiont slide pales by comparison with some landslides for which evidence appears in the geologic record.

Avalanches

In the winter of 1999, an unusual weather system passed over the Austrian Alps. First it snowed; then the temperature warmed and the snow began to melt. But then the weather turned cold again, and the melted snow froze into a hard, icy crust. This cold snap ushered in a blizzard that blanketed the ice crust with tens of centimeters (1–2 ft.) of snow, and at the mountain tops the wind built the snow into huge overhanging drifts, called cornices. Skiers delighted in the bounty of white, but not for long. Who knows how it started—perhaps a gust of wind or a sudden noise was enough—but the world witnessed the aftermath. With the frozen snow layer underneath acting as a failure surface, the heavy layer of new snow began to slide down the mountain, accelerating as it moved and then disintegrating and mixing with air. It became a roaring cloud—an avalanche—traveling at hurricane speeds, and it flattened everything in its path. Trees and ski lodges toppled like toothpicks before the mass finally reached the valley floor and came to a halt (▶Fig. 16.7a, b). Unfortunately, many of those who survived the impact succumbed to suffocation in the minutes that followed, and the avalanche blocked roads into the region, tragically slowing rescue efforts.

Avalanches are turbulent clouds of debris mixed with air that rush down steep hill slopes at high velocity. If the debris consists of snow, like the Austrian avalanche, it's a snow avalanche. If it consists of fragments of rock and dust, it's a debris avalanche. The moving air-debris mass is denser than clear air and thus hugs the ground and acts like an extremely strong and viscous wind that can knock down and blow away anything in its path. As illustrated by the Austrian example, snow avalanches pose a particular threat when frozen snow layers get buried and thus can act as a failure surface for the overlying snow. Typically, avalanches happen again and again in the same area, creating pathways, called "avalanche chutes," in which no mature trees grow.

Not all snow avalanches are the same—some are clouds of light powder traveling at up to 250 km per hour, while some are much slower flows of wet snow. The snow in an avalanche can break off a cornice, or it can start as a huge slab that detaches from its substrate in response to little more than the weight of a single skier. In some cases, avalanches occur because of failure of a glacier. We've already mentioned the example of Yungay, Peru. A more re-

FIGURE 16.8 Successive rock falls have littered the base of this sandstone cliff with boulders. Note the talus at the base of the cliff.

cent example occurred in the Caucasus Mountains of Russia. In September 2002, a three-million-ton portion of the Maili Glacier detached and charged 16 km (10 miles) down a gorge. It buried the village of Karmadon with up to 170 m (500 ft.) of ice blocks and other debris.

Rock Falls and Debris Falls

Rock falls and **debris falls,** as their names suggest, occur when a mass free-falls from a steep (vertical) cliff (▶Fig. 16.8). Friction and collision with other rocks may bring some blocks to a halt before they reach the bottom of the slope; these blocks pile up to form a **talus,** a sloping apron of rocks along the base of the cliff (Fig. 16.8). Rock or debris that has fallen a long way can reach speeds of 300 km per hour, and may have so much momentum that it keeps going when it touches bottom and triggers a debris avalanche that can cross a valley floor and rise up the other side. As with avalanches, large, fast rock falls push the air in front of them, creating a short blast of hurricane-like wind. For example, the wind alone from a 1996 rock fall in Yosemite National Park flattened over 2,000 trees. Commonly, rock falls happen when a rock separates from a cliff face along a joint.

Most rock falls involve only a few blocks detaching from a cliff face and dropping into the talus. But some falls dislodge immense quantities of rock. In September 1881, a 600-m-high crag of slate, undermined by quarrying, suddenly collapsed onto the Swiss town of Elm in a valley of the Swiss Alps. Over 10 million cubic meters of rock fell to the valley floor, burying Elm and its 115 inhabitants to a depth of 10 to 20 m.

Rock falls typically take place along steep highway road cuts, leading to the posting of "falling-rock zone" signs.

FIGURE 16.7 (a) Aftermath of the 1999 avalanches in the Austrian Alps. (b) Trees that were flattened by an avalanche, now exposed after the snow melted.

(a)

(b)

Such rock falls occur with increasing frequency as the road cut ages, because frost wedging and/or root wedging pries fragments loose, and water infiltrates the outcrop and weakens clay-rich layers.

Submarine Mass Movements

So far, we've focused on mass movements that occur subaerially, for these are the ones we can see and are affected by most. But mass wasting also happens under water. The sedimentary record contains abundant evidence of submarine mass movements, because after they take place, they tend to be buried by younger sediments and are preserved.

Geologists distinguish three types of submarine mass movements, according to whether the mass remains coherent or disintegrates as it moves. The degree to which the mass comes apart during movement typically depends on the amount of water mixed with the sediment in the moving mass. In **submarine slumps,** semicoherent blocks (olistostromes) slip downslope on weak mud detachments. (▶Fig.

FIGURE 16.9 Submarine mass movements. (a) Slump blocks remain semicoherent as they move. (b) In a debris flow, the mass becomes a viscous slurry of chunks floating in a mud matrix. (c) In a turbidity current, sediment is suspended like a cloud in water. Submarine slumping and debris flow can generate tsunamis.

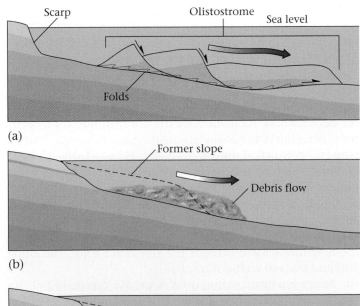

(a)

(b)

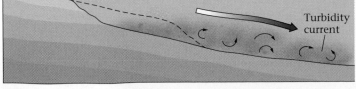

(c)

16.9a-c). In some cases, the layers constituting the blocks become contorted as they move, like a tablecloth that has slid off a table. In **submarine debris flows,** the moving mass breaks apart to form a slurry containing larger clasts (pebbles to boulders) suspended in a mud matrix. And in **turbidity currents,** sediment disperses in water to create a turbulent cloud of suspended sediment that avalanches downslope. As the turbidity current slows, sediment settles out in sequence from coarse to fine, creating graded beds (see Chapter 7).

In recent years, marine geologists have used a sensitive type of sonar, called GLORIA (geologic long-range inclined asdic [sonar]), to map the sea floor. The instrument "sees" sideways and can map a 60-km-wide swath of the ocean floor all at once. GLORIA maps have documented huge slumps along the margins of Hawaii. These have substantially modified the shape of the island and have created a broad apron of hummocky sea floor around them (▶Fig. 16.10a, b). Major slumping events, which seem to recur (on average) every 100,000 years, may generate catastrophic tsunamis in the Pacific basin. Marine geologists have mapped similar slumps along the edge of continental shelves suggesting that tsunamis generated by submarine slumps and debris flows could be a major hazard worldwide (see Box 16.1).

16.3 SETTING THE STAGE FOR MASS MOVEMENTS

We've seen that mass movements travel at a range of different velocities, from the slowest (creep) to the faster (slumps, mud and debris flows, and rock and debris slides), to the fastest (snow and debris avalanches, and rock and debris falls—see art on pp. 516–17). The velocity in turn depends on the steepness of the slope and the water or air content of the mass. In order for these movements to take place, the stage must be set by the following phenomena: fracturing and weathering, which weaken materials at Earth's surface so that they cannot hold up against the pull of gravity; and the development of relief, which creates slopes down which masses move.

Fragmentation and Weathering: Weakening the Surface

If the Earth's surface were covered by intact (unbroken) rock, mass movements would be of little concern, for intact rock has great strength and could form stalwart mountain faces that would never tumble, even if they were vertical. But the rock of Earth's upper crust has been affected by jointing and faulting, and in many locations the surface has a cover of regolith created by the weathering of rock in Earth's corrosive atmosphere. Fragmented

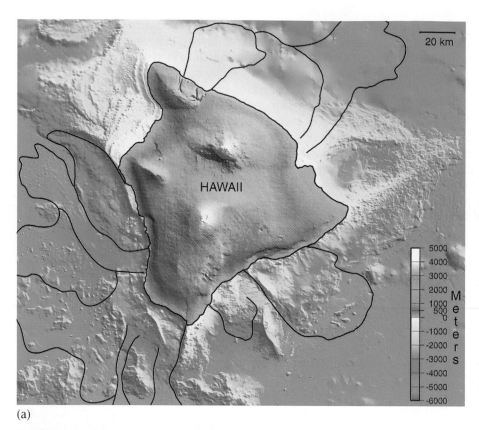

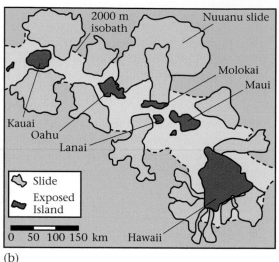

(a)

(b)

FIGURE 16.10 (a) The irregular sea-floor surface surrounding Hawaii consists of a succession of huge slumps, here outlined; each slump is a mass of rock and sediment that slipped seaward. (b) Regional map of the Hawaiian Islands showing some of the larger slides. The longest is 300 km long.

rock and regolith are much weaker than intact rock and can indeed collapse in response to Earth's gravitational pull (▶Fig. 16.12). Thus, jointing, faulting, and weathering make mass movements possible.

Why are regolith and fractured rock so much weaker than intact rock? Intact rock is held together by the strong chemical bonds within mineral crystals, by mineral cement, or by the interlocking of grains. In contrast, a joint-bounded or fault-bounded block is held in place only by friction between the block and its surroundings. Regolith is unconsolidated; that is, it consists of unattached grains. Dry regolith holds together because of friction between adjacent grains and/or because weak electrical charges cause grains to attract each other. Slightly wet regolith holds together because of water's surface tension. Surface tension, the phenomenon that makes water form drops, exists because water molecules have a positively charged side and a negatively charged side, so the molecules bond to mineral surfaces and attract each other. (Because of surface tension, damp sand holds together to form a sand castle, while dry sand collapses into a shapeless pile.)

Slope Stability: The Battle between Downslope Force and Resistance Force

Mass movements do not take place on all slopes, and even on slopes where such movements are possible, they occur only occasionally. Geologists distinguish between **stable slopes,** on which sliding is unlikely, and **unstable slopes,** on which sliding will likely happen. When material starts moving on an unstable slope, we say that slope failure has occurred. Whether a slope fails or not depends on the balance between two forces—the downslope force, caused by gravity, and the resistance force, which inhibits sliding. If the downslope force exceeds the resistance force, the slope fails and mass movement results.

Imagine a block sitting on a slope. We can represent the gravitational attraction between this block and the Earth by an arrow (a vector) that points straight down, toward the Earth's center of gravity. This arrow can be separated into two components, the downslope force parallel to the slope and the normal force perpendicular to the slope. The resistance force can be represented by an arrow pointing uphill. If the downslope force is larger than the resistance force, then the block moves; otherwise, it stays in place (▶Fig. 16.13a, b). Note that for a given mass, the magnitude of the downslope force increases as the slope angle increases, so downslope forces are larger on steeper slopes.

What causes the resistance force? As we saw above, chemical bonds in mineral crystals, cement, and the jigsaw-puzzle-like interlocking of crystals hold intact rock in place,

friction holds an unattached block in place, electrical charges and friction hold dry regolith in place, and surface tension holds slightly wet regolith in place. Because of resistance force, granular debris tends to pile up and create the steepest slope it can without collapsing. The angle of this slope is called the **angle of repose,** and for most dry unconsolidated materials (such as dry sand) it typically has a

The Storegga Slide and the North Sea Tsunamis

BOX 16.1
GEOLOGIC CASE STUDY

The Firth of Forth, a long inlet of the North Sea, forms the waterfront of Edinburgh, Scotland. At its western end, it merges with a broad plain in which mud and peat have been accumulating since the last ice age. Around 1865, geologists investigating this sediment discovered an unusual layer of sand containing bashed seashells, marine plankton, and torn-up fragments of substrate. This sand layer lies sandwiched between mud layers at an elevation of up to 4 m above the high-tide limit and 80 km inland from the shore. How could the sand layer have been deposited? The mystery simmered for many decades. During this time, layers of shelly sand, similar to the one found in Scotland, were discovered at many other localities along both sides of the North Sea (▶Fig. 16.11). In some cases, the layer occurred 20 m above the high-tide limit. Geologists studying the coast also found locations where coastal cliffs appeared to have been eroded by wave action at elevations well out of reach of normal storm waves.

While land-based geologists puzzled about these unusual sand beds and coastal erosional features, marine geologists investigating the continental shelf off the western coast of Norway discovered a region of very irregular sea floor underlaid with a jumble of chaotic blocks, some of which are 10 km by 30 km across and 200 m thick. When mapped out, they indicate that a 290-km-long sector of the continental shelf had collapsed in a series of at least three submarine slides that together constitute about 5,580 km³ of debris—the overall feature is called the Storegga Slide. Further studies show that the Storegga Slide formed during three movement events: one occurred 30,000 years ago, the second about 7,950 years ago, and the third about 6,000 years ago. Now the pieces of the puzzle were in place—slides the size of those constituting the Storegga Slide had displaced enough ocean water to create tsunamis (see Fig. 16.10).

Tsunamis produced by the second Storegga slide may coincide with the disappearance of Stone Age tribes along the North Sea coast, suggesting the tribes simply washed away. If such a calamity happened in the past, could it happen again, with submarine-slide-generated tsunamis inundating coastal cities with large populations? Geologists now realize that submarine landslides can trigger tsunamis every bit as devastating as earthquake- and volcano-generated tsunamis. A slide in 1929 along the coast of Newfoundland, for example, not only created a turbidity current that broke the trans-Atlantic telephone cable, but also generated tsunamis that washed away houses and boats along the coast of Newfoundland at eleva-

tions of up to 27 m above sea level. With a bit of looking, geologists have found huge boulders flung by tsunamis onto the land, layers of sand and gravel deposited well above the high-tide limit, and erosional features high up on shoreline cliffs along many coastal areas, even in areas (such as the Bahamas and southeastern Australia) far from seismic or volcanic regions. And submarine mapping shows that many large slumps occur all along continental shelves. It's no wonder that Edward Bryant, in his recent book on tsunamis, calls them "The Underrated Hazard."

FIGURE 16.11 Map of the North Sea region showing the location of sites (blue dots) where marine sand layers occur significantly above the high-tide limit. These were caused by tsunamis generated by movement of the Storegga Slide. The map shows the estimated position of the tsunamis at 2 hours, 4 hours, and 6 hours.

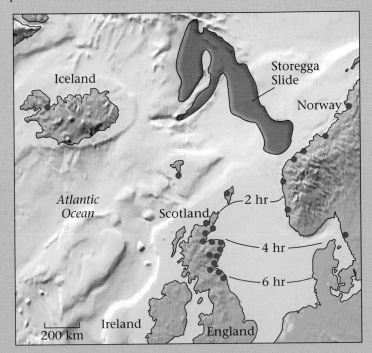

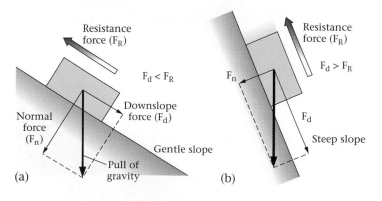

FIGURE 16.12 Perfectly intact rock is rare at the surface of the Earth. Most outcrops, like this one on the western coast of Ireland, are highly jointed.

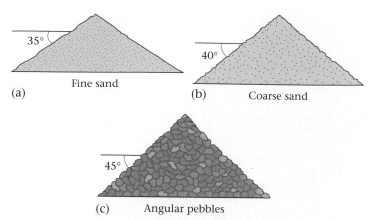

FIGURE 16.14 The angle of repose is the steepest slope that a pile of unconsolidated sediment can have and remain stable. Angles of repose depend on the size and shape of grains. (a) Fine, well-rounded sand has a small angle of repose. (b) Coarse, angular sand has a larger angle. (c) Large, irregularly shaped pebbles have a large angle of repose.

value of between 30° and 37°. The angle depends partly on the shape and size of grains, which determine the amount of friction across boundaries. For example, larger angles of repose (up to 45°) tend to form on slopes composed of large, irregularly shaped grains, for these grains interlock with one another (▶Fig. 16.14a–c).

In many locations, the resistance force is less than might be expected because a weak surface exists at some depth

FIGURE 16.13 (a) Gravity, represented by the black arrow, pulls a block toward the center of the Earth. The gravitational force has two components, the downslope force parallel to the slope and the normal force perpendicular to the slope. On gentle slopes, the normal force is larger than the downslope force. The resistance force, caused by friction and represented by an arrow pointing upslope, is larger in this example than the downslope force. (b) If the slope angle increases, the normal force becomes smaller than the downslope force. If the downslope force then becomes greater than the resistance force, then the block starts to move.

below ground level. This weak surface separates unstable rock and debris above from the substrate below. If downslope movement begins on the weak surface, we say that "failure" has occurred and that the weak surface has become a "failure surface." Geologists recognize several different kinds of weak surfaces that are prone to become failure surfaces (▶Fig. 16.15a–c). These include: wet clay layers; wet, unconsolidated sand layers; surface-parallel joints (also known as exfoliation joints); weak bedding planes (shale beds and evaporite beds are particularly weak); and metamorphic foliation planes.

Failure surfaces that dip parallel to the slope are particularly likely to fail because the downslope force is parallel to the surface. As an example, consider the 1959 landslide that occurred in Madison Canyon, in southwestern Montana. On August 17 of that year, shock waves from a strong earthquake jarred the region. The southern wall of the canyon is underlaid with metamorphic rock with a strong foliation that provided a plane of weakness, and when the ground vibrated, rock detached along a foliation plane and tumbled downslope. Unfortunately, twenty-eight campers lay sleeping on the valley floor. They were awakened by the hurricane-like winds blasting in front of the moving mass, but seconds later were buried under 45 m of rubble.

Fingers on the Trigger: Factors Causing Slope Failure

What triggers an individual mass-wasting event? In other words, what causes the balance of forces to change so that the downslope force exceeds the resistance force, and a slope suddenly fails? Here, we look at various phenomena—natural and human-made—that trigger slope failure.

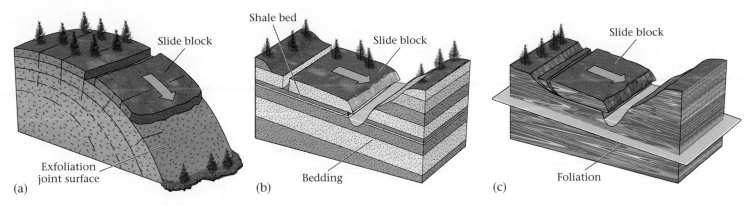

FIGURE 16.15 Different kinds of surfaces become failure surfaces in different geologic settings. (a) In exfoliated massive granite, exfoliation joints become failure surfaces. (b) In sedimentary rock, bedding planes become failure surfaces. (c) In metamorphic rock, foliation planes, especially schistosity (the parallel alignment of mica flakes), become failure surfaces.

Shocks, Vibrations, and Liquefaction

Earthquake tremors, the passing of large trucks, or blasting in construction sites may cause a mass that was on the verge of moving to actually start moving. For example, an earthquake-triggered slide dumped debris into Lituya Bay, in southeastern Alaska, in 1958. The debris displaced the water in the bay, creating a 300-m-high splash that washed the slope on the opposing side of the bay clean of their forest and carried fishing boats many kilometers out to sea. The vibrations of an earthquake break bonds that hold a mass in place and/or cause the mass and the slope to separate slightly, thereby decreasing friction. As a consequence, the resistance force decreases, and the downslope force sets the mass in motion.

Shaking produces a unique effect in certain types of clay, called **quick clay.** Quick clay, which consists of damp clay flakes, behaves like a solid when still, for surface tension holds water-coated flakes together. But shaking separates the flakes from one another and suspends them in the water, thereby transforming the clay into a slurry that flows like a fluid (▶Fig. 16.16a, b).

Shaking can also cause **liquefaction** of wet sand—the shaking causes the sand grains to try to fit together more tightly, which in turn increases the water pressure in the pores between grains, destroying the cohesion between the grains. Without cohesion, the mixture of sand and water turns into a weak slurry.

Changing Slope Angles, Slope Loads, and Slope Support

Factors that make a slope steeper or heavier may cause the slope to fail (▶Fig. 16.17a–c). For example, when a river eats away at the base of the slope, or a contractor excavates at the base of a slope, the slope becomes steeper and the downslope force increases. But while this happens, the resistance force stays the same. Therefore, if the excavation continues, the downslope force will eventually exceed the resistance force and the slope will fail. The same phenomenon occurs when it rains heavily, for addition of water to regolith not only makes the regolith heavier, thereby increasing the downslope force, but at the same time weakens failure surfaces, thereby decreasing resistance force. So, as trucks dump loads of waste on the side of a tailing pile, the pile gradually becomes steeper than the angle of repose, and when this happens, collapse becomes inevitable. The largest observed landslide in U.S. history, the Gros Ventre slide, which took place in 1925 on the flank of Sheep Mountain, near Jackson Hole, Wyoming, illustrates this phenomenon (▶Fig. 16.18a–c). Almost 40 million cubic meters of rock, soil, and forest detached from the side of the mountain and slid 600 m down a slope, filling the valley and creating a 75-m-high natural dam

FIGURE 16.16 (a) In a quick clay, before shaking, the grains stick together. (b) During shaking, the grains become suspended in water, and the formerly solid mass becomes a movable slurry.

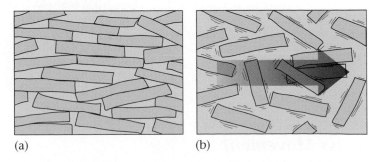

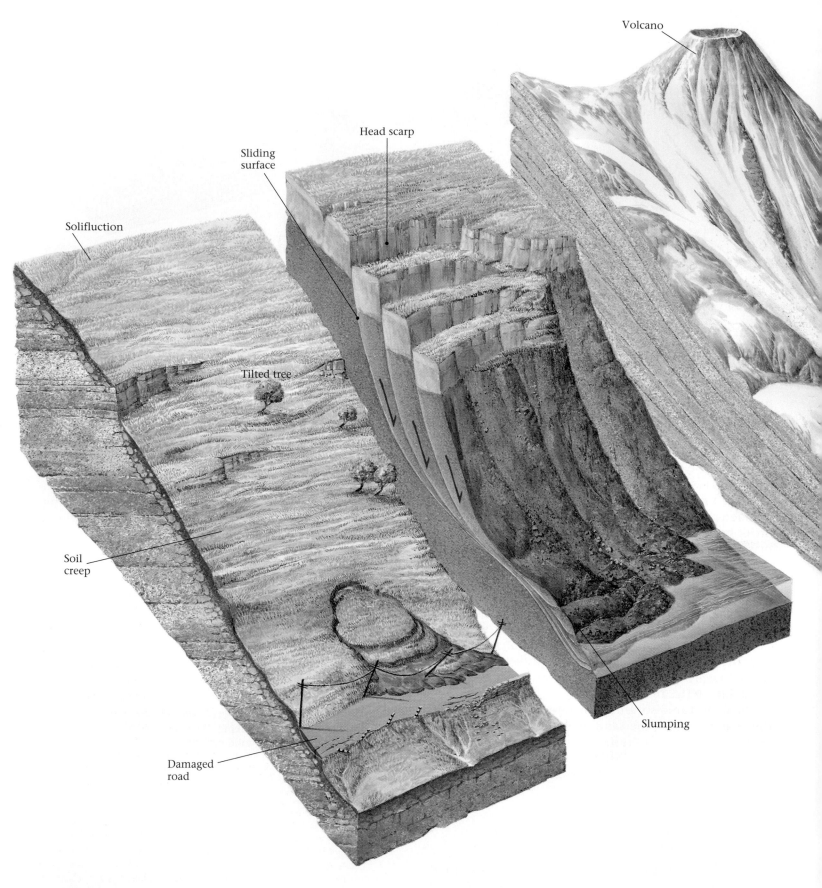

Volcano

Head scarp

Sliding surface

Solifluction

Tilted tree

Soil creep

Damaged road

Slumping

Mass Movement

In Earth's gravity field, what goes up must come down—
sometimes with disastrous consequences. Rock and regolith are
not infinitely strong, so every now and then slopes or cliffs give
way in response to gravity, and materials slide, tumble, or career

downslope. This downslope movement, called mass movement,
or mass wasting, is the first step in the process of erosion and
sediment formation. The resulting debris may eventually be
carried away by water, ice, or wind.

The kind of mass wasting that takes place at a given location
reflects the composition of the slope (is it composed of weak soil,

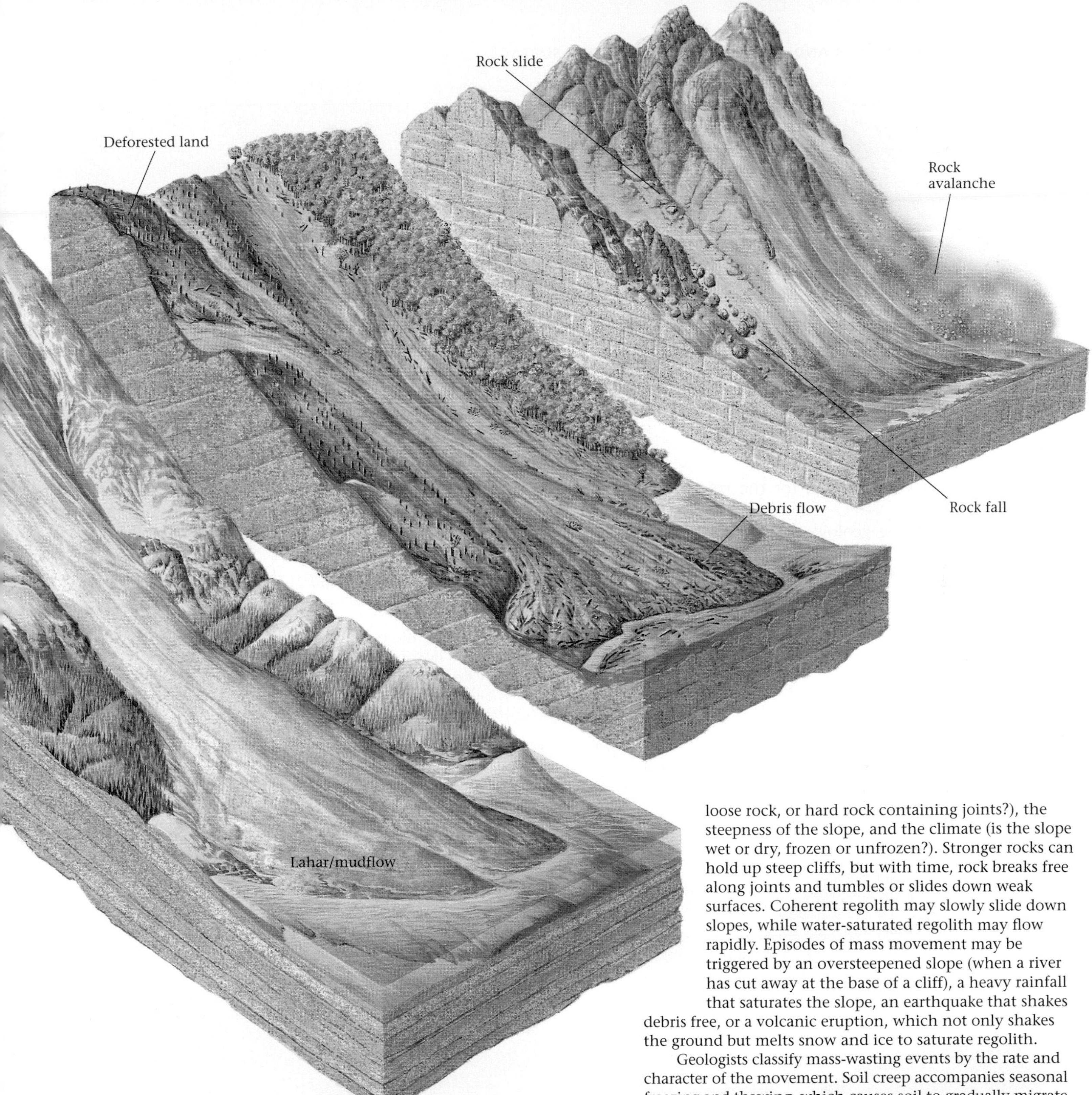

Deforested land

Rock slide

Rock avalanche

Rock fall

Debris flow

Lahar/mudflow

loose rock, or hard rock containing joints?), the steepness of the slope, and the climate (is the slope wet or dry, frozen or unfrozen?). Stronger rocks can hold up steep cliffs, but with time, rock breaks free along joints and tumbles or slides down weak surfaces. Coherent regolith may slowly slide down slopes, while water-saturated regolith may flow rapidly. Episodes of mass movement may be triggered by an oversteepened slope (when a river has cut away at the base of a cliff), a heavy rainfall that saturates the slope, an earthquake that shakes debris free, or a volcanic eruption, which not only shakes the ground but melts snow and ice to saturate regolith.

Geologists classify mass-wasting events by the rate and character of the movement. Soil creep accompanies seasonal freezing and thawing, which causes soil to gradually migrate downslope; if it creeps over a frozen substrate, it's called solifluction. Slumping involves semicoherent slices of earth that move slowly down spoon-shaped sliding surfaces, leaving behind a head scarp. Mudflows and debris flows happen where regolith has become saturated with water and moves downslope as a slurry. When volcanoes erupt and melt ice and snow at their summit, or if heavy rains fall during an eruption, water mixes with ash, creating a fast-moving lahar. Steep, rocky cliffs may suddenly give way in rock falls. If the rock breaks up into a cloud of debris that rushes downslope at high velocity, it is a rock avalanche. Snow avalanches are similar, but the debris consists only of snow.

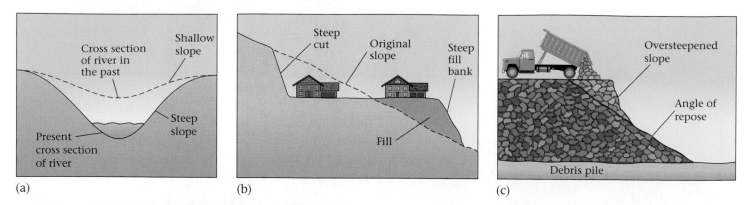

FIGURE 16.17 Slope angles may become steeper, making the slopes unstable. (a) A river can cut into the base of a slope, steepening the sides of the valley. (b) Cutting terraces in a hill slope creates a steeper slope. (c) Adding debris to the top of an unconsolidated sediment pile may cause the angle of repose to be exceeded.

across the Gros Ventre River, for the river had removed support.

In retrospect, the geology of the slide area made this landslide almost inevitable. The flank of Sheep Mountain is a **dip slope,** meaning that bedding parallels the face of the mountain. The Tensleep Formation, the stratigraphic unit exposed at the surface, consists of interbedded sandstone and shale. In the past, a thick sandstone layer spanned the valley and propped up sandstone farther up the side of the mountain. But river erosion cut down through the sand layer to a

FIGURE 16.18 (a) The huge Gros Ventre slide took place after heavy rains had seeped into the ground, weakening the Amsden Shale and making the overlying Tensleep Formation heavier. The slope was already unstable because the Gros Ventre River had cut down to the shale, and the bedding planes dipped parallel to the slope. (b) After the slide moved, it filled the river valley and dammed the river, creating a lake. A huge landslide scar formed on the hill slope. (c) The Gros Ventre slide.

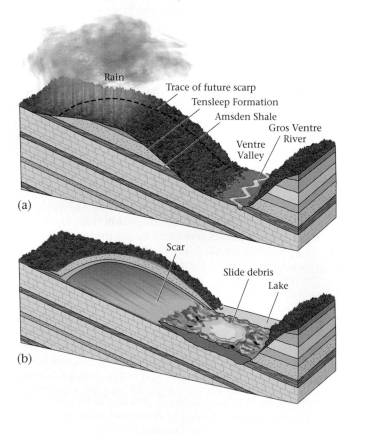

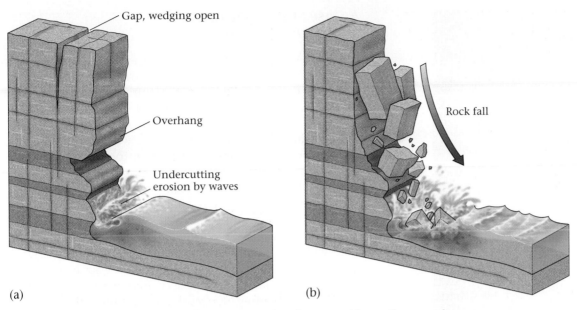

FIGURE 16.19 (a) Undercutting by waves removes the support beneath an overhang. (b) Eventually, the overhang breaks off along joints, and a rock fall takes place.

weak shale (the Amsden Shale) beneath. Rainfall in the weeks before the landslide made the ground heavier than usual and increased the amount of water seeping into the shale layer, weakening it further. By June, the downslope force exceeded the restraining force, and a huge slab of rock upslope broke off and raced downhill, with the wet shale layer acting as a failure surface.

In some cases, excavation results in the formation of an overhang. When such **undercutting** has occurred, rock making up the overhang eventually breaks away from the slope and falls. Overhangs commonly develop above a weak horizontal layer that erodes back preferentially, or along seacoasts and rivers where the water cuts into a fairly strong slope (▶Fig. 16.19a, b).

Changing the Slope Strength: The Effects of Weathering, Vegetation, and Water

The stability of a slope depends on the strength of the material constituting it. If the material weakens with time, the slope becomes weaker and eventually collapses. Three factors influence the strength of slopes: weathering, vegetation cover, and water.

With time, chemical weathering produces weaker minerals, and physical weathering breaks rocks apart. Thus, a formerly intact rock composed of strong minerals is transformed into a weaker rock or into regolith.

We've seen that thin films of water create cohesion between grains. Water in larger quantities, though, decreases cohesion, because it fills pore spaces entirely and keeps grains

apart. Though slightly damp sand makes a better sand castle than dry sand, a slurry of sand and water can't make a castle at all. Likewise, the saturation of regolith with water during a torrential rainstorm weakens the regolith so much that it may begin to move downslope as a slurry. If the water weakens a specific subsurface layer, then the layer becomes a failure surface. Similarly, if the water table (the top surface of the groundwater layer) rises above a weak failure surface after water has sunk into the ground, overlying rock or regolith may start to slide over the further weakened failure surface.

Water infiltration has a particularly notable effect in regions underlain by **swelling clays.** These clays possess a mineral structure that allows them to absorb water—water molecules form sheets between layers of silica tetrahedra, which cause clay flakes to swell to several times their original size. Such swelling pushes up the ground surface, making it crack, and weakens the upper layer of the ground, making it susceptible to creep or slip. When the clay dries, it shrinks, and the ground surface subsides. This up-and-down movement is enough to wrinkle road surfaces and crack foundations.

In the case of slopes underlaid with regolith, vegetation tends to strengthen the slope, because the roots hold otherwise unconsolidated grains together. Also, plants absorb water from the ground, thus keeping it from turning into slippery mud. The removal of vegetation therefore has the net result of making slopes more susceptible to downslope mass movement. In 2003, terrifying wildfires, stoked by strong winds, destroyed the ground-covering vegetation in many areas of California. When heavy rains followed, the barren ground of this hilly region became saturated with water

FIGURE 16.20 Deforestation makes slopes more susceptible to mass movement, as shown in this example from Puebla, Mexico. The slide destroyed the small village of Acalama, killing all but 30 of its 150–200 residents.

and turned into mud, which then flowed downslope, damaging and destroying many homes and roads. Deforestation in tropical rain forests, similarly, leads to catastrophic mass wasting of the forest's substrate (▶Fig. 16.20).

16.4 PLATE TECTONICS AND MASS MOVEMENTS

The Importance of the Tectonic Setting

Most unstable ground on Earth ultimately owes its existence to the activity of plate tectonics. As we've seen, plate tectonics causes uplift, generating relief, and causes faulting, which fragments the crust. And, of course, earthquakes on plate boundaries trigger devastating landslides. Spend a day along the steep slopes of the Alpine Fault, a plate boundary that transects New Zealand, and you can hear mass movement in progress: during heavy rains, rockfalls and landslides clatter with astounding frequency, as if the mountains were falling down around you. To see the interplay of plate tectonics and other factors, let's consider the example of mass movements in southern California.

A Case Study: Slumping in Southern California

Some Californians pay immense prices for the privilege of building homes on cliffs overlooking the Pacific. The sunset views from their backyard patios are spectacular. But the landscape is not ideal from the standpoint of stability (see Box 16.2). Slumps and mudflows on coastal cliffs have consumed many homes over the years, with a cost to their owners (or insurance companies) of untold millions of dollars (see Fig. 10.30a).

What is special about the region that makes it so prone to mass wasting? California is an active plate boundary. The coast borders the San Andreas Fault, a transform plate boundary accommodating the northward movement of the Pacific Plate with respect to North America. Faulting has shattered the rock of California's crust. The fractures act as planes of weakness, and allow water to seep into the rock and cause chemical weathering; the resulting clay and other slippery minerals further weaken the rock. Also, the rocks in many areas are weak to begin with, because they formed as part of an accretionary prism, a chaotic mass of clay-rich sediment that was scraped off subducting oceanic lithosphere during the Mesozoic Era. Though most of the movement between the North American and Pacific plates involves strike-slip displacement on the San Andreas Fault, there is a component of compression across the fault and this compression leads to uplift which creates slopes. Since the uplifted region borders the coast, wave erosion steepens and in some places undercuts cliffs. Finally, because of the plate motion, numerous earthquakes rock the region, thus shaking regolith loose.

Another reason for southern California's susceptibility to mass movements is its climate. In general, the region is hot and dry and thus supports only semidesert flora. Brush fires remove much of this cover, leaving large areas with no dense vegetation. But since the region lies on the West Coast, it endures occasional heavy winter rains. The water sinks quickly into the sparsely vegetated ground, adds weight to the mass on the slope, and weakens failure surfaces.

Development of cities and suburbs serve as the final factors that trigger mass movements in southern California. Development has oversteepened slopes, overloaded slopes, and has caused the water content of regolith to change. The consequences can be seen in an event called the "Portuguese Bend slide" of the Palos Verdes area near Los Angeles. The Portuguese Bend region borders the Pacific coast and is underlaid with a thick, seaward-dipping layer of weak volcanic ash (now altered to weak clay) resting on the shale. The land slopes down to the sea. To provide a foundation for homes and roads, developers deposited a 23-m-thick layer of fill over the ground surface. New residents living in the area began to water their lawns and use septic tanks that were prone to leaking—the water seeped into the ground and decreased the strength of the ash layer. Because of the decrease in strength, the added weight, and the erosion of the toe of the hill by the sea, the upper 30 m of land began to move. Between 1956 and 1985, the Portuguese Bend slide moved between 0.3 and 2.5 cm per day, and in the process it destroyed over 150 homes (▶Fig. 16.21).

16.5 HOW CAN WE PROTECT AGAINST MASS-MOVEMENT DISASTERS?

Identifying Regions at Risk

Clearly, landslides, mudflows, and slumps are natural hazards we cannot ignore. Too many of us live in areas where mass wasting has the potential to kill people and destroy property. In many cases, the best solution is avoidance: don't build, live, or work in an area where mass movement will take place. But avoidance is only possible if we know where the hazards are.

To pinpoint dangerous regions, geologists look for landforms known to result from mass movements, for where these movements have happened in the past, they might happen again in the future. For example, the Portuguese Bend slide occurred on top of at least two other slides that had happened in the past several thousand years. Features such as slump head scarps, swaths of forest in which trees have been flattened and point downslope, piles of loose debris at the base of hills, and hummocky land surfaces all indicate recent mass wasting.

Geologists may also be able to detect regions that are beginning to move (▶Fig. 16.22). For example, roads, buildings, and pipes begin to crack over unstable ground. Power lines may be too tight or too loose because the poles to which they are attached move together or apart. Visible

---

Los Angeles, a Mobile Society

BOX 16.2
THE HUMAN ANGLE

During a year of abundant sliding in southern California, Art Buchwald wrote the following newspaper column.

I came to Los Angeles last week for rest and recreation, only to discover that it had become a rain forest.

I didn't realize how bad it was until I went to dinner at a friend's house. I had the right address, but when I arrived, there was nothing there. I went to a neighboring house where I found a man bailing out his swimming pool.

I beg your pardon, I said. Could you tell me where the Cables live?

"They used to live above us on the hill. Then, about two years ago, their house slid down in the mud, and they lived next door to us. I think it was last Monday, during the storm, that their house slid again, and now they live two streets below us, down there. We were sorry to see them go—they were really nice neighbors."

I thanked him and slid straight down the hill to the new location of the Cables' house. Cable was clearing out the mud from his car. He apologized for not giving me the new address and explained, "Frankly, I didn't know until this morning whether the house would stay here or continue sliding down a few more blocks."

Cable, I said, you and your wife are intelligent people, why do you build your house on the top of a canyon, when you know that during a rainstorm it has a good chance of sliding away?

"We did it for the view. It really was fantastic on a clear night up there. We could sit in our Jacuzzi and see all of Los Angeles, except of course when there were brush fires. Even when our house slid down two years ago, we still had a great sight of the airport. Now I'm not too sure what kind of view we'll have because of the house in front of us, which slid down with ours at the same time."

But why don't you move to safe ground so that you don't have to worry about rainstorms?

"We've thought about it. But once you live high in a canyon, it's hard to move to the plains. Besides, this house is built solid and has about three more good mudslides in it."

Still, it must be kind of hairy to sit in your home during a deluge and wonder where you'll wind up next. Don't you ever have the desire to just settle down in one place?

"It's hard for people who don't live in California to understand how we people out here think. Sure we have floods, and fire and drought, but that's the price you have to pay for living the good life. When Esther and I saw this house, we knew it was a dream come true. It was located right on the tippy top of the hill, way up there. We would wake up in the morning and listen to the birds, and eat breakfast out on the patio and look down on all the smog.

"Then, after the first mudslide, we found ourselves living next to people. It was an entirely different experience. But by that time we were ready for a change. Now we've slid again and we're in a whole new neighborhood. You can't do that if you live on solid ground. Once you move into a house below Sunset Boulevard, you're stuck there for the rest of your life.

"When you live on the side of a hill in Los Angeles, you at least know it's not going to last forever."

Then, in spite of what's happened, you don't plan to move out?

"Are you crazy? You couldn't replace a house like this in L.A. for $500,000."

What happens if it keeps raining and you slide down the hill again?

"It's no problem. Esther and I figure if we slide down too far, we'll just pick up and go back to the top of the hill, and start all over again; that is, if the hill is still there after the earthquake."

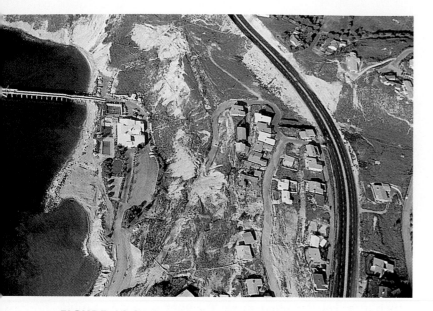

cracks form on the ground at the potential head of a slump, while the ground may bulge up at the toe of the slump. Subsurface cracks may drain the water from an area and kill off vegetation, while another area may sink and form a swamp. Slow movements cause trees to develop pronounced curves at their base. In some cases, the activity of land masses moving too slowly to be perceptible to people can be documented with sensitive surveying techniques that can detect a subtle tilt of the ground or changes in distance between nearby points.

If various clues indicate that a land mass is beginning to move, and if conditions make accelerating movement likely (e.g., persistent rain, rising floodwaters, or continuing earthquake aftershocks), then officials may order an evacuation. Evacuations have saved lives, and ignored warnings have cost lives. But unfortunately, some mass movements happen without any warning, and some evacuations prove costly but unnecessary.

FIGURE 16.21 The Portuguese Bend slide viewed from the air.

FIGURE 16.22 The features shown here indicate that a large slump is beginning to develop. Note the cracks at the site of the growing head scarp, which drain water and kill trees. Power-line poles crossing the unstable ground bend, and the lines become overtight. Fences and roads that straddle the scarp begin to break up. Houses that straddle the scarp begin to crack, and their foundations sink.

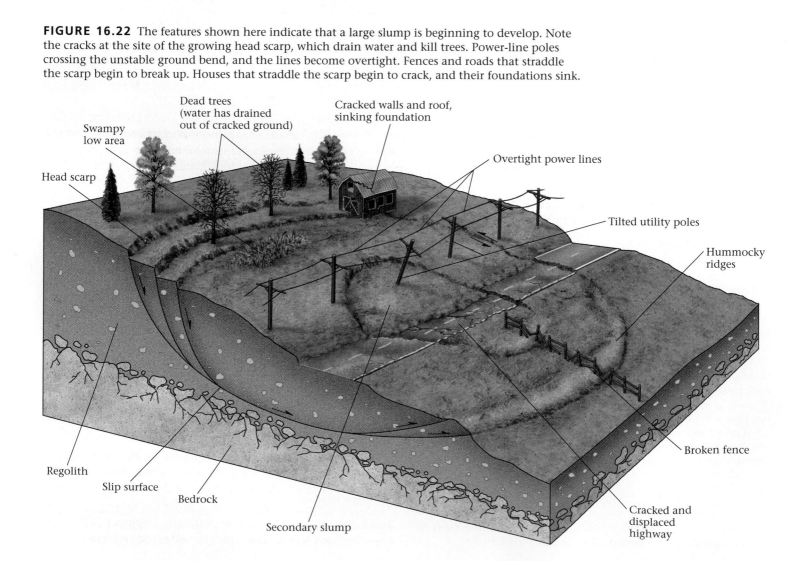

Even if there is no evidence of recent movement, a danger may still exist, for just because a steep slope hasn't collapsed in the recent past doesn't mean it won't in the future. In recent years, geologists have begun to identify such potential hazards (by using computer programs that evaluate factors that trigger mass wasting) and create maps that portray the degree of risk for a certain location. These factors include the following: slope steepness; strength of substrate; degree of water saturation; orientation of bedding, joints, or foliation relative to the slope; nature of vegetation cover; potential for heavy rains; potential for undercutting to occur; and likelihood of earthquakes. From such hazard-assessment studies, geologists compile **landslide-potential maps,** which rank regions according to the likelihood that a mass movement will occur. In any case, common sense suggests that you should avoid building on or below particularly dangerous slide-prone slopes. In Japan, regulations on where to build in regions prone to mass wasting, careful monitoring of ground movements, and well-designed evacuation plans have drastically reduced property damage and the number of fatalities.

Preventing Mass Movements

In areas where a hazard exists, people can take certain steps to remediate the problem and stabilize the slope (▶Fig. 16.23a–i).

- *Revegetation:* Since bare ground is much more prone to downslope movement than vegetated ground, stability in deforested areas will be greatly enhanced if owners replant the region with vegetation that sends down deep roots.

- *Regrading:* An oversteepened slope can be regraded so that it does not exceed the angle of repose.

- *Reducing subsurface water:* Because water weakens material beneath a slope and adds weight to the slope, an unstable situation may be remedied either by improving drainage so that water does not enter the subsurface in the first place, or by removing water from the ground.

- *Preventing undercutting:* In places where a river undercuts a cliff face, engineers can divert the river. Similarly, along coastal regions they may build an offshore breakwater or pile **riprap** (loose boulders or concrete) along the beach to absorb wave energy before it strikes the cliff face.

- *Constructing safety structures:* In some cases, the best way to prevent mass wasting is to build a structure that stabilizes a potentially unstable slope or protects a region downslope from debris if a mass movement does occur. For example, civil engineers can build retaining walls or bolt loose slabs of rock to more coherent masses in the substrate in order to stabilize highway embankments. The danger from rock falls can be decreased by covering a road cut with chainlink fencing or by spraying road cuts with "shotcrete," a cement that coats the wall and prevents water infiltration and consequent freezing and thawing. Highways at the base of an avalanche chute can be covered by an avalanche shed, whose roof keeps debris off the road.

- *Controlled blasting of unstable slopes:* When it is clear that unstable ground threatens a particular region, the best solution may be to blast the unstable ground or snow loose at a time when its movement can do no harm.

Clearly, the cost of preventing mass-wasting calamities is high, and people might not always be willing to pay the price. In such cases, they have a choice of avoiding the risky area, taking the chance that a calamity will not happen while they are around, buying appropriate insurance, or counting on relief agencies to help if disaster does strike. Once again, geology and society cross paths.

CHAPTER SUMMARY

- Rock or regolith on unstable slopes has the potential to move downslope under the influence of gravity. This process, called mass movement, or mass wasting, plays an important role in the erosion of hills and mountains.

- Slow mass movement, caused by the freezing and thawing of regolith, is called creep. In places where slopes are underlaid with permafrost, solufluction causes a melted layer of regolith to flow down slopes. During slumping, a semicoherent mass of material moves down a spoon-shaped failure surface. Mudflows and debris flows occur where regolith has become saturated with water and moves downslope as a slurry.

- Landslides (rock and debris slides) move very rapidly down a slope; the rock or debris breaks apart and tumbles. During avalanches, debris mixes with air and moves downslope as a turbulent cloud. And in a debris fall or rock fall, the material free-falls down a vertical cliff.

- Intact, fresh rock is too strong to undergo mass movement. Thus, for mass movement to be possible, rock must be weakened by fracturing (joint formation) or weathering.

- Unstable slopes start to move when the downslope force exceeds the resistance force that holds material in place. The steepest angle at which a slope of unconsolidated material can remain without collapsing is the angle of repose.

- Downslope movement can be triggered by shocks and vibrations, a change in the steepness of a slope, a change in the strength of a slope, deforestation, weathering, or heavy rain.

- Geologists produce landslide-potential maps to identify areas susceptible to mass movement. Engineers can help prevent mass movements using a variety of techniques.

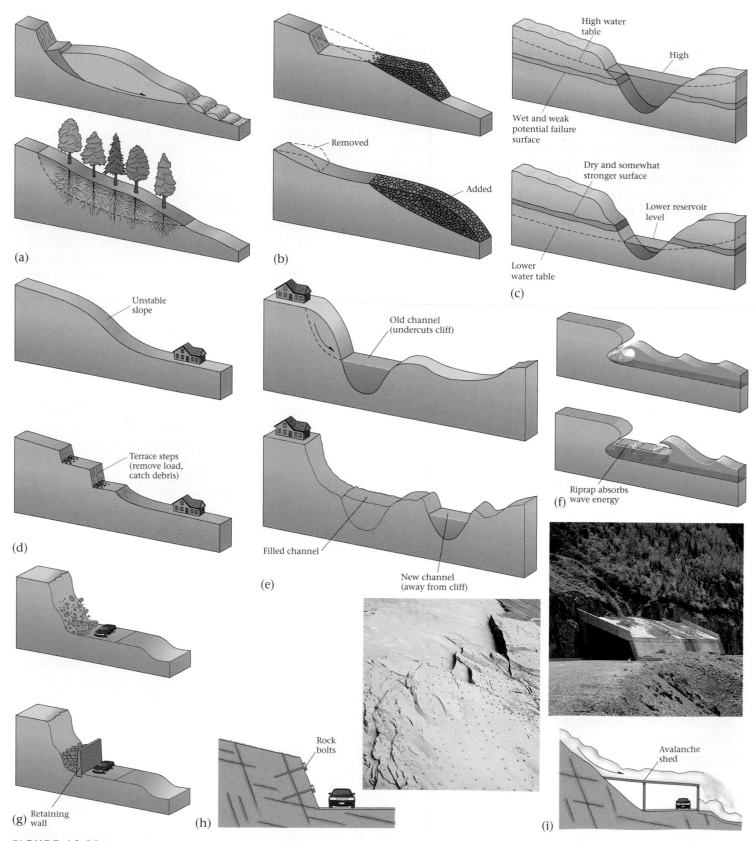

FIGURE 16.23 A variety of remedial steps can stabilize unstable ground. (a) Revegetation removes water, and tree roots bind regolith. (b) Redistributing the mass on a slope eases the load where necessary, adds support where necessary, and decreases slope angles. (c) Lowering the level of groundwater (the water table) may allow a failure surface to dry out. (d) Terracing a steep slope may decrease the load and provide benches to catch debris. (e) Relocating a river channel stops undercutting, and filling the old channel adds support. (f) Riprap absorbs wave energy along the coast. (g) A retaining wall traps falling rock. (h) Bolting rock to a steep cliff face holds loose blocks in place. (i) An avalanche shed diverts avalanche debris over a roadway.

KEY TERMS

angle of repose (p. 513)
avalanche (p. 509)
creep (p. 506)
debris fall (p. 510)
debris flow (p. 508)
debris slide (p. 509)
dip slope (p. 518)
failure surface (p. 507)
head scarp (p. 507)
lahar (p. 508)
landslide (p. 509)
landslide-potential maps
   (p. 523)
liquefaction (p. 515)
mass movement (wasting)
   (p. 505)
mudflow (p. 508)
natural hazard (p. 505)

permafrost (p. 506)
quick clay (p. 515)
riprap (p. 523)
rock fall (p. 510)
rock glacier (p. 506)
rock slide (p. 509)
slumping (p. 507)
solifluction (p. 506)
stable slopes (p. 512)
submarine debris flow
   (p. 511)
submarine slump (p. 511)
swelling clays (p. 519)
talus (p. 510)
turbidity current (p. 511)
undercutting (p. 519)
unstable slopes (p. 512)

REVIEW QUESTIONS

1. What factors distinguish the various types of mass movement?

2. How does a slump differ from creep? How does it differ from a mudflow or debris flow?

3. How does a rock or debris slide differ from a slump? What conditions trigger a snow avalanche?

4. How are submarine slumps similar to those above water? How might they be related to tsunamis?

5. How does a small amount of water between grains hold material together? How does this change when the sediment is oversaturated?

6. What force is responsible for downslope movement? What force helps resist that movement?

7. How does the angle of repose change with grain size? How does it change with water content?

8. What factors trigger downslope movement?

9. How do geologists predict whether an area is prone to mass wasting?

10. What steps can people take to reduce the risk of mass wasting?

SUGGESTED READING

Brabb, E. E., and B. L. Harrod, eds. 1989. *Landslides: Extent and Economic Significance.* Brookfield, Va.: Balkema.

Costa, J. E., and G. F. Wieczorek. 1987. *Reviews in Engineering Geology.* Vol. 7, *Debris Flows, Avalanches: Process, Recognition, and Mitigation.* Boulder, Colo.: Geological Society of America.

Crozier, M. J. 1986. *Landslides: Causes, Consequences, and Environment.* Dover, N.H.: Croom Helm.

Dikau, R., D. Brunsden, and L. Schrott, eds. 1996. *Landslide Recognition: Identification, Movement, and Causes.* Chichester, England: Wiley.

Slosson, J. E., A. G. Keene, and J. A. Johnson, eds. 1993. *Reviews in Engineering Geology.* Vol. 9, *Landslides/Landslide Mitigation.* Boulder, Colo.: Geological Society of America.

Voight, B., ed. 1978. *Rockslides and Avalanches.* Vol. 1, *Natural Processes.* New York: Elsevier.

Zaruba, Q., and V. Mencl. 1969. *Landslides and Their Control.* New York: Elsevier.

CHAPTER 17

Streams and Floods: The Geology of Running Water

17.1 INTRODUCTION

By the 1880s, Johnstown, built along the Conemaugh River in scenic western Pennsylvania, had become a significant industrial town with numerous steel-making factories. Recognizing the attraction of the region as a summer retreat from the heat and pollution of nearby Pittsburgh, speculators built a mud and gravel dam across the river, upstream of Johnstown, to trap a pleasant reservoir of cool water. A group of industrialists and bankers bought the reservoir and established the exclusive South Fork Hunting and Fishing Club, a cluster of lavish fifteen-room "cottages" on the shore. Unfortunately, the dam had been poorly designed, and debris blocked its spillway (a passageway for surplus water), setting the stage for a monumental tragedy. On May 31, 1889, torrential rain drenched Pennsylvania, and the reservoir surface rose until water flowed over the dam and down its face. Despite frantic attempts to strengthen the dam, the soggy structure abruptly collapsed, and the reservoir emptied into the Conemaugh River Valley. A 20-m-high wall of water roared downstream and slammed into Johnstown, transforming bridges and buildings into twisted wreckage (▶Fig. 17.1). When the water subsided, 2,300 people lay dead, and Johnstown became the focus of national sympathy. Clara Barton mobilized the recently

Water that falls on land drains back to the sea via rivers. The character of an individual river depends on a variety of factors, such as slope, water volume, and sediment load. The Genesee River drains the hills of Central New York State and carries the water over waterfalls, down into Lake Ontario. The energy of flowing water has cut a canyon into beds of shale and siltstone.

FIGURE 17.1 During the disastrous 1889 flood in Johnstown, Pennsylvania, the force of the water was able to move and tumble sturdy buildings.

founded Red Cross, which set to work building dormitories, and citizens nationwide donated everything from clothes to beds. Nevertheless, it took years for the town to recover, and many residents simply picked up and left. Despite several lawsuits, no one payed a penny of restitution, but the South Fork Hunting and Fishing Club abandoned its property.

The unlucky inhabitants of Johnstown experienced one of the more destructive consequences of running water, or **runoff,** water that flows on the land surface. The Conemaugh River overflowed when the dam broke, causing a **flood,** an event during which the volume of water in a stream becomes so great that it covers areas outside the stream's normal limits. But even when not in flood, running water relentlessly digs into and scrapes away at the surface of the Earth, carrying away debris and depositing it in other locations. Without running water, our planet would look very different.

Much of the work of running water takes place in **streams,** ribbons of water that flow down **channels,** or troughs, cut into the land. Streams remove, or drain, excess water (runoff) from the landscape and carry it eventually to the sea just like culverts drain water from parking lots, and in the process they have modified much of Earth's landscape. (Note that geologists may call any channelized body of flowing water a stream, regardless of its size. In common usage, though, a large stream is a "river." Any segment of the channel, as measured along the length of the stream, is called a "reach" of the stream, and a broad curving reach is a "meander.") In the process of draining the land, streams transport sediment and nutrients, and provide a home for living organisms. And for human society, streams supply avenues for commerce, water for agriculture, and sources for power.

Earth is the only planet in the solar system that currently has streams of running water. (Mars may have had them in the past; see Interlude E.) In this chapter, we see how streams operate in the Earth System. We first learn about the origin of running water and the architecture of streams and how running water ultimately drops sediment. We then look at the landforms that erosion and/or deposition generate in response to stream flow, and conclude with a consideration of flooding and its effects on human society.

17.2 DRAINING THE LAND

Runoff, in the Hydrologic Cycle

Nothing that is can pause or stay—
  The moon will wax, the moon will wane,
  The mist and cloud will turn to rain,
  The rain to mist and cloud again,
  Tomorrow be today.
        —HENRY WADSWORTH LONGFELLOW (1807–1882)

Perhaps without realizing it, Longfellow, an American poet fascinated with reincarnation, provided an accurate if somewhat romantic image of the hydrologic cycle (see Interlude E). Solar energy evaporates water from the Earth's surface; when the vapor rises, it cools and condenses to form rain or snow that falls back to Earth; this water is called "meteoric water." Recall that some of this meteoric water infiltrates (seeps into) the ground and either dampens the regolith or rock near the surface, or sinks deeper to a realm in which water fills pores and cracks. The boundary defining the top of the subsurface realm in which water fills pores and cracks completely is called the **water table;** above the water table, pores and cracks contain some air (see Chapter 19). Water that does not sink into the ground remains at the surface either in the form of snow or ice, or as liquid. All water that lies on the land surface is "surface water," whereas the water held below the ground is "groundwater."

On horizontal ground, surface water collects in puddles or swamps, but on sloping ground, the water flows downslope in response to gravity, following the steepest route possible; this flowing water eventually collects in streams. Meltwater (from snow and ice) and rain contribute to the volume of water in streams (▶Fig. 17.2). Also, when new meteoric water infiltrates into the ground higher on a slope, it pushes some of the water that had been underground, but above the water table, out and back to the surface, lower on the slope. Springs provide conduits for even deep groundwater to return to the surface. Thus, much of the water flowing in a stream, in a temperate climate, resided underground prior to entering the stream.

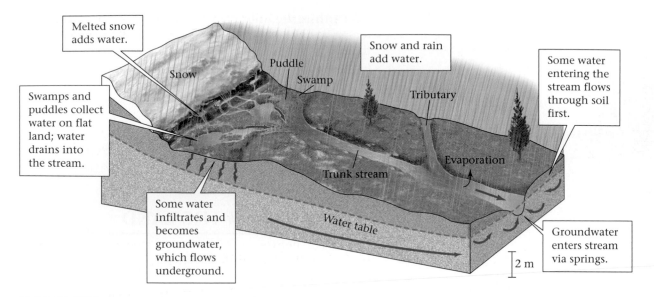

FIGURE 17.2 Excess surface water comes from rain, melting ice or snow, and groundwater springs. Where the ground is flat, the water accumulates in puddles or swamps, but on sloping ground, it flows downslope, collecting in natural troughs called streams.

Water flowing in streams may pause temporarily in lakes or reservoirs. But all *fresh* standing bodies of water have an outlet through which water escapes and continues to the sea. On a global basis, 36,000 cubic km of water becomes runoff every year, about 10% of the total volume that passes through the hydrologic cycle.

Forming Streams and Drainage Networks

Running water begins its downslope journey as **sheetwash,** a film of water less than a few mm thick that covers the surface of the ground. You've seen sheetwash if you've watched water flowing down a sloping street during a rainstorm. Like any flowing fluid, sheetwash erodes its substrate (the material it flows over). The efficiency of such erosion depends on the velocity of the flow—faster flows erode more effectively. In nature, the ground is not perfectly planar, not all substrate has the same resistance to erosion, and the amount of vegetation that covers and protects the ground varies with location. Thus, the velocity of sheetwash also varies with location. Where the flow happens to be a bit faster, or the substrate a little weaker, erosion scours (digs) a channel into the substrate (▶Fig. 17.3a, b). This channel is lower than the surrounding ground, so sheetwash in adjacent areas starts to head toward it. With time, the extra flow deepens the channel relative to its surroundings, a process called **downcutting,** and creates a stream.

As its flow increases, a stream channel begins to lengthen up its slope, a process called **headward ero-sion** (▶Fig. 17.3c, d). Headward erosion occurs because the flow is more intense at the entry to the channel (upslope) than in the surrounding sheetwashed areas. At the same time, new channels form nearby; these merge with the main channel, because once a channel forms, the surrounding land slopes into it. An array of linked streams evolves, with the smaller streams, or **tributaries,** flowing into a single larger stream, or **trunk stream.** The array of interconnecting streams together constitute the **drainage network.**

Like transportation networks, drainage networks reach into all corners of a region to provide conduits for the removal of runoff. The configuration of tributaries and trunk streams defines the map pattern of a drainage network. This pattern depends on the shape of the landscape and the composition of the substrate. Geologists recognize several types of networks, based on their map pattern:

- *Dendritic:* When rivers flow over a fairly uniform substrate with a fairly uniform initial slope, they develop a dendritic network, which looks like the pattern of branches connecting to the trunk of a deciduous tree (▶Fig. 17.4a). In fact, the word "dendritic" comes from the Greek *dendros,* meaning "tree."

- *Radial:* Drainage networks forming on the surface of a cone-shaped mountain flow outward from the mountain peak, like spokes on a wheel. Such a pattern defines a radial network (▶Fig. 17.4b).

- *Rectangular:* In places where a rectangular grid of fractures (vertical joints) breaks up the ground, channels form

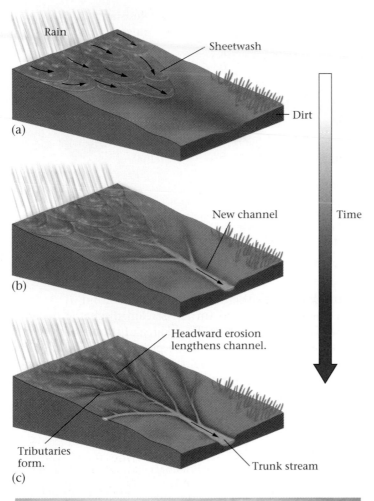

FIGURE 17.3 (a) Drainage on a slope first occurs when sheetwash, overlapping films or sheets of water, moves downslope. (b) Where the sheetwash happens to move a little faster, it scours a channel. (c) The channel grows upslope, a process called headward erosion, and new tributary channels form. The interconnecting streams comprise a drainage network. (d) Headward erosion in Canyonlands National Park, Utah, where the main canyon and its tributaries are cutting upstream slowly.

along the preexisting fractures, and streams join each other at right angles, creating a rectangular network (▶Fig. 17.4c).

• *Trellis:* In places where a drainage network develops across a landscape of parallel valleys and ridges, major tributaries flow down a valley and join a trunk stream that cuts across the ridges; the place where a trunk stream cuts across a resistant ridge is a "water gap." The resulting map pattern resembles a garden trellis, so the arrangement of streams constitutes a trellis network (▶Fig. 17.4d).

Drainage Basins and Divides

A drainage network collects water from a broad region, variously called a **drainage basin,** catchment, or **watershed,** and feeds it into the trunk stream, which carries the water away. The highland, or ridge, that separates one watershed from another is a **drainage divide** (▶Fig. 17.5). A continental divide separates drainage that flows into one ocean from drainage that flows into another. For example, if you straddle the North American continental divide and pour a cup of water out of each hand, the water in one hand flows to the Atlantic, and the water in the other flows to the Pacific. This continental divide is not, however, the only important divide in North America. A divide runs along the crest of the Appalachians, separating Atlantic Ocean drainage from Gulf of Mexico drainage, and another one runs just south of the Canada-U.S. border, separating Gulf of Mexico drainage from Hudson Bay (Arctic Ocean) drainage. These three divides bound the Mississippi drainage basin, which drains the interior of the United States (▶Fig. 17.6).

Streams That Last, Streams That Don't: Permanent and Ephemeral Streams

Some streams flow all year long, while others flow for only part of the year; in fact, some flow only for a brief time after a heavy rain. The character of a stream depends on the depth of the water table. If the bed, or floor, of a stream lies below the water table, then the stream flows year-round (▶Fig. 17.7a). In such **permanent streams,** found in humid or temperate climates, water comes not only from upstream or from surface runoff, but also from springs through which groundwater seeps. But if the bed of a stream lies above the water table, then water flows only when the rate at which water enters the stream channel exceeds the rate at which water infiltrates the ground below (▶Fig. 17.7b). Such streams can be permanent only if supplied by abundant water from upstream. In dry climates with intermittent rainfall and high evaporation rates, water entirely sinks into the ground, and the stream dries up when the supply of water stops. Streams which do not flow all year are called **ephemeral streams.**

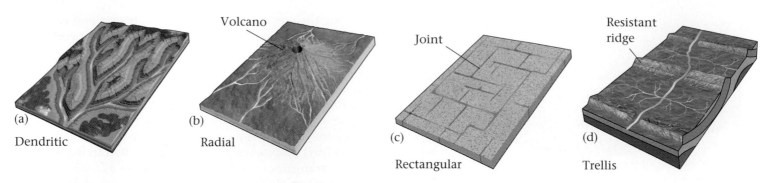

FIGURE 17.4 (a) Dendritic patterns resemble the branches of a tree and form on land with a uniform substrate. (b) A radial network drains a conical mountain; in this case, a volcano. (c) Rectangular patterns develop where a gridlike array of vertical joints controls drainage. (d) Trellis patterns (resembling a garden trellis) form where drainage networks cross a landscape in which ridges of hard rock separate valleys of soft rock. In this example, the alternation is due to folding of the rock layers.

Ephemeral streams only flow during rainstorms or after spring thaws. A dry ephemeral stream bed (channel floor) is called a **dry wash,** or wadi.

17.3 DISCHARGE AND TURBULENCE

Geologists and engineers describe the amount of water a stream carries by its **discharge,** the volume of water passing through an imaginary cross section drawn across the stream perpendicular to the bank, in a unit of time. We can specify stream discharge either in cubic feet per second (ft^3/s) or in cubic meters per second (m^3/s). Stream discharge depends on two factors: the cross-sectional area of the stream (A_c; the area measured in a vertical plane perpendicular to the flow direction) and the average velocity at which water moves in the downstream direction (v_a). Thus, we can calculate stream discharge by using the simple formula $D = A_c \times v_a$. Stream discharge can be determined at a stream-gauging sta-

FIGURE 17.5 A drainage divide is a relatively high ridge that separates one drainage basin from another.

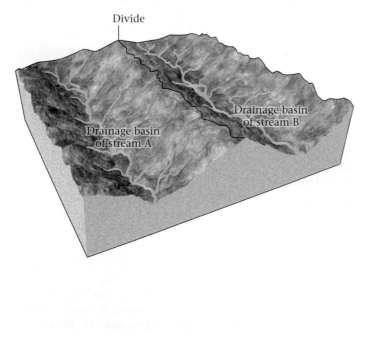

FIGURE 17.6 The Mississippi drainage basin is one of several drainage basins in North America. The continental divide separates basins that drain into the Atlantic (and waters connected to the Atlantic) from basins that drain into the Pacific.

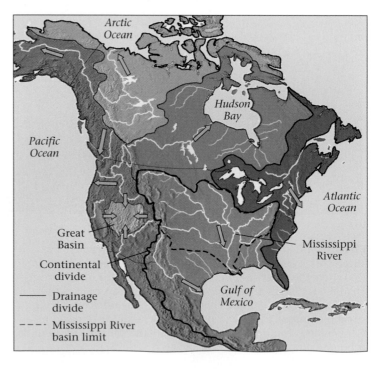

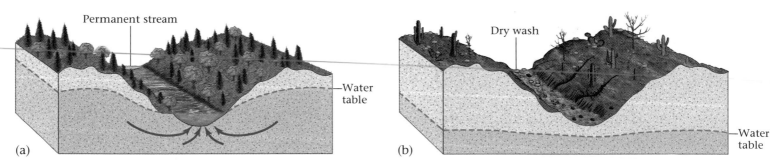

(a) (b)

FIGURE 17.7 (a) If the bed of a stream channel lies below the water table, then springs add water to the stream, and the stream contains water even during periods when there is no rainfall. Such streams are permanent. (b) If the stream bed lies above the water table, then the stream flows only during rainfall or spring thaws, when water enters the stream faster than it can infiltrate into the ground. Such streams are ephemeral. In desert regions, a dry stream bed is called a dry wash.

tion, where instruments measure the velocity and depth of the water (▶Fig. 17.8).

Different streams have different average discharges. For example, the Amazon River has the largest average discharge in the world—about 200,000 m³/s, or 15% of the total amount of runoff on Earth. The next-largest stream, the Congo River, has an average discharge of 40,000 m³/s, while the "mighty" Mississippi's is only 17,000 m³/s. Also, the discharge of a given stream varies along its length. For example, the discharge in a temperate region *increases* in the downstream direction, because each tributary that enters the stream adds more water, while the discharge in an arid region may *decrease* downstream, as progressively more water

seeps into the ground or evaporates. Discharge can also be affected by human activity, if people divert the river's water for irrigation, the river's discharge decreases downstream. Finally, the discharge at a given location can vary with time: in a temperate climate, a stream's discharge during the spring may be double or triple the amount during a hot summer, and a flood may increase the discharge to more than a hundred times normal.

The average velocity of stream water (v_a) can be difficult to calculate, because the water doesn't all travel at the same velocity for two reasons. First, friction along the sides and floor of the stream slows the flow. Thus, water near the channel walls or the stream bed (the floor of the stream) moves more slowly than water in the middle of the flow, and the fastest-moving part of the stream flow lies near the surface in the center of the channel. In a curved channel, the fastest flow shifts toward the outside curve, somewhat like a car swerves to the outer edge of a curve on a highway. Therefore, the deepest part of a channel, its **thalweg,** lies near the outside curve. In fact, as the water flows toward the outside wall of a curving channel, it begins a spiral motion; because water near the surface can flow faster toward the outer bank, water deeper down must flow toward the inner bank to replace the surface water. The amount by which friction slows the flow depends both on the roughness of the walls and bed and on the channel shape. A wide, shallow stream channel has a larger "wetted perimeter" (the area in which water touches the channel walls) than does a semicircular channel, so water flows more slowly in the former than in the latter (▶Fig. 17.9a–c). Second, **turbulence,** or turbulent flow, the twisting, swirling motion that, at a large scale, can create eddies (whirlpools) in which water curves and actually flows upstream, or circles in place (▶Fig. 17.10). Turbulence develops in part because the shearing motion of one volume against its neighbor causes the neighbor to spin, and in part because obstacles like boulders deflect volumes, forcing them to move in a different direction.

FIGURE 17.8 Geologists obtain information needed to calculate discharge at a stream-gauging station. First, they make a survey of the channel so they know its shape. Then they measure its depth, using a well, and its velocity, using a current meter (either a propeller that spins in response to moving water as shown or, more recently, Doppler radar). The current meter takes measurements at various points in the stream, for velocity changes with location.

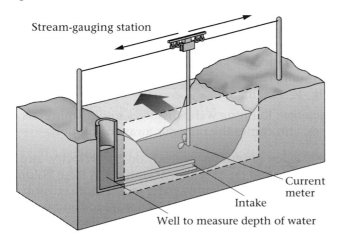

Stream-gauging station

Current meter

Intake

Well to measure depth of water

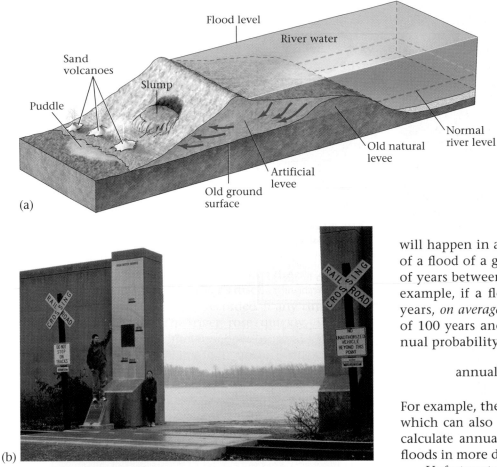

(a)

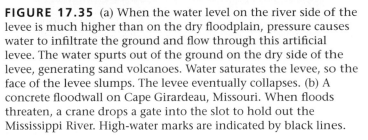

(b)

FIGURE 17.35 (a) When the water level on the river side of the levee is much higher than on the dry floodplain, pressure causes water to infiltrate the ground and flow through this artificial levee. The water spurts out of the ground on the dry side of the levee, generating sand volcanoes. Water saturates the levee, so the face of the levee slumps. The levee eventually collapses. (b) A concrete floodwall on Cape Girardeau, Missouri. When floods threaten, a crane drops a gate into the slot to hold out the Mississippi River. High-water marks are indicated by black lines.

floodplains back into natural wetlands helps prevent floods, for wetlands absorb water like a sponge. A solution to some flooding in some cases may lie in the removal, rather than the construction, of levees. Property may also be kept safe by defining **floodways,** regions likely to be flooded, and then by moving or abandoning buildings located there. Even the simple act of moving levees farther away from the river and creating natural habitats in the resulting floodways would decrease flooding damage immensely (▶Fig. 17.36a, b).

When making decisions about investing in flood-control measures, mortgages, or insurance, planners need a basis for defining the hazard or risk posed by flooding. If floodwaters submerge a locality every year, a bank officer would be ill-advised to approve a loan that would pro-

mote building there. But if floodwaters submerge the locality very rarely, then the loan may be worth the risk. Geologists characterize the risk of flooding in two ways. The **annual probability** (more formally known as the "annual exceedance probability") of flooding indicates the likelihood that a flood of a given size or larger will happen at a specified locality during any given year. For example, if we say that a flood of a given size has an annual probability of 1%, then we mean that there is a 1 in 100 chance that a flood of at least this size will happen in any given year. The **recurrence interval** of a flood of a given size is defined as the *average* number of years between successive floods of at least this size. For example, if a flood of a given size happens once in 100 years, *on average,* then it is assigned a recurrence interval of 100 years and is called a 100-year-flood. Note that annual probability and recurrence interval are related:

$$\text{annual probability} = \frac{1}{\text{recurrence interval}}.$$

For example, the annual probability of a 50-year-flood is $\frac{1}{50}$, which can also be written as 0.02 or 2%. To learn how to calculate annual probabilities and recurrence intervals of floods in more detail, see Box 17.1.

Unfortunately, some people become misled by the meaning of recurrence interval, and think that they do not face a flooding hazard if they buy a home built within an area submerged by 100-year floods just after such a flood has occurred. Their confidence comes from making the incorrect assumption that because such flooding just happened, it

FIGURE 17.36 Concept of a floodway. (a) Building artificial levees directly on natural levees creates a larger channel for a river. (b) Building artificial levees at a distance from the river creates a floodway on either side of the river, an even larger channel than in (a), and a surface of wetland that can absorb floodwaters.

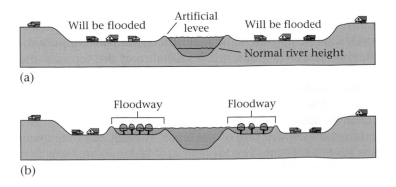

can't happen again until "long after I'm gone." They may regret their decision because two 100-year floods can occur in consecutive years or even in the same year (alternatively, the interval between such floods could be, say, 210 years). Because the term "recurrence interval" can lead to confusion, it may be better to report risk in terms of annual probability.

Knowing the discharge during a flood of a specified annual probability, and knowing the shape of the river channel and the elevation of the land bordering the river, hydrologists can predict the extent of land that will be submerged by such a flood. (▶Fig. 17.37). Such data, in turn, permits hydrologists to produce "flood-hazard maps." In the United States, the Federal Emergency Management Agency (FEMA) produces Flood Insurance Rate Maps that show the 1% annual probability (100-year) flood area and the 0.2% annual probability (500-year) flood risk zones (▶Fig. 17.38).

17.9 RIVERS: A VANISHING RESOURCE?

As *Homo sapiens* evolved from hunter-gatherers into farmers, areas along rivers became attractive places to settle. Rivers serve as avenues for transportation and as sources for food, irrigation water, drinking water, power, recreation, and (unfortunately) waste disposal. Further, their floodplains provide particularly fertile soil for fields, replenished annually by seasonal floods. Considering the multitudinous resources that rivers provide, it's no coincidence that early civilizations gathered in river valleys and on floodplains: Mesopotamia arose around the Tigris and Euphrates Rivers, Egypt around the Nile, India in the Indus Valley, and China

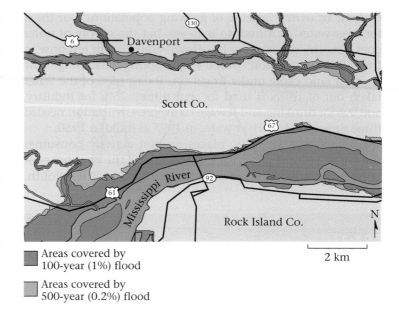

Areas covered by 100-year (1%) flood

Areas covered by 500-year (0.2%) flood

2 km

FIGURE 17.38 A flood hazard map for a region near Davenport, Iowa, as prepared by FEMA. It shows areas likely to be flooded.

along the Hwang (Yellow) River. Over the millennia, rivers have killed millions of people in floods, but they have been the lifeblood for hundreds of millions more. Nevertheless, over time, humans have increasingly tended to abuse or overuse the Earth's rivers. Here we note four pressing environmental issues.

Pollution. The capacity of some rivers to carry pollutants has long been exceeded, transforming them into deadly cesspools. Pollutants include raw sewage and storm drainage from urban areas, spilled oil, toxic chemicals from industrial sites, and excess fertilizer and animal waste from agricultural fields. Some pollutants directly poison aquatic life, some feed algae blooms that strip water of its oxygen, and some settle out to be buried along with sediments. River pollution has become overwhelming in developing countries, where there are few waste-treatment facilities.

Dam construction. In 1950, there were about 5,000 large (over 15 m high) dams worldwide, but today there are over 38,000. Damming rivers has both positive and negative results. Reservoirs provide irrigation water and hydroelectric power, and they trap some floodwaters and create popular recreation areas. But sometimes their construction destroys "wild rivers" (the whitewater streams of hilly and mountainous areas) and alters the ecosystem of a drainage network by providing barriers to migrating fish, decreasing the nutrient supply to organisms downstream, by removing the source of sediment for the delta, and by eliminating seasonal floods that replenish nutrients in the landscape.

FIGURE 17.37 A 100-year flood covers a larger area than a 2-year flood.

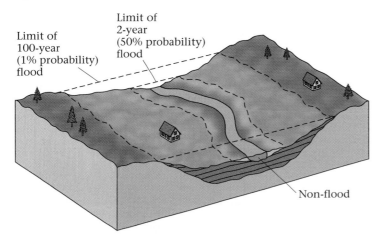

18.3 OCEAN WATER

Composition

If you've ever had a chance to swim in the ocean, you may have noticed that you float much more easily in ocean water than you do in freshwater. That's because ocean water contains an average of 3.5% dissolved salt ions (▶Fig. 18.7); in contrast, typical freshwater contains only 0.02% salt. The dissolved ions fit between water molecules without changing the volume of the water, so adding salt to water increases the water's density, and you float higher in a denser liquid.

Leonardo da Vinci, a Renaissance artist and scientist, speculated that sea salt came from rivers passing through salt mines, but modern studies demonstrate that most cations in sea salt—sodium (Na^+), potassium (K^+), calcium (Ca^{2+}), and magnesium (Mg^{2+})—come from the chemical weathering of rocks, and the anions, chloride (Cl^-) and sulfate (SO_4^{-2}), from volcanic gases. Still, da Vinci was right in believing that dissolved ions get carried to the sea by flowing groundwater and river water—rivers deliver over 2.5 billion tons of salt every year.

There's so much salt in the ocean that if all the water suddenly evaporated, a 60-m-thick layer of salt would coat the ocean floor. This layer would consist of about 75% halite (NaCl), with lesser amounts of gypsum ($CaSO_4 \cdot H_2O$), anhydrite ($CaSO_4$), and other salts. Oceanographers refer to the concentration of salt in water as "salinity." Although ocean salinity averages 3.5%, measurements from around the world demonstrate that salinity varies with location, ranging from about 1.0% to about 4.1% (▶Fig. 18.8a). Salinity reflects the balance between the addition of freshwater by rivers or rain and the removal of freshwater by evaporation,

for when seawater evaporates, salt stays behind; salinity may also depend on water temperature, for warmer water can hold more salt in solution than can cold water.

Notably, the salinity of the ocean changes with depth. A graph of the variation in salinity with depth (▶Fig. 18.8b) indicates that such differences in salinity are only found in seawater down to a depth of about 1 km. Deeper water tends to be more homogenous. Oceanographers refer to the gradational boundary between surface-water salinities and deep-water salinities as the "halocline."

Temperature

When the *Titanic* sank after striking an iceberg in the North Atlantic, most of the unlucky passengers and crew who jumped or fell into the sea died within minutes, because the seawater temperature at the site of the tragedy approached freezing, and cold water removes heat from a body very rapidly. Yet swimmers can play for hours in the Caribbean, where sea-surface temperatures reach 28°C (83°F). Though the *average* global sea-surface temperature hovers around 17°C, it ranges between freezing near the poles to almost 35°C in restricted tropical seas (▶Fig. 18.8c). The correlation of average temperature with latitude exists because the intensity of solar radiation varies with latitude.

The intensity of solar radiation also varies with the season, so surface seawater temperature varies with the season. But the difference is only around 2° in the tropics, 8° in the temperate latitudes, and 4° near the poles. (By contrast, the seasonal temperature change on land can be much greater—in central Illinois, for example, temperatures may reach 40°C [104°F] in the summer and drop to −32°C [−25°F] in the winter.) The seasonal seawater temperature change remains in a narrow range because water can absorb or release large amounts of heat without changing temperature very much. Thus, the ocean regulates the temperatures of coastal regions; air temperature in Vancouver, on the Pacific coast of Canada, rarely drops below freezing even though it lies farther north than Illinois.

Water temperature in the ocean varies markedly with depth (▶Fig. 18.8d). Waters warmed by the Sun are less dense and tend to remain at the surface. An abrupt "thermocline," below which water temperatures decrease sharply, reaching near freezing at the sea floor, appears at a depth of about 300 m, in the tropics. There is no pronounced thermocline in polar seas, since surface waters there are already so cold.

FIGURE 18.7 The composition of average seawater. The expanded part of the graph shows the proportions of ions in the salt of seawater.

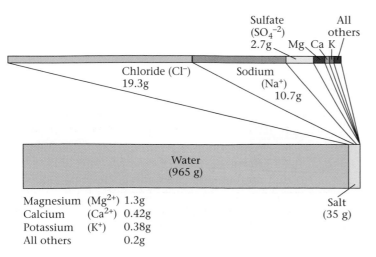

18.4 CURRENTS: RIVERS IN THE SEA

Since first setting sail on the open ocean, people have known that the water of the ocean does not stand still, but rather flows or circulates at velocities of up to several kilometers

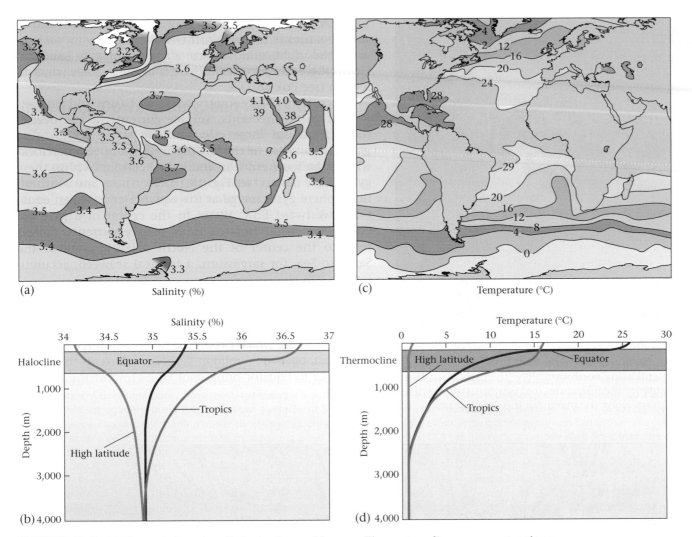

FIGURE 18.8 (a) The variations in salinity in the world ocean. The contour lines represent regions of different salinity (numbers are percentages). (b) The variation of salinity with depth in the ocean. (c) The variation in temperature with latitude. Contours are given in degrees Celsius. (d) The variation in temperature with depth.

per hour in fairly well-defined streams called **currents.** Oceanographic studies made since the *Challenger* expedition demonstrate that circulation in the sea occurs at two levels: surface currents affect the upper hundred meters of water, and deep currents keep even water at the bottom of the sea in motion.

Surface Currents: A Consequence of the Wind

When the skippers of sailing ships planned their routes from Europe to North America, they paid close attention to the directions of surface currents, for sailing against a current slowed down the voyage substantially. If they headed due west at a high latitude, they would find themselves battling an eastward-flowing surface current, the

Gulf Stream. Further, they found that the water moving in a surface current does not flow smoothly but displays some turbulence. Isolated swirls or ring-shape currents of water, called eddies, form along the margins of currents (▶Fig. 18.9).

Surface currents occcur in all the world's oceans (▶Fig. 18.10). They result from interaction between the sea surface and the wind—as moving air molecules shear across the surface of the water, the friction between air and water drags the water along. If we look at a map that shows global wind patterns along with oceanic currents, we can see this relationship (see Fig. 18.10 inset). But the movement of water resulting from wind shear does not exactly parallel the movement of the wind. This is a consequence of Earth's rotation, which generates the **Coriolis effect**

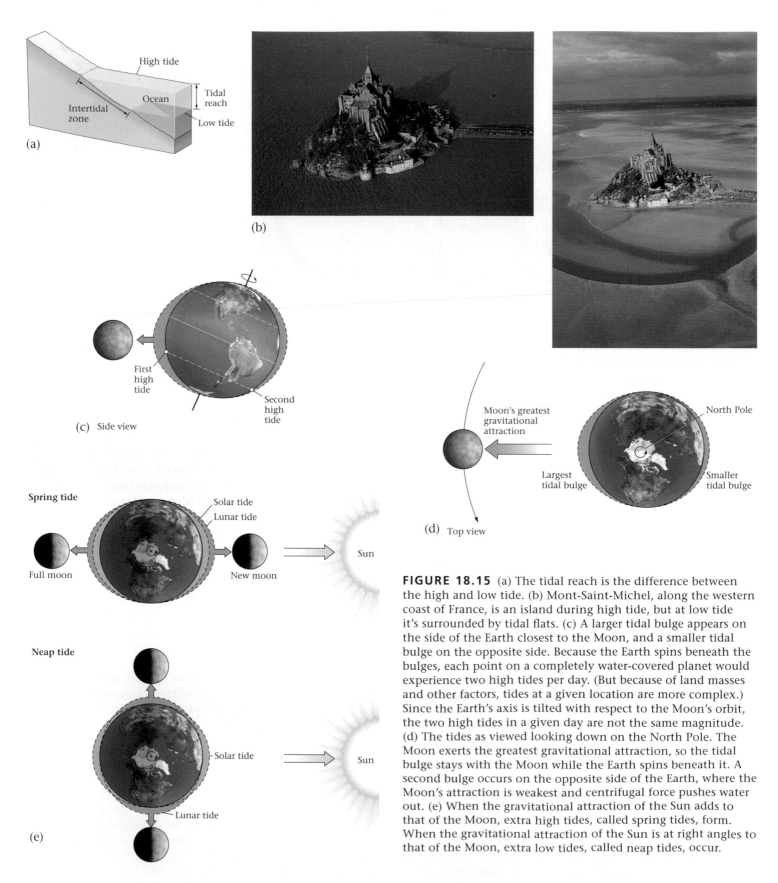

FIGURE 18.15 (a) The tidal reach is the difference between the high and low tide. (b) Mont-Saint-Michel, along the western coast of France, is an island during high tide, but at low tide it's surrounded by tidal flats. (c) A larger tidal bulge appears on the side of the Earth closest to the Moon, and a smaller tidal bulge on the opposite side. Because the Earth spins beneath the bulges, each point on a completely water-covered planet would experience two high tides per day. (But because of land masses and other factors, tides at a given location are more complex.) Since the Earth's axis is tilted with respect to the Moon's orbit, the two high tides in a given day are not the same magnitude. (d) The tides as viewed looking down on the North Pole. The Moon exerts the greatest gravitational attraction, so the tidal bulge stays with the Moon while the Earth spins beneath it. A second bulge occurs on the opposite side of the Earth, where the Moon's attraction is weakest and centrifugal force pushes water out. (e) When the gravitational attraction of the Sun adds to that of the Moon, extra high tides, called spring tides, form. When the gravitational attraction of the Sun is at right angles to that of the Moon, extra low tides, called neap tides, occur.

The Forces Causing Tides

BOX 18.2
SCIENCE TOOLBOX

Fundamentally, tides result from interaction between two forces: gravitational attraction exerted by the Moon and Sun on the Earth, and centrifugal force caused by the revolution of the Earth around the center of mass of the Earth-Moon system. To explain this statement, we must review some key terms from physics.

- **Gravitational pull** is the attractive force that one mass exerts on another. The magnitude of gravitational pull depends on the amount of mass in each object, and on the distance between the two masses.

- **Centrifugal force** is the apparent outward-directed ("center-fleeing") force that material on or in an object feels when the object moves in orbit around a point. Note that centrifugal force differs from *centripetal force,* the "center-seeking" force; this distinction can be confusing. To picture centripetal force, tie a ball to a string and swing it around your head. The string exerts an inward-directed centripetal force on the ball—if the string breaks, the centripetal force ceases to exist and the ball heads off in a straight-line path. To picture centrifugal force, imagine that the ball is hollow and that you've placed a marble inside. As you twirl the ball around your head, the marble moves to the outer edge of the ball. The apparent force pushing the marble outward is the centrifugal force. But as such, centrifugal force is not a real force—it is simply a manifestation of inertia, and it *only* exists from the perspective, or "reference frame," of the orbiting object. (A physics book explains this contrast in greater detail.)

- **Earth-Moon system** refers to this pair of objects viewed as a unit, as they move together through space.

- **Center of mass** is the point within an object, or a group of objects, about which mass is evenly distributed; put another way, it is the location of the average position, or the balance point, of the total mass in a single object or a group of objects. Because the Earth is 81 times more massive than the Moon, the center of mass of the Earth-Moon system actually lies 1,700 km below the surface of the Earth.

First, let's consider the origin of centrifugal force in the Earth-Moon system. To do this, we must consider the way in which the Earth-Moon system moves. The center of the Earth itself does not follow a simple orbit around the Sun. Rather, it is the *center of mass* of the Earth-Moon system that follows this trajectory; the Earth actually spirals around this trajectory as it speeds around the Sun. To picture this motion, imagine that the Earth-Moon system is a pair of dancers, one of whom is much heavier than the other. The dancers face each other, hold hands, and whirl in a circle as they drift across the dance floor (▶Fig. 18.16a). Each dancer's head orbits the center of mass.

Revolution of the Earth around the Earth-Moon system's center of mass generates centrifugal forces on both the Earth and the Moon that would cause the Earth and the Moon to fly away from each other, were it not for the gravitational attraction holding them together. We can see this by looking again at our dancer analogy (▶Fig. 18.16b)—the centrifugal force acting on each dancer points outward, away from his or her partner, and is the *same* for all points on each dancer. We can represent the direction and magnitude of this centrifugal force by arrows called vectors. (A vector is a number that has magnitude and direction.) In this case, the length of the arrow represents the magnitude of the force, and the orientation of the arrow indicates the direction of the force. If we think of the dancers as the Earth and the Moon, then centrifugal force vectors at all points on the surface of the Earth point away from the Moon (▶Fig. 18.17a). On the Earth, therefore, centrifugal force causes the surface of the ocean to bulge outward, away from the center of mass of the Earth-Moon system, on the far side of the Earth.

Now, let's consider how the force of gravity comes into play in causing tides. To simplify this discussion, we only examine the effect of the Moon's gravity on Earth.

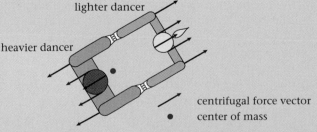

dancer's trajectory across the floor

dancer's head orbits the center of mass

(a)

lighter dancer

heavier dancer

centrifugal force vector
• center of mass

(b)   centrifugal force vectors point outwards; they are the same magnitude for all points on a dancer.

FIGURE 18.16 (a) To picture the Earth-Moon system, imagine two dancers spinning around each other as they move along a straight-line trajectory. The center of mass of the two-dancer system lies closer to the heavier dancer. Each dancer orbits the center of mass. (b) Each point on each dancer feels a centrifugal force (represented by a vector) that points outward.

Vectors representing the magnitude and direction of the Moon's gravitational pull at any point on the surface of the Earth all point toward the center of the Moon. Because the magnitude of gravity depends on distance, the Moon exerts more attraction on the near side of the Earth than at the Earth's center, and less attraction on the far side of the Earth than at the Earth's center. Gravity, therefore, causes the surface of the ocean on the near side of the Earth to bulge toward the Moon.

In the Earth-Moon system, both centrifugal force and gravitational pull operate at the same time. How do they interact? If we draw vectors representing both centrifugal force and gravitational force at various points on or in the Earth, we see that the vectors representing centrifugal force do not have the same length as those representing gravitational attraction, except at the Earth's center. Moreover, the vectors representing centrifugal force do not point in the same direction as the vectors representing gravitational attraction. The force that the ocean water feels is the sum of the two forces acting on the water. You can determine the sum of two vectors by drawing the vectors so they touch head to tail—the sum is the vector that completes the triangle. This sum is the *tide-generating force*, and its magnitude and direction vary with location on the Earth.

Let's look at the tide-generating force a little more closely. On the side of the Earth closer to the Moon, gravitational vectors are larger than centrifugal force vectors, so adding the two leaves a net tide-generating force that pulls the sea surface to bulge toward the Moon. On the side of the Earth further from the Moon, the centrifugal force vectors are larger, so centrifugal force caused by the orbiting of the Earth-Moon system around the center of mass causes the surface of the sea to bulge outward, away from the Moon (▶Fig. 18.17b). Thus, the ocean has two tidal bulges—one on the side close to the Moon, and one on the opposite side of the Earth. The bulge closer to the Moon is larger. Note that tidal bulges have nothing to do with the spinning of the Earth on its axis—this spin has no measurable effect on the sea surface.

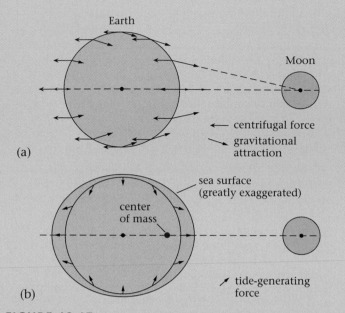

FIGURE 18.17 (a) Each point on the surface of the Earth feels the same centrifugal force (due to the spin of the Earth-Moon system around its center of mass), but feels a different gravitational attraction (due to the pull of the Moon). Centrifugal force vectors pointing away from the Moon are all the same magnitude, but gravitational force vectors pointing toward the center of the Moon, and their magnitude varies with distance from the Moon. (b) Tide-generating force is the sum of the centrifugal force vector and the gravitational force vector. Vectors on the side of the Earth close to the Moon are dominated by gravitational force and thus point toward the Moon, creating a tidal bulge. Vectors on the other side of the Earth are dominated by centrifugal force and point away from the Moon, creating a second tidal bulge.

so the two high tides at the given point are not the same size (Fig. 18.15c).

- *The Moon's orbit:* The moon progresses in its 28-day orbit around the Earth in the same direction as the Earth rotates. High tides arrive 50 minutes later each day because of the difference between the time it takes for Earth to spin on its axis, and the time it takes for the Moon to orbit the Earth.

- *The Sun's gravity:* When the angle between the direction to the Moon and the direction to the Sun is 90°, we experience extra low tides (neap tides) because the Sun's gravitational attraction counteracts the Moon's. When the Sun is on the same side as the Moon, we experience extra high tides (spring tides) because the Sun's attraction *adds* to the Moon's (▶Fig. 18.15e).

- *Focusing effect of bays:* In the open ocean, the maximum tidal reach is only a few meters. But in the Bay of Fundy,

along the eastern coast of Canada, the tidal reach approaches 20 m. In a bay that narrows to a point, like the Bay of Fundy, the flood tide brings a large volume of water into a small area, so the point experiences an especially large high tide.

- *Basin shape:* The shape of the basin containing a portion of the sea influences the sloshing of water back and forth within the basin as tides rise and fall. Depending on the timing and magnitude of this sloshing, this effect can locally add to the global tidal bulge or subtract from it, and thus can affect the rhythm of tides. In some locations, the net effect is to cancel one of the daily tides entirely, so that the locality experiences only one high tide and one low tide in a day.

- *Air pressure:* The effects of air pressure on tides can contribute to disaster. For example, during a hurricane the air pressure drops radically so the sea surface rises;

if the hurricane coincides with a high tide, the "storm surge" (water driven landward by the wind) can inundate the coast.

Because of the complexity of factors contributing to tides, the timing and magnitude of tides vary significantly along the coast. Nevertheless, at a given location the tides are periodic and can be predicted. Tides gave early civilizations a rudimentary way to tell time. In fact, in some languages the word for "tide" is the same as the word for "time."

Notably, friction between ocean water and the ocean floor causes the movement of the tidal bulge to lag slightly behind the movement of the Moon across the Earth. The Moon, therefore, exerts a slight pull on the side of the bulge. This pull acts like a brake and slows the Earth's spin, so that days grow longer at a rate of about 0.002 seconds per century. Over geologic time, the seconds add up; a day was only 21.9 hours long in the Middle Devonian Period (390 Ma). As the spinning Earth slows, the Moon moves farther away. During the Archean Eon (3.8 Ga), the Moon was 15,000 km closer, so the tidal reach on Earth was larger.

18.6 WAVE ACTION

Waves make the ocean surface restless, an ever-changing vista. They develop because of the shear between the molecules of air in the wind and the molecules of water at the surface of the sea. It may seem surprising that so much friction can arise between two fluids, but it can, as Benjamin Franklin demonstrated. Franklin noted that oily waste spilled from ships on a windy day made the water surface smoother. He proposed that the oily coating on the water decreased frictional shear between the air and the water, and therefore prevented waves from forming.

When you watch a wave travel across the open ocean, you may get the impression that the whole mass of water constituting the wave moves with the wave. But drop a cork overboard and watch it bob up and down and back and forth; it does not move along with a wave. Within a wave, away from shore, a particle of water moves in a circular motion, as viewed in cross section. The diameter of the circle is greatest at the ocean's surface, where it equals the amplitude of the wave. With increasing depth, though, the diameter of the circle decreases until, at a depth equal to about half the "wavelength" (the horizontal distance between two wave troughs), there is no wave movement at all (▶Fig. 18.18). Submarines traveling below this **wave base** cruise through smooth water, while ships toss about above.

The character of waves in the open ocean depends on the strength of the wind (how fast the air moves) and on the "fetch" of the wind (over how long a distance it blows). When the wind first begins to blow, it creates rip-

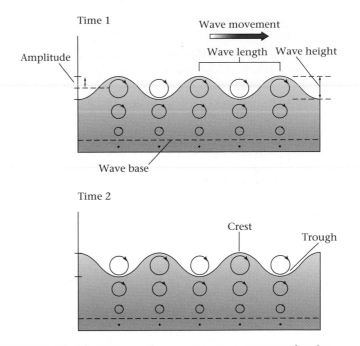

FIGURE 18.18 Within a deep-ocean wave, water molecules follow a circular path. The diameter of the circle decreases with depth to the wave base, below which the wave has no effect. When a wave passes, the shape of the water surface changes, but water does not move as a mass. Note that the amplitude is one-half the wave height.

ples in the water surface, pointed waves whose "amplitude" (the height from rest to crest or rest to trough) and wavelength are small. With continued blowing over a long fetch, swells, larger waves with amplitudes of 2–10 m and wavelengths of 40–500 m, begin to build. Hurricane wave amplitudes may grow to over 25 m. The largest documented swell in the open ocean was 35 m high.

Swells may travel for thousands of kilometers across the ocean, well out of the region where they were created, and thus may interact with waves formed elsewhere. The overlap, or "interference," between comparable-sized swells leads to the formation of larger waves, and some of these can be truly mountainous. One of the most disastrous instances of wave interference took place in 1979 off the coast of Ireland, where over 300 yachts were competing in a race. Waves generated by an east-blowing gale collided with waves generated by a west-blowing gale and built up to an amplitude of over 15 m, capsizing twenty-three yachts and killing fifteen sailors.

Waves have no effect on the ocean floor, as long as the floor lies below the wave base. However, near the shore, where the wave base just touches the floor, it causes a slight back-and-forth motion of sediment. Closer to shore, as the water gets shallower, friction between the wave and the sea floor slows the deeper part of the wave, and the motion in the wave becomes more elliptical. Eventually, water at the top of the wave curves over the base, and the wave becomes a breaker, ready for surfers to ride. Breakers crash onto the shore in the surf zone, sending a surge of water up the beach. This upward surge, or **swash,** continues until friction brings motion to a halt. Then gravity draws the water back down the beach as **backwash** (▶Fig. 18.19).

Waves may make a large angle with the shoreline as they're coming in, but they bend as they approach the shore, a phenomenon called **wave refraction;** right at the shore, their crests make no more than about a 5° angle with the shoreline (▶Fig. 18.20a). To see why this happens, imagine a wave approaching the shore so that its crest makes an angle of 45° with the shoreline. The end of the wave closest to the shore touches bottom first and slows down because of friction, while the end farther offshore still continues to move at its original velocity, swinging the whole wave around so that it's more parallel with the shoreline.

Though refraction decreases the angle at which a wave rolls onto shore, the wave may still arrive at an angle. When the water returns seaward in the backwash, however, it must flow straight down the slope of the beach in response to gravity. Overall, this sawtooth-like flow results in a **longshore current,** which flows parallel to the beach (Fig. 18.20a). Also because of wave refraction, wave energy is focused on headlands (places where higher land protrudes into the sea), and is weaker in embayments (places set back from the sea).

Thus, erosion happens at headlands, forming a cliff, while deposition takes place in embayments, forming a beach (▶Fig. 18.20b, c).

Waves pile water up on the shore incessantly. As the excess water moves back to the sea, it creates a strong, localized seaward flow perpendicular to the beach called a "rip current" (▶Fig. 18.21). Rip currents are the cause of many drownings every year along beaches, because they suddenly carry unsuspecting swimmers away from the beach.

18.7 WHERE LAND MEETS SEA: COASTAL LANDFORMS

Tourists along the Amalfi coast of Italy thrill to the sound of waves crashing on rocky shores. But on the Gulf Coast of Florida, sunbathers lie on seemingly endless white sand beaches next to calm seas. Large dome-like mountains rise directly from the sea in Rio de Janeiro, Brazil, but a 100-m-high vertical cliff marks the boundary between the Nullarbor Plain of southern Australia and the Great Southern Ocean (▶Fig. 18.22a–d). And New Orleans sprawls over a vast plain of former swamp. As these examples illustrate, coasts, the belt of land bordering the sea, vary dramatically in terms of topography and associated landforms (▶Fig. 18.23a–g).

FIGURE 18.19 This profile shows the various landforms of a beach, as well as a cross section of a barrier island. As a wave approaches the shore, it touches the bottom of the sea, at a depth of about half the wavelength. Because of friction, the wave slows down and the wavelength decreases, so the wave height must increase. As the bottom of the wave moves slower than the top, the wave builds up into a breaker that carries water up onto the beach, with the top of the wave falling over the bottom. The water washing up on the beach is swash, and the water rushing back is backwash (indicated by arrows).

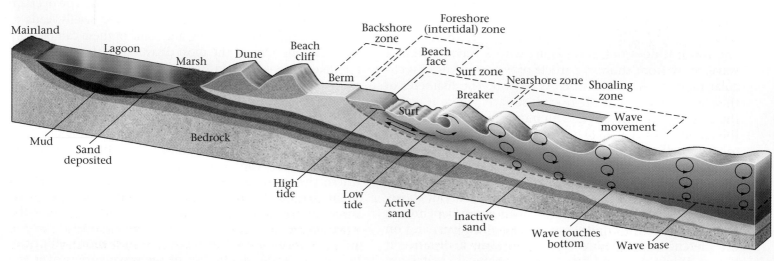

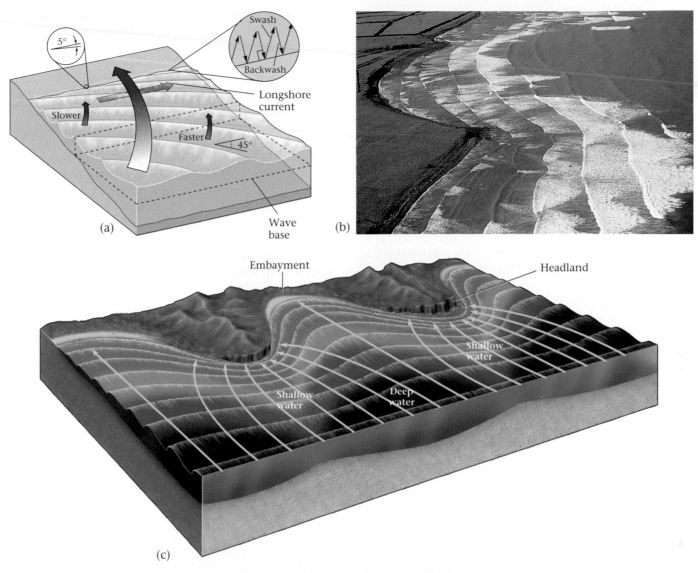

FIGURE 18.20 (a) Wave refraction occurs when waves approach the shore at an angle. The part of the wave that touches bottom first slows down, then the rest of the wave catches up. As a result, the wave bends so that it's nearly parallel with the shore. However, because the wave hits the shore at an angle, water moving parallel to the shore creates a longshore current. (b) Wave refraction on a beach. (c) Like a lens, wave refraction focuses wave energy on a headland, so erosion occurs; and it disperses wave energy in embayments, so deposition occurs.

Beaches and Tidal Flats

For millions of vacationers, the ideal holiday includes a trip to a **beach,** a gently sloping fringe of sediment along the shore. Some beaches consist of pebbles or boulders, whereas others consist of sand grains (▶Fig. 18.24a, b). This is no accident, for waves winnow out finer sediment like silt and mud and carry it to quieter water, where it settles. Storm waves, which can smash cobbles against one another with enough force to shatter them, have little effect on sand, for sand grains can't collide with enough energy to crack. Thus, cobble beaches exist only where nearby cliffs continuously supply large rock fragments.

The composition of sand itself varies from beach to beach, because different sands come from different sources. Sands derived from the weathering and erosion of silicic-to-intermediate rocks consist mainly of quartz; other minerals in these rocks chemically weather to form clay, which

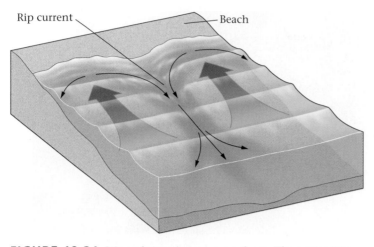

FIGURE 18.21 Waves bring water up on shore. The water may return to sea in a narrow rip current perpendicular to the shore.

washes away in waves. Beaches made from the erosion of limestone or of recent corals and shell beds consist of carbonate sand, including lots of sand-sized chips of shells. And beaches derived by the recent erosion of basalt may have black sand, made of tiny basalt grains.

A "beach profile," a cross section drawn perpendicular to the shore, illustrates the shape of a beach (Fig. 18.19). Starting from the sea and moving landward, a beach consists of a foreshore zone, or intertidal zone, across which the tide rises and falls. The **beach face,** a steeper, concave part of the foreshore zone, forms where the swash of the waves actively scours the sand. The backshore zone extends from a small step, or escarpment, cut by high-tide swash to the front of the dunes or cliffs that lie farther inshore. The backshore zone includes one or more **berms,** horizontal to landward-sloping terraces that received sediment during a storm.

FIGURE 18.22 (a) A rocky shore in eastern Italy. (b) A flat, sandy beach along the Gulf Coast of Florida. (c) The sugar loafs (rounded mountains) rising out of the sea at Rio de Janeiro, Brazil. (d) The abrupt edge of the Nullarbor Plain in South Australia.

(a)

(c)

(b)

(d)

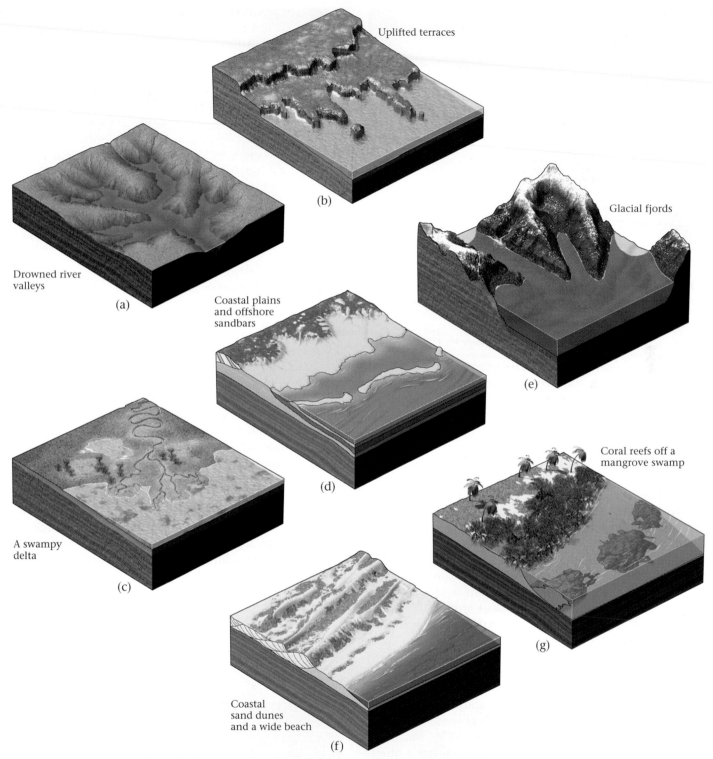

FIGURE 18.23 A wide variety of coastal landforms have developed on Earth. (a) Drowned river valleys, formed where sea level rises and floods valleys, create complex, irregular coastlines. (b) Uplifted terraces develop where the coastline rises relative to sea level, and creates escarpments. (c) Swampy deltas form where a sediment-laden stream deposits sediment along the coast. (d) Along sandy coastal plains, large beaches and offshore bars appear. (e) Glacial fjords develop where sea level rises and floods a glacially carved valley. (f) Coastal dunes form where there is a large sand supply and strong wind. (g) In tropical environments, mangrove swamps grow along the shore, protected from wave action by offshore coral reefs.

(a) (b)

FIGURE 18.24 (a) A pebble and cobble beach, Olympic Peninsula, Washington. The clasts were derived from nearby cliffs. (b) A sand beach, western coast of Puerto Rico.

Geologists commonly refer to beaches as "rivers of sand," to emphasize that beach sand moves along the coast over time—it is not a permanent substrate. Wave action at the shore moves an active sand layer on the sea floor on a daily basis. Inactive sand, buried below this layer, moves only during severe storms or not at all. Where waves hit the beach at an angle, the swash of each successive wave moves active sand up the beach at an angle to the shoreline, but the backwash moves this sand down the beach parallel to the slope of the shore. This sawtooth motion causes sand to gradually migrate along the beaches, a process called **beach drift** (Fig. 18.20a). Beach drift, which happens in association with the long-shore drift of water, can transport sand hundreds of kilometers along a coast in a matter of centuries. Where the coastline indents landward, beach drift stretches beaches out into open water to create a **sand spit.** Some sand spits grow across the opening of a bay, to form a bay-mouth bar (▶Fig. 18.25).

The scouring action of waves piles sand up in a narrow ridge away from the shore called an offshore bar, which parallels the shoreline. In regions with an abundant sand supply, offshore bars rise above the mean high-water level and become **barrier islands** (▶Fig. 18.26a). The water between a barrier island and the mainland becomes a quiet-water **lagoon,** a body of shallow seawater separated from the open ocean.

Though developers have covered some barrier islands with expensive resorts, in the time frame of centuries to millennia barrier islands are temporary features. For example, wind and waves pick up sand from the ocean side of the barrier island and drop it on the lagoon side, causing the island to migrate landward. Storms may breach barrier islands and create an inlet (a narrow passage of water). Finally, beach drift gradually transports the sand of barrier islands and modifies their shape.

Tidal flats, regions of mud and silt exposed or nearly exposed at low tide but totally submerged at high tide, develop in regions protected from strong wave action (Fig. 18.15b; ▶Fig. 18.26b). They are typically found along the margins of lagoons or on shores protected by barrier islands. Here, mud and silt accumulate to form thick, sticky layers. In tidal flats that provide a home for burrowing organisms like clams and worms, **bioturbation** ("stirred by life") mixes sediments together.

FIGURE 18.25 Beach drift can generate sand spits and bay-mouth bars. Sedimentation fills in the region behind a baymouth bar. As a result, the shoreline gets smoother with time.

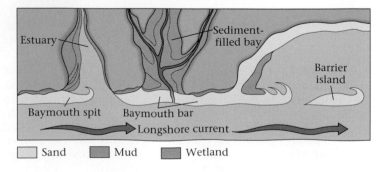

(a)

(b)

FIGURE 18.26 (a) The barrier islands of the Outer Banks off the coast of North Carolina, as viewed by *Apollo 9* astronauts. The white dots are clouds. (b) Tidal flats are broad muddy areas submerged only at high tide. At low tide, boats at anchor rest in the mud of this tidal flat along the coast of Wales.

FIGURE 18.27 The sediment budget along a coast. Sediment is brought into the system by rivers, by the erosion of cliffs and moraines, and by wind. Sediment moves along the coast as a result of beach drift. And sediment leaves the system by being blown off the beach, by sinking into deeper water, or by being carried out by the longshore current.

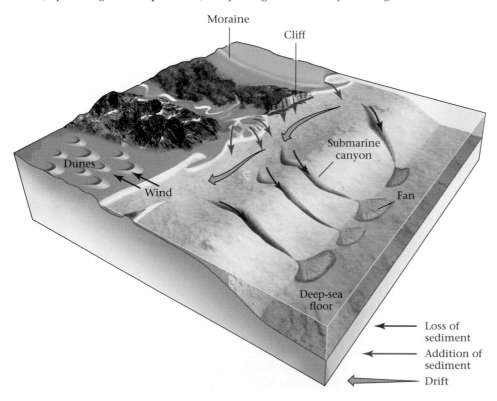

Because of the movement of sediment, the "sediment budget" (the difference between sand supplied and sand removed) plays an important role in determining the long-term evolution of a beach. Let's look at how the budget works for a small segment of beach (▶Fig. 18.27). Sand may be supplied to the segment from local rivers or by wind from nearby dune fields; it may also be brought from just offshore by waves or from far away by beach drift. (In fact, the large quantity of sand along beaches of the southeastern United States may have originated in Pleistocene glacial outwash far to the north.) Some of the sand from a stretch of beach may be removed by beach drift, while some gets carried offshore by waves, where it either settles locally or tumbles down a submarine canyon into the deep sea. If the lost sand cannot be replaced, the beach segment grows narrower, whereas if the supply of sand exceeds the amount that washes away, the beach becomes wider.

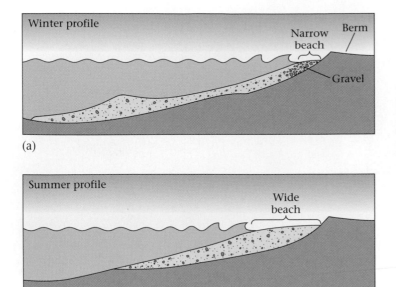

FIGURE 18.28 (a) In the winter, when waters are stormier, sand moves offshore, and the beach narrows and may become stonier. (b) During the summer, waves bring sand back to replenish the beach.

In temperate climates, winter storms tend to be stronger and more frequent than summer ones. The larger, shorter-wavelength waves of winter storms wash beach sand into deeper water and thus make the beach narrower, while the smaller, longer-wavelength summer waves bring sand in from offshore and deposit it on the beach (▶Fig. 18.28a, b).

Rocky Coasts

More than one ship has met its end smashed and splintered in the spray and thunderous surf of a rocky coast, where bedrock cliffs rise directly from the sea (Fig. 18.22a; ▶Fig. 18.29a). Lacking the protection of a beach, rocky coasts feel the full impact of ocean breakers. The water pressure generated during the impact of a breaker can pick up boulders and smash them together until they shatter, and it can squeeze air into cracks, creating enough force to widen them. Further, because of its turbulence, the water hitting a cliff face carries suspended sand, and thus can abrade the cliff. The combined effects of shattering, wedging, and abrading, together called wave erosion, gradually undercut a cliff face and make a **wave-cut notch** (▶Fig. 18.29b, c). Undercutting continues until the overhang becomes unstable and breaks away at a joint, creating a pile of rubble at the base of the cliff that waves immediately attack and break up. By this process, wave erosion cuts away at a rocky coast, so that the cliff gradually migrates inland. Such cliff retreat leaves behind a **wave-cut bench,** or platform, which becomes visible at low tide (▶Fig. 18.29d).

Other processes besides wave erosion break up the rocks along coasts. For example, salt spray coats the cliff face above the waves and infiltrates into pores. When the water evaporates, salt crystals grow and push apart the grains, thereby weakening the rock. Biological processes also contribute to erosion, for plants and animals in the intertidal zone bore into the rocks and gradually break them up.

Many rocky coasts start out with an irregular coastline, with headlands protruding into the sea and embayments set back from the sea. Such irregular coastlines tend to be temporary features in the context of geologic time: wave energy focuses on headlands and disperses in embayments (a result of wave refraction), so that erosion removes debris at headlands and sediment accumulates in embayments (Fig. 18.21c); thus, over time the shoreline becomes less irregular.

A headland erodes in stages (▶Fig. 18.30a–c). Because of refraction, waves curve and attack the sides of a headland, slowly eating through it to create a **sea arch** connected to the mainland by a narrow bridge (▶Fig. 18.31a). Eventually the arch collapses, leaving isolated **sea stacks** just offshore (▶Fig. 18.31b). Once formed, a sea stack protects the adjacent shore from waves. Therefore, sand collects in the lee of the stack, slowly building a **tombolo,** a narrow ridge of sand that links the sea stack to the mainland.

Coastal Wetlands

Let's move now from the crashing waves of rocky coasts to the gentlest type of shore, the **coastal wetland,** a vegetated flat-lying stretch of coast that floods with shallow water but does not feel the impact of strong waves. In temperate climates, coastal wetlands include swamps (wetlands dominated by trees), marshes (wetlands dominated by grasses; ▶Fig. 18.32a), and bogs (wetlands dominated by moss and shrubs). So many marine species use wetlands to spawn that despite their relatively small area when compared with the oceans as a whole, wetlands account for 10–30% of marine organic productivity.

In tropical or semitropical climates (between 30° north and 30° south of the equator), mangrove swamps thrive in wetlands (▶Fig. 18.32b). Mangrove tree roots can filter salt out of water, so the trees have evolved the ability to survive in freshwater or saltwater. Some mangrove species form a broad network of roots above the water surface, making the plant look like an octopus standing on its tentacles, and some send up small protrusions from roots that rise above the water and allow the plant to breathe. Dense stands of

FIGURE 18.29 (a) The major landforms of a rocky shore include cliffs, sea caves, wave-cut notches, sea stacks, sea arches, wave-cut benches, and tombolos. Beaches tend to collect in embayments, while erosion happens at headlands. (b) Erosion by waves creates a wave-cut notch. Eventually, the overhanging rock collapses into the sea to form gravel on the wave-cut bench. (c) A wave-cut notch exposed along a rocky shore. (d) A wave-cut bench at the foot of the cliffs at Etrétat, France.

mangroves counter the effects of stormy weather and thus prevent coastal erosion.

Estuaries

Along some coastlines, a relative rise in sea level causes the sea to flood river valleys that merge with the coast, resulting in **estuaries,** where seawater and river water mix. You can recognize an estuary on a map by the dendritic pattern of its river-carved coastline (▶Fig. 18.33). Oceanic and fluvial water interact in two ways within an estuary. In quiet estuaries, protected from wave action or river turbulence, the water becomes stratified, with denser oceanic saltwater flowing upstream as a wedge beneath less dense fluvial freshwater. Such saltwater wedges migrate about 100 km up the Hudson River in New York, and about 40 km up the Columbia River in Oregon. In turbulent estuaries, like Chesapeake Bay, oceanic and fluvial water combine to create nutrient-rich brackish water with a salinity between that of oceans and rivers. Estuaries are complex ecosystems inhabited by unique species of shrimp, clams, oysters, worms, and fish that can tolerate large changes in salinity.

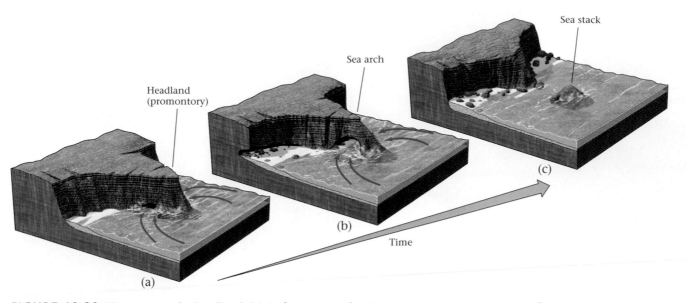

FIGURE 18.30 The erosion of a headland. (a) At first, wave refraction causes wave energy to attack the sides of a promontory, making a sea cave on either side. (b) Gradually erosion breaks through the promontory to create a sea arch. (c) The arch finally collapses, leaving a sea stack.

Fjords

During the last ice age, glaciers carved deep valleys in coastal mountain ranges. When the ice age came to a close, the glaciers melted away, leaving deep, U-shaped valleys (see Chapter 22). The water stored in the glaciers, along with the water within the vast ice sheets that covered continents during the ice age, flowed back into the sea and caused sea level to rise.

The rising sea filled the deep valleys, creating **fjords,** or flooded glacial valleys. Coastal fjords are fingers of the sea surrounded by mountains; because of their deep-blue water and steep walls of polished rock, they are distinctively beautiful (▶Fig. 18.34a, b). Some of the world's most spectacular fjords decorate the western coasts of Norway, British Columbia, and New Zealand. Smaller examples appear along the coast of Maine and southeastern Canada.

FIGURE 18.31 (a) A sea arch exposed along a rocky coast of southern Australia. Another arch once bridged the gap to the cliff on the left, but it collapsed, stranding several tourists. (b) These sea stacks along the southern coast of Australia, together with eight others, comprise a tourist attraction called the Twelve Apostles.

(a)

(b)

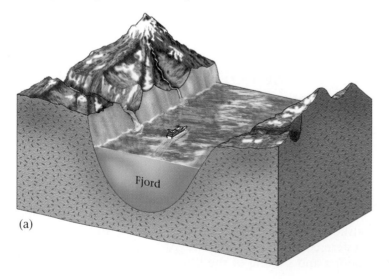

FIGURE 18.32 Examples of coastal wetlands: (a) a salt marsh, (b) a mangrove swamp.

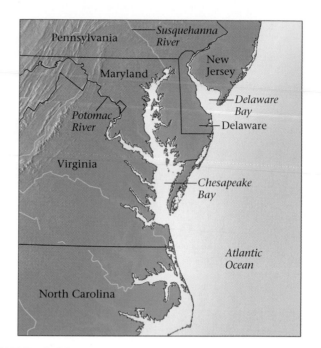

FIGURE 18.33 Chesapeake Bay, a large estuary along the East Coast of the United States, formed when sea level rose and flooded the Potomac and Susquehanna river valleys and the mouths of their tributaries.

Coral Reefs

In the Undersea National Park of the Virgin Islands, visitors swim through colorful growths of living coral (▶Fig. 18.35a). Some corals look like brains, others like elk antlers, still others like delicate fans. Sea anemones, sponges, and clams grow on and around the coral. Though at first glance coral looks like a plant, it is actually a colony of tiny invertebrates

FIGURE 18.34 (a) The subsurface shape of a fjord, a drowned U-shaped glacial valley. (b) Fjords in Norway have spectacular scenery.

(a)

(b)

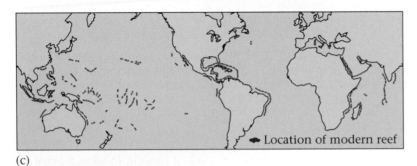

(c)

FIGURE 18.35 (a) Corals and other organisms make up a reef. (b) A coral reef bordering an island in the western Pacific. (c) The distribution of coral reefs on Earth today.

related to jellyfish. An individual coral animal, or polyp, has a tube-like body with a head of tentacles. Corals obtain part of their livelihood by filtering nutrients out of seawater; the remainder comes from algae that live on the corals' tissue. Corals have a symbiotic (mutually beneficial) relationship with the algae, in that the algae photosynthesize and provide nutrients and oxygen to the corals, while the corals provide carbon dioxide and nutrients for the algae.

Coral polyps secrete calcite shells, which gradually build into a mound of solid limestone, whose top surface lies from just below the low-tide level down to a depth of about 60 m. At any given time, only the surface of the mound lives—the mound's interior consists of shells from previous generations of coral. The realm of shallow water underlaid by coral mounds, associated organisms, and debris comprises a **coral reef** (▶Fig. 18.35b). Reefs absorb wave energy and thus serve as a living buffer zone that protects coasts from erosion. Corals need clear, well-lit, warm (18°–30°C) water with normal oceanic salinity, so coral reefs only grow along clean coasts at latitudes of less than about 30° (▶Fig. 18.35c).

Marine geologists distinguish three different kinds of coral reef, based on their geometry (▶Fig. 18.36a–c). A fringing reef forms directly along the coast, a barrier reef develops offshore (separated from the coast by a lagoon), and an atoll

makes a circular ring surrounding a lagoon. As Charles Darwin first recognized back in 1859, coral reefs associated with islands in the Pacific start out as fringing reefs and then later become barrier reefs and finally atolls. Darwin suggested, correctly, that this progression reflects the continued growth of the reef as the island around which it formed gradually sinks. Eventually, the reef itself sinks too far below sea level to remain alive and becomes the cap of a guyot.

18.8 CAUSES OF COASTAL VARIABILITY

Plate Tectonic Setting

The tectonic setting of a coast plays a role in determining whether the coast has steep-sided mountain slopes or a broad plain that borders the sea (see art on pp. 588–589). Along an active margin, compression squeezes the crust and pushes it up, creating mountains like the Andes along the western coast of South America. Along a passive margin, the cooling and sinking of the lithosphere may create a broad

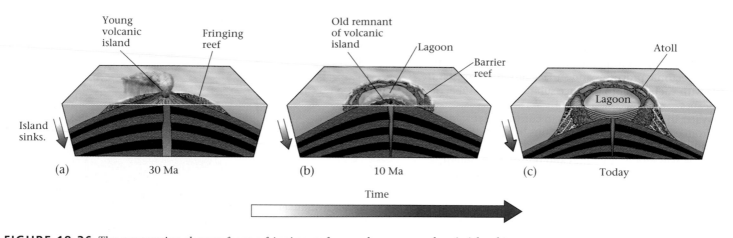

FIGURE 18.36 The progressive change from a fringing reef around a young volcanic island to a ring-shaped atoll. (a) The reef begins to grow around the volcano. (b) The volcano subsides as the sea floor under it ages, so the reef is now a ring, separated from a small island (the peak of the volcano) by a lagoon. (c) The volcano has subsided completely, so that only an atoll surrounding a lagoon remains. When the lagoon fills with debris, and together with the atoll finally sinks below sea level, the result is a guyot.

coastal plain, a flat land that merges with the continental shelf, as exists along the Gulf Coast and southeastern Atlantic coast of the United States. But not all passive margins have coastal plains. At some, the margin of the rift that gave birth to the passive margin remains at a high elevation, even tens of millions of years after rifting ceased. For example, highlands formed during recent rifting border the Red Sea, while highlands formed during Cretaceous rifting persist along portions of the Brazilian coast (Fig. 18.22c).

Relative Sea-Level Changes (Emergent and Submergent Coasts)

Sea level, relative to the land surface, changes during geologic time. Some changes develop due to vertical movement of the land. These may reflect plate tectonic processes or the addition or removal of a load (such as a glacier) on the crust. Locally, changes in sea level reflect human activity—when people pump out groundwater, for example, the pores between grains in the sediment beneath the ground collapse, and the land surface sinks (see Chapter 19). Some relative sea-level changes, however, are due to a *global* rise or fall of the ocean surface. Such **eustatic sea-level changes** may reflect changes in the volume of mid-ocean ridges—an increase in the number or size of ridges, for example, displaces water and causes sea level to rise. Eustatic sea-level changes may also reflect changes in the volume of glaciers, for glaciers store water on land (▶Fig. 18.37).

Geologists refer to coasts where the land is rising or rose relative to sea level as **emergent coasts.** At emergent coasts, steep slopes typically border the shore. A series of step-like terraces form along some emergent coasts (▶Fig. 18.38a, b). These terraces reflect episodic changes in relative sea level.

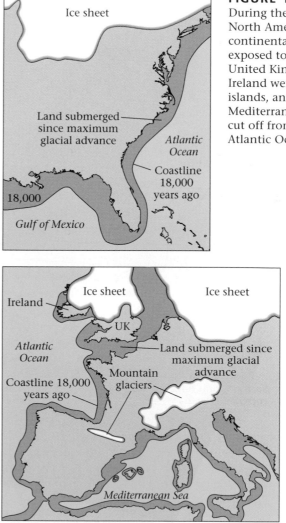

FIGURE 18.37 During the last ice age, North America's continental shelf lay exposed to the air, the United Kingdom and Ireland were not islands, and the Mediterranean Sea was cut off from the Atlantic Ocean.

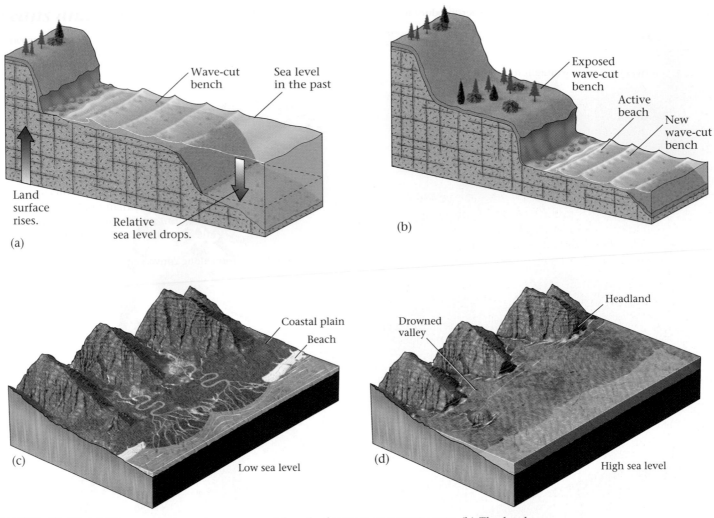

FIGURE 18.38 (a) Wave erosion creates a wave-cut bench along an emergent coast. (b) The land rises, and the bench becomes a terrace. (c) A coast before sea level rises. Rivers drain valleys onto a coastal plain. (d) As a submergent coast forms, sea level rises and floods the valleys, and waves erode the headlands.

Those coasts at which the land sinks relative to sea level become **submergent coasts** (▶Fig. 18.38c, d). At submergent coasts, landforms include estuaries and fjords that developed when the sea flooded coastal valleys. Many of the coastal landforms of eastern North America represent the consequences of submergence.

Sediment Supply and Climate

The quantity and character of sediment supplied to a shore affects its character. That is, coastlines where the sea washes sediment away faster than it can be supplied (erosional coasts) recede landward and may become rocky, while coastlines that receive more sediment than erodes away (accretionary coasts) grow seaward and develop broad beaches.

Climate also affects the character of a coast. Shores that enjoy generally calm weather erode less rapidly than those constantly subjected to ravaging storms. A sediment supply large enough to generate an accretionary coast in a calm environment may be insufficient to prevent the development of an erosional coast in a stormy environment. The climate also affects biological activity along coasts. For example, in the warm water of tropical climates, mangrove swamps flourish along the shore, and coral reefs form offshore. The reefs may build into a broad carbonate platform such as appears in the Bahamas today. In cooler climates, salt marshes develop, while in arctic regions, the coast may be a stark environment of lichen-covered rock and barren sediment.

18.9 COASTAL PROBLEMS AND SOLUTIONS

Contemporary Sea-Level Changes

People tend to view a shoreline as a permanent entity. But in fact, shorelines are ephemeral geologic features. On a time scale of hundreds to thousands of years, a shoreline moves inland or seaward depending on whether relative sea level rises or falls. In places where sea level is rising today, shoreline towns will eventually be submerged. For example, the Persian Gulf now covers about twice the area that it did 4,000 years ago. And if present rates of sea-level rise along the East Coast of the United States continue, major coastal cities like Washington, New York, Miami, and Philadelphia may be inundated within the next millennium (▶Fig. 18.39).

Beach Destruction—Beach Protection?

In a matter of hours, a hurricane can radically alter a landscape that took centuries or millennia to form. The backwash of storm waves sweeps vast quantities of sand seaward, leaving the beach a skeleton of its former self. The surf submerges barrier islands and shifts them toward the lagoon. Waves and wind together rip out mangrove swamps and salt marshes and fragment coral reefs, thereby destroying the organic buffer that normally protects the coast and leaving it vulnerable to erosion for years to come. Of course, major storms also destroy human constructions: erosion undermines shoreside buildings, causing them to collapse into the sea; wave impacts smash buildings to bits; and the storm surge—very high water levels created when storm winds push water toward the shore—floats buildings off their foundations (▶Fig. 18.40a, b).

But even less dramatic events, such as the loss of river sediment, a gradual rise in sea level, a change in the shape of a shoreline, or the destruction of coastal vegetation, can alter the balance between sediment accumulation and sediment removal from a beach, leading to "beach erosion." In some places, beaches retreat landward at rates of 1–2 m per year, forcing homeowners to pick up and move their houses. Even large lighthouses have been moved to keep them from washing away or tumbling down eroded headlands.

In many parts of the world, beachfront property has great value; but if a hotel loses its beach sand, it probably won't stay in business. Thus, property owners often construct artificial barriers to "protect" their stretch of coastline, or to shelter the mouth of a harbor from waves. These barriers alter the natural movement of sand in the beach system and thus change the shape of the beach, sometimes with undesirable results. For example, people may build "groins," concrete or stone walls protruding perpendicular to the shore, to prevent beach drift from removing sand

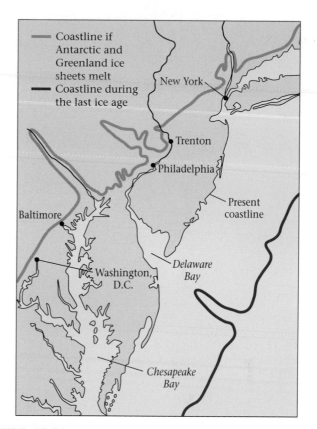

FIGURE 18.39 A possible sea-level rise in the future may flood major cities of the northeastern United States. The Washington–New York corridor would lie underwater.

(▶Fig. 18.41a). Sand accumulates on the updrift side of the groin, forming a long triangular wedge, but sand erodes away on the downdrift side. Needless to say, the property owner on the downdrift side doesn't appreciate this process.

A pair of walls called "jetties" may protect the entrance to a harbor (▶Fig. 18.41b). But jetties erected at the mouth of a river channel effectively extend the river into deeper water, and thus may lead to the deposition of an offshore sandbar. Engineers may also build an offshore wall called a "breakwater," parallel or at an angle to the beach, to prevent the full force of waves from reaching a harbor. With time, however, sand builds up in the lee of the breakwater and the beach grows seaward, clogging the harbor (▶Fig. 18.41c). And to protect expensive shoreside homes, people build "seawalls," out of riprap (large stone or concrete blocks) or reinforced concrete, on the landward side of the backshore zone. Seawalls reflect wave energy that crosses the beach back to sea. Unfortunately, this process increases the rate of erosion at the foot of the seawall, and thus during a large storm the seawall may be undermined so that it collapses (▶Fig. 18.42).

In some places, people have given up trying to decrease the rate of beach erosion, and instead have worked

(a)

(b)

FIGURE 18.40 (a) Damaged beachfront homes after a hurricane in Florida. (b) Hurricane Isabel (September 2003) eroded away the foundation of this house at Kitty Hawk, North Carolina, causing it to collapse.

FIGURE 18.41 (a) The construction of groins creates a sawtooth beach. (b) Jetties extend a river farther into the sea, but may result in the deposition of a sandbar at the jetties' ends. (c) A breakwater causes the beach to build out in the lee.

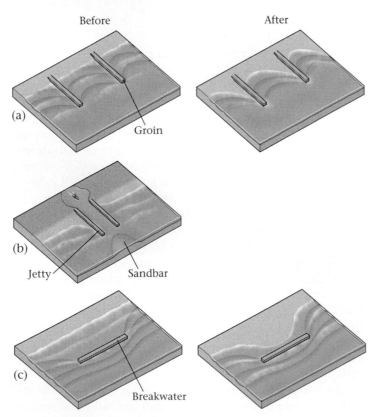

to increase the rate of sediment supply. To do this, they truck or ship in vast quantities of sand to replenish a beach. This procedure, called **beach nourishment,** can be hugely expensive and at best provides only a temporary fix, for the backwash and beach drift that removed the sand in the first place continue unabated as long as the wind blows and the waves break. Clearly, beach management remains a controversial issue, for beachfront properties are expensive, but the shore is, geologically speaking, a temporary feature whose shape can change radically with the next storm.

Pollution and the Destruction of Organic Coasts

Bad cases of beach pollution create headlines. Because of beach drift, garbage dumped in the sea in an urban area may drift along the shore and be deposited on a tourist beach far from its point of introduction. For example, hospital waste from New York City has washed up on beaches tens of kilometers to the south. Oil spills—most commonly from ships that flush their bilges but also from tankers that have run aground or foundered in stormy seas—have contaminated shorelines at several places around the world.

Coasts in which living organisms control landforms along the shore are called **organic coasts.** These coasts, a manifestation of interaction between the physical and biological components of the Earth System, are particularly susceptible to changes in the environment. The loss of

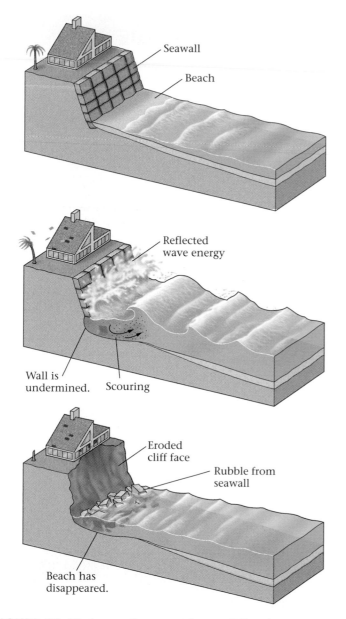

FIGURE 18.42 A seawall protects the sea cliff under most conditions, but during a severe storm the wave energy reflected by the seawall helps scour the beach. As a result, the wall may be undermined and collapse.

such landforms can increase a coast's vulnerability to erosion and, considering that they provide spawning grounds for marine organisms, can upset the food chain of the global ocean.

In wetlands and estuaries, sewage, chemical pollutants, and agricultural runoff cause havoc. Toxins settle along with clay and concentrate in the sediments, where they contaminate burrowing marine life and then move up the food chain. Fertilizers and sewage that enter the sea with runoff increase the nutrient content of water, creating algae blooms that absorb oxygen and therefore kill animal and plant life. Coastal wetlands face destruction by development—they have been filled or drained to be converted into farmland or suburbia, and have been used as garbage dumps. In most parts of the world, between 20 and 70% of coastal wetlands were destroyed in the last century.

Reefs, which depend on the health of delicate coral polyps, can be devastated by even slight changes in the environment. Pollutants and hydrocarbons, for example, will poison them. Organic sewage fosters algae blooms that rob water of dissolved oxygen and suffocate the coral. And agricultural runoff or suspended sediment introduced to coastal water during beach-nourishment projects reduces the light, killing the algae that live in the coral, and clogs the pores that coral polyps use to filter water. Changes in water temperature or salinity caused by dumping waste water from power plants into the sea or by global warming of the atmosphere also destroy reefs, for reef-building organisms

THE VIEW FROM SPACE The Great Barrier Reef fringes the northeast coast of Australia. It has been built from the shells of corals and other marine animals. This reef is the largest in the world—it forms the living coast of a continent.

are very sensitive to temperature changes. People can destroy reefs directly by dragging anchors across reef surfaces, by touching reef organisms, or by quarrying reefs to obtain construction materials. In the last decade, marine biologists have noticed that reefs around the world have lost their color and died. This process, called **reef bleaching,** may be due to the removal or death of symbiotic algae, in response to the warming of seawater triggered by El Niño, or may be a result of the dust carried by winds from desert or agricultural areas.

CHAPTER SUMMARY

- The landscape of the sea floor depends on the character of the underlying crust. Wide continental shelves form over passive-margin basins, while narrow continental shelves form over accretionary prisms. Continental shelves are cut by submarine canyons. Abyssal plains develop on old, cool oceanic lithosphere. Seamounts and guyots form above hot spots.

- The salinity, temperature, and density of seawater vary with location and depth.

- Water in the oceans circulates in currents. Surface currents are driven by the wind and are deflected in their path by the Coriolis effect. The vertical upwelling and downwelling of water create deep currents. Some of this movement is thermohaline circulation, a consequence of variations in temperature and salinity.

- Tides—the daily rise and fall of sea level—are caused by a tide-generating force, mostly driven by the gravitational pull of the Moon.

- Waves are caused by friction where the wind shears across the surface of the ocean. Water particles follow a circular motion, in a vertical plane, as a wave passes. Waves refract (bend) when they approach the shore because of frictional drag with the sea floor.

- Sand on beaches moves with the swash and backwash of waves. If there is a longshore current, the sand gradually moves along the beach and may extend outward from headlands to form sand spits.

- At rocky coasts, waves grind away at rocks, yielding such features as wave-cut beaches and sea stacks. Some shores are wetlands, where marshes or mangrove swamps grow. Coral reefs grow along coasts in warm, clear water.

- The differences in coasts reflect their tectonic setting, whether sea level is rising or falling, the sediment supply, and the climate.

- To protect beach property, people build groins, jetties, breakwaters, and seawalls.

- Human activities have led to the pollution of coasts. Reef bleaching has become dangerously widespread.

KEY TERMS

abyssal plain (p. 563)
active continental margin (p. 564)
backwash (p. 576)
barrier island (p. 580)
bathymetry (p. 562)
beach (p. 577)
beach drift (p. 580)
beach face (p. 578)
beach nourishment (p. 592)
berm (p. 578)
bioturbation (p. 580)
center of mass (p. 573)
centrifugal force (p. 573)
coast (p. 562)
coastal plain (p. 587)
coastal wetland (p. 582)
continental rise (p. 563)
continental shelf (p. 562)
continental slope (p. 563)
coral reef (p. 586)
Coriolis effect (p. 567)
currents (p. 567)
Earth-Moon system (p. 573)
emergent coasts (p. 587)
estuaries (p. 583)
eustatic sea-level change (p. 587)
fjord (p. 584)
gravitational pull (p. 573)

guyot (p. 565)
gyre (p. 568)
intertidal zone (p. 571)
lagoon (p. 580)
longshore current (p. 576)
organic coasts (p. 592)
passive continental margin (p. 563)
pelagic sediment (p. 565)
reef bleaching (p. 594)
sand spit (p. 580)
sea arch (p. 582)
sea stack (p. 582)
seamount (p. 565)
submarine canyons (p. 564)
submarine fan (p. 565)
submergent coasts (p. 590)
swash (p. 576)
thermohaline circulation (p. 570)
tidal flat (p. 571)
tidal reach (p. 571)
tide (p. 570)
tide-generating force (p. 571)
tombolo (p. 582)
wave base (p. 575)
wave refraction (p. 576)
wave-cut bench (platform) (p. 582)
wave-cut notch (p. 582)

REVIEW QUESTIONS

1. How much of the Earth's surface is covered by oceans? How much of the world's population lives near a coast?

2. Describe the typical topography of a passive continental margin, from the shoreline to the abyssal plain.

3. How do the shelf and slope of an active continental margin differ from those of a passive margin?

4. Where does the salt in the ocean come from? How does the salinity in the ocean vary?

5. What factors control the direction of surface currents in the ocean? What is the Coriolis effect, and how does it affect oceanic circulation? Explain thermohaline circulation.

6. What causes the tides?

7. Describe the motion of water molecules in a wave. How does wave refraction cause longshore currents?

8. Describe the components of a beach profile.

9. How does beach sand migrate as a result of longshore currents? Explain the sediment budget of the coast.

10. Describe how waves affect a rocky coast, and how such coasts evolved.

11. What is an estuary? Why is it such a delicate ecosytem? What is the difference between an estuary and a fjord?

12. Discuss the different types of coastal wetlands. Describe the different kinds of reefs, and how a reef surrounding an oceanic island changes with time.

13. How do plate tectonics, sea-level changes, sediment supply, and climate change affect the shape of a coastline? Explain the difference between emergent and submergent coasts.

14. In what ways do people try to modify or "stabilize" coasts? How do the actions of people threaten the natural systems of coastal areas?

SUGGESTED READING

Ballard, R. D., and W. Hively. 2002. *The Eternal Darkness: A Personal History of Deep-Sea Exploration.* Princeton, N.J.: Princeton University Press.

Bird, E. C. 2001. *Coastal Geomorphology: An Introduction.* New York: Wiley.

Davis, R. A. 1997. *The Evolving Coast.* New York: Holt.

Dean, R. G., and R. A. Dalrymple. 2001. *Coastal Processes with Engineering Applications.* Cambridge, England: Cambridge University Press.

Erickson, J. 2003. *Marine Geology.* London: Facts on File.

Garrison, T. 2002. *Oceanography: An Invitation to Marine Science,* 4th ed. Pacific Grove, Calif.: Wadsworth/Thomson.

Komar, P. D. 1997. *Beach Processes and Sedimentation.* 2nd ed. Upper Saddle River, N.J.: Pearson.

Kunzig, R. 1999. *The Restless Sea: Exploring the World beneath the Waves.* New York: Norton.

Seibold, E., and W. H. Berger. 1995. *The Sea Floor: An Introduction to Marine Geology.* 3rd ed. New York: Springer-Verlag.

Sverdrup, K. A., A. B. Duxbury, and A. C. Duxbury. 2002. *An Introduction to the World's Oceans.* 7th ed. New York: McGraw-Hill.

Viles, H., and T. Spencer. 1995. *Coastal Problems: Geomorphology, Ecology and Society at the Coast.* New York: Wiley.

Woodroffe, C. D. 2002. *Coasts: Form, Process and Evolution.* Cambridge, England: Cambridge University Press.

CHAPTER 19

A Hidden Reserve: Groundwater

19.1 INTRODUCTION

Imagine Rosa May Owen's surprise when, on May 8, 1981, she looked out her window and discovered that a large sycamore tree in the backyard of her Winter Park, Florida, home had suddenly disappeared. It wasn't a particularly windy day, so the tree hadn't blown over—it had just vanished! When Owen went outside to investigate, she found that more than the tree had disappeared. Her whole backyard had become a deep, gaping hole. The hole continued to grow for a few days until finally it swallowed Owen's house and six other buildings, as well as the deep end of the municipal swimming pool (and all of the pool's water), part of a road, and the stock of expensive Porsches in a car dealer's lot (▶Fig. 19.1a).

What had happened in Winter Park? The bedrock beneath the town consists of limestone, a fairly soluble rock. **Groundwater,** the water that resides under the surface of the Earth, had gradually dissolved the limestone, carving open rooms, or caverns, underground. On May 8, the roof of a cavern underneath Owen's backyard began to collapse, forming a circular depression called a **sinkhole.** The sycamore tree and the rest of the neighborhood simply dropped down into the sinkhole. It would have taken too much effort to fill in the hole with soil, so the commu-

Groundwater can turn a desert green, as shown by these circular irrigated fields sprouting in the sands of Jordan. A water well lies at the center of each circle.

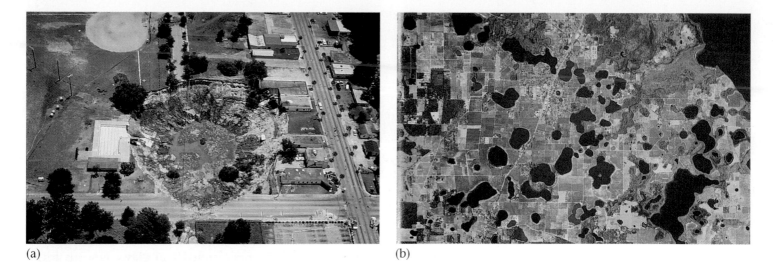

(a)   (b)

FIGURE 19.1 (a) The Winter Park, Florida, sinkhole as seen from a helicopter; (b) numerous sinkhole lakes dot central Florida, as seen from high altitude.

nity allowed it to fill with water, and now it's a circular lake, the centerpiece of a pleasant municipal park. Similar lakes appear throughout central Florida (▶Fig. 19.1b).

The Winter Park sinkhole serves as one of the more dramatic reminders that significant quantities of water reside underground. Whereas we can easily see Earth's surface water (in lakes, rivers, streams, marshes, and oceans) and atmospheric water (in clouds and rain), groundwater lies hidden beneath the surface in the pores and cracks found within sediment or rock. Nevertheless, groundwater has increasingly become a major supply of water for homes, agriculture, and industry. This chapter provides a basic picture of where groundwater comes from, how it flows, and how it interacts with the rock and sediment it flows through.

19.2 WHERE DOES GROUNDWATER RESIDE?

The Underground Reservoir of the Hydrologic Cycle

As we saw in Interlude E, water moves among various reservoirs (the ocean, the atmosphere, rivers and lakes, groundwater, living organisms and soil, and glaciers) during the hydrologic cycle. Of the water that falls on land, some evaporates directly back into the atmosphere, some gets trapped in glaciers, and some becomes runoff that enters a network of streams and lakes that drains to the sea. The remainder sinks, via the process of **infiltration,** into the ground; in effect, the upper part of the crust behaves like a giant sponge that can soak up water. How much water infiltrates into the crust at a certain location depends on the

vegetation cover (vegetation absorbs water) and on the composition of materials making up the surface layer of the Earth—loose sand lets water sink in easily, while hard-packed clay, concrete, or dense, unfractured rock does not.

Of the water that does infiltrate, some descends only into the soil and wets the surfaces of grains and organic material making up the soil. This water, called **soil moisture,** later evaporates back into the atmosphere or gets sucked up by the roots of plants and then transpires back into the atmosphere. But some water sinks deeper and fills available spaces in cracks and between grains of sediment or rock; this water, as well as water trapped in rock at the time the rock formed, comprises groundwater. Groundwater slowly flows underground for anywhere from a few months to tens of thousands of years before returning to the surface to pass once again into other reservoirs of the hydrologic cycle. At any given time, groundwater accounts for about two-thirds of the world's freshwater supply.

Porosity: The Home of Groundwater

Contrary to popular belief, only a small proportion of groundwater flows freely in the underground lakes and streams of cavern networks. Most groundwater resides within the pore space of what might at first glance look like solid rock or sediment. Generally speaking, a **pore** is any open space (as opposed to solid material) within a body of sediment or rock, and **porosity** refers to the total volume of empty space in a material, usually expressed as a percentage. For example, if we say that a piece of sandstone has 30% porosity, then 30% of a block of what looks like solid sandstone actually consists of open space.

Geologists further distinguish between primary and secondary porosity. **Primary porosity** consists of space that

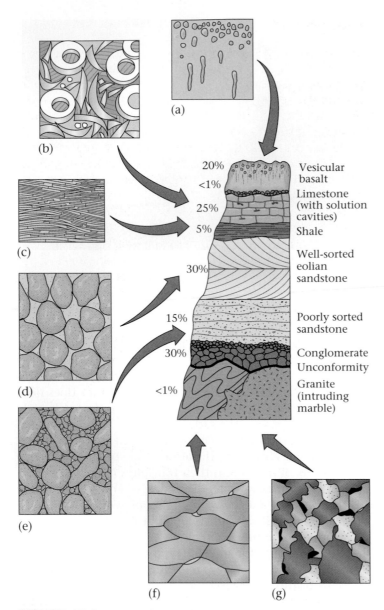

FIGURE 19.2 Various kinds of primary porosity in rock. Porosities are indicated as percentages. (a) Vesicles create variable porosity in basalt. (b) Uncemented spaces create moderate porosity in fossiliferous limestone. (c) Shale has packed-together clay flakes, resulting in very low porosity. (d) Well-sorted sandstone has high porosity. (e) Poorly sorted sandstone has lower porosity. (f) Metamorphic rocks (such as marble) have very low porosity. (g) Igneous rocks (such as granite) have very low porosity.

remains between solid grains or crystals after sediment accumulates or rocks form. To see why solid pieces do not fit together perfectly, fill a glass with marbles. The marbles contact one another at many points, but no matter how you shake the glass to encourage the marbles to settle together, open spaces still exist between the marbles except at contact points.

Let's look at different types of primary porosity (▶Fig. 19.2a–g). In unlithified sediment, like sand or gravel, primary

porosity exists for the simple reason that rounded clasts can't fit together tightly. In such sediment, porosity depends on the size and sorting of clasts; the overall porosity of a poorly sorted sediment is less than that of a well-sorted sediment, because in a poorly sorted sediment smaller grains fill in spaces between larger grains. The primary porosity of a sediment tends to decrease with greater burial depth, because the weight of overburden pushes the sediment grains together. During the transition from sediment to sedimentary rock, primary porosity decreases still further because some space between sediment grains fills with cement. Nevertheless, a significant amount of primary porosity typically remains in sedimentary rock.

In most crystalline igneous and metamorphic rocks, grains interlock as they grow, so the amount of primary porosity is small. In fine-grained and glassy igneous rocks, vesicles (gas bubbles trapped when the rock freezes from lava), if they exist, provide primary porosity.

Secondary porosity refers to new pore space in rocks, produced some time after the rock first formed. For example, when rocks fracture, the opposing walls of the fracture do not fit together tightly, so narrow spaces remain in between. Thus, joints and faults may provide openings for water (▶Fig. 19.3). Faulting may also produce breccia, a jumble of angular fragments, with the space between the fragments providing another type of secondary pore space. Finally, as groundwater passes through rock, it may dissolve and remove some minerals, creating solution cavities.

19.3 PERMEABILITY: THE EASE OF FLOW

If solid rock completely surrounds a pore, the water in the pore cannot flow to another location. For groundwater to flow, therefore, pores must be linked by conduits (openings). The ability of a material to allow fluids to pass through an

FIGURE 19.3 Fractures in a rock provide secondary porosity.

interconnected network of pores is a characteristic known as **permeability** (►Fig. 19.4a, b). Water flows easily through a permeable material. In contrast, water flows slowly or not at all through an impermeable material. The permeability of a material depends on several factors.

- *Number of available conduits:* As the number of conduits increases, the permeability increases.

- *Size of the conduits:* More fluids can travel through wider conduits than through narrower ones.

- *Straightness of the conduits:* Water flows more rapidly through straight conduits than it does through crooked ones. In crooked channels, the distance a water molecule actually travels may be many times the straight-line distance between the two end points.

Note that the factors that control permeability in rock or sediment resemble those that control the ease with which traffic moves through a city. Traffic can flow quickly through cities with many straight, multilane boulevards, whereas it flows slowly through cities with only a few narrow, crooked streets.

Porosity and permeability are not the same. A material whose pores are isolated from one another can have high porosity but low permeability. For example, if you plug the bottom of a water-filled tube with a cork, the water remains in the tube even though cork is very porous; this is because woody cell walls separate adjacent pores in the cork and prevent communication between them, making cork impermeable.

FIGURE 19.4 (a) Isolated, nonconnected pores in an impermeable material; (b) pores connected to one another by a network of conduits in permeable material.

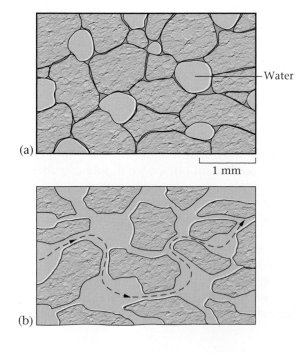

(a)

|___ 1 mm ___|

(b)

Aquifers and Aquitards

With the concept of permeability in mind, hydrogeologists (geologists who study groundwater) distinguish between **aquifers,** sediment or rocks that transmit water easily, and **aquitards,** sediment or rocks that do not transmit water easily and therefore retard the motion of water. Aquicludes do not transmit water at all. When hydrogeologists talk about the "principal aquifer" of a region, they are referring to the geologic unit that serves as the primary source of groundwater in that region. For example, a Cretaceous stratigraphic formation called the Dakota Sandstone serves as the principal aquifer for much of the high plains region of the United States. Gravel deposited during the last ice age provides the principal aquifers for the northern Midwest. Aquifers that intersect the surface of the Earth are called "unconfined aquifers," because water can percolate directly from the surface down into the aquifer, and water in the aquifer can rise to the surface. Aquifers that are separated from the surface by an aquitard are called "confined aquifers," as the water they contain is isolated from the surface (►Fig. 19.5).

19.4 THE WATER TABLE

We've seen that groundwater resides in subsurface pore space. But are pores filled with water right up to the ground surface everywhere? The answer is no. Geologists define a boundary, called the **water table,** above which pore spaces contain mostly air and below which they contain only water (►Fig. 19.6a). In technical jargon, the region of

FIGURE 19.5 An aquifer is a high-porosity, high-permeability rock. If it has access to the ground surface, it's an unconfined aquifer. If it's trapped below an aquitard, a rock with low permeability, it's a confined aquifer.

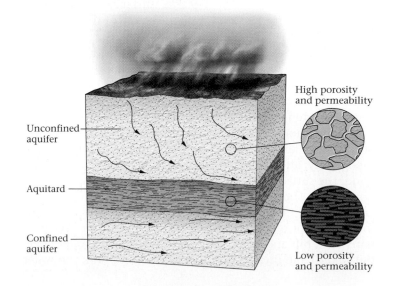

the subsurface above the water table is called the **unsaturated zone** (or the vadose zone) and the region below the water table is the **saturated zone** (or the phreatic zone). Material above the water table can be damp, but pores are not full. Typically, surface tension, the electrostatic attraction of water molecules to mineral surfaces, causes water to seep up from the water table (just as water rises in a thin straw), filling pores in the **capillary fringe,** a thin layer between the saturated and unsaturated zones.

The depth of the water table in the subsurface varies greatly with location. The surface of a permanent stream, lake, or marsh, for example, defines the water table at that location, for water saturates the soil or rock below (▶Fig. 19.6b, c). Elsewhere, the water table lies hidden below the ground surface: in humid regions, it typically lies within a few meters of the ground surface, whereas in arid regions, it may lie tens of meters to over 200 m below the surface.

Rainfall affects the water-table depth—the level sinks during a dry season (▶Fig. 19.6d). If the water table drops below the floor of a river or lake, the river or lake dries up, because the water it contains infiltrates into the ground. In arid regions, there are no permanent streams, as the water table lies below the stream bed. Streams in such regions flow only during storms, when rain falls at a faster rate than the water can infiltrate.

Note that "water table" refers to the position of the *top* of the groundwater reservoirs beneath land. Defining the bottom of the groundwater reservoirs proves to be more controversial. Below about 15 to 20 km, rocks are too weak to keep pore spaces open (minerals plastically flow and close up any space), and water becomes involved in metamorphic reactions. Thus, water does exist in the deeper crust, but not in the form found at shallower depths.

Topography of the Water Table

In hilly regions, if the ground has relatively low permeability, the water table is not a planar surface, but rather its shape mimics, in a subdued way, the shape of the overlying topography (▶Fig. 19.7). This means that the water table lies at a higher elevation beneath hills than it does beneath valleys. But the relief (the vertical distance between the highest and lowest elevations) of the water table is not as great as that of the overlying land, so the surface of the water table tends to be smoother than that of the landscape.

At first thought, it may seem surprising that the elevation of the water table varies as a consequence of ground-surface topography; after all, when you pour a bucket of water into a pond, the surface of the pond immediately adjusts to remain horizontal. The elevation of the water table varies because groundwater moves so slowly through rock and sediment that it cannot quickly assume a horizontal surface. When it rains on a hill and water infiltrates down to the water table, the water table rises a little. When it doesn't rain, the water

table sinks slowly, but so slowly that rain will fall again and make the water table rise before the water table has had time to sink very far.

Perched Water Tables

In some locations, layers of strata are discontinuous, meaning that they pinch out at their sides. As a result, lens-shaped layers of impermeable rock (such as shale) may lie within a thick aquifer. A mound of groundwater accumulates above this aquitard. The result is a **perched water table,** a quantity of groundwater that lies above the regional water table because an underlying lens of impermeable rock or sediment prevents the water from sinking down to the regional water table (▶Fig. 19.8).

19.5 GROUNDWATER FLOW

Groundwater Flow Paths

What happens to water that has infiltrated down into the ground? Does it just sit, unmoving, like the water in a stagnant puddle, or does it flow and eventually find its way back to the surface? Countless measurements confirm that groundwater enjoys the latter fate—groundwater indeed flows, and in some cases it moves great distances underground. In this section we examine factors that drive groundwater flow and determine the path that this flow follows. Then, in the next section, we examine the rate (velocity) at which groundwater moves.

In the zone of aeration, the region between the ground surface and the water table, water percolates straight down, like the water passing through a drip coffee maker, for this water moves only in response to the downward pull of gravity. But in the zone of saturation, the region below the water table, water flow is more complex, for in addition to the downward pull of gravity, water responds to differences in pressure. Pressure may cause groundwater to flow sideways, or even upward. (If you've ever watched water spray up from a fountain, you've seen how pressure can push water upward.) Thus, to understand the nature of groundwater flow, we must first understand the origin of pressure in groundwater. For simplicity, we'll only consider the case of groundwater in an unconfined aquifer.

Pressure in groundwater at a specific point underground is caused by the weight of all the overlying water from that point up to the water table. (The weight of overlying rock does not contribute to the pressure pressing on groundwater, for the contact points between mineral grains bear the rock's weight.) Thus, a point at a greater depth below the water table feels more pressure than does a point at lesser depth. If the water table is horizontal, the pressure acting on an imaginary

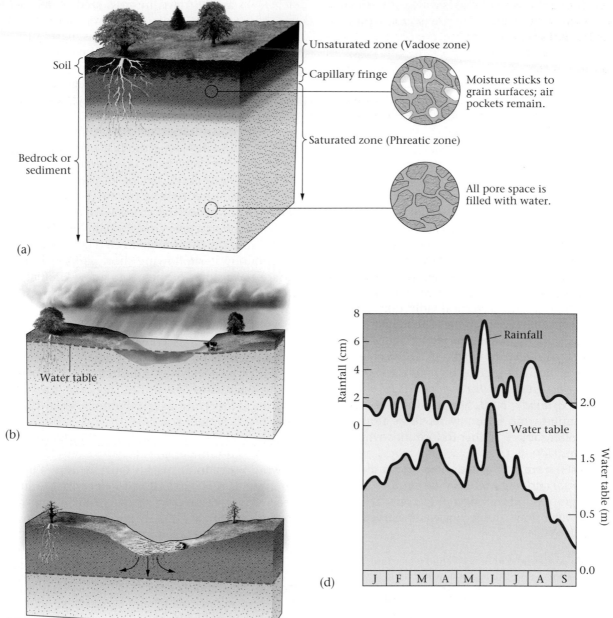

FIGURE 19.6 (a) The geometry of the water table, illustrating the saturated zone, the unsaturated zone, and the capillary fringe. (b) The surface of a permanent pond is the water table. (c) During the dry season, the water table can drop substantially, causing the pond to dry up. (d) The graph shows the relation between the height of the water table and rainfall between January and September, in a temperate region.

horizontal reference plane at a specified depth below the water table is the same everywhere. But if the water table is not horizontal, as shown in Figure 19.7, the pressure at points on a horizontal reference plane at depth changes with location. For example, the pressure acting at point p_1, which lies below the hill in Figure 19.7, is greater than the pressure acting at point p_2, which lies below the valley, even though both p_1 and p_2 are at the same elevation (sea level, in this case).

Both the elevation of a volume of groundwater and the pressure within the water provide energy that, if given the chance, will cause the water to flow—physicists refer to such stored energy as "potential energy" (see Appendix A). (To understand why elevation provides potential energy, imagine a bucket of water high on a hill; the water has potential energy, due to Earth's gravity, which will cause it to flow downslope if the bucket were suddenly to rupture. To understand why

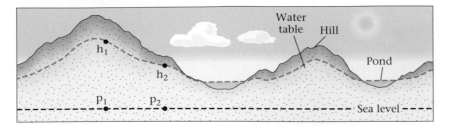

FIGURE 19.7 The shape of a water table beneath hilly topography. Note that the water table can be above sea level. Point h_1, on the water table is higher than point h_2, relative to a reference elevation (sea level).

pressure provides potential energy, imagine a water-filled plastic bag sitting on a table; if you puncture the bag and then squeeze the bag to exert pressure, water spurts out.) The potential energy available to drive the flow of a given volume of groundwater at a location is called the **hydraulic head.** To measure the hydraulic head at a point in an aquifer, hydrogeologists drill a vertical hole down to the point and then insert a pipe in the hole. The height above a reference elevation (e.g., sea level) to which water rises in the pipe represents the hydraulic head—water rises higher in the pipe where the head is higher. As a rule, *groundwater flows from regions where it has higher hydraulic head to regions where it has lower hydraulic head.* Simplistically, this statement generally implies that groundwater flows from locations where the water table is higher to locations where the water table is lower.

Hydrogeologists have calculated how hydraulic head changes with location underground, by taking into account both the effect of gravity and the effect of pressure. They conclude that groundwater flows along concave-up curved paths,

as illustrated in cross section (▶Fig. 19.9). (Specialized books on hydrogeology provide the details of why flow paths have such a specific shape.) These curved paths eventually take groundwater from regions where the water table is high (under a hill) to regions where the water table is low (below a valley), but because of flow-path shape, some groundwater may flow deep down into the crust—in some cases, to depths of 10 km—along the first part of its path, and then may flow back up, toward the ground surface, along final part of its path. The location where water enters the ground (i.e., where the flow direction has a downward trajectory) is called the **recharge area**, and the location where groundwater flows back up to the surface is called the **discharge area** (Fig. 19.9).

Groundwater following short paths close to the Earth's surface travels tens of meters to a few kilometers before returning to the surface; such movement constitutes "local flow" and keeps water underground for only hours to weeks. Groundwater following paths of several kilometers to tens of kilometers constitutes "intermediate flow" and stays underground for weeks to years. Groundwater following paths that carry it hundreds of kilometers across a sedimentary basin constitutes "regional flow" and stays underground for centuries to millennia (▶Fig. 19.10).

Rates of Groundwater Flow: Darcy's Law

Flowing water in an ocean current moves at up to 3 km per hour (over 26,000,000 m per year), and water in a steep river channel can reach speeds of up to 30 km per hour (over 260,000,000 m per year). In contrast, ground-

FIGURE 19.8 The configuration of a perched water table. A lens of groundwater lies above, and the regional water table lies at greater depth.

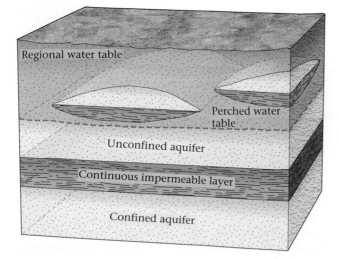

FIGURE 19.9 The flow lines from the recharge area to the discharge area curve through the substrate. In fact, some groundwater descends to great depth and then rises back to the surface.

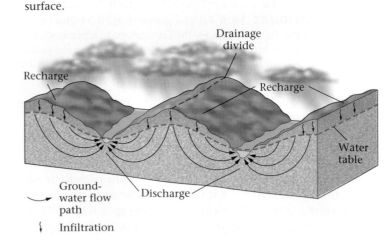

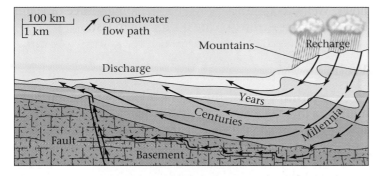

FIGURE 19.10 Cross section showing regional-scale groundwater flow in a sedimentary basin. The arrows indicate the flow paths of groundwater. Note that recharge occurs in highlands on one side of the basin, and discharge occurs on the far side of the basin. Some of the water flows through fractures deep in the basement.

water moves at a snail's pace—typical rates range between 0.01 and 1.4 m per day (about 4 to 500 m per year). Groundwater moves much more slowly than surface water for two reasons. First, groundwater moves by percolating through a complex, crooked network of tiny conduits; it must travel a much greater distance than it would if it could follow a straight path. Second, friction and/or electrostatic attraction between groundwater and conduit walls slows down the water flow. Hydrogeologists may measure the rate (velocity) of groundwater flow in a region by "tagging" water with dye or a radioactive element; they inject the tracer in one well and time how long it takes for the tracer to reach another well.

The rate at which groundwater flows, at a given location, depends on the permeability of the material containing the groundwater; groundwater flows faster in material with greater permeability than in material with lesser permeability. Rate also depends on the **hydraulic gradient,** the change in hydraulic head per unit of distance between two locations as measured along the flow path. Recall that groundwater flows from a region where the hydraulic head is higher (due to higher elevation and/or greater pressure) to a region where the hydraulic head is lower. If there is a large difference in the hydraulic head over a given distance, then there is a greater amount of energy driving the flow, so the flow is faster.

To calculate the hydraulic gradient, we simply divide the difference in hydraulic head between two points by the distance between the two points as measured along the flow path. This can be written as a formula:

$$\text{hydraulic gradient} = \Delta h/j,$$

where Δh is the difference in head (given in meters or feet, because head can be represented as an elevation), and j is the distance between the two points (▶Fig. 19.11). A hydraulic gradient exists anywhere that the water table has a

slope; in an unconfined aquifer the hydraulic gradient is roughly equivalent to the slope of the water table, if the slope is gentle.

With an understanding of the controls on the rate of groundwater flow, we can consider the practical problem of determining the volume of water that will pass through an area underground during a specified time. In 1856, a French engineer named Henry Darcy addressed this problem because he wanted to find out whether the city of Dijon in central France could supply its water needs from groundwater. Darcy carried out a series of experiments in which he measured the rate of water flow through sediment-filled tubes tilted at varying angles. Darcy found that the volume of water that passed through a specified area in a given time, a quantity he called the *discharge* (Q), can be calculated from the equation

$$Q = K(\Delta h/j)A,$$

where $\Delta h/j$ is the hydraulic gradient, A is the area through which the water is passing (see Fig. 19.11), and K is a number called the "hydraulic conductivity." The hydraulic conductivity takes into account the permeability as well as the fluid's viscosity ("stickiness"). Hydrogeologists refer to this equation as **Darcy's law.**

For quick estimates, geologists sometimes simplify Darcy's Law, by writing it as follows: discharge ≅ (slope of the water table) × (permeability). Thus, Darcy's law states that groundwater flow is faster through very permeable rocks than it is through impermeable rocks, and that it is faster where the water table has a steep slope than where the water table has a shallow slope. It is no surprise, therefore, that groundwater moves very slowly (0.5 to 3.0 cm per day) through a well-cemented bed of the Dakota Sandstone, a major aquifer beneath the gently sloping Great Plains central

FIGURE 19.11 A hydraulic gradient (HG) is the change in hydraulic head per unit of distance between two points along the flow path.

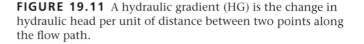
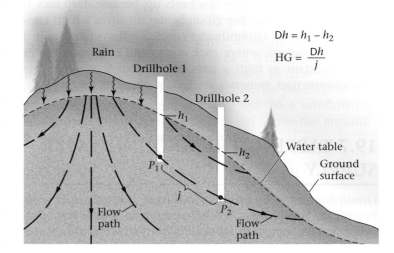

(a)

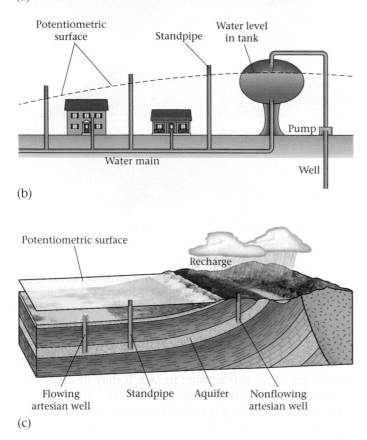

(b)

(c)

FIGURE 19.14 (a) A flowing artesian well. (b) The configuration of a city water supply. Water will rise in vertical pipes up to the level of the potentiometric surface. (c) The configuration of an artesian system. Artesian wells flow if the potentiometric surface lies above the ground surface. Nonflowing artesian wells form where the potentiometric surface lies below the ground.

drinking or irrigation, without the expense of drilling or digging. Springs form under a variety of conditions:

• Where the ground surface intersects the water table in a discharge area (▶Fig. 19.15a): such springs typically occur in valley floors where they may add water to lakes or streams.

• Where downward-percolating water runs into an impermeable layer and migrates along the top surface of the layer to a hillslope (▶Fig. 19.15b).

• Where a particularly permeable layer or zone intersects the surface of a hill (▶Fig. 19.15c): water percolates down through the hill and then migrates along the permeable layer to the hill face.

• Where a network of interconnected fractures channels groundwater to the surface of a hill (▶Fig. 19.15d).

• Where flowing groundwater collides with a steep impermeable barrier, and pressure pushes it up to the ground along the barrier (▶Fig. 19.15e): faulting can create such barriers by juxtaposing impermeable rock against permeable rock.

• **Artesian springs** form if the ground surface intersects a natural fracture (joint) that taps a confined aquifer in which the pressure is sufficient to drive the water to the surface (▶Fig. 19.15f and Box 19.1).

• Where a perched water table intersects the surface of a hill (▶Fig. 19.15g, h).

19.8 HOT SPRINGS AND GEYSERS

Hot springs, springs that emit water ranging in temperature from about 30° to 104°C, are found in two geologic settings. First, they occur where very deep groundwater, heated in warm bedrock at depth, flows up to the ground surface. This water brings heat with it as it rises. Such hot springs form in places where faults or fractures provide a high-permeability conduit for deep water, or where the water emitted in a discharge region followed a trajectory that first carried it deep into the crust. Second, hot springs develop in **geothermal regions,** places where volcanism currently takes place or has occurred recently, so that magma and/or very hot rock resides close to the Earth's surface. In hot springs, groundwater is a steaming tea of water and dissolved minerals. Hot groundwater contains more dissolved minerals because water becomes a more effective solvent when hot. People use the water emitted at hot springs to fill relaxing mineral baths (▶Fig. 19.17).

Numerous distinctive geologic features form in geothermal regions as a result of the eruption of hot water (▶Fig. 19.18a–d). In places where the hot water rises into soils rich in volcanic ash and clay, a viscous slurry forms and fills bub-

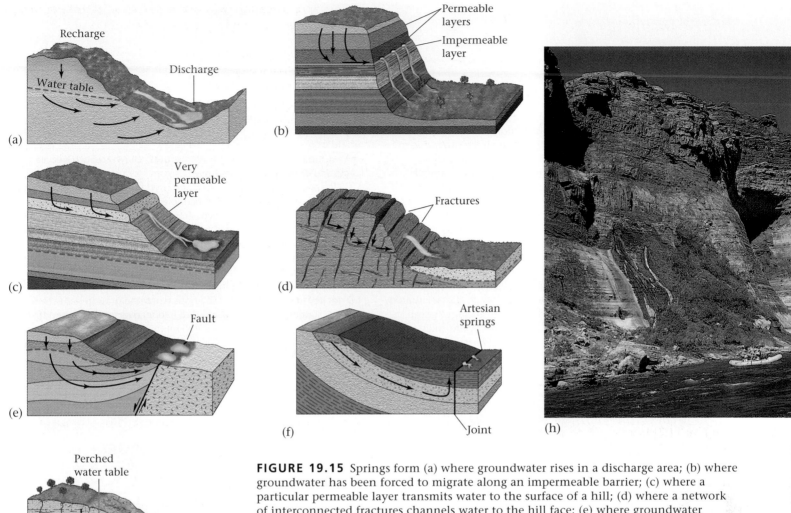

FIGURE 19.15 Springs form (a) where groundwater rises in a discharge area; (b) where groundwater has been forced to migrate along an impermeable barrier; (c) where a particular permeable layer transmits water to the surface of a hill; (d) where a network of interconnected fractures channels water to the hill face; (e) where groundwater collides with a steep impermeable barrier, and pressure pushes it up to the ground along the barrier. (f) An artesian spring forms where water from a confined aquifer migrates up a joint; (g) springs also form where a perched water table intersects the surface of a hill; (h) photo of a spring arising from a perched water table intersecting a wall of the Grand Canyon.

bling mud pots. Bubbles of steam rising through the slurry cause it to splatter about in goopy drops. Where geothermal waters spill out of natural springs and then cool, dissolved minerals in the water precipitate, forming colorful mounds or terraces of travertine and other minerals. Geothermal waters may accumulate in brightly colored pools—the gaudy greens, blues, and oranges of these pools come from thermophyllic (heat-loving) bacteria and archaea that thrive in hot water and metabolize the sulfur-containing minerals dissolved in the groundwater.

The most spectacular consequence of geothermal waters is a **geyser** (from the Icelandic word for "gush"), a fountain of steam and hot water that erupts episodically from a vent in the ground (▶Fig. 19.19a). To understand why a geyser erupts, we first need a picture of its underground plumbing. Beneath a geyser lies a network of irregular fractures in very hot rock; groundwater sinks and fills these fractures. Adjacent hot rock then "superheats" the water: it raises the temperature above the temperature at which water at a pressure of 1 atmosphere will boil. Eventually, this superhot water rises through a conduit to the surface. When some of this water transforms into steam, the resulting expansion causes water higher up to spill out of the conduit at the ground surface. When this spill happens, pressure in the

BOX 19.1
THE HUMAN ANGLE

Oases

The Sahara Desert of northern Africa is now one of the most barren and desolate places on Earth, for it lies in a climatic belt where rain seldom falls. But it wasn't always that way. During the last ice age, when glaciers covered parts of northern Europe on the other side of the Mediterranean, the Sahara enjoyed a more temperate climate, and the water table was high enough that permanent streams dissected the landscape. In fact, using ground-penetrating radar, geologists can detect these streams today—they look like ghostly valleys beneath the sand (▶Fig. 19.16a). When the climate warmed after the ice age, rainfall diminished, the water table sank below the floor of most stream beds, and the streams dried up.

Today, the water of the Sahara region lies locked in a vast underground aquifer composed of porous sandstone. Recharge into this aquifer comes from highlands bordering the desert, from occasional downpours, and from the Nile River (particularly Lake Nassar, created by the Aswan High Dam). In general, the water of the aquifer can only be obtained by drilling deep wells, but locally, water spills out at the surface—either because folding brings the aquifer particularly close to the ground so that valley floors intersect the water table, or

because artesian pressure pushes groundwater up along joints of faults (▶Fig. 19.16b). In either case, the aquifer feeds springs that quench the thirst of desert and tropical plants and create an **oasis,** an island of green in the sand sea (▶Fig. 19.16c). Oases became important stopping points along caravan routes, allowing both people and camels to replenish water supplies.

In some oases, people settled and used the groundwater to irrigate date palms and other crops. For example, the Bahariya Oasis, about 400 km southwest of Cairo, Egypt, hosted a town of perhaps 30,000 between 300 B.C.E. and 300 C.E. During that time, the water table lay only 5 m below the ground and could be easily accessed by shallow wells. Today, as a result of changing climates and centuries of use, the water table lies 1,500 m below the ground, almost out of reach. Bahariya's glorious past came to light in 1996, quite by accident. A man was riding his donkey in the desert near the oasis when the ground beneath the donkey suddenly caved in. The guard had inadvertently opened the roof into a tomb filled with over 150 mummies, along with thousands of well-preserved artifacts. The site has since come to be known as the Valley of the Mummies.

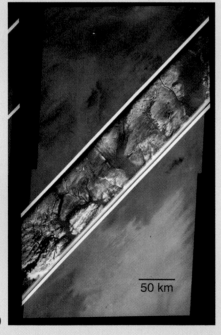

(a)

50 km

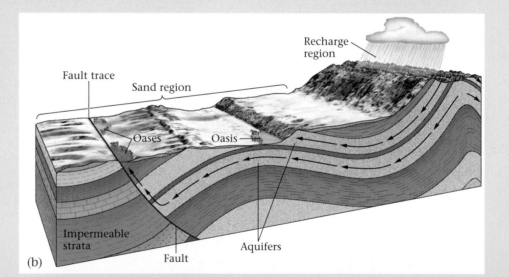

(b)

(c)

FIGURE 19.16 (a) A satellite image, taken using ground-penetrating radar, showing a long-abandoned drainage network now buried by Saharan sand. The gray strip shows darker channels and lighter high areas under the sand. The orange area is the sand-surface that you see without radar. (b) This subsurface configuration of aquifers leads to the formation of an oasis, where groundwater reaches the surface. (c) An oasis in the Sahara Desert.

FIGURE 19.17 In the city of Bath, England, the Romans built an elaborate spa around an artesian hot spring. The spring formed along a fault that tapped a supply of deep groundwater that still flows today. The baths of Bath have remained popular through the ages, and are currently undergoing renovation.

(a)

(b)

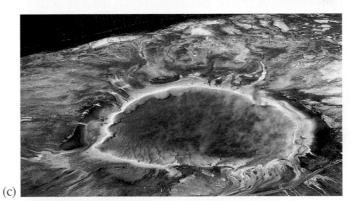

(c)

(d)

FIGURE 19.18 Features of geothermal regions. (a) Mud pots in Yellowstone National Park; (b) mounds of siliceous minerals precipitated around the Grotto Geyser, in Yellowstone; (c) colorful bacteria- and archaea-laden pools, Yellowstone; (d) Old Faithful geyser at Yellowstone. The geyser erupts somewhat predictability.

conduit, from the weight of overlying water, suddenly decreases. A sudden drop in pressure causes the superhot water at depth to instantly turn to steam, and this steam quickly rises, ejecting all the water and steam above it out of the conduit in a geyser eruption. Once the conduit empties, the eruption ceases, and the conduit fills once again with water that gradually heats up, starting the eruptive cycle all over again.

Hot springs are found in many localities around the world: at Hot Springs, Arkansas, where deep groundwater rises to the surface; in Yellowstone National Park, above the magma chamber of a continental hot spot; around the Salton Sea in southern California, where the mid-ocean ridge of the Gulf of California merges with the San Andreas Fault; in the Geysers Geothermal Field of California, an area of significant geothermal power generation formed above a felsic magma intrusion; in Iceland, which has grown on top of an oceanic hot spot along the Mid-Atlantic Ridge; and in Rotorua, New Zealand, which lies in an active volcanic field above a subduction zone.

People do live in some geothermal regions, though the areas have inherent natural hazards. In Rotorua, signs along the road warn of steam, which can obscure visibility, and steam indeed spills out of holes in backyards and parking lots (▶Fig. 19.19b). But all this hot water does offer a benefit: in Rotorua, waters circulate through pipes to provide home heating, and in geothermal areas worldwide, steam provides a relatively inexpensive means of generating electricity.

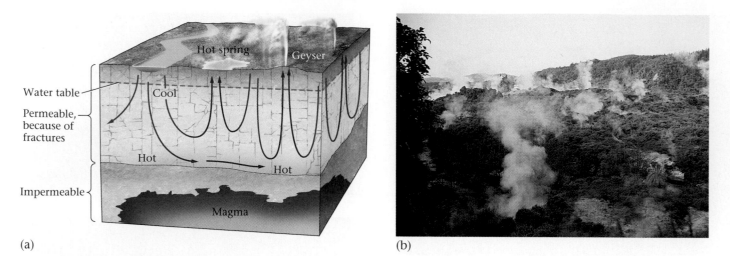

(a)   (b)

FIGURE 19.19 (a) Geysers and hot springs form where groundwater, heated at depth, rises to the surface. (b) Geysers in the geothermal region of Rotorua, New Zealand.

19.9 GROUNDWATER USAGE PROBLEMS

Since prehistoric times, groundwater has been an important resource that people have relied on for drinking, irrigation, and industry. Groundwater feeds the lushness of desert oases in the Sahara Desert, the amber grain in the North American high plains, and the growing cities of arid regions. Agricultural and industrial usage accounts for about 93% of all water usage, so as once-empty land comes under cultivation and countries become increasingly industrialized, demands on the groundwater supply soar. Globally, groundwater provides only about 20% of the water we use, but this number has increased as surface-water resources decrease. And locally, groundwater is the sole water source.

Though groundwater accounts for about 95% of the liquid freshwater on the planet, accessible groundwater cannot be replenished quickly in important locations, and this leads to shortages. In the twentieth century, the problem was exacerbated by the contamination of existing groundwater. Such pollution, caused when toxic wastes and other impurities infiltrate down to the water table, may be invisible to us but may ruin a water supply for generations to come. In this section, we'll take a look at problems associated with the use of groundwater supplies.

Depletion of Groundwater Supplies

Is groundwater a renewable resource? In a time frame of 10,000 years, the answer is yes, for the hydrologic cycle will eventually resupply depleted reserves. But in a time frame of 100 to 1,000 years—the span of a human lifetime or a

civilization—groundwater in many regions may be a nonrenewable resource. By pumping water out of the ground at a rate faster than nature replaces it, people are effectively "mining" the groundwater supply. In fact, in portions of the desert "Sunbelt" region of the United States, supplies of young groundwater have already been exhausted, and deep wells now extract 10,000-year-old groundwater. But such ancient water has been in rock so long that some of it has become too mineralized to be usable. A number of other problems accompany the depletion of groundwater.

- *Lowering the water table:* When we extract groundwater from wells at a rate faster than it can be resupplied, the water table drops. First, a cone of depression forms locally around the well; then the water table gradually becomes lower in a broad region. As a consequence, existing wells, springs, and rivers dry up (▶Fig. 19.20a, b). To continue tapping into the water supply, we must drill progressively deeper.

    The water table can also drop when people divert surface water from the recharge area. Such a problem has developed in the Everglades of southern Florida, a huge swamp where, before the expansion of Miami and the development of agriculture, the water table lay at the ground surface (▶Fig. 19.20c, d). Diversion of water from the Everglades' recharge area into canals has significantly lowered the water table, causing parts of the Everglades to dry up.

- *Reversing the flow direction of groundwater:* The cone of depression that develops around a well creates a local slope to the water table. The resulting hydraulic gradient may be large enough to reverse the flow direction of nearby groundwater (▶Fig. 19.21a, b). Such reversals can

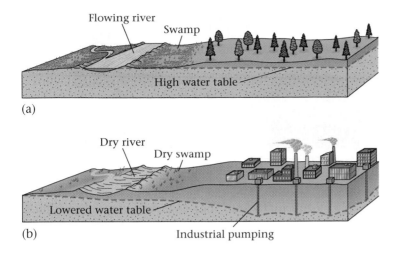

(a)

(b)

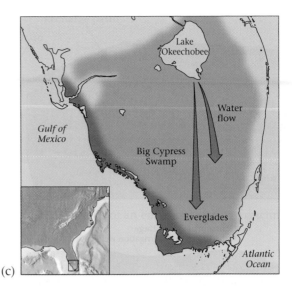

(c)

FIGURE 19.20 (a) Before a water table is lowered, a large swamp exists. (b) Pumping by a nearby city causes the water table to sink, so the swamp dries up. (c) The Everglades, in Florida, before the advent of urban growth and intensive agriculture. Water flowed south from Lake Okeechobee, creating a fast swamp in which the water table lay just above the ground surface. In effect, the Everglades is a "river of grass." (d) Channelization and urbanization have removed water from the recharge area, disrupting the groundwater flow path in the Everglades. Since less water enters the Everglades, the water table has dropped, so locally the swamp has dried up. The decrease in the supply of freshwater has also led to saltwater (saline) intrusion along the coast.

(d)

lead to the contamination of a well, from pollutants seeping out of a septic tank.

- *Saline intrusion:* In coastal areas, fresh groundwater lies in a layer above saltwater that entered the aquifer from the adjacent ocean (▶Fig. 19.21c, d). (Saltwater is denser than freshwater, so the fresh groundwater floats above it.) If people pump water out of a well too quickly, the boundary between the saline water and the fresh groundwater rises. And if this boundary rises above the base of the well, then the well will start to yield useless saline water.

- *Pore collapse and land subsidence:* When groundwater fills the pore space of a rock, it holds the grains of the rock or regolith apart, for water cannot be compressed. The

extraction of water from a pore eliminates the support holding the grains apart, because the air that replaces the water *can* be compressed. As a result, the grains pack more closely together. Such **pore collapse** permanently decreases the porosity and permeability of a rock, and thus lessens its value as an aquifer (▶Fig. 19.21e, f).

Pore collapse also decreases the volume of the aquifer, with the result that the ground above the aquifer sinks. Such **land subsidence** may cause fissures at the surface to develop and the ground to tilt. Buildings constructed over regions undergoing land subsidence may themselves tilt, or their foundations may crack. The Leaning Tower of Pisa, in Italy, tilts because the removal of groundwater caused its foundation to subside (▶Fig. 19.23g). In the San Joaquin Valley of California, the land surface

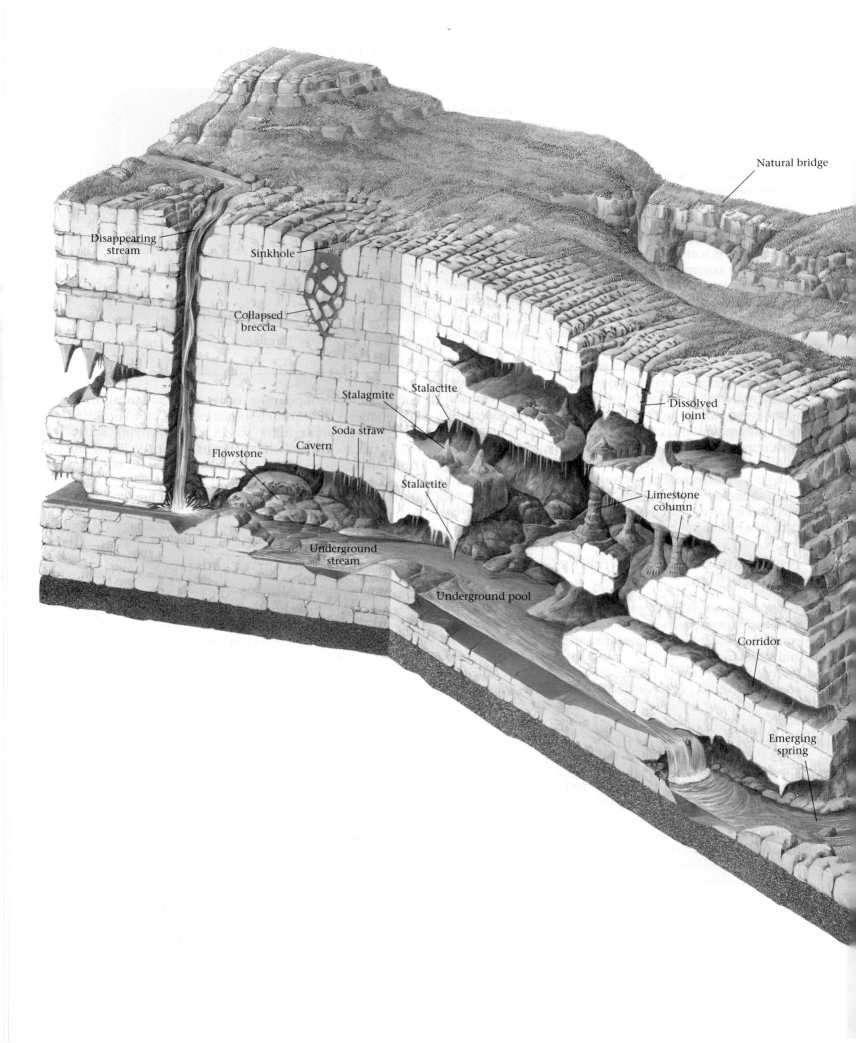

Natural bridge

Disappearing stream

Sinkhole

Collapsed breccia

Stalagmite

Stalactite

Dissolved joint

Soda straw

Flowstone

Cavern

Stalactite

Limestone column

Underground stream

Underground pool

Corridor

Emerging spring

Caves and Karst Landscapes

Limestone, a sedimentary rock made of the mineral calcite, is soluble in acidic water. Much of the water that falls to the ground as rain, or seeps through the ground as groundwater, tends to be acidic, so in regions of the Earth where bedrock consists of limestone, there are signs of dissolution. Underground openings that develop by dissolution are called caves or caverns. Some of these may be large open rooms, while others are long, narrow passages. Underground lakes and streams may form on the floor. A cave's location depends on the orientation of bedding and joints, for these features localize the flow of groundwater.

Caves originally form at or near the water table (the subsurface boundary between rock or sediment in which pores contain air, and rock or sediment in which pores contain water). As the water table drops, caves empty of most water and become filled with air. In many locations, groundwater drips from the ceiling of a cave or flows along its walls. As the water evaporates and thus loses its acidity (because of the evaporation of dissolved carbon dioxide), new calcite precipitates. Over time, this calcite builds into cave formations, or speleothems, such as stalactites, stalagmites, columns, and flowstone.

Distinctive landscapes, called karst landscapes, develop at the Earth's surface over limestone bedrock. In such regions, the ground may be rough where rock has dissolved along joints, and where the roofs of caves collapse, sinkholes develop. If a surface stream flows through an open joint into a cave network, we say that the stream is "disappearing." The water from such streams may flow underground for a ways and then reemerge elsewhere as a spring. In some places, the collapse of subsurface openings leaves behind natural bridges.

and the landscape is pockmarked by deep circular-to-elongate depressions, or "sinkholes," which, as we have seen, can form where the roof of a cave collapses. Such sinkholes are called "collapse sinkholes." (Sinkholes can also form where acidic water accumulates in a pool at the surface, seeps down, and dissolves bedrock below. Eventually, a solution pit develops in the bedrock, and the overlying soil sinks downward, forming a depression called a "solution sinkhole.") The sinkholes of the Kras Plateau are separated from one another by hills or walls of bedrock (▶Fig. 19.29a–d). Locally, where most of a cave collapsed, a **natural bridge** spans the cave remnant. Where the water table rises above the floor of a sinkhole, the sinkhole fills to become a lake. Where surface streams intersect cracks or holes that link to the caves below, the water disappears into the subsurface and becomes an underground stream. Such **disappearing streams** reemerge from a cave entrance downstream.

Geologists refer to landscapes such as the Kras Plateau in which surface features reflect the dissolution of bed rock below as **karst landscapes,** from the Germanized version of "kras." Karst landscapes form in a series of stages (▶Fig. 19.30a–c).

- *The establishment of a water table in limestone:* The story of a karst landscape begins after the formation of a thick interval of limestone. Limestone forms in seawater, and thus initially lies below sea level. If relative sea level drops, a water table can develop in the limestone below the ground surface. If, however, the limestone gets buried deeply, it must first be uplifted and exposed by erosion before it can contain the water table.

- *The formation of a cave network:* Once the water table has been established, dissolution begins and a cave network develops.

- *A drop in the water table:* If the water table later becomes lower, either because of a decrease in rainfall or because nearby rivers cut down through the landscape and drain the region, newly formed caves dry out. Downward-percolating groundwater emerges from the roofs of the caves; dripstone and flowstone precipitate.

FIGURE 19.29 (a) Numerous sinkholes of a karst landscape. (b) The Arecibo Radio telescope in Puerto Rico was built in a sinkhole. (c) Natural Bridge, Virginia. (d) A disappearing stream.

(a)

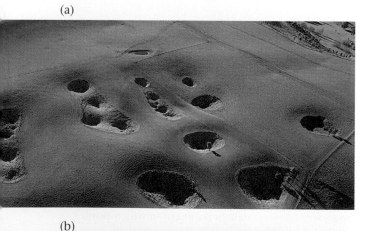

(b)

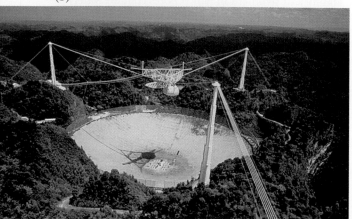

(c)

(d)

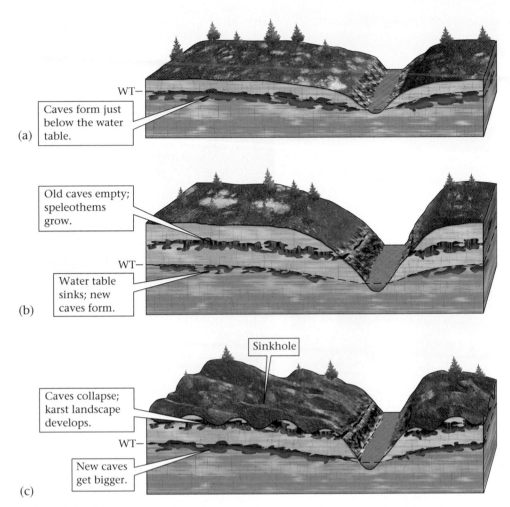

(a) | Caves form just below the water table.

(b) | Old caves empty; speleothems grow.
Water table sinks; new caves form.

(c) | Sinkhole
Caves collapse; karst landscape develops.
New caves get bigger.

FIGURE 19.30 The formation of caves and a karst landscape. (a) Dissolution takes place near the water table (WT) in an uplifted sequence of limestone. (b) Downcutting by an adjacent river lowers the water table, and the caves empty. Speleothems grow on the cave walls. (c) After roof collapse, the landscape becomes pockmarked by sinkholes.

• *Roof collapse:* If rocks fall off the roof of a cave for a long time, the roof eventually collapses. Such collapse creates sinkholes and troughs, leaving behind hills, ridges, and natural bridges of limestone.

Some karst landscapes contain many round sinkholes separated by hills. The giant Arecibo Radio telescope in Puerto Rico, for example, consists of a dish formed by smoothing the surface of a 300-m-wide round sinkhole (Fig. 19.29b). In regions where vertical joints control roof collapse, steep-sided residual bedrock towers remain between sinkholes. A karst landscape with such spires is called "tower karst." The surreal collection of pinnacles constituting the tower karst landscape in the Guilin region of China has inspired generations of artists to portray them on scroll paintings (▶Fig. 19.31a, b).

Life in Caves

Despite their lack of light, caves are not sterile, lifeless environments. Caves that are open to the air provide a refuge for bats as well as for various insects and spiders. Similarly, fish and crustaceans enter caves where streams flow in or out. Species living in caves have evolved some unusual characteristics. For example, cave fish lose their pigment and in some cases their eyes (▶Fig. 19.32a). Recently, explorers discovered caves in Mexico in which warm, mineral-rich groundwater currently flows. Colonies of bacteria metabolize sulfur-containing minerals in this water, and create thick mats of living ooze in the complete darkness of the cave. Long gobs of this bacteria slowly drip from the ceiling. Because of the mucus-like texture of these drips, they have come to be known as "snotites" (▶Fig. 19.32b).

(a)

(b)

FIGURE 19.31 (a) Tower karst in China; (b) painting of tower karst by an unknown Chinese artist.

FIGURE 19.32 Life in caves: (a) blind fish and (b) snotites (gobs of bacteria).

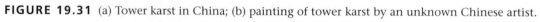

(a)

(b)

CHAPTER SUMMARY

• During the hydrologic cycle, water infiltrates the ground and fills the pores and cracks in rock and sediment. This sub-surface water is called groundwater. The amount of open space in rock or sediment is its porosity, and the degree to which pores are interconnected, so that water can flow through, defines its permeability.

• Geologists classify rock and sediment according to their permeability. Aquifers are relatively permeable, and aquitards are relatively impermeable.

• The water table is the surface in the ground above which pores contain mostly air, and below which pores are filled with water. The shape of a water table is a subdued im-itation of the shape of the overlying land surface.

• Groundwater flows wherever the water table has a hy-draulic gradient, and moves slowly from recharge areas to discharge areas. Darcy's law shows that this rate depends on permeability and on the hydraulic gradient.

• Groundwater contains dissolved ions. These ions may come out of solution to form the cement of sedimentary rocks or to fill veins.

• Groundwater can be extracted in wells. An ordinary well penetrates below the water table, but in an artesian well, water rises on its own. Pumping water out of a well too fast causes drawdown, yielding a cone of depression. At a spring, groundwater exits the ground on its own.

• Hot springs and geysers release hot water to the Earth's surface. This water may have been heated by residing very deep in the crust, or by the proximity of a magma chamber or recently formed volcanic rock.

• Groundwater is a precious resource, used for municipal water supplies, industry, and agriculture. In recent years, some regions have lost their groundwater supply because of overuse or contamination.

• When limestone dissolves just below the water table, underground caves are the result. Soluble beds and joints determine the location and orientation of caves. If the water table drops, caves empty out. Limestone precipitates out of water dripping from cave roofs, and creates speleo-thems (such as stalagmites and stalactites). Regions that contain abundant caves, some of which have collapsed to form sinkholes, are called karst landscapes.

KEY TERMS

aquifer (p. 599)
aquitard (p. 599)
artesian springs (p. 606)
artesian well (p. 604)

bioremediation (p. 615)
capillary fringe (p. 600)
cone of depression (p. 604)
contaminant plume (p. 615)

Darcy's law (p. 603)
disappearing streams (p. 620)
discharge area (p. 602)
geothermal region (p. 606)
geyser (p. 607)
groundwater (p. 596)
groundwater contamination (p. 613)
hard water (p. 613)
hot springs (p. 606)
hydraulic gradient (p. 603)
hydraulic head (p. 602)
infiltration (p. 597)
karst landscape (p. 620)
land subsidence (p. 611)
limestone column (p. 617)
natural bridge (p. 620)
ordinary well (p. 604)
perched water table (p. 600)

permeability (p. 599)
pore (p. 597)
pore collapse (p. 611)
porosity (p. 597)
potentiometric surface (p. 604)
primary porosity (p. 597)
recharge area (p. 602)
saturated zone (phreatic zone) (p. 600)
secondary porosity (p. 598)
sinkhole (p. 596)
soil moisture (p. 597)
speleothem (p. 617)
springs (p. 604)
stalactite (p. 617)
stalagmite (p. 617)
unsaturated zone (vadose zone) (p. 600)
water table (p. 599)
wells (p. 604)

REVIEW QUESTIONS

1. How do porosity and permeability differ? Give examples of substances with high porosity but low permeability.

2. What factors affect the level of the water table? What factors affect the flow direction of the water below the water table?

3. How does the rate of groundwater flow compare with that of moving ocean water or river currents?

4. What does Darcy's law tell us about how the hydraulic gradient and permeability affect discharge?

5. How does the chemical composition of groundwater change with time? Why is "hard water" hard?

6. How does excessive pumping affect the local water table?

7. How is an artesian well different from an ordinary well?

8. Explain why hot springs form and what makes a geyser erupt.

9. Is groundwater a renewable or nonrenewable resource? Explain how the difference in time frame changes this answer.

10. Describe some of the ways in which human activities can adversely affect the water table.

11. What are some sources of groundwater contamination? How can it be prevented?

12. Describe the process leading to the formation of caves and the speleotherms within caves.

SUGGESTED READING

Alley, W. M., et al. 1999. *Sustainability of Ground-Water Resources.* U.S. Geological Survey Circular 1186. Denver: U.S. Government Printing Office.

Deutsch, W. J. 1997. *Groundwater Geochemistry: Fundamentals and Applications to Contamination.* Boca Raton, Fla.: Lewis Publishers.

Domenico, P. A., and F. W. Schwartz. 1997. *Physical and Chemical Hydrogeology.* 2nd ed. New York: Wiley.

Freeze, R. A., and J. A. Cherry. 1979. *Groundwater.* Englewood Cliffs, N.J.: Prentice-Hall.

Glennon, R. J. 2002. *Water Follies: Groundwater Pumping and the Fate of America's Fresh Waters:* Washington, D.C.: Island Press.

Gunn, J. 2004. *Encyclopedia of Caves and Karst Science.* London: Fitzroy Dearborn.

Klimchouk, A. B. 2000. *Speleogenesis: Evolution of Karst Aquifers.* Huntsville, Ala.: National Speleological Society.

Schwartz, F. W., and H. Zhang. 2002. *Introduction to Groundwater Hydrology.* New York: Wiley.

White, W. B. 1997. *Geomorphology and Hydrology of Karst Terrains.* Oxford, England: Oxford University Press.

CHAPTER 20

An Envelope of Gas: Earth's Atmosphere and Climate

20.1 INTRODUCTION

On March 21, 1999, Bertrand Piccard, a Swiss psychiatrist, and Brian Jones, a British balloon instructor, became the first people to circle the globe non-stop in a balloon (▶Fig. 20.1). They began their flight on March 1, following more than twenty unsuccessful attempts by various balloonists during the previous two decades. Their air-tight gondola, which could float in case they had to ditch in the sea, contained a heater, bottled air, food and water, and instruments for navigation and communication. The nature of their equipment hints at the challenges the balloonists faced, and why it took so many years before anyone finally succeeded in circling the globe.

Balloons can rise from the Earth only because an **atmosphere,** a layer consisting of a mixture of gases called **air,** surrounds our planet. Any object placed in such a fluid feels a buoyancy force, and if the object is less dense than the fluid, then the buoyancy force can lift it off the ground. Balloons rise because the gas (either helium or hot air) in a balloon is less dense than air. Balloonists control their vertical movements by changing either the buoyancy of the balloon or the weight of the payload, but they cannot directly control their horizontal motions—balloons float with the **wind,** the flow of air from one place to another. In order to

Fed by the warm waters of the tropical Atlantic, Hurricane Hugo's spiraling winds flattened buildings, eroded coastal islands, and triggered floods in 1989. Hugo was particularly destructive because it generated a 6-m-high storm surge that arrived on top of very high tides. This image shows the storm striking the southeastern coast of the United States.

FIGURE 20.1 The balloon and gondola used by Piccard and Jones during their successful attempt to circle the globe in March 1999.

reach their destination before running out of supplies, long-distance balloonists must find a fast wind flowing in the correct direction. Thus, round-the-world balloonists study the **weather,** the physical conditions (the temperature, pressure, moisture content, and wind velocity and direction) of the atmosphere at a given time and location, in great detail to decide when and where to take off and how high to fly. Different winds blow at different elevations, so balloonists vary their elevation to catch the best wind. On leaving their launching point in the Swiss Alps, Piccard and Jones entered a strong wind flowing from west to east. Despite a few problems, fuel supplies, and heaters, the balloonists circled the globe and touched down in Egypt.

In this chapter, we explore the envelope of air—the atmosphere—through which Piccard and Jones traveled. We begin by learning where the gases came from and how the atmosphere evolved in the context of the Earth System. Then we look at the structure of the atmosphere, and the global-scale and local-scale circulation of the lowermost layer; this circulation ultimately controls the weather, and can lead to the growth of storms. We conclude by exploring **climate,** the average weather conditions during the year.

20.2 THE FORMATION OF THE ATMOSPHERE

When the Earth formed about 4.57 Ga, it was initially surrounded by gas molecules gravitationally attracted to its surface from the gas and dust rings surrounding the newborn Sun (see Chapter 1). This "primary atmosphere," which consisted mostly of hydrogen and helium and traces of other gases, survived only a short time. Heat from the Sun caused the light atoms (H and He) in it to move about so rapidly that they eventually escaped the attraction of Earth's gravity. Effectively, the primary atmosphere leaked into space and was then blown away by the solar wind.

Even as the primary atmosphere was disappearing, volcanic activity on Earth released new gases that accumulated to form a "secondary atmosphere." The elements in these gases had been bonded to minerals inside the Earth, but when melting released them they bubbled out of volcanoes. Volcanic gas consists of about 70% to 90% water (H_2O), with lesser amounts of carbon dioxide (CO_2), and sulfur dioxide (SO_2), along with traces of other gases including nitrogen (N_2) and ammonia (NH_3). The secondary atmosphere consisted of these volcanic gases plus, according to some researchers, other gases brought to Earth by comets.

The original secondary atmosphere of the Earth has undergone major evolutionary changes throughout geologic time. Specifically, when the Earth cooled sufficiently for water to condense, probably by 3.9 to 3.8 Ga (though possibly earlier), rains extracted most of the water from the atmosphere. This water accumulated on the surface and either filled oceans, lakes, and streams or sank underground to become groundwater, so the proportion of water in the atmosphere decreased. Once liquid water existed at Earth's surface, the concentration of CO_2 in the atmosphere began to decrease. This occurred because CO_2 dissolves in oceans and then combines with calcium to form solid carbonate minerals that precipitate to the sea floor. CO_2 also reacts with rocks exposed on the surface of continents to produce solid chemical weathering products (see Chapter 7). Also over time, ultraviolet radiation from the Sun split apart molecules of NH_3 to produce nitrogen and hydrogen atoms. The light-weight hydrogen atoms escaped into space, but the nitrogen atoms combined to form N_2 molecules. Molecular nitrogen is a stable gas that does not chemically react with rocks, so once it has formed it remains in the air for a long time. Because of the accumulation of new N_2 molecules, and the loss of atmospheric H_2O and CO_2, the proportion of nitrogen in the atmosphere gradually increased.

It is noteworthy that if the Earth's surface had been too warm for liquid water to exist on it, then CO_2 would not have been removed from the atmosphere, and Earth's atmosphere today would resemble the present atmosphere of Venus. Venus's atmosphere, currently contains 96.5% CO_2, whereas Earth's contains only 0.033%. CO_2 is a greenhouse gas, meaning that it traps heat in the atmosphere. The high concentration of CO_2 makes Venus's atmosphere so hot that lead can melt at the planet's surface.

If you were suddenly to travel back through time and appear on Earth 3 billion years ago, you would instantly suffocate, for the atmosphere back then contained virtually no molecular oxygen (O_2). It took the appearance of life on Earth to add significant quantities of oxygen (O_2) to the air, for O_2 is produced by photosynthesis. The first photosynthetic organisms, cyanobacteria ("blue-green algae"), appeared on Earth between 3.8 and 3.5 Ga and began to add O_2 to the atmosphere. By 2 Ga, Earth's atmosphere contained about 1% of its present oxygen level. Oxygen concentrations increased very slowly. At about 1.2 Ga, there may have been a boost in the production of O_2 with the appearance of photosynthetic algae. Only around 0.6 Ga did oxygen levels reach 10% of their present level. As more species of photosynthetic organisms appeared, still more oxygen entered the atmosphere, with oxygen approaching present values by 400 to 250 Ma (▶Fig. 20.2).

Oxygen is important not only because it allows complex multicellular organisms to breathe, but also because it supplies the raw components for the production of **ozone** (O_3), a gas that absorbs harmful ultraviolet (short-wavelength) radiation from the Sun. Ozone, which accumulates primarily at an elevation of about 30 km, forms by a two-step reaction:

(1) O_2 + energy (from the Sun) → 2O (2 oxygen atoms that are not bonded together);

(2) $O_2 + O \rightarrow O_3$.

Only when enough ozone had accumulated in the atmosphere could life leave the protective blanket of seawater, which also absorbs ultraviolet radiation. When this happened, terrestrial plants and animals could evolve. These organisms have themselves interacted with and modified the atmosphere ever since.

In sum, Earth's atmosphere today consists mostly of volcanic gas modified by interactions with the land and life. It must be maintained, though—if all life were to vanish and all volcanic activity to shut off, lighter gases would leak into space in only a few million years.

20.3 THE ATMOSPHERE IN PERSPECTIVE

The Components of Air

Completely dry air consists of 78% nitrogen and 21% oxygen. The remaining 1% includes several gases in trace amounts. Trace gases are important in the Earth System. They include carbon dioxide (CO_2) and methane (CH_4), which are greenhouse gases that regulate Earth's atmospheric temperature; greenhouse gases allow solar radiation

FIGURE 20.2 Stages in the evolution of the atmosphere.

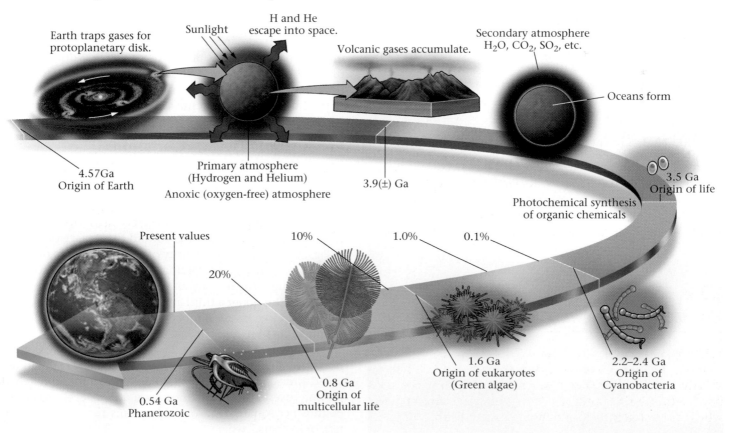

from the Sun to pass through, but trap infrared radiation rising from the Earth's surface. Trace gases also include ozone (O_3), which protects the surface from ultraviolet radiation. Not all trace gases are good, though. Radon, produced by the natural decay of uranium in rocks, is radioactive and can cause cancer; it can seep from the Earth and collect in dangerous concentrations in the basements of houses.

In addition to gases, the air contains trace amounts of **aerosols.** These tiny particles (less than 1 micrometer) of liquid or solid material are so small that they remain suspended in the air, just as fine mud remains suspended in river water. Aerosols include tiny droplets of water and acid and microscopic particles of sea salt, volcanic ash, clay, soot, and pollen (▶Fig. 20.3).

Atmospheric Pollutants

During the past two centuries, human activity has added substantial amounts of pollutants (both gases and aerosols) to the air, primarily through the burning of fossil fuels and through industrial operations. Pollutants include sulfate (SO_4^{-2}) and nitrate (NO_3^-) molecules, which react with water to make a weak acid that then falls from the sky as **acid rain.** Where acid rain falls, lakes, streams, and the ground become more acidic and thus toxic to fish and vegetation (particularly coniferous trees). The burning of fossil fuels has significantly increased the amount of CO_2 in the atmosphere. Because CO_2 is a greenhouse gas, this increase might contribute to "global warming," a rise in the average atmospheric temperature. Recently, atmospheric scientists have discovered that certain pollutants, notably chlorofluorocarbons (CFCs), react with ultraviolet light from the Sun to release chlorine atoms, which in turn react with ozone

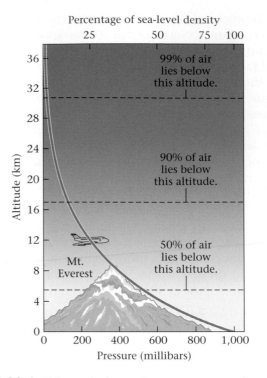

FIGURE 20.4 This graph shows air pressure versus elevation on the Earth. Note that climbers on top of Mt. Everest breathe an atmosphere that contains only about 33% of the atmospheric gases at sea level. A commercial airliner flies through air that is only 20% as dense as the air at sea level.

and break it down. These reactions appear to happen mainly in high clouds above polar regions during certain times of the year, thus preferentially removing ozone from these regions to create an "ozone hole." We'll look further at the human impact on the atmosphere in Chapter 23.

Pressure and Density Variations

Air is not uniformly distributed in the atmosphere. In the Earth's gravity field, the weight of air at higher elevations presses down on and compresses air at lower elevations. **Air pressure,** the push that air can exert on its surroundings, and air density, therefore increases toward the surface of the Earth (▶Fig. 20.4). Because the density of a gas reflects the number of gas molecules in a given volume, a gulp of air on the top of Mt. Everest, where the air pressure is about one-third that at sea level, contains about a third as many O_2 molecules as air at sea level. Therefore, most climbers seeking to reach Mt. Everest's summit must breathe bottled oxygen. Similarly, the cabin of an airliner must be pressurized to provide adequate oxygen for normal breathing (Fig. 20.4). We measure air pressure in units called atmospheres (atm), where one atm is approximately the pressure exerted by the atmosphere at sea level (about 14.7 pounds per square inch, or 1,035 grams per square centimeter), or in "bars," where 1 bar is about 0.986 atm.

FIGURE 20.3 A recent forest fire produces smoke that adds aerosols (e.g., soot), as well as CO_2 gas, to the atmosphere.

Because of the decrease in air density with elevation, 50% of the atmosphere's molecules lie below an elevation of 5.6 km, 90% lie below 16 km, and 99.99997% lie below 100 km. Thus, even though the outer edge of the atmosphere, a vague boundary where the gas density becomes the same as that of interplanetary space, lies as far as 10,000 km from the Earth's surface, most of the atmosphere's molecules lie within a shell only 0.5% as wide as the solid Earth. Though thin, the atmospheric shell contains sufficient gas to turn the sky blue (see Box 20.1).

Heat and Temperature

The molecules that constitute the atmosphere, or any gas, are not standing still but are constantly moving. We refer to the *total* kinetic energy (energy of motion; see Appendix A) resulting from the movement of molecules in a gas as its thermal energy, or heat. Note that heat and temperature are not the same—a gas's temperature is a measure of the *average* kinetic energy of its molecules. A volume of gas with a small number of rapidly moving molecules has a higher temperature but may contain less heat than a volume with a large number of slowly moving molecules. If we add heat to a gas, its molecules move faster and its temperature rises, and the gas will try to expand to occupy a larger volume.

Relations between Pressure and Temperature

When air moves from a region of higher pressure to a region of lower pressure, without adding or subtracting heat, it expands. When this happens, the air temperature decreases. Such a process is called **adiabatic cooling** (from the Greek *adiabatos,* meaning "impassable"; air cools at 6°–10°C per kilometer that it rises). The reverse is also the case: if air moves from a region of lower pressure to a region of higher pressure, without adding or subtracting heat, it contracts, and the air temperature increases. Such a process is called **adiabatic heating.** Adiabatic cooling and heating are important processes in the atmosphere, because pressure changes with elevation—when air near the ground surface (where pressure is higher) flows up to higher elevations (where pressure is less), it undergoes adiabatic cooling, but when air from high elevations flows down and compresses, it undergoes adiabatic heating.

Water in the Air: Relative Humidity and Latent Heat

Earlier, we examined the percentages of gases in completely dry (water-free) air. In reality, however, air contains variable amounts of water—from 0.3% above a hot desert to 4% in a rainforest during a heavy downpour.

Meteorologists, scientists who study the weather, specify the water content of air by a number called the **relative humidity,** the ratio between the measured water content and

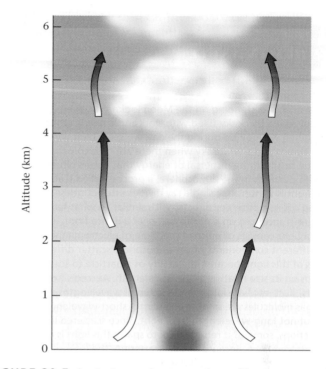

FIGURE 20.5 As air rises and enters regions of lower pressure, it expands and adiabatically cools. If the air contains moisture, the moisture condenses (it turns into water droplets or ice crystals) after the air has risen high enough. Here, we see moist air rising and becoming less dense; its moisture condenses at elevations above 3 km and produces a cloud.

the maximum possible amount of water the air could hold, expressed as a percentage.* The maximum possible amount varies with temperature—warmer air can hold more water than colder air. When air contains as much water as possible, it is saturated, while air with less water is undersaturated. If we say that air at a given temperature has a relatively humidity of 20%, we mean the air contains only 20% of the water that it could hold at that temperature when saturated. Such air feels dry. Air with a relative humidity of 100% is saturated and feels very humid, or damp.

Because cold air can't hold as much water as warm air, air that is undersaturated when warm may become saturated when cooled, without the addition of any new water. The temperature at which the air becomes saturated is called the **dewpoint temperature;** dew forms when undersaturated air cools at night and becomes saturated, so that water condenses on surfaces. When the dewpoint temperature is below freezing, frost develops. And when moist air rises and adiabatically cools, its moisture condenses to form a **cloud,** a mist of tiny droplets (▶Fig. 20.5).

---

*"Relative humidity" differs from "absolute humidity." The latter term refers to the mass of water in a volume of air. Absolute humidity is given in grams per cubic meter.

three layers has essentially the same proportion of different gases regardless of location. For this reason, atmospheric scientists refer to the troposphere, stratosphere, and mesosphere together as the "homosphere." In contrast, atoms and molecules in the low-density thermosphere collide so infrequently that this layer does not homogenize. Rather, gases separate into distinct layers based on composition, with the heaviest (nitrogen) on the bottom, followed in succession by oxygen, helium, and at the top, hydrogen, the lightest atom. To emphasize this composition, atmospheric scientists refer to the thermosphere as the "heterosphere."

So far, we've distinguished atmospheric layers according to their thermal structure (troposphere, stratosphere, mesosphere, and thermosphere) and according to the degree their gases mix (homosphere and heterosphere). We need to add one more "sphere" to our discussion. The **ionosphere** is the interval between 60 and 400 km, and thus includes most of the mesophere and the lower part of the thermosphere. It was given its name because in this layer, short-wavelength solar energy strips nitrogen molecules and oxygen atoms of their electrons and transforms them into positive ions. The ionosphere plays an important role in modern communication in that, like a mirror, it reflects radio transmissions from Earth so that they can be received over great distances.

The ionosphere also hosts a spectacular atmospheric phenomenon, the auroras (**aurora borealis** in the Northern Hemisphere and **aurora australis** in the Southern), which look like undulating, ghostly curtains of varicolored light in the night sky (►Fig. 20.8). They appear when charged particles (protons and electrons) ejected from the Sun, especially when solar flares (erupt), reach the Earth and interact with the ions in the ionosphere, making them release energy. Auroras occur primarily at high latitudes because Earth's magnetic field traps solar particles and carries them to the poles.

FIGURE 20.8 The splendor of an aurora borealis lights up the night sky in Arctic Canada. The colors result when particles emitted from the Sun interact with atoms in the thermosphere of the Earth.

20.4 WIND AND GLOBAL CIRCULATION IN THE ATMOSPHERE

A gusty breeze on a summer day, the steady "trades" that once blew clipper ships across the oceans, and a fierce hurricane all are examples of the wind, the movement of air from one place to another. We can feel the wind, because of the impacts of air molecules as they strike us. The existence of wind illustrates that the lower part of the atmosphere is in constant motion, swirling and overturning at rates between a fraction of a kilometer and a few hundred kilometers per hour. This circulation happens on two scales, local and global. "Local circulation" refers to the movement of air over a distance of tens to a thousand kilometers. "Global circula-

tion" refers to the movement of volumes of air in paths that ultimately carry it around the entire planet. (We can picture local circulation as eddies in global-scale "rivers" of air.) To understand both kinds of circulation, we must first see what drives air from one place to another, and examine energy inputs into the atmosphere.

Lateral Pressure Changes and the Cause of Wind

The air pressure of the atmosphere not only changes vertically, it also changes horizontally at a given elevation. The rate of pressure change over a given horizontal distance, called a pressure gradient, can be represented by the slope of a line on a graph plotting pressure on the vertical axis (specified in bars) and distance on the horizontal axis (►Fig. 20.9a). Winds form wherever a pressure gradient exists. Air always flows from a high-pressure region to a low-pressure region; in other words, it flows "down" a pressure gradient. To see why, step on one end of a long balloon filled with air; you momentarily increase the pressure at that end, so the air flows toward the other end (►Fig. 20.9b).

We can use a map to represent air pressure at a given elevation. A line on a map along which the air has a specified pressure is called an **isobar** (►Fig. 20.9c). In other words,

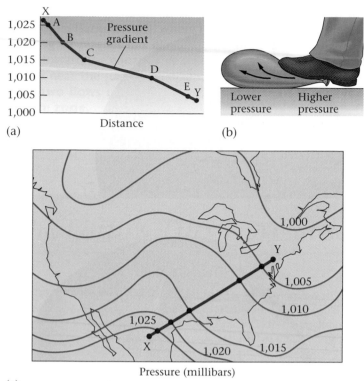

(a)

(b) Lower pressure   Higher pressure

(c) Pressure (millibars)

FIGURE 20.9 (a) This graph shows a profile from location X to location Y. Notice that the pressure gradient, the slope of the line, is greater between X and C than between C and D. (b) Air flows from a high-pressure region to a low-pressure region to cause wind. If you step on one end of a balloon, for example, you increase the pressure there, so that the air flows toward the unsqueezed end. (c) Isobars on a map are lines of equal pressure. Every place along the 1,005 isobar is experiencing a pressure of 1,005 millibars (1 millibar = 0.001 bar). By moving in any direction not parallel to the isobars, you will feel a pressure change. For example, if you walk from X to Y, you will experience a decrease in pressure.

the pressure is the same all along an isobar. Isobars can never touch, because they represent different values of pressure. A difference in pressure exists between one isobar and the next, so the air starts flowing perpendicular to these lines. As we will see, however, the Coriolis effect modifies wind direction.

Energy Input into the Atmosphere: Convection

Ultimately, air circulation, like the movement of water in a heated pot, results from convection: heated air expands and becomes less dense, so it rises to be replaced by sinking cooler, denser air. In the case of Earth's atmosphere, the energy comes from solar radiation. Solar energy constantly bathes the Earth. Of this energy, 30% reflects back to space (off of clouds, water, and land); air or clouds absorb 19%; and land and water absorb 51%. The energy absorbed by land and water later reradiates as infrared radiation and thus bakes the atmosphere from below. As noted earlier, greenhouse gases absorb part of this reradiated energy before it can return to space.

FIGURE 20.10 (a) A flashlight beam aimed straight down produces a narrower and more intense beam than a flashlight aimed obliquely. Thus, the area under the straight beam heats up more than the area under the oblique beam. (b) Sunlight hitting the Earth near the equator provides more heat per unit area of surface than sunlight hitting the Earth at a polar latitude. That is why the poles are colder.

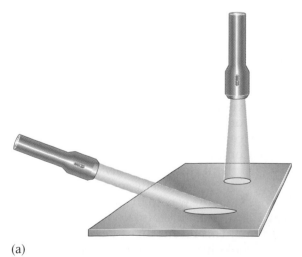

(a)

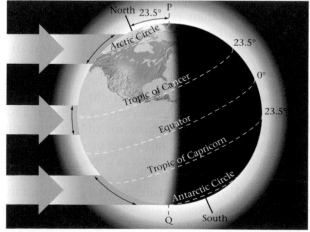

(b)

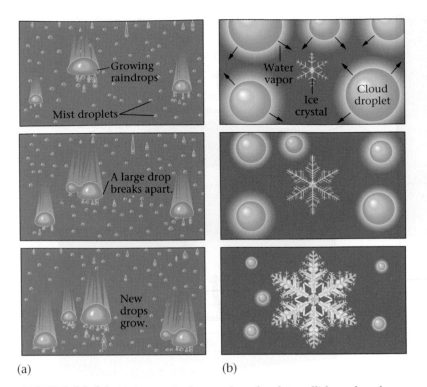

(a)   (b)

FIGURE 20.21 (a) Some rain forms when droplets collide and coalesce. Once a drop becomes large enough to fall, it incorporates more drops on the way down. Note that drops are flattened at their base because of air resistance (real raindrops are not teardrop-shaped). When a drop gets too big, it splits in two. (b) During the Bergeron process, water drops evaporate, releasing vapor that attaches to growing snowflakes, which then fall. If the air below a cloud is warm enough, the snowflakes melt and turn to rain before hitting the ground.

before it hits the ground. This kind of precipitation, involving the growth of ice crystals in a cloud at the expense of water droplets, is called the **Bergeron process,** after Tor Bergeron, the Swedish meteorologist who discovered it (▶Fig. 20.21b).

Many kinds of clouds form in the troposphere. It wasn't until 1803, however, that Luke Howard, a British naturalist, proposed a simple terminology for describing clouds (▶Fig. 20.22). First, we divide clouds into types based on their shape: puffy, cotton-ball- or cauliflower-shaped clouds are **cumulus** (from the Latin word for "stacking"). Clouds that occur in relatively thin, stable layers and thus have a sheet-like or layered shape are called **stratus.** Clouds that have a wispy shape and taper into delicate, feather-like curls are called **cirrus.** We can then add a prefix to the name of a cloud to indicate its elevation: high-altitude clouds (above about 7 km) take the prefix "cirro," mid-altitude clouds take the prefix "alto," and low-altitude clouds (below 2 km) do not have a prefix. Finally, we add the suffix "nimbus" or the prefix "nimbo" if the cloud produces rain.

Applying this cloud terminology, we see that a nimbostratus is a layered, sheet-like raincloud, and a "cumulonimbus" is a rain-producing puffy cloud. Cumulonimbus clouds can be

truly immense, with their base lying at less than 1 km high and their top butting up against the tropopause at an elevation of over 14 km. These clouds are "vertically developed," because they grow across altitude divisions. Large cumulonimbus clouds spread laterally at the tropopause to form broad, flat-topped clouds called "anvil clouds."

The differences in cloud types depend on whether the clouds develop in stable or unstable air. "Stable air" does not have a tendency to rise, because it is colder than its surroundings. "Unstable air" has a tendency to rise, because it is warmer than its surroundings. Cumulus clouds, which in time-lapse photography look like they're boiling, form in unstable air. They billow because of updrafts (upward-moving air) and downdrafts (downward-moving air)—plane flights through these clouds will be rather bumpy. In contrast, stratus clouds indicate stable air.

20.6 STORMS: NATURE'S FURY

The rain from an overcast sky may be inconvenient for a picnic, but it won't threaten life or property and will be appreciated by farmers. In contrast, a storm poses a threat. A **storm** is an episode of severe weather, when winds, rainfall, snowfall, and in some cases lightning become strong enough to be bothersome and even dangerous (▶Fig. 20.23). Storms form where large pressure gradients develop (as may exist across a front), for pressure gradients produce strong winds. Storms also form where local conditions cause warm, moist air to rise. Rising moist air can trigger a storm because when the air reaches higher elevation, it condenses and, as we have seen, releases latent heat. This heat warms the air, makes it more buoyant, and thus causes it to rise still further, until it becomes cool enough to produce clouds. Meanwhile, at ground level, new moist air flows in beneath the clouds to replace the air that has already risen—this new air then immediately starts to rise and causes the clouds to build upwards. Warm moist air effectively feeds the storm. Once the clouds become thick enough to start producing heavy rain, and/or the wind becomes strong enough to be troublesome, we can say that a storm has been born. We'll now look at various types of storms.

Thunderstorms

A "thunderstorm" is a local episode of intense rain accompanied by strong, gusty winds, and by lightning. Over 2,000 are occurring at any given time somewhere on the Earth, and over 100,000 take place in the United States every year. Thunderstorms form where a cold front moves into a region of particularly warm, moist air (as happens, for example, in

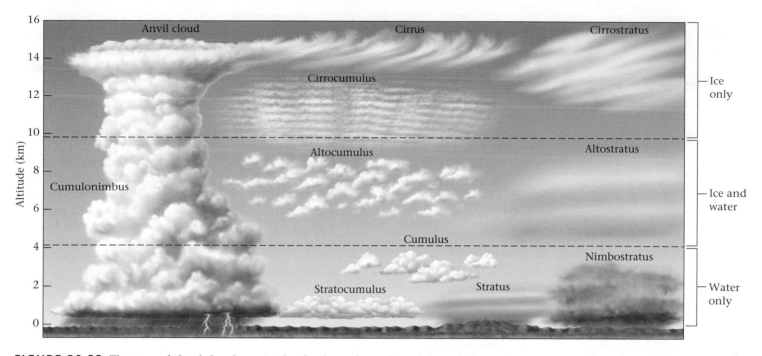

FIGURE 20.22 The type of cloud that forms in the sky depends on the stability of the air, the elevation at which moisture condenses, and the wind speed. Note that cumulonimbus clouds develop vertically, in that they grow across elevation boundaries.

North America's mid-latitudes during the summer where cold polar air masses collide with warm gulf air masses); where convective lifting is driven by solar radiation in a region with an immense supply of moisture (as occurs over tropical rain forests); or where orographic lifting causes clouds to form over mountains.

A typical thunderstorm has a relatively short life, lasting from under an hour to a few hours (▶Fig. 20.24). The

FIGURE 20.23 A thunderstorm can drench an area with rain and attack it with lightning.

storm begins when a cumulonimbus cloud, fed by a steady supply of warm, moist air, grows large. The rising hot air, kept warm by the addition of energy from the latent heat of condensation, creates updrafts that cause the cloud to stack, or billow upward, toward the tropopause. When this air adiabatically cools, precipitation begins.

If updrafts in the cloud are strong enough, ice crystallizes in the higher levels of the cloud, where temperatures are below freezing, building into ice balls known as **hail** (or hailstones). A discrete mass of hail may fall from a cloud over a few minutes to form a hail streak on the ground, typically 2 by 10 km and elongated in the direction that the storm moves. Though most hailstones are pea-sized, the largest recorded hailstone reached a diameter of 14 cm and weighed 0.7 kilograms.

Once precipitation begins, a thunderstorm has reached its "mature stage." Now, falling rain pulls air down with it, creating strong downdrafts. By this stage, the top of the cloud has reached the top of the troposphere and begins to spread laterally to form an anvil cloud. Because of the simultaneous occurrence of updrafts and downdrafts, a mature thunderstorm produces gusty winds and the greatest propensity for lightning. Eventually, downdrafts become the overwhelming wind; their cool air cuts off the supply of warm, moist air, so the thunderstorm dissipates.

Lightning accompanies thunderstorms because electrical charges separate in a storm cloud (▶Fig. 20.25a). Surprisingly, meteorologists still aren't sure why this happens, but they

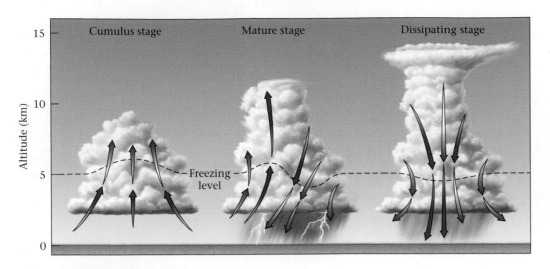

FIGURE 20.24 A thunderstorm evolves in three stages. First, warm, unstable moist air rises and builds a cumulus cloud. Second, updrafts, fueled by more warm air at the surface and the heat released by condensation, cause the cloud to billow higher. Heavy rain falls, and lightning flashes occur and the downpour creates cold downdrafts. Finally, the cloud reaches the top of the troposphere, but downdrafts cut off the supply of warm air at the ground surface, and without this fuel, the thunderstorm dissipates.

can become very large until a giant spark or pulse of current, a **lightning flash,** jumps across the gap—essentially, lightning is like a giant short circuit across which a huge (30-million-volt) pulse of electricity flows.

Lightning flashes can jump from one part of a cloud to another, or from a cloud to the ground. In the case of cloud-to-ground lightning, the flash begins when electrons leak from the negatively charged base of the cloud incrementally downward across the insulating air gap, creating a conductive path, or leader. While this happens, positive charges flow upward to the cloud through conducting materials, like trees or buildings (▶Fig. 20.25b). The instant that the charge flows connect, a

speculate that the rubbing of air and water molecules together in a cloud creates positively charged hydrogen ions and/or ice crystals that drift to the top of the cloud, and negatively charged OH⁻ ions and/or water droplets that sink to the base. The negative ions at the base of the cloud repel negative ions on the ground below, creating a zone of positive charge on the ground. Air is a good insulator, so the charge separation strong current carries positive charges up into the cloud; this upward-flowing current is called the "return stroke" (▶Fig. 20.25c).

We hear **thunder,** the cracking or rumbling noise that accompanies lightning, because the immense energy of a flash almost instantaneously heats the surrounding air to a temperature of 8,000 to 33,000°C, and this abrupt expansion,

FIGURE 20.25 (a) Lightning flashes when a charge separation develops in a cloud, with a negative charge at the base and a positive charge at the top. The negative charge repels negative charges on the ground, so positive charges develop on the ground. A leader begins to descend from the cloud. (b) As the leader grows downward, positive charges begin to flow upward from an object on the ground. (c) When the connection is complete, the return stroke carries positive charges rapidly from the ground to the cloud, creating the main part of the flash.

(a)   (b)   (c)

like an explosion, creates sound waves that travel through the air to our ears. But because sound travels so much more slowly than light, we hear thunder after we see lightning. A five-second time delay between the two means that the lightning flashed about 1.6 km (1 mile) away.

Over eighty people a year die from lightning strikes in the United States alone, and many more are seriously burned or shocked. Lightning that strikes trees heats the sap so quickly that the trees literally explode. Lightning can spark devastating forest fires and set buildings on fire. You can reduce the hazard to buildings by installing lightning rods, upward-pointing iron spikes that conduct electricity directly to the ground so that it doesn't pass through the building.

Tornadoes

Some thunderstorms grow to be quite violent, and spawn one or more tornadoes. A **tornado** is a near-vertical, funnel-shaped cloud in which air rotates extremely rapidly around the axis (center line) of the funnel (▶Fig. 20.26a). In other words, a tornado is a vortex beneath a severe thunderstorm. The word probably comes either from the Spanish *tonar,* meaning "to turn," or *tronar,* meaning "thunder" (perhaps referring to the loud noise generated by a tornado).

In the mid-latitudes of the Northern Hemisphere, where most tornadoes form, air in the funnel rotates counterclockwise around the center and spirals upward. Air in the fiercest tornadoes probably moves at speeds of up to 500 km per hour (about 300 mph). The diameter of the base of the funnel in a small tornado may be only 5 m across, while in the largest tornadoes it may be as wide as 1,500 m across. Because of the upward movement of air, air pressure within a tornado drops.

In North America, tornadoes drift with a thunderstorm from southwest to northeast, because of the prevailing wind direction, traveling at speeds of 0 to 100 km per hour. They tend to hopscotch across the landscape, touching down for a stretch, then rising up into the air for a while before touching down again. This characteristic leads to a bizarre incidence of damage—one house may be blasted off its foundation while its next-door neighbor remains virtually unscathed (▶Fig. 20.26b). Small tornadoes may cut a swath less than a kilometer long, but large tornadoes raze the ground for tens of kilometers, and the largest have left a path of destruction up to 500 km long (▶Fig. 20.27a, b). In 1925, for example, one of the most enormous tornadoes on record ripped across Missouri, Illinois, and Indiana, killing 689 people before it dissipated. In some cases, two or three tornadoes may erupt from a single thunderstorm. Massive thunderstorm fronts may produce a tornado swarm, dozens of tornadoes out of the same storm. In April 1974, a single thunderstorm system generated a swarm of at least 148 individual tornadoes, which killed 307 people over eleven states. On November 11, 2002, a chain of thunderstorms covering a belt from Ohio to Alabama spawned 66 tornadoes that left 66 people dead. The

(a)

(b)

FIGURE 20.26 (a) A tornado touches down. (b) The intense winds of a tornado rip through houses and scatter debris.

death toll would have been higher were it not for warnings broadcast by the U.S. National Weather Service, which sent people scrambling for safety.

Tornadoes cause damage both because of the force of their rapidly moving wind and because of their low air pressure. The wind lifts trucks and tumbles them for hundreds of meters, uproots trees, and flattens buildings (Fig. 20.26). Particularly large tornadoes can even rip asphalt off a highway. The low air pressure around a tornado can make windows pop out of buildings and may cause airtight buildings

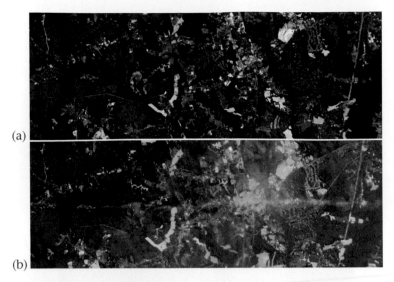

(a)

(b)

FIGURE 20.27 Two satellite images showing the swath cut by an F5 tornado that passed through Maryland on April 28, 2002. The field of view is 6 × 17.8 km. (a) Before the tornado; (b) after the tornado. Red is vegetation. The turquoise strip shows where the tornado stripped vegetation away. The swath here is about 150 m wide and 5 km long.

to explode, as the air inside suddenly expands relative to the air outside. In some cases, tornadoes cause strange kinds of damage: they have been known to drive straw through wood, lift cows and carry them unharmed for hundreds of meters, and raise railroad cars right off the ground.

Because of the range of damage a tornado can cause, T. T. Fujita, of the University of Chicago, proposed a scale that distinguishes among tornadoes on the basis of wind speed, path dimensions, and possible damage (see Table 20.2). The wind speeds in the **Fujita scale** are estimates,

based on the damage assessment, as no one has yet succeeded in measuring winds directly in a tornado.

Tornadoes in North America form where strong westerlies exist at high altitudes while strong southeast winds develop near the ground surface (▶Fig. 20.28a–c). These opposing winds shear the air between them, so that the air begins to rotate in a horizontal cylinder. Tornado watchers search for the resulting funnel cloud, as a precursor to a true tornado. As the associated thunderstorm matures, updrafts tilt one end of the cylinder up and downdrafts push the other end down, until the air in the cylinder starts spiraling inward and upward, gaining speed like spinning figure skaters who pull their arms inward. At this stage, the cloud consists only of rotating moisture, and may look greyish white. But if the process continues, the low end of the funnel touches ground, and at the instant of contact dirt and debris get sucked into the tornado, giving it a dark color.

The special weather conditions that spawn tornadoes in the Midwestern United States and Florida develop when cold polar air from Canada collides with warm tropical air from the Gulf of Mexico. These conditions happen most frequently during the months of March–September. So many tornadoes occur during the summer in a belt from Texas to Indiana that this region has the unwelcome nickname "tornado alley" (▶Fig. 20.29). During a thirty-year span, the number of reported tornadoes per year ranged between about 420 and 1,100 in the United States (with an average of 770 per year; fewer than twenty strike Canada annually). On average, about eighty people a year die in tornadoes. But a single F5 event may kill hundreds.

Because of the threat tornadoes pose to life and property, meteorologists have worked hard to be able to forecast them. First they search for appropriate weather conditions. If these conditions exist, meteorologists issue a "tornado watch." If

TABLE 20.2 Fujita Scale for Tornadoes

Scale	Category	Wind Speed km/h (mph)	Path Length; Path Width	Typical Damage
F0	Weak	64–116 (40–72)	0–1.6 km; 0–17 m	Branches and windows broken.
F1	Moderate	117–180 (73–112)	1.6–5.0 km; 18–55 m	Trees broken; shingles peeled off; mobile homes moved off their foundations.
F2	Strong	181–253 (113–157)	5–16 km; 56–175 m	Large trees broken; mobile homes destroyed; roofs torn off.
F3	Severe	254–332 (158–206)	16–50 km; 176–556 m	Trees uprooted; cars overturned; well-constructed roofs and walls removed.
F4	Devastating	333–418 (207–260)	50–160 km; .56–1.5 km	Strong houses destroyed; buildings torn off foundations; cars thrown; trees carried away.
F5	Incredible	419–512 (261–318)	160–500 km; 1.5–5.0 km	Cars and trucks carried more than 90 m; strong houses disintegrated; bark stripped off trees; asphalt peeled off roads.

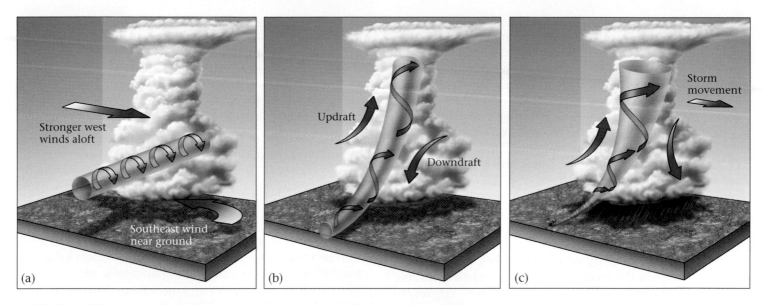

FIGURE 20.28 (a) Tornadoes initiate at intense fronts where high-altitude westerlies flow over low-altitude southeasterlies. The resulting shear creates a horizontal cylinder of rotating air in a cloud. (b) Updrafts and downdrafts in the cloud eventually tilt the cylinder, creating a tornado. (c) When the tornado touches down, destruction follows. Note that the spiraling winds of the funnel also circulate inside the main cloud; the tornado we see protruding from the base of the cloud is only the tip of a much larger flow.

observers spot an actual tornado forming, they issue a "tornado warning" for the region in its general path (the exact path can't be predicted). If you hear a warning, it's best to take cover immediately in a basement, or at least in an interior room away from windows. With the invention of Doppler radar, which uses the Doppler effect (see Chapter 1)

FIGURE 20.29 North American tornadoes are most common in "tornado alley," a band extending from Texas to Indiana, where the polar mass collides with the Gulf Coast maritime tropical air mass. The storm systems are urged eastward by the jet stream and related high-altitude westerlies.

Number of tornadoes per year (per 26,000 sq. km, for a 27-year period)

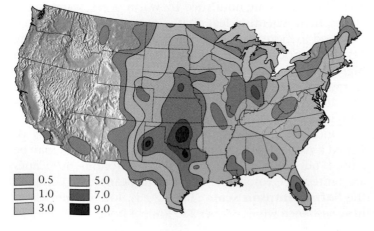

0.5	5.0
1.0	7.0
3.0	9.0

to identify rain moving in strong winds, meteorologists may detect tornadoes without even going outside.

Nor'easters

In some cases, large mid-latitude (wave) cyclones of North America affect the Atlantic coast. Because the cold, counterclockwise winds of these cyclones come out of the northeast, they are called **nor'easters.** Some nor'easters are truly phenomenal storms. One of the strongest occurred at the end of October 1991 (the "perfect storm" made famous by the book and movie of that name). During this storm, winds were not as strong as those of a hurricane, but they covered such a large area that waves in the open ocean built to a height of over 11 m, a disaster for ships. When the waves reached shore, they eroded huge stretches of beach. The rainfall from the storm caused extensive flooding inland.

Hurricanes

During the summer and early fall, cyclonic wind systems called "tropical disturbances" develop off the western coast of Africa, near the Cape Verde Islands (latitude 20°N). In these low-pressure regions, air converges and rises and, because of the Coriolis effect, begins to circulate counterclockwise. But because these storms form over the warm tropical waters of the central Atlantic, the air that rises within them is particularly warm and moist. As it rises, it cools, and moisture condenses, releasing latent heat of condensation.

This latent heat is a fuel that provides energy to the storm—it causes the air to rise still higher, creating even lower pressure at the Earth's surface, which in turn sucks up even more warm, moist air. Thus, the storm grows in strength until it becomes a "tropical depression," in which winds may reach 61 km per hour.

As long as there continues to be a supply of warm, moist air, the storm continues to grow broader, and air within begins to rotate still faster. When the sustained wind speed exceeds 119 km per hour, a hurricane has been born. In other words, a **hurricane** (named for the Carib god of evil) is a huge rotating storm, resembling a giant spiral in map view, in which sustained winds are greater than 119 km (74 miles) per hour (▶Fig. 20.30a). Within a hurricane, air pressure becomes much lower, because of the upward flow of air. Before the days of satellite forecasting, mariners closely watched their barometers, knowing that an extreme drop signaled the approach of a hurricane.

Traditionally, the designation "hurricane" applies to storms born in the east-central Atlantic that drift first westward and then northward, with the prevailing winds. The path, or "hurricane track," that they follow allows them to inflict damage to land regions in the Caribbean and the Gulf Coast and East Coast of North America (▶Fig. 20.30b). Occasionally, hurricanes make it across Central America and then thrash the Pacific coast of Mexico and the western United States; some may even drift northwestward to Hawaii. Because hurricanes require warm water (warmer than 27°C), they develop only at latitudes south of about 20°. They do not form close to the equator, because there is not enough atmospheric motion or Coriolis effect at that latitude. Similar storms that form at latitude 20°N and 20°S in the western Pacific are called "typhoons;" those at latitude 20°N in the Indian Ocean are called "cyclones" (a second use of the word; ▶Fig. 20.30c).

A typical hurricane consists of several spiral arms, called rain bands, extending inward to a central zone of relative calm known as the hurricane's eye. A rotating vertical cylinder of clouds, called the eye wall, surrounds the eye (▶Fig. 20.31). The entire width of a hurricane ranges from 100 to 1,500 km, with the average around 600 km. Winds spiral toward the eye, and thus their angular momentum, and therefore speed, increases toward the center; as a result, the greatest rotary wind velocity occurs in the eye wall.

Hurricanes as a whole move along their track because of the prevailing motion of the atmosphere. Typically, a hurricane's velocity along its track, the storm-center velocity, ranges from 0 to 60 km per hour (on rare occasions, as fast as 100 km per hour). Because of the rotary motion of wind around the eye, winds on one side of the eye move in the opposite direction to those on the other side. On the side of the hurricane where winds move in the same direction as the storm's center, the storm-center velocity adds to the rotary motion, making surface winds particularly fast. On the other side of the hurricane, the storm-center velocity subtracts from the rotary velocity, so the winds are slower.

Hurricanes cause damage in several ways (▶Fig. 20.32).

- *Wind:* Hurricane-force winds may reach such intensity that buildings cannot stand up to them. Hurricanes can tear off branches, uproot trees, rip off roofs, collapse walls, force vehicles off the road, knock trains from their tracks, and blast mobile homes off their foundations.

- *Waves:* The force of hurricane winds shearing across the sea surface generates huge waves, sometimes tens of meters high. Out in the ocean, these waves can swamp or capsize even large ships. Near shore, they batter beachside property, rip up anchored boats and carry them inland, and strip beaches of their sand.

- *Storm surge:* Coastal areas may be severely flooded by a **storm surge,** excess water that is carried landward by the hurricane. In the portion of a hurricane where winds blow onshore, water piles up and becomes deeper over a region of 60 to 80 km, allowing waves to break at a higher elevation than they otherwise would. Storm surges during hurricanes are exacerbated by the very low air pressure in a hurricane, which allows the sea surface to rise still higher—if a hurricane hits at high tide, the damage may be even worse. A storm surge during the 1900 hurricane that hit Galveston, Texas, flooded the city and killed 6,000 people. Storm surges in Bangladesh cause horrendous loss of life because the Indian Ocean submerges the low-lying delta plain of the Ganges. During one 1970 cyclone, the death toll reached 500,000.

- *Rainfall:* The intense rainfall during a hurricane (in some cases, half a meter of rain has fallen in a single day) causes streams far inland to flood. The flooding itself can submerge towns and cause death. Intense rains can also weaken the soil of steep slopes, especially in deforested areas, and thus can trigger mudslides.

On average, five Atlantic hurricanes happen every year. But in 1995, there were eleven—in fact, three or four tropical depressions and hurricanes existed at a single time, forming a deadly procession that marched across the Atlantic. Because hurricanes are nourished by warm ocean water, some atmospheric scientists fear that a warming of the climate may lead to more and fiercer hurricanes in the future. Many Atlantic hurricanes veer northward when they bump into high-pressure air masses over North America. This sends the storms up the eastern seaboard. Hurricanes die out when they lose their source of warm, moist air—when they cross onto land or move over the cold waters of high latitudes.

Once a storm reaches tropical-depression status, it is assigned a name. Names for less significant hurricanes may be reused, but the names of particularly notorious hurricanes are retired. Meteorologists classify hurricanes according to the **Saffir-Simpson scale** (Table 20.3). Intense hurricanes have sustained winds of over 230 km per hour.

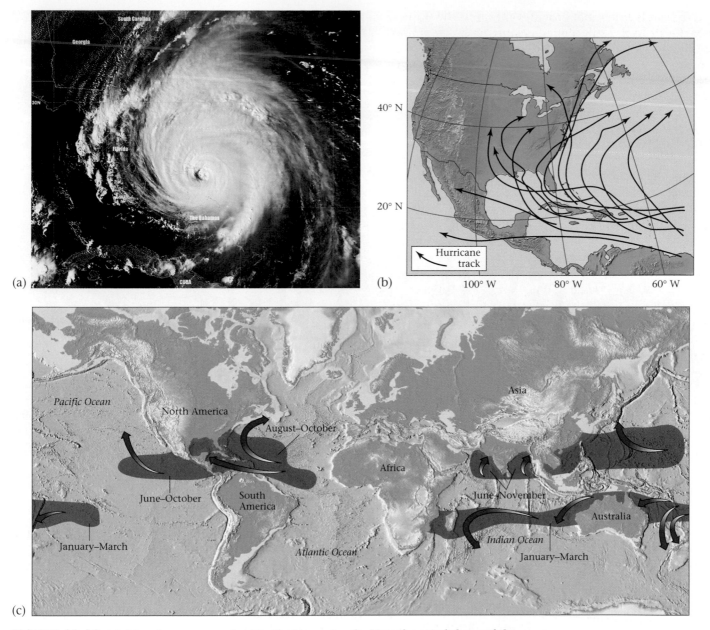

FIGURE 20.30 (a) A hurricane approaches the Florida peninsula. Note the spiral shape of the storm and the relatively small eye. (b) The tracks of several important Atlantic hurricanes show how most begin at latitudes of 15°–20° off the western coast of Africa, then drift westward and northward. (c) Some Atlantic hurricanes make it across Central America and lash the eastern Pacific. Similar storms occur in the western Pacific, where they are known as typhoons, and in the Indian Ocean, where they are known as cyclones; these storms also originate at latitudes of about 20°.

20.7 GLOBAL CLIMATE

When we talk about the weather, we're referring to the atmospheric conditions at a certain location at a specified time. But when we speak of the characteristic weather conditions, the typical range of conditions, the nature of seasons, and the possible weather extremes of that region over a long time (say, 30 years), we use the term "climate." For example,

on a given summer day in Winnipeg, central Canada, it may be sunny, hot, and humid, and on a given winter day in southern Florida, it may drop below freezing. But averaged over a year, the weather in Winnipeg is more likely to be cooler and drier than the weather in southern Florida. The variables that characterize the climate of a region include its temperature (both the yearly average and the yearly range), its humidity, its precipitation (both the yearly amount and

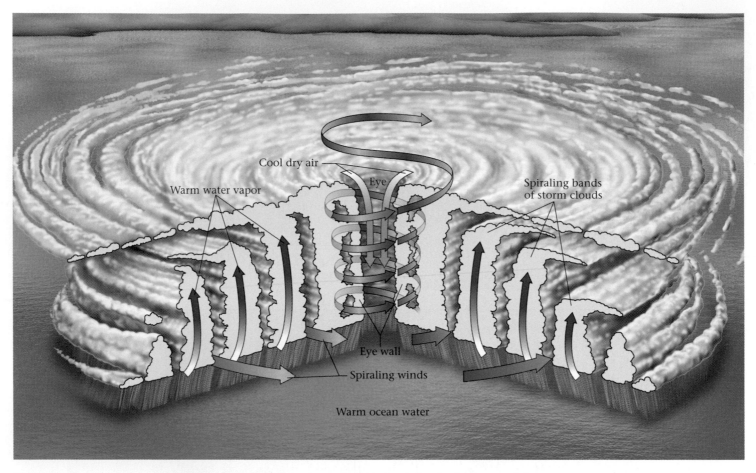

Cool dry air

Warm water vapor

Eye

Spiraling bands
of storm clouds

Eye wall

Spiraling winds

Warm ocean water

FIGURE 20.31 This cutaway diagram of a hurricane shows the spirals of clouds, the eye, and the eye wall. Dry air descends in the eye, creating a small region of calm weather.

FIGURE 20.32 In 1992, Hurricane Andrew caused immense damage in southern Florida before crossing the Gulf of Mexico and slamming into Louisiana.

the distribution during the year), its wind conditions, and the character of its storms.

Climate Controls and Belts

Climatologists, scientists who study the Earth's climate, suggest that several distinct factors control the climate of a region.

- *Latitude:* This is perhaps the most significant factor, for the latitude determines the amount of solar energy a region receives, as well as the contrasts between seasons. Polar regions, which receive much less solar radiation over the year, have colder climates than equatorial regions. And the contrast between winter and summer is greater in mid-latitudes than at the poles or the equator. We can easily see the influence of latitude by examining the global distribution of temperature, represented on a map by **isotherms,** lines along which the temperature is exactly the same (▶Fig. 20.33). Because land and sea do not heat up at the same rate, because the distribution of clouds is not uniform, and because ocean currents transfer heat across latitudes, isotherms are not perfect circles.

TABLE 20.3 Saffir-Simpson Scale for Hurricanes

Scale	Category	Wind Speed (km/h)	Air Pressure in Eye (millibars)	Damage
1	Minimal	119–153	980 or more	Branches broken; unanchored mobile homes damaged; some flooding of coastal areas; no damage to buildings; storm surge of 1.2–1.5 m.
2	Moderate	154–177	965–979	Some roofs, doors, and windows damaged; mobile homes seriously damaged; some trees blown down; small boat moorings broken; storm surge of 1.6–2.4 m.
3	Extensive	178–209	945–964	Some structural damage to small buildings; large trees blown down; mobile homes destroyed; structures along coastal areas destroyed by flooding and battering; storm surge of 2.5–3.6 m.
4	Extreme	210–250	920–944	Some roofs completely destroyed; extensive window and door damage; major damage and flooding along coast; storm surge of 3.7–5.4 m. Widespread evacuation of regions within up to 10 km of the coast may be necessary.
5	Catastrophic	over 250	less than 920	Many roofs and buildings completely destroyed; extensive flooding; storm surge greater than 5.4 m. Widespread evacuation of regions within up to 16 km of the coast may be necessary.

- *Altitude:* Because temperature decreases with elevation, cold climates exist at high elevations even at the equator. Hiking from the base of a high mountain at the Andes to its summit takes you through the same range of climate belts you would pass through on a hike from the equator to the pole.

- *Proximity of water:* Land and water have very different heat capacities (ability to absorb and hold heat). Land absorbs or loses heat quickly, while water absorbs or loses heat slowly. Also, water can absorb and hold on to more heat than land can because water is semi-transparent; sunlight heats water down to a depth of up to 100 m, while sunlight heats land down to a depth of only a few centimeters. Thus, the proximity of the sea tempers the climate of a region: as a rule, locations in the interior of a continent experience a much greater range of weather conditions than regions along the coast.

- *Proximity to ocean currents:* Where a warm current flows, it may warm the overlying air, and where a cold current flows, it may cool down the overlying air. For example, the Gulf Stream brings warm water north from the Gulf of Mexico and keeps Ireland, the United Kingdom, and Scandinavia much warmer than they would be otherwise.

- *Proximity to orographic barriers:* An **orographic barrier** is a landform (such as a mountain range) that diverts air flow upward or laterally. This diversion affects the amount of precipitation and wind a region receives.

- *Proximity to high- or low-pressure zones:* Zones of high and low pressure, roughly parallel to the equator, encircle the planet (Fig. 20.13). Because land and sea have different

FIGURE 20.33 Isotherms of January roughly parallel lines of latitude, owing to contrasts in amounts of insolation, except where they are distorted by oceanic currents. The Gulf Stream in the Atlantic Ocean deflects isotherms northward, granting the United Kingdom, Ireland, and Scandinavia milder climates than they might otherwise have. The pattern of isotherms is different in July.

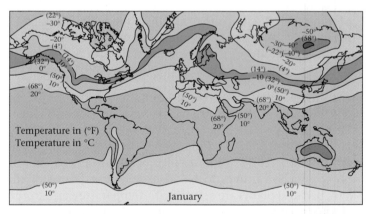

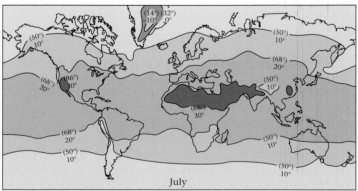

TABLE 20.4 Climate Types of the Earth

Climate Type	Regions and Characteristics
Tropical rainy	Tropical rainforests lie at equatorial latitudes and experience rain throughout the year. Rain commonly falls during afternoon thunderstorms. Tropical savanna (grasslands with brush and drought-resistant trees), which lie on either side of a rainforest, have a rainy season and a dry season. Rainforests and savannas may receive tropical monsoons.
Dry	Dry regions include deserts (regions with very little moisture or vegetation cover; vegetation that does exist has adapted to long periods without moisture) and steppes. Steppe regions (vast grassy plains with no forest) border the desert and have somewhat more precipitation. Some steppe regions occur at high elevations, in latitudes where the climates would otherwise be more humid.
Humid mesothermal	This category includes humid subtropical climates, with moist air and warm temperatures for much of the year, in which mixed deciduous-coniferous forest thrives; Mediterranean climates, coastal regions with most rainfall in the winter, very hot summers, and scrub forests; and marine west-coast climates, where the sea tempers the climate and may create a coastal temperate rainforest.
Humid microthermal	These higher-latitude temperate climates, which occur only in the Northern Hemisphere, include humid continental regions, with long summers (as in the U.S. Midwest and mid-Atlantic states), in which deciduous forest thrives; humid continental regions with short summers, characterized by mixed deciduous-coniferous forest or coniferous-only forest; and subarctic climates, with very short, cool summers and coniferous forest that becomes lower and scrubbier at higher latitudes.
Polar	These cold climates include tundra and ice caps. Tundra are regions with no summer and an extremely cold winter, in which only low, cold-resistant plants (moss, lichen, and grass) can survive. Much of the ground in tundra is permafrost (permanently frozen ground). In ice-cap regions, near the poles, the climate is sub-freezing year-round and, any land not covered by ice has essentially no vegetation cover. Highlands are regions that lie at lower (non-polar) latitudes but have such a high elevation that they have polar-like climates. When you enter a region above the treeline, you have entered a highland polar climate.

heat capacities, they modify the zones, so that high-pressure zones tend to be narrower over land. Meteorologists refer to the resulting somewhat elliptical regions of high or low pressure as "semipermanent pressure cells" (▶Fig. 20.34). These influence prevailing wind direction and relative humidity.

Climatologists who have studied the distribution of climatic conditions around the globe have developed a classification scheme for climates, based on such factors as the average monthly and annual temperatures and the total monthly and yearly amounts of precipitation. The vegetation of a region proves to be an excellent indicator of climate, because plants are sensitive to temperature and to the amount and distribution of rainfall. Table 20.4 lists the principal types of climate belts (▶Fig. 20.35).

Climate Variability: Monsoons and El Niño

The climate at a certain location may change during the course of one or more years. Here, we look at two important examples of climate variability that affect human populations in notoriously significant ways.

A **monsoon** is a major reversal in the wind direction that causes a shift from a very dry season to a very rainy season. In southern Asia, home to about half the world's population, people depend on monsoonal rains to bring moisture for their crops.

The Asian monsoon develops primarily because Asia is so large that it includes vast tracts of land far from the sea. Further, a substantial part of this land, the Tibet Plateau, lies

FIGURE 20.34 Because of land masses, atmospheric pressure belts vary in width to create lens-shaped, semipermanent high- and low-pressure cells. Compare this figure with Figure 20.13.

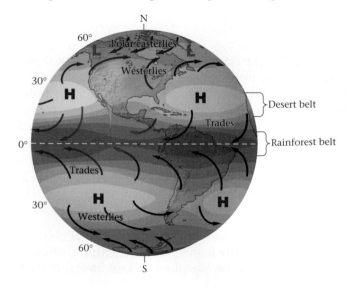

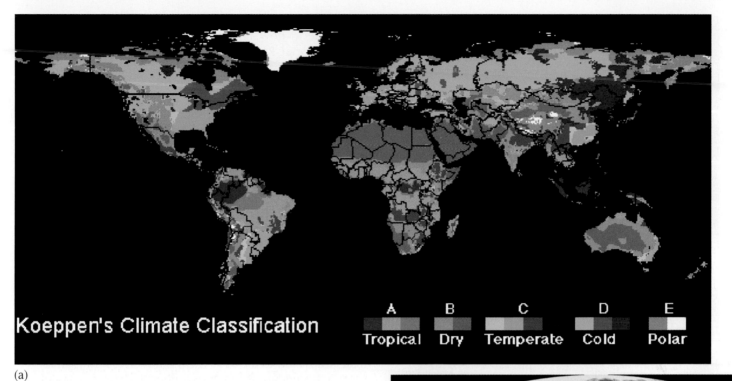

(a)

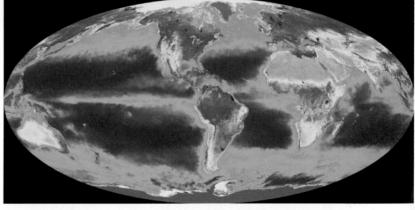

(b)

FIGURE 20.35 (a) The basic climate belts on Earth as originally defined by W. Koeppen (1846–1940). Koeppen was the father-in-law of Alfred Wegener (see Chapter 3). On this map, darker colors imply more extreme conditions. Koeppen's scheme is but one approach to defining climate. Note that highland areas (above snow line) are considered to have polar climates. (b) Satellite image showing the global biosphere as represented by vegetation on land and chlorophyll production in the sea. The distribution of vegetation is indicative of climate.

at a high elevation. During the winter, central Asia becomes very cold, much colder than coastal regions to the south. This coldness creates a stable high-pressure cell over central Asia. Dry air sinks and spreads outward from this cell and flows southward over southern Asia, pushing the intertropical convergence zone (ITCZ) out over the Indian Ocean, south of Asia (▶Fig. 20.36a). Thus, during the winter, southern Asia experiences a dry season. During the summer, central Asia warms up dramatically. As warm air rises over central Asia, a pronounced low-pressure cell develops, and the intertropical convergence zone moves north. When this happens, warm air flows northward from the Indian Ocean, bringing with it substantial moisture, and the summer rains begin. Rainfalls are especially heavy on the southern slope of the Himalayas, because orographic lifting leads to the production of huge cumulonimbus clouds (▶Fig. 20. 36b).

Long before the modern science of meteorology became established, fishermen from Peru and Ecuador who ventured into the coastal waters west of South America knew that in late December the fish population that provided their livelihood diminished. Because of the timing of this event, it came to be known as **El Niño,** Spanish for "the Christ child." Why did the fish vanish? Fish are near the top of a food chain that begins with plankton, which live off nutrients in the water. These nutrients increase when cold water upwells from the deep along the coast of South America. During El Niño, warm water currents flow eastward from the central Pacific, and the cold, nutrient-rich water that supports the marine food chain remains at depth. With fewer nutrients, there are fewer plankton, and without the plankton, the fish migrate elsewhere.

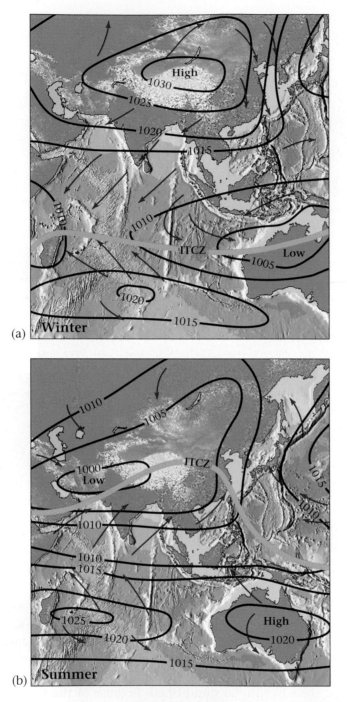

(a) **Winter**

(b) **Summer**

FIGURE 20.36 In the monsoonal climate of Asia, each year can be divided into a dry season and a wet season. (a) During winter, the dry season, a large high-pressure cell develops over central Asia, and the intertropical convergence zone lies well south of Asia. (b) During summer, the wet season, a low-pressure cell develops over Asia, so warm, moisture-laden water from the Indian Ocean flows landward. Orographic lifting along the Himalayas leads to cloud formation and intense rainfall.

To understand why El Niño occurs, we need to look at atmospheric flow and related surface ocean currents in the equatorial Pacific (▶Fig. 20.37a, b). When El Niño is not in progress, a major equatorial low-pressure cell exists in the western Pacific over Indonesia and Papua New Guinea, while a high-pressure cell forms over the eastern Pacific, along the coast of equatorial South America. This geometry means that air rises in the western Pacific, flows east, sinks in the eastern Pacific, and then flows west at the surface. The easterly surface winds blow warm surface water westward, so that it pools in the western Pacific. Cold water from the deep ocean rises along South America, to replace the warm water that moved west. It is this rising cold water that brings nutrients to the surface. During El Niño, the low-pressure cell moves eastward over the central Pacific, and a high-pressure cell develops over

FIGURE 20.37 El Niño exists because of a change in winds and currents in the central Pacific. (a) During the times between El Niño, a low-pressure cell lies over the western Pacific, and surface trade winds blow to the west. These winds drive warm surface water westward, so cold water rises along the western coast of South America to replace it. (b) During El Niño, the low-pressure cell moves eastward, and the westward flow stops, so that cold water no longer upwells.

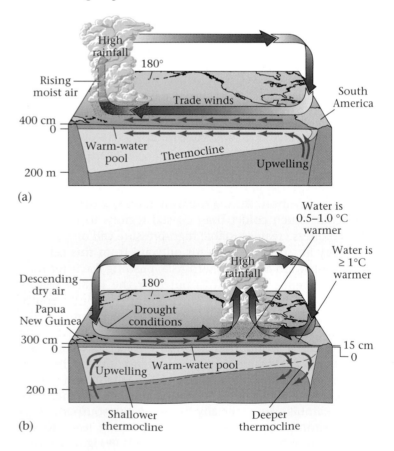

Indonesia; so two convective cells develop. As a result, surface winds start to blow east in the western Pacific, driving warm surface water back to South America. This warm surface water prevents deep cold water from rising, sending the fish away. In effect, pressure cells oscillate back and forth across the Pacific, an event now called the **southern oscillation.**

El Niño gained world notoriety in late 1982 and early 1983 when a particularly large low-pressure cell developed in the eastern part of the Pacific. As a result, the jet streams stayed farther north than is typical. In effect, El Niño caused a temporary climate change worldwide. Drought conditions persisted in the normally rainy western Pacific, while unusually heavy rains drenched western South America. In North America, rains swamped the southern United States and storms battered California, winters were warmer than usual in Canada, and snowfalls were heavier in the Sierra Nevada, leading to spring floods.

Climatologists have been working intensely to understand the periodicity of El Niño. It is clear that strong El Niños take place around once every four years, with even stronger ones possibly happening at other intervals.

CHAPTER SUMMARY

•   The early atmosphere of the Earth contained high concentrations of water, carbon dioxide, and sulfur dioxide, gases erupted by volcanoes.

•   After the oceans formed, much of the carbon dioxide was removed from the atmosphere. When photosynthetic organisms evolved, they produced oxygen, and the concentration of this gas gradually increased.

•   Air consists mostly of nitrogen (78%) and oxygen (21%). Several other gases occur in trace amounts. The atmosphere also contains aerosols. Air pressure decreases with elevation. Thus, 90% of the air in the atmosphere occurs below an elevation of 16 km.

•   When air rises, it expands and cools, a process called adiabatic cooling. If air is compressed, it heats up, a process called adiabatic heating.

•   Air generally contains water. The ratio between the measured water content and the maximum possible amount of water that the air can hold is its relative humidity.

•   The atmosphere is divided into layers, separated from each other by pauses. In the lowest layer, the troposphere, temperature decreases with elevation. The troposphere convects—its air movement causes weather. The other layers are the stratosphere, the mesosphere, and, at the top, the thermosphere.

•   Air circulates on two scales, local and global. Winds blow because of pressure gradients: air moves from regions of higher pressure to regions of lower pressure.

•   High latitudes receive less solar energy than low latitudes. This contrast initiates convection in the atmosphere. Because of the Coriolis effect, air moving north from the equator to the pole deflects to the east, and in each hemisphere, three convection cells develop (the Hadley, Ferrel, and polar cells).

•   Prevailing surface winds—the northeast tradewinds, the surface westerlies, and the polar easterlies—develop because of circulation in global convection cells.

•   Air pressure at a given latitude decreases from the equator to the pole, causing a poleward flow of air. In the Northern Hemisphere, the Coriolis effect deflects this flow to generate high-altitude westerlies. These winds, where particularly strong, are known as jet streams.

•   "Weather" refers to the temperature, air pressure, wind speed, and relative humidity at a given location and time. Weather reflects the interaction of air masses. The boundary between two air masses is a front.

•   Air sinks in high-pressure air masses and rises in low-pressure air masses. Because of the Coriolis effect, the air begins to rotate around the center of the mass as a consequence, generating cyclones or anti-cyclones.

•   Clouds, which consist of tiny droplets of water or tiny crystals of ice, form when the air is saturated with water and contains condensation nuclei on which water condenses.

•   Thunderstorms begin when cumulonimbus clouds grow large. Friction between air and water molecules separates positive and negative charges. Lightning flashes when a giant spark jumps across the charge separation.

•   Tornadoes, rapidly rotating funnel-shaped clouds, develop in violent thunderstorms. Nor'easters are large storms associated with wave cyclones. Hurricanes, huge rotating storms, originate over oceans where the water temperature exceeds 27°C.

•   "Climate" refers to the typical range of conditions, the nature of seasons, and the possible weather extremes of a region over a long time (say 30 years). Climate is controlled by latitude, altitude, proximity to water, ocean currents, orographic barriers, and high- or low-pressure zones. Climate classes can be recognized by the vegetation they support.

•   Monsoonal climates occur where there is a seasonal shift in the wind direction. El Niño is a temporary shift in weather conditions triggered by shifts in the position of high- and low-pressure cells in the Pacific.

KEY TERMS

acid rain (p. 628)
adiabatic cooling, heating (p. 629)
aerosols (p. 628)
air (p. 625)
air mass (p. 638)
air pressure (p. 628)
anticyclone (p. 639)
atmosphere (p. 625)
aurora (p. 632)
Bergeron process (p. 642)
cirrus (p. 642)
climate (p. 626)
cloud (p. 629)
collision and coalescence (p. 641)
condensation nuclei (p. 640)
convergence zone (p. 635)
cumulus (p. 642)
cyclone (p. 639)
dewpoint temperature (p. 629)
divergence zone (p. 635)
doldrums (p. 636)
El Niño (p. 653)
Ferrel cell (p. 636)
fog (p. 640)
front (p. 639)
Fujita scale (p. 646)
Hadley cell (p. 636)
hail (hailstones) (p. 643)
hurricane (p. 648)

insolation (p. 634)
ionosphere (p. 632)
isobar (p. 632)
isotherm (p. 650)
jet stream (p. 637)
lifting mechanism (p. 640)
lightning flash (p. 644)
mesosphere (p. 631)
monsoon (p. 652)
nor'easter (p. 647)
orographic barrier (p. 651)
ozone (p. 627)
pauses (p. 631)
polar cell (p. 636)
polar front (p. 635)
prevailing winds (p. 636)
relative humidity (p. 629)
Saffir-Simpson scale (p. 648)
southern oscillation (p. 655)
storm (p. 642)
storm surge (p. 648)
stratosphere (p. 631)
stratus (p. 642)
thermosphere (p. 631)
thunder (p. 644)
tornado (p. 645)
tradewinds (p. 636)
troposphere (p. 631)
wave cyclone (p. 639)
weather (p. 626)
weather system (p. 638)
wind (p. 625)

REVIEW QUESTIONS

1. Describe the stages in the formulation and evolution of Earth's atmosphere. Where does the ozone in the atmosphere come from and why is it important?

2. Describe the composition of air (considering both its gases and its aerosols). Why are trace gases important?

3. How does air pressure change with elevation? Does the density of the atmosphere also change with elevation? Explain why or why not.

4. Describe the atmosphere's structure from base to top. What characteristics define the boundaries between layers?

5. What is the relative humidity of the atmosphere? What is the latent heat of condensation, and what is its relevance to a thunderstorm or hurricane?

6. Explain the relation between the wind and variations in air pressure.

7. Why do changes in atmospheric temperature depend on latitude and the seasons? Why does global circulation break into three distinct convection belts in each hemisphere, separated by high-pressure or low-pressure belts?

8. Why do prevailing winds develop at the Earth's surface? Why do the jet streams form?

9. Explain the origin of cyclones and anticyclones, and note their relationship to high-pressure and low-pressure air masses. What is a mid-latitude cyclone?

10. How does a cold front differ from a warm front and from an occluded front?

11. Why do clouds form? (Include a discussion of lifting mechanisms.) What are the basic categories of clouds?

12. Under what conditions do thunderstorms develop? What provides the energy that drives clouds to the top of the troposphere? How do meteorologists explain lightning?

13. What conditions lead to the formation of a tornado? Where do most tornadoes appear?

14. Describe the stages in the development of a hurricane. Describe a hurricane's basic geometry.

15. What factors control the climate of a region? What special conditions cause monsoons? El Niño?

SUGGESTED READING

Barry, R. G., R. J. Chorley, and N. J. Yokoi. 2003. *Atmosphere, Weather and Climate.* 8th ed. New York: Routledge.

Bluestein, H. B. 1999. *Tornado Alley: Monster Storms of the Great Plains.* New York: Oxford University Press.

Burroughs, W. J. 1999. *The Climate Revealed.* Cambridge, England: Cambridge University Press.

Davies, P. 2000. *Inside the Hurricane.* New York: Holt.

Graedel, T. E., and P. J. Crutzen. 1993. *Atmospheric Change: An Earth System Perspective.* New York: Freeman.

Grazulis, T. P., and D. Flores. 2003. *The Tornado: Nature's Ultimate Windstorm.* Norman: University of Oklahoma Press.

Hartmann, D. 1994. *Global Physical Climatology.* New York: Academic Press.

Holton, J. 2004. *An Introduction to Dynamic Meteorology.* 4th ed. New York: Academic Press.

Larson, E. 2001. *Isaac's Storm: A Man, a Time, and the Deadliest Hurricane in History.* New York: Crown, Random House.

Lutgens, F. K., E. J. Tarbuck, and D. Tasa. 2003. *The Atmosphere: An Introduction to Meteorology.* 9th ed. Englewood Cliffs, N.J.: Prentice-Hall.

CHAPTER 21

Dry Regions: The Geology of Deserts

The bare hills are cut out with sharp gorges, and over their stone skeletons scanty earth clings . . . A white light beat down, dispelling the last tract of shadow, and above hung the burnished shield of hard, pitiless sky.
—CLARENCE KING (1842–1901; 1ST DIRECTOR OF THE U.S. GEOLOGICAL SOCIETY, describing A Desert)

21.1 INTRODUCTION

For generations, nomadic traders have used camels to traverse the Sahara Desert in northern Africa (▶Fig. 21.1) The Sahara, the world's largest desert, receives so little rainfall that it has little if any surface water or vegetation. So camels must be able to walk for up to three weeks without drinking or eating. They can survive these journeys because they sweat relatively little, thereby conserving their internal water supply; they have the ability to metabolize their own body fat (up to 20 kg of fat make up the animal's hump alone) to produce new water; and they can withstand severe dehydration (loss of water). Most mammals die after losing only 10–15% of their body fluid, for their blood plasma dries up, but camels can survive 30% dehydration with no ill effects. Camels do get thirsty, though. After a marathon trek across the desert, a camel may guzzle up to 100 liters of water in less than ten minutes.

The survival challenges faced by a camel emphasize that deserts are lands of extremes—extreme dryness, heat, cold, and, in some places, beauty. Desert vistas include everything from sand seas to sagebrush plains, cactus-covered hills to endless stony pavements. While less populated than other regions on Earth, deserts cover a significant percentage

In a desert, there is very little vegetation because it rarely rains. But not all deserts are seas of sand. Here, in the Sonoran Desert of northwestern Mexico, a saguaro cactus stands guard over a landscape of stony plains and barren rock cliffs.

FIGURE 21.1 Camels, as in this Sahara Desert caravan in Mauritania, survive harsh desert conditions by storing water in fat reserves and by sweating little, if at all.

FIGURE 21.2 The shimmering in this desert mirage may look like water, but it's not. Mirages result from the interaction of light with a thin layer of hot air just above the ground surface.

(about 25%) of the land surface, and thus constitute an important component of the Earth System. In this chapter, we take a look at the desert landscape. We learn why deserts occur where they do, and how erosion and deposition shape their surface. We conclude by exploring life in the desert and by examining the problem of desertification, the gradual transformation of temperate lands into desert.

21.2 WHAT IS A DESERT?

Formally defined, a **desert** is a region that is so **arid** (dry) that it contains no permanent streams, except for rivers that bring water in from temperate regions elsewhere, and supports vegetation on no more than 15% of its surface. In general, desert conditions exist where less than 25 cm per year (10 inches per year) of rain falls, on average. But rainfall alone does not determine the aridity of a region. Aridity also depends on rates of evaporation and on whether rainfall occurs only sporadically or more continuously during the year. If all the rain in a region falls during isolated downpours once every few years, the region becomes a desert, because the intervals of drought last so long that plants and permanent streams cannot survive. Similarly, if high temperatures and dry air cause evaporation rates from the ground to exceed the rate at which rainfall wets the ground, then the region becomes a desert even if there is more than 25 cm per year of rain.

Note that the definition of a desert depends on a region's aridity, not on its temperature. Geologists distinguish between cold deserts, where temperatures generally stay below about 20°C, and hot deserts, where summer daytime temperatures exceed 35°C. Cold deserts exist at high latitudes where the Sun's rays strike the Earth obliquely and thus don't provide much energy, at high elevations where the air is too thin to hold much heat, or in lands adjacent to cold oceans, where the cold water absorbs heat from the air above. Hot deserts are found at low latitudes where the Sun's rays strike the desert at a high angle, at low elevations where dense air can hold a lot of heat, and in regions distant from the cooling effect of cold ocean currents. The hottest recorded temperatures on Earth occur in low-latitude, low-elevation deserts—58°C (136°F) in Libya and 56°C (133°F) in Death Valley, California.

Notably, the ground surface absorbs so much heat in hot deserts that a layer of very hot air (up to 77°C, or 170°F) forms just above the ground. This layer refracts sunlight, creating a **mirage,** a wavering pool of light, on the ground. Mirages make the dry sand of a desert wasteland look like a shimmering lake and distant mountains look like islands (▶Fig. 21.2). Heat also contributes to aridity by increasing the rate of evaporation. In fact, evaporation rates in hot deserts may be so great that even when it rains, the ground stays dry because raindrops evaporate in mid-air. But even the hottest of hot deserts become cold at night: because of the dryness of the air, the lack of cloud cover, and the lack of foliage, deserts re-radiate their heat back into space at night. As a consequence, the air temperature at the ground surface in a desert may change by as much as 80°C in a single day.

The aridity of deserts causes weathering, erosion, and depositional processes to be different from those of temperate or tropical regions. Without plant cover, rain and wind batter and scour the ground, and during particularly heavy rains

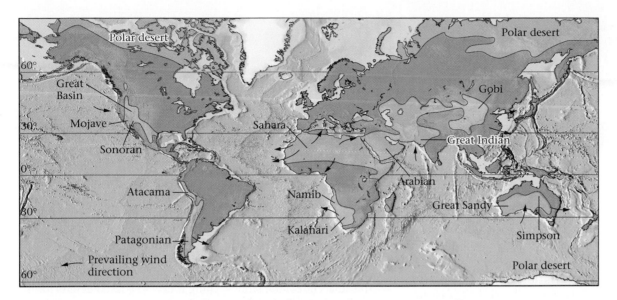

FIGURE 21.3 The global distribution of deserts. Note that the largest lie in the subtropical belts.

water accumulates into flash floods of immense power. Rocks and sediment do not undergo rapid chemical weathering, and humus (organic matter) does not collect on the ground surface. Thus, the desert land surface consists of any of the following: exposed bedrock, accumulations of clasts, relatively unweathered sediment, precipitated salt, or windblown sand. Overall, therefore, desert landscapes tend to be harsher and more rugged than temperate or tropical ones. If eastern North America were a desert, the Appalachians would not be gentle, forested hills but rather would consist of stark, rocky ridges.

21.3 TYPES OF DESERTS

Each desert on Earth has unique characteristics of landscape and vegetation that distinguish it from others. Geologists group deserts into five different classes, based on the environment in which the desert forms (▶Fig. 21.3).

*   *Subtropical deserts:* Subtropical deserts (e.g., the Sahara, Arabian, Kalahari, and Australian) form because of the pattern of convection cells in the atmosphere (▶Fig. 21.4; see also Chapter 20). At the equator, the air becomes warm and humid, for sunlight is intense and water rapidly evaporates from the ocean. The hot, moisture-laden air rises to great heights above the equator. As this air rises, it expands and cools, and can no longer hold as much moisture. Water condenses and falls in downpours that feed the lushness of the equatorial rainforest. The now-dry air high in the

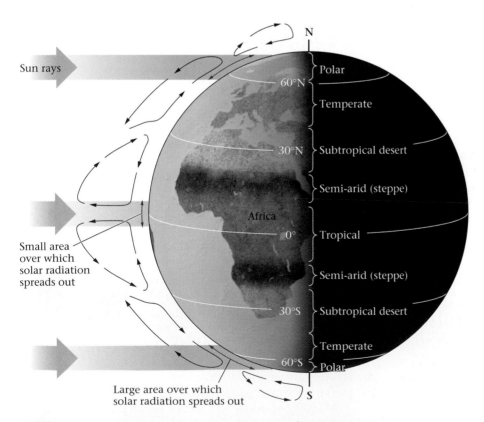

FIGURE 21.4 Rising air at the equator loses its moisture by raining over rainforests. When the air sinks over the subtropics, it warms and absorbs water. Thus, rainfall rarely occurs in the subtropics.

FIGURE 21.5 Moist air, when forced to rise by mountains, cools. As this happens, the moisture condenses and rain falls, nourishing coastal rain forests, so by the time the air reaches the inland side of the mountains, it no longer holds enough moisture to rain. Deserts form in the rain shadow of mountains.

troposphere spreads laterally north or south. When this air reaches latitudes of 20° to 30°, a region called the subtropics, it has become cold and dense enough to sink. Because the air is dry, no clouds form, and intense solar radiation strikes the Earth's surface. The sinking dry air condenses and heats up, soaking up any moisture present. In the regions swept by this hot air on its journey back to the equator, evaporation rates greatly exceed rainfall rates.

In subtropical deserts that border the sea, episodes of high sea level flood coastal areas. The eventual evaporation of stranded seawater leaves broad regions, called "sabkhas," where a salt crust covers a mire of organic-rich mud.

- *Deserts formed in rain shadows:* As air flows over the sea toward a coastal mountain range, the air must rise (▶Fig. 21.5). As the air rises, it expands and cools. The water it contains condenses and falls as rain on the seaward flank of the mountains, nourishing a coastal rain forest. When the air finally reaches the inland side of the mountains, it has lost all its moisture and can no longer provide rain. As a consequence, a **rain shadow** forms, and the land beneath the rain shadow becomes a desert. A rain-shadow desert can be found east of the Cascade Mountains in Washington.

- *Coastal deserts formed along cold ocean currents:* Cool ocean water cools the overlying air by absorbing heat, and decreases the capacity of the air to hold moisture. For example, the cold Humboldt Current, which carries water northward from Antarctica to the western coast of South America, absorbs water from the breezes that blow east, over the coast. Thus, rain rarely falls on the coastal areas of Chile and Peru. As a result, this region hosts a desert landscape, including one of the driest deserts in the world, the Atacama (▶Fig. 21.6a–c). Portions of this narrow (less than 200 km wide) desert, which lies between the Pacific coast on the west and the Andes on the east, received no rain at all between 1570 and 1971.

- *Deserts formed in the interiors of continents:* As air masses move across a continent, they lose moisture by dropping

FIGURE 21.6 (a) Currents bringing cold water up from the Antarctic cool the air along the southwestern coasts of South America and Africa. (b) The cool, dry air absorbs moisture from the adjacent coastal land, keeping it dry, so coastal deserts form. (c) The Atacama Desert of South America is the driest place in the world.

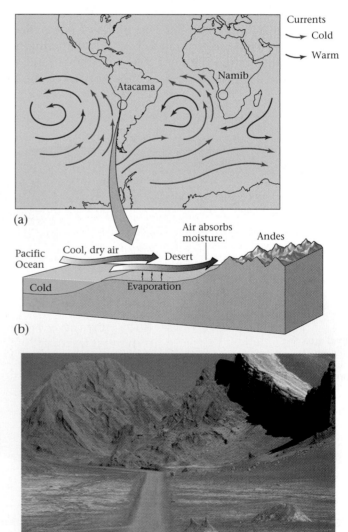

rain, even in the absence of a coastal mountain range. Thus, when an air mass reaches the interior of a particularly large continent like Asia, it has grown quite dry, so the land beneath becomes arid. The largest example of such a continental-interior desert, the Gobi, lies in central Asia, over 2,000 km away from the nearest ocean.

• *Deserts of the polar regions:* So little precipitation falls in Earth's polar regions (north of the Arctic Circle, at 66°30′N, and south of the Antarctic Circle, at 66°30′S) that these areas are, in fact, arid. Polar regions are dry, in part, the same reason that the subtropics are dry (the global pattern of air circulation means that the air flowing over these regions is dry) and, in part, for the same reason that coastal areas along cold currents are dry (cold air holds little moisture).

The distribution of deserts around the world through geologic time reflects the process of plate tectonics, for plate movements determine the latitude of land masses, the position of land masses relative to the coast, and the proximity of land masses to a mountain range. Because of continental drift, some regions that were deserts in the past are temperate regions now, and vice versa.

21.4 WEATHERING AND EROSIONAL PROCESSES IN DESERTS

Without the protection of foliage to catch rainfall and slow the wind, and without roots to hold regolith in place, rain and wind can attack and erode the land surface of deserts. The result, as we have noted, is that hillslopes are typically bare, and plains can be covered with stony debris or drifting sand.

Weathering and Soil Formation in Deserts

In the desert, as in temperate climates, physical weathering happens primarily when joints (natural fractures) split rock into pieces. Joint-bounded blocks eventually break free of bedrock and tumble down slopes, fragmenting into smaller pieces as they fall. In temperate climates, thick soil forms over bedrock, so it lies buried beneath the surface. In deserts, however, the jointed bedrock commonly remains exposed at the surface of hillslopes, creating rugged, rocky escarpments.

Chemical weathering happens more slowly in deserts than in temperate or tropical climates, because there is less water available to react with rock. Still, rain or dew provides enough moisture for *some* weathering to occur. This water seeps into rock and leaches (dissolves and carries away) calcite, quartz, and various salts. Leaching effectively

rots the rock, by transforming it into a poorly cemented aggregate. Over time, the rock will crumble and form a pile of unconsolidated sediment, susceptible to transport by water or wind. If water seeping into the rock contains dissolved salts, salt crystals may grow in the rock when the water dries out. The growth of such crystals pushes neighboring crystals apart and can also weaken the rock.

Although enough rain falls in deserts to leach chemicals out of rock, there is insufficient water to flush the dissolved minerals entirely away. Thus, when the water percolates down and dries up, minerals precipitate in regolith below the surface. If calcite precipitates, it cements loose grains together, forming solid "calcrete." The growth of hard masses like calcrete can occur rapidly enough to incorporate abandoned tools from prospectors.

Shiny **desert varnish,** a dark, rusty brown coating of iron oxide, manganese oxide, and clay, covers the surface of most rock varieties in deserts. Desert varnish was once thought to form when water from rain or dew seeped into a rock, dissolved iron and magnesium ions, and carried the ions back to the surface of the rock by capillary action. More recent studies, however, have shown that desert varnish is not derived from the rock below, but actually forms when wind-borne dust settles on the surface of the rock; in the presence of moisture, microorganisms (bacteria) extract elements from the dust and transform it into iron or manganese oxide. The oxides bind together clay flakes. Such varnish won't form in humid climates, because rain washes the ions away too fast.

Desert varnish takes a long time to form. In fact, the thickness of a desert varnish layer provides an approximate estimate of how long a rock has been exposed at the ground surface. In past centuries, Native Americans used desert-varnished rock as a medium for art: by chipping away the varnish to reveal the underlying lighter-colored rock, they were able to create to figures or symbols on a dark background. The resulting drawings are called **petroglyphs** (▶Fig. 21.7a).

Because of the lack of plant cover in deserts, variations in bedrock color stand out. Locally derived soils typically retain the color of the bedrock from which they were derived. Slight variations in the concentration of iron, or in the degree of iron oxidation, in adjacent beds result in spectacular color bands in rock layers and the thin soils derived from them. The Painted Desert of northern Arizona earned its name from the brilliant and varied hues of oxidized iron (▶Fig. 21.7b).

Water Erosion

Although rain rarely falls in deserts, when it does come, it can radically alter a landscape in a matter of minutes. Since deserts lack plant cover, rainfall, sheetwash, and stream flow all are extremely effective agents of erosion. It may seem surprising, but water generally causes more erosion than does wind in most deserts.

(a)

(b)

FIGURE 21.7 (a) Desert varnish, made from iron oxide, manganese oxide, and clay, is a dark coating on desert rock surfaces. Native American artists created petroglyphs, images on the rock surface, by chipping through the varnish to reveal the lighter rock beneath. (b) In the Painted Desert of Arizona, the different colors of the rock layers are due, in part, to the oxidation state of iron.

Water erosion begins with the impact of raindrops, which eject sediment into the air. On a hill, the ejected sediment lands downslope, and thus during a rain, sediment gradually migrates to lower elevations. The ground quickly becomes saturated with water during a heavy rain, so water starts flowing across the surface, carrying the loose sediment with it. Within minutes after a heavy downpour begins, dry stream channels fill with a turbulent mixture of water and sediment, which rushes downstream as a flash flood. When the rain stops, the water sinks into the stream bed's gravel and disappears—such streams are called intermittent, or ephemeral streams (see Chapter 17). Because of the relatively high viscosity of the water (owing to its load of suspended sediment) and the velocity and turbulence of the flow, flash floods in deserts cause intense erosion—they undercut cliffs and transport huge boulders downstream. As rocks roll and tumble along, they strike one another and shatter, creating smaller pieces that can be carried still farther. Between floods, the stream floor consists of gravel, littered with boulders.

Flash floods carve steep-sided channels into the ground. Scouring of bedrock walls by sand-laden water may polish the walls and create grooves. Dry stream channels in desert regions of the western United States are called **dry washes,** or **arroyos,** and in the Middle East and North Africa they are called wadis (▶Fig. 21.8).

Wind Erosion

In temperate and humid regions, plant cover protects the ground surface from the wind, but in deserts, the wind has direct access to the ground. In hot deserts, gusts of hot air feel like blasts from a furnace. Wind, just like flowing water, can carry sediment both as suspended load and as bed load. **Suspended load** (fine-grained sediment such as dust and silt held in suspension) floats in the air and moves with it (▶Fig. 21.9). The suspended sediment can be carried so high into the atmosphere (up to several kilometers above the Earth's surface) and so far downwind (tens to hundreds of kilometers) that it may move completely out of its source region. In some cases, tiny vortices can churn up dust. In general, these vortices are very

FIGURE 21.8 Gravel and sand are left behind on the floor of a dry wash after a flash flood. The wash has steep walls because downcutting happens so fast.

FIGURE 21.9 Dust clouds form in deserts when turbulent air carries very fine sediment in suspension.

small (centimeters to meters high). But in some cases, they become "dust devils" up to 100 m high, and look like miniature tornadoes. Cars driving down dirt roads in deserts have the same effect; they break dust free from the ground and generate plumes of suspended sediment.

Moderate to strong winds can roll and bounce sand grains along the ground, a process called **saltation** (▶Fig. 21.10). Saltating sand constitutes the wind's **surface load.** Saltation begins when turbulence caused by wind shearing along the ground surface lifts sand grains. The grains move downwind, following an asymmetric, arch-like trajectory, but eventually return to the ground, where they strike other sand grains, causing the new grains to bounce up and drift or roll downwind. The collisions between sand grains make the grains rounded and frosted. Saltating grains generally rise no more than 0.5 m. But where sand bounces on bedrock during a desert sandstorm, they may rise 2 m, and can strip the paint off a car.

FIGURE 21.10 During saltation, sand grains roll and bounce along the ground surface. As they bounce, they follow parabolic paths.

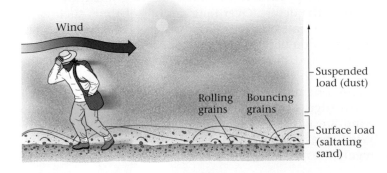

The size of clasts that wind can carry depends on the wind velocity. Wind, therefore, does an effective job of sorting sediment, sending dust-sized particles skyward and sand-sized particles bouncing along the ground, while pebbles and larger grains remain behind. In some cases, wind carries away so much fine sediment that pebbles and cobbles become concentrated at the ground surface (▶Fig. 21.11). An accumulation of coarser sediment left behind when fine-grained sediment blows away is called a **lag deposit.**

In many locations, the desert surface resembles a tile mosaic in that it consists of separate stones that fit together tightly, forming a fairly smooth surface layer above a soil composed of silt and clay. Such natural mosaics constitute **desert pavement** (▶Fig. 21.12a, b). Typically, desert varnish coats the top surfaces of the stones forming desert pavement. Geologists have come up with several explanations for the origin of desert pavements: (1) They are lag deposits, formed when wind blows away the fine sediment between clasts, so that the clasts can settle down and fit together. (2) They form when sheetwash during heavy rainfalls washes away the fine sediment between larger clasts. (3) They form in response to wetting and drying of desert soils. According to this hypothesis, clays in the soil absorb water and expand during heavy rains. The expansion pushes larger stones upward. When the soils dry and shrink, the stones settle down and fit together while the finer-grained sediment sinks between the stones and disappears. (4) Bubbles formed by the metabolism of microorganisms in the desert soil gradually buoy stones upward. (5) Pavements form as wind-blown dust slowly sifts down onto the stones, and then washes down between the stones. In this model, the pavement is "born at the surface," meaning that the stones forming the pavement were never buried, but have been progressively lifted up as sediment collects and builds up beneath

FIGURE 21.11 The progressive development of a lag deposit. Pebbles in a desert are distributed through a matrix of finer sediment. With time, wind blows the finer sediment away, and the pebbles concentrate on the ground surface and may settle together to create a lag deposit. The deposit acts like armor, protecting the substrate from further erosion.

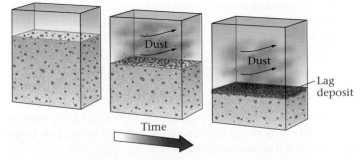

FIGURE 21.12 (a) A well-developed desert pavement in Arizona. (b) Photo of a trench dug in the Sonoran Desert of Arizona, showing how the desert pavement lies on top of a fine-grained sediment. This pavement was lifted up from the surface of a now-buried alluvial fan. (c) Desert pavement forms in stages. First, loose pebbles and cobbles collect at the surface. These can be formed by mechanical weathering, as shown, or by deposition of alluvium. (d) Dust settles among stones and builds up a layer of soil beneath the stone layer. Stones crack into smaller pieces. (e) A durable, mosaic-like pavement has formed.

(►Fig. 21.12c–e). Through time, the sediment beneath the stones is transformed into a pavement. Recent studies lend most support to the last model.

Desert pavements are remarkably durable and can last for hundreds of thousands of years if they are left alone. But like many features of the desert, they can be disrupted by human activity. For example, people driving vehicles across the pavement indent and crack its surface, making it susceptible to erosion. In parts of Arizona, vast desert pavements have turned into parking lots for campers in motor homes who migrate to the desert for the winter season hoping to escape the snows of the North.

Just as sand blasting cleans the grime off the surface of a building, windblown sand and dust grind away at surfaces in the desert. Over long periods of time, such wind abrasion creates smooth faces, or facets, on pebbles, cobbles, and boulders. If a rock rolls or tips relative to the prevailing wind direction after it has been faceted on one side, or if the wind shifts direction, a new facet with a different orientation forms, and the two facets join at a sharp edge. Rocks whose surface has been faceted by the wind are faceted rocks, or **ventifacts** (►Fig. 21.13a–d). Wind abrasion also gradually polishes and bevels down irregularities on a desert pavement and polishes the sur-

(a)   (b)   (c) Old facet New facet   (d)

FIGURE 21.13 The progressive development of a ventifact. (a) Wind-blown sand and dust abrade the face of a rock. (b) Slowly, erosion carves a smooth surface called a facet. (c) Later, the wind shifts direction, and a new facet forms. The two facets join at a sharp edge. (d) Ventifacts formed by strong winds in Antarctica.

faces of desert-varnished outcrops, giving them a reflective sheen.

In places where a resistant layer of rock overlies a softer layer of rock, wind abrasion may create mushroom-like columns with a resistant block perched on an eroding column of softer rock. These unusual features are called **yardangs** (▶Fig. 21.14). If a strong wind blows in only one direction, the yardangs become elongate, aligned with the wind direction.

Over time, in regions where the substrate consists of soft sediment, wind picks up and removes so much sediment that the land surface sinks. The process of lowering the land surface by wind erosion is called **deflation.** Shrubs can stabilize a small patch of sediment with their roots, so after deflation a forlorn shrub with its residual

pedestal of soil stands isolated above a lowered ground surface. In some places, the shape of the land surface twists the wind into a turbulent vortex that causes enough deflation to scour a deep, bowl-like depression called a **blowout** (▶Fig. 21.15).

Satellite exploration in the last two decades shows that wind affects the desert-like surface of Mars just as it does the deserts on Earth. In fact, during the Martian fall and spring, when differences in temperature between the ice-covered

FIGURE 21.15 A blowout, made where wind erosion has hollowed out a topographic depression.

FIGURE 21.14 Yardangs are small landforms sculpted by the wind, where a resistant rock layer overlies a softer rock layer. They may be elongate, aligned with the wind direction. These yardangs formed in the Sahara Desert of Egypt.

poles and the ice-free equator are large, strong winds blow from the poles to the equator, transporting so much dust that the planet's surface, as seen from Earth, visibly changes. At times, the entire planet becomes enveloped in a cloud of dust. Close-up images taken by spacecraft that have landed on Mars show that rocks have been abraded by saltating sand and that sand has accumulated on the lee side of the rocks (▶Fig. 21.16).

21.5 DEPOSITIONAL ENVIRONMENTS IN DESERTS

We've seen that erosion relentlessly eats away at bedrock and sediment in deserts. Where does the debris go? Below, we examine the various desert settings in which sediment accumulates.

Talus Aprons

With time, joint-bounded blocks of rock break off rock ledges and cliffs on the sides of hills. Under the influence of gravity, the resulting debris tumbles downslope and accumulates as a **talus apron** at the base of a hill. Talus aprons can survive for a long time in desert climates, so we typically see them fringing the base of cliffs in deserts (▶Fig. 21.17). The angular clasts constituting talus aprons gradually become coated in desert varnish.

Alluvial Fans

Flash floods can carry sediment downstream in an ephemeral stream channel. When the turbulent water flows out into a plain at the mouth of a canyon, it spreads out over a broader

FIGURE 21.17 This talus apron along the base of a desert cliff formed from rocks that broke off and tumbled down the cliff.

surface and slows. As a consequence, sediment in the water settles out. In some cases, debris flows also emerge from the canyon and spread out. The resulting lenses of sediment cause the channel that has emerged from the mountains to subdivide into a number of subchannels (distributaries) that diverge outward in a broad fan. The fan of distributaries spreads the sediment, or alluvium, out into a broad **alluvial fan,** a wedge- or apron-shaped pile of sediment (▶Fig. 21.18). Alluvial fans emerging from adjacent valleys may merge and overlap along the front of a mountain range, creating an elongate wedge of sediment called a **bajada.** Over long periods, the sediment of bajadas fills in adjacent valleys to depths of several kilometers.

FIGURE 21.16 This NASA photo shows the wind abrasion of clasts and small sand dunes on the surface of Mars. Clearly, the surface of the "red planet" now experiences desert conditions.

FIGURE 21.18 The sediment constituting this alluvial fan in Death Valley was carried to the mountain front during flash floods. The water drops the sediment at the mountain front, because the water slows down.

Playas and Salt Lakes

Water from a flash flood may make it out to the center of an alluvium-filled basin, but if the supply of water is relatively small, it quickly sinks into the permeable alluvium without accumulating as a standing body of water. During a particularly large storm or an unusually wet spring, however, a temporary lake may develop over the low part of a basin. During drier times, such desert lakes evaporate entirely, leaving behind a dry flat lake bed known as a **playa** (▶Fig. 21.19a). Over time, a smooth crust of clay and various salts (halite, gypsum, borax, and other minerals) accumulates on the surface of playas—some of these minerals have industrial uses and thus have been mined. Notably, when it rains slightly, clay-covered playa surfaces become very slippery. Racetrack Playa, in California, becomes so slippery, in fact, that the wind sends stones sliding out across the surface, leaving grooves behind them to mark their path (▶Fig. 21.19b).

Where sufficient water flows into a desert basin, it creates a permanent lake. If the basin is an **interior basin,** with no outlet to the sea, the lake becomes very salty, because although its water escapes by evaporation in the desert sun, its salt cannot. The Great Salt Lake, in Utah, exemplifies this process. Even though the streams feeding the lake are fresh enough to drink, their water contains trace amounts of dissolved salt ions. Because the lake has no outlet, these ions have become concentrated in the lake over time, making it even saltier than the ocean.

Deposition from the Wind

As mentioned earlier, wind carries two kinds of sediment loads—a suspended load of dust-sized particles and a surface load of sand. Much of the dust is carried out of the desert and accumulates elsewhere, forming layers of fine-grained sediment called **loess.** Sand, however, cannot travel far, and accumulates within the desert in piles called **dunes,** ranging in size from less than a meter to over 300 m high. In favorable locations, dunes accumulate to form vast sand seas hundreds of meters thick. We'll look at dunes in more detail later in the chapter.

(a)

(b)

FIGURE 21.19 (a) A playa, a basin covered with clay and salt, develops where a shallow desert lake dries up. (b) Rocks slide along the slippery clay surface of California's Racetrack Playa when strong winds blow.

21.6 DESERT LANDSCAPES

The popular media commonly portray deserts as endless seas of sand, piled into dunes that hide the occasional palm-studded oasis. In reality, vast sand seas are merely one type of desert landscape. Some deserts are vast rocky plains, others sport a stubble of cacti and other hardy desert plants, and still others contain intricate rock formations that look like medieval castles. Explorers of the Sahara, for example, traditionally distinguished among hamada (barren, rocky highlands), reg (vast stony plains), and erg (sand seas in which large dunes form). In this section, we'll see how the erosional and depositional processes described above lead to the formation of such contrasting landscapes.

(a) Massive sandstone · B · A · Scarp · Thin-bedded siltstone and shale · Weak shale · Before · Time · After

(b) Sandstone (strong) · Joint · Shale (weak) · Stair-step slope · Cliff · Talus apron · Slope · Cliff · Bajada · Stony plain · Rubble cover · Playa · Desert

(c) Small escarpments of resistant rock · Slump · Soil cover · Temperate

(d) Plateau · Time · Isolated buttes

(e) · (f)

FIGURE 21.20 (a) The process of cliff retreat in a desert. Rocks break off the cliff along joints that run parallel to the cliff face. The cliff face therefore maintains the same shape and orientation, even though erosion causes the cliff to retreat from point A to point B. (b) Stair-step cliffs appear where beds of strong rock are interlayered with thin beds of rock. Joints are more closely spaced in the thin layers. (c) If the sequence of rocks shown in (b) were to occur in a temperate or humid climate, the slope would instead be smooth and underlain by thick soil. (d) Because of cliff retreat, a once-continuous layer of rock evolves into a series of isolated remnants. (e) These buttes were carved from massive red sandstone layers in Monument Valley, Arizona. (f) These hoodoos, in Bryce Canyon, Utah, were made from the erosion of multicolored layers of sandstone, siltstone, and shale.

Rocky Cliffs and Mesas

In hilly desert regions, the lack of soil exposes rocky ridges and cliffs. As noted earlier, cliffs erode when rocks split away along vertical joints. When this happens, the cliff face retreats but retains roughly the same form. The process, commonly referred to as **cliff retreat,** or scarp retreat, occurs in fits and starts—a cliff may remain unchanged for decades or centuries, and then suddenly a block of rock falls off and crumbles into rubble at the foot of the cliff (▶Fig. 21.20a). Cliff height depends on bed thickness: in places where particularly thick resistant beds crop out, tall cliffs develop. This is because large, widely spaced joints form in thick beds, so the collapse of a portion of the wall generates huge blocks. In thinly bedded shales, joints are small and closely spaced, so shale beds erode to make an overall gradual slope, consisting of many tiny stair steps. Thus, cliffs formed from stratified rocks (such as beds of sandstone and shale) develop a step-like shape; strong layers (sandstone or limestone) become vertical cliffs, and weak layers (shale) become rubble-covered slopes (▶Fig. 21.20b). (This landscape contrasts with landscapes in humid climates, where thick soils form; ▶Fig. 21.20c.).

With continued erosion and cliff retreat, a plateau of rock slowly evolves into a cluster of isolated hills, ridges, or columns (▶Fig. 21.20d). Flat-lying strata or flat-lying layers of volcanic rocks erode to make flat-topped hills. These go by different names, depending on their size. Large examples (with a top surface area of several square km) are **mesas,** from the Spanish word for "table." Medium-sized examples are **buttes** (▶Fig. 21.20e). Small examples, whose height greatly exceeds their top surface area, are **chimneys.** Erosion of strata has resulted in the skyscraper-like buttes of Monument Valley, Arizona, and the stark cliffs of Canyonlands National Park. Bryce Canyon National Park in Utah contains countless chimneys of brightly colored shale and sandstone—locally, these chim-

neys are called "hoodoos" (▶Fig. 21.20f). **Natural arches,** such as those of Arches National Monument, form when erosion along joints leaves narrow walls of rock. When the lower part of the wall erodes while the upper part remains, an arch results. (See art on pp. 672–673.)

In places where bedding dips at an angle to horizontal, flat-topped mesas and buttes don't form; rather, asymmetric ridges called **cuestas** develop. A joint-controlled cliff forms the steep front side of a cuesta, while the tilted top surface of a resistant bed forms the gradual slope on the backside (▶Fig. 21.21). Because the angle of the gradual slope is the same as the dip angle of the bed (the angle the bed surface makes with respect to horizontal), it is called a "dip slope." If the bedding dip is steep to near vertical, a narrow symmetrical ridge, called a **hogback,** forms. If desert hills consist of homogeneous rock like granite, rather than stratified rock, they typically erode to make a pile of rounded blocks (see Fig. 7.10b).

With progressive cliff retreat on all sides of a hill, finally all that remains of the hill is a relatively small island of rock, surrounded by alluvium filled-basins. Geologists refer to such islands of rock by the German word **inselberg** ("island mountain"; ▶Fig. 21.22). Depending on the rock type or the orientation of stratification in the rock, and on rates of erosion, inselbergs may be sharp-crested, plateau-like, or loaf-shaped (steep sides and a rounded crest). Inselbergs with a loaf geometry, as exemplified by Uluru (Ayers Rock) in central Australia (Box 21.1), are also known as "bornhardts."

FIGURE 21.21 Asymmetric ridges called cuestas appear where the strata in a region are not horizontal. A joint-controlled cliff forms the steep side, while a dip slope makes up the gentle side. The surface of a dip slope, by definition, is parallel to the bedding of strata beneath.

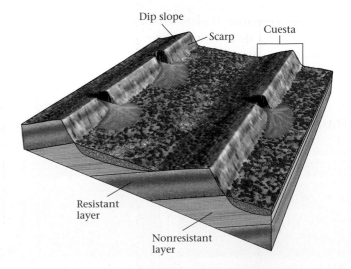

Dip slope

Scarp

Cuesta

Resistant layer

Nonresistant layer

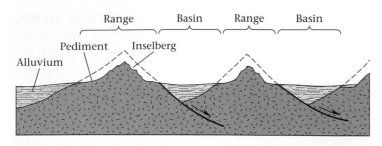

FIGURE 21.22 Inselbergs are small islands of rock surrounded by pediments. Alluvium gradually covers the pediments.

Stony Plains

The coarse sediment eroded from desert mountains and ridges washes into the lowlands and builds out to form gently sloping alluvial fans. The surfaces of these gravelly piles are strewn with pebbles, cobbles, and boulders, and are dissected by dry washes (wadis or arroyos). Portions of these stony plains evolve into desert pavements.

Pediments

When travelers began trudging through the desert southwest of the United States during the nineteenth century, they found that in many locations the wheels of their wagons were rolling over flat or gently sloping bedrock surfaces. These bedrock surfaces extended outward like ramps from the steep cliffs of a mountain range on one side, to alluvium-filled valleys on the other (Fig. 21.22). Geologists now refer to such surfaces as **pediments.** A pediment is a consequence of erosion, left behind as a mountain front gradually retreats. Pediments develop when sheetwash during floods carries sediment away from the mountain front—as it moves, the sediment grinds away the bedrock that it tumbles over. Between erosional events, weathering weakens the surface of the pediment. Alluvium that is washed off pediments accumulates further downslope, and may eventually build up sufficiently to bury the pediments.

Seas of Sand: The Geometry of Dunes

A "sand dune" is a pile of sand deposited by a moving current. Dunes in deserts form because of the wind. They start to form where sand becomes trapped on the windward side of an obstacle, such as a rock or a shrub. (Dunes formed around shrubs are known as "coppice dunes.") Gradually the sand builds downwind into the lee of the obstacle. Once initiated, the dune itself affects the wind flow, and sand accumulates on the lee (downwind) side of the dune. Here, sand slides down the lee surface of the dune, so this surface is aptly named the "slip face" (▶Fig. 21.23).

In places where abundant sand accumulates, sand seas (ergs) bury the landscape. The wind builds the sand in these ergs into dunes that display a variety of shapes and sizes, depending on the character of the wind and the sand supply (▶Fig. 21.25). Where the sand is relatively scarce and the wind blows steadily in one direction, beautiful crescents called "barchan dunes" develop, with the tips of the crescents pointing downwind. If the wind shifts direction frequently, a group of crescents pointing in different directions overlap one another, creating a constantly changing "star dune." Where enough sand accumulates to bury the ground surface completely, and only moderate winds blow, sand piles into simple, wavelike shapes called "transverse dunes." The crests of transverse dunes lie perpendicular to the wind direction. Strong winds may break through transverse dunes and change them into "parabolic dunes," whose ends point in the upwind direction. Finally, if there is abundant sand and a strong, steady wind, the sand streams into "longitudinal dunes" (also called seif dunes, after the Arabic word for "sword"), whose axis lies parallel to the wind direction. In the southern third of the Arabian Peninsula, a region called the Empty Quarter because of its total lack of population, a vast erg called the Rub al Khali contains seif dunes that stretch for almost 200 km and reach heights of over 300 m.

In a sand dune, sand saltates up the windward side of the dune, blows over the crest of the dune, and then settles on the steeper, lee face of the dune. The slope of this face at-

FIGURE 21.23 Progressive stages in the growth of a small sand dune.

Blowing sand

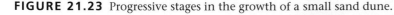

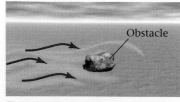

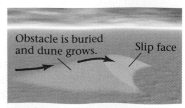

Time 1   Time 2   Time 3   Time 4

Uluru (Ayers Rock)

BOX 21.1
THE REST OF THE STORY

Much of central Australia comprises an immense desert. Parts are barren and sandy, but much of the land is covered with scrub brush, which provides meager grazing for cattle and kangaroo. Uluru, also known by its English name, Ayers Rock, is a huge bornhardt that towers 360 m above the desert plain (▶Fig. 21.24a, b). This rock mass, 3.6 km long, and 2 km wide, consists of nearly vertical dipping sandstone beds. It makes up one limb of a huge regional syncline. The other limb is also a bornhardt, known locally as The Olgas, and the entire area in between is buried in alluvium. Uluru formed because its sandstone resisted erosion, while adjacent rock formations did not.

Thus, through geologic time, alluvium buried the surrounding landscape, but Uluru remained high. Because its strata dip vertically, it has not developed the stair-step shape of mesas and buttes.

Because of its grandeur, Uluru serves a sacred role in Australian Aboriginal traditions. In the dreamtime (time of creation) legends of the Aboriginal people, erosional features on the surface of the rock are scars from a fierce battle between ancient clans. In recent years, the rock has attracted tourists from around the world, who risk life and limb to climb to its top. Plaques at the base of the rock record the names of those who slipped and fell from its steep side.

(a) Time 1 / Time 2 / The Olgas Uluru Alluvium Time 3 (b)

FIGURE 21.24 (a) Originally Uluru (Ayers Rock), in Australia, was part of a large syncline beneath a mountain. Erosion has beveled the mountain away, until all that's left now above a sediment-covered plain are the resistant sandstone limbs of the syncline; softer units are covered by sediment. Uluru is the vertical limb, while The Olgas are the gently dipping limb. (b) Uluru is a dramatic landform.

tains the angle of repose, the slope angle of a freestanding pile of sand. As sand collects on this surface, it may become unstable and slide down the slope—as we've noted, geologists refer to the lee side of a dune as the slip face. As progressively more sand accumulates on the slip face, the crest of the dune migrates downwind, and former slip faces become preserved inside the dune. In cross section, these slip faces appear as cross beds (▶Fig. 21.26a–d). The surfaces of dunes are not, in general, smooth surfaces, but rather are covered with delicate ripples.

With the exception of star and longitudinal dunes, sand dunes migrate downwind as the wind continuously picks up sand from the gently dipping windward slope and drops it onto the leeward side, or slip face. Rates of

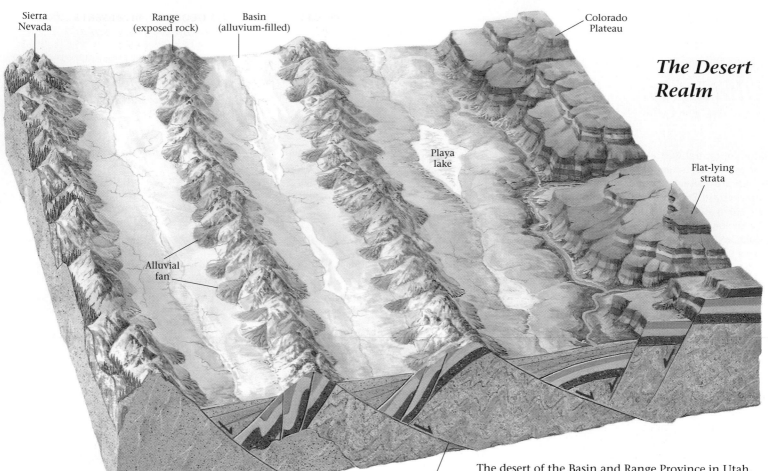

Sierra
Nevada

Range
(exposed rock)

Basin
(alluvium-filled)

Colorado
Plateau

The Desert Realm

Playa
lake

Flat-lying
strata

Alluvial
fan

Normal fault

Granite

The desert of the Basin and Range Province in Utah, Nevada, and Arizona consists of alternating basins (grabens or half-grabens) separated by narrow ranges (tilted fault blocks). The Sierra Nevada, underlain largely by granite, border the western edge of the province, while the Colorado Plateau, underlain by flat-lying sedimentary strata, borders the eastern edge. The overall climate of the region is dry. Because of the great variety of elevations and rock types, the region hosts different desert landscapes.

Except in the case of large rivers, like the Nile or Colorado, which bring water into a desert region from a more temperate region, streams in deserts fill with water only after heavy rains. At other times, the stream channels are dry. These channels are called dry washes, arroyos, or wadis. When there is a heavy rain, water cannot be absorbed into the ground fast enough, so runoff enters dry washes and fills them very quickly, creating a flash flood. The turbulent, muddy water of a flash flood can transport even large boulders. This flash flood is rushing down a stream in the Sonoran Desert of Arizona.

Barchan dune

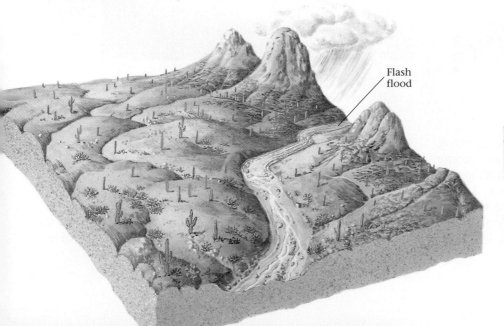

Flash
flood

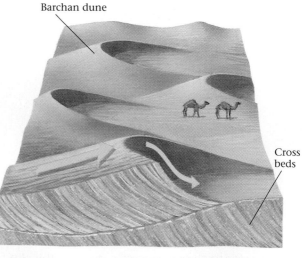

Cross
beds

Where there is a large supply of sand, a variety of sand dunes develop. The geometry of a particular sand dune (e.g., barchan, longitudinal, or star) depends on the sand supply and the wind. Inside sand dunes, we find cross beds.

Inselberg

Pediment

Alluvial apron with dry channels

Alluvial apron

Pediment

Pediment

In places where ranges consist of granitic rock, they tend to be bordered by pediments. The isolated mountains that remain become inselbergs. Sediment derived by the erosion of the mountains fills the basins between the mountains. Dry washes (arroyos) channel both the pediments and the alluvium.

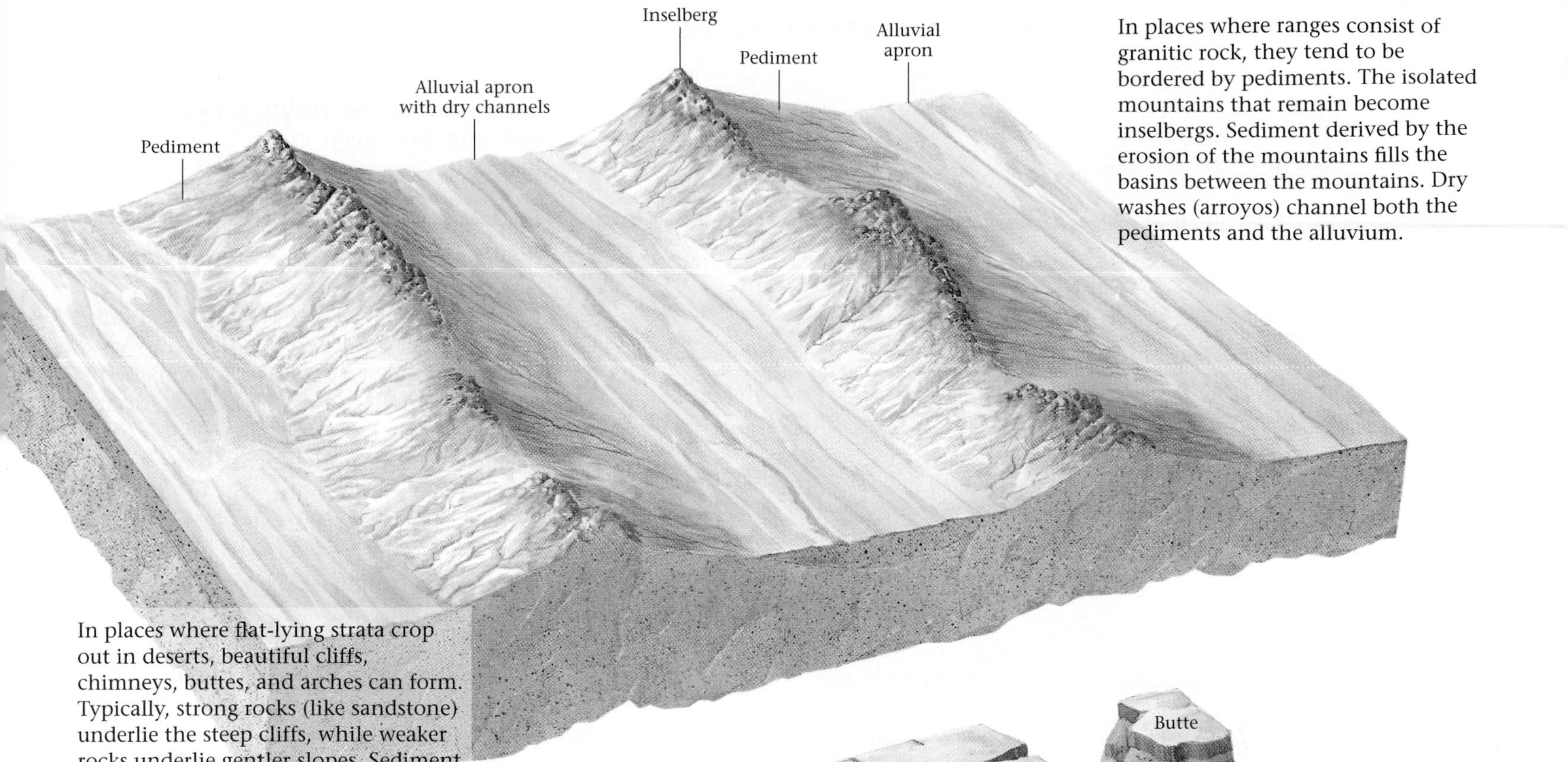

In places where flat-lying strata crop out in deserts, beautiful cliffs, chimneys, buttes, and arches can form. Typically, strong rocks (like sandstone) underlie the steep cliffs, while weaker rocks underlie gentler slopes. Sediment is washed out of valleys during floods to create alluvial fans. Some of the debris carried out of the highlands breaks up and settles together to form desert pavements (stony plains).

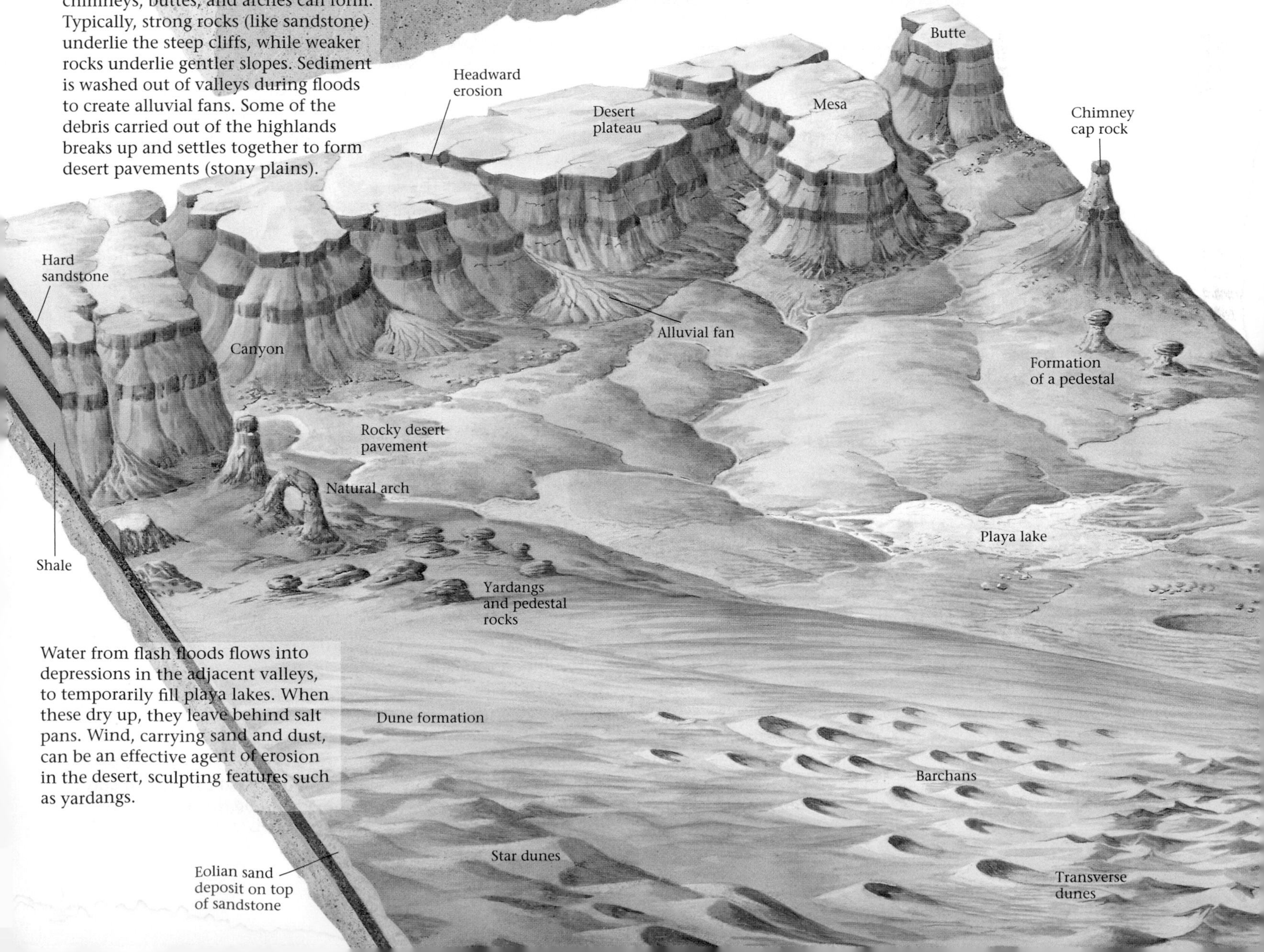

Headward erosion

Desert plateau

Mesa

Butte

Chimney cap rock

Hard sandstone

Canyon

Alluvial fan

Formation of a pedestal

Rocky desert pavement

Natural arch

Shale

Playa lake

Yardangs and pedestal rocks

Water from flash floods flows into depressions in the adjacent valleys, to temporarily fill playa lakes. When these dry up, they leave behind salt pans. Wind, carrying sand and dust, can be an effective agent of erosion in the desert, sculpting features such as yardangs.

Dune formation

Barchans

Star dunes

Transverse dunes

Eolian sand deposit on top of sandstone

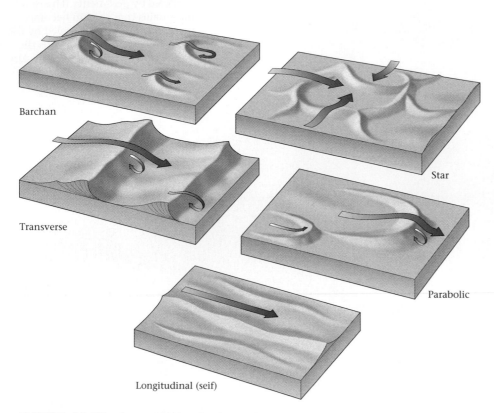

Barchan

Transverse

Star

Parabolic

Longitudinal (seif)

FIGURE 21.25 The various kinds of sand dunes.

migration can exceed 25 m per year. Because of moving sand on an active dune, vegetation can't grow there. If a change in climate brings more rain, however, plant cover may grow and stabilize the dunes. At the end of the last ice age, for example, the "sand hills region" of western Nebraska was a vast dune field, but in the past 11,000 years it has been covered by grasslands, and the dunes have become stabilized.

21.7 LIFE IN THE DESERT

In the midst of a large erg, there seems to be nothing growing or moving at all. But most desert landscapes do include plants and animals. These organisms must possess special characteristics to enable them to survive in the desert: they must be able to withstand extremes of temperature—oppressive heat during the day and chilling cold at night—and to survive without abundant water.

Plants have evolved a number of different means to survive desert conditions. Some produce thick-skinned seeds that last until a heavy rainfall, then quickly germinate, grow, and create new seeds only while water is available. The new generation of seeds then waits until the next rainfall to start the cycle over again. Other plants

have evolved the ability to send roots down to find deep groundwater; these plants have very long taproot systems. Still others have shallow root systems that spread over a broad area so that they can efficiently soak up water when it does rain.

Many desert plants have thick, fleshy stems and leaves. These plants, known as **succulents,** can store water for long periods of time (▶Fig. 21.27). Because succulents may be the only source of water during a time of drought, they have developed threatening thorns or needles to keep away thirsty animals. Plant life is much more diverse in desert "oases," the verdant islands that crop up where natural springs spill groundwater onto the surface (see Chapter 19). The nearly year-round supply of water in an oasis nourishes a variety of palms and other nonsucculent plants.

Animal life in the desert includes scavengers, hunters, and plant eaters. Animals face the same challenge as plants: to retain water and survive extreme temperatures. To accomplish these goals, desert animals have also evolved numerous strategies. Frogs, for example, burrow beneath the ground and remain dormant for months, waiting until the next rain. Reptiles escape the midday heat by crawling into dark cracks between rocks. Rodents forage for food only during the cool night. And kit foxes, jack rabbits, and mule deer have disproportionately large ears through which they efficiently lose body heat. Many desert mammals, such as camels, retain body water by not sweating.

Humans are not meant to live in the desert. The loss of body moisture in extreme heat can be so rapid that a person will die in less than 24 hours unless shaded from the Sun and supplied with at least 8 liters of water per day. Nevertheless, all but the most barren deserts are inhabited, though sparsely. Before civilization provided water wells, pipelines, and mechanized transportation, desert peoples lived in small nomadic groups, spaced far enough apart that they could live off the land; deserts have too little water to sustain agriculture or husbandry (▶Fig. 21.28). Australian Aboriginal groups, for example, rarely included more than a dozen individuals. In the past, nomadic desert dwellers either built temporary shelters out of local materials or traveled with tents. Locally, people carved underground dwellings in sediment or soft rock, for rock is such a good insulator that a few meters below the ground surface it stays close to the region's mean temperature year-round.

(a)

(c)

(b)

(d)

FIGURE 21.26 (a) A sand dune with surface ripples; (b) cross bedding inside a dune; (c) cross beds exposed in the Mesozoic sandstone of Zion National Park, in Utah; (d) close-up of the contact between the top of one set of cross beds and the bottom of another.

FIGURE 21.27 These plants in the Sonoran Desert, Arizona, are well adapted for dry conditions.

21.8 DESERT PROBLEMS

The modern era has seen a remarkable change in desert margins. Natural droughts (periods of unusually low rainfall), aggravated by overpopulation, overgrazing, careless agriculture, and diversion of water supplies, have transformed semi-arid grasslands into true deserts, leading to tragic famines that have killed millions of people. **Desertification,** the process of transforming nondesert areas to desert, has accelerated.

The consequences of desertification have devastated the Sahel, the belt of semi-arid land that fringes the southern margin of the Sahara Desert. In the past, the Sahel provided sufficient vegetation to support a small population of nomadic people and animals. But during the second half of the twentieth century, large numbers of people migrated into the Sahel to escape overcrowding in central Africa. The immigrants began farming, and maintained large herds of cattle and goats. Plowing and overgrazing removed soil-preserving grass and caused the soil to dry out. In addition, the trampling of animal hooves

FIGURE 21.28 Before the modern era, inhabitants of Australia's central desert (the "Red Center," named for its red soil and rocks) lived in small nomadic groups. There was not enough water to supply permanent communities until the advent of technology.

compacted the ground so it could no longer soak up water. In the 1960s and again in the 1980s, a series of natural droughts hit the region, bringing catastrophe (▶Fig. 21.29a, b). Wind erosion stripped off the remaining topsoil. Without vegetation, the air grew drier, and the semi-arid grassland of the Sahel became desert, with mass starvation as the result.

Other regions on Earth are developing the same problem. The Aral Sea in Kazakhstan, for example, has started to dry up. Diversion of the rivers that once fed the sea have so diminished its water supply that the area of the sea has shrunk dramatically. Boats that once plied its waters now lie as rusting hulks in a sea of dust (▶Fig. 21.29c, d).

Desertification does not only happen in less industrialized nations. People in the western Great Plains of the United States and Canada suffered from the problem beginning in 1933, the fourth year of the Great Depression. Banks had failed, workers had lost their jobs, the stock market had crashed, and hardship burdened all. No one needed yet another disaster—but that year, even nature turned hostile. All through the fall, so little rain fell in the plains of Texas and Oklahoma that the region's grasslands and croplands browned and withered, and the topsoil turned to powdery dust. Then, on November 12 and 13, strong storms blew eastward across the plains. Without vegetation to protect it, the wind lapped at the ground, stripped off the topsoil, and sent it skyward to form rolling black clouds that literally blotted out the sun (▶Fig.

21.29e). People caught in the resulting **dust storm** found themselves choking and gasping for breath. When the dust finally settled, it had buried houses and roads under huge drifts, and dirtied every nook and cranny. The dust blew east as far as New England, where it turned the snow brown. What had once been a rich farmland in the southwestern plains turned into a wasteland that soon acquired a nickname, the "Dust Bowl."

For several more years the drought persisted, leading to starvation and bankruptcy. Many of the region's residents were forced to move to wherever they could find work. Hundreds of "Okies," as the natives of Oklahoma were then called, piled into jalopies and drove on old Route 66 out to California, looking for jobs in the state's still-green agricultural regions. Many were subjected to exploitation and persecution once they arrived in California. John Steinbeck dramatized this staggering human tragedy, which came to symbolize the Depression, in his novel *The Grapes of Wrath*.

Why did the fertile soils of the southern Great Plains suddenly dry up? The causes were complex; some were natural and some were human-induced. Drought episodically visits certain regions, but drought alone doesn't inevitably lead to dust-bowl conditions. They may result from human activity. Typically, the Great Plains region has a semi-arid climate in which only thin soil develops. But the plains were settled in the 1880s and 1890s, unusually wet years. Not realizing its true character, far more people moved into the region than it could sustain, and the land was farmed too intensively. When farmers used steel plows, they destroyed the fragile grassland root systems that held the thin soil in place. And when the drought of the 1930s came, it was catastrophic.

Desertification can be reversed, but at a price. Planting and irrigation may transform desert into farm fields, orchards, forests, or lawns. But water to nourish the plants has to come from somewhere, and people obtain it by diverting rivers or by pumping groundwater, activities that create their own set of problems. River diversion robs regions downstream of their water supply, and pumping out too much groundwater lowers the water table so substantially that the pore space in aquifers collapses and the ground surface subsides. In short, people will need to rethink land-use policies in semi-arid lands to avoid catastrophe.

The Dust Bowl of the 1930s and the fate of the Sahel in Africa remind us of how fragile the Earth's green blanket of vegetation really is. Global climate changes can shift climatic belts sufficiently to transform agricultural regions into deserts. Some 5,000 years ago, the swath of land between the Nile Valley in Egypt and the Tigris-Euphrates Valley of Mesopotamia was known as the "fertile crescent"; here, people first abandoned their nomadic ways and settled in agricultural communities. Yet the original

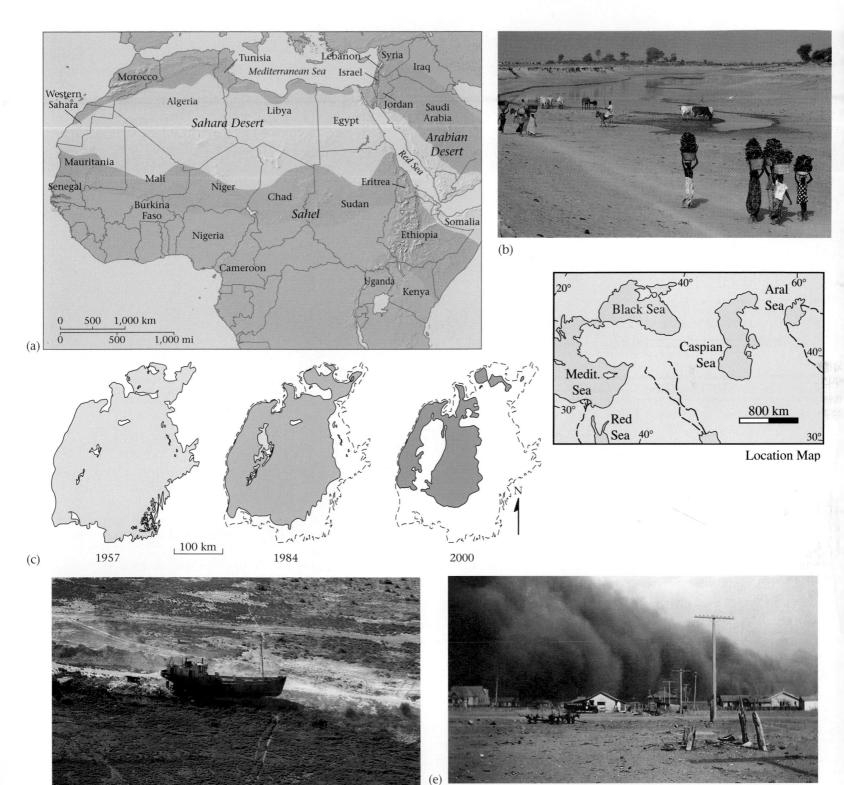

FIGURE 21.29 (a) The Sahel is the semi-arid land along the southern edge of the Sahara Desert. As a result of grazing and agriculture, the vegetation in this region has vanished, and large parts have undergone desertification. (b) Drought in the Sahel has brought deadly consequences. Here, residents seek water from a dwindling pond. (c) The Aral Sea in central Asia has started to dry up because of diversion of rivers that once flowed into it. What was once the floor of the sea is now a parched land. The inset shows the location. (d) Fishing vessels now lie stranded in the Aral Sea area. (e) During the "Dust Bowl" days in central Oklahoma (ca. 1930s), dust storms stripped valuable topsoil off the land.

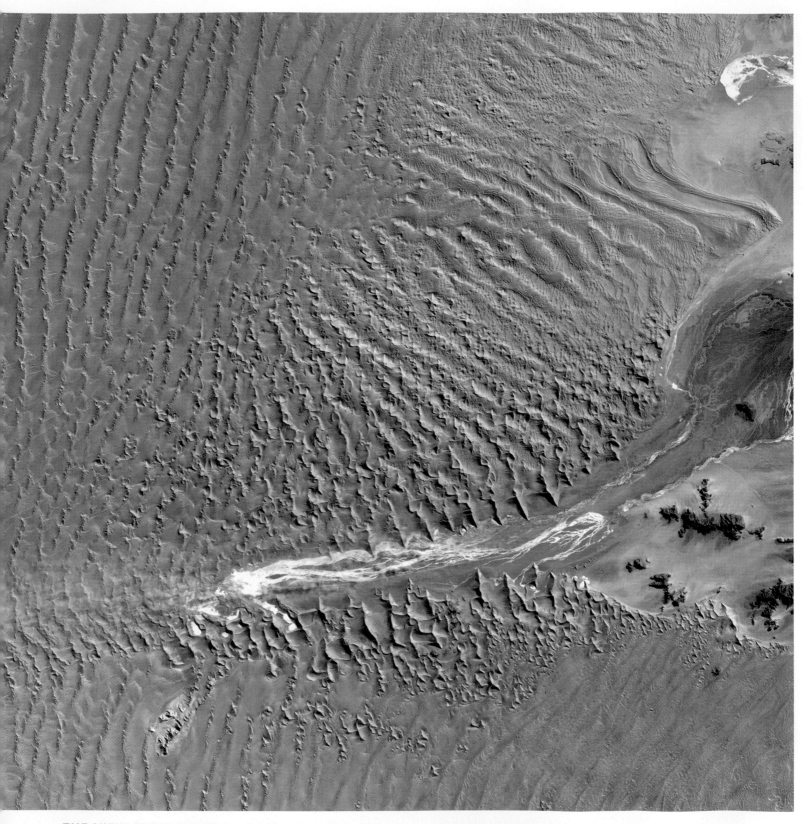

THE VIEW FROM SPACE Erosion of rock produces sand, which, under appropriate conditions in arid climates, can build into large dunes. These dunes, some of which are 300 m high, developed in the Namib Desert of Namibia, southwestern Africa.

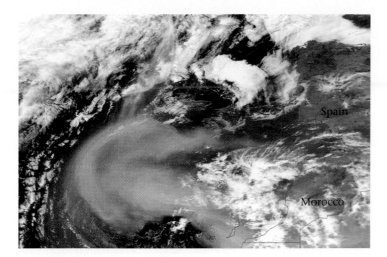

FIGURE 21.30 In this satellite image, we see a huge dust cloud blowing across the Atlantic. The dust originated in the Sahara.

land of "milk and honey" has become a barren desert, in need of intensive irrigation to maintain any agriculture. The change in landscape reflects a change in climate—the beginning of Western civilization occurred during the warmest and wettest period of global climate since before the last ice age. So much water drenched the Middle East and North Africa that rivers flowed where Saharan sands now blow. Unfortunately, if the current trend in climate change continues, our present agricultural belts could someday become new Saharas.

Desertification has a potentially dangerous side effect—blowing dust. In recent years, satellite images have revealed that wind-blown dust from deserts can travel across oceans and affect regions on the other side. For example, dust blown off the Sahara can traverse the Atlantic and settle over the Caribbean (▶Fig. 21.30). Geologists are concerned that this dust, along with the fungi and microbes that it carries, may infect corals with disease or in some other way inhibit their life processes. Thus, wind-blown dust can contribute to the destruction of coral reefs. As desert areas expand in response to desertification, wind-blown dust becomes more of a problem—not only do winds have larger areas of dry, dusty land to churn, but the dust generated from lands that were once agricultural and are now desert may contain harmful chemicals (e.g., residues of herbicides and pesticides) that can themselves become windborne.

CHAPTER SUMMARY

• Deserts generally receive less than 25 cm of rain per year. Vegetation covers no more than 15% of their surface.

• Subtropical deserts form between 20° and 30° latitude, rain-shadow deserts are found on the inland side of mountain ranges, coastal deserts are located on the land adjacent to cold ocean currents, continental-interior deserts exist in land-locked regions far from the ocean, and polar deserts form at high latitude.

• In deserts, chemical weathering happens slowly, so rock bodies tend to erode primarily by physical weathering. Desert varnish forms on rock surfaces, and soils tend to accumulate soluble minerals.

• Water causes significant erosion in deserts, mostly during heavy downpours. Flash floods carry large quantities of sediments down ephemeral streams. When the rain stops, these streams dry up, leaving steep-sided washes.

• Wind causes significant erosion in deserts, for it picks up dust and silt and carries them as suspended load, and causes sand to saltate. Where wind blows away finer sediment, a lag deposit remains. Wind-blown sediment abrades the ground, creating a variety of features such as ventifacts and yardangs.

• Desert pavements are mosaics of varnished stones armoring the surface of the ground.

• Talus aprons form when rock fragments accumulate at the base of a slope. Alluvial fans form at a mountain front where water in ephemeral streams deposits sediment. When temporary desert lakes dry up, they leave playas.

• In some desert landscapes, erosion causes cliff retreat, eventually resulting in the formation of inselbergs. Pediments of nearly flat or gently sloping bedrock surround some inselbergs. The erosion of stratified rock yields such landforms as buttes.

• Where there is abundant sand, the wind builds it into dunes. Common types include barchan, star, transverse, parabolic, and longitudinal (seif) dunes.

• Deserts contain a great variety of plant and animal life. All are adapted to survive extremes in temperature and without abundant water.

• Changing climates and land abuse may cause desertification, the transformation of semi-arid land into deserts. Wind-blown dust from deserts may waft across oceans.

KEY TERMS

alluvial fan (p. 666)
arid (p. 658)
arroyos (p. 662)
bajada (p. 666)
blowout (p. 665)
butte (p. 669)
chimney (p. 669)
cliff retreat (p. 669)

cuesta (p. 669)
deflation (p. 665)
desert (p. 658)
desert pavement (p. 663)
desert varnish (p. 661)
desertification (p. 675)
dry wash (p. 662)
dunes (p. 667)

dust storm (p. 676)
hogback (p. 669)
inselberg (p. 669)
interior basin (p. 667)
lag deposit (p. 663)
loess (p. 667)
mesa (p. 669)
mirage (p. 658)
natural arches (p. 669)
pediment (p. 670)

petroglyph (p. 661)
playa (p. 667)
rain shadow (p. 660)
saltation (p. 663)
succulents (p. 674)
surface load (p. 663)
suspended load (p. 662)
talus apron (p. 666)
ventifact (p. 664)
yardang (p. 665)

REVIEW QUESTIONS

1. What factors determine whether a region can be classified as a desert?

2. Explain why deserts form.

3. Have today's deserts always been deserts? Keep in mind the consequences of plate tectonics.

4. How do weathering processes in deserts differ from those in temperate or humid climates?

5. Describe how water modifies the landscape of a desert. Be sure to discuss both erosional and depositional landforms.

6. Explain the ways in which desert winds transport sediment.

7. Explain how the following features form: (a) desert varnish; (b) desert pavement; (c) ventifacts; (d) yardangs.

8. Describe the process of formation of the different types of depositional landforms that develop in deserts.

9. Describe the process of cliff (scarp) retreat and the landforms that result from it.

10. What are the various types of sand dunes? What factors determine which type of dune develops?

11. Discuss various adaptations that life forms have evolved in order to be able to survive in desert climates.

12. What is the process of desertification, and what causes it? How can desertification in Africa affect the Caribbean?

SUGGESTED READING

Abrahams, A. D., and A. J. Parsons, eds. 1994. *Geomorphology of Desert Environments*. London: Chapman and Hall.

Dregne, H. E. 1983. *Desertification of Arid Lands*. Chur, Switzerland: Harwood Academic.

Livingston, I., S. Stokes, and A. S. Goudie, eds. 2000. *Aeolian Environments, Sediment and Landforms*. New York: Wiley.

Tchakerian, V. P., ed. 1995. *Desert Aeolian Processes*. London: Chapman and Hall.

CHAPTER 22

Amazing Ice: Glaciers and Ice Ages

22.1 INTRODUCTION

There's nothing like a good mystery, and one of the most puzzling in the annals of geology came to light in northern Europe early in the nineteenth century. When farmers of the region prepared their land for spring planting, they occasionally broke their plows by running them into large boulders buried randomly through otherwise fine-grained sediment. Many of these boulders did not consist of local bedrock, but rather came from outcrops hundreds of kilometers away. Because the boulders had apparently traveled so far, they came to be known as **erratics** (from the Latin *errare,* "to wander").

The mystery of the wandering boulders became a subject of great interest to early nineteenth-century geologists, who realized that deposits of *un*sorted sediment (containing a variety of different clast sizes) could not be examples of typical stream alluvium, for running water sorts sediment by size. Most attributed the deposits to a vast flood that they imagined had somehow spread a slurry of boulders, sand, and mud across the continent. In 1837, however, a young Swiss geologist named Louis Agassiz proposed a radically different interpretation. Agassiz often hiked among **glaciers** (slowly flowing masses of ice that survive the summer melt) in the Alps near his

Glaciers are rivers or sheets of ice. They carve beautiful landscapes and deposit hills of sediment. During an ice age, glaciers covered vast areas of continents.

FIGURE 22.1 During the ice age, glaciers covered much of North America and Europe, so that these continents would have resembled present-day Antarctica, where glacial ice extends from the foot of the Transantarctic Mountains (seen here on the horizon) to the far side of Antarctica, over 3,000 km away.

home. He observed that glacial ice could carry enormous boulders as well as sand and mud, because ice is solid and has the strength to support the weight of rock. Agassiz realized that because solid ice does not sort sediment as it flows, glaciers leave behind unsorted sediment when they melt. Based on these observations, he proposed that the mysterious sediment and erratics of Europe represented deposits left by **ice sheets,** vast glaciers that had once covered much of the continent. In Agassiz's mind, Europe had at one time been in the grip of an **ice age,** a time when the climate was significantly colder and glaciers grew (▶Fig. 22.1).

Agassiz's radical proposal faced intense criticism for the next two decades. But he didn't back down, and instead challenged his opponents to visit the Alps and examine the sedimentary deposits that glaciers had left behind. By the late 1850s, most doubters had changed their minds, and the geological community concluded that the notion that Europe once had had Arctic-like climates was correct.

Later in life, Agassiz traveled to the United States and documented many glacier-related features in North America's landscape, proving that an ice age had affected vast areas of the planet.

Glaciers, which have many forms, cover only about 10% of the land on Earth today (▶Fig. 22.2a, b), but during the most recent ice age, which ended only about 11,000 years ago, as much as 30% of the continents had a coating of ice. New York City, Montreal, and many of the great cities of Europe now occupy land that once lay beneath hundreds of meters to a few kilometers of ice.

The work of Louis Agassiz brought the subject of glaciers and ice ages into the realm of geologic study and led people to recognize that major climate changes happen in Earth history. In this chapter, after looking at the nature of ice, we see how glaciers form and why they move. Next, we consider how glaciers modify landscapes by erosion and deposition. A substantial portion of the chapter concerns the Pleistocene ice ages, for their impact on the landscape can still be seen today. We briefly inroduce ice ages that happened earlier in Earth history too. We conclude by considering hypotheses to explain why ice ages happen.

FIGURE 22.2 (a) A piedmont glacier near the coast of Greenland. (b) A large valley glacier in Pakistan.

(a)

(b)

22.2 ICE: A ROCK MADE FROM WATER

Ice consists of solid water, formed when liquid water cools below its freezing point. We can consider a single ice crystal to be a mineral specimen: it is a naturally occurring, inorganic solid, with a definite chemical composition (H_2O) and a regular crystal structure. Ice crystals have a hexagonal form, so snowflakes grow into six-pointed stars (▶Fig. 22.3a). We can think of a layer of fresh snow as a layer of sediment, and a layer of snow that has been compacted so that the grains stick together as a layer of sedimentary rock (▶Fig. 22.3b). We can also think of the ice that appears on the surface of a pond as an igneous rock, for it forms when "molten ice" (liquid water) solidifies (▶Fig. 22.3c). Glacier ice, in effect, is a metamorphic rock. It develops when preexisting ice recrystallizes in the solid state, meaning that the molecules in solid water rearrange to form new crystals (▶Fig. 22.3d).

Pure ice, has the transparency of glass, but if ice contains tiny air bubbles or cracks that disperse light, it becomes milky white. Like glass, ice has a high **albedo,** meaning that it reflects light well—so well, in fact, that if you walk on ice without eye protection, you risk blindness from the glare. Ice differs from most other familiar materials in that its solid form is less dense than its liquid form—the architecture of an ice crystal holds water molecules apart. Ice, therefore, floats on water. This unusual characteristic has benefits; if ice didn't float, ice in oceans would sink, leaving room for new ice to form until the oceans froze solid.

22.3 THE NATURE OF GLACIERS

Categories

Glaciers are streams or sheets of recrystallized ice that last all year long and flow under the influence of gravity. Today, they highlight coastal and mountain scenery in Alaska, western North America, the Alps of Europe, the Southern Alps of New

FIGURE 22.3 (a) The hexagonal shape of snowflakes. No two are alike. (b) Snowdrifts carved by the wind in Antarctica. Note the person for scale. (c) Frost on a glass window, showing the large crystals of water. (d) Thin section of glacial ice, showing the texture of ice crystals and air bubbles.

(a)

(b)

(c)

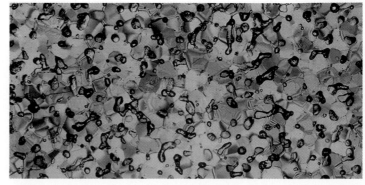

(d)

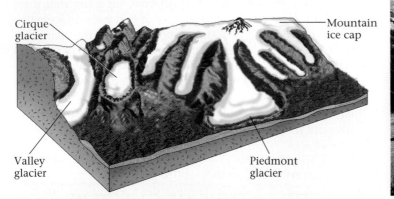

Cirque glacier

Mountain ice cap

Valley glacier

Piedmont glacier

(a)

(b)

(d)

(c)

FIGURE 22.4 (a) The various kinds of mountain glaciers. (b) A valley glacier in Alaska. (c) A mountain ice cap in Greenland. Note the ice flowing out of the ice cap in a big valley glacier. (d) The Malaspina Glacier, a piedmont glacier along the coast of Alaska, as viewed by satellite. This is a "false color" image; red indicates rocky debris on the glacier surface, and gold indicates forest. Lines of debris trace out folds in the ice. The glacier is several kilometers across.

Zealand, the Himalayas of Asia, and the Andes of South America, and they cover most of Greenland and Antarctica. Geologists distinguish between two main categories, mountain glaciers and continental glaciers.

Mountain glaciers (also called alpine glaciers) exist in or adjacent to mountainous regions (▶Fig. 22.4a). Topographical features of the mountains control their shape; overall, mountain glaciers flow from higher elevations to lower elevations. Mountain glaciers include "cirque glaciers", which fill bowl-shaped depressions, or **cirques,** on the flank of a mountain; "valley glaciers", rivers of ice that flow down valleys (▶Fig. 22.4b and Fig. 22.2b); mountain "ice caps", mounds of ice that submerge peaks and ridges at the crest of a mountain range (▶Fig. 22.4c); and piedmont glaciers (Fig. 22.2a), fans or lobes of ice that form where a valley glacier emerges from a valley and spreads out into the adjacent plain

(▶Fig. 22.4). Mountain glaciers range in size from a few hundred meters to a few hundred kilometers long.

Continental glaciers are vast ice sheets that spread over thousands of square kilometers of continental crust. Continental glaciers now exist only on Antarctica and Greenland (▶Fig. 22.5a, b). (Remember that Antarctica is a continent, so the ice beneath the South Pole rests on solid ground; in contrast, the ice beneath the North Pole comprises part of a thin sheet of sea ice floating on the Arctic Ocean.) Continental glaciers flow outward from their thickest point (up to 3.5 km thick) and thin toward their margins, where they may be only a few hundred meters thick. The front edge of the glacier may divide into several tongue-shaped lobes, because not all of the glacier flows at the same speed.

Geologists also find it valuable to distinguish between types of glaciers on the basis of the thermal conditions in

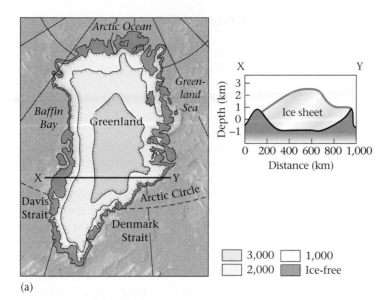

(a)

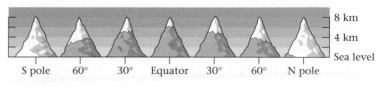

FIGURE 22.6 The snow line depends on latitude.

X ———————— Y

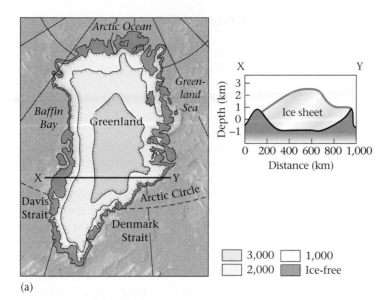

(b)

FIGURE 22.5 (a) A map and cross section of the Greenland ice sheet. (b) A map and cross section of the Antarctic ice sheet. Note that the Transantarctic Mountains divide the ice sheet in two. The Ross Ice Shelf lies between the East Antarctic and West Antarctic sheets. Valley glaciers carry ice from the ice sheets down to the shelf.

which the glaciers exist. **Temperate glaciers** occur in regions where atmospheric temperatures become high enough during a substantial portion of the year for the glacial ice to be at or near its melting temperature throughout much of the year. **Polar glaciers** occur in regions where atmospheric temperatures stay so low all year long that the glacial ice remains well below melting temperature throughout the entire year.

Forming a Glacier

In order for a glacier to form, three conditions must be met: first, the local climate must be sufficiently cold that winter snow does not melt entirely away during the summer; second, there must be, or must have been sufficient snowfall for a large amount to accumulate; and third, the slope of the surface on which the snow accumulates must be gentle enough that the snow does not slide away in avalanches, and must be protected enough that the snow doesn't blow away.

Glaciers develop in polar regions because even though relatively little snow falls today, temperatures remain so low that most ice and snow survive all year. Glaciers develop in mountains, even at low latitudes, because temperature decreases with elevation; at high elevations, the mean temperature stays low enough for ice and snow to survive all year. Since the temperature of a region depends on latitude, the specific elevation at which mountain glaciers form also depends on latitude. In Earth's present-day climate, glaciers form only at elevations above 5 km between 0° and 30° latitude and down to sea level at 60° to 90° latitude (▶Fig. 22.6). Thus, you can see high latitude glaciers from a cruise ship, but you have to climb the highest mountains of the Andes to find glaciers at the equator. Mountain glaciers tend to develop on the side of mountains that receives less wind, and on the side that receives less sunlight. Glaciers do not exist on slopes greater than about 30°, because avalanches clear such slopes.

The transformation of snow to glacier ice takes place slowly, as younger snow progressively buries older snow. Freshly fallen snow consists of delicate hexagonal crystals with sharp points. The crystals do not fit together tightly, so this snow contains about 90% air. With time, the points of the snowflakes become blunt, because they either **sublimate** (evaporate directly into vapor) or melt, and the snow packs more tightly. As snow becomes buried, the weight of the overlying snow increases pressure, which causes remaining points of contact between snowflakes to melt. This process of melting at points of contact where the pressure is greatest is another example of pressure solution (see chapter 11). Gradually, the snow transforms into a packed granular material called **firn,** which contains only about 25% air (▶Fig. 22.7a, b). Melting of firn grains at contact points produces water that crystallizes in the spaces between grains until eventually the firn transforms into a solid mass of glacial ice composed of interlocking ice crystals. Such glacial ice, which may still contain up to 20% air trapped in bubbles, tends to absorb red light and thus has a bluish color. The transformation of fresh snow to glacier ice can take as little as tens of years in regions with abundant snowfall, or as long as thousands of years in regions with little snowfall.

FIGURE 22.7 (a) Snow compacts and melts to form firn, which recrystallizes to make ice. (b) The size of ice crystals increases with depth in a glacier, where some crystals grow at the expense of others.

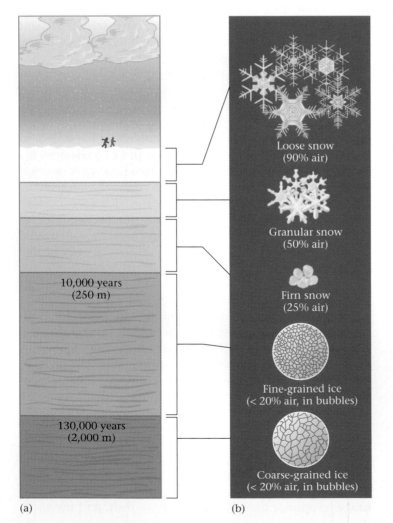

(a)   (b)

The Movement of Glacial Ice

When Louis Agassiz became fascinated by glaciers, he decided to find out how fast they flowed, so he hammered stakes into an Alpine glacier and watched the stakes move during the year. More recently, researchers have observed glacial movement with the aid of time-lapse photography, which shows the evolution of a glacier over several years in a movie that lasts a few minutes. In such movies, you can actually see a glacier flow across the screen, making its nickname "river of ice" seem perfectly appropriate.

How do glaciers move? Some move when meltwater accumulates at their bases, so that the mass of the glacier slides partially on a layer of water or on a slurry of water and sediment. During this kind of movement, known as "basal sliding", the water or wet slurry holds the glacier above the underlying rock and reduces friction between the glacier and its substrate. Basal sliding is the dominant style of movement for **wet-bottom glaciers** (▶Fig. 22.8a). The water forms at the base of such glaciers because of pressure-induced melting, or because the glaciers exist in a climate that is warm enough for the ice to be at or near its melting temperature, or because the glacier ice acts like an insulating blanket and traps heat rising from the Earth below. Temperate glaciers are generally wet bottomed due primarily to melting by heat from the ground.

Glaciers also move by means of internal flow, during which the mass of ice slowly changes shape internally without breaking apart or completely melting. By studying ice deformation with a microscope, geologists have determined that internal flow involves two processes. First, ice crystals become plastically deformed, by the rearrangement of water molecules within the crystal lattice (during this process, ice crystals change shape, old crystals disappear, and new ones form); second, if conditions allow very thin films of water to form on the surfaces of ice crystals, the crystals can slide past one another. Internal flow is the dominant style of movement for **dry-bottom glaciers,** those that are so cold that their base remains frozen to their substrate (▶Fig. 22.8b). Polar glaciers tend to be dry bottomed.

Note that the plastic deformation of ice in glaciers resembles the plastic deformation of metamorphic rock in mountain belts. However, metamorphic rock deforms plastically at depths greater than 10–15 km in the crust, while ice deforms plastically at depths below about 60 m (▶Fig. 22.8c). In other words, the "brittle-plastic transition" in ice lies at a depth of only about 60 m—above this depth, large cracks can form in ice, while below this depth, cracks cannot form because ice is too plastic. A large crack that develops by brittle deformation in a glacier is called a **crevasse.** In large glaciers, crevasses can be hundreds of meters long and tens of meters deep (▶Fig. 22.9). Tragically, several explorers have met their death by falling into crevasses.

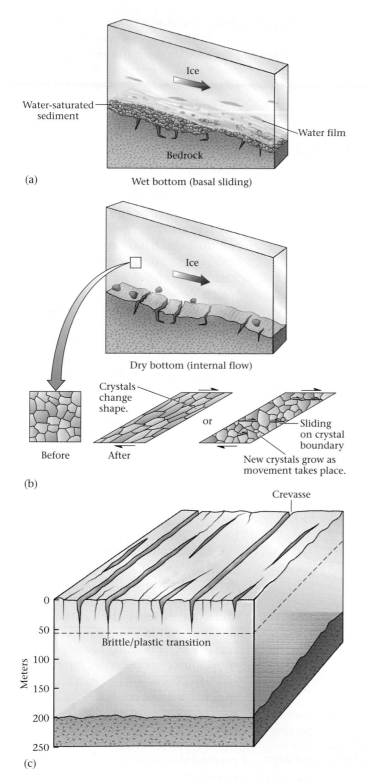

FIGURE 22.8 (a) Wet-bottom glaciers move by means of basal sliding. (b) Dry-bottom glaciers flow in the solid state. In some cases, the ice crystals stretch and rotate, while in other cases, the ice recrystallizes and some crystals slide past each other, especially if there are thin films of water between grains. (c) Crevasses form in the brittle ice above the brittle-plastic transition in glaciers.

Why do glaciers move? Ultimately, because the pull of gravity can cause weak ice to flow (▶Fig. 22.10a). Glaciers flow in the direction in which their surfaces slope. Thus, valley glaciers flow down their valleys, and continental ice sheets "spread" outward from their thickest point. To picture the movement of a continental ice sheet, imagine that a thick pile of ice builds up, gravity causes the top of the pile to push down on the ice at the base. Eventually, the basal ice can no longer support the weight of the overlying ice and begins to deform plastically. When this happens, the basal ice starts squeezing out to the side, carrying the overlying ice with it. The greater the volume of ice that builds up, the wider the sheet of ice can become. You've seen a similar process of "gravitational spreading" if you've ever poured honey onto a plate. The honey can't build up into a narrow column because it's too weak; rather, it flows laterally away from the point where it lands to form a wide, thin layer (▶Fig. 22.10b, c).

Glaciers generally flow at rates of between 10 and 300 m per year—far slower than a river, but far faster than a silicate rock even under high-grade metamorphic conditions. The velocity of a particular glacier depends, in part, on the magnitude of the force driving its motion. For example, a glacier whose surface slopes steeply moves faster than one with a gently sloping surface. Flow velocity also depends on whether liquid water exists at places along the base of the glacier; wet-bottom glaciers tend to move faster than dry-bottom glaciers.

Not all parts of a glacier move at the same rate. For example, friction between rock and ice slows a glacier, so the center of a valley glacier moves faster than its margins, and the top of a glacier moves faster than its base (▶Fig. 22.11a, b). And because water at the base of a glacier

FIGURE 22.9 Crevasses in an Antarctic glacier. The crevasses are up to 10 m wide.

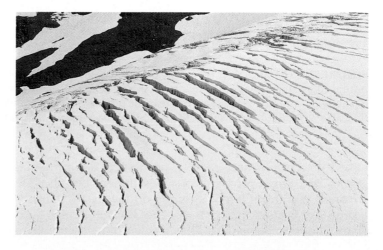

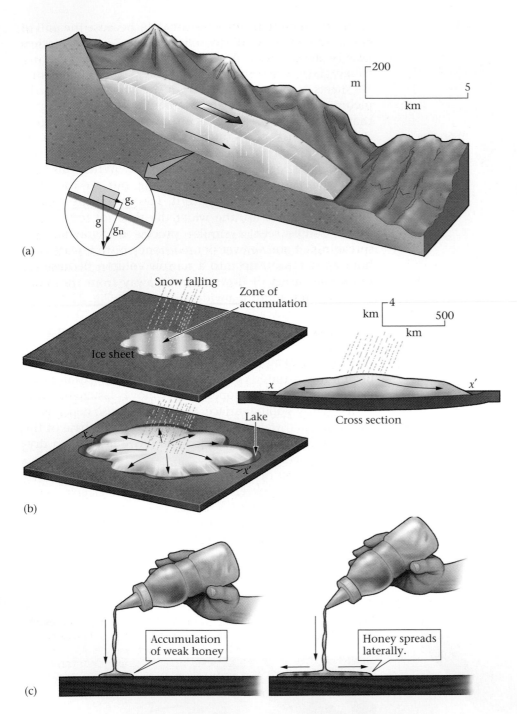

FIGURE 22.10 (a) Gravity drives the downslope motion of glaciers. Gravity can be portrayed by an arrow (vector) that points straight down. One component (g_s) is parallel to the slope, and drives the flow, whereas the other (g_n) is perpendicular to the slope. (b, c) The gravitational spreading of an ice sheet resembles honey spreading across a table.

it lifts the glacier off its substrate, the glacier undergoes a **surge** and flows much faster for a limited time (rarely more than a few months), until the water escapes. During surges, glaciers have been clocked at speeds of 20–110 m per day.

Glacial Advance and Retreat

Glaciers resemble bank accounts: snowfall adds to the account, while **ablation**—the removal of ice by sublimation (the evaporation of ice into water vapor), melting (the transformation of ice into liquid water, which flows away), and calving (the breaking off of chunks of ice at the edge of the glacier)—subtracts from the account (▶Fig. 22.12). Snowfall adds to the glacier in the **zone of accumulation,** while ablation subtracts in the **zone of ablation;** the boundary between these two zones is the **equilibrium line.** The zone of accumulation occurs where the temperature remains cold enough year round so that winter snow does not melt or sublimate away entirely during the summer. Therefore, elevation and latitude control the position of the equilibrium line.

The leading edge or margin of a glacier is called its **toe,** or terminus (▶Fig. 22.13a). If the rate at which ice accumulates in the zone of accumulation exceeds the rate at which ablation occurs below the equilibrium line, then the toe moves forward into previously unglaciated regions. Such a change is called a **glacial advance** (▶Fig. 22.13b). In the case of mountain glaciers, the position of a toe moves downslope during an advance, and in the case of continental glaciers, the toe moves outward, away from the glacier's origin. If the rate of ablation below the

allows it to travel more rapidly, portions of a continental glacier that flow over water or wet sediment become "ice streams" that travel ten to a hundred times faster than adjacent dry-bottom portions of the glacier. Similarly, if water builds up beneath a valley glacier to the point where equilibrium line equals the rate of accumulation, then the position of the toe remains fixed. But if the rate of ablation exceeds the rate of accumulation, then the position of the toe moves back toward the origin of the glacier; such a change is called a **glacial retreat** (▶Fig. 22.13c).

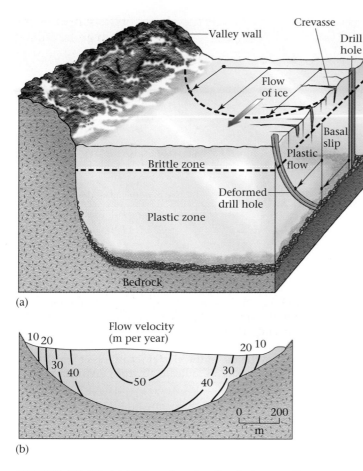

(a)

Valley wall

Crevasse

Drill hole

Flow of ice

Basal slip

Brittle zone

Plastic flow

Deformed drill hole

Plastic zone

Bedrock

Flow velocity (m per year)

10 20

30

40

50

40

30

20 10

0   200
m

(b)

FIGURE 22.11 (a) Different parts of a glacier may flow at different velocities. The vector lengths indicate the velocity of flow. (b) This cross section of a glacier shows measured velocities of flow.

During a mountain glacier's retreat, the position of the toe moves upslope. It's important to realize that when a glacier retreats, it's only the *position* of the toe that moves back toward the origin, for ice continues to flow toward the toe. Glacial ice cannot flow back toward the glacier's origin.

One final point before we leave the subject of glacial flow: beneath the zone of accumulation, a crystal of ice gradually moves down toward the base of the glacier as new ice accumulates above it, while beneath the zone of ablation, a crystal of ice gradually moves up toward the surface of the glacier, as overlying ice ablates. Thus, as a glacier flows, ice crystals follow curved trajectories (Fig. 22.13 a–c). For this reason, rocks picked up by ice at the base of the glacier may slowly move to the surface. The upward flow of ice where the Antarctic ice sheet collides with the Transantarctic Mountains brings up meteorites long buried in the ice (▶Fig. 22.14a, b).

Ice in the Sea

On the moonless night of April 14, 1912, the great ocean liner *Titanic* plowed through the calm but frigid waters of the North Atlantic on her maiden voyage from Southampton, England, to New York. Although radio broadcasts from other ships warned that "icebergs," large blocks of ice floating in the water, had been sighted in the area and might pose a hazard, the ship sailed on, its crew convinced that they could see and avoid the biggest bergs, and that smaller ones would not be a problem for the steel hull of

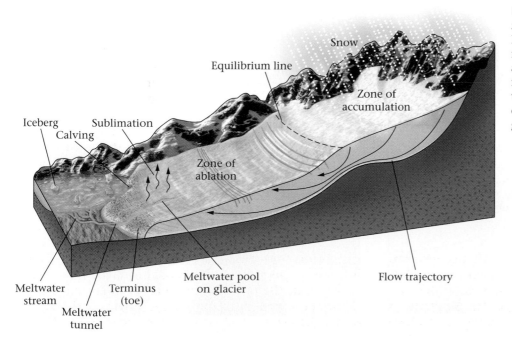

FIGURE 22.12 The zone of ablation, zone of accumulation, and equilibrium line. The arrows illustrate how a grain of ice flows down in the zone of accumulation and up in the zone of ablation, so that it follows a curved trajectory. Note that ice at the base of the glacier can flow up and over obstacles as long as the surface of the glacier has a downhill slope.

Snow

Equilibrium line

Zone of accumulation

Iceberg

Calving

Sublimation

Zone of ablation

Flow trajectory

Meltwater stream

Meltwater tunnel

Terminus (toe)

Meltwater pool on glacier

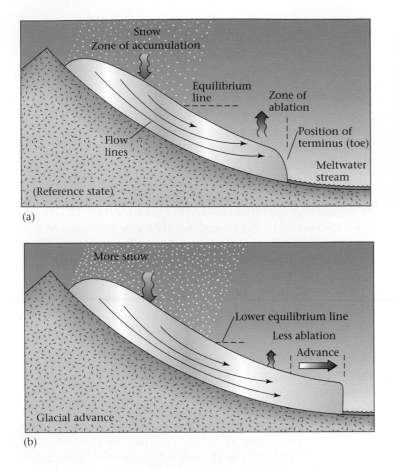

(a)

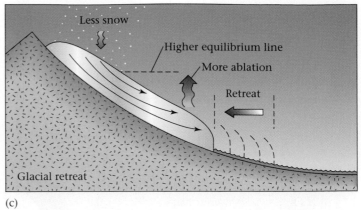

(c)

FIGURE 22.13 (a) The position of a glacier's toe represents the balance between the amount of ice that forms beneath the zone of accumulation and the amount of ice lost in the zone of ablation. (b) Glacial advance and (c) glacial retreat. Notice that ice always flows toward the toe of the glacier regardless of whether the toe advances or retreats, as indicated by the curving flow line.

(b)

this "unsinkable" vessel. But, in a story now retold countless times, their confidence was fatally wrong. At 11:40 P.M., while the first-class passengers danced, the *Titanic* struck a large iceberg. Lookouts had seen the ghostly mass of frozen water only minutes earlier and had alerted the ship's pilot, but the ship had been unable to turn fast enough to avoid

disaster. The force of the blow split the steel hull that spanned five of the ship's sixteen water-tight compartments—the ship could stay afloat if four compartments flooded, but the flooding of five meant it would sink. At about 2:15 A.M., the bow disappeared below the water, and the stern rose until the ship protruded nearly vertically

FIGURE 22.14 (a) Beneath the zone of ablation, ice crystals move up to the surface. Long-buried meteorites also collect at the surface. (b) A meteorite sitting on the surface of a glacier.

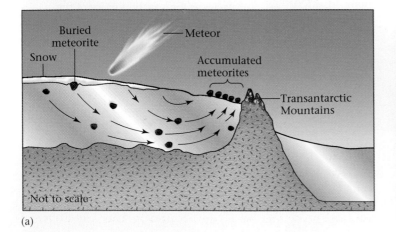

(a)

(b)

Ice tongue

Sea ice

(a)

(b) Land   Ice shelf   Water

FIGURE 22.15 (a) An ice tongue protruding into the Ross Sea along the coast of Antarctica. (b) The Larson ice shelf along the coast of Antarctica, as seen from a satellite in January 2002. Since then, much of the shelf has disintegrated.

from the water. Without water to support its weight, the hull buckled and split in two—the stern section fell back down onto the water and momentarily bobbed horizontally before following the bow, settling downward through over 3.5 km of water to the silent sea floor below. Because of an inadequate number of lifeboats and the inability of the crew to load passengers efficiently, only 705 passengers survived; 1,500 expired in the frigid waters of the

Atlantic. The *Titanic* remained lost until 1985, when a team of oceanographers led by Robert Ballard located the sunken hull and photographed its eerie form where it lies, upright in the mud.

Where do icebergs, like the one responsible for the *Titanic*'s demise, originate? In high latitudes, mountain glaciers and continental ice sheets flow down to the sea, and they either stop at the shore or flow into the sea. Valley glaciers whose terminus lies in the water are "tidewater glaciers." Some of these may protrude into the ocean to become ice tongues (▶Fig. 22.15a), continental glaciers entering the sea become broad, flat sheets called **ice shelves** (▶Fig. 22.15b). Four-fifths of the ice lies below the sea surface, so in shallow water, it remains grounded (▶Fig. 22.16). But where the water becomes deep enough, the ice floats. At the boundary between glacier and ocean, blocks of ice calve off and tumble into the water with an impressive splash. If a free-floating chunk rises 6 m above the water and is at least 15 m long, it is formally called an **iceberg.** Smaller pieces, formed when ice blocks fragment before entering the water or after icebergs have had time to melt, include "bergy bits," rising 1 to 5 m above the water and covering an area of 100–300 m², and "growlers," rising less than 1 m above the water and covering an area of about 20 m²—still big enough to damage a ship. Growlers get their name because of the sound they make as they bob in the sea and grind together.

Most large icebergs form along the western coast of Greenland or along the coast of Antarctica. Icebergs that calve off valley glaciers tend to be irregularly shaped with pointed peaks rising upward—such glaciers are called "castle bergs" or "pinnacle bergs". One of the largest on record protruded about 180 m above the sea. Since four-fifths of the ice lies below the surface of the sea, the base of a large iceberg may actually be a few hundred meters below the surface (▶Fig. 22.17). Icebergs that originate in Greenland float into the "iceberg alley" region of the North Atlantic. These are the bergs that threaten ships, although the danger has diminished in modern times because of less ice and because of ice patrols that report the locations of floating ice. Blocks that calve off the vast ice shelves of Antarctica tend to have flat tops and nearly vertical sides—such glaciers are called "tabular bergs". Some of the tabular bergs in the Antarctic are truly immense; air photos have revealed bergs over 160 km across.

Not all ice floating in the sea originates as glaciers on land. In polar climates, the surface of the sea itself freezes, forming **sea ice** (▶Fig. 22.18). Some sea ice, such as that covering the interior of the Arctic Ocean, floats freely, but some protrudes outward from the shore. Ice breakers can crunch through sea ice that is up to 2.5 m thick; the ice breaker rides up on the ice, then crushes it. Vast areas of ice shelves and of sea ice have been disintegrating in recent years, perhaps because of global warming. For example, open regions have been found in the Arctic Ocean

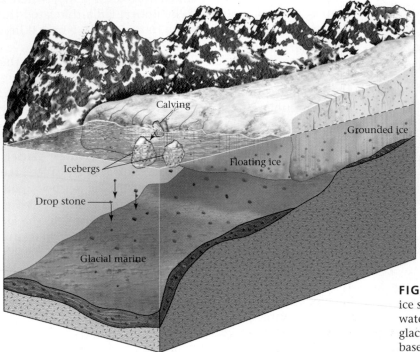

Calving

Grounded ice

Icebergs

Floating ice

Drop stone

Glacial marine

FIGURE 22.16 Where a glacier reaches the sea, the ice stays grounded in shallow water and floats in deep water. Icebergs calve off the front of such tidewater glaciers. Sediment known as drop stones falls off the base of icebergs and collects on the sea floor.

FIGURE 22.17 An artist's rendition of an iceberg. This image is a composite of photographs spliced together. It conveys a sense of the relative proportions of ice above and below the surfce of the sea.

FIGURE 22.18 Broken-up sea ice off the coast of Antarctica.

during the summers, and the ice shelf in Antarctica has been decreasing rapidly in area. In some locations, large openings known as "polynyas", have developed in the sea ice of Antarctica. Some sea ice forms in winter and melts away in summer. But at high latitudes, sea ice may last for several years. For example, in the Arctic Ocean, sea ice may last long enough to make a 7- to 10-year voyage around the Arctic Ocean; this movement occurs in response to currents.

The existence of icebergs leaves a record in the stratigraphy of the sea floor, for icebergs carry ice-rafted sediment (▶Fig. 22.19). Larger rocks that drop from the ice to the sea floor are called **drop stones** (Fig. 22.16). In ancient glacial deposits, drop stones appear as isolated blocks surrounded by mud. Icebergs and smaller fragments also drop sand and gravel, derived by the erosion of continents, onto the sea floor. In cores extracted by drilling into sea-floor sediment, horizons of such land-derived sediment, sandwiched between layers of sediment formed from marine plankton shells, indicate times in Earth history when glaciers were breaking up and icebergs became particularly abundant.

22.4 CARVING AND CARRYING BY ICE

Glacial Erosion and Its Products

The Sierra Nevada mountains of California consist largely of granite that formed during the Mesozoic Era in the crust beneath a volcanic arc. During the past 10–20 million years, the land surface slowly rose and erosion stripped away overlying rock to expose the granite. The style of erosion formed rounded domes. Then, during the last ice age, valley glaciers cut deep, steep-sided valleys into the range. In the process, some of the domes were cut in half, leaving a rounded surface on one side and a steep cliff on the other. Half Dome, in Yosemite National Park, formed in this way (▶Fig. 22.20a); its steep cliff has challenged many rock climbers. Such glacial erosion also produces the knife-edge ridges and pointed spires of high mountains (▶Fig. 22.20b), and the broad expanses where rock outcrops have been stripped of overlying sediment and polished smooth (▶Fig. 22.20c). Glacial erosion in the mountains can lower a valley floor by over 100 m, and continental glaciation during the last ice age stripped up to 30 m of material off the land in northern Canada.

Glaciers erode their substrate in several ways (▶Fig. 22.21a, b). During "glacial incorporation", ice surrounds and incorporates debris. During "glacial plucking" (or glacial quarrying), a glacier breaks off and then carries away fragments of bedrock. Plucking occurs when ice freezes onto rock that has already started to crack and separate from its substrate; movement of the ice lifts off pieces of the rock. At the toe of a glacier, ice may actually bulldoze sediment and trees slightly before flowing over them.

As glaciers flow, clasts embedded in ice act like the teeth of a giant rasp and grind away the substrate. This process, "glacial abrasion", produces very fine sediment called "rock flour", just as sanding wood produces sawdust. Rasping by embedded sand yields shiny, **glacially polished surfaces.** Rasping by large clasts produces long gouges, grooves, or scratches (1 cm to 1 m across) called **glacial striations** (▶Fig. 22.21c). As you might expect, striations run parallel to the flow direction of the ice. When boulders entrained in the base of the ice impact bedrock below, as the ice moves, the impact may break off asymmetric wedges of bedrock, leaving behind indentations called "chatter marks" (▶Fig. 22.21d). In regions of wet-bottom glaciers, sediment-laden water rushes through tunnels at the base of the glaciers and can carve substantial channels.

Let's now look more closely at the erosional features associated with a mountain glacier (▶Fig. 22.22a–f). Freezing and thawing during the fall and spring help fracture the rock bordering the head of the glacier (the ice edge high in the mountains). This rock falls on the ice or gets picked up at the base of the ice, and moves downslope with the glaciers. As a consequence, a bowl-shaped depression, or cirque, develops on the side of the mountain. If the ice later melts, a lake called a **tarn** may form at the base of the cirque. An **arête** (French for "ridge"), a residual knife-edge ridge of rock, separates two adjacent cirques. A pointed mountain peak surrounded by at least three cirques is called a **horn.** The Matterhorn, a peak in Switzerland, serves as a particularly beautiful example of a horn (Fig. 22.22e).

FIGURE 22.19 A 7-m-long growler in Alaska containing now tilted gravel-rich layers. The gravel fell on the ice and was buried during successive snowfalls when the ice was part of a glacier. The murky color of water is due to suspended glacial sediment.

(a)

(b)

(c)

FIGURE 22.20 Products of glacial erosion. (a) Half Dome, in Yosemite National Park, California. Before glaciation, the mountain was a complete dome; then glacial erosion truncated one side. (b) Glaciated mountains with sharp peaks. (c) A glacially polished surface along the shore of Lake Superior. The bedrock is so smooth that the girl can slide down it.

FIGURE 22.21 The various kinds of glacial erosion. (a) Plowing and incorporation; (b) plucking and abrasion; (c) glacial striations in downtown Victoria, British Columbia (Canada); (d) chatter marks on a glacially eroded surface in Switzerland.

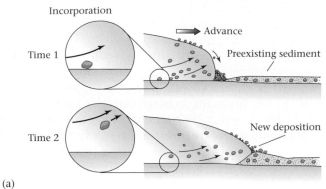

(a)

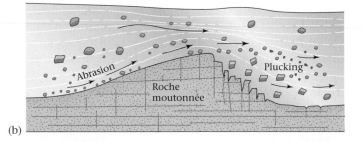

(b)

(c)

(d)

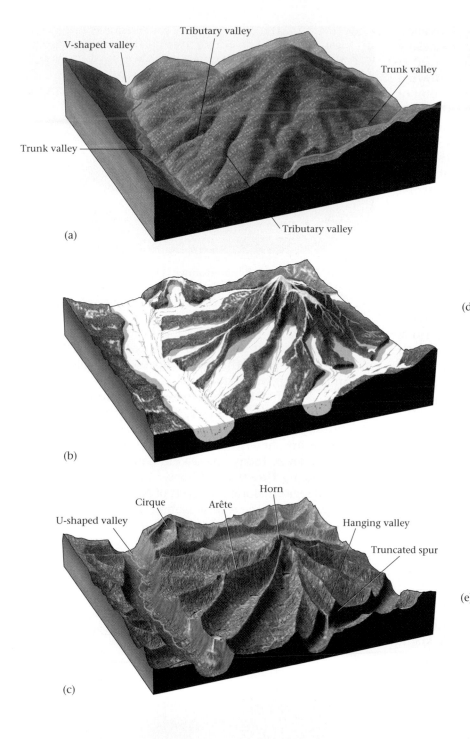

(a)

(b)

(c)

(d)

(e)

(f)

FIGURE 22.22 (a) A mountain landscape before glaciation. The V-shaped valleys are the result of river erosion; the floor of the tributary valleys are at the same elevation as the trunk valley where they intersect. (b) During glaciation, the valleys fill with ice. (c) After glaciation, the region contains U-shaped valleys, hanging valleys, truncated spurs, and horns. (d) A U-shaped glacial valley. (e) The Matterhorn in Switzerland. (f) A waterfall spilling out of a hanging valley. Truncated spurs occur on both sides of the waterfall.

Glacial erosion severely modifies the shape of a valley. To see how, compare a river-eroded valley with a glacially eroded valley. If you look along the length of a river in unglaciated mountains, and you'll see that it flows down a V-shaped valley, with the river channel forming the point of the V. The V develops because river erosion occurs only in the channel, and mass wasting causes the valley slopes to approach the angle of repose. But if you look down the length of a glacially eroded valley, you'll see that it resembles a U, with steep walls. A **U-shaped valley,** forms because the combined processes of glacial abrasion and plucking not only lower the floor of the valley, they also bevel its sides.

Glacial erosion in mountains also modifies the intersections between tributaries and the trunk valley. In a river system, tributaries cut side valleys that merge with the trunk valley, such that the mouths of the tributary valleys lie at the same elevation as the trunk valley. The ridges ("spurs") between valleys taper to a point when they join the trunk valley floor. During glaciation, tributary glaciers flow down side valleys into a trunk glacier. But the trunk glacier cuts the floor of its valley down to a depth that far exceeds the depth cut by the tributary glaciers. Thus, when the glaciers melt away, the mouths of the tributary valleys perch at a higher elevation than the floor of the trunk valley. Such side valleys are called **hanging valleys.** The water in post-glacial streams that flow down a hanging valley must cascade over a spectacular waterfall to reach the post-glacial trunk stream (Fig. 22.22f). Trunk glaciers, as they erode, also chop off the ends of spurs, creating truncated spurs.

Now let's look at the erosional features produced by continental ice sheets. To a large extent, these depend on the nature of the preglacial landscape. Where an ice sheet spreads over a region of low relief, like the Canadian Shield, glacial erosion creates a vast region of polished, flat striated surfaces. Where an ice sheet spreads over a hilly area, it deepens valleys and smoothes hills. In central New York, for example, continental glaciers carved the deep valleys that now cradle the Finger Lakes, and in Maine glaciers smoothed and streamlined the granite and metamorphic rock hills of Acadia National Park. Notably, glacially eroded hills end up being elongate in the direction of flow and are asymmetric; glacial rasping smoothes and bevels the upstream part of the hill, creating a gentle slope, while glacial plucking eats away at the downstream part, making a steep slope. Ultimately, the hill's profile resembles that of a sheep lying in a meadow—such hills are called **roche moutonnée,** from the French for "sheep rock" (▶Fig. 22.23).

Fjords: Submerged Glacial Valleys

As noted earlier, where a valley glacier meets the sea, the glacier's base remains in contact with the ground until the

FIGURE 22.23 A roche moutonnée. The glacier flowed from right to left.

water depth exceeds about four-fifths of the glacier's thickness, at which point the glacier floats. Thus, glaciers can continue carving a U-shaped valley even below sea level. In addition, during an ice age, water extracted from the sea becomes locked in the ice sheets on land, so sea level drops significantly. Therefore, the floors of valleys cut by coastal glaciers during the last ice age were cut much deeper than present sea level. Today, the sea has flooded these deep valleys, producing **fjords** (see Chapter 18). In the spectacular fjord-land regions along the coasts of Norway, New Zealand, Chile, and Alaska, the walls of submerged U-shaped valleys rise straight from the sea as vertical cliffs up to 1,000 m high (▶Fig. 22.24). Fjords also develop where a glacial valley fills to become a lake.

FIGURE 22.24 One of the many spectacular fjords of Norway.

The Glacial Conveyor: The Transport of Sediment by Ice

Glaciers can carry sediment of any size and, like a conveyor belt, transport it in the direction of flow (i.e., toward the toe). The sediment load either falls onto the surface of the glacier from bordering cliffs or gets plucked and lifted from the substrate and incorporated into the moving ice. Because of the curving flow lines of glacial ice, rocks plucked off the floor may eventually reach the surface.

Sediment dropped on the glacier's surface moves with the ice and becomes a stripe of debris. Stripes formed along the side edges of the glacier are **lateral moraines.** The word **moraine** was a local term used by Alpine farmers and shepherds for piles of rock and dirt. When the glacier finally melts, lateral moraines lie stranded along the side of the glacially carved valley, like a bathtub ring. If flowing water runs along the edge of the glacier and sorts the sedi-

ment of a lateral moraine, a stratified sequence of sediment called a **kame** forms. Where two valley glaciers merge, the debris constituting two lateral moraines merges to become a **medial moraine,** running as a stripe down the interior of the composite glacier (▶Fig. 22.25a, b). Trunk glaciers created by the merging of tributary glaciers contain several medial moraines. Glaciers passing through ranges with extremely high erosion rates may be completely buried by rocky debris. In some cases, a glacier incorporates so much rock that geologists refer to it as a "rock glacier." Sediment transported to a glacier's toe by the glacial "conveyor" accumulates in a pile at the toe and builds up to form an **end moraine** (▶Fig. 22.25c).

So far, we've emphasized sediment moved by ice. But in temperate glacial environments the flowing water at the base of the glacier, moving through channels, transports much of the sediment load, eventually depositing it beyond the terminus of the glacier.

FIGURE 22.25 (a) Lateral and medial moraines on a glacier. (b) Medial moraines on a glacier near Alyeska, Alaska. The source of the moraine on the right is off the image. (c) The surface load plus the internal load accumulates at the toe of the glacier to constitute an end moraine; in effect, glaciers act like conveyor belts, constantly transporting more sediment toward the toe, regardless of whether the glacier advances or retreats.

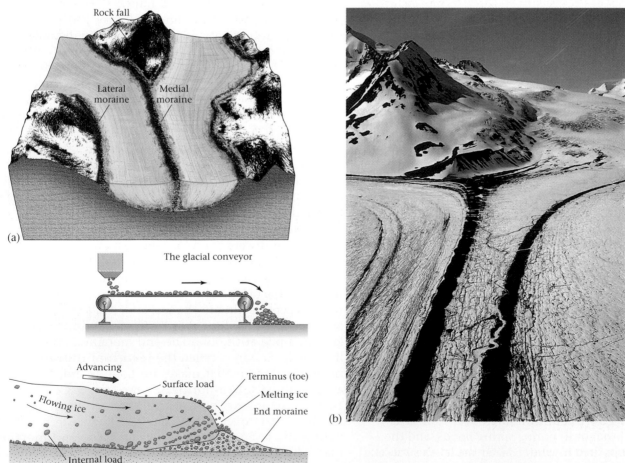

22.5 DEPOSITION ASSOCIATED WITH GLACIATION

Types of Glacial Sedimentary Deposits

If you drill through the soil throughout much of the upper midwestern and northeastern United States and adjacent parts of Canada, the drill penetrates a layer of sediment deposited during the last ice age. A similar story holds for much of northern Europe. Thus, many of the world's richest agricultural regions rely on soil derived from sediment deposited by glaciers during the ice age. Notably, this sediment buries a pre–ice age landscape, as frosting fills the irregularities on a cake. Preglacial valleys may be completely filled with sediment.

Several different types of sediment can be deposited in glacial environments; all of these types together constitute **glacial drift.** (The term dates from pre-Agassiz studies of glacial deposits, when geologists thought that the sediment had "drifted" into place during an immense flood.) Specifically, glacial drift includes the following:

- *Till:* Sediment transported by ice and deposited beneath or at the toe of a glacier comprises **glacial till.** Glacial till is unsorted (it is a type of distinction), because the solid ice of glaciers can carry clasts of all sizes (▶Fig. 22.26a).

- *Erratics:* Glacial *erratics* are cobbles and boulders that have been dropped by a glacier. Some protrude from till piles, and others rest on glacially polished surfaces.

- *Glacial marine:* Where a sediment-laden glacier flows into the sea, icebergs calve off the toe and raft clasts out to sea. As the icebergs melt, they drop the clasts, which settle into the muddy sediment on the sea floor. Pebble- and larger-size clasts deposited in this way, as we have seen, are "drop stones". Sediment consisting of ice-rafted clasts mixed with marine sediment makes up **glacial marine.** Glacial marine can also consist of sediment carried into the sea by water flowing at the base of a glacier.

- *Glacial outwash:* Till deposited by a glacier at its toe may be picked up and transported by meltwater streams that sort the sediment. The clasts are deposited by a braided stream network in a broad area of gravel and sand bars called an outwash plain. This sediment is known as **glacial outwash** (▶Fig. 22.26b).

- *Glacial lake-bed sediment:* Streams transport fine clasts, including rock flour, away from the glacial front. This sediment eventually settles in meltwater lakes, forming a thick layer of **glacial lake-bed sediment.** This sediment commonly contains varves. A **varve** is a pair of thin layers deposited during a single year. One layer consists of silt brought in during spring floods, and the other of clay deposited in winter when the lake's surface freezes over and the water is still (▶Fig. 22.26c).

- *Loess:* When the warmer air above ice-free land beyond the toe of a glacier rises, the cold, denser air from above the glacier rushes in to take its place; a strong wind called "catabatic wind" therefore blows at the margin of a glacier. This wind picks up fine clay and silt and transports it away from the glacier's toe. Where the winds die down, the sediment settles and forms a thick layer. This sediment, called **loess,** sticks together because of the electrical charges on clay flakes; thus steep escarpments develop by erosion of loess deposits (▶Fig. 22.26d).

Till, which contains no layering, is sometimes called "unstratified drift", while glacial sediments that have been redistributed by flowing water are called "stratified drift."

Depositional Landforms of Glacial Environments

Picture a hunter, dressed in deerskin, standing at the toe of a continental glacier in what is now southern Canada, waiting for an unwary woolly mammoth to wander by. It's a sunny summer day 12,000 years ago, and milky, sediment-laden streams gush from tunnels and channels at the base of the glacier and pour off the top as the ice melts. No mammoths venture by today, so the bored hunter climbs to the top of the glacier for a view. The climb isn't easy, partly because of the incessant catabatic wind, and partly because deep crevasses interrupt his path. Reaching the top of the ice sheet, the hunter looks northward, and the glare almost blinds him. Squinting, he sees the white of snow, and where the snow has blown away, he sees the rippled, glassy surface of bluish ice (Fig. 22.1). Here and there, a rock protrudes from the ice. Now looking southward, he surveys a stark landscape of low, sinuous ridges separated by hummocky (bumpy) plains. Braided streams, which carry meltwater out across this landscape, flow through the hummocky plains and supply a number of lakes. Dust fills the air because of the wind.

All the landscape features observed by the hunter as he looks southward form by deposition in glacial environments, both mountain and continental (▶Fig. 22.27a, b). The low, sinuous ridges, called "end moraines," develop when the toe of a glacier stalls in one position for a while; the ice keeps flowing to the toe, and, like a giant conveyor belt, transports sediment with it; the sediment accumulates in a pile at the toe. The end moraine at the farthest limit of glaciation is called the **terminal moraine.** (The ridge of sediment that makes up Long Island, New York, and continues east-northeast into Cape Cod, Massachusetts, represents part of the terminal moraine of the ice sheet that covered New England and eastern Canada during the last ice age; ▶Fig. 22.28.) The end moraines that form when a glacier stalls for a while as it recedes are **recessional moraines.**

Till that has been released at the base of a flowing glacier and remains after the glacier has melted away comprises "lodgment till". Clasts in lodgment till may be aligned and scratched during their movement or as ice flows over them. The flow of the glacier may mold till and other subglacial sediment into streamlined, elongate hills called **drumlins** (from the Gaelic word for "hills"). Drumlins tend to be asymmetric along their length, with a gentle downstream slope, tapered in the direction of flow, and a steeper up-stream slope (▶Fig. 22.29a–c). The till left behind during rapid recession forms a thin, hummocky layer on the land surface; this till, together with lodgment till, forms a land-scape feature known as **ground moraine.**

The hummocky surface of moraines reflects partly the variations in the amount of sediment supplied by the glacier, and partly the occurrence of **kettle holes,** circular depressions made when blocks of ice calve off the toe of the glacier, become buried by till, and then melt (▶Fig. 22.30a). A land surface with many kettle holes separated by round hills of till displays "knob-and-kettle topography" (▶Fig. 22.30b, c).

FIGURE 22.26 (a) The unsorted sediment constituting glacial till in Ireland. (b) Braided streams choked with glacial outwash near the toe of a glacier in Alaska. (c) Note the alternating light and dark layers in this 20 cm-thick cross section of varved glacial lake bed sediment. (d) A small escarpment cut into glacial loess deposits of Illinois.

(a)

(b)

(c)

(d)

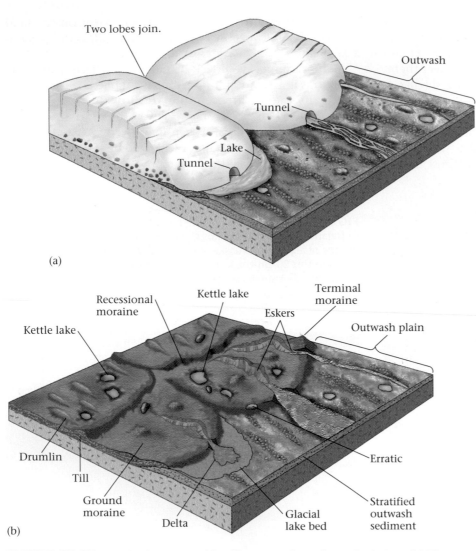

FIGURE 22.27 (a) The depositional landforms resulting from glaciation. (b) The setting in which various types of moraines form.

FIGURE 22.28 The moraines that constitute Long Island and Cape Cod.

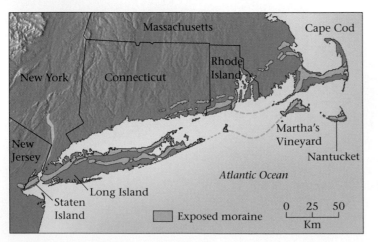

Sediment-choked water pours out of tunnels in the ice. This water feeds large braided streams that sort till and redeposit it as glacial outwash, stratified layers and lenses of gravel and sand. Thus, outwash plains form between recessional moraines and beyond the terminal moraine. Some of the melt water pools in lowlands between recessional moraines, or in ice-margin lakes between the glacier's toe and the nearest recessional moraine. Even long after the glacier has melted away, the lowlands between recessional moraines persist as lakes or swamps. Glacial lake beds provide particularly fertile soil for agriculture. When a glacier eventually melts away, ridges of sorted sand and gravel, deposited in subglacial meltwater tunnels, snake across the ground moraine. These ridges are called **eskers** (▶Fig. 22.31). Sediments in glacial outwash plains and in eskers serve as important sources of sand and gravel for road building and construction.

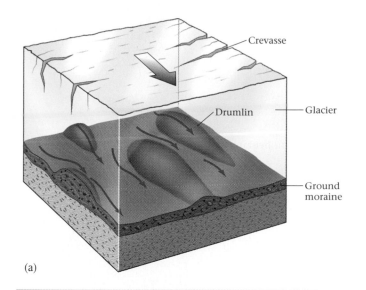

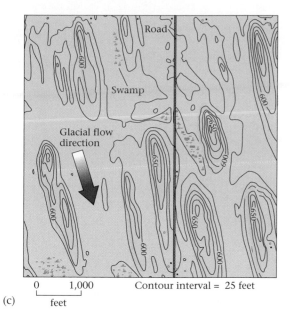

FIGURE 22.29 (a) The formation of a drumlin beneath a glacier. (b) Drumlins near Rochester, New York. (c) Topographic map emphasizing the shape of drumlins in New York State.

FIGURE 22.30 (a) When this ice block melts away, a kettle hole will form. (b, c) Knob-and-kettle topography, the hummocky surface of a moraine.

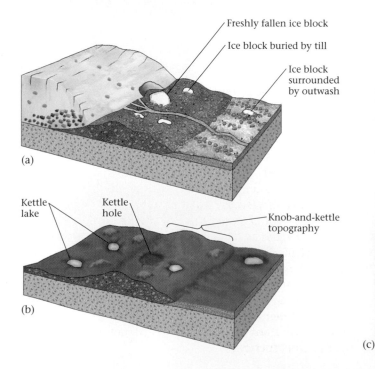

Glaciers and Glacial Landforms

Continental ice sheet

Crevasses

Ice shelf

Lower sea level

Drop stones

Iceberg

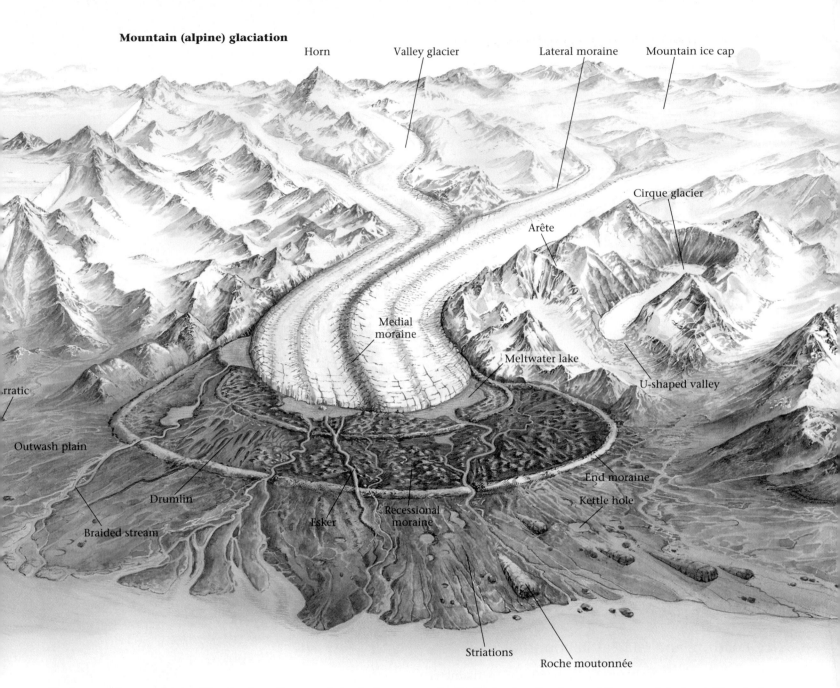

Mountain (alpine) glaciation

Horn · Valley glacier · Lateral moraine · Mountain ice cap

Cirque glacier

Arête

Medial moraine

Meltwater lake

U-shaped valley

rratic

Outwash plain

Drumlin

Esker

Recessional moraine

End moraine

Kettle hole

Braided stream

Striations

Roche moutonnée

Glaciers are rivers or sheets of ice that last all year and slowly flow. Continental glaciers, vast sheets of ice up to a few kilometers thick, covered extensive areas of land during times when Earth had a colder climate. Continental glaciers form when snow accumulates at high latitudes, then, when buried deeply enough, packs together and recrystallizes to make glacial ice. Ice, though solid, is weak, and thus ice sheets spread over the landscape, like syrup over a pancake. At the peak of the last ice age, ice sheets covered almost all of Canada, much of the United States, northern Europe, and parts of Russia.

The upper part of a sheet is brittle and may crack to form crevasses. Because ice sheets store so much of the Earth's water, sea level becomes lower during an ice age. When a glacier reaches the sea, it becomes an ice shelf. Rock that has been plucked up by the glacier along the way is carried out to sea with the ice; when the ice melts, the rocks fall to the sea floor as drop stones. At the edge of the shelf, icebergs calve off and float away.

A second class of glaciers, called mountain, or alpine, glaciers, exist in mountainous areas because snow can last all year at high elevations. During the ice age, mountain glaciers grew and flowed out onto the land surface beyond the mountain front. The glacier at the right has started to recede after formerly advancing and covering more of the land. Glacial recession may happen when the climate warms, so ice melts away faster at the toe (terminus) of the glacier than can be added at the source. In front of the glacier, you can find consequences of glacial erosion such as striations on bedrock and roche moutonnée.

When the glacier pauses, till (unsorted glacial sediment) accumulates to form an end moraine. Meltwater lakes gather at the toe. Streams carry sediment and deposit it as glacial outwash. Sediment that accumulates in ice tunnels, exposed when the glacier melts, make up sinuous ridges called eskers. Even when the toe remains fixed in position for a while, the ice continues to flow, and thus molds underlying sediment into drumlins. Ice blocks buried in till melt to form kettle holes. (Though the examples shown here were left after the melting away of a piedmont glacier, most actually form during continental glaciation.)

In the mountains, the glacier is confined to a valley. Sediment falling on it from the mountains creates lateral and medial moraines. Glaciers carve distinct landforms in the mountains, like cirques, arêtes, horns, and U-shaped valleys.

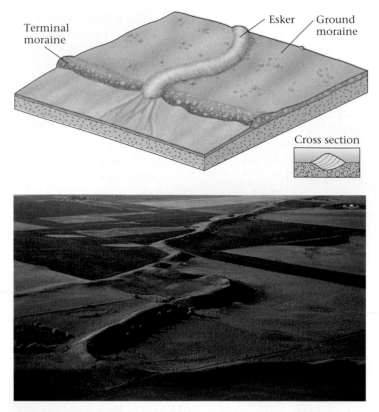

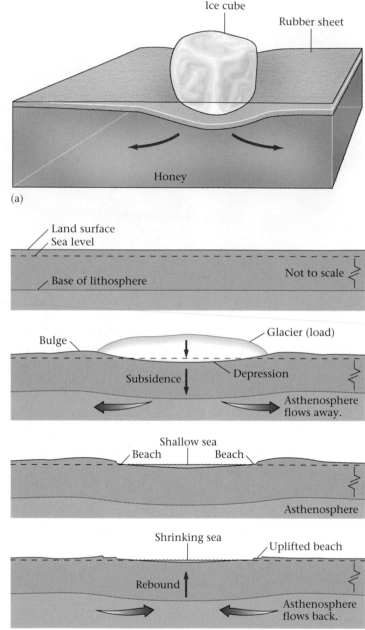

FIGURE 22.31 Eskers are snake-like ridges of sand and gravel that form when sediment fills meltwater tunnels at the base of a glacier.

22.6 OTHER CONSEQUENCES OF CONTINENTAL GLACIATION

Ice Loading and Glacial Rebound

When a large ice sheet (more than 50 km in diameter) grows on a continent, its weight causes the surface of the lithosphere to sink. In other words, ice loading causes **glacial subsidence.** Lithosphere, the relatively rigid outer shell of the Earth, can sink because the underlying asthenosphere is soft enough to flow slowly out of the way. You can see how this works by conducting a simple experiment. First, fill a bowl with honey and then place a thin rubber sheet over the honey (▶Fig. 22.32a). The rubber represents the lithosphere, and the honey represents the asthenosphere. If you place an ice cube on the rubber sheet, the sheet sinks because the weight of the ice pushes it down; the honey flows out of the way to make room. Because of ice loading, much of Antarctica and Greenland now lie below sea level (Fig. 22.5), so if their ice were instantly to melt away, these continents would be flooded by a shallow sea.

FIGURE 22.32 (a) An ice cube placed on the surface of a rubber sheet floating on honey illustrates the concept of glacial loading. (b) Cross sections illustrating the concept of glacial rebound. Before glaciation, the surface of the lithosphere is flat. The weight of the glacier pushes the lithosphere down below sea level. The asthenosphere flows out of the way below, and lithosphere on either side bulges up. When the glacier melts, the depression fills with water. During glacial rebound, the floor of the shallow sea rises, and beaches along its shore rise.

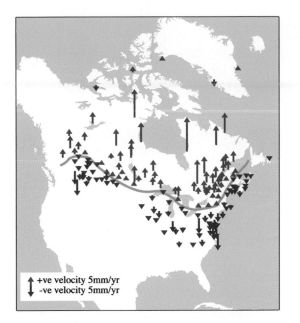

FIGURE 22.33 Rates of isostatic movement still taking place in response to the melting of the pleistocene ice sheet, as indicated by satellite data. Red bars indicate upward movement; blue bars indicate downward movement.

What happens when continental ice sheets do melt away? Gradually, the surface of the underlying continent rises back up, a process called **glacial rebound,** and the asthenosphere flows back underneath to fill the space (▶Fig. 22.32b). In the honey and rubber example, when you remove the ice cube, the rubber sheet slowly returns to its original shape. This process doesn't take place instantly, because the honey can only flow slowly. Similarly, because the asthenosphere flows *so* slowly (at rates of a few millimeters per year), it takes thousands of years for ice-depressed continents to rebound. Thus, glacial rebound is still taking place in some regions that were burdened by ice during the last ice age. Recently, researchers in North America have documented this movement by using GPS measurements (▶Fig. 22.33). Regions north of a line passing through the Great Lakes are rising, whereas regions south of this line are sinking. Where rebound affects coastal areas, beaches along the shoreline rise several meters above sea level and become terraces. Glacial rebound can also take place as mountain glaciers melt away.

Sea-Level Changes: The Glacial Reservoir in the Hydrologic Cycle

More of the Earth's surface and near-surface freshwater is stored in glacial ice than in any other reservoir. In fact, glacial ice accounts for 2.15% of Earth's total water supply, while lakes, rivers, soil, and the atmosphere together contain only 0.03%. During the last ice age, when glaciers covered almost three times as much land area as they do today,

they held significantly more water (70 million cubic km, as opposed to 25 million cubic km today). In effect, water from the ocean reservoir transferred to the glacial reservoir and remained trapped on land. As a consequence, sea level dropped by as much as 100 m, and extensive areas of continental shelves became exposed as the coastline migrated seaward, in places by more than 100 km (▶Fig. 22.34a–c). People and animals migrated into the newly exposed coastal plains; in fact, fishermen dragging their nets along the Atlantic Ocean floor off New England today occasionally recover artifacts. The drop in sea level also created land bridges across the Bering Strait between North America and northeastern Asia and between Australia and Indonesia, providing convenient migration routes.

If today's ice sheets in Antarctica and Greenland were to melt, the crust of these continents would undergo glacial rebound, global sea level would rise substantially, and low-lying areas of other continents would undergo flooding. In the United States, large areas of the coastal plain along the East Coast and Gulf Coast would flood, and cities like Miami, Houston, New York, and Philadelphia would disappear beneath the waves. In Canada, substantial parts of the shield would flood.

Ice Dams, Drainage Reversals, and Lakes

When ice freezes over a sewer opening in a street, neither meltwater nor rain can enter the drain, and the street floods. Ice sheets play a similar role in glaciated environments. The ice may block the course of a river, leading to the formation of a lake. In addition, the weight of a glacier changes the tilt of the land surface and therefore the gradient of streams, and glacial sediment may fill preexisting valleys. In sum, continental glaciation destroys preexisting drainage networks. While the glacier exists, streams find different routes and carve out new valleys; by the time the glacier melts away, these new streams have become so well established that old river courses may remain abandoned.

Glaciation during the last ice age profoundly modified North America's drainage. Before the ice age, several major rivers drained much of the interior of the continent to the north, into the Arctic Ocean (▶Fig. 22.35a, b). The ice sheet buried this drainage network and diverted the flow into the Mississippi-Missouri network, which became larger. Then after the ice receded, regions covered by knob-and-kettle topography became spotted with thousands of small lakes, as now occur in central and southern Minnesota, while regions of the Midwest covered by a smooth frosting of till became swampland, with little or no drainage at all. In the Canadian Shield, scouring left innumerable depressions that have now become lakes.

Inevitably, the ice dams that held back large ice-margin lakes melted and broke. In a matter of hours to days, the

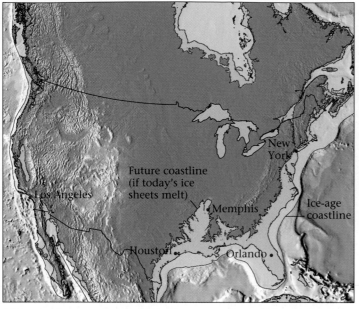

(a)

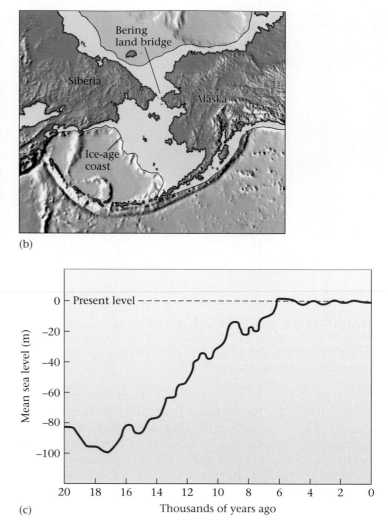

(b)

(c)

FIGURE 22.34 (a) This map shows the coastline of North America during the last ice age, and the coastline should present-day ice sheets melt. Note that much of the continental shelf was exposed during the last ice age. (b) A land bridge existed across the Bering Strait between Asia and North America during the last ice age. (c) The graph shows the change in sea level during the past 20,000 years.

contents of the lakes drained, creating immense floodwaters that stripped the land of soil and left behind huge ripple marks. Glacial Lake Missoula, in Montana, filled when glaciers advanced and blocked the outlet of a large valley. When the glaciers retreated, the ice dam broke, releasing immense torrents that scoured eastern Washington, creating a barren, soil-free landscape called the channeled scablands (see Chapter 17). Recent evidence suggests that this process occurred many times.

Pluvial Features

During ice ages, regions to the south of continental glaciers were wetter than they are today. Fed by enhanced rainfall, lakes accumulated in low-lying land even at a great distance from the ice front. The largest of these **pluvial lakes** (from the Latin *pluvia,* for "rain") in North America flooded interior basins of the Basin and Range Province in Utah and Nevada (▶Fig. 22.36a). These basins received drainage from the adjacent ranges but had no outlet to the sea, so they filled with water. The largest pluvial lake, Lake

Bonneville, covered almost a third of western Utah (▶Fig. 22.36b). When this lake suddenly drained after a natural dam holding it back broke, it left a bathtub ring of shorelines rimming the Wasatch Mountains near Salt Lake City. Today's Great Salt Lake itself is but a small remnant of Lake Bonneville.

22.7 PERIGLACIAL ENVIRONMENTS

In polar latitudes today, and in regions adjacent to the front of continental glaciers during the last ice age, the mean annual temperature stays low enough (below –5°C) that soil moisture and groundwater freeze and, except in the upper few meters, stay solid all year. Such permanently frozen ground, or **permafrost,** may extend to depths of 1,500 m below the ground surface. Regions with widespread permafrost but without a blanket of snow or

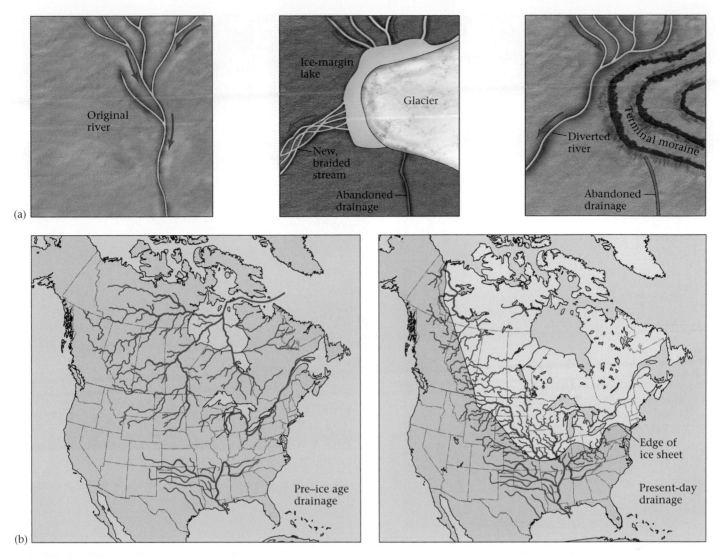

FIGURE 22.35 (a) When a glacier advances on the course of a river, the glacier blocks the drainage and causes a new stream to form. After the glacier melts away, the river remains diverted. (b) In North America, the major river systems that flowed northward before the last ice age have been destroyed.

ice are called periglacial environments (the Greek *peri* means "around," or "encircling"; periglacial environments appear around the edges of glacial environments) (▶Fig. 22.37a).

The upper few meters of permafrost may melt during the summer months, only to refreeze again when winter comes. As a consequence of the freeze-thaw process, the ground splits into pentagonal or hexagonal shapes, creating a landscape called **patterned ground** (▶Fig. 22.37b). Water fills the gaps between the cracks and freezes to create wedge-shaped walls of ice. In some places, freeze and thaw cycles in permafrost gradually push cobbles and pebbles up from the subsurface. Because the expansion of the ground is not even, the stones gradually collect between adjacent

bulges to form stone rings (▶Fig. 22.37c). Some stone rings may also form when mud at depth pushes up from beneath a permafrost layer and forces stones aside.

Permafrost presents a unique challenge to people who live in polar regions or who work to extract resources from these regions. For example, heat from a building may warm and melt underlying permafrost, creating a mire into which the building settles. For this reason, buildings in permafrost regions must be placed on stilts, so that cold air can circulate beneath them to keep the ground frozen. When geologists discovered oil on the northern coast of Alaska, oil companies faced the challenge of shipping the oil to markets outside of Alaska. After much debate over the environmental impact, the trans-Alaska pipeline was built, and now

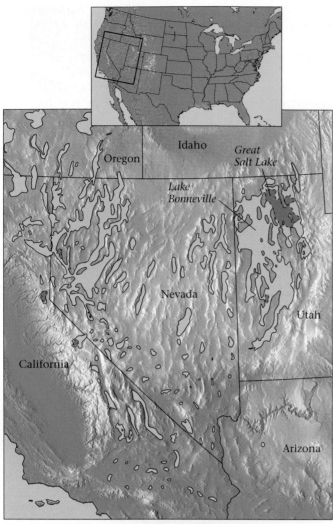

(a)

(b)

FIGURE 22.36 (a) The distribution of pluvial lakes in the Basin and Range Province during the last ice age. (b) The shoreline of Lake Bonneville along a mountain near Salt Lake City. The Great Salt Lake is a small remnant of this once-huge pluvial lake.

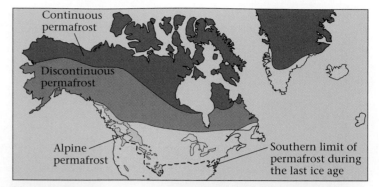

(a)

(b)

(c)

FIGURE 22.37 (a) The present-day distribution of periglacial environments in the Northern Hemisphere. (b) Patterned ground in the Northwest Territories, Canada, as viewed from a low-flying plane. The polygons are about 10 m across. The straight lines on the left side of the photo are caribou tracks. (c) Stone circles that are about 2 m across.

it carries oil for 1,000 km to a seaport in southern Alaska (see Chapter 14). The oil must be warm during transport, or it would be too viscous to flow; thus, to prevent the warm pipeline from melting underlying permafrost, it had to be built on a frame that holds it above the ground for its entire length.

22.8 THE PLEISTOCENE ICE AGES

The Pleistocene Glaciers

Today, most of the land surface in New York City lies hidden beneath concrete and steel, but in Central Park it's still possible to see land in a seminatural state. If you stroll through the park and study the rock outcrops, you'll find that their top surfaces are smooth and polished, and in places have been grooved and scratched (▶Fig. 22.38). You are seeing evidence that an ice sheet once scraped along this ground, rasping and gouging the bedrock as it moved. Geologists estimate that the ice sheet that overrode the New York City area may have been 250 m thick, enough to bury the Empire State Building up to the 75th floor.

Glacial features like those on display in Central Park first led Louis Agassiz to propose the idea that vast continental glaciers advanced over substantial portions of North America, Europe, and Asia during a great "ice age." Since Agassiz's day, thousands of geologists, by mapping out the distribution of glacial deposits and landforms, have gradually defined the extent of ice-age glaciers and a history of their movement.

The fact that these glacial features decorate the surface of the Earth today means that the most recent ice age occurred fairly recently during Earth history. This ice age, responsible for the glacial landforms of North America and Eurasia, happened mostly during the Pleistocene Epoch, which began 1.8 million years ago (see Chapter 13), so it is commonly known as the "Pleistocene ice age." This traditional title is a bit of a misnomer. Recent studies demonstrate that the glaciations of this ice age actually began between 3.0 and 2.5 million years ago, during the Pliocene Epoch, and continued through the Pleistocene. Further, there was not just a single ice advance, but, as we will see, there were many—probably over twenty. (Thus, all of these events together might better be called the Pleistocene "ice ages.") Geologists use the name Holocene to refer to the time since the last glaciation (i.e., to the last 11,000 years).

Based on their mapping of glacial striations and deposits, geologists have determined where the great Pleistocene ice sheets originated and flowed. In North America, the Laurentide ice sheet started to grow over northeastern Canada, then merged with the Keewatin ice sheet, which originated in northwestern Canada. Together, these ice sheets eventually covered all of Canada east of the Rocky Mountains and extended southward across the border as far as southern Illinois (▶Fig. 22.39a, b). At their maximum, they attained a thickness of 2–3 km; each thinned toward their toe. In Northeastern Canada, the ice sheet eroded the land surface. Further south and west, it deposited sediment (▶Fig. 22.39c) These ice sheets also eventually merged with the Greenland ice sheet to the northeast and the Cordilleran ice sheet to the west; the Cordilleran covered the mountains of western Canada, as well as the southern third of Alaska.

During the Pleistocene ice age, mountain ice caps and valley glaciers also grew in the Rocky Mountains and the Sierra Nevada and Cascade Mountains. In Eurasia, a large ice sheet formed in northernmost Europe and adjacent Asia, and gradually covered all of Scandinavia and northern Russia. This ice sheet flowed southward across France until it reached the Alps and merged with Alpine mountain glaciers; it also covered all of Ireland and almost all of the United Kingdom. A smaller ice sheet grew in eastern Siberia and expanded in the mountains of central Asia. In the Southern Hemisphere, Antarctica remained ice-covered, and mountain ice caps expanded in the Andes, but there were no continental glaciers in South America, Africa, or Australia.

In addition to continental ice sheets, sea ice in the Northern Hemisphere expanded to cover all of the Arctic Ocean and parts of the North Atlantic. Sea ice surrounded Iceland and approached Scotland, and also fringed most of western Canada and southeastern Alaska.

Life and Climate in the Pleistocene World

During the Pleistocene ice age, all climatic belts shifted southward (▶Fig. 22.40a, b). Geologists can document this

FIGURE 22.38 A glacially polished surface in Central Park, New York City.

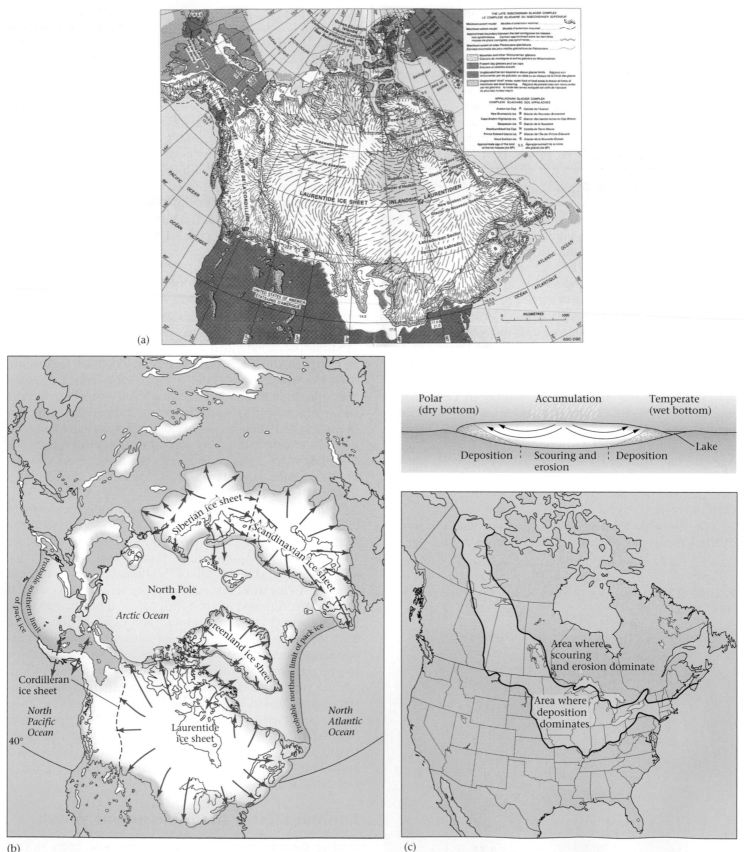

FIGURE 22.39 (a) A map of the major ice sheets in North America, as first compiled by V. K. Prest. The thin lines indicate the flow direction of the ice. Note that one major sheet originated to the west of Hudson Bay and another to the east. (b) A simplified map of the distribution of major ice sheets during the Pleistocene Epoch. The arrows indicate the flow trajectories of the ice sheets. (c) Top: A continental glacier scours and erodes the land surface beneath its center, while at the margins it deposits sediment. Map: The Laurentide ice sheet scoured and eroded the land in northern and eastern Canada, and deposited sediment in western Canada and the midwestern United States.

shift by examining fossil pollen, which can survive for thousands of years, preserved in the sediment of bogs. Presently, the southern boundary of North America's tundra, a treeless region supporting only low shrubs, moss, and lichen capable of living on permafrost, lies at a latitude of 68°N; during the Pleistocene ice age, it moved down to 48°N. Much of the interior of the United States, which now has temperate, deciduous forest, harbored cold-weather spruce and pine forest. Ice-age climates also changed the distribution of rainfall on the planet: increased rainfall in North America led to the filling of pluvial lakes in Utah and Nevada, while decreased rainfall in equatorial regions led to shrinkage of the rain forest. Overall, the contrast between colder glaciated regions and warmer unglaciated regions created windier conditions worldwide. These winds sent glacial rock flour skyward, creating a dusty atmosphere (and, presumably, spectacular sunsets). The dust settled to create extensive deposits of loess. And because glaciers trapped so much water, as we have seen, sea level dropped.

Numerous species of now-extinct large mammals inhabited the Pleistocene world (▶Fig. 22.41). Giant mammoths and mastodons, relatives of the elephant, along with woolly rhinos, musk oxen, reindeer, giant ground sloths, bison, lions, saber-toothed cats, giant cave bears, and hye-

nas wandered forests and tundra in North America. Early human-like species were already foraging in the woods by the beginning of the Pleistocene Epoch, and by the end modern *Homo sapiens* lived on every continent except Antarctica, and had discovered fire and invented tools. Rapidly changing climates may have triggered a global migration of early humans, who gained access to the Americas, Indonesia, and Australia via land bridges that became exposed when sea level dropped.

Chronology of the Pleistocene Ice Ages

Louis Agassiz assumed that only one ice age had affected the planet. But close examination of the stratigraphy of glacial deposits on land revealed that "paleosol" (ancient soil preserved in the stratigraphic record), as well as beds containing fossils of warmer-weather animals and plants, separated distinct layers of glacial sediment. This observation suggested that between episodes of glacial deposition, glaciers receded and temperate climates prevailed. In the second half of the twentieth century, when modern methods for dating geological materials became available, the difference in ages between the different layers of glacial sediment could be confirmed. Clearly, ice-age glaciers had advanced and then retreated

---

So You Want to See Glaciation?

BOX 22.1
THE HUMAN ANGLE

Though the last of the Pleistocene continental ice sheet that once covered much of North America vanished about 6,000 years ago, you can find evidence of its power quite easily. The Great Lakes, along the U.S.-Canadian border, the Finger Lakes and drumlins of New York, the low-lying moraines and outwash plains of Illinois, and the polished outcrops of southern Canada all formed in response to the movement of this glacier. But if you want to see continental glaciers in action today, you must trek to Greenland or Antarctica.

Mountain glaciers are easier to reach. A trip to the mountains of western North America (including Alaska), the Alps of France or Switzerland, the Andes of South America, or the mountains of southern New Zealand will bring you in contact with live glaciers. You can even spot glaciers from the comfort of a cruise ship.

Some of the most spectacular glacial landscapes in North America formed during the Pleistocene Epoch, when mountain glaciers were more widespread. These are now on display in national parks.

- *Glacier National Park (Montana):* This park, which borders Waterton Lakes National Park in Canada, displays giant cirques, U-shaped valleys, hanging valleys, terminal moraines, and fifty small relicts of formerly larger glaciers, all in a mountainous terrain that reaches elevations of over 3 km. Unfortunately, these glaciers are melting away quickly!

- *Yosemite National Park (California):* A huge U-shaped valley, carved into the Sierra Nevada granite batholith, makes up the centerpiece of this park. Waterfalls spill out of hanging valleys bordering the valley.

- *Voyageurs National Park (Minnesota):* This park lacks the high peaks of mountainous parks, but shows the dramatic consequences of glacial scouring and deposition on the Canadian Shield. The low-lying landscape, dotted with lakes, contains abundant polished surfaces, glacial striations, and erratics, along with moraines, glacial lake beds, and outwash plains.

- *Acadia National Park (Maine):* During the last ice age, the continental ice sheet overrode low bedrock hills and flowed into the sea along the coast of Maine. This park provides some of the best examples of the consequences. Its hills were scoured and shaped into large roches moutonnées by glacial flow. Some of the deeper valleys have now become small fjords.

- *Glacier Bay National Park (Alaska):* In Glacier Bay, huge tidewater glaciers fringe the sea, creating immense ice cliffs from which icebergs calve off. Cruise ships bring tourists up to the toes of these glaciers. More adventurous visitors can climb the coastal peaks and observe lateral and medial moraines, crevasses, and the erosional and depositional consequences of glaciers that have already retreated up the valley.

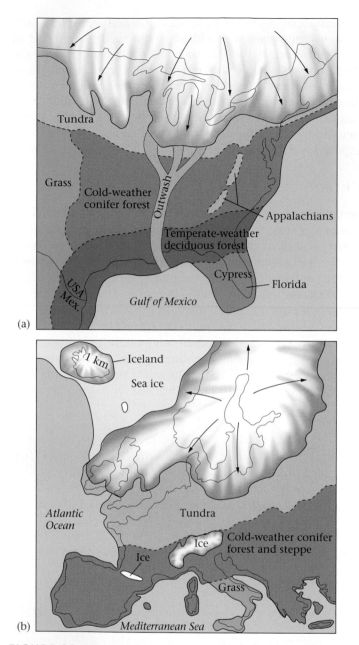

FIGURE 22.40 (a) Pleistocene climatic belts in North America; (b) Pleistocene climatic belts in Europe.

FIGURE 22.41 Examples of now-extinct large mammals that roamed the countryside during the Pleistocene Epoch.

▶Fig. 22.42). Since the mid-1980s, geologists no longer recognize Nebraskan and Kansan; they are lumped together as "pre-Illionoian." With the advent of radiometric dating in the mid-twentieth century, the ages of the younger glaciations were determined by dating wood trapped in glacial deposits. Geologists estimate the ages of the older glaciations by identifying fossils in the deposits. Because of their greater age, these deposits have been thoroughly weathered and dissected.

The four-stage chronology of North American glaciation was turned on its head in the 1960s, when geologists began to study submarine sediment containing the fossilized shells of microscopic marine plankton. Because the assemblage of plankton species living in warm water is not the same as the assemblage living in cold water, geologists

FIGURE 22.42 Pleistocene deposits in the United States.

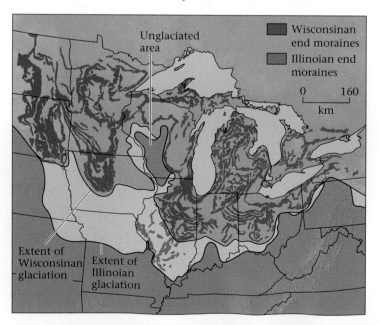

more than once. Times during which the glaciers grew and covered substantial areas of the continents are called glacial periods, or **glaciations,** and times between glacial periods are called interglacial periods, or **interglacials.**

Using the on-land sedimentary record, geologists recognized five Pleistocene glaciations in Europe (named, in order of increasing age: Würm, Riss, Mindel, Gunz, and Donau) and, traditionally four in the midwestern United States (Wisconsinan, Illinoian, Kansan, and Nebraskan, named after the southernmost states in which their till was deposited;

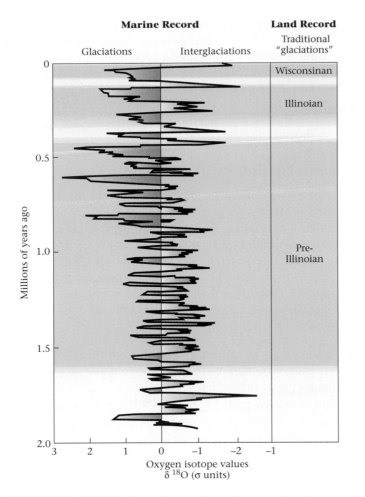

Marine Record **Land Record**

Traditional "glaciations"

Glaciations Interglaciations

FIGURE 22.43 This time column shows the variations in oxygen-isotope ratios from marine sediment that define twenty to thirty glaciations and interglacials during the Pleistocene Epoch. The green bands represent the approximate boundaries of the principle glacial stages recognized on land in the Midwestern USA. Note that the traditional names "Kansan" and "Nebraskan" are no longer used, and have been replaced by "Pre-Illinoian."

can track changes in the temperature of the ocean by studying plankton fossils. Researchers found that in sediment of the last 2 million years, assuming that cold water indicates a glacial period and warm water an interglacial period, there were twenty to thirty different glacial advances during the Pleistocene Epoch. The four traditionally recognized glaciations probably represent only the largest of these—sediments deposited on land by other glaciations were eroded and redistributed during subsequent glaciations, or were eroded by streams and wind during interglacials.

Geologists refined their conclusions about the frequency of Pleistocene glaciations by examining the composition of fossil shells. Shells of many plankton species consist of calcite ($CaCO_3$). The oxygen in the shells includes two isotopes, a heavier one (^{18}O) and a lighter one (^{16}O). The ratio of these isotopes tells us about the water temperature in which the plankton grew; this is because as water gets colder, plankton incorporate a higher proportion of ^{18}O into their shells (see Chapter 23). Thus, intervals in the stratigraphic record during which plankton shells have a large ratio of $^{18}O/^{16}O$ define times when Earth had a colder, glacial climate. The record also indicates that twenty to thirty of these events occurred during the last 3 million years (▶Fig. 22.43).

Older Ice Ages During Earth History

So far, we've focused on the Pleistocene ice age, because of its importance in developing Earth's present landscape. Was this the only ice age during Earth history, or do ice ages happen frequently? To answer such questions, geologists study the stratigraphic record and search for ancient glacial deposits that have hardened into rock. These deposits, called **tillites,** consist of larger clasts distributed throughout a matrix of sandstone and mudstone (▶Fig. 22.44a). In many cases, tillites are deposited on glacially polished surfaces.

FIGURE 22.44 (a) This time column shows pre-Pleistocene glaciations during Earth history. (b) The distribution of Permian glacial features on a reconstruction of Gondwana, the southern part of the supercontinent that existed at the time.

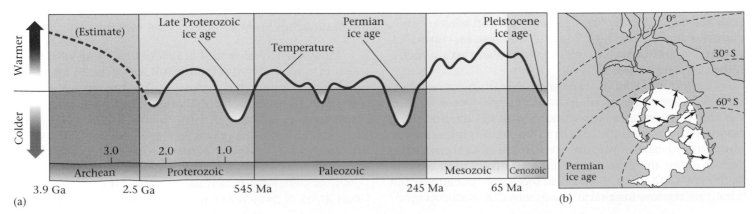

By using the stratigraphic principles described in Chapter 12, geologists have determined that tillites were deposited about 280 million years ago (in Permian time; these are the deposits Alfred Wegener studied when he argued in favor of continental drift; ▶Fig. 22.44b), about 600 to 700 million years ago (at the end of the Proterozoic Eon), about 2.2 billion years ago (near the beginning of the Proterozoic), and perhaps about 2.7 billion years ago (in the Archean Eon). Strata deposited at other times in Earth history do not contain tillites. Thus, it appears that glacial advances and retreats have not occurred steadily throughout Earth history, but rather are restricted to specific time intervals, or ice ages, of which there are four or five: Pleistocene, Permian, late Proterozoic, early Proterozoic, and perhaps Archean. Of particular note, some tillites of the late Proterozoic event were deposited at equatorial latitudes, suggesting that for at least a short time the continents worldwide were largely glaciated, and the sea may have been covered worldwide by ice. Geologists refer to the ice encrusted planet as **snowball Earth.**

22.9 THE CAUSES OF ICE AGES

Ice ages occur only during restricted intervals of Earth history, hundreds of millions of years apart. But within an ice age glaciers advance and retreat with a frequency measured in tens of thousands to hundreds of thousands of years. Thus, there must be both long-term and short-term controls on glaciation. The nature of these controls emphasizes the complexity of interactions among components of the Earth System.

Long-Term Causes

Plate tectonics probably provides some long-term controls on glaciation. First, for an ice age to occur, substantial areas of continents must have drifted to high latitudes; if all continents sat along the equator, the land would be too warm for snow to accumulate. Second, glaciations can only take place when most continents lie well above sea level; sea-level changes may be controlled, in part, by changes in rates of sea-floor spreading. Finally, ice ages can't develop when oceanic currents carry heat to high latitudes; currents are controlled, in part, by positions of continents and volcanic arcs, as determined by plate motions.

The concentration of carbon dioxide in the atmosphere may also determine whether an ice age can occur. Carbon dioxide is a greenhouse gas—it traps infrared radiation rising from the Earth—so if the concentration of CO_2 increases, the atmosphere becomes warmer. Ice sheets cannot form during periods when the atmosphere has a relatively high concentration of CO_2, even if other factors favor glaciation. But what might cause long-term changes in CO_2 concentration?

Possibilities include changes in the number of marine organisms that extract CO_2 to make shells, changes in the amount of chemical weathering on land (determined by the abundance of mountain ranges) for weathering absorbs CO_2, and changes in the amount of volcanic activity. Major stages in evolution may also affect CO_2 concentration. For example, the appearance of coal swamps at the end of Paleozoic may have removed CO_2, for plants incorporate CO_2.

Short-Term Causes

Now we've seen how the stage could be set for an ice age to occur, but why do glaciers advance and retreat periodically *during* an ice age? In 1920, Milutin Milankovitch, a Serbian astronomer and geophysicist, came up with an explanation. Milankovitch studied how the Earth's orbit changes shape and how its axis changes orientation through time, and he calculated the frequency of these changes. In particular, he evaluated three aspects of Earth's movement around the Sun.

- *Orbital eccentricity:* Milankovitch showed that the Earth's orbit gradually changes from a more circular shape to a more elliptical shape. This "eccentricity cycle" takes around 100,000 years (▶Fig. 22.45a).

- *Tilt of Earth's axis:* We have seasons because the Earth's axis is not perpendicular to the plane of its orbit. Milankovitch calculated that over time, the tilt angle varies between 22.5° and 24.5°, with a frequency of 41,000 years (▶Fig. 22.45b).

- *Precession of Earth's axis:* If you've ever set a top spinning, you've probably noticed that its axis gradually traces a conical path. This motion, or wobble, is called precession (▶Fig. 22.45c). Milankovitch determined that the Earth's axis wobbles over the course of about 23,000 years; right now, the Earth's axis points toward Polaris, making it the north star, but 12,000 years ago the axis pointed to Vega. Precession determines the relationship between the timing of the seasons and the position of Earth along its orbit around the Sun. If summer happens when Earth is closer to the Sun, then we have a warm summer, but if it happens when Earth is farther away from the Sun, then we have a cool summer.

Milankovitch showed that precession, along with variations in orbital eccentricity and tilt, combine to affect the total annual amount of "insolation" (exposure to the Sun's rays) and the seasonal distribution of insolation that the Earth receives at the high latitudes (such as 65°N). For example, high-latitude regions receive more insolation when the Earth's axis is almost perpendicular to its orbital plane than when its axis is greatly tilted. According to Milankovitch, glaciers tend to advance during times of cool summers at 65°N, which occur roughly 100,000, 40,000, and

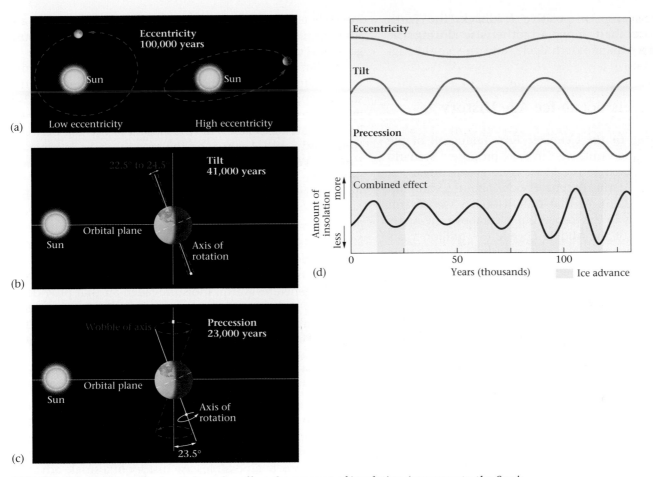

FIGURE 22.45 The Milankovitch cycles affect the amount of insolation (exposure to the Sun's rays) at high latitudes. (a) Variations in insolation caused by changes in orbital shape; (b) Variations caused by changes in the tilt angle of Earth's axis; (c) variations caused by the precession of Earth's axis. (d) Eccentricity, tilt, and precession all affect the amount of insolation, but with different periodicities. When the effects are combined, we see that there are distinct warm and cold periods; cold periods occur when there is less insolation.

20,000 years apart. When geologists began to study the climate record, they found climate cycles with the frequency predicted by Milankovitch. These climate cycles are now called **Milankovitch cycles** (▶Fig. 22.45d).

The discovery of Milankovitch cycles in the geologic record strongly supports the contention that changes in the Earth's orbit and tilt help trigger short-term advances and retreats during an ice age. But orbit and tilt changes cannot be the whole story, because they could cause only about a 4°C temperature decrease (relative to today's temperature), and during glaciations the temperature decreased 5°–7°C along coasts and 10°–13°C inland. Geologists suggest that several other factors may come into play in order to trigger a glacial advance.

- *A changing albedo:* When snow remains on land throughout the year, or clouds form in the sky, the albedo (reflectivity) of the Earth increases, so Earth's surface reflects incoming sunlight and thus becomes even cooler.

- *Interrupting the global heat conveyor:* As the climate cools, evaporation rates from the sea decrease, so seawater does not become as salty. And decreasing salinity might stop the system of thermohaline currents that brings warm water to high latitudes (see Chapter 18). Thus, the high latitudes become even colder than they would otherwise.

- *Biological processes that change CO_2 concentration:* Several kinds of biological processes may have amplified climate changes by altering the concentration of carbon dioxide in the atmosphere. For example, a greater amount of plankton growing in the oceans could absorb more carbon dioxide and thus remove it from the atmosphere.

The three processes described above are called "positive-feedback" mechanisms: they enhance the process that causes them. Because of positive feedback, the Earth could cool more than it would otherwise during the cooler stage of a Milankovitch cycle, and this could trigger a glacial advance.

A Model for Pleistocene-Ice-Age History

Long-term cooling in the Cenozoic Era. Taking all of the above causes into account, we can now propose a scenario for the events that led to the Pleistocene glacial advances. Our story begins in the Eocene Epoch, about 55 million years ago (▶Fig. 22.46). At that time, climates were warm and balmy, not only in the tropics but even above the Arctic Circle. At the end of the Middle Eocene (37 million years ago), the climate began to cool, and by Early Oligocene time (33 million years ago), Antarctica became glaciated. The Antarctic ice sheet came and went until the middle of the Miocene Epoch (15 million years ago), when an ice sheet formed that has lasted ever since. Ice sheets did not appear in the Arctic, however, until 2–3 million years ago, when the Pleistocene ice age began.

These long-term climate changes may have been caused, in part, by changes in the pattern of oceanic currents that happened, in turn, because of plate tectonics. For example, in the Eocene, the collision of India with Asia cut off warm equatorial currents that had been flowing in the Tethys Sea. And in the Miocene and Oligocene, Australia and South America drifted away from Antarctica, allowing the cold Circum-Antarctic current to develop. This new current prevented warm, southward-flowing currents from reaching Antarctica, allowing ice to form and survive in the region. The climate of Antarctica overall, with the loss of the warm currents, underwent cooling. Changes to atmospheric circulation and temperature may also have happened at this time—the uplift of the Himalayas and Tibet diverted winds in a way that, models suggest, cooled climate. Further, this uplift exposed more rock to chemical weathering, perhaps leading to extraction of CO_2

FIGURE 22.46 The graph shows the gradual cooling of Earth's atmosphere since the Cretaceous Period.

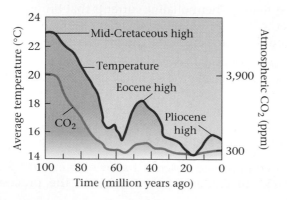

from the atmosphere (for chemical weathering reactions absorb CO_2); a decrease in the concentration of this greenhouse gas would contribute to atmospheric cooling.

So far, we've examined hypotheses that explain long-term cooling since 55 million years ago, but what caused the sudden appearance of the Laurentide ice sheet about 2–3 million years ago? This event coincides with another well-known plate-tectonic event, the closing of the gap between North and South America by the growth of the Isthmus of Panama. When this land bridge formed, it separated the waters of the Caribbean from those of the tropical Pacific for the first time, and when this happened, warm currents that previously flowed out the Caribbean into the Pacific were blocked and diverted northward to merge with the Gulf Stream. This current transfers warm water from the Caribbean up the Atlantic Coast of North America and ultimately to the British Isles. As the warm water moves up the Atlantic Coast, it generates warm, moisture-laden air that provides a source for the snow that falls over New England, eastern Canada, and Greenland. In other words, the Arctic has long been cold enough for ice caps, but until the Gulf Stream was diverted northward by the growth of Panama, there was no source of moisture to make abundant snow and ice.

Short-term advances and retreats in the Pleistocene Epoch. Once the Earth's climate had cooled overall, short-term processes such as the Milankovitch cycles led to periodic advances and retreats of the glaciers. To understand how, let's look at a possible case history of a single advance and retreat of the Laurentide ice sheet. (Note that such models remain the subject of vigorous debate.)

- *Stage 1:* During the overall cooler climates of the late Cenozoic Era, the Earth reaches a point in the Milankovitch cycle when the average mean temperature in temperate latitudes drops. Because of glacial rebound, the ice-free surface of northern Canada has risen to an altitude of several hundred meters above sea level. With lower temperatures and higher elevations, not all of winter's snow melts away during the summer. Eventually, snow covers the entire region of northern Canada even during the summer. Because of the snow's high albedo, it reflects sunlight, so the region grows still colder (a positive-feedback effect) and even more snow accumulates. Precipitation rates are high, because evaporation off the Gulf Stream provides moisture. Finally, the snow at the base of the pile turns to ice, and the ice begins to spread outward under its own weight. A new continental glacier has been born.

- *Stage 2:* The ice sheet continues to grow as more snow piles up in the zone of accumulation. And as the ice sheet grows, the atmosphere continues to cool because of the albedo effect. But now, the weight of the ice loads the continent and makes it sink, so the elevation of the glacier decreases, and its surface approaches the equilibrium line. Also, the temperature becomes cold enough that the Atlantic Ocean in high latitudes begins to freeze. As the sea ice covers the ocean, the amount of evaporation decreases, so the source of snow is cut off and the amount of snowfall diminishes. The glacial advance pretty much chokes on its own success. The decrease in the glacier's elevation (leading to warmer summer temperatures) on the ice surface, as well as the decrease in snowfall, causes ablation to occur faster than accumulation, and the glacier begins to retreat.

- *Stage 3:* As the glacier retreats, temperatures gradually increase, and the sea ice begins to melt. The supply of water to the atmosphere from evaporation increases once again, but with the warmer temperatures and lower elevations, this water precipitates as rain during the summer. The rain drastically accelerates the rate of ice melting, and the retreat progresses quite rapidly.

22.10 WILL THERE BE ANOTHER GLACIAL ADVANCE?

What does the future hold? Considering the periodicity of glacial advances and retreats during the Pleistocene Epoch, we may be living in an interglacial period. Pleistocene interglacials lasted about 10,000 years, and since the present interglacial began about 11,000 years ago, the time seems ripe for a new glaciation. If a glacier on the scale of the Laurentide ice sheet were to develop, major cities and agricultural belts would be overrun by ice, and their populations would have to migrate southward. Long before the ice front arrived, though, the climate would become so hostile that the cities would already be abandoned.

The Earth actually had a brush with ice-age conditions between the 1300s and the mid-1800s, when average annual temperatures in the Northern Hemisphere fell sufficiently for mountain glaciers to advance significantly. During this period, now known as the **little ice age,** sea ice surrounded Iceland, and canals froze in the Netherlands, leading to that country's tradition of skating (▶Fig. 22.47). Some researchers speculate that the depopulation of the western hemisphere, in the wake of European conquest, caused temporary reforestation, for without inhabitants, farmlands went untended. The new forests absorbed CO_2, and caused atmospheric concentrations of CO_2 to decrease, leading to the cooler conditions that triggered the little ice age. Others speculate that the change reflects increased cloud cover, not a change in CO_2 concentration. Researchers will likely propose additional ideas, as work on this problem continues. During the past 150 years, temperatures have warmed, and most mountain glaciers have retreated significantly (▶Fig. 22.48). We no longer see the

FIGURE 22.47 Skaters (c. 1600) on the frozen canals of the Netherlands during the little ice age.

FIGURE 22.48 A couple of hundred years ago, glacial ice filled the cirque in the background and all of the valley up to the height of the lateral moraine in the foreground. In this 2003 photo, most of the ice of this Alaskan glacier has vanished.

icebergs that once threatened Atlantic shipping lanes and sank the *Titanic,* and large slabs frequently calve off the Antarctic ice sheet. Some researchers suggest that this global-warming trend is due to the addition of carbon dioxide to the atmosphere from the burning of forests and the use of fossil fuels (see Chapter 23). Global warming could conceivably cause a "super-interglacial."

If the climate were to become significantly warmer than it is today, the ice sheet of West Antarctica might begin to float and then break up rapidly. If all of today's ice caps melted, global sea level would rise by 70 m (230 feet), extensive areas of coastal plains would be flooded, and major coastal cities like New York, Miami, and London would be submerged (Fig. 22.36). Instead of protruding from ice, the tip of the Empire State Building would protrude from the sea. Ice house or greenhouse? We may not know which scenario will play out in the future until it happens. However, researchers have voiced concern that, at least in the near term, glacial melting will be the order of the day as global temperatures seem to be rising.

THE VIEW FROM SPACE Glaciers and glacially carved features dominate the landscape of southeastern Alaska, as seen in this infrared image. Here we see Hubbard glacier as it enters Yakutat Bay. Where the glacier meets the sea, large blocks of ice calve off and float away. Many tributary glaciers can still be seen, but some glaciers have melted away, leaving behind long fjords.

CHAPTER SUMMARY

• Glaciers are streams or sheets of recrystallized ice that survive for the entire year and flow in response to gravity. Mountain glaciers exist in high regions, and fill cirques and valleys. Continental glaciers (ice sheets) spread over substantial areas of the continents.

• Glaciers form when snow accumulates over a long period of time. With progressive burial, the snow first turns to firn, and then to ice.

• Wet-bottom glaciers move by basal sliding over water or wet sediment. Dry-bottom glaciers move by internal flow. In general, glaciers move tens of meters per year.

• Glaciers move because of gravitational pull, as long as the glaciers have a surface slope.

• Whether the toe of a glacier stays fixed in position, advances farther from the glacier's origin, or retreats back toward the origin depends on the balance between the rate at which snow builds up in the zone of accumulation and the rate at which glaciers melt or sublimate in the zone of ablation.

• Icebergs break off glaciers that flow into the sea. Continental glaciers that flow out into the sea along a coast make ice shelves. Sea ice forms where the ocean surface freezes.

• Glacial ice can flow over sediment or incorporate sediment. The clasts embedded in glacial ice act like a rasp that abrades the substrate.

• Mountain glaciers carve numerous landforms, including cirques, arêtes, horns, U-shaped valleys, hanging valleys, and truncated spurs. Glacially carved valleys that fill with water when sea level rises after an ice age are fjords.

• Glaciers can transport sediment of all sizes. Glacial drift includes till, glacial marine, glacial outwash, lake-bed mud, and loess. Lateral moraines accumulate along the sides of valley glaciers, and medial moraines form down the middle of a glacier. End moraines accumulate at a glacier's toe.

• Glacial depositional landforms include moraines, knob-and-kettle topography, drumlins, kames, eskers, meltwater lakes, and outwash plains.

• Continental crust subsides as a result of ice loading. When the glacier melts away, the crust rebounds.

• When water is stored in continental glaciers, sea level drops. When glaciers melt, sea level rises.

• During past ice ages, the climate in regions south of the continental glaciers was wetter, and pluvial lakes formed. Permafrost (permanently frozen ground) exists in periglacial environments.

• During the Pleistocene ice age, large continental glaciers covered much of North America, Europe, and Asia.

• The stratigraphy of Pleistocene glacial deposits preserved on land records five European and four U.S. glaciations, times during which ice sheets advanced. The record preserved in marine sediments records twenty to thirty such events. The land record, therefore, is incomplete.

• Long-term causes of ice ages include plate tectonics and changes in the concentration of CO_2 in the atmosphere. Short-term causes include the Milankovitch cycles (caused by periodic changes in Earth's orbit and tilt).

KEY TERMS

ablation (p. 688)
albedo (p. 683)
arête (p. 693)
cirque (p. 684)
continental glacier (ice sheet) (p. 684)
crevasse (p. 686)
drop stones (p. 693)
drumlin (p. 699)
dry-bottom glaciers (p. 686)
equilibrium line (p. 688)
end moraine (p. 697)
erratics (p. 681)
esker (p. 700)
firn (p. 686)
fjord (p. 696)
glacial advance (p. 688)
glacial drift (p. 698)
glacial lake-bed sediment (p. 698)
glacial marine (p. 698)
glacial outwash (p. 698)
glacial rebound (p. 705)
glacial retreat (p. 688)
glacial striations (p. 693)
glacial subsidence (p. 704)
glacial till (p. 698)
glacially polished surface (p. 693)
glaciations (p. 712)
glacier (p. 681)
ground moraine (p. 699)
hanging valleys (p. 696)
horn (p. 693)
ice age (p. 682)
ice sheets (p. 682)

ice shelf (p. 691)
icebergs (p. 691)
interglacials (p. 712)
kame (p. 697)
kettle hole (p. 699)
lateral moraine (p. 697)
little ice age (p. 717)
loess (p. 698)
medial moraine (p. 697)
Milankovitch cycles (p. 715)
moraine (p. 697)
mountain (alpine) glacier (p. 684)
patterned ground (p. 707)
permafrost (p. 706)
pluvial lakes (p. 706)
polar glacier (p. 685)
recessional moraine (p. 698)
roche moutonnée (p. 696)
sea ice (p. 691)
Snowball Earth (p. 714)
sublimation (p. 685)
surge (p. 688)
tarn (p. 693)
temperate glaciers (p. 685)
terminal moraine (p. 698)
tillite (p. 713)
toe (p. 688)
U-shaped valley (p. 696)
varve (p. 698)
wet-bottom glaciers (p. 686)
zone of ablation (p. 688)
zone of accumulation (p. 688)

REVIEW QUESTIONS

1. What evidence did Louis Agassiz offer to support the idea of an ice age?

2. How do mountain glaciers and continental glaciers differ in terms of dimensions, thickness, and patterns of movement?

3. Describe the transformation from snow to ice.

4. Explain how arêtes, cirques, and horns form.

5. Describe the mechanisms that enable glaciers to move, and explain why they move.

6. How fast do glaciers normally move? How fast can they move during a surge?

7. Explain how the balance between ablation and accumulation determines whether a glacier advances or retreats.

8. How can a glacier continue to flow toward its toe even though its toe is retreating?

9. How does a glacier transform a V-shaped river valley into a U-shaped valley? Discuss how hanging valleys develop.

10. Describe the various kinds of glacial deposits. Be sure to note the materials from which the deposits are made and the landforms that result from deposition.

11. How do the crust and mantle respond to the weight of glacial ice?

12. How was the world different during the glacial advances of the Pleistocene ice ages? Be sure to mention the relation between glaciations and sea level.

13. How was the standard four-stage chronology of U.S. glaciations developed? Why was it so incomplete? How was it modified with the study of marine sediment?

14. Were there ice ages before the Pleistocene? If so, when?

15. What are some of the long-term causes that lead to ice ages? What are the short-term causes that trigger glaciations and interglacials?

SUGGESTED READING

Alley, R. B. 2002. *The Two-Mile Time Machine: Ice Cores, Abrupt Climate Change, and Our Future.* Princeton, N.J.: Princeton University Press.

Anderson, B. G., and H. W. Borns, Jr. 1994. *The Ice Age World.* Oslo-Copenhagen-Stockholm: Scandinavian University Press.

Bennett, M. R., and N. F. Glasser. 1996. *Glacial Geology: Ice Sheets and Landforms.* New York: Wiley.

Dawson, A. G. 1992. *Ice Age Earth: Late Quaternary Geology and Climate.* New York: Routledge, Chapman, and Hall.

Erickson, J. 1996. *Glacial Geology: How Ice Shapes the Land.* New York: Facts on File.

Fagan, B. 2002. *The Little Ice Age: How Climate Made History, 1350–1850.* New York: Basic Books.

Hambrey, M. J., and J. Alean. 1992. *Glaciers.* Cambridge, England: Cambridge University Press.

Menzies, J. 2002. *Modern and Past Glacial Environments.* Woburn, Mass.: Butterworth-Heinemann.

Paterson, W. S. B. 1999. *Physics of Glaciers.* Woburn, Mass.: Butterworth-Heinemann.

Post, A., and E. R. Lachapelle. 2000. *Glacier Ice.* Seattle: University of Washington Press.

CHAPTER 23

Global Change in the Earth System

All we in one long caravan
are journeying since the world began,
we know not whither, we know . . . all must go.
—BHARTRIHARI (INDIAN POET, C. 500 C.E.)

23.1 INTRODUCTION

Would the Earth's surface have looked the same in the Jurassic Period as it did to the *Apollo* astronauts? Definitely not—the Earth of 200 million years ago differed from that of today in many ways. In the Jurassic, the North Atlantic Ocean was a narrow sea and the South Atlantic Ocean didn't exist at all, so most dry land connected to form a single vast continent (▶Fig. 23.1a). Today, both parts of the Atlantic are wide oceans, and the Earth has seven separate continents (▶Fig. 23.1b). Moreover, in the Jurassic, the call of the wild rumbled from the throats of dinosaurs, while today, the largest land animals are mammals. In essence, what we see of the Earth today is just a snapshot, an instant in the life story of a constantly changing planet. This idea arguably stands as geology's greatest philosophical contribution to humanity's understanding of its surroundings.

Why does the Earth change so much through time? Ultimately, it's because the Earth's asthenosphere is warm and soft enough to flow, because the Sun is close enough to heat the Earth's surface, and because gravity causes heavy objects to fall and buoyant ones to rise. If the asthenosphere could not flow and gravitational force did not exist, internal processes such as sea-floor spreading, subduction, and continental drift would not occur.

This view, from an airplane landing at Chicago's O'Hare Airport, emphasizes the extent to which the Earth's surface has changed. Fifteen thousand years ago, the view would have been the surface of a glacier. Five hundred years ago, it would have been a vast tall-grass prairie. One hundred and fifty years ago, it would have been a checkerboard of farm fields. Today, most of the land has been covered by a layer of concrete.

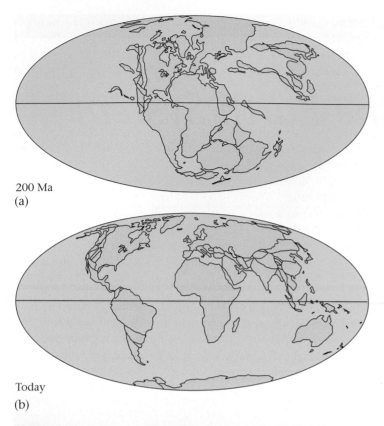

200 Ma
(a)

Today
(b)

FIGURE 23.1 A comparison of (a) a map of the Earth 200 million years ago with (b) a map of today's Earth emphasizing the change that has resulted from continental drift.

Without the warmth of the Sun, there could be no liquid water, advanced life, wind, rivers, or glaciers. Without gravity, the wind, rivers, and glaciers would not move. And without wind, rivers, or glaciers, processes like weathering and erosion would not occur and thus landscapes would not evolve and the rock cycle would not take place. No other object in our solar system has a mobile asthenosphere and a surface whose temperature straddles the freezing point of water (see Box 23.1 and ▶Fig. 23.2), so no other object of our solar system undergoes the kinds of changes that Earth does. The Moon, for example, changes so little through time that it looked essentially the same to a Jurassic dinosaur as it does to you.

Changes that take place on Earth also reflect complex interactions among geologic and biological phenomena. For example, photosynthetic organisms affect the composition of the atmosphere by providing oxygen, and atmospheric composition in turn determines the nature of chemical weathering in rocks. For purposes of discussion, we refer to the global interconnecting web of physical and biological phenomena on Earth as the **Earth System,** and we define **global change** as the transformations or modifications of both physical and biological components of the Earth System through time.

Geologists distinguish among different types of global change, based on the rate or way in which change progresses with time. "Gradual change" takes place over long periods of geologic time (millions to billions of years), while "catastrophic change" takes place relatively rapidly in the context of geologic time (seconds to millennia). "Unidirectional change" involves transformations that never repeat, while "cyclic change" repeats the same steps over and over, though not necessarily with the same results. Some types of cyclic change are "periodic," in that the cycles happen with a definable frequency.

In this chapter, we begin by reviewing examples of global change, both unidirectional and cyclic, involving phenomena discussed earlier in the book. Then we look at an example of a **biogeochemical cycle,** the exchange of chemicals among living and nonliving reservoirs; some kinds of global change are due to changes in the proportions of chemicals held in different reservoirs through time. Finally, we focus on **global climate change** (transformations or modifications in Earth's climate through time) and on anthropogenic (human-caused) contributions to global change. We conclude this chapter, and the book, by considering hypotheses that describe the ultimate global change—the end of the Earth.

23.2 UNIDIRECTIONAL CHANGES

The Evolution of the Solid Earth

Recall from Chapter 1 that Earth began as a fairly homogenous mass, formed by the coalescence of planetesimals. But the homogeneous proto-Earth did not last long—within about one hundred million years of its birth the planet began to melt, yielding a liquid iron alloy that sank rapidly to the center to form the core (▶Fig. 23.3a, b). This process of "differentiation" represents major unidirectional change: it produced a layered, onion-like planet, with an iron alloy core surrounded by a rocky mantle.

Soon after differentiation, a Mars-sized proto-planet appears to have collided with the newborn Earth. This collision caused a catastrophic change—a significant portion of the Earth and the colliding object fragmented and vaporized, creating a ring of debris that coalesced to form the Moon (▶Fig. 23.3c–e). After the collision, the Earth's mantle was probably partially molten, and its surface became a sea of magma. The Earth continued to endure intense bombardment by asteroids and comets until about 3.9 Ga, so any crust that had formed prior to 3.9 Ga was largely pulverized or melted. Eventually, however, bombardment ceased and our planet gradually cooled, permitting a crust to form at its surface and plate tectonics to begin operating. Subduction, and/or the rise of mantle plumes, produced relatively low-density rocks (e.g., granite). These rocks could not be subducted and thus remained as buoyant blocks at the Earth's surface. Plate motion eventually caused these buoyant blocks to collide and suture together, forming the first continents. Overall, therefore, the transition from the Hadean Eon to the Archean Eon saw remarkable unidirectional change in the nature of the Earth—by early Archean time, our planet had distinct continents and ocean basins, and thus looked radically different from the other terrestrial planets (see chapter 13).

The Evolution of the Atmosphere and Oceans

Like its surface, the Earth's atmosphere has also changed through time. Partial melting in the mantle produced magma and also released large quantities of gases that belched from volcanoes. More gases may have arrived when comets collided with our new planet. Eventually, Earth accumulated an "early atmosphere" composed dominantly of carbon dioxide (CO_2) and water (H_2O). Other gases, such as nitrogen (N_2), composed only a minor proportion of the early atmosphere. When the Earth's surface cooled, however, water condensed and fell as rain, collecting in low areas to form oceans. This may have happened before 4.0 Ga, but had certainly happened by 3.8 Ga. Gradually, CO_2 dissolved in the oceans and was absorbed by chemical-weathering reactions on land, so its concentration in the atmosphere decreased. Nitrogen, which doesn't react with other chemicals, was left behind. Thus, the atmosphere's composition changed to become dominated by nitrogen. Photosynthetic organisms appeared early in the Archean. But it probably wasn't until between 2.5 and 2.0 Ga (the early part of the Proterozoic) that oxygen (O_2) became a significant proportion of the atmosphere. Present concentrations of oxygen may have only existed for the past 400 million years.

The Goldilocks Effect

BOX 23.1
THE REST OF THE STORY

Like Baby Bear's porridge in the tale of *Goldilocks and the Three Bears,* Earth is not too hot, and it's not too cold . . . it's just right, as far as complex life is concerned (▶Fig. 23.2a–c). This condition is known as the **Goldilocks effect.**

Two factors play the key roles in determining Earth's surface temperature: the distance between the Earth and the Sun, and the concentration of carbon dioxide (CO_2) and other greenhouse gases in the atmosphere. If Earth were just 13% closer to the Sun, the heat would be so intense that liquid water could not exist. Without liquid water,

CO_2 would not be stored as limestone and coal but rather would remain in the atmosphere. And without uplift and exposure of continental rocks, CO_2-absorbing chemical-weathering reactions could not take place. As a result, the atmosphere would contain so much CO_2 that it would become hot enough to melt lead, a condition that now exists on Venus. On the other hand, if Earth were significantly farther from the Sun so that sunlight was weaker, or if the atmosphere contained less CO_2 with which to trap heat, our planet's surface would be so cool that the oceans would freeze solid and complex life would die.

FIGURE 23.2 (a) Venus is too hot, (b) Mars is too cold, and (c) Earth is just right.

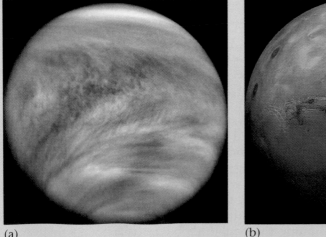
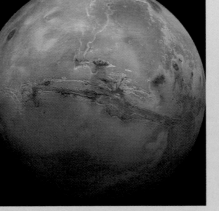

(a)　　　　　　　　　　　　　　　(b)　　　　　　　　　　　　　　　(c)

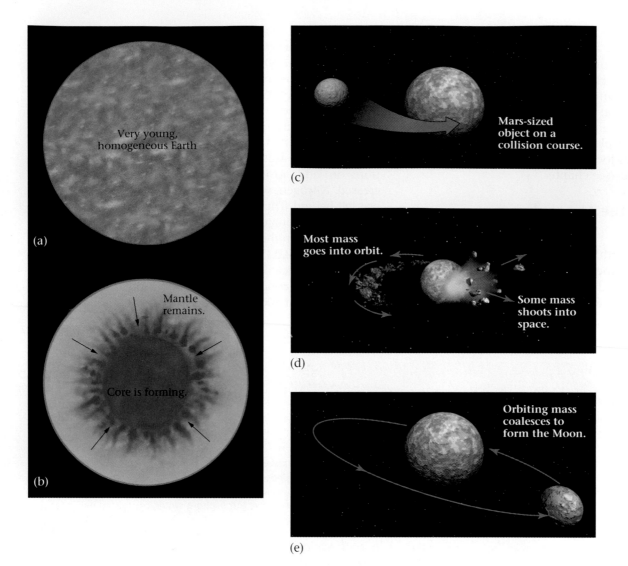

FIGURE 23.3 (a) When it first formed, the Earth was probably homogeneous. (b) Soon thereafter, the iron in the Earth melted and sank to the center. When this differention, a unidirectional change, was complete, the Earth had a distinct core and mantle. (c–e) Then, a Mars-sized body collided with Earth, sending off fragments that coalesced to form the Moon. All these phenomena radically changed the Earth.

The Evolution of Life

During most of the Hadean Eon, Earth's surface was probably lifeless, for carbon-based organisms could not survive the high temperatures of the time. The fossil record indicates that life had appeared at least by 3.8 billion years ago and has undergone unidirectional change (evolution) in fits and starts ever since (see Interlude D). Though simple organisms like archaea and bacteria still exist, life evolution during the late Proterozoic and early Phanerozoic yielded multicellular plants and animals (▶Fig. 23.4). Life now inhabits regions from a few kilometers below the surface to a few kilometers above, yielding a diverse biosphere.

23.3 PHYSICAL CYCLES

The Supercontinent Cycle

During Earth history, the map of the planet's surface has constantly changed. At times, almost all continental crust merged to form a single supercontinent, but usually the crust is distributed among several smaller continents. The process of change during which supercontinents form and later break apart is the **supercontinent cycle** (▶Fig. 23.5). Geologists have found evidence that at least three or four times during the past three billion years of Earth history,

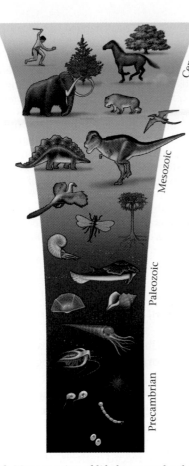

FIGURE 23.4 New species of life have evolved through geologic time. Though some of the simplest still exist, more complex organisms have appeared more recently.

supercontinents existed. The most recent one, Pangaea, formed at the end of the Paleozoic Era. Others likely coalesced at 1.1 Ga, 2.1 Ga, and perhaps 2.7 Ga.

Plates move only 1–15 cm per year, so one passage through the supercontinent cycle takes at least a few hundred million years. Note that ocean basins do not simply open and close like accordions. In reality, plate motions are more complex, so the land never rearranges exactly the same way through two supercontinent cycles.

The Sea-Level Cycle

Global sea level has risen and fallen by as much as 300 m during the Phanerozoic, and likely did the same in the Precambrian. When sea level rises, the shoreline migrates inland, and low-lying plains in the continents become submerged. During periods of particularly high sea level, more than half of Earth's continental area can be covered by shallow seas; at such times, sediment buries continental regions, thereby changing their surface (▶Fig. 23.6a). When sea level falls, the continents become dry again, and regional unconformities develop. For example, the sedimentary strata of the midwestern United States record at least six continent-wide advances and retreats of the sea, each of which left behind a blanket of sediment called a **sedimentary sequence;** unconformities define the boundaries between the sequences (▶Fig. 23.6b). Of note, the sequence deposited during the Pennsylvanian contains at least thirty shorter repeated intervals, called "cyclothems", each of which contains a specific succession of sedimentary beds. At the base of each cyclothem you'll find sandstone, and in the middle you'll find coal (▶Fig. 23.6c).

FIGURE 23.5 During the supercontinent cycle, smaller continents coalesce to form a supercontinent, which then later rifts and breaks apart, only to recombine later on. Since continents drift around the surface of the Earth, collisions do not necessarily bring previously adjacent continents back together again.

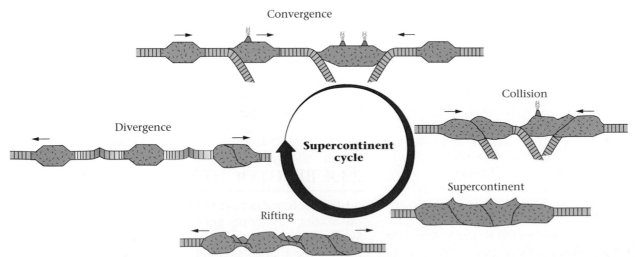

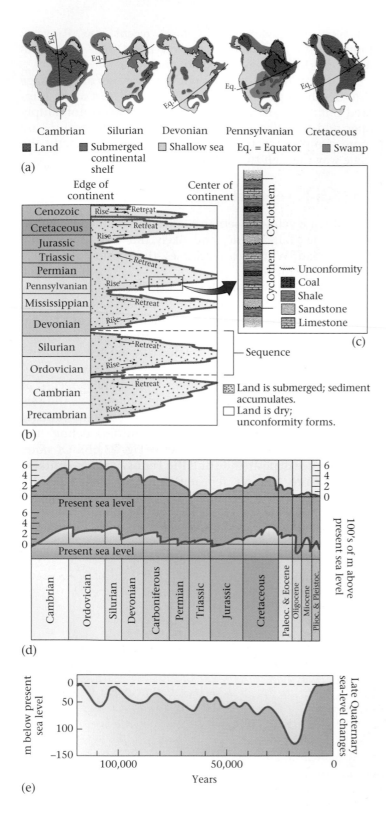

(a) Cambrian   Silurian   Devonian   Pennsylvanian   Cretaceous

■ Land   ■ Submerged continental shelf   □ Shallow sea   Eq. = Equator   ■ Swamp

(b)

Edge of continent / Center of continent

Cenozoic · Rise ← Retreat
Cretaceous · Retreat
Jurassic · Rise
Triassic
Permian · Retreat
Pennsylvanian · Rise →
Mississippian · Retreat
Devonian · Rise →
Silurian · Retreat
Ordovician · Rise →
Cambrian · Retreat
Precambrian · Rise →

▦ Land is submerged; sediment accumulates.
□ Land is dry; unconformity forms.

Sequence

(c)

Cyclothem | Cyclothem

〜〜 Unconformity
■ Coal
■ Shale
□ Sandstone
▦ Limestone

(d)

Cambrian Ordovician Silurian Devonian Carboniferous Permian Triassic Jurassic Cretaceous Paleoc. & Eocene Oligocene Miocene Plioc. & Pleistoc.

Present sea level

Present sea level

100's of m above present sea level

(e)

m below present sea level — 0, 50, 100, −150

100,000   50,000   0
Years

Late Quaternary sea-level changes

FIGURE 23.6 (a) Sea level has changed significantly over geologic time. For example, large parts of North America were once submerged by shallow seas. (The continental outline is shown for reference only—the continent did not have its present shape in the past.) (b) The stratigraphic record shows that sedimentary sequences are deposited during long-term transgressions (sea-level rise) and regressions (sea-level retreat). Transgression starts near the edge of a continent and then moves inland. (c) During the Pennsylvanian, short-term transgressions and regressions created cyclothems. (d) To some extent, the transgressions and regressions indicated by this sedimentary sequence chart reflect global (eustatic) sea-level change. Here, two versions of the sequence chart are shown, each produced by a different author—the shape of the curve is still a subject for research. (e) At a finer time scale, there are many short-term ups and downs in sea level. This graph shows sea-level changes during the last 150,000 years.

Cyclothems represent short-term cycles of sea-level rise and fall.

After studying sedimentary sequences around the world, geologists at Exxon Corporation pieced together a chart defining the succession of global transgressions and regressions during the Phanerozoic Eon. The global **sedimentary cycle chart** may largely reflect the cycles of **eustatic** (worldwide) **sea-level change** (▶Fig. 23.6d, e). However, the chart probably does not give us an exact image of sea-level change, because the sedimentary record reflects other factors as well, such changes in sediment supply. Eustatic sea-level changes may be due to a variety of factors, including advances and retreats of continental glaciers, changes in the volume of mid-ocean ridge systems, and changes in continental elevation and area.

The Rock Cycle

We learned early in this book that the crust of the Earth consists of three rock types: igneous, sedimentary, and metamorphic. Atoms making up the minerals of one rock type may later become part of another rock type. In effect, rocks are simply reservoirs of atoms, and the atoms move from reservoir to reservoir through time. As we learned in Interlude B, this process is the "rock cycle". Each stage in the rock cycle changes the Earth by redistributing and modifying material.

23.4 BIOGEOCHEMICAL CYCLES

A **biogeochemical cycle** involves the passage of a chemical among nonliving and living reservoirs in the Earth System, mostly on or near the surface. Nonliving reservoirs include the atmosphere, the crust, and the ocean, while liv-

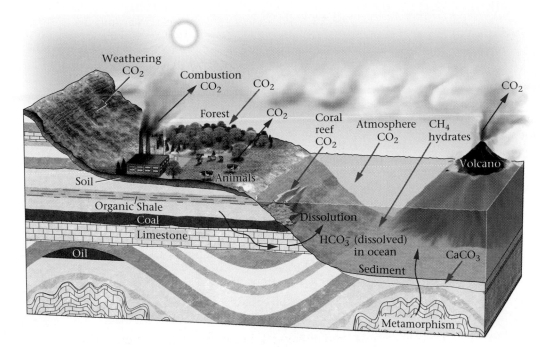
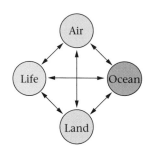

FIGURE 23.7 In the carbon cycle, carbon transfers among various reservoirs at or near the Earth's surface. Red arrows indicate release to the air, and green arrows indicate absorbtion from air.

ing reservoirs include plants, animals, and microbes. Although a great variety of chemicals (water, carbon, oxygen, sulfur, ammonia, phosphorous, and nitrogen) participate in biogeochemical cycles, here we look at only two: water (H_2O) and carbon (C).

Some stages in a biogeochemical cycle may take only hours, some may take thousands of years, and others may take millions of years. Because chemicals can cycle rapidly, the transfer of a chemical from reservoir to reservoir during these cycles doesn't really seem like a "change" in the Earth in the way that the movement of continents or the metamorphism of rock seems like a change. In fact, for intervals of time, biogeochemical cycles attain a **steady-state condition,** meaning that the proportions of a chemical in different reservoirs remain fairly constant even though there is a constant flux (flow) of the chemical among reservoirs. When we speak of global change in a biogeochemical cycle, we mean a change in the relative proportions of a chemical held in different reservoirs at a given time—in other words, a change in the steady-state condition.

The Hydrologic Cycle

As we learned in Interlude E, the hydrologic cycle involves the movement of water from reservoir to reservoir on or near the surface of the Earth. The hydrologic cycle is an example of a biogeochemical cycle, in that a chemi-

cal (H_2O) passes through both nonliving and living entities—the oceans, the atmosphere, surface water, groundwater, glaciers, soil, and living organisms. Global change in the hydrologic cycle occurs when a change in global climate alters the ratio between the amount of water held in the ocean and the amount held in continental ice sheets. For example, during an ice age, water that had been stored in oceans moves into glacial reservoirs. Thus, the continents become covered with ice, and sea level drops. When the climate warms, water returns to the oceans, and sea level rises.

The Carbon Cycle (the Movement of a Greenhouse Gas)

Most carbon in the near-surface realm of Earth originally bubbled out of the mantle in the form of CO_2 gas released at volcanoes (▶Fig. 23.7). Once it enters the atmosphere, it can be removed in various ways. Some dissolves in seawater to form bicarbonate (HCO_3^-) ions, whereas some is absorbed by photosynthetic organisms that convert it into sugar and other organic chemicals. This carbon enters the food chain and ultimately makes up the flesh, fat, and sinew of animals. In fact, about 63 billion tons of carbon move from the atmosphere into life forms every year. Some of the reactions that take place when rock undergoes chemical weathering incorporate atmospheric CO_2, and thus also remove carbon from the atmosphere.

The Earth System

External energy

Sun

Thunderhead

Lightning

Rain and snow

Mountain uplift

Continental glacier

City

Ocean

Rocky coastline

Desert

Valley

Arid mountains

Mining

Lakes

Deciduous forest

Forested mountains

Beach

The Earth's surface represents the interface among the solid Earth (lithosphere); the ice and liquid water of oceans, lakes, streams, groundwater, and glaciers (the hydrosphere); and the planet's gaseous envelope (the atmosphere). Countless species of life, ranging from nearly invisible bacteria to giant whales and trees, make up the complex ecosystems of Earth's biosphere. All of these components—the lithosphere, hydrosphere, atmosphere, and biosphere—interact with one another. These components and the interactions among them constitute the Earth System.

Various materials cycle among living and nonliving components of the Earth System. In the hydrologic cycle, for example, water evaporates from the sea, rains on the land, and eventually flows back to the sea. During this process, water may be temporarily trapped in living organisms, clouds, subsurface pores, or ice sheets. Carbon dioxide can be stored in the air, dissolved in water, or trapped in plants, coal, or limestone. Some limestone forms when coral extracts ions from water. Meanwhile, the atoms that make up minerals, over the vastness of geologic time, pass through the rock cycle. New elements from the mantle may enter the cycles of the Earth System at volcanoes or black smokers. Elements at the surface may be carried back into the mantle at subduction zones. Some atoms escape from the atmosphere into space.

Tropical rain forest

Coral reef

Shark

Internal energy

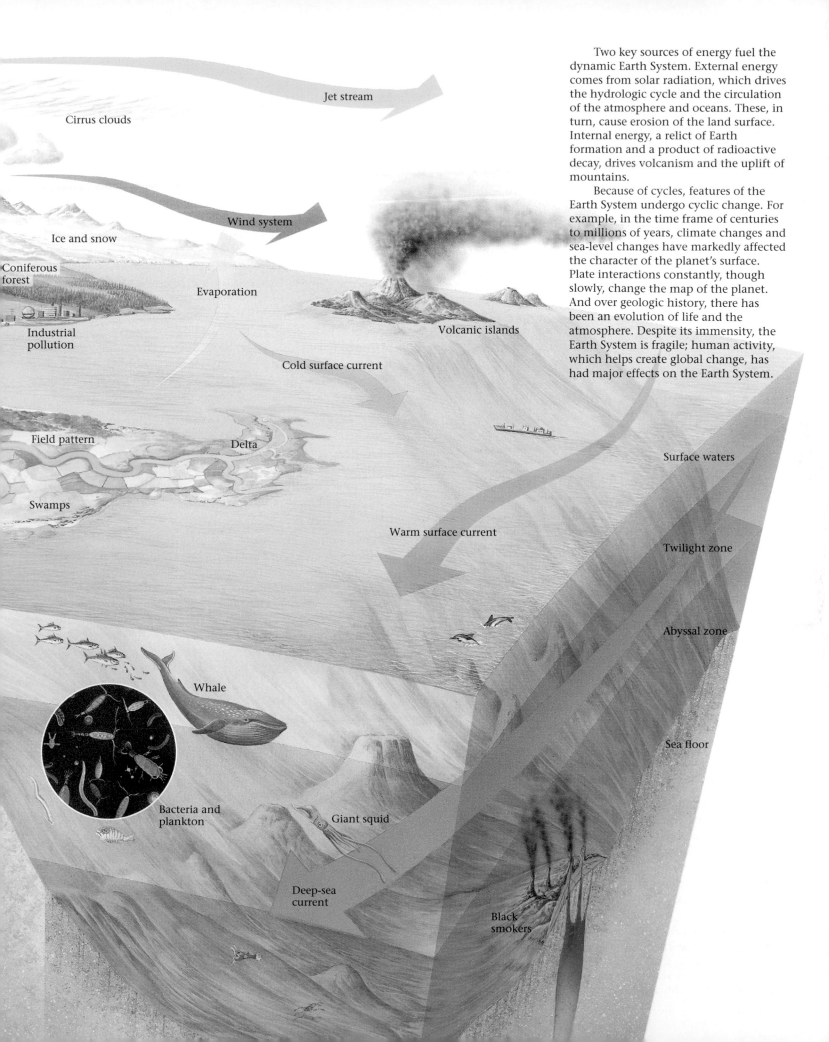

Jet stream

Cirrus clouds

Wind system

Ice and snow

Coniferous forest

Evaporation

Volcanic islands

Industrial pollution

Cold surface current

Field pattern

Delta

Surface waters

Swamps

Warm surface current

Twilight zone

Abyssal zone

Whale

Sea floor

Bacteria and plankton

Giant squid

Deep-sea current

Black smokers

Two key sources of energy fuel the dynamic Earth System. External energy comes from solar radiation, which drives the hydrologic cycle and the circulation of the atmosphere and oceans. These, in turn, cause erosion of the land surface. Internal energy, a relict of Earth formation and a product of radioactive decay, drives volcanism and the uplift of mountains.

Because of cycles, features of the Earth System undergo cyclic change. For example, in the time frame of centuries to millions of years, climate changes and sea-level changes have markedly affected the character of the planet's surface. Plate interactions constantly, though slowly, change the map of the planet. And over geologic history, there has been an evolution of life and the atmosphere. Despite its immensity, the Earth System is fragile; human activity, which helps create global change, has had major effects on the Earth System.

Some carbon returns directly to the atmosphere by the respiration of animals (again as CO_2), by the flatulence of animals (as methane [CH_4]), or by the decay of dead organisms. But some can be stored for long periods of time in fossil fuels (oil and coal), in organic shale, in methane hydrates (see Chapter 14), or in limestone. Fossil fuel deposits, limestone, methane hydrates, and the organic portion of shale contain most of the carbon in the near-surface realm of Earth and can hold on to it for long periods of time. But this carbon either returns to the atmosphere in the form of CO_2, as a result of the burning of fossil fuels and the metamorphism of rocks containing carbonate, or returns to the sea after undergoing chemical weathering followed by dissolution as HCO_3^- in river water or groundwater. Melting of methane hydrates releases methane to the atmosphere.

The concentration of carbon dioxide and methane in the atmosphere play an essential role in controlling Earth's climate because, as we saw in Chapter 20, these gases, along with several other trace gases (such as water), are greenhouse gases—an increase in their concentration warms the atmosphere, while a decrease cools it down.

23.5 GLOBAL CLIMATE CHANGE

How often have you seen a newspaper proclaim "Record High Temperatures!" In August 2003, such a claim became reality for much of Europe, where thermometers registered weeks of temperatures as much as 8°C above "normal." The *New York Times* ran the headline "Europe Sizzles and Suffers in a Summer of Merciless Heat." Does this mean that the climate, the average range of weather conditions for a given region, is changing? Is it something we should be worried about? Perhaps. "Global climate change," the transformations or modifications in Earth's climate through time, is indeed important because it affects sea level and the distribution and character of climatic belts and, therefore, the distribution of habitats, agricultural lands, and landscapes. A rise in the average global near-surface atmospheric temperature of only a few degrees might melt enough polar ice to cause a sea-level rise that would in turn flood coastal population centers and also cause devastating droughts.

For purposes of discussion, we distinguish between "long-term climate change", which takes place over millions to tens of millions of years, and "short-term climate change", which takes place over tens to hundreds of thousands of years. If the average atmospheric and sea-surface temperature rises, we have **global warming**, while if it falls, we have **global cooling**. Some changes are great enough to cause oceanic islands and large regions of continents to be submerged by shallow seas, or to be covered by ice, while others are subtle, creating only a slight latitudinal shift in vegetation belts and a sea-level change measured in meters or less.

Methods of Study

Geologists and climatologists are working hard to define the kinds of climate changes that can occur, the rates at which these changes take place, and the effects they may have on society. There are two basic approaches to studying global climate change: (1) researchers measure past climate change, as indicated by the stratigraphic record, to document the magnitude of change that is possible and the rate at which such change occurred; (2) researchers make computer programs to calculate how factors like atmospheric composition, topography, ocean currents, and Earth's orbit affect the climate. The resulting **climate-change models** provide insight into when and why changes took place in the past and whether they will happen in the future.

Let's look first at how geologists study the **paleoclimate** (past climate), so as to document climate changes throughout Earth history. Any feature whose character depends on the climate and whose age can be determined serves as a clue to defining paleoclimate.

- *The stratigraphic record:* The nature of sedimentary strata deposited at a certain location reflects the climate at that location. For example, an outcrop exposing cross-bedded sandstone, overlain successively by coal and glacial till, indicates that the site of the outcrop has endured different climates (desert, then tropical, then glacial) through time.

- *Paleontological evidence:* Different assemblages of species survive in different climatic belts. Thus, the succession of species in a sedimentary sequence provides clues to the changes in climate at that site. For example, a record of short-term climate change can be obtained by studying the succession of plankton fossils in sea-floor sediments, for cold-water species of plankton are different from warm-water species. Fossil pollen also yields clues to the paleoclimate. Pollen, tiny grains involved in plant reproduction, looks like dust to the unaided eye. But under a microscope, each grain has a distinctive structure, and grains of one species look different from grains of another species (▶Fig. 23.8a, b). Further, pollen grains have a tough coating and can survive burial. By studying pollen in sediment, palynologists (scientists who study pollen) can determine whether the sediment accumulated in a cold-climate coniferous forest or in a warm-climate deciduous forest. And by recording changes in the pollen assemblage found in successive layers of sediment, palynologists can track the movement of climate belts over the landscape (▶Fig. 23.8c). For example, studies of spruce pollen preserved in the mud of bogs show that spruce forests, indicative of cool climates, have slowly migrated north since the ice age (▶Fig. 23.8d, e).

- ***Oxygen-isotope ratios:*** Two isotopes of an element have the same atomic number but different atomic

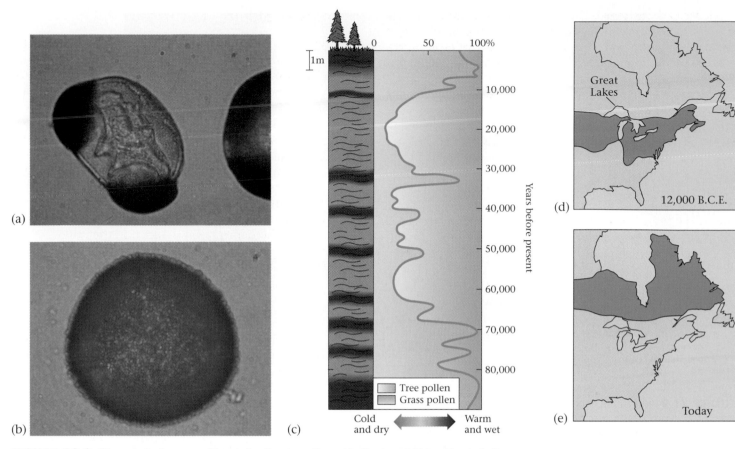

FIGURE 23.8 Changes in the assemblage of pollen in sediment indicate a shift in climate belts. (a) Spruce pollen from a cold-climate coniferous forest. (b) Hemlock pollen from a warm-climate deciduous forest. (c) This model shows how the proportion of tree pollen relative to grass pollen can change in a sedimentary sequence through time. Tree pollen indicates cooler and drier conditions, while grass pollen indicates warmer and wetter conditions. (d) Pollen data suggest that about 12,000 years ago, spruce forests (green areas) lay south of the Great Lakes. (e) Today, they are found north of the Great Lakes.

weights (see Appendix A). For instance, oxygen occurs as ^{16}O (8 protons and 8 neutrons) and ^{18}O (8 protons and 10 neutrons). Geologists have found that the ratio of $^{18}O/^{16}O$ in glacial ice indicates the atmospheric temperature in which the snow that made up the ice formed: the ratio is larger in snow that forms in warmer air, while the ratio is smaller in snow that forms in colder air. Because of this relationship, the isotope ratio measured in a succession of ice layers in a glacier indicates temperature change through time. Researchers have now obtained ice cores down to a depth of almost 3 km in Antarctica and in Greenland; this record spans almost 400,000 years (▶Fig. 23.9a). For a number of reasons, the $^{18}O/^{16}O$ ratio in the $CaCO_3$ making up plankton shells also gives geologists an indication of past temperatures. Thus, measurement of oxygen-isotope ratios in drill cores of marine sediment extends the record of temperature change back through millions of years (▶Fig. 23.9b, c).

- *Bubbles in ice:* Bubbles in ice trap the air present at the time the ice forms. By analyzing these bubbles, geologists can measure the concentration of CO_2 in the atmosphere back through time. This information can be used to correlate CO_2 concentration with past atmospheric temperature. The CO_2 record has been extended back through 240,000 years.

- *Growth rings:* If you've ever looked at a tree stump, you'll have noticed the concentric rings visible in the wood. Each ring represents one year of growth, and the thickness of the ring indicates the rate of growth in a given year. Trees grow faster during warmer, wetter years and more slowly during cold, dry years (▶Fig. 23.10a); thus, the succession of ring widths provides an easily calibrated record of climate during the lifetime of the tree. Bristlecone pines supply a record back through 4,000 years. To go further into the past, dendrochronologists (scientists who study tree rings) look at the record of rings in logs dated by the

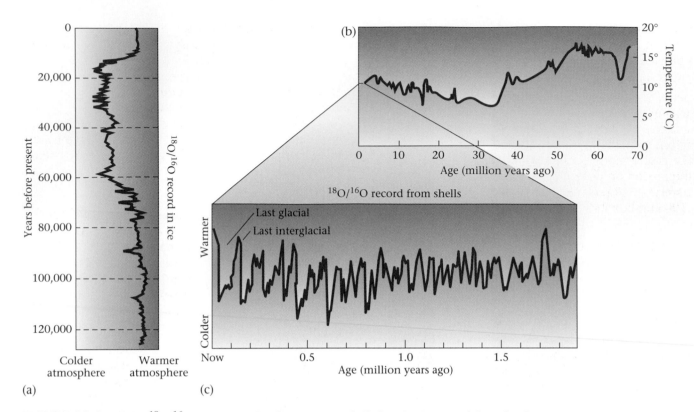

(a)   (c)

FIGURE 23.9 (a) The $^{18}O/^{16}O$ ratio in a 2-km-long ice core drilled in the ice cap of Greenland varies with depth in the core, indicating that atmospheric temperature varies over time. Smaller ratios mean a colder atmosphere. (b) The ratio of $^{18}O/^{16}O$ in the calcite of fossil plankton shells of deep-marine sediment shows variations in temperature for the past 70 million years. (c) The detailed plankton record of temperature for the past 2 million years. The decreases in the $^{18}O/^{16}O$ ratio correspond with glacial advances.

FIGURE 23.10 (a) Tree rings provide a record of climate, for more growth happens in wet years than in dry years. (b) The stories of great floods recorded by artists and historians provide clues to the timing of climate changes. This woodcut depicts a flood in medieval England.

(a)   (b)

radiocarbon technique or in logs whose ages overlap with that of the oldest living tree. Growth rings in corals and shells can provide similar information.

- *Human history:* Researchers have made careful, direct measurements of climate changes only for the past few decades. This record is not long enough to display long-term climate change. But history, both written and archaeological, contains important clues to climates at times in the past. Periods of unusual cold or drought leave an impression on people, who record them in paintings, stories, and records of crop success or failure (▶Fig. 23.10b and Box 23.2).

Long-Term Climate Change

Using the variety of techniques described above, geologists have reconstructed an approximate record of global climate, represented by mean temperature and rainfall, for geologic time. The record shows that at some times in the past, the Earth's atmosphere was significantly warmer than it is today, whereas at other times it was significantly cooler. The warmer periods have come to be known as **greenhouse** (or hothouse) **periods** and the colder as **ice-house periods.** (The more familiar term, "ice age," refers to the times during an ice-house period when the Earth was cold enough for ice sheets to advance and cover substantial areas of the continents.) As the chart in ▶Figure 23.11a shows, there have been at least five major ice-house periods during geologic history.

Let's look a little more closely at the climate record of the last 100 million years, for this time interval includes the transition between a greenhouse and an ice-house period. Paleontological and other data suggest that the climate of the Mesozoic Era, the Age of Dinosaurs, was much warmer than the climate of today. At the equator, average annual temperatures may have been 2°–6°C warmer, while at the poles, temperatures may have been 20°–60°C warmer. In fact, during the Cretaceous Period, dinosaurs were able to live at high latitudes, and there were no polar ice caps on Earth. But starting about 80 million years ago, the Earth's atmosphere began to cool. We entered an ice-house period about 33 million years ago, and the climate reached its coldest condition about 2 million years ago, during the Pleistocene ice age (see Chapter 22).

What caused long-term global climate change? The answer may lie in the complex relationships among the various geologic and biogeochemical cycles of the Earth system, as described earlier. Following are some likely influences on long-term global change:

- *Positions of continents:* Continental drift influences the climate by controlling the pattern of oceanic currents, which redistribute heat around the planet's surface (▶Fig. 23.11b). Drift also determines whether the land is at high or low latitudes (and thus how much solar radiation strikes it), and whether or not there are large continental interior regions where extremely cold winter temperatures can develop.

- *Volcanic activity:* A long-term global increase in volcanic activity may contribute to long-term global warming, because it increases the concentration of greenhouse gases in the atmosphere. For example, during the Cretaceous Period when Pangaea broke up, numerous rifts formed, and sea-floor-spreading rates were particularly high, so volcanoes were more abundant than they are today; volcanic activity may thus have triggered Cretaceous greenhouse conditions. The eruption of large igneous provinces (LIPs; see Chapter 6) at other times may have also caused cooling.

- *The uplift of land surfaces:* Tectonic events that lead to the long-term uplift of the land affect atmospheric CO_2 concentration, because such events expose land to weathering, and chemical-weathering reactions absorb CO_2. Thus, uplift decreases the greenhouse effect and causes global cooling. For example, uplift of the Himalayas and Tibet may have triggered Cenozoic ice-house

Global Climate Change and the Birth of Legends

BOX 23.2
THE HUMAN ANGLE

Some geologists argue that myths passed down from the early days of civilization may have their roots in global climate change. For example, recent evidence suggests that earlier than 7,600 years ago, the region that is now the Black Sea contained a much smaller freshwater lake surrounded by settlements. Subsequent to the most recent ice-age glacial advance, the ice sheets melted and sea level rose, and the Mediterranean eventually broke through a dam at the site of the present Bosporous Strait. Researchers suggest that seawater from the Mediterranean spilled into the Black Sea basin via a waterfall two hundred times larger than Niagara Falls. This influx of water caused the lake level to rise by perhaps 10 cm per day, and within a year 155,000 square km (60,000 square miles) of populated land had become submerged beneath hundreds of meters of water. This flooding presumably led to a huge human migration, and its timing has suggested to some researchers that it may have inspired the Babylonian *Epic of Gilgamesh* (ca. 2000 B.C.E.) and, later, the biblical epic of Noah's Ark.

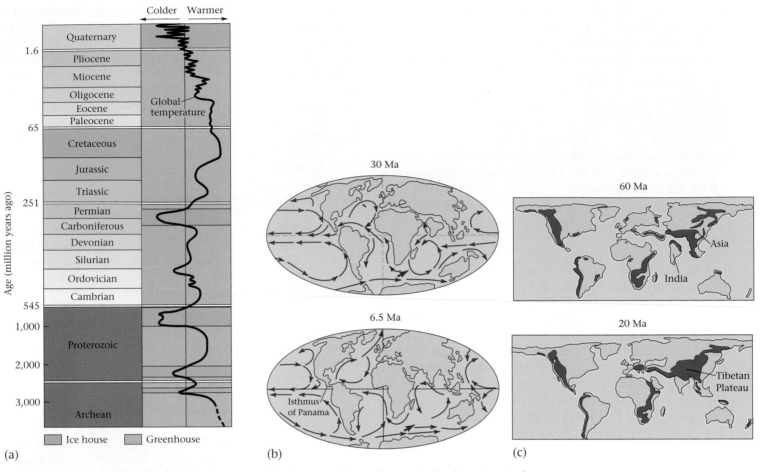

FIGURE 23.11 (a) The chart shows the timing of ice-house and greenhouse (or hot-house) periods during Earth history. (b) Continental drift and the growth of volcanic arcs affect the pattern of currents in the ocean. Here, we see that currents changed when the Isthmus of Panama (a volcanic arc) developed. (c) The maps show that the proportion of the Earth's surface at high elevations, where it is exposed to chemical weathering, has changed through time as a consequence of mountain building. For example, the collision of India with Asia about 40 Ma created a broad high plateau.

conditions (▶Fig. 23.11c). Such uplift will also affect atmospheric circulation and rainfall rates; see Chapter 20.

• *The formation of coal, oil, or organic shale:* At various times during Earth history, environments suitable for coal or oil formation have been particularly widespread. Such formation removes CO_2 from the atmosphere and stores it underground. For example, global cooling in the Carboniferous may correlate with the development of vast coal swamps on Pangaea.

• *Life evolution:* The appearance of or extinction of certain life forms may have impacted climate significantly. For example, some researchers speculate that the appearance of lichens in the Neoproterozoic may have decreased atmospheric CO_2 concentration, and thus could have triggered Neoproterozoic ice house conditions. Similarly, the appearance of grass about 30-35 Ma may have triggered Cenozoic ice house conditions.

Note that some of the effects described above add or subtract CO_2 from the atmosphere. Some researchers argue that change in the distribution of CO_2 among various reservoirs in the carbon cycle plays a particularly influential role in controlling climate because CO_2 is an important greenhouse gas—factors that add CO_2 lead to warming, while those that decrease CO_2 lead to cooling. Others suggest that the greenhouse effect of CO_2 serves only to amplify the effects caused by other phenomena, such as changes in cloud cover. The relative importance of these different factors remains a focus of research.

It is important to note that feedback among components of the Earth System helps regulate the amount of CO_2 in the atmosphere. "Negative feedback" slows a process down or even reverses it. For example, as global temperature rises because of an increase in CO_2, rates of evaporation and therefore amounts of precipitation increase. As a consequence, weathering rates (which absorb CO_2) increase, so the concentration of CO_2 then goes down. "Positive feedback," on the other hand, makes a process continue or even accelerate. For example, positive feedback on Venus may have led to a **runaway greenhouse effect,** through the following steps:

1. Because Venus orbits closer to the Sun than does Earth, it receives more solar radiation. In the past, the high heat caused any surface water to evaporate until the planet became enveloped in clouds.

2. The water vapor, a greenhouse gas, did not allow infrared radiation to escape, so the atmosphere became still hotter, approaching 1,500°C. At these extremely high temperatures, water molecules break apart, forming hydrogen and oxygen gas. The hydrogen escaped to space, and the oxygen reacted with surface rocks to oxidize (rust) them.

3. Without water, CO_2-absorbing chemical weathering ceased, so CO_2 built up in the atmosphere, making the temperature rise even more. This dense hot atmosphere persists on Venus today.

Short-Term Climate Change

The record of the past million years gives a sense of the magnitude and duration of short-term climate change. During this period, there have been about five major and twenty to thirty minor episodes of glaciation, separated by interglacial periods. If we focus on the last 15,000 years, we see that overall the temperature has increased, but there are still notable ups and downs (▶Fig. 23.12a). In fact, by studying evidence for icebergs in the ocean, researchers have identified cycles in this period that lasted for centuries to millennia. Some evidence suggests that significant shifts in climate can occur in only ten years.

As a result of the warming that began 15,000 years ago, the glaciers retreated for about 4,500 years. Then there was a return to colder conditions for a few more thousand years. This interval of cooler temperature is named the "Younger Dryas," after an Arctic flower that became widespread during the time. The climate then warmed, reaching a peak at 5,000–6,000 years ago, a period called the "Holocene maximum," when average temperatures were about 2°C above temperatures of today. This warming peak led to increased evaporation and therefore precipitation, making the Middle East unusually wet and fertile—conditions that may partially account for the rise of civilization in this part of the world.

The temperature dipped to a low about 3,000 years ago, before returning to a high during the Middle Ages, a time called the "Medieval warm period." During this time, when temperatures were 0.5 to 0.8° above those of today, Vikings established self-supporting agricultural settlements along the coast of Greenland (▶Fig. 23.12b). The temperature dropped again from 1500 to about 1800, a period known as the "little ice age," when Alpine glaciers advanced and the canals of the Netherlands froze over in winter (▶Fig. 23.12c, d; see Fig. 22.47). The climate, overall, has warmed since the end of the little ice age, and today it is as warm as it was during the Medieval warm period.

Geologists have focused on four factors to explain short-term climate change.

• *Fluctuations in solar radiation and cosmic rays:* The amount of energy produced by the Sun varies with the **sunspot cycle.** This cycle involves the appearance of large numbers of sunspots (black spots thought to be magnetic storms on the Sun's surface) about every 9 to 11.5 years (▶Fig. 23.13a, b). There may be longer-term cycles that have not yet been identified. This variation in energy may affect the climate. Recently, some researchers have speculated that changes in the rate of influx of cosmic rays may affect climate, perhaps by generating clouds. Specifically, recent research suggests that cosmic rays

The Faint Young Sun Paradox

BOX 23.3
THE REST OF THE STORY

The intensity of sunlight striking the Earth has likely changed substantially during the history of the solar system, because the composition of the Sun changes with time. The Sun's energy comes primarily from the fusion reaction that bonds four hydrogen atoms together to form one helium atom. Through time, the proportion of helium in the Sun increases. Calculations suggest that because one helium atom takes up less space than four separate hydrogen atoms, the production of helium should allow the inside of the Sun to contract. This contraction raises the Sun's temperature, which in turn increases the rate of fusion reactions and therefore the amount of energy the Sun produces. Overall, the Sun may be 30% brighter today than it was when the Earth first formed.

If this is the case, the Earth's average temperature *should have been* over 20°C less in the Archean Eon than it is today, and Archean oceans *should have been* frozen solid. But fossil evidence suggests that liquid water has existed on our planet's surface almost continuously since at least 3.8 billion years ago, so the Earth couldn't have been so cold. This apparent conflict between calculation and observation is called the "faint young Sun paradox".

Most researchers agree that the paradox can be explained by remembering that earlier in Earth history, before the widespread appearance of photosynthetic life, the atmosphere contained more carbon dioxide than it does today. The greenhouse effect caused by this excess CO_2 could have increased the temperature of the Earth sufficiently to counter the effect of the faint young Sun, so that surface temperatures remained above freezing.

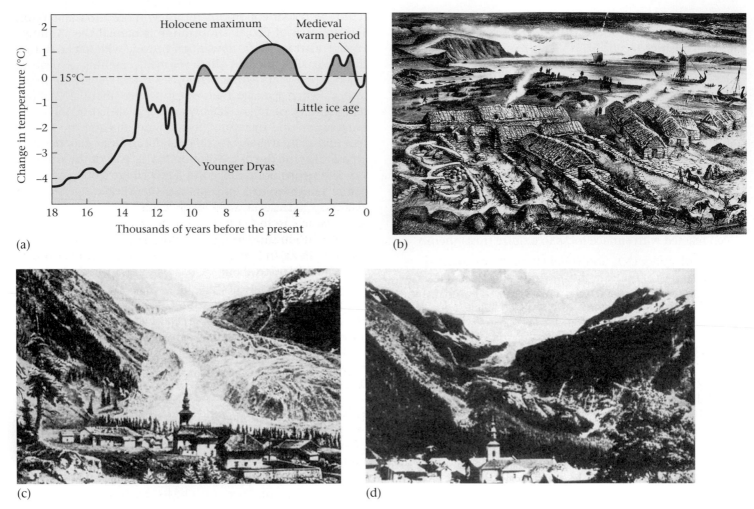

FIGURE 23.12 (a) The past 15,000 years (the Holocene Epoch) experienced several periods of warming and cooling. (b) Coastal Greenland was settled by the Vikings during one of the warmer periods, when the region could support agriculture. (c) Glaciers that formed during the little ice age persisted into the nineteenth century. Here we see a glacier in the French Alps as it appeared in 1850. (d) By the second half of the twentieth century (1966), the glacier had almost disappeared.

striking the atmosphere produce clusters of ions that serve as condensation nuclei around which water molecules congregate, thus forming the mist droplets making up clouds. But, how cloud formation changes climate remains uncertain. High-elevation clouds could reflect incoming solar radiaton and would cool the planet, whereas low-elevation clouds could absorb infrared rays rising from the Earth's surface and would warm the planet.

- *Changes in Earth's orbit and tilt:* As Milankovitch first recognized in 1920, the change in the tilt of Earth's axis over a period of 41,000 years, the Earth's 23,000-year precession cycle, and changes in the eccentricity of its orbit over a period of 100,000 years together cause the amount of summer heat in high latitudes to vary, and cause the overall amount of heat reaching Earth to vary

(see Chapter 22). These changes correlate with observed ups and downs in atmospheric and oceanic temperature.

- *Changes in volcanic emissions:* Not all of the sunlight that reaches the Earth penetrates its atmosphere and warms the ground. Some is reflected by the atmosphere. The degree of reflectivity, or **albedo,** of the atmosphere increases not only if cloud cover increases, as we have seen, but also if the concentration of volcanic aerosols in the atmosphere increases.

The short-term effect of volcanism on global temperature is abundantly clear. For example, the year following the 1815 eruption of Mt. Tambora in the western Pacific became known as the "year without a summer," for aerosols that erupted encircled the Earth and blocked the Sun—snow fell in Europe throughout the spring, and the entire summer was cold.

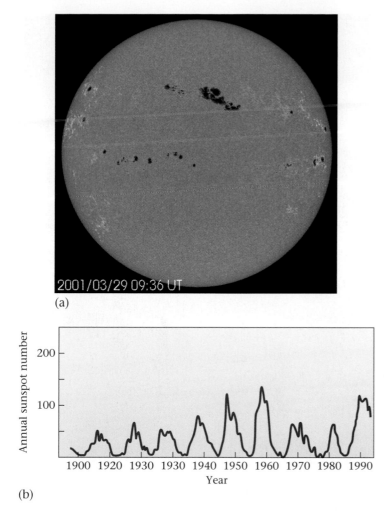

(a)

(b)

FIGURE 23.13 (a) The appearance of sunspots correlates with increased energy production on the Sun. (b) There may be cycles in sunspot activity that could influence the climate.

- *Changes in ocean currents:* Recent studies suggest that the configuration of currents can change quite quickly, and that this configuration affects the climate. The Younger Dryas may have resulted when a layer of freshwater from melting glaciers spread out over the North Atlantic and prevented thermohaline circulation in the ocean, thereby shutting off the Gulf Stream (see Chapter 18).

- *Changes in surface albedo:* Regional-scale changes in the nature of continental vegetation cover, and/or in the proportion of snow and ice on our planet's surface, would affect our planet's albedo. Increasing albedo causes cooling, whereas decreasing albedo causes warming.

- *Abrupt changes in concentrations of greenhouse gases:* A sudden change in greenhouse gas concentration in the atmosphere could affect climate. One such change might happen if sea temperature warmed or sea level dropped, causing some of the methane hydrate that crystallized in

sediment of the sea floor to suddenly melt—such melting would release CH_4 to the atmosphere. Algal blooms and reforestation (or deforestation) conceivably could change CO_2 concentrations.

Catastrophic Climate Change and Mass-Extinction Events

The changes that happen on Earth almost instantaneously are called *catastrophic changes.* For example, a volcanic eruption, an earthquake, a tsunami, or a landslide can change a local landscape in seconds or minutes. But such events affect only relatively small areas. Can such catastrophes happen on a global scale? In the past decades, geoscientists have come to the conclusion that the answer is yes. The stratigraphic record shows that Earth history includes several **mass-extinction events,** when large numbers of species abruptly vanished (▶Fig. 23.14a). Some of these define boundaries between geologic periods. A mass-extinction event decreases the **biodiversity** (the number of different species that exist at a given time) of life on Earth (▶Fig. 23.14b). It takes millions of years after a mass-extinction event for biodiversity to increase again, and the new species that appear differ from the ones that vanished, for evolution is unidirectional.

Geologists speculate that some mass-extinction events reflect a catastrophic change in the planet's climate, brought about by unusually voluminous volcanic eruptions or by the impact of a comet or asteroid with the Earth (▶Fig. 23.14c). Either of these events could eject enough debris into the atmosphere to block sunlight. Without the warmth of the Sun, winter-like or night-like conditions would last for weeks to years, long enough to disrupt the food chain. Either event, in addition, could eject aerosols that would turn into global acid rain, scatter hot debris that would ignite forest fires, or give off chemicals that, when dissolved in the ocean, would make the ocean either toxic or so nutritious that oxygen-consuming algae could thrive.

Let's examine possible causes for two of the more profound mass-extinction events in Earth history. The first event marks the boundary between the Permian and Triassic periods (i.e., between the Paleozoic and Mesozoic eras). During the Permian-Triassic extinction event, some 90% of the species on Earth became extinct—in fact, this boundary was first defined, in the nineteenth century, precisely because the assemblage of fossils from rocks below the boundary differs so markedly from the assemblage in rocks above. Radiometric dating suggests that the extinction event roughly coincided with the eruption of vast quantities of basalt in Siberia—this basalt covered over 2.5 million km^2 of continental crust. So much basalt erupted that geologists attribute their source to a superplume, a mantle plume many times larger than the one currently beneath Hawaii. Because of the correlation between the time of the basalt eruptions and the time of the mass extinction, geologists suggest that the former caused the latter.

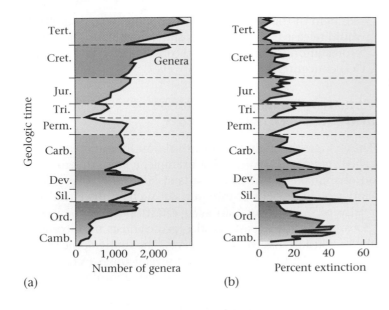

(a)

(b)

But recently, researchers have found evidence suggesting that a large asteroid collided with the Earth (perhaps at a site that now lies off the northern coast of Australia) at the Permian-Triassic boundary. Thus, the mass extinction may have resulted, instead, from this collision.

The second event, called the "K-T boundary event", caused the mass extinction that marks the boundary between the Cretaceous and Tertiary Periods (i.e., the boundary between the Mesozoic and Cenozoic Eras). As discussed in Chapter 13, the timing of this event correlates very well with the time at which an asteroid collided with the Earth at a site now called the Chicxulub crater in Yucatán, Mexico. Thus, most geologists suggest that the mass extinction is the aftermath of this collision. But it is interesting to note that the extinction is comparable in age to the eruption of extensive basalt flows in India, so there remains a possibility that volcanic activity played a role. Some geologists are even speculating that the eruptions are related to the impact of the asteroid. This debate remains active.

23.6 ANTHROPOGENIC CHANGES IN THE EARTH SYSTEM

When Stone Age artists decorated the caves of Lascaux, in France (▶Fig. 23.15a), the human population worldwide had not yet topped 10 million. By the dawn of civilization, 4000 B.C.E., it was still, at most, a few tens of millions. But by the beginning of the nineteenth century, revolutions in industrial methods, agriculture, medicine, and hygiene had substantially lowered death rates and raised living standards, so that the population began to grow at accelerating rates—it took tens of thousands of years to grow from Stone Age populations to

(c)

FIGURE 23.14 (a) During a mass-extinction event, biodiversity on Earth, as indicated by the number of fossil species (grouped together in genera), suddenly decreases. (b) Paleontologists can calculate the percentage of species that went extinct during a given event. Notice that most existing species went extinct at the boundary between the Permian and Triassic periods. (c) The K-T (Cretaceous-Tertiary) boundary event may have been caused by the impact of a large comet or asteroid. Here, we see an artist's rendition of the impact and its aftermath.

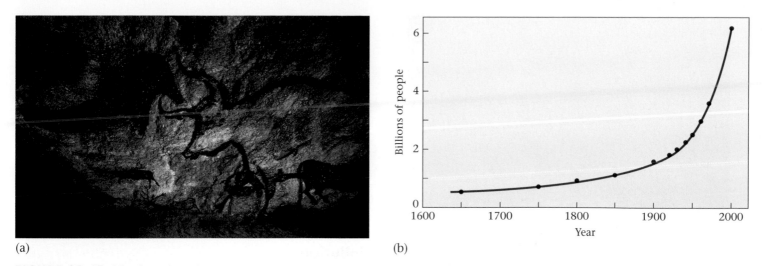

(a)

(b)

FIGURE 23.15 (a) When these Stone Age paintings were drawn on cave walls of Lascaux in France, the world's human population was less than 10 million. (b) The population has increased dramatically during the past two centuries. Currently, it doubles every forty-four years.

1 billion people worldwide in 1850, but it took only eighty years to double again, reaching 2 billion in 1930. The growth rate increased during the twentieth century, with the population reaching 4 billion in 1975. Now, the doubling time is only forty-four years, and the population passed the 6 billion mark just before the year 2000 (▶Fig. 23.15b).

As the population grows and the standard of living improves, *per capita* usage of resources increases; we use land for agriculture and grazing, forests for wood, rock and dirt for construction, oil and coal for energy or plastics, and ores for metals. Without a doubt, our usage of resources has affected the Earth System profoundly, and thus humanity has become a major agent of global change. Here, we examine some of these anthropogenic impacts.

The Modification of Landscapes

Every time we pick up a shovel and move a pile of dirt, we redistribute a portion of the Earth's crust, an activity that prior to humanity was only accomplished by rivers, the wind, rodents, and worms. In the last century, the pace of Earth movement has accelerated, for now we have shovels in coal mines that can move 300 cubic meters of coal in a single scoop, trucks that can carry 200 tons of ore in a single load, and tankers that can transport 50 million liters of oil during a single journey. In North America, human activity now moves more sediment each year than rivers do. The extraction of rock during mining, the building of levees and dams along rivers or of sea walls along the coast, and the construction of highways and cities all involve the redistribution of Earth materials (▶Fig. 23.16a). In addition, people clear and plow fields, drain and fill wetlands, and pave over the land surface (▶Fig. 23.16b; see also chapter-opening photo). All these activities change the landscape.

Landscape modification has side effects. For example, it may make the ground unstable and susceptible to landslides. And it may expose the land to erosion, thereby changing the volume of sediment transported by natural agents (such as running water and wind). Locally, flood-control projects may diminish the sediment supply downstream, also with unfortunate consequences; for example, the damming of the Nile by the Aswan High Dam has cut off the sediment supply to the Nile Delta, so ocean waves along the Mediterranean coast of the delta have begun to erode the coastline by more than 1 m per year.

The Modification of Ecosystems

In undisturbed areas, the **ecosystem** of a region (an interconnected network of organisms and the physical environment in which they live) is the product of evolution for an extended period of time. The ecosystem's flora (plant life) includes species that have adapted to living together in that particular climate and on the substrate available, while its fauna (animal life) can survive local climate conditions and utilize local food supplies. Human-caused deforestation, overgrazing, agriculture, and urbanization disrupt ecosystems and lead to a decrease in biodiversity.

Archaeological studies have found that the earliest example of human modification of an ecosystem occurred in the Stone Age, when hunters played a major role in causing the mass extinction of many species of large mammals (mammoths, giant sloths, giant bears). Today, less than 5% of Europe retains its original habitats. The same number can be applied to the eastern United States, which lost its original forest and prairie. Tropical rain forests cover less than about half the area worldwide that they covered before the dawn of civilization, and they are disappearing at a rate of

(a)

(b)

FIGURE 23.16 (a) A giant, power-driven shovel can move more dirt and rock in a day than a stream can move in a decade. (b) Agriculture and development radically change the landscape of a region. Prairies of the midwestern United States have been replaced by giant farm fields.

about 1.8% per year (►Fig. 23.17a, b). Much of this loss comes from slash-and-burn agriculture during which farmers and ranchers destroy forest to make open land for farming and grazing (►Fig. 23.17c). Unfortunately, the heavy rainfall of tropical regions removes nutrients from the soil, making the soil useless in just a few years. Overgrazing by

domesticated animals can remove vegetation so completely that some grasslands have undergone desertification. And urbanization replaces the natural land surface with concrete or asphalt, a process that completely destroys an ecosystem and radically changes the amount of rain that infiltrates the land surface to become groundwater.

Human-caused changes to ecosystems affect the broader Earth System, because they modify biogeochemical cycles and Earth's albedo (surface reflectivity). For example, deforestation increases the CO_2 concentration in the atmosphere, for much of the carbon that was stored in trees is burned and rises. And the replacement of forest cover with concrete or fields increases Earth's albedo.

Pollution

The environment has always contained contaminants such as soot, dust, and the byproducts of organisms. But when human populations grew, urbanization, industrial and agricultural activity, the production of electricity, and modern modes of transportation greatly increased both the quantity and diversity of contaminants that entered the air, surface water, and groundwater. These contaminants, or **pollution,** include both natural and synthetic materials (in liquid, solid, or gaseous form), and have become a problem because they are produced at a higher rate than can be naturally absorbed or modified by the Earth System. For example, while small quantities of sewage can be absorbed by clay minerals in the soil or destroyed by bacterial metabolism, large quantities overwhelm natural controls and can accumulate into destructive concentrations. Further, because many contaminants are not produced in nature, they are not easily removed by natural processes. Pollution of the Earth System is a type of global change, because it represents a redistribution and reformulation of materials. Some key problems associated with this change include the following.

- *Smog:* The term was originally coined to refer to the dank, dark air that resulted when smoke from the burning of coal mixed with fog in London and other industrial cities. Another kind of smog, called **photochemical smog,** is the ozone-rich brown haze that blankets cities when exhaust from cars and trucks reacts with air in the presence of sunlight.

- *Water contamination:* We dump a great variety of chemicals into surface water and groundwater, including gasoline, other organic chemicals, radioactive waste, acids, fertilizers—the list could go on for pages. These chemicals affect biodiversity.

- *Acid runoff:* Dissolution of sulfide-containing minerals in ores or coal by groundwater or stream water makes the water acidic (it increases the concentration of hydrogen ions in the water) and toxic to life forms.

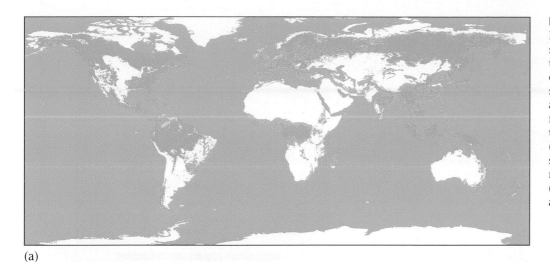

(a)

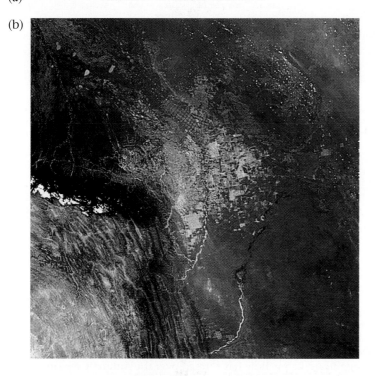

(b)

(c)

FIGURE 23.17 (a) The area of the Earth covered by rain forest shrank steadily during the past century. On this map, the dark green areas indicate existing forest (including high-latitude scrub forest), while the light green areas indicate regions that were forested 8,000 years ago. Note how tropical rain forests are shrinking. (b) A satellite image of Bolivia showing forest (dark green) being replaced by fields (light yellow). (c) Much of the loss is due to slash-and-burn agriculture.

• *Acid rain:* When rain passes through air that contains sulfur-containing aerosols (emitted from power plants), the water dissolves the sulfur, creating sulfuric acid, or **acid rain.** Wind can carry aerosols far from a power plant, so acid rain can damage a broad region (▶Fig. 23.18).

• *Radioactive materials:* Nuclear weapons, nuclear energy, and medical waste transfer radioactive materials from rock to Earth's surface environment. Also, human-caused nuclear reactions produce new, nonnatural radioactive isotopes, some of which have relatively short half-lives. Thus, society has changed the distribution and composition of radioactive material worldwide.

• *Ozone depletion:* Human-produced chemicals, most notably chlorofluorocarbons (CFCs), when emitted into the atmosphere react with ozone in the stratosphere. This reaction, which happens most rapidly on the surfaces of tiny ice crystals in polar stratopheric clouds, destroys ozone molecules, thus creating an **ozone hole** over high-latitude regions, particularly during the spring (▶Fig. 23.19a, b). Note that the "hole" is not really an area where no ozone is present, but rather is a region where atmospheric ozone has been reduced substantially. The ozone hole is more prominent in the Antarctic than in the Arctic, because a current of air circulates around the land mass of Antarctica and traps the air above the continent, with its CFCs, from mixing with air from elsewhere.

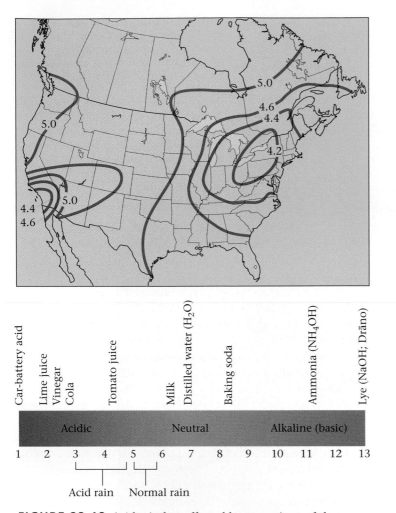

FIGURE 23.18 Acid rain has affected large portions of the United States. The map contours pH numbers, indicating the concentration of hydrogen ions in a solution (according to the formula pH = −log[H⁺]; the square brackets mean "concentration"). Note that very acidic rain falls in the Northeast and Southwest. The scale gives a sense of what the numbers mean. A solution with a pH of 7 is neutral; acidic solutions have a pH less than 7, while alkaline solutions have a pH greater than 7.

Ozone holes have dangerous consequences, for they affect the ability of the atmosphere to shield the Earth's surface from harmful ultraviolet radiation. In 1987, a summit conference in Montreal proposed a global reduction of CFC emissions that destroy the ozone layer. Reduction of such emissions may substantially reduce ozone depletion.

Effects on the Climate: The Global-Warming Issue

During the past two centuries, industry, energy production, and agriculture significantly altered the rate at which greenhouse gases, such as carbon dioxide (CO_2) and methane

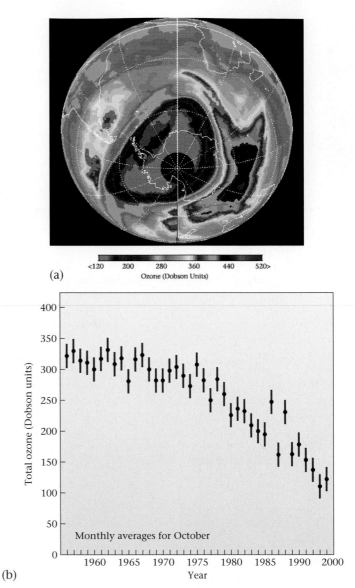

FIGURE 23.19 (a) The colors on this map show the quantity of ozone in the atmosphere when ozone levels are at a minimum. A significant ozone hole has formed over the Antarctic region, where the amount of ozone has decreased by as much as 50%. The shape and size of the hole vary over time; in 2002, for example, the hole divided into two pieces. (b) The graph shows the decrease in ozone above Antarctica in October, specified in Dobson units (named for a scientist who helped identify the ozone hole). One Dobson unit is the amount of ozone needed to make a 0.01-mm-thick layer at the surface of the Earth.

(CH_4), have been added to the atmosphere. In fact, the rate of addition has exceeded the rate at which these gases can be absorbed by dissolution in the ocean, by incorporation in plants, or by chemical-weathering reactions. Effectively, by burning fossil fuels at the rate that we do, we transfer CO_2 from underground reservoirs (oil and coal deposits)

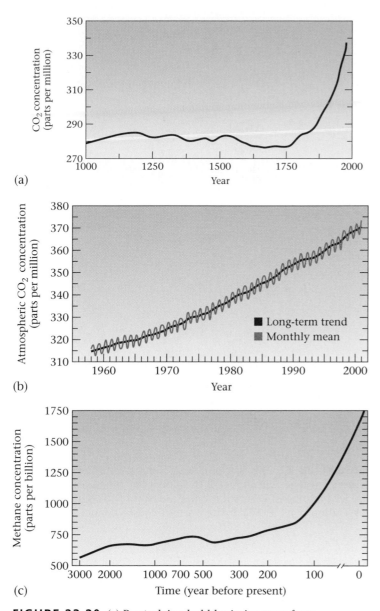

(a)

(b)

(c)

FIGURE 23.20 (a) By studying bubbles in ice cores from Antarctica, the record of CO_2 in the atmosphere can be traced back in time. Note that there has been a significant increase since the start of the Industrial Revolution. (b) Direct measurements of the concentration of CO_2 in the atmosphere, taken since 1958 at Mauna Loa Observatory, Hawaii, show that the concentration changes with the seasons and that the average concentration has increased steadily during the past several decades. (c) The concentration of methane (CH_4) has increased steadily during the past few centuries.

back into the atmosphere. In 1800, the mean concentration of CO_2 in the atmosphere was 280 parts per million (ppm), in 1900 it was 295 ppm, and by 2000 it had reached 370 ppm (▶Fig. 23.20a, b). At the same time, the decay of organic material in rice paddies and the flatulence of cows have released enough methane ito measurably change the

concentration of this organic chemical in the atmosphere (▶Fig. 23.20c).

We might expect this increase in greenhouse gases to cause global warming, and indeed global mean temperature appears to have risen by about 0.9° between 1880 and 2000 (▶Fig. 23.21a, b). This may not seem like much, but by comparison, the magnitude of change between the last ice age and now was only a few degrees. Similarly, mean ocean temperature has been rising: in 2000, temperatures at the North Pole were the warmest in four centuries, and a 15-km-by-5-km patch of open water appeared at the pole. Though the majority of researchers accept global warming as fact, not all do. Some argue that it would have happened anyway, in the context of natural cycles of short-term climate change. But of note, a recent study of ice-core data characterizing Earth's climate during the past few hundred thousand years suggests that the Earth's climate started on a cooling trend about 10,000 years ago and that, if only solar and volcanic phenomena controlled climate, we should now be heading into another ice age. Instead, according to this study, temperatures began to deviate from the cooling trend about 8,000 years ago, and we have followed an overall warming trend ever since. The timing of this warming trend suggests it results from human activity that produce greenhouse gases. Specifically, deforestation, rice-paddy farming, and the burning of fossil fuels all began within the last 8,000 years. Brief cooling episodes during the past 8,000 years (e.g., the little ice age) may correspond to times when human population decreased abruptly, due to pandemics, so that activities producing greenhouse gases decreased.

There is even more disagreement about whether global warming will continue in the future and what its consequences might be, because predictions are dependent on computer models, and not all researchers agree on how to represent the factors that affect climate in these models. In the worst-case scenario, global warming will continue into the future at the present rate, so that by 2050—within the lifetime of many readers of this book—the average annual temperature will have increased in some parts of the world by 1.5° to 2.0°C (▶Fig. 23.21c). At these rates, by 2150 global temperatures may be 5° to 11° warmer than present—the warmest since the Eocene Epoch, 40 million years ago. The effects of such a change are controversial, but according to some climate models, the following events might happen.

• *A shift in climate belts:* Temperate climates would move to higher latitudes, and vegetation belts would follow this trend. As a result, desert regions would expand, and the soil would dry out in present agricultural regions (▶Fig. 23.22a, b).

• *Stronger storms:* An increase in average ocean temperatures would mean that more of the ocean could evaporate when a tropical depression passed over. This evaporation would nourish stronger hurricanes. In nondesert areas,

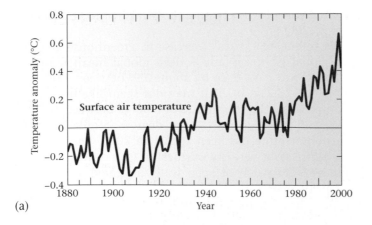

(a)

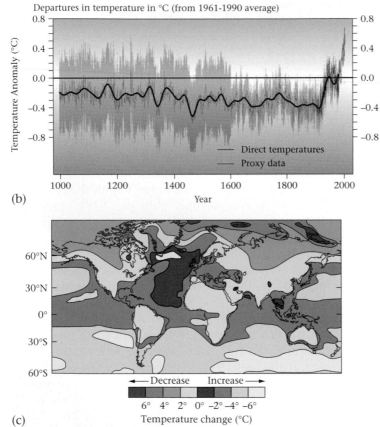

(b)

(c)

FIGURE 23.21 (a) The average global surface atmospheric temperature varies year by year, but overall, there has been a noticeable increase since about 1920. Here, the difference in global temperature, relative to an arbitrary reference value, is plotted as a function of time. (b) A reconstruction of Northern Hemisphere temperature for the past 1,000 years, based on measurements of tree rings and ice cores. The 0° reference line is the 1920–1980 mean. Note that climate seemed to be cooling slowly between 1000 and 1900, but since 1900 it has increased substantially. (c) This map shows the results of a complicated computer model predicting the annual mean temperature change (°C) between 1795 and 2050 if the concentration of greenhouse gases continues to increase. Notice that some regions of the world will cool, but others will warm. The warming effect is most pronounced in the Northern Hemisphere.

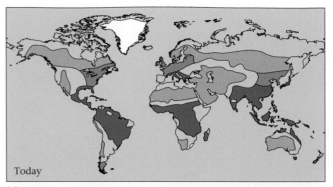

(a)

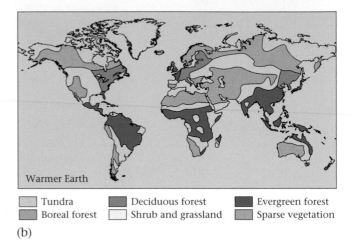

(b)

| Tundra | Deciduous forest | Evergreen forest |
| Boreal forest | Shrub and grassland | Sparse vegetation |

(b)

FIGURE 23.22 The distribution of climate belts, indicated by vegetation type, will change if the global temperature rises. (a) Distribution of vegetation types today. (b) Distribution of vegetation types if the climate warms by just a couple of degrees.

there might be more precipitation and, therefore, flooding.

- *A rise in sea level:* The melting of ice sheets in polar regions and the expansion of water in the sea as the water warms would make sea level rise enough to flood coastal wetlands and communities and damage deltas. Measurements suggest that there has been a rise of almost 12 cm in the past century (▶Fig. 23.23a). Sea-level rise is already taking its toll on islands in the South Pacific, where some islands have already become submerged. Even a change of only 1 m, which some models say may happen in little more than a century, could inundate regions of the world where 20% of the population currently lives (▶Fig. 23.23b).

- *An interruption of the oceanic heat conveyor:* Oceanic currents play a major role in transferring heat across latitudes. If global warming melts polar ice, the resulting freshwater would dilute surface ocean water at high latitudes. This water could not sink, and thus

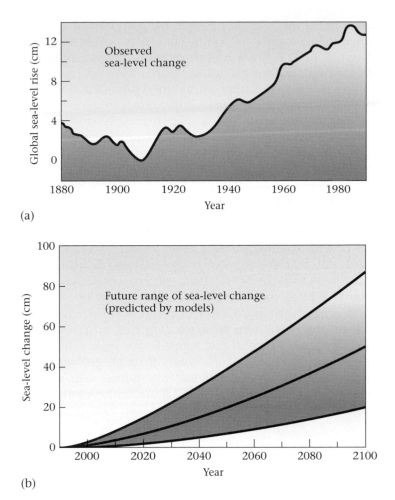

(a)

(b)

FIGURE 23.23 (a) Estimates of the average global sea level between 1880 and 1990 suggest that there has been a rise of about 12 cm. (b) All models of future sea-level change predict that there will be a rise. In the worst-case model, sea level could rise by almost 1 m in the next century.

thermohaline circulation would be shut off (see Chapter 18), preventing it from conveying heat.

Models suggesting that unchecked addition of greenhouse gases will cause global warming imply that governments must work to require modifications in industrial procedures or life-styles. Because of the importance of this issue, the United Nations and the World Meteorological Association have established the Intergovernmental Panel on Climate Change (IPCC), which issues a report every five years. There have also been meetings designed to address issues of future climate change. Global warming was discussed at the 1992 Earth Summit held in Rio de Janeiro, and again at a 1997 summit in Kyoto, Japan, where 160 nations signed a treaty intended to slow the input of CO_2 into the atmosphere by reducing emissions to levels below those of 1990.

23.7 THE FUTURE OF THE EARTH: A SCENARIO

Most of the discussion in this book has focused on the past, for the geologic record preserved in rocks tells us of earlier times. Let's now bring this book to a close by facing in the opposite direction and speculating what the world might look like in geologic time to come.

In the geologic near term, the future of the world depends largely on human activities. Whether the Earth System undergoes a major disruption and shifts to a new equilibrium, whether a catastrophic mass-extinction event takes place, or whether society achieves **sustainable growth** (meaning an ability to prosper within the constraints of the Earth System) will depend on our own foresight and ingenuity. Projecting thousands of years into the future, we might well wonder if the Earth will return to ice-age conditions, with glaciers growing over major cities and the continental shelf becoming dry land, or if the ice age is over for good because of global warming. No one really knows for sure.

If we project millions of years into the future, it is clear the map of the planet will change significantly because of the continuing activity of plate tectonics. For example, during the next 50 million years or so, the Atlantic Ocean will probably become bigger, the Pacific Ocean will shrink, and the western part of California will migrate northward. Eventually, Australia will crush against the southern margin of Asia, and the islands of Indonesia will be flattened in between.

Predicting the map of the Earth beyond that is hard, because we don't know for sure where new subduction zones will develop. Most likely, subduction of the Pacific Ocean will lead to the collision of the Americas with Asia, to produce a supercontinent ("Amasia"). A subduction zone eventually will form on one side of the Atlantic Ocean, and the ocean will be consumed. As a consequence, the eastern margin of the Americas will collide with the western margin of Europe and Africa. The sites of major cities—New York, Miami, Rio de Janeiro, Buenos Aires, London—will be incorporated in a collisional mountain belt, and likely will be subjected to metamorphism and igneous intrusion before being uplifted and eroded. Shallow seas may once again cover the interiors of continents and then later retreat, and glaciers may once again cover the continents—it happened in the past, so it could happen again. And if the past is the key to the future, we *Homo sapiens* might not be around to watch our cities enter the rock cycle, for biological evolution may have introduced new species to the biosphere, and there is no way to predict what these species will be like.

And what of the end of the Earth? Geologic catastrophes resulting from asteroid and comet collisions will undoubtedly

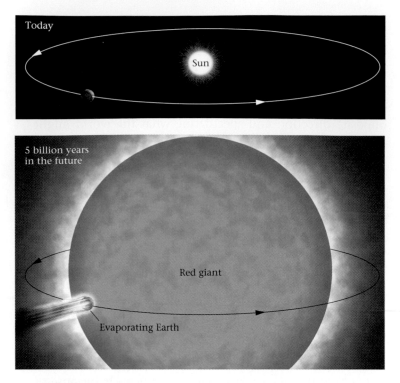

FIGURE 23.24 In about 5 billion years, the Sun will grow to become a red giant. When that happens, the Earth will first dry out, then vaporize. Initially, it may look like a giant comet. Perhaps some of its atoms will become part of a nebula that someday condenses to form a new Sun and planetary system.

occur in the future as they have in the past. We can't predict when the next strike will come, but unless the object can be diverted, Earth is in for another radical readjustment of surface conditions. But it's not likely that such collisions will destroy our planet. Rather, astronomers predict that the end of the Earth will occur some 5 billion years from now, when the Sun begins to run out of nuclear fuel. When this happens, thermal pressure caused by fusion reactions will no longer be able to prevent the Sun from collapsing inward, because of the immense gravitational pull of its mass. Were the Sun a few times larger than it is, the collapse would trigger a supernova explosion that would blast matter of the Universe out into space to form a new nebula, perhaps surrounding a black hole. But since the Sun is not that large, the thermal energy generated when its interior collapses inward will heat the gases of its outer layers sufficiently to cause them to expand. As a result, the Sun will become a **red giant,** a huge star whose radius would grow beyond the orbit of Earth (▶Fig. 23.24). Our planet will then vaporize, and its atoms will join an expanding ring of gas—the ultimate global change. If this happens, the atoms that once made Earth and all its inhabitants through geologic time may eventually be incorporated in a future solar system, where the cycle of planetary formation and evolution will begin anew.

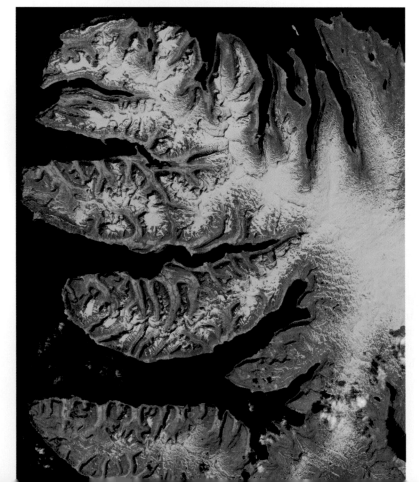

THE VIEW FROM SPACE Numerous fjords make western Iceland's coast resemble the shape of a leaf. The glaciers that carved these fjords vanished over 10,000 years ago. Their growth and demise is one manifestation of climate change on Earth.

CHAPTER SUMMARY

• We refer to the global interconnecting web of physical and biological phenomena on Earth as the Earth System. Global change involves the transformations or modifications of physical and biological components of the Earth System through time. Unidirectional change results in transformations that never repeat, while cyclic change involves repetition of the same steps over and over.

• Examples of unidirectional change include the gradual evolution of the solid Earth from a homogeneous collection of planetesimals to a layered planet; the formation of the oceans and the gradual change in the composition of the atmosphere; and the evolution of life.

• Examples of physical cycles that take place on Earth include the supercontinent cycle, the sea-level cycle, and the rock cycle.

• A biogeochemical cycle involves the passage of a chemical among nonliving and living reservoirs. Examples include the hydrologic cycle and the carbon cycle. Global change occurs when factors change the relative proportions of the chemical in different reservoirs.

• Tools for documenting global climate change include the stratigraphic record, paleontology, oxygen-isotope ratios, bubbles in ice, growth rings, and human history.

• Studies of long-term climate change show that at times in the past the Earth experienced greenhouse (warmer) periods, while at other times there were ice-house (cooler) periods. Factors leading to long-term climate change include the positions of continents, volcanic activity, the uplift of land, and the formation of materials that remove CO_2, an important greenhouse gas.

• Short-term climate change can be seen in the record of the last million years. In fact, during only the past 15,000 years, we see that the climate has warmed and cooled a few times. Causes of short-term climate change include fluctuations in solar radiation and cosmic rays, changes in Earth's orbit and tilt, changes in reflectivity, and changes in ocean currents.

• Mass extinction, a catastrophic change in biodiversity, may be caused by the impact of a comet or asteroid or by intense volcanic activity associated with a superplume.

• During the last two centuries, humans have changed landscapes, modified ecosystems, and added pollutants to the land, air, and water at rates faster than the Earth System can process.

• The addition of CO_2 and CH_4 to the atmosphere may be causing global warming, which could shift climate belts and lead to a rise in sea level.

• In the future, in addition to climate change, the Earth will witness a continued rearrangement of continents resulting from plate tectonics, and will likely suffer the impact of asteroids and comets. The end of the Earth may come when the Sun runs out of fuel in about 5 billion years and becomes a red giant.

KEY TERMS

acid rain (p. 741)
albedo (p. 736)
biodiversity (p. 737)
biogeochemical cycle
 (pp. 722, 726)
climate-change models (p. 730)
Earth System (p. 722)
ecosystem (p. 739)
eustatic sea-level change (p. 726)
global change (p. 722)
global climate change
 (p. 722)
global cooling (p. 730)
global warming (p. 730)
Goldilocks effect (p. 723)
greenhouse (hot-house) period
 (p. 733)

ice-house period (p. 733)
mass-extinction events (p. 737)
oxygen-isotope ratio (p. 730)
ozone hole (p. 741)
paleoclimate (p. 730)
photochemical smog (p. 740)
pollution (p. 740)
red giant (p. 746)
runaway greenhouse effect
 (p. 735)
sedimentary cycle chart (p. 726)
sedimentary sequence (p. 725)
steady-state condition (p. 727)
sunspot cycle (p. 735)
supercontinent cycle (p. 724)
sustainable growth (p. 745)

REVIEW QUESTIONS

1. Why do we use the term "Earth System" to describe the processes operating on this planet?

2. How have the Earth's crust and atmosphere changed since they first formed?

3. What processes control the rise and fall of sea level on Earth?

4. How does carbon cycle through the various Earth systems?

5. How do paleoclimatologists study ancient climate change?

6. Contrast ice-house and greenhouse conditions.

7. What are the possible causes of long-term climatic change?

8. What factors explain short-term climatic change?

9. Give some examples of events that cause catastrophic change.

10. Give some examples of how humans have changed the Earth.

11. What is the ozone hole, and how does it affect us?

12. Describe how carbon dioxide–induced global warming takes place, and how humans may be responsible. What effects might global warming have on the Earth System?

13. What are some of the likely scenarios for the long-term future of the Earth?

SUGGESTED READING

Alvarez, W. 1997. *T. Rex and the Crater of Doom.* Princeton, N.J.: Princeton University Press.

Burroughs, W. J. 2001. *Climate Change: A Multidisciplinary Approach.* Cambridge, England: Cambridge University Press.

Collier, M., and R. H. Webb. 2002. *Floods, Droughts, and Climate Change.* Tucson: University of Arizona Press.

Harvey, D. 1999. *Global Warming: The Hard Science.* Upper Saddle River, N.J.: Prentice-Hall.

Holland, H. D., and U. Petersen. 1995. *Living Dangerously: The Earth, Its Resources, and the Environment.* Princeton, N.J.: Princeton University Press.

Houghton, J. T., et al., eds. 2001. *Climate Change 2001: The Scientific Basis.* A report of the Intergovernmental Panel on Climate Change. Cambridge, England: Cambridge University Press.

Kump, L. R., J. F. Kasting, and R. G. Crane. 1999. *The Earth System.* Upper Saddle River, N.J.: Prentice-Hall.

Leggett, J. K. 2001. *The Carbon War: Global Warming and the End of the Oil Era.* New York: Routledge.

Mackay, A., et al., eds. 2003. *Global Change in the Holocene.* London: Edward Arnold.

MacKenzie, F. T. 1998. *Our Changing Planet: An Introduction to Earth System Science and Global Environmental Change.* Upper Saddle River, N.J.: Prentice-Hall.

Officer, C., and J. Page. 1993. *Tales of the Earth: Paroxysms and Perturbations of the Blue Planet.* New York: Oxford University Press.

Ruddiman, W. F. 2003. The anthropogenic greenhouse era began thousands of years ago: *Climate Change,* v. 61, p. 261–293.

Speth, J. G. 2004. *Red Sky at Morning: America and the Crisis of the Global Environment.* New Haven: Yale University Press.

Turco, R. P. 1997. *Earth under Siege: From Air Pollution to Global Change.* Oxford, England: Oxford University Press.

Van Andel, T. H. 1994. *New Views on an Old Planet: A History of Global Change.* Cambridge, England: Cambridge University Press.

Weart, S. R. 2003. *The Discovery of Global Warming.* Cambridge, Mass.: Harvard University Press.

Metric Conversion Chart

Length

1 kilometer (km) = 0.6214 mile (mi)
1 meter (m) = 1.094 yards = 3.281 feet
1 centimeter (cm) = 0.3937 inch
1 millimeter (mm) = 0.0394 inch
1 mile (mi) = 1.609 kilometers (km)
1 yard = 0.9144 meter (m)
1 foot = 0.3048 meter (m)
1 inch = 2.54 centimeters (cm)

Area

1 square kilometer (km^2) = 0.386 square mile (mi^2)
1 square meter (m^2) = 1.196 square yards (yd^2)
= 10.764 square feet (ft^2)
1 square centimeter (cm^2) = 0.155 square inch (in^2)
1 square mile (mi^2) = 2.59 square kilometers (km^2)
1 square yard (yd^2) = 0.836 square meter (m^2)
1 square foot (ft^2) = 0.0929 square meter (m^2)
1 square inch (in^2) = 6.4516 square centimeters (cm^2)

Volume

1 cubic kilometer (km^3) = 0.24 cubic mile (mi^3)
1 cubic meter (m^3) = 264.2 gallons
= 35.314 cubic feet (ft^3)
1 liter (1) = 1.057 quarts
= 33.815 fluid ounces
1 cubic centimeter (cm^3) = 0.0610 cubic inch (in^3)
1 cubic mile (mi^3) = 4.168 cubic kilometers (km^3)
1 cubic yard (yd^3) = 0.7646 cubic meter (m^3)
1 cubic foot (ft^3) = 0.0283 cubic meter (m^3)
1 cubic inch (in^3) = 16.39 cubic centimeters (cm^3)

Mass

1 metric ton = 2,205 pounds
1 kilogram (kg) = 2.205 pounds
1 gram (g) = 0.03527 ounce
1 pound (lb) = 0.4536 kilogram (kg)
1 ounce (oz) = 28.35 grams (g)

Pressure

1 kilogram per square
centimeter (kg/cm^2)* = 0.96784 atmosphere (atm)
= 0.98066 bar
= 9.8067×10^4 pascals (Pa)
1 bar = 10 megapascals (Mpa)
= 1.0×10^5 pascals (Pa)
= 29.53 inches of mercury (in a barometer)
= 0.98692 atmosphere (atm)
= 1.02 kilograms per square centimeter (kg/cm^2)
1 pascal (Pa) = 1 kg/m/s^2
1 pound per square inch = 0.06895 bars
= 6.895×10^3 pascals (Pa)
= 0.0703 kilogram per square centimeter

Temperature

To change from Fahrenheit (F) to Celsius (C):
$$°C = \frac{(°F - 32°)}{1.8}$$

To change from Celsius (C) to Fahrenheit (F):
$$°F = (°C \times 1.8) + 32°$$

To change from Celsius (C) to Kelvin (K):
$$K = °C + 273.15$$

To change from Fahrenheit (F) to Kelvin (K):
$$K = \frac{(°F - 32°)}{1.8} + 273.15$$

---

*Note: Because kilograms are a measure of mass whereas pounds are a unit of weight, pressure units incorporating kilograms assume a given gravitational constant (g) for Earth. In reality, the gravitational for Earth varies slightly with location.

Scientific Background: Matter and Energy

a.I INTRODUCTION

In order to understand the formation and evolution of the Universe, as well as descriptions of materials that constitute the Earth and processes that shape the Earth, we must first understand some basic facts about matter, energy, and heat (▶Fig. a.1). The following synopsis highlights key topics from physics and chemistry that serve as an essential background to the rest of the book.

a.1 DISCOVERING THE NATURE OF MATTER

Matter takes up space—you can feel it and see it. We use the word **matter** to refer to any material making up the universe. The amount of matter in an object is its **mass.** An object with a greater mass contains more matter—for example, a large tree contains more matter than a blade of grass. There's a subtle but important distinction between mass and weight. **Weight** depends on the amount of an object's mass but also on the strength of gravity. An astronaut has the same mass on both the Earth and the Moon, but weighs a different amount in each place. Since the amount of gravitational pull exerted by an object depends on its mass, and the Moon has about one-sixth the mass of the Earth, an astronaut weighing 68 kilograms (150 pounds) on Earth would weigh only about 11 kilograms (25 pounds) on the Moon. Thus, lunar explorers have no trouble jumping great distances even when burdened by a space suit and oxygen tanks.

What does matter consist of? Early philosophers deduced that the Earth and the plants and animals on it must be formed from simpler components, just as bread is made from a measured mixture of primary ingredients. They initially thought that the "primary ingredients" making up matter included only earth, air, fire, and water. Then a philosopher named Democritus (ca. 460–370 B.C.E.) argued that if you were able to keep dividing matter into progressively smaller pieces, you would eventually end up with nothing, and since it doesn't seem possible to make something out of nothing, there must be a smallest piece of matter that can't be subdivided further. He proposed the name **atom** for these

FIGURE a.1 A lightning storm over mountains. The solid rock, the air, and the clouds consist of matter. The lightning, a rapidly moving current of electrons, is a manifestation of energy.

smallest pieces, based on the Greek word *atomos,* which means "indivisible."

Our modern understanding of matter didn't become established until the seventeenth century, when chemists such as Robert Boyle (1627–1691) recognized that certain substances, like hydrogen, oxygen, carbon, and sulfur, cannot break down into other substances, while others, like water and salt, *can* break down into other substances

FIGURE a.2 A compound, salt, can be subdivided to form two elements, sodium metal and chlorine gas. Neither sodium nor chlorine can be divided further.

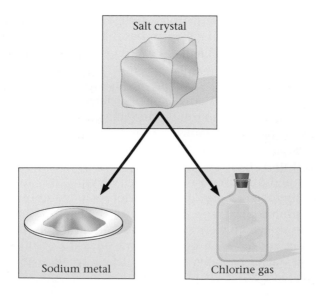

(►Fig. a.2). For example, water breaks down into oxygen and hydrogen. Substances that can't be broken down came to be known as **elements,** while those that can be broken down came to be known as **compounds.** An English schoolteacher, John Dalton (1766–1844), adopted the word "atom" for the smallest piece of an element that maintains the property of the element, and suggested that compounds consisted of combinations of different atoms. Then in 1869, a Russian chemist named Dmitri Mendeléev (1834–1907) realized that groups of elements share similar characteristics. Mendeléev organized the elements into a chart we now call the **periodic table of the elements** (►Fig. a.3). In the figure, elements within each column of the table behave similarly; for example, all elements in the right-hand column are inert gases, meaning that they can't combine with other elements to form compounds. But Mendeléev and his contemporaries didn't know what caused the similarities and differences. An understanding of the cause would have to wait until twentieth-century physicists discovered the internal structure of the atom.

a.2 A MODERN VIEW OF MATTER

Atoms

In modern terminology, an element is a substance composed only of atoms of the same kind. Ninety-two different elements occur naturally on Earth, but physicists have created more than a dozen new elements in the laboratory. Each element has a name (e.g., nitrogen, hydrogen, sulfur) and a **symbol,** an abbreviation of its English or Latin name (N = nitrogen, H = hydrogen, Fe = iron, Ag = silver).

Work in the late 1800s and early 1900s demonstrated that, contrary to the view of Democritus, atoms actually can be divided. Ernest Rutherford, a British physicist, made this key discovery in 1910 when he shot a beam of atoms at a gold foil and found, to his amazement, that only a tiny fraction of the atoms bounced back; most of the mass in the beam passed through the foil as if it were invisible. This result could mean only one thing. Most of the mass in an atom clusters in a dense ball at the atom's center, and this ball is surrounded by a cloud that contains very little mass, so an atom as a whole consists mostly of empty space.

Physicists now refer to the dense ball at the center of the atom as the **nucleus,** and the low-density cloud surrounding the nucleus as the **electron cloud.** Further study in the first half of the twentieth century led to the conclusion that the nucleus contains two types of subatomic particles: **neutrons,** which have a neutral electrical charge, and **protons,** which have a positive electrical charge (►Fig. a.4a–d). The electron cloud consists of negatively charged

Alkali metals

Symbol → He 2 ← Atomic number
Helium ← Name
4.002 ← Atomic weight

Inert gases

Nonmetals

Transition elements (metals)

H 1 Hydrogen 1.007																	He 2 Helium 4.002
Li 3 Lithium 6.941	Be 4 Beryllium 9.0121											B 5 Boron 10.811	C 6 Carbon 12.011	N 7 Nitrogen 14.006	O 8 Oxygen 15.999	F 9 Fluorine 18.998	Ne 10 Neon 20.179
Na 11 Sodium 22.989	Mg 12 Magnesium 24.305											Al 13 Aluminum 26.981	Si 14 Silicon 28.085	P 15 Phosphorus 30.973	S 16 Sulfur 32.066	Cl 17 Chlorine 35.452	Ar 18 Argon 39.948
K 19 Potassium 39.098	Ca 20 Calcium 40.078	Sc 21 Scandium 44.955	Ti 22 Titanium 47.88	V 23 Vanadium 50.941	Cr 24 Chromium 51.996	Mn 25 Manganese 54.938	Fe 26 Iron 55.847	Co 27 Cobalt 58.933	Ni 28 Nickel 58.693	Cu 29 Copper 63.546	Zn 30 Zinc 65.39	Ga 31 Gallium 69.723	Ge 32 Germanium 72.61	As 33 Arsenic 74.921	Se 34 Selenium 78.96	Br 35 Bromine 79.904	Kr 36 Krypton 83.80
Rb 37 Rubidium 85.467	Sr 38 Strontium 87.62	Y 39 Yttrium 88.905	Zr 40 Zirconium 91.224	Nb 41 Niobium 92.906	Mo 42 Molybdenum 95.94	Tc 43 Technetium 98.907	Ru 44 Ruthenium 101.07	Rh 45 Rhodium 102.905	Pd 46 Palladium 106.42	Ag 47 Silver 107.868	Cd 48 Cadmium 112.411	In 49 Indium 114.82	Sn 50 Tin 118.710	Sb 51 Antimony 121.757	Te 52 Tellurium 127.60	I 53 Iodine 126.904	Xe 54 Xenon 131.29
Cs 55 Cesium 132.905	Ba 56 Barium 137.327	La 57 Lanthanum 138.905	Hf 72 Hafnium 178.49	Ta 73 Tantalum 180.947	W 74 Tungsten 183.85	Re 75 Rhenium 186.207	Os 76 Osmium 190.2	Ir 77 Iridium 192.22	Pt 78 Platinum 195.08	Au 79 Gold 196.966	Hg 80 Mercury 200.59	Tl 81 Thallium 204.383	Pb 82 Lead 207.2	Bi 83 Bismuth 208.980	Po 84 Polonium 208.982	At 85 Astatine 209.987	Rn 86 Radon 222.017
Fr 87 Francium 223.019	Ra 88 Radium 226.025	Ac 89 Actinium 227.027															

Ce 58 Cerium 140.115	Pr 59 Praseodymium 140.907	Nd 60 Neodymium 144.24	Pm 61 Promethium 144.912	Sm 62 Samarium 150.36	Eu 63 Europium 151.965	Gd 64 Gadolinium 157.25	Tb 65 Terbium 158.925	Dy 66 Dysprosium 162.50	Ho 67 Holmium 164.930	Er 68 Erbium 167.26	Tm 69 Thulium 168.934	Yb 70 Ytterbium 173.04	Lu 71 Lutetium 174.967
Th 90 Thorium 232.038	Pa 91 Protactinium 231.035	U 92 Uranium 238.028	Np 93 Neptunium 237.048	Pu 94 Plutonium 244.064	Am 95 Americium 243.061	Cm 96 Curium 247.070	Bk 97 Berkelium 247.070	Cf 98 Californium 251.079	Es 99 Einsteinium 252.083	Fm 100 Fermium 257.095	Md 101 Mendelevium 258.10	No 102 Nobelium 259.100	Lr 103 Lawrencium 262.11

FIGURE a.3 The modern periodic table of the elements. The columns group elements with related properties. For example, inert gases are listed in the column on the right. Metals are found in the central and left parts of the chart.

particles, **electrons,** which are only about ¹⁄₁,₈₃₆ times as massive as protons. Remember that opposite charges attract; the positive charge on the nucleus attracts the negative charge of the electrons, so the nucleus holds on to the electron cloud. (For simplicity, think of a positive charge as the "+" end of a battery and a negative charge as the "−" end.) The mass of a neutron approximately equals the sum of the mass of a proton and the mass of an electron. In the past few decades, physicists have found that protons and neutrons are made up of a myriad of even smaller particles, the smallest of which is called a **quark.**

Electron clouds have a complex internal structure. Electron are grouped in intervals called **orbitals,** energy levels, or shells. Some shells have a spherical shape, whereas others resemble dumbbells, rings, or groups of balls—for simplicity, we portray shells as circles, in cross section (▶Fig. a.5). Electrons in "inner" shells are concentrated near the nucleus, whereas those of "outer" shells predominate farther from the nucleus. Successively higher shells lie at progressively greater distances from the nucleus. Each shell can only contain a specific number of electrons: the lowest shell can contain only two electrons, while the next several shells each can hold 8. Electrons fill the lowest shells first, so that atoms with a small number of electrons only have the innermost shells; the outer shells do not exist unless there are electrons to fill them.

The outermost shell of electrons, in effect, defines the outer edge of an atom, so an atom with many occupied electron shells is larger than one with few occupied shells (argon, for example, with 18 electrons, contains more occupied electron shells than helium, with 2 electrons, so an argon atom is bigger than a helium atom). If we picture the nucleus of a carbon atom as an orange, the electrons of the outermost shell would lie at a distance of over 1 km from the orange.

Atoms are so small that they can't be seen with even the strongest light microscopes. (However, by using some clever techniques, scientists have been able, in recent years, to create images of very large atoms; ▶Fig. a.6.) Atoms are so small, in fact, that 1 gram (0.036 ounces) of helium contains 6.02×10^{23} atoms; in other words, a quantity of helium weighing little more than a postage stamp contains 602,000,000,000,000,000,000,000 atoms! The number of atoms of helium in a small balloon is approximately the same as the number of balloons it would take to replace the entirety of Earth's atmosphere.

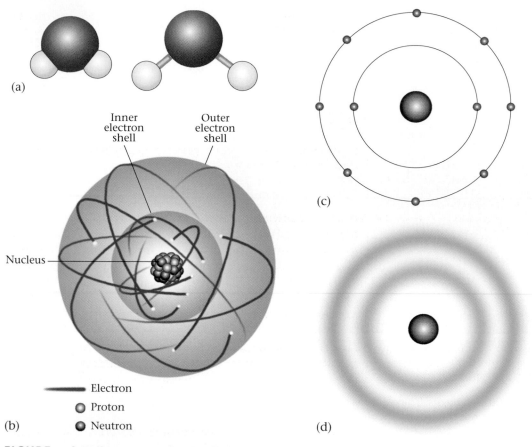

FIGURE a.4 Different ways of portraying atoms. (a) Two ways of portraying a water molecule. The large ball is oxygen, and the small ones are hydrogen. The "sticks" represent chemical bonds. (b) An image of an atom with a nucleus surrounded by electrons. (c) This diagram shows the number of electrons in the inner shells. (d) An alternative depiction of electron shells, implying that the electrons constitute a cloud. In reality, electrons do not follow simple circular orbits.

FIGURE a.5 This schematic drawing of a neon atom shows the two complete electron shells. The inner shell contains two electrons, the outer one eight. The "shells" merely represent the most likely location for an election to be.

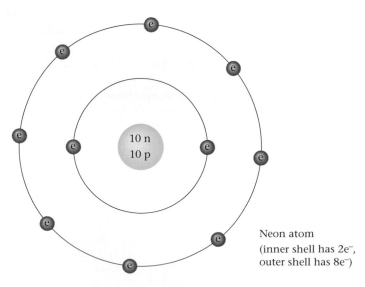

Neon atom
(inner shell has 2e⁻,
outer shell has 8e⁻)

Atomic Number, Atomic Mass, and Isotopes

We distinguish atoms of different elements from one another by their **atomic number,** the number of protons in their nucleus. For example, hydrogen's atomic number is 1, oxygen's is 8, lead's is 82, and uranium's is 92. We write the atomic number as a subscript to the left of the element's symbol (e.g., $_1H$, $_8O$, $_{82}Pb$, $_{92}U$). With the exception of the most common hydrogen nuclei, all atomic nuclei also contain neutrons. In smaller atoms, the number of neutrons equals the number of protons, but in larger atoms, the number of neutrons exceeds the number of protons. Subatomic particles are held together in a nucleus by **nuclear bonds.**

Atomic weight (or **atomic mass**) defines the amount of matter in a single atom. For a given element, the atomic weight *approximately* equals the number of protons plus the number of neutrons. (Precise atomic weights are actually slightly greater than this sum, because neutrons have slightly more mass than protons, and because the electrons have mass.) For example, helium contains 2 protons and

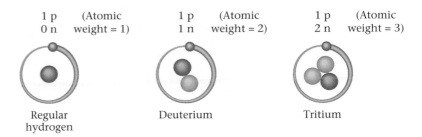

FIGURE a.7 The three isotopes of hydrogen: regular hydrogen, deuterium, and tritium.

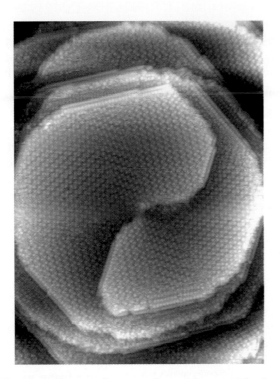

FIGURE a.6 An image of atoms taken with a special microscope that uses electrons to illuminate the surface.

2 neutrons, and thus has an atomic weight of about 4, whereas oxygen contains 8 protons and 8 neutrons and thus has an atomic weight of about 16. We indicate the weight of an atom by a superscript to the left of the element's symbol (^{4}He, ^{16}O).

Some elements occur in more than one form, and these differ in atomic weight. For example, uranium 235 ($^{235}_{92}$U) contains 92 protons and 143 neutrons, while uranium 238 ($^{238}_{92}$U) contains 92 protons and 146 neutrons. Note that both forms of uranium have the *same* atomic number—they must, if they are to be considered the same element. But they have different quantities of neutrons and thus differ in atomic weight. Multiple versions of the same element, which differ from one another in atomic weight, are called **isotopes** of the element (▶Fig. a.7).

Ions: Atoms with a Charge

If the number of electrons (negatively charged particles) exactly equals the number of protons (positively charged particles) in an atom, then the atom is electrically neutral. But if the numbers aren't equal, then the atom has a net electrical charge and is called an **ion.** For example, if an atom has two fewer electrons than protons, then we say it has a charge of +2, and if it has two more electrons than protons, it has a charge of −2. We write the charge as a superscript to the right of the symbol (e.g., Na$^+$). Ions with a negative charge are **anions** (pronounced ANN-eye-ons), and ions with a positive charge are **cations** (pronounced CAT-eye-ons). Ions form because atoms "prefer" to have a complete outer electron shell. Thus, an atom may give up or take on electrons in order for its outer shell to contain the proper number of electrons. As is the case with neutral atoms, the size of an ion depends on the number of shells containing electrons. Note that oxygen ions are larger than silicon ions, even though silicon has a larger atomic number, because oxygen has gained electrons and thus uses more electron shells than does silicon, for silicon has lost electrons (▶Fig. a.8).

FIGURE a.8 Relative sizes of common ions making up materials in rocks at the Earth's surface.

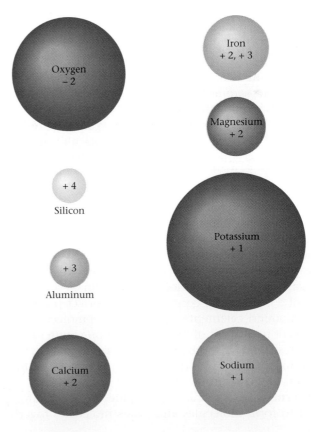

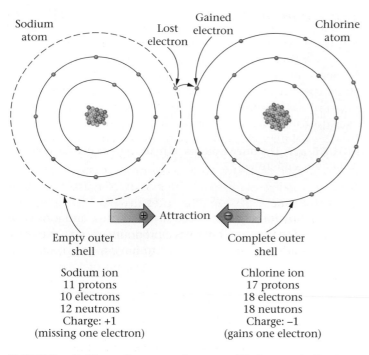

FIGURE a.9 The sodium atom has an unfilled outer shell—the shell has room for eight electrons but only has one. The chlorine atom also has an unfilled outer shell—it's missing one electron. The sodium atom gives up its outer electron, and thus has one more proton than electron (i.e., a net positive charge), while the chlorine gains an electron and thus has one more electron than proton (a net negative charge). The two ions attract each other.

Molecules and Chemical Bonds

Most of the materials we deal with in everyday life—oxygen, water, plastic—are not composed of isolated atoms. Rather, most atoms tend to stick, or "bond," to other atoms; two or more atoms stuck together constitute a **molecule.** Hydrogen gas, for example, consists of H_2 molecules; note that a hydrogen molecule consists of two of the same kind of atom. But many materials contain different kinds of atoms bonded together. As we noted earlier, such materials are called "compounds." For example, common table salt is a compound containing sodium and chlorine atoms bonded together. Compounds generally differ markedly from the elements that make them up—salt bears no resemblance at all to pure sodium (a shiny metal) or pure chlorine (a noxious gas) (Fig. a.2). A molecule is the smallest identifiable piece of a compound, containing the correct proportion of the compound's elements. For example, a molecule of water consists of two atoms of hydrogen and one atom of oxygen. We represent water by the **chemical formula** H_2O, a concise recipe that indicates the relative proportions of different elements in the molecule (Fig. a.4a).

Chemical bonds act as the glue that holds atoms together to form molecules and holds molecules together to form larger pieces of a material. Chemical bonding results from the interaction among the electrons of nearby atoms, and can take place in four different ways. (Note that chemical bonds are not the same as the nuclear bonds that hold together protons and neutrons in a nucleus.)

- *Ionic bonds:* As an inviolate rule of nature, "like" electrical charges repel (two positive charges push each other away), while "unlike" electrical charges attract (a negative charge sticks to a positive charge). Bonds that form in this way are called **ionic bonds** (▶Fig. a.9). For example, in a molecule of salt, positively charged sodium ions (Na^+) attract negatively charged chloride (Cl^-) ions. ("Chloride" is the name given to ions of chlorine.)

- *Covalent bonds:* The atoms of carbon making up a diamond do not transfer electrons to one another, but rather share electrons. Bonding that involves the sharing of electrons is called **covalent bonding** (▶Fig. a.10). Because of the sharing, the electron shells of all the carbon atoms in a diamond are complete, and all the carbon atoms have a neutral charge. Water molecules also exist because of covalent bonding: in a water molecule, two hydrogen atoms are covalently bonded to one oxygen atom.

- *Metallic bonds:* In metals, electrons of the outer shells move easily from atom to atom and bind the atoms to each other. We call this type of bonding **metallic bonding** (▶Fig. a.11). Because outer-shell electrons move so freely, metals conduct electricity easily—when you connect a metal wire to an electrical circuit, a current of electrons flows through the metal.

- *Bonds resulting from the polarity of atoms or molecules:* Chemists long recognized that some materials break or split so easily that they must be held together by

FIGURE a.10 Covalent bonding. Carbon only has four electrons in its outer shell, which has a capacity of eight. Thus, the carbon shares electrons with four other carbon atoms—it is covalently bonded to the other atoms.

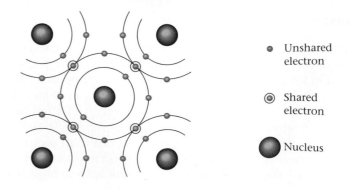

particularly weak chemical bonds. Eventually, they realized that these bonds are due to the permanent or temporary polarity of molecules. **Polarity** means that the molecule has a positive charge on one side and a negative charge on the other. Polarity-related bonds form because the negative side of one molecule attracts the positive side of another.

Bonds resulting from the polarity of molecules are important in many geologic materials. For example, liquid water consists of molecules in which two hydrogen atoms covalently bond to one oxygen atom. The hydrogen atoms both lie on the same side of the oxygen atom. The oxygen nucleus, which contains 8 protons, exerts more attraction to the shared electrons than do the hydrogen nuclei, each of which contains only one proton. As a consequence, the shared electrons concentrate closer to the oxygen side of the molecule, making the oxygen side more negative than the hydrogen side. Thus, the water molecules are polar, and they attract each other. This attraction is called a **hydrogen bond** (▶Fig. a.12). The polarity of water molecules makes water a good solvent (meaning, it can dissolve substances), because the polar water molecules attract and surround ions of soluble materials and pull them apart.

Johannes van der Waals (1837–1923), a Dutch physicist, discovered another type of weak chemical bonding that depends on polarity. This type, now known as **van der Waals bonding,** links one covalently bonded molecule to another. The bonds exist because electrons temporarily cluster on one side of each molecule, giving it a polarity.

FIGURE a.11 In a metallically bonded material, nuclei and their inner shells of electrons float in a "sea" of free electrons. Sometimes the electrons orbit the nuclei, but at other times they stream through the metal.

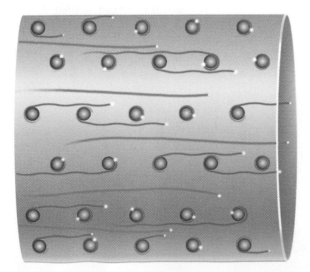

The Forms of Matter

Matter exists in one of four states: **solids,** which can maintain their shape for a long time without the restraint of a container; **liquids,** which flow fairly quickly and assume the shape of the container they fill (it's possible for a liquid to fill only part of its container); **gases,** which expand to fill the entire container they have been placed in and will disperse in all directions when not restrained; and **plasma,** an unfamiliar, gas-like mixture of positive ions and free electrons that exists only at very high temperatures (▶Fig. a.13a–d). Which state of matter exists at a location depends on the pressure and temperature (▶Fig. a.14).

To understand the differences at an atomic scale among solids, liquids, and gases—the common forms of matter on

FIGURE a.12 Water is a polar molecule, because the two hydrogen atoms lie on the same side of the oxygen atom. Thus, water molecules tend to attract one another (this attraction creates surface tension, and causes water to form drops). Similarly, the polar molecules surround chlorine and sodium ions when salt dissolves in water.

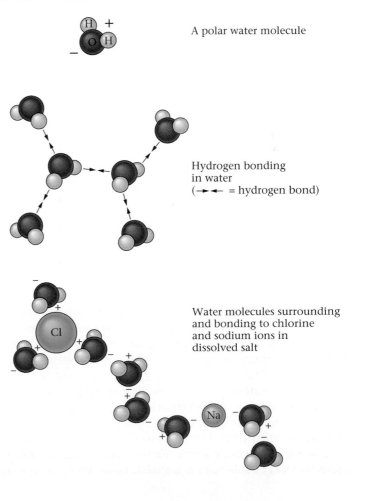

A polar water molecule

Hydrogen bonding in water
(→►◄─ = hydrogen bond)

Water molecules surrounding and bonding to chlorine and sodium ions in dissolved salt

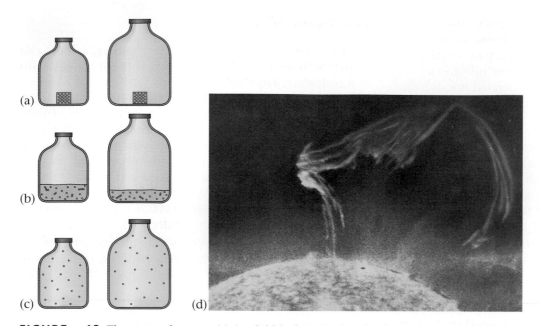

FIGURE a.13 The states of matter. (a) A solid block sitting in a bottle retains its shape regardless of the size of the container. (b) A liquid conforms to the shape of the container. If the container changes shape, the liquid also changes shape, so that it stays constant in density (mass per unit volume). (c) A gas will expand to fill whatever volume it occupies, and thus can change density if the volume changes. (d) The Sun contains plasma, a gas-like material that is so hot that electrons have been stripped from the nuclei.

Earth—imagine a large group of students getting ready to take a 9:00 A.M. exam in a windowless auditorium. The students represent atoms. It's 8 o'clock, and some students are in their rooms, some are in the library, some are walking to the auditorium, and some are at the cafeteria. In other words, everyone is scattered all over the place; there is no order to the distribution of people, and they are too far apart to interact and bond to one another. Such a distribution is characteristic of a gas.

At ten minutes before nine, most students have arrived at the auditorium. Some are already in their seats, but many are walking in or out of the room. Such a distribution characterizes the state of a liquid, in which most of the particles are in the container, except for a slow exchange with the atmosphere (by evaporation or condensation). There is short-range order, represented by little clumps of students in their seats or chatting in the aisles, but overall there is still a lot of motion and no overall organization, or long-range order.

Suddenly the lights go off, when someone bumps against the switch, and everyone stops in place to avoid stumbling but people continue to fidget in place. Even though the migration of people has stopped, there is still no long-range order in the room. Such a situation characterizes a special kind of solid called a **glass.** Glass has a disordered

arrangement of atoms, like a liquid, but cannot flow and thus behaves like solid.

The lights go on again, and the examiner finally enters. Everyone takes a seat according to a specified plan—alternate rows and alternate seats. Even though individual students are not motionless, because they are still fidget-

FIGURE a.14 A material such as water can "change state" (from gas to liquid, liquid to solid, or solid to liquid), depending on the pressure and temperature. Water can exist simultaneously in all three states at a unique pressure and temperature called the triple point.

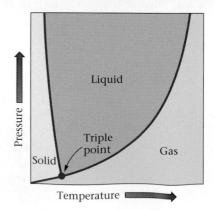

ing in place, they are no longer moving out of an orderly arrangement. The situation now corresponds to that in a solid.

a.3 THE FORCES OF NATURE

In our everyday experience, we constantly see or feel the effect of force. Forces can squash objects, speed them up and slow them down, tear, stretch, spin, and twist them, and make them float or sink. Isaac Newton, the great British scientist who effectively founded the field of physics, was the first to describe the way forces work. He defined a **force** as simply the push or pull that causes the velocity (speed) of a mass to change in magnitude and/or direction.

We can distinguish between two categories of force. The first, which includes force applied by the movement of a mass (a hand, a hammer, the wind, waves), is called **mechanical force,** or contact force. When you push a block across the floor, you are applying a mechanical force. The second category, which includes force resulting from the action of an invisible agent, is called a **field force,** or noncontact force. If you drop a book, the invisible force of gravity pulls the book toward the floor.

Physicists further recognize four types of field forces: gravity, electromagnetic, strong nuclear, and weak nuclear. **Gravity** is a force of attraction between any two masses, and can act over large distances. The magnitude of gravitational attraction depends on the size of the masses and the distance between them. We feel a much stronger gravitational pull to the planet we walk on than we do to a baseball. **Electromagnetic force,** the force associated with electricity and magnetism, is stronger than gravity, but only operates between materials that have electrical charges or are magnetic, and only operates over short distances. Like gravity, electromagnetic force depends on the distance between objects, but unlike gravity it can be either attractive or repulsive—as mentioned previously, like charges repel and unlike charges attract. Particles within atoms are subject to two kinds of nuclear force. We'll leave the discussion of nuclear forces to a physics text.

a.4 ENERGY

To a physicist, **energy** is the ability to do work, or, in other words, the ability to apply a force that moves a mass by some distance (**work** = force × distance). According to this definition, gasoline serves as an energy source because we can burn gasoline to move heavy cars and trucks (a type of work).

Kinetic and Potential Energy

Formally, we classify energy into two basic types: **kinetic energy,** the energy of motion, and **potential energy,** the energy stored in a material. A boulder sitting at the top of a hill has potential energy, because the force of gravity will do work on it as the boulder topples down. If the boulder falls down, part of its potential energy converts into kinetic energy (►Fig. a.15). Some of this kinetic energy can be transferred to a second boulder if the original boulder strikes another boulder, causing a contact force that starts the second boulder moving. Similarly, the gas in a car's tank has potential energy, which converts to kinetic energy when the gas burns and molecules start moving quickly. This kinetic energy starts the cylinders in the engine moving up and down. Now we briefly look at where energy comes from.

Energy from Chemical Reactions

A **chemical reaction** is a process whereby one or more compounds (the reactants) come together or break apart to form new compounds and/or ions (the products). Many chemical reactions produce energy. In the case of burning gasoline, gasoline molecules react with oxygen molecules in the air to form carbon dioxide molecules and water molecules, plus energy. We can describe this reaction in a chemical formula, that uses symbols for the molecules:

$$2C_8H_{18} + 25O_2 \rightarrow 16CO_2 + 18H_2O + energy.$$

Note that *reactants* appear on the left side of the formula (C_8H_{18} is a typical compound in gasoline), and *products* on the right side. The arrow indicates the direction in which

FIGURE a.15 A boulder sitting on a ledge in a gravitational field has potential energy. When the boulder starts to roll, this potential energy converts to kinetic energy.

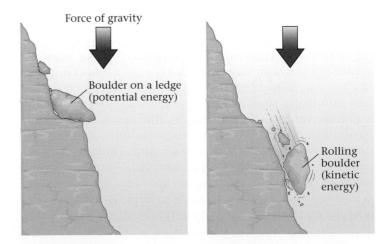

Force of gravity

Boulder on a ledge (potential energy)

Rolling boulder (kinetic energy)

the reaction proceeds. The energy comes from the breaking of chemical bonds—more energy is stored in the chemical bonds of gasoline and oxygen than in the chemical bonds of carbon dioxide and water.

Energy from Field Forces

When an object you place in a field force, the force acts on the object, and the object, unless it is restrained, moves. In a gravity field, as described earlier, a boulder resting on top of a hill has potential energy (it is restrained by the ground), but when the boulder starts rolling, it has kinetic energy. Similarly, an iron bar taped to a table near a magnet has potential energy, but an unrestrained bar moving toward the magnet has kinetic energy.

Fission and Fusion: Energy from Breaking or Making Atoms

During **nuclear fission,** a large atomic nucleus splits into two or more smaller nuclei and other by-products, including free neutrons (▶Fig. a.16). Such fission is one form of "nuclear decay". Other forms include: (1) the ejection of an electron from one of the neutrons in the nucleus, transforming the neutron into a proton; (2) the ejection of a couple of protons or neutrons from the nucleus. Nuclear decay produces new atoms with different atomic numbers from that of the original atom. In other words, one element transforms into another. In effect, the ultimate goal of medieval alchemists, to turn lead into gold, can now be achieved in modern atom smashers, though at a significant cost.

When fission or other nuclear decay reactions occur, some of the matter constituting the original atom transforms into a huge amount of energy (heat and electromagnetic radiation), as defined by Einstein's famous equation: $E = mc^2$, where E is energy, m is mass, and c is the speed of light. The realization that fission releases vast quantities of energy led to the rush to build atomic weapons during and after World War II. Nuclear fission provides the energy in atomic bombs, nuclear power plants, and nuclear submarines.

Nuclear fusion results when two nuclei slam together at such high velocity that they get close enough for nuclear forces to bind them together, thereby creating a new and larger atom of a different element (▶Fig. a.17). In a manner of speaking, fusion is the opposite of fission. Fusion reactions also produce huge amounts of energy, because during fusion some of the matter converts into energy, according to Einstein's equation. Fusion reactions generate the heat in stars; in the Sun, for example, hydrogen atoms fuse together to form helium. Fusion reactions also generate the explosive energy of a hydrogen bomb

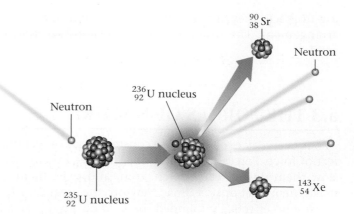

FIGURE a.16 A uranium atom splits during nuclear fission.

(a "thermonuclear device"). Such reactions can only take place at extremely high temperatures (over 1,000,000°C). In fact, in order to get a hydrogen bomb to blow up, you need to use an atomic bomb as the trigger, because only an atomic explosion can generate high enough temperatures to start fusion. So far, no one has figured out how to economically sustain a controlled fusion reaction on Earth for the purpose of generating electricity.

FIGURE a.17 (a) In the Sun, four hydrogen atoms fuse in sequence to form a helium atom. (b) In a hydrogen bomb, a deuterium and a tritium atom fuse to form a helium atom plus a free neutron.

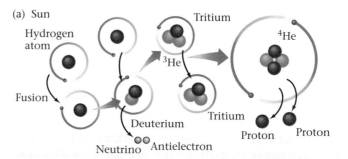

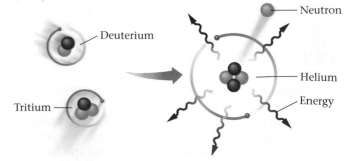

a.5 HOT AND COLD

Thermal Energy and Temperature

The atoms and molecules that make up an object do not stay rigidly fixed in place, but rather jiggle and jostle with respect to one another. This vibration creates **thermal energy**—the faster the atoms move, the greater the thermal energy and the hotter the object. In simple terms, the thermal energy in a substance represents the sum of the kinetic energy of all the substance's atoms (this includes the back-and-forth displacements that an atom makes as it vibrates as well as the movement of an atom from one place to another).

When we say that one object is hotter or colder than another, we are describing its temperature. It represents the average kinetic energy of atoms in the material. **Temperature** is a measure of warmth relative to some standard. In everyday life, we generally use the freezing or boiling point of water at sea level as the standard. When using the **Celsius (centigrade) scale,** we arbitrarily set the freezing point of water (at sea level) as 0°C and the boiling point as 100°C; whereas in the **Fahrenheit scale,** we set the freezing point as 32°F and the boiling point as 212°F.

The coldest a substance can be is the temperature at which its atoms or molecules stand still. We call this temperature **absolute zero,** or 0K (pronounced "zero kay"), where **K** stands for **Kelvin** (after Lord Kelvin [1824–1907], a British physicist), another unit of temperature. You simply can't get colder than absolute zero, meaning that you can't extract any thermal energy from a substance at 0K (–273.15°C). Degrees in the Kelvin scale are the same increment as degrees in the Celsius scale. On the Kelvin scale, ice melts at 273.15K and water boils at 373.15K (▶Figure a.18).

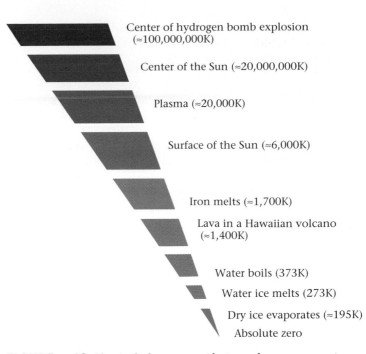

Center of hydrogen bomb explosion (≈100,000,000K)

Center of the Sun (≈20,000,000K)

Plasma (≈20,000K)

Surface of the Sun (≈6,000K)

Iron melts (≈1,700K)

Lava in a Hawaiian volcano (≈1,400K)

Water boils (373K)

Water ice melts (273K)

Dry ice evaporates (≈195K)

Absolute zero

FIGURE a.18 Physical phenomena that we observe occur at a wide range of temperatures (here, as specified by the Kelvin scale).

Heat and Heat Transfer

Heat is the thermal energy transferred from one object to another. Heat can be measured in **calories.** One thousand calories can heat 1 kilogram of water by 1°C. There are four ways in which heat transfer takes place in the Earth System: radiation, conduction, convection, and advection.

Radiation is the process by which electromagnetic waves transmit heat into a body or out of a body (▶Fig. a.19a). For example, when the Sun heats the ground during the day, radiative heating takes place. Similarly, when heat rises from the ground at night, radiative heating is occurring—in the opposite direction.

Conduction takes place when you stick the end of an iron bar in a fire (▶Fig. a.19b). The iron atoms at the fire-licked end of the bar start to vibrate more energetically;

they gradually incite atoms farther from the flame to start jiggling, and these atoms in turn set atoms even farther along in motion. In this way, heat slowly flows down the bar until you feel it with your hand. Note that even though the iron atoms vibrate more as the bar heats up, the atoms remain locked in their position within the solid. Thus, conduction does not involve actual movement of atoms from one place to another. If you place two bars of different temperatures in contact with each other, heat conducts from the hot bar into the cold one until both end up at the same temperature.

Convection takes place when you set a pot of water on a stove (▶Fig. a.19c). The heat from the stove warms the water at the base of the pot (it makes the molecules of water vibrate faster and move around more). As a consequence, the density of the water at the base of the pot decreases, for as you heat a liquid, the atoms move away from each other and the liquid expands. For a time, cold water remains at the top of the pot, but eventually the warm, less dense water becomes buoyant relative to the cold, dense water. In a gravitational field, a buoyant material rises (like a styrofoam ball in a pool of water) if the material above is weak enough to flow out of the way. Since liquid water can easily flow, hot water rises. When this happens, cold water sinks to take its place. The new volume of cold water then heats up and then rises itself. Thus, during convection, by the actual flow of the material itself carries heat. The trajectory of

FIGURE a.19 The four processes of heat transfer. (a) Radiation occurs when sunlight strikes the ground. (b) Conduction occurs when you heat the end of an iron bar in a flame. Heat flows from the hot region toward the cold region, as vibrating atoms cause their neighbors to vibrate. (c) Convection takes place when moving fluid carries heat with it. Hot water at the bottom of the pot rises, while cool water sinks, setting up a convective cell. (d) During advection, a hot liquid (such as molten rock) rises into cooler material. Heat conducts from the hot liquid into the cooler material.

flow defines **convective cells;** in a convective cell, water follows a loop, with warm water rising and cold water sinking. Convection occurs in the atmosphere, the ocean, and the interior of the Earth.

Advection, a less familiar process, happens when heat moves with a fluid, flowing through cracks and pores within a solid material (▶Fig. a.19d). The heat brought by the fluid conductively heats up the solid that the fluid passes through. Advection takes place, for example, if you pass hot water through a sponge and the sponge itself gets hot. In the Earth, advection occurs where molten rock rises through the crust beneath a volcano and heats up the crust in the process or where hot water circulates through cracks in rocks beneath geysers and hot springs.

Additional Maps and Charts

This Appendix contains several maps and charts for general reference. We list the purpose of each below.

Mineral Identification Flow Charts: Geologists use these charts to identify unknown mineral specimens (▶Fig. b.1a, b). A mineral flow chart is simply an organized series of questions concerning the mineral's physical properties. The questions are arranged in a sequence such that appropriate answers ultimately lead you on a path to a specific mineral. To understand this concept, let's imagine that we are trying to identify a shiny, bronze-colored, metallic-looking mineral specimen. We start by observing the specimen's luster. It is metallic, so we follow the path on the chart for metallic-luster minerals (Fig. b.1a). Next, we determine if the mineral is magnetic or nonmagnetic. If it is nonmagnetic, we follow the path for non-magnetic minerals. Then, we look at the mineral's color. Since it is bronze-colored, our path ends at pyrite.

Notice that one of the flow chart questions in Figure b.1b asks about the reaction of the specimen with hydrochloric acid (HCl). Only calcite and dolomite react, so the question allows definitive identification of these minerals. Another question pertains to striations, faint parallel lines on cleavage planes. Only plagioclase has striations.

World Magnetic Declination Map: This map shows the variation of magnetic declination with location on the surface of the Earth. Declination exists because the position of the Earth's magnetic pole does not coincide exactly with that of the geographic pole. In fact, the magnetic pole location constantly moves, currently at a rate of about 20 km per year. It now lies off the north coast of Canada, and in the not too distant future, it may lie along the coast of Siberia. In order for a compass to give an accurate indication of direction, it must be adjusted to accommodate for the declination at the location of measurement.

US Magnetic Declination Map: This map shows the magnetic declination for the United States.

Volcanoes of the Past Few Million Years: This map shows the location of volcanoes that have erupted in relatively recent time. The map also shows the location of earthquakes.

Earthquake Epicenter Map: This map shows the positions of epicenters for earthquakes that were large enough to be detected at seismic stations over a broad region. An epicenter is the point on the surface of the Earth above the earthquake's hypocenter (focus). The hypocenter is the place where the earthquake energy is generated. Different colors indicate the different depths of the earthquake hypocenters.

World Soils Map: This map displays areas of different types of soils, using one of the standard classification schemes for soils.

A Satellite Image of the Earth at Night: This image shows light sources on the surface of the planet. Light sources indicate both the density of human habitation, and the degree of industrialization. Note, for example, that most of the lights in Australia occur in coastal cities.

FIGURE b.1a   Mineral-Identification Flow Chart (metallic or dark-colored nonmetallic)

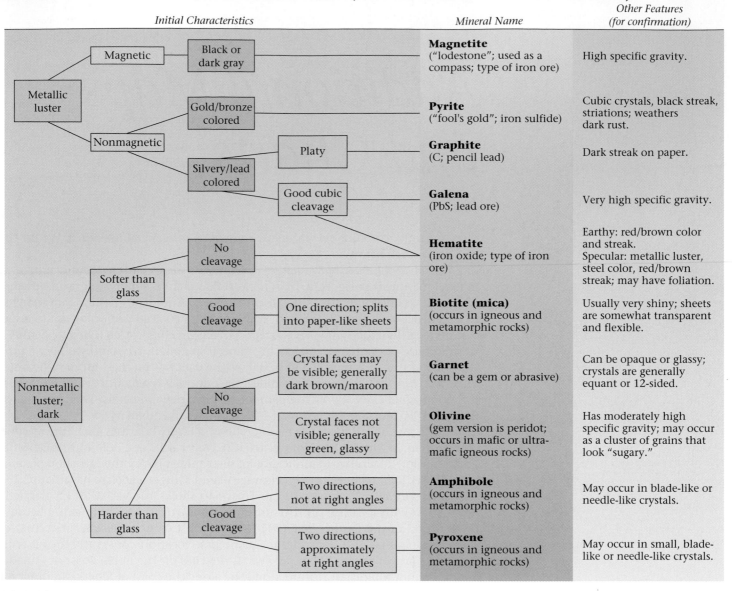

FIGURE b.1b   **Mineral-Identification Flow Chart (light-colored nonmetallic)**

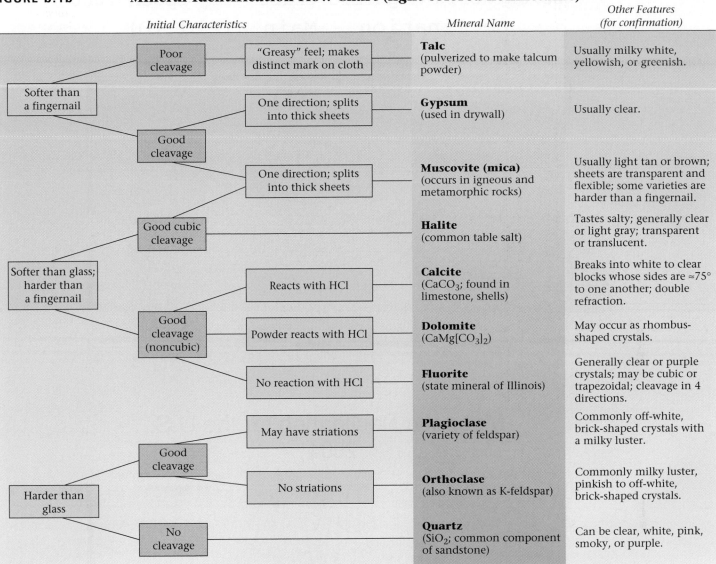

Initial Characteristics   *Mineral Name*   *Other Features (for confirmation)*

Softer than a fingernail → Poor cleavage → "Greasy" feel; makes distinct mark on cloth → **Talc** (pulverized to make talcum powder) — Usually milky white, yellowish, or greenish.

Good cleavage → One direction; splits into thick sheets → **Gypsum** (used in drywall) — Usually clear.

Softer than glass; harder than a fingernail → Good cleavage → One direction; splits into thick sheets → **Muscovite (mica)** (occurs in igneous and metamorphic rocks) — Usually light tan or brown; sheets are transparent and flexible; some varieties are harder than a fingernail.

Good cubic cleavage → **Halite** (common table salt) — Tastes salty; generally clear or light gray; transparent or translucent.

Good cleavage (noncubic) → Reacts with HCl → **Calcite** ($CaCO_3$; found in limestone, shells) — Breaks into white to clear blocks whose sides are ≈75° to one another; double refraction.

Powder reacts with HCl → **Dolomite** ($CaMg[CO_3]_2$) — May occur as rhombus-shaped crystals.

No reaction with HCl → **Fluorite** (state mineral of Illinois) — Generally clear or purple crystals; may be cubic or trapezoidal; cleavage in 4 directions.

Harder than glass → Good cleavage → May have striations → **Plagioclase** (variety of feldspar) — Commonly off-white, brick-shaped crystals with a milky luster.

No striations → **Orthoclase** (also known as K-feldspar) — Commonly milky luster, pinkish to off-white, brick-shaped crystals.

No cleavage → **Quartz** (SiO_2; common component of sandstone) — Can be clear, white, pink, smoky, or purple.

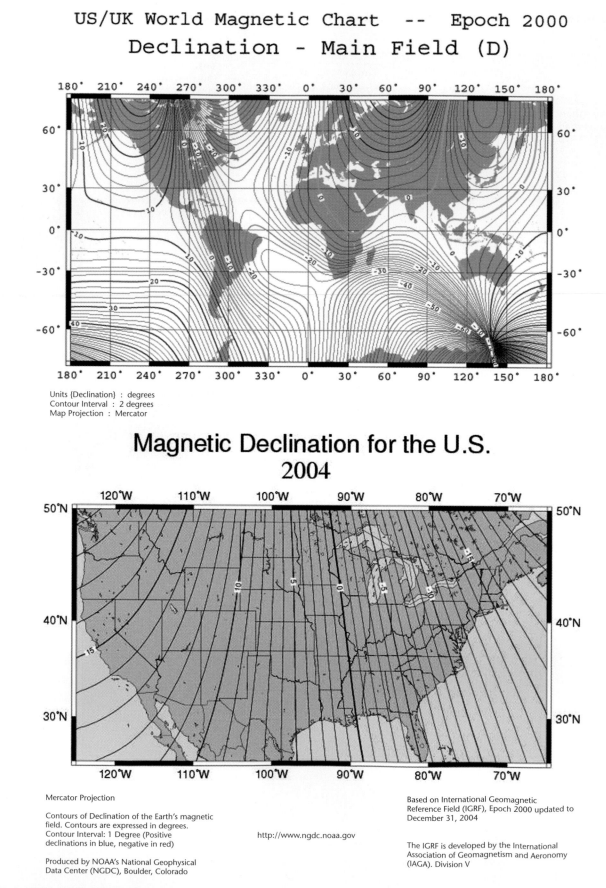

FIGURE b.2 (a) World magnetic declination chart. (b) Magnetic declination chart for the United States.

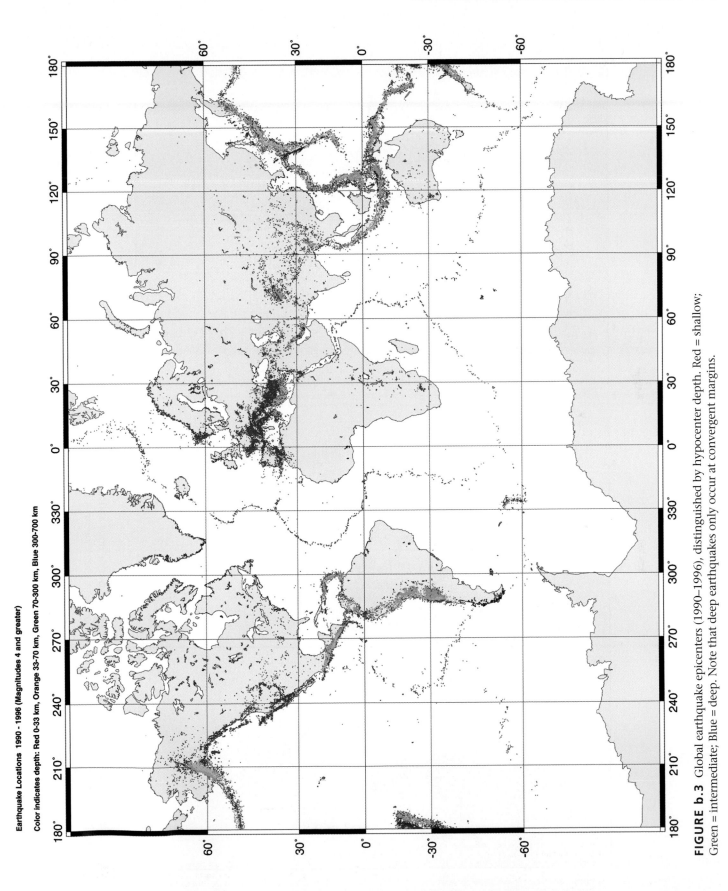

Earthquake Locations 1990 - 1996 (Magnitudes 4 and greater)

Color indicates depth: Red 0-33 km, Orange 33-70 km, Green 70-300 km, Blue 300-700 km

FIGURE b.3 Global earthquake epicenters (1990–1996), distinguished by hypocenter depth. Red = shallow; Green = intermediate; Blue = deep. Note that deep earthquakes only occur at convergent margins.

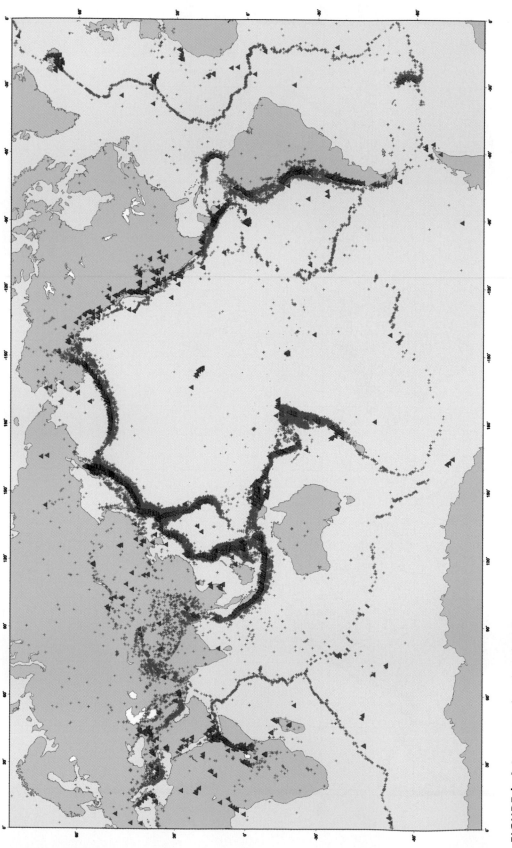
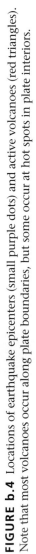

FIGURE b.4 Locations of earthquake epicenters (small purple dots) and active volcanoes (red triangles). Note that most volcanoes occur along plate boundaries, but some occur at hot spots in plate interiors.

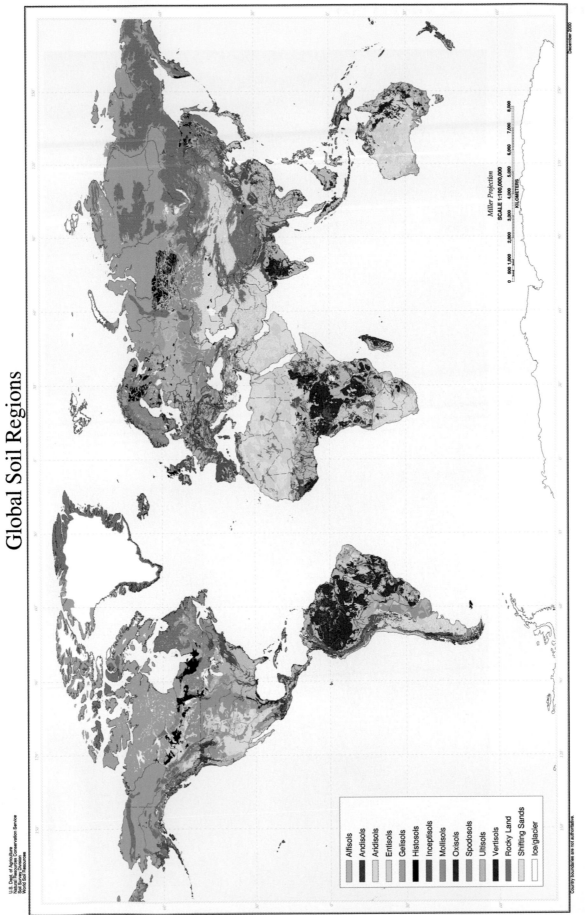

Global Soil Regions

U.S. Dept. of Agriculture
Natural Resources Conservation Service
Soil Survey Division
World Soil Resources

Alfisols
Andisols
Aridisols
Entisols
Gelisols
Histosols
Inceptisols
Mollisols
Oxisols
Spodosols
Ultisols
Vertisols
Rocky Land
Shifting Sands
Ice/glacier

Miller Projection

SCALE 1:100,000,000

0 500 1,000   2,000   3,000   4,000   5,000   6,000   7,000   8,000

KILOMETERS

Country boundaries are not authoritative.

December 2000

FIGURE b.5 Map of soil types around the world. The different colors indicate areas in which a particular soil type dominates.

FIGURE b.6 A composite satellite image portraying the distribution of lights at night on the Earth. Light density represents a combination of population density and level of industrialization.

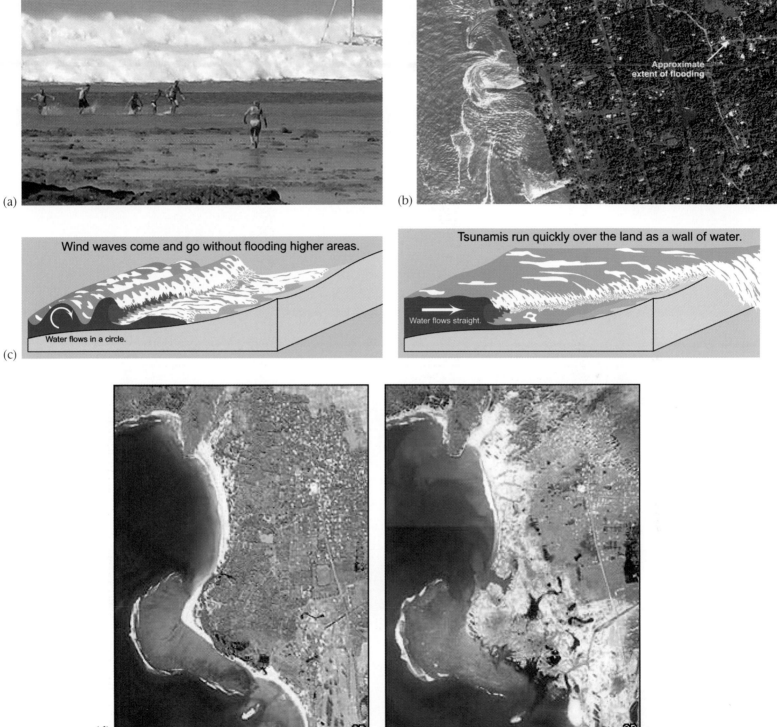

FIGURE b.7 The Indian Ocean Tsunami of 2004 (a) A photograph showing the wave front approaching. (b) A satellite photograph showing the water extending 500 m inshore at Kalutara, on the west coast of Sri Lanka. (c) The difference between wind waves and tsunami. (d) Satellite photos taken of the Indonesian province of Aceh before and after the tsunami hit.

Glossary

aa   A lava flow with a rubbly surface.

abandoned meander   A meander that dries out after it was cut off.

ablation   The removal of ice at the toe of a glacier by melting, sublimation (the evaporation of ice into water vapor), and/or calving.

abrasion   The process in which one material (such as sand-laden water) grinds away at another (such as a stream channel's floor and walls).

absolute age   Numerical age (the age specified in years).

absolute plate velocity   The movement of a plate relative to a fixed point in the mantle.

abyssal plain   A broad, relatively flat region of the ocean that lies at least 4.5 km below sea level.

Acadian orogeny   A convergent mountain-building event, that occurred around 400 million years ago, during which continental slivers accreted to the eastern edge of the North American continent.

accreted terrane   A block of crust that collided with a continent at a convergent margin and stayed attached to the continent.

accretionary coast   A coastline that receives more sediment than erodes away.

accretionary orogen   An orogen formed by the attachment of numerous buoyant slivers of crust to an older, larger continental block.

accretionary prism   A wedge-shaped mass of sediment and rock scraped off the top of a downgoing plate and accreted onto the overriding plate at a convergent plate margin.

acid mine runoff   A dilute solution of sulfuric acid, produced when sulfur-bearing minerals in mines react with rainwater, that flows out of a mine.

acid rain   Precipitation in which air pollutants react with water to make a weak acid that then falls from the sky.

active continental margin   A continental margin that coincides with a plate boundary.

active fault   A fault that has moved recently or is likely to move in the future.

active sand   The top layer of beach sand, which moves daily because of wave action.

active volcano   A volcano that has erupted within the past few centuries and will likely erupt again.

adiabatic cooling   The cooling of a body of air or matter without the addition or subtraction of thermal energy (heat).

adiabatic heating   The warming of a body of air or matter without the addition or subtraction of heat.

aerosols   Tiny solid particles or liquid droplets that remain suspended in the atmosphere for a long time.

aftershocks   The series of smaller earthquakes that follow a major earthquake.

air   The mixture of gases that make up the Earth's atmosphere.

air-fall tuff   Tuff formed when ash settles gently from the air.

air mass   A body of air, about 1,500 km across, that has recognizable physical characteristics.

air pressure   The push that air exerts on its surroundings.

albedo   The reflectivity of a surface.

Alleghenian orogeny   The orogenic event that occurred about 270 million years ago when Africa collided with North America.

alloy   A metal containing more than one type of metal atom.

alluvial fan   A gently sloping apron of sediment dropped by an ephemeral stream at the base of a mountain in arid or semi-arid regions.

alluvium   Sorted sediment deposited by a stream.

alluvium-filled valley   A valley whose floor fills with sediment.

amber   Hardened (fossilized) ancient sap or resin.

amphibolite facies   A set of metamorphic mineral assemblages formed under intermediate pressures and temperatures.

amplitude   The height of a wave from crest to trough.

Ancestral Rockies   The late Paleozoic uplifts of the Rocky Mountain region; they eroded away long before the present Rocky Mountains formed.

angiosperm   A flowering plant.

angle of repose   The angle of the steepest slope that a pile of uncemented material can attain without collapsing from the pull of gravity.

angularity   The degree to which grains have sharp or rounded edges or corners.

angular unconformity   An unconformity in which the strata below were tilted or folded before the unconformity developed; strata below the unconformity therefore have a different tilt than strata above.

anhedral grains   Crystalline mineral grains without well-formed crystal faces.

Antarctic bottom water mass   The mass of cold, dense water that sinks along the coast of Antarctica.

antecedent stream   A stream that cuts across an uplifted mountain range; the stream must have existed before the range uplifted and must then have been able to downcut as fast as the land was rising.

anthracite coal   Shiny black coal formed at temperatures between 200° and 300°C. A high-rank coal.

anticline   A fold with an arch-like shape in which the limbs dip away from the hinge.

anticyclone   The clockwise flow of air around a high-pressure mass.

Antler orogeny   The Late Devonian mountain-building event in which slices of deep-marine strata were pushed eastward, up and over the shallow-water strata on the western coast of North America.

anvil cloud   A large cumulonimbus cloud that spreads laterally at the tropopause to form a broad, flat top.

aphanitic   A textural term for fine-grained igneous rock.

apparent polar-wander path   A path on the globe along which a magnetic pole appears to have wandered over time; in fact, the continents drift, while the magnetic pole stays fairly fixed.

aquiclude   Sediment or rock that transmits no water.

aquifer   Sediment or rock that transmits water easily.

aquitard   Sediment or rock that does not transmit water easily and therefore retards the motion of the water.

archaea   A kingdom of "old bacteria," now commonly found in extreme environments like hot springs. (Also called "archaeobacteria.")

Archean The middle Precambrian Eon.

Archimedes' principle The mass of the water displaced by a block of material equals the mass of the whole block of material.

arête A residual knife-edge ridge of rock that separates two adjacent cirques.

argillaceous sedimentary rock Sedimentary rock that contains abundant clay.

arroyo The channel of an ephemeral stream; dry wash; wadi.

artesian well A well in which water rises on its own.

ash fall Ash that falls to the ground out of an ash cloud.

ash flow An avalanche of ash that tumbles down the side of an explosively erupting volcano.

assimilation The process of magma contamination in which blocks of wall rock fall into a magma chamber and dissolve.

asthenosphere The layer of the mantle that lies between 100–150 km and 350 km deep; the asthenosphere is relatively soft and can flow when acted on by force.

atm A unit of air pressure that approximates the pressure exerted by the atmosphere at sea level.

atmosphere A layer of gases that surrounds a planet.

atoll A coral reef that develops around a circular reef surrounding a lagoon.

atomic number The number of protons in the nucleus of a given element.

atomic weight The number of protons plus the number of neutrons in the nucleus of a given element. (Also known as atomic mass.)

aurora australis The same phenomenon as the aurora borealis, but in the Southern Hemisphere.

aurora borealis A ghostly curtain of varicolored light that appears across the night sky in the Northern Hemisphere when charged particles from the Sun interact with the ions in the ionosphere.

avalanche A turbulent cloud of debris mixed with air that rushes down a steep hill slope at high velocity; the debris can be rock and/or snow.

avalanche chute A downslope hillside pathway along which avalanches repeatedly fall, consequently clearing the pathway of mature trees.

avulsion The process in which a river overflows a natural levee and begins to flow in a new direction.

axial plane The imaginary surface that encompasses the hinges of successive layers of a fold.

axial trough A narrow depression that runs along a mid-ocean ridge axis.

backscattered light Atmospheric scattered sunlight that returns back to space.

backshore zone The zone of beach that extends from a small step cut by high-tide swash to the front of the dunes or cliffs that lie farther inshore.

backswamp The low marshy region between the bluffs and the natural levees of a floodplain.

backwash The gravity-driven flow of water back down the slope of a beach.

bajada An elongate wedge of sediment formed by the overlap of several alluvial fans emerging from adjacent valleys.

Baltica A Paleozoic continent that included crust that is now part of today's Europe.

banded-iron formation (BIF) Iron-rich sedimentary layers consisting of alternating gray beds of iron oxide and red beds of iron-rich chert.

bar (1) A sheet or elongate lens or mound of alluvium; (2) a unit of air-pressure measurement approximately equal to 1 atm.

barchan dune A crescent-shaped dune whose tips point downwind.

barrier island An offshore sand bar that rises above the mean high-water level, forming an island.

barrier reef A coral reef that develops offshore, separated from the coast by a lagoon.

basal sliding The phenomenon in which meltwater accumulates at the base of a glacier, so that the mass of the glacier slides on a layer of water or on a slurry of water and sediment.

basalt A fine-grained mafic igneous rock.

base level The lowest elevation a stream channel's floor can reach at a given locality.

basement Older igneous and metamorphic rocks making up the Earth's crust beneath sedimentary cover.

basement uplift Uplift of basement rock by faults that penetrate deep into the continental crust.

base metals Metals that are mined but not considered precious. Examples include copper, lead, zinc, and tin.

basin A fold or depression shaped like a right-side-up bowl.

Basin and Range Province A broad, Cenozoic continental rift that has affected a portion of the western United States in Nevada, Utah, and Arizona; in this province, tilted fault blocks form ranges, and alluvium-filled valleys are basins.

batholith A vast composite, intrusive, igneous rock body up to several hundred km long and 100 km wide, formed by the intrusion of numerous plutons in the same region.

bathymetric map A map illustrating the shape of the ocean floor.

bathymetric profile A cross section showing ocean depth plotted against location.

bathymetry Variation in depth.

bauxite A residual mineral deposit rich in aluminum.

baymouth bar A sandspit that grows across the opening of a bay.

beach drift The gradual migration of sand along a beach.

beach erosion The removal of beach sand caused by wave action and longshore currents.

beach face A steeply concave part of the foreshore zone formed where the swash of the waves actively scours the sand.

bedding Layering or stratification in sedimentary rocks.

bed load Large particles, such as sand, pebbles, or cobbles, that bounce or roll along a stream bed.

bedrock Rock still attached to the Earth's crust.

Bergeron process Precipitation involving the growth of ice crystals in a cloud at the expense of water droplets.

berm A horizontal or landward-sloping terrace in the backshore zone of a beach that receives sediment during a storm.

big bang A cataclysmic explosion that scientists suggest represents the formation of the Universe; before this event, all matter and all energy were packed into one volumeless point.

biochemical sedimentary rock Sedimentary rock formed from material (such as shells) produced by living organisms.

biodiversity The number of different species that exist at a given time.

biogeochemical cycle The exchange of chemicals between living and nonliving reservoirs in the Earth System.

bioremediation The injection of oxygen and nutrients into a contaminated aquifer to foster the growth of bacteria that will ingest or break down contaminants.

biosphere The region of the Earth and atmosphere inhabited by life; this region stretches from a few km below the Earth's surface to a few km above.

bioturbation The mixing of sediment by burrowing animals such as clams and worms.

bituminous coal Dull, black intermediate-rank coal formed at temperatures between 100° and 200°C.

black-lung disease Lung disease contracted by miners from the inhalation of too much coal dust.

black smoker The cloud of suspended minerals formed where hot water spews out of a vent along a mid-ocean ridge; the dissolved sulfide components of the hot water instantly precipitate when the water mixes with seawater and cools.

blind fault A fault that does not intersect the ground surface.

blocking temperature The temperature below which isotopes in a mineral are no longer free to move, so the radiometric clock starts.

blowout A deep, bowl-like depression scoured out of desert terrain by a turbulent vortex of wind.

blue shift The phenomenon in which a source of light moving toward you appears to have a higher frequency.

body waves Seismic waves that pass through the interior of the Earth.

bog A wetland dominated by moss and shrubs.

bolide A solid extraterrestrial object such as a meteorite, comet, or asteroid that explodes in the atmosphere.

bornhardt An inselberg with a loaf geometry, like that of Uluru (Ayers Rock) in central Australia.

Bowen's reaction series The sequence in which different silicate minerals crystallize during the progressive cooling of a melt.

braided stream A sediment-choked stream consisting of entwined subchannels.

breaker A water wave in which water at the top of the wave curves over the base of the wave.

breakwater An offshore wall, built parallel or at an angle to the beach, that prevents the full force of waves from reaching a harbor.

breccia Coarse sedimentary rock consisting of angular fragments; or rock broken into angular fragments by faulting.

breeder reactor A nuclear reactor that produces its own fuel.

brine Water that is not fresh but is less salty than seawater; brine may be found in estuaries.

brittle deformation The cracking and fracturing of a material subjected to stress.

brittle-ductile transition (brittle-plastic transition) The depth above which materials behave brittlely and below which materials behave ductilely (plastically); this transition typically lies between a depth of 10 and 15 km in continental crustal rock, and 60 m deep in glacial ice.

buoyancy The upward force acting on a less dense object immersed or floating in denser material.

butte A medium-size, flat-topped hill in an arid region.

caldera A large circular depression with steep walls and a fairly flat floor, formed after an eruption as the center of the volcano collapses into the drained magma chamber below.

caliche A solid mass created where calcite cements the soil together (also called calcrete).

calving The breaking off of chunks of ice at the edge of a glacier.

Cambrian explosion of life The remarkable diversification of life, indicated by the fossil record, that occurred at the beginning of the Cambrian Period.

Canadian Shield A broad, low-lying region of exposed Precambrian rock in the Canadian interior.

canyon A trough or valley with steeply sloping walls, cut into the land by a stream.

capillary fringe The thin subsurface layer in which water molecules seep up from the water table by capillary action to fill pores.

carbonate rocks Rocks containing calcite and/or dolomite.

carbon-14 dating A radiometric dating process that can tell us the age of organic material containing carbon originally extracted from the atmosphere.

cast Sediment that preserves the shape of a shell it once filled before the shell dissolved or mechanically weathered away.

catabatic winds Strong winds that form at the margin of a glacier where the warmer air above ice-free land rises and the cold, denser air from above the glaciers rushes in to take its place.

catastrophic change Change that takes place either instantaneously or rapidly in geologic time.

catchment *Drainage network.*

cement Mineral material that precipitates from water and fills the spaces between grains, holding the grains together.

cementation The phase of lithification in which cement, consisting of minerals that precipitate from groundwater, partially or completely fills the spaces between clasts and attaches each grain to its neighbor.

Cenozoic The most recent era of the Phanerozoic Eon, lasting from 65 Ma up until the present.

chalk Very fine-grained limestone consisting of weakly cemented plankton shells.

change of state The process in which a material changes from one phase (liquid, gas, or solid) to another.

channel A trough dug into the ground surface by flowing water.

channeled scablands A barren, soil-free landscape in eastern Washington, scoured clean by a flood unleashed when a large glacial lake drained.

chatter marks Wedge-shaped indentations left on rock surfaces by glacial plucking.

chemical sedimentary rocks Sedimentary rocks made up of minerals that precipitate directly from water solution.

chemical weathering The process in which chemical reactions alter or destroy minerals when rock comes in contact with water solutions and/or air.

chert A sedimentary rock composed of very fine-grained silica (cryptocrystalline quartz).

Chicxulub crater A circular excavation buried beneath younger sediment on the Yucután peninsula; geologists suggest that a meteorite landed there 65 Ma.

chimney (1) A conduit in a magma chamber in the shape of a long vertical pipe through which magma rises and erupts at the surface; (2) an isolated column of strata in an arid region.

cinder cone A subaerial volcano consisting of a cone-shaped pile of tephra whose slope approaches the angle of repose for tephra.

cinders Fragments of glassy rock ejected from a volcano.

cirque A bowl-shaped depression carved by a glacier on the side of a mountain.

cirrus cloud A wispy cloud that tapers into delicate, feather-like curls.

clastic (detrital) sedimentary rock Sedimentary rock consisting of cemented-together detritus derived from the weathering of preexisting rock.

cleavage (1) The tendency of a mineral to break along preferred planes; (2) a type of foliation in low-grade metamorphic rock.

cleavage planes A series of surfaces on a crystal that form parallel to the weakest bonds holding the atoms of the crystal together.

cliff (or scarp) retreat The change in the position of a cliff face caused by erosion.

climate The average weather conditions, along with the range of conditions, of a region over a year.

cloud A mist of tiny water droplets in the sky.

coal rank A measurement of the carbon content of coal; higher-rank coal forms at higher temperatures.

coal reserve The quantities of discovered, but not yet mined, coal in sedimentary rock of the continents.

coal swamp A swamp whose oxygen-poor water allows thick piles of woody debris to accumulate; this debris transforms into coal upon deep burial.

coastal plain Low-relief regions of land adjacent to the coast.

cold front The boundary at which a cold air mass pushes underneath a warm air mass.

collision The process of two buoyant pieces of lithosphere converging and squashing together.

columnar jointing A type of fracturing that yields roughly hexagonal columns of basalt; columnar joints form when a dike, sill, or lava flow cools.

comet A ball of ice and dust, probably remaining from the formation of the solar system, that orbits the Sun.

compaction The phase of lithification in which the pressure of the overburden on the buried rock squeezes out water and air that was trapped between clasts, and the clasts press tightly together.

composite volcano *Stratovolcano.*

compositional banding A type of metamorphic foliation, found in gneiss, defined by alternating bands of light and dark minerals.

compressibility The degree to which a material's volume changes in response to squashing.

compression A push or squeezing felt by a body.

compressional waves Waves in which particles of material move back and forth parallel to the direction in which the wave itself moves.

conchoidal fractures Smoothly curving, clamshell-shaped surfaces along which materials with no cleavage planes tend to break.

condensation The process of gas molecules linking together to form a liquid.

condensation nuclei Preexisting solid or liquid particles, such as aerosols, onto which water condenses during cloud formation.

cone of depression The downward-pointing, cone-shaped surface of the water table in a location where the water table is experiencing drawdown because of pumping at a well.

confined aquifer An aquifer that is separated from the Earth's surface by an overlying aquitard.

conglomerate Very coarse-grained sedimentary rock consisting of rounded clasts.

consuming boundary *Convergent plate boundary.*

contact The boundary surface between two rock bodies (as between two stratigraphic formations, between an igneous intrusion and adjacent rock, between two igneous rock bodies, or between rocks juxtaposed by a fault).

contact metamorphism *Thermal metamorphism.*

contaminant plume A cloud of contaminated groundwater that moves away from the source of the contamination.

continental crust The crust beneath the continents.

continental divide A highland separating drainage that flows into one ocean from drainage that flows into another.

continental-drift hypothesis The idea that continents have moved and are still moving slowly across the Earth's surface.

continental glacier A vast sheet of ice that spreads over thousands of square km of continental crust.

continental-interior desert An inland desert that develops because by the time air masses reach the continental interior, they have lost all of their moisture.

continental lithosphere Lithosphere topped by continental crust; this lithosphere reaches a thickness of 150 km.

continental margin A continent's coastline.

continental rift A linear belt along which continental lithosphere stretches and pulls apart.

continental rifting The process by which a continent stretches and splits along a belt; if it is successful, rifting separates a larger continent into two smaller continents separated by a divergent boundary.

continental rise The sloping sea floor that extends from the lower part of the continental slope to the abyssal plain.

continental shelf A broad, shallowly submerged region of a continent along a passive margin.

continental slope The slope at the edge of a continental shelf, leading down to the deep sea floor.

continental volcanic arc A long curving chain of subaerial volcanoes on the margin of a continent adjacent to a convergent plate boundary.

contour lines Lines on a map along which a parameter has a constant value; for example, all points along a contour line on a topographic map are at the same elevation.

control rod Rods that absorb neutrons in a nuclear reactor and thus decrease the number of collisions between neutrons and radioactive atoms.

convection Heat transfer that results when warmer, less dense material rises while cooler, denser material sinks.

convergence zone A place where two surface air flows meet so that air has to rise.

convergent margin *Convergent plate boundary.*

convergent plate boundary A boundary at which two plates move toward each other so that one plate sinks (subducts) beneath the other; only oceanic lithosphere can subduct.

coral reef A mound of coral and coral debris forming a region of shallow water.

core The dense, iron-rich center of the Earth.

core-mantle boundary An interface 2,900 km below the Earth's surface separating the mantle and core.

Coriolis effect The deflection of objects, winds, and currents on the surface of the Earth owing to the planet's rotation.

cornice A huge, overhanging drift of snow built up by strong winds at the crest of a mountain ridge.

correlation The process of defining the age relations between the strata at one locality and the strata at another.

cosmic rays Nuclei of hydrogen and other elements that bombard the Earth from deep space.

cosmology The study of the overall structure of the Universe.

country rock (wall rock) The preexisting rock into which magma intrudes.

crater (1) A circular depression at the top of a volcanic mound; (2) a depression formed by the impact of a meteorite.

craton A long-lived block of durable continental crust commonly found in the stable interior of a continent.

cratonic platform A province in the interior of a continent in which Phanerozoic strata bury most of the underlying Precambrian rock.

creep The gradual downslope movement of regolith.

crevasse A large crack that develops by brittle deformation in the top 60 m of a glacier.

critical mass A sufficiently dense and large mass of radioactive atoms in which a chain reaction happens so quickly that the mass explodes.

cross section A diagram depicting the geometry of materials underground as they would appear on an imaginary vertical slice through the Earth.

crude oil Oil extracted directly from the ground.

crust The rock that makes up the outermost layer of the Earth.

crustal root Low-density crustal rock that protrudes downward beneath a mountain range.

crystal A single, continuous piece of a mineral bounded by flat surfaces that formed naturally as the mineral grew.

crystal form The geometric shape of a crystal, defined by the arrangement of crystal faces.

crystal habit The general shape of a crystal or cluster of crystals that grew unimpeded.

crystal lattice The orderly framework within which the atoms or ions of a mineral are fixed.

crystalline Containing a crystal lattice.

cuesta An asymmetric ridge formed by tilted layers of rock, with a steep cliff on one side cutting across the layers and a gentle slope on the other side; the gentle slope is parallel to the layering.

cumulonimbus cloud A rain-producing puffy cloud.

cumulus cloud A puffy, cotton-ball-shaped cloud.

current (1) A well-defined stream of ocean water; (2) the moving flow of water in a stream.

cut bank The outside bank of the channel wall of a meander, which is continually undergoing erosion.

cutoff A straight reach in a stream that develops when erosion eats through a meander neck.

cyanobacteria Blue-green algae; a type of archaea.

cycle A series of interrelated events or steps that occur in succession and can be repeated, perhaps indefinitely.

cyclone (1) The counterclockwise flow of air around a low-pressure mass; (2) the equivalent of a hurricane in the Indian Ocean.

cyclothem A repeated interval within a sedimentary sequence that contains a specific succession of sedimentary beds.

Darcy's law A mathematical equation stating that a volume of water, passing through a specified area of material at a given time, depends on the material's permeability and hydraulic gradient.

daughter isotope The decay product of radioactive decay.

day The time it takes for the Earth to spin once on its axis.

debris avalanche An avalanche in which the falling debris consists of rock fragments and dust.

debris flow A downslope movement of mud mixed with larger rock fragments.

debris slide A sudden downslope movement of material consisting only of regolith.

decompression melting The kind of melting that occurs when hot mantle rock rises to shallower depths in the Earth so that pressure decreases while the temperature remains unchanged.

deep current An ocean current at a depth greater than 100 m.

deep-focus earthquake An earthquake that occurs at a depth between 300 and 670 km; below 670 km, earthquakes do not happen.

deflation The process of lowering the land surface by wind abrasion.

deformation A change in the shape, position, or orientation of a material, by bending, breaking, or flowing.

dehydration Loss of water.

delta A wedge of sediment formed at a river mouth when the running water of the stream enters standing water, the current slows, the stream loses competence, and sediment settles out.

delta plain The low, swampy land on the surface of a delta.

delta-plain flood A flood in which water submerges a delta plain.

dendritic network A drainage network whose interconnecting streams resemble the pattern of branches connecting to a deciduous tree.

dendrochronologist A scientist who analyzes tree rings to determine the geologic age of features.

density Mass per unit volume.

denudation The removal of rock and regolith from the Earth's surface.

deposition The process by which sediment settles out of a transporting medium.

depositional landform A landform resulting from the deposition of sediment where the medium carrying the sediment evaporates, slows down, or melts.

desert A region so arid that it contains no permanent streams except for those that bring water in from elsewhere, and has very sparse vegetation cover.

desertification The process of transforming nondesert areas into desert.

desert pavement A mosaic-like stone surface forming the ground in a desert.

desert varnish A dark, rusty-brown coating of iron oxide and magnesium oxide that accumulates on the surface of the rock.

detachment fault A nearly horizontal fault at the base of a fault system.

detritus The chunks and smaller grains of rock broken off outcrops by physical weathering.

dewpoint temperature The temperature at which air becomes saturated so that dew can form.

differential stress A condition causing a material to experience a push or pull in one direction of a greater magnitude than the push or pull in another direction; in some cases, differential stress can result in shearing.

differential weathering What happens when different rocks in an outcrop undergo weathering at different rates.

diffraction The splitting of light into many tiny beams that interfere with one another.

dike A tabular (wall-shaped) intrusion of rock that cuts across the layering of country rock.

dimension stone An intact block of granite or marble to be used for architectural purposes.

dipole A magnetic field with a north and south pole, like that of a bar magnet.

dipole field (for Earth) The part of the Earth's magnetic field, caused by the flow of liquid iron alloy in the outer core, that can be represented by an imaginary bar magnet with a north and south pole.

dip-slip fault A fault in which sliding occurs up or down the slope (dip) of the fault.

dip slope A hill slope underlain by bedding parallel to the slope.

disappearing stream A stream that intersects a crack or sinkhole leading to an underground cavern, so that the water disappears into the subsurface and becomes an underground stream.

discharge The volume of water in a conduit or channel passing a point in one second.

discharge area A location where groundwater flows back up to the surface, and may emerge at springs.

disconformity An unconformity parallel to the two sedimentary sequences it separates.

displacement (or **offset**) The amount of movement or slip across a fault plane.

disseminated deposit A hydrothermal ore deposit in which ore minerals are dispersed throughout a body of rock.

dissolution A process during which materials dissolve in water.

dissolved load Ions dissolved in a stream's water.

distillation column A vertical pipe in which crude oil is separated into several components.

distributaries The fan of small streams formed where a river spreads out over its delta.

divergence zone A place where sinking air separates into two flows that move in opposite directions.

divergent plate boundary A boundary at which two lithosphere plates move apart from each other; they are marked by mid-ocean ridges.

diversification The development of many different species.

DNA (deoxyribonucleic acid) The complex molecule, shaped like a double helix, containing the code that guides the growth and development of an organism.

doldrums A belt with very slow winds along the equator.

dome Folded or arched layers with the shape of an overturned bowl.

Doppler effect The phenomenon in which the frequency of wave energy appears to change when a moving source of wave energy passes an observer.

dormant volcano A volcano that has not erupted for hundreds to thousands of years but does have the potential to erupt again in the future.

downcutting The process in which water flowing through a channel cuts into the substrate and deepens the channel relative to its surroundings.

downdraft Downward-moving air.

downgoing plate (or **slab**) A lithosphere plate that has been subducted at a convergent margin.

downslope force The component of the force of gravity acting in the downslope direction.

downslope movement The tumbling or sliding of rock and sediment from higher elevations to lower ones.

downwelling zone A place where near-surface water sinks.

drag fold A fold that develops in layers of rock adjacent to a fault during or just before slip.

drainage divide A highland or ridge that separates one watershed from another.

drainage network (or **basin**) An array of interconnecting streams that together drain an area.

drawdown The phenomenon in which the water table around a well drops because the users are pumping water out of the well faster than it flows in from the surrounding aquifer.

drilling mud A slurry of water mixed with clay that oil drillers use to cool a drill bit and flush rock cuttings up and out of the hole.

dripstone Limestone (travertine in a cave) formed by the precipitation of calcium carbonate out of groundwater.

drop stone A rock that drops to the sea floor once the iceberg that was carrying the rock melts.

drumlin A streamlined, elongate hill formed when a glacier overrides glacial till.

dry-bottom (polar) glacier A glacier so cold that its base remains frozen to the substrate.

dry wash The channel of an ephemeral stream when empty of water.

dry well (1) A well that does not supply water because the well has been drilled into an aquitard or into rock that lies above the water table; (2) a well that does not yield oil, even though it has been drilled into an anticipated reservoir.

ductile (plastic) deformation The bending and flowing of a material (without cracking and breaking) subjected to stress.

dune A pile of sand generally formed by deposition from the wind.

dust storm An event in which strong winds hit unvegetated land, strip off the topsoil, and send it skyward to form rolling dark clouds that block out the Sun.

dynamic metamorphism Metamorphism that occurs as a consequence of shearing alone, with no change in temperature or pressure.

dynamo A power plant generator in which water or wind power spins an electrical conductor around a permanent magnet.

dynamothermal metamorphism Metamorphism that involves heat, pressure, and shearing.

earthquake A vibration caused by the sudden breaking or frictional sliding of rock in the Earth.

earthquake belt A relatively narrow, distinct belt of earthquakes that defines the position of a plate boundary.

earthquake engineering The design of buildings that can withstand shaking.

earthquake zoning The determination of where land is relatively stable and where it might collapse because of seismicity.

Earth System The global interconnecting web of physical and biological phenomena involving the solid Earth, the hydrosphere, and the atmosphere.

ebb tide The falling tide.

eccentricity cycle The cycle of the gradual change of the Earth's orbit from a more circular to a more elliptical shape; the cycle takes around 100,000 years.

ecliptic The plane defined by a planet's orbit.

ecosystem An environment and its inhabitants.

eddy An isolated, ring-shaped current of water.

effusive eruption An eruption that yields mostly lava, not ash.

Ekman spiral The change in flow direction of water with depth, caused by the Coriolis effect.

Ekman transport The overall movement of a mass of water, resulting from the Eckman spiral, in a direction 90° to the wind direction.

elastic strain A change in shape of a material that disappears instantly when stress is removed.

electromagnet An electrical device that produces a magnetic field.

electron microprobe A laboratory instrument that can focus a beam of electrons on a small part of a mineral grain in order to create a signal that defines its chemical composition.

El Niño The flow of warm water eastward from the Pacific Ocean that reverses the upwelling of cold water along the western coast of South America and causes significant global changes in weather patterns.

embayment A low area of coastal land.

emergent coast A coast where the land is rising relative to sea level or sea level is falling relative to the land.

end moraine (terminal moraine) A low, sinuous ridge of till that develops when the terminus (toe) of a glacier stalls in one position for a while.

energy The capacity to do work.

energy resource Something that can be used to produce work; in a geologic context, a material (such as oil, coal, wind, flowing water) that can be used to produce energy.

eon The largest subdivision of geologic time.

epeirogenic movement The gradual uplift or subsidence of a broad region of the Earth's surface.

epeirogeny An event of epeirogenic movement; the term is usually used in reference to the formation of broad mid-continent domes and basins.

ephemeral (intermittent) stream A stream whose bed lies above the water table, so that the stream flows only when the rate at which water enters the stream from rainfall or meltwater exceeds the rate at which water infiltrates the ground below.

epicenter The point on the surface of the Earth directly above the focus of an earthquake.

epicontinental sea A shallow sea overlying a continent.

epoch An interval of geologic time representing the largest subdivision of a period.

equant A term for a grain that has the same dimensions in all directions.

equatorial low The area of low pressure that develops over the equator because of the intertropical convergence zone.

equinox One of two days out of the year (September 22 and March 21) in which the Sun is directly overhead at noon at the equator.

era An interval of geologic time representing the largest subdivision of the Phanerozoic Eon.

erg Sand seas formed by the accumulation of dunes in a desert.

erosion The grinding away and removal of Earth's surface materials by moving water, air, or ice.

erosional coast A coastline where sediment is not accumulating and wave action grinds away at the shore.

erosional landform A landform that results from the breakdown and removal of rock or sediment.

erratic A boulder or cobble that was picked up by a glacier and deposited hundreds of kilometers away from the outcrop from which it detached.

esker A ridge of sorted sand and gravel that snakes across a ground moraine; the sediment of an esker was deposited in subglacial meltwater tunnels.

estuary An inlet in which seawater and river water mix, created when a coastal valley is flooded because of either rising sea level or land subsidence.

Eubacteria The kingdom of "true bacteria."

euhedral crystal A crystal whose faces are well formed and whose shape reflects crystal form.

eukaryotic cell A cell with a complex internal structure, capable of building multicellular organisms.

eustatic sea-level change A global rising or falling of the ocean surface.

evaporate To change from liquid to vapor.

evapotranspiration The sum of evaporation from bodies of water and the ground surface and transpiration from plants and animals.

exfoliation The process by which an outcrop of rock splits apart into onion-like sheets along joints that lie parallel to the ground surface.

exhumation The process (involving uplift and erosion) that returns deeply buried rocks to the surface.

exotic terrane A block of land that collided with a continent along a convergent margin and attached to the continent; the term "exotic" implies that the land was not originally part of the continent to which it is now attached.

expanding Universe theory The theory that the whole Universe must be expanding because galaxies in every direction seem to be moving away from us.

explosive eruptions Violent volcanic eruptions that produce clouds and avalanches of pyroclastic debris.

external process A geomorphologic process—such as downslope movement, erosion, or deposition—that is the consequence of gravity or of the interaction between the solid Earth and its fluid envelope (air and water). Energy for these processes comes from gravity and sunlight.

extinction The death of the last members of a species so that there are no parents to pass on their genetic traits to offspring.

extinct volcano A volcano that was active in the past but has now shut off entirely and will not erupt in the future.

extraordinary fossil A rare fossilized relict, or trace, of the soft part of an organism.

extrusive igneous rock Rock that forms by the freezing of lava above ground, after it flows or explodes out (extrudes) onto the surface and comes into contact with the atmosphere or ocean.

eye The relative calm in the center of a hurricane.

eye wall A rotating vertical cylinder of clouds surrounding the eye of a hurricane.

facies (1) Sedimentary: a group of rocks and primary structures indicative of a given depositional environment; (2) Metamorphic: a set of metamorphic mineral assemblages formed under a given range of pressures and temperatures.

fault A fracture on which one body of rock slides past another.

fault-block mountains An outdated term for a narrow, elongate range of mountains that develops in a continental-rift setting as normal faulting drops down blocks of crust, or tilts blocks.

fault breccia Fragmented rock in which angular fragments were formed by brittle fault movement; fault breccia occurs along a fault.

fault creep Gradual movement along a fault that occurs in the absence of an earthquake.

fault gouge Pulverized rock consisting of fine powder that lies along fault surfaces; gouge forms by crushing and grinding.

faulting Slip events along a fault.

fault scarp A small step on the ground surface where one side of a fault has moved vertically with respect to the other.

fault system A grouping of numerous related faults.

fault trace (or line) The intersection between a fault and the ground surface.

felsic An adjective used in reference to igneous rocks that are rich in elements forming feldspar and quartz.

Ferrel cells The name given to the middle-latitude convection cells in the atmosphere.

fetch The distance across a body of water along which a wind blows to build waves.

fine-grained A textural term for rock consisting of many fine grains or clasts.

firn Compacted granular ice (derived from snow) that forms where snow is deeply buried; if buried more deeply, firn turns into glacial ice.

fission track A line of damage formed in the crystal lattice of a mineral by the impact of an atomic particle ejected during the decay of a radioactive isotope.

fissure A conduit in a magma chamber in the shape of a long crack through which magma rises and erupts at the surface.

fjord A deep, glacially carved, U-shaped valley flooded by rising sea level.

flank eruption An eruption that occurs when a secondary chimney, or fissure, breaks through the flank of a volcano.

flash flood A flood that occurs during unusually intense rainfall or as the result of a dam collapse, during which the floodwaters rise very fast.

flexing The process of folding in which a succession of rock layers bends and slip occurs between the layers.

flocculation The clumping together of clay suspended in river water into bunches that are large enough to settle out.

flood An event during which the volume of water in a stream becomes so great that it covers areas outside the stream's normal channel.

flood basalt Vast sheets of basalt that spread from a volcanic vent over an extensive surface of land; they may form where a rift develops above a continental hot spot, and where lava is particularly hot and has low viscosity.

floodplain The flat land on either side of a stream that becomes covered with water during a flood.

floodplain flood A flood during which a floodplain is submerged.

flood stage The stage when water reaches the top of a stream channel.

flood tide The rising tide.

floodway A mapped region likely to be flooded, in which people avoid constructing buildings.

flow fold A fold that forms when the rock is so soft that it behaves like weak plastic.

flowstone A sheet of limestone that forms along the wall of a cave when groundwater flows along the surface of the wall.

fluvial deposit Sediment deposited in a stream channel, along a stream bank, or on a floodplain.

flux Flow.

focus The location where a fault slips during an earthquake (hypocenter).

fog A cloud that forms at ground level.

fold A bend or wrinkle of rock layers or foliation; folds form as a consequence of ductile deformation.

fold axis An imaginary line that, when moved parallel to itself, can trace out the shape of a folded surface.

fold-thrust belt An assemblage of folds and related thrust faults that develop above a detachment fault.

foliation Layering formed as a consequence of the alignment of mineral grains, or of compositional banding in a metamorphic rock.

foraminifera Microscopic plankton with calcitic shells, components of some limestones.

foreland sedimentary basin A basin located under the plains adjacent to a mountain front, which develops as the weight of the mountains pushes the crust down, creating a depression that traps sediment.

foreshocks The series of smaller earthquakes that precede a major earthquake.

foreshore zone The zone of beach regularly covered and uncovered by rising and falling tides.

formation *Stratigraphic formation.*

fossil The remnant, or trace, of an ancient living organism that has been preserved in rock or sediment.

fossil assemblage A group of fossil species found in a specific sequence of sedimentary rock.

fossil correlation A determination of the stratigraphic relation between two sedimentary rock units, reached by studying fossils.

fossil fuel An energy resource such as oil or coal that comes from organisms that lived long ago, and thus stores solar energy that reached the Earth then.

fossiliferous limestone Limestone consisting of abundant fossil shells and shell fragments.

fossilization The process of forming a fossil.

fractional crystallization The process by which a magma becomes progressively more silicic as it cools, because early-formed crystals settle out.

fracture zone A narrow band of vertical fractures in the ocean floor; fracture zones lie roughly at right angles to a mid-ocean ridge, and the actively slipping part of a fracture zone is a transform fault.

fresh rock Rock whose mineral grains have their original composition and shape.

friction Resistance to sliding on a surface.

fringing reef A coral reef that forms directly along the coast.

front The boundary between two air masses.

frost wedging The process in which water trapped in a joint freezes, forces the joint open, and may cause the joint to grow.

fuel rod A metal tube that holds the nuclear fuel in a nuclear reactor.

Fujita scale A scale that distinguishes among tornadoes on the basis of wind speed, path dimensions, and possible damage.

Ga Billions of years ago (abbreviation).

gabbro A coarse-grained, intrusive mafic igneous rock.

Gaia The term used for the Earth System, with the implication that it resembles a complex living entity.

galaxy An immense system of hundreds of billions of stars.

gene An individual component of the DNA code that guides the growth and development of an organism.

genetics The study of genes and how they transmit information.

geocentric Universe concept An ancient Greek idea suggesting that the Earth sat motionless in the center of the Universe while stars and other planets and the Sun orbited around it.

geochronology The science of dating geologic events in years.

geode A cavity in which euhedral crystals precipitate out of water solutions passing through a rock.

geographical pole The locations (north and south) where the Earth's rotational axis intersects the planet's surface.

geologic column A composite stratigraphic chart that represents the entirety of the Earth's history.

geologic history The sequence of geologic events that has taken place in a region.

geologic map A map showing the distribution of rock units and structures across a region.

geologic time The span of time since the formation of the Earth.

geologic time scale A scale that describes the intervals of geologic time.

geology The study of the Earth, including our planet's composition, behavior, and history.

geotherm The change in temperature with depth in the Earth.

geothermal energy Heat and electricity produced by using the internal heat of the Earth.

geothermal gradient The rate of change in temperature with depth.

geothermal region A region of current or recent volcanism in which magma or very hot rock heats up groundwater, which may discharge at the surface in the form of hot springs and/or geysers.

geyser A fountain of steam and hot water that erupts periodically from a vent in the ground in a geothermal region.

glacial abrasion The process by which clasts embedded in the base of a glacier grind away at the substrate as the glacier flows.

glacial advance The forward movement of a glacier's toe when the supply of snow exceeds the rate of ablation.

glacial drift Sediment deposited in glacial environments.

glacial incorporation The process by which flowing ice surrounds and incorporates debris.

glacial marine Sediment consisting of ice-rafted clasts mixed with marine sediment.

glacial outwash Coarse sediment deposited on a glacial outwash plain by meltwater streams.

glacially polished surface A polished rock surface created by the glacial abrasion of the underlying substrate.

glacial plucking (or quarrying) The process by which a glacier breaks off and carries away fragments of bedrock.

glacial rebound The process by which the surface of a continent rises back up after an overlying continental ice sheet melts away and the weight of the ice is removed.

glacial retreat The movement of a glacier's toe back toward the glacier's origin; glacial retreat occurs if the rate of ablation exceeds the rate of supply.

glacial subsidence The sinking of the surface of a continent caused by the weight of an overlying glacial ice sheet.

glacial till Sediment transported by flowing ice and deposited beneath a glacier or at its toe.

glaciation A period of time during which glaciers grew and covered substantial areas of the continents.

glacier A river or sheet of ice that slowly flows across the land surface and lasts all year long.

glass A solid in which atoms are not arranged in an orderly pattern.

glassy igneous rock Igneous rock consisting entirely of glass, or of tiny crystals surrounded by a glass matrix.

glide horizon The surface along which a slump slips.

global change The transformations or modifications of both physical and biological components of the Earth System through time.

global circulation The movement of volumes of air in paths that ultimately take it around the planet.

global climate change Transformations or modifications in Earth's climate over time.

global cooling A fall in the average atmospheric temperature.

global positioning system (GPS) A satellite system people can use to measure rates of movement of the Earth's crust relative to one another, or simply to locate their position on the Earth's surface.

global warming A rise in the average atmospheric temperature.

gneiss A compositionally banded metamorphic rock typically composed of alternating dark- and light-colored layers.

Gondwana A supercontinent that consisted of today's South America, Africa, Antarctica, India, and Australia. Also called Gondwanaland.

graben A down-dropped crustal block bounded on either side by a normal fault dipping toward the basin.

gradualism The theory that evolution happens at a constant, slow rate.

grain A fragment of a mineral crystal or of a rock.

grain rotation The process by which rigid, inequant mineral grains distributed through a soft matrix may rotate into parallelism as the rock changes shape owing to differential stress.

granite A coarse-grained intrusive silicic igneous rock.

granulite facies A set of metamorphic mineral assemblages formed at very high pressures and temperatures.

gravitational spreading A process of lateral spreading that occurs in a material because of the weakness of the material; gravitational spreading causes continental glaciers to grow and mountain belts to undergo orogenic collapse.

graywacke An informal term used for sedimentary rock consisting of sand-size up to small-pebble-size grains of quartz and rock fragments all mixed together in a muddy matrix; typically, graywacke occurs at the base of a graded bed.

greenhouse conditions (greenhouse period) Relatively warm global climate leading to the rising of sea level for an interval of geologic time.

greenhouse effect The trapping of heat in the Earth's atmosphere by carbon dioxide and other greenhouse gases, which absorb infrared radiation; somewhat analogous to the effect of glass in a greenhouse.

greenhouse gases Atmospheric gases, such as carbon dioxide and methane, that regulate the Earth's atmospheric temperature by absorbing infrared radiation.

greenschist facies A set of metamorphic mineral assemblages formed under relatively low pressures and temperatures.

greenstone A low-grade metamorphic rock formed from basalt; if foliated, the rock is called greenschist.

Greenwich mean time (GMT) The time at the astronomical observatory in Greenwich, England; time in all other time zones is set in relation to GMT.

Grenville orogeny The orogeny that occurred about 1 billion years ago and yielded the belt of deformed and metamorphosed rocks that underlie the eastern fifth of the North American continent.

groin A concrete or stone wall built perpendicular to a shoreline in order to prevent beach drift from removing sand.

ground moraine A thin, hummocky layer of till left behind on the land surface during a rapid glacial recession.

groundwater Water that resides under the surface of the Earth, mostly in pores or cracks of rock or sediment.

group A succession of stratigraphic formations that have been lumped together, making a single, thicker stratigraphic entity.

growth ring A rhythmic layering that develops in trees, travertine deposits, and shelly organisms as a consequence of seasonal changes.

gusher A fountain of oil formed when underground pressure causes the oil to rise on its own out of a drilled hole.

guyot A seamount that had a coral reef growing on top of it, so that it is now flat-crested.

gymnosperm A plant whose seeds are "naked," not surrounded by a fruit.

gyre A large circular flow pattern of ocean surface currents.

Hadean The oldest of the Precambrian eons; the time between Earth's origin and the formation of the first rocks that have been preserved.

Hadley cells The name given to the low-latitude convection cells in the atmosphere.

hail Falling ice balls from the sky, formed when ice crystallizes in turbulent storm clouds.

hail streak An approximately 2-by-10-km stretch of ground, elongate in the direction of a storm, onto which hail has fallen.

half-graben A wedge-shaped basin in cross section that develops as the hanging-wall block above a normal fault slides down and rotates; the basin develops between the fault surface and the top surface of the rotated block.

half-life The time it takes for half of a group of a radioactive element's isotopes to decay.

halocline The boundary in the ocean between surface-water and deep-water salinities.

hamada Barren rocky highlands in a desert.

hanging valley A glacially carved tributary valley whose floor lies at a higher elevation than the floor of the trunk valley.

hanging wall The rock or sediment above an inclined fault plane.

hard water Groundwater that contains dissolved calcium and magnesium, usually after passing through limestone or dolomite.

head (1) The elevation of the water table above a reference horizon; (2) the edge of ice at the origin of a glacier.

headland A place where a hill or cliff protrudes into the sea.

head scarp The distinct step along the upslope edge of a slump where the regolith detached.

headward erosion The process by which a stream channel lengthens up its slope as the flow of water increases.

headwaters The beginning point of a stream.

heat Thermal energy resulting from the movement of molecules.

heat capacity A measure of the amount of heat that must be added to a material to change its temperature.

heat flow The rate at which heat rises from the Earth's interior up to the surface.

heat-transfer melting Melting that results from the transfer of heat from a hotter magma to a cooler rock.

heliocentric Universe concept An idea proposed by Greek philosophers around 250 B.C.E. suggesting that all heavenly objects including the Earth orbited the Sun.

Hercynian orogen The late Paleozoic orogen that affected parts of Europe; a continuation of the Alleghenian orogen.

heterosphere A term for the upper portion of the atmosphere in which gases separate into distinct layers on the basis of composition.

hiatus The interval of time between deposition of the youngest rock below an unconformity and deposition of the oldest rock above the unconformity.

high-altitude westerlies Westerly winds at the top of the troposphere.

high-grade metamorphic rocks Rocks that metamorphose under relatively high temperatures.

high-level waste Nuclear waste containing greater than 1 million times the safe level of radioactivity.

hinge The portion of a fold where curvature is greatest.

hogback A steep-sided ridge of steeply dipping strata.

Holocene The period of geologic time since the last glaciation.

Holocene climatic maximum The period from 5,000 to 6,000 years ago, when Holocene temperatures reached a peak.

homosphere The lower part of the atmosphere, in which the gases have stirred into a homogenous mixture.

hoodoo The local name for the brightly colored shale and sandstone chimneys found in Bryce Canyon National Park in Utah.

horn A pointed mountain peak surrounded by at least three cirques.

hornfels Rock that undergoes metamorphism simply because of a change in temperature, without being subjected to differential stress.

horse latitudes The region of the subtropical high in which winds are weak.

horst The high block between two grabens.

hot spot A location at the base of the lithosphere, at the top of a mantle plume, where temperatures can cause melting.

hot-spot track A chain of now-dead volcanoes transported off the hot spot by the movement of a lithosphere plate.

hot-spot volcano An isolated volcano not caused by movement at a plate boundary, but rather by the melting of a mantle plume.

hot spring A spring that emits water ranging in temperature from about 30° to 104°C.

hummocky surface An irregular and lumpy ground surface.

hurricane A huge rotating storm, resembling a giant spiral in map view, in which sustained winds blow over 119 km per hour.

hurricane track The path a hurricane follows.

hyaloclastite A rubbly extrusive rock consisting of glassy debris formed in a submarine or sub-ice eruption.

hydration The absorption of water into the crystal structure of minerals; a type of chemical weathering.

hydraulic conductivity The coefficient K in Darcy's law; hydraulic conductivity takes into account the permeability of the sediment or rock as well as the fluid's viscosity.

hydraulic gradient The slope of the water table.

hydrocarbon A chain-like or ring-like molecule made of hydrogen and carbon atoms; petroleum and natural gas are hydrocarbons.

hydrocarbon system The association of source rock, migration pathway, reservoir rock, seal, and trap geometry that leads to the occurrence of a hydrocarbon reserve.

hydrologic cycle The continual passage of water from reservoir to reservoir in the Earth System.

hydrolysis The process in which water chemically reacts with minerals and breaks them down.

hydrosphere The Earth's water, including surface water (lakes, rivers, and oceans), groundwater, and liquid water in the atmosphere.

hydrothermal deposit An accumulation of ore minerals precipitated from hot-water solutions circulating through a magma or through the rocks surrounding an igneous intrusion.

hypsometric curve A graph that plots surface elevation on the vertical axis and the percentage of the Earth's surface on the horizontal axis.

ice age An interval of time in which the climate was colder than it is today, glaciers occasionally advanced to cover large areas of the continents, and mountain glaciers grew; an ice age can include many glacials and interglacials.

iceberg A large block of ice that calves off the front of a glacier and drops into the sea.

icehouse period A period of time when the Earth's temperature was cooler than it is today and ice ages could occur.

ice-margin lake A meltwater lake formed along the edge of a glacier.

ice-rafted sediment Sediment carried out to sea by icebergs.

ice sheet A vast glacier that covers the landscape.

ice shelf A broad, flat region of ice along the edge of a continent formed where a continental glacier flowed into the sea.

ice stream A portion of a glacier that travels much more quickly than adjacent portions of the glacier.

ice tongue The portion of a valley glacier that has flowed out into the sea.

igneous rock Rock that forms when hot molten rock (magma or lava) cools and freezes solid.

ignimbrite Rock formed when deposits of pyroclactic flows solidify.

inactive fault A fault that last moved in the distant past and probably won't move again in the near future, yet is still recognizable because of displacement across the fault plane.

inactive sand The sand along a coast that is buried beneath a layer of active sand and moves only during severe storms or not at all.

incised meander A meander that lies at the bottom of a steep-walled canyon.

index minerals Minerals that serve as good indicators of metamorphic grade.

induced seismicity Seismic events caused by the actions of people (e.g. filling a reservoir, that lies over a fault, with water).

industrial minerals Minerals that serve as the raw materials for manufacturing chemicals, concrete, and wallboard, among other products.

inequant A term for a mineral grain whose length and width are not the same.

inertia The tendency of an object at rest to remain at rest.

infiltrate Seep down into.

injection well A well in which a liquid is pumped down into the ground under pressure so that it passes from the well back into the pore space of the rock or regolith.

inner core The inner section of the core 5,155 km deep to the Earth's center at 6,371 km, and consisting of solid iron alloy.

inselberg An isolated mountain or hill in a desert landscape created by progressive cliff retreat, so that the hill is surrounded by a pediment or an alluvial fan.

insolation Exposure to the Sun's rays.

interglacial A period of time between two glaciations.

interior basin A basin with no outlet to the sea.

interlocking texture The texture of crystalline rocks in which mineral grains fit together like pieces of a jigsaw puzzle.

internal process A process in the Earth System, such as plate motion, mountain building, or volcanism, ultimately caused by Earth's internal heat.

intertidal zone The area of coastal land across which the tide rises and falls.

intertropical convergence zone The equatorial convergence zone in the atmosphere.

intraplate earthquakes Earthquakes that occur away from plate boundaries.

intrusive contact The boundary between country rock and an intrusive igneous rock.

intrusive igneous rock Rock formed by the freezing of magma underground.

ionosphere The interval of Earth's atmosphere, at an elevation between 50 and 400 km, containing abundant positive ions.

iron catastrophe The proposed event very early in Earth history when the Earth partly melted and molten iron sank to the center to form the core.

isobar A line on a map along which the air has a specified pressure.

isograd (1) A line on a pressure-temperature graph along which all points are taken to be at the same metamorphic grade; (2) A line on a map making the first appearance of a metamorphic index mineral.

isostasy (or isostatic equilibrium) The condition that exists when the buoyancy force pushing lithosphere up equals the gravitational force pulling lithosphere down.

isostatic compensation The process in which the surface of the crust slowly rises or falls to reestablish isostatic equilibrium after a geologic event changes the density or thickness of the lithosphere.

isotherm Lines on a map or cross section along which the temperature is constant.

isotopes Different versions of a given element that have the same atomic number but different atomic weights.

jet stream A fast-moving current of air that flows at high elevations.

jetty A manmade wall that protects the entrance to a harbor.

joints Naturally formed cracks in rocks.

joint set A group of systematic joints.

Jovian A term used to describe the outer gassy, Jupiter-like planets (gas-giant planets).

kame A stratified sequence of lateral-moraine sediment that's sorted by water flowing along the edge of a glacier.

karst landscape A region underlain by caves in limestone bedrock; the collapse of the caves creates a landscape of sinkholes separated by higher topography, or of limestone spires separated by low areas.

kerogen The waxy molecules into which the organic material in shale transforms on reaching about 100°C. At higher temperatures, kerogen transforms into oil.

kettle hole A circular depression in the ground made when a block of ice calves off the toe of a glacier, becomes buried by till, and later melts.

knob-and-kettle topography A land surface with many kettle holes separated by round hills of glacial till.

K-T boundary event The mass extinction that happened at the end of the Cretaceous Period, 65 million years ago, possibly because of the collision of an asteroid with the Earth.

lag deposit The coarse sediment left behind in a desert after wind erosion removes the finer sediment.

lagoon A body of shallow seawater separated from the open ocean by a barrier island.

lahar A thick slurry formed when volcanic ash and debris mix with water, either in rivers or from rain or melting snow and ice on the flank of a volcano.

landslide A sudden movement of rock and debris down a nonvertical slope.

landslide-potential map A map on which regions are ranked according to the likelihood that a mass movement will occur.

land subsidence Sinking elevation of the ground surface; the process may occur over an aquifer that is slowly draining and decreasing in volume because of pore collapse.

La Niña Years in which the El Niño event is not strong.

lapilli Marble-to-plum-sized fragments of pyroclastic debris.

Laramide orogeny The mountain-building event that lasted from about 80 Ma to 40 Ma, in western North America; in the United States, it formed the Rocky Mountains as a result of basement uplift and the warping of the younger overlying strata into large monoclines.

latent heat of condensation The heat released during condensation, which comes only from a change in state.

lateral moraine A strip of debris along the side margins of a glacier.

laterite soil Soil formed over iron-rich rock in a tropical environment, consisting primarily of a dark-red mass of insoluble iron and/or aluminum oxide.

Laurentia A continent in the early Paleozoic Era composed of today's North America and Greenland.

Laurentide ice sheet An ice sheet that spread over northeastern Canada during the Pleistocene ice age(s).

lava Molten rock that has flowed out onto the Earth's surface.

lava dome A dome-like mass of rhyolitic lava that accumulates above the eruption vent.

lava flows Sheets or mounds of lava that flow onto the ground surface or sea floor in molten form and then solidify.

lava lake A large pool of lava produced around a vent when lava fountains spew forth large amounts of lava in a short period of time.

lava tube The empty space left when a lava tunnel drains; this happens when the surface of a lava flow solidifies while the inner part of the flow continues to stream downslope.

leach To dissolve and carry away.

leader A conductive path stretching from a cloud toward the ground, along which electrons leak from the base of the cloud, and which provides the start for a lightning flash to the ground.

lightning flash A giant spark or pulse of current that jumps across a gap of charge separation.

light year The distance that light travels in one Earth year (about 6 trillion miles or 9.5 trillion km).

lignite Low-rank coal that consists of 50% carbon.

limb The side of a fold, showing less curvature than at the hinge.

limestone Sedimentary rock composed of calcite.

liquification The process by which wet sediment becomes a slurry; liquification may be triggered by earthquake vibrations.

lithification The transformation of loose sediment into solid rock through compaction and cementation.

lithologic correlation A correlation based on similarities in rock type.

lithosphere The relatively rigid, nonflowable, outer 100- to 150-km-thick layer of the Earth; constituting the crust and the top part of the mantle.

little ice age A period of cooler temperatures, between 1500 and 1800 C.E., during which many glaciers advanced.

local base level A base level upstream from a drainage network's mouth.

lodgment till A flat layer of till smeared out over the ground when a glacier overrides an end moraine as it advances.

loess Layers of fine-grained sediments deposited from the wind; large deposits of loess formed from fine-grained glacial sediment blown off outwash plains.

longitudinal (seif) dune A dune formed when there is abundant sand and a strong, steady wind, and whose axis lies parallel to the wind direction.

longitudinal profile A cross-sectional image showing the variation in elevation along the length of a river.

longshore current A current that flows parallel to a beach.

lower mantle The deepest section of the mantle, stretching from 670 km down to the core-mantle boundary.

low-grade metamorphic rocks Rocks that underwent metamorphism at relatively low temperatures.

low-velocity zone The asthenosphere underlying oceanic lithosphere in which seismic waves travel more slowly, probably because rock has partially melted.

luster The way a mineral surface scatters light.

L-waves (*love waves*) Surface seismic waves that cause the ground to ripple back and forth, creating a snake-like movement.

Ma Millions of years ago (abbreviation).

macrofossil A fossil large enough to be seen with the naked eye.

mafic A term used in reference to magmas or igneous rocks that are relatively poor in silica and rich in iron and magnesium.

magma Molten rock beneath the Earth's surface.

magma chamber A space below ground filled with magma.

magma contamination The process in which flowing magma incorporates components of the country rock through which it passes.

magmatic deposit An ore deposit formed when sulfide ore minerals accumulate at the bottom of a magma chamber.

magnetic anomaly The difference between the expected strength of the Earth's magnetic field at a certain location and the actual measured strength of the field at that location.

magnetic declination The angle between the direction a compass needle points at a given location and the direction of true north.

magnetic field The region affected by the force emanating from a magnet.

magnetic field lines The trajectories along which magnetic particles would align, or charged particles would flow, if placed in a magnetic field.

magnetic force The push or pull exerted by a magnet.

magnetic inclination The angle between a magnetic needle free to pivot on a horizontal axis and a horizontal plane parallel to the Earth's surface.

magnetic reversal The change of the Earth's magnetic polarity; when a reversal occurs, the field flips from normal to reversed polarity, or vice versa.

magnetic-reversal chronology The history of magnetic reversals through geologic time.

magnetization The degree to which a material can exert a magnetic force.

magnetometer An instrument that measures the strength of the Earth's magnetic field.

magnetosphere The region protected from the electrically charged particles of the solar winds by Earth's magnetic field.

magnetostratigraphy The comparison of the pattern of magnetic reversals in a sequence of strata, with a reference column showing the succession of reversals through time.

manganese nodules Lumpy accumulations of manganese-oxide minerals precipitated onto the sea floor.

mantle The thick layer of rock below the Earth's crust and above the core.

mantle plume A column of very hot rock rising up through the mantle.

marble A metamorphic rock composed of calcite and transformed from a protolith of limestone.

mare The broad darker areas on the Moon's surface, which consist of flood basalts that erupted over 3 billion years ago and spread out across the Moon's lowlands.

marginal sea A small ocean basin created when sea-floor spreading occurs behind an island arc.

maritime tropical air mass A mass of air that originates over tropical or subtropical oceanic regions.

marsh A wetland dominated by grasses.

mass-extinction event A time when vast numbers of species abruptly vanish.

mass movement (or **mass wasting**) The gravitationally caused downslope transport of rock, regolith, snow, or ice.

matrix Finer-grained material surrounding larger grains in a rock.

meander A snake-like curve along a stream's course.

meandering stream A reach of stream containing many meanders (snake-like curves).

meander neck A narrow isthmus of land separating two adjacent meanders.

mean sea level The average level between the high and low tide over a year at a given point.

mechanical weathering *Physical weathering.*

medial moraine A strip of sediment in the interior of a glacier, parallel to the flow direction of the glacier, formed by the lateral moraines of two merging glaciers.

Medieval warm period A period of high temperatures in the Middle Ages.

melt Molten (liquid) rock.

meltdown The melting of the fuel rods in a nuclear reactor that occurs if the rate of fission becomes too fast and the fuel rods become too hot.

melting curve The line defining the range of temperatures and pressures at which a rock melts.

melting temperature The temperature at which the thermal vibration of the atoms or ions in the lattice of a mineral is sufficient to break the chemical bonds holding them to the lattice, so a material transforms into a liquid.

meltwater lake A lake fed by glacial meltwater.

Mercalli intensity scale An earthquake characterization scale based on the amount of damage that the earthquake causes.

mesa A large, flat-topped hill (with a surface area of several square km) in an arid region.

mesopause The boundary that marks the top of the mesosphere of Earth's atmosphere.

mesosphere The cooler layer of atmosphere overlying the stratosphere.

Mesozoic The middle of the three Phanerozoic eras; it lasted from 245 Ma to 65 Ma.

metal A solid composed almost entirely of atoms of metallic elements; it is generally opaque, shiny, smooth, and malleable, and can conduct electricity.

metallic bond A chemical bond in which the outer atoms are attached to each other in such a way that electrons flow easily from atom to atom.

metamorphic aureole The region around a pluton, stretching tens to hundreds of meters out, in which heat transferred into the country rock and metamorphosed the country rock.

metamorphic facies A set of metamorphic mineral assemblages indicative of metamorphism under a specific range of pressures and temperatures.

metamorphic foliation A fabric defined by parallel surfaces or layers that develop in a rock as a result of metamorphism; schistocity and gneissic layering are examples.

metamorphic mineral assemblage A group of minerals that form in a rock as a result of metamorphism.

metamorphic rock Rock that forms when preexisting rock changes into new rock as a result of an increase in pressure and temperature and/or shearing under elevated temperatures; metamorphism occurs without the rock first becoming a melt or a sediment.

metamorphic zone The region between two metamorphic isograds, typically named after an index mineral found within the region.

metamorphism The process by which one kind of rock transforms into a different kind of rock.

metasomatism The process by which a rock's overall chemical composition changes during metamorphism because of reactions with hot water that bring in or remove elements.

meteoric water Water that falls to Earth from the atmosphere as either rain or snow.

meteorite A piece of rock or metal alloy that fell from space and landed on Earth.

micrite Limestone consisting of lime mud (i.e., very fine-grained limestone).

microfossil A fossil that can be seen only with a microscope or an electron microscope.

mid-latitude (wave) cyclone The circulation of air around large, low-pressure masses.

mid-ocean ridge A 2-km-high submarine mountain belt that forms along a divergent oceanic plate boundary.

migmatite A rock formed when gneiss is heated high enough so that it begins to partially melt, creating layers, or lenses, of new igneous rock that mix with layers of the relict gneiss.

Milankovitch cycles Climate cycles that occur over tens to hundreds of thousands of years, because of changes in Earth's orbit and tilt.

mine A site at which ore is extracted from the ground.

mineral A homogenous, naturally occurring, solid inorganic substance with a definable chemical composition and an internal structure characterized by an orderly arrangement of atoms, ions, or molecules in a lattice. Most minerals are inorganic.

mineral classes Groups of minerals distinguished from each other on the basis of chemical composition.

mineral resources The minerals extracted from the Earth's upper crust for practical purposes.

Mississippi Valley–type (MVT) ore An ore deposit, typically in dolostone, containing lead- and zinc-bearing minerals that precipitated from groundwater that had moved up from several km depth in the upper crust; such deposits occur in the upper Mississippi Valley.

Moho The seismic-velocity discontinuity that defines the boundary between the Earth's crust and mantle.

Mohs hardness scale A list of ten minerals in a sequence of relative hardness, with which other minerals can be compared.

mold A cavity in sedimentary rock left behind when a shell that once filled the space weathers out.

monocline A fold in the land surface whose shape resembles that of a carpet draped over a stair step.

monsoon A seasonal reversal in wind direction that causes a shift from a very dry season to a very rainy season in some regions of the world.

moraine A sediment pile composed of till deposited by a glacier.

mountain front The boundary between a mountain range and adjacent plains.

mountain (alpine) glacier A glacier that exists in or adjacent to a mountainous region.

mountain ice cap A mound of ice that submerges peaks and ridges at the crest of a mountain range.

mouth The outlet of a stream where it discharges into another stream, a lake, or a sea.

mudflow A downslope movement of mud at slow to moderate speed.

mud pot A viscous slurry that forms in a geothermal region when hot water or steam rises into soils rich in volcanic ash and clay.

mudstone Very fine-grained sedimentary rock that will not easily split into sheets.

mylonite Rock formed during dynamic metamorphism and characterized by foliation that lies roughly parallel to the fault (shear zone) involved in the shearing process; mylonites have very fine grains formed by the nonbrittle subdivision of larger grains.

native metal A naturally occurring pure mass of a single metal in an ore deposit.

natural arch An arch that forms when erosion along joints leaves narrow walls of rock; when the lower part of the wall erodes while the upper part remains, an arch results.

natural levees A pair of low ridges that appear on either side of a stream and develop as a result of the accumulation of sediment deposited naturally during flooding.

natural selection The process by which the fittest organisms survive to pass on their characteristics to the next generation.

neap tide An especially low tide that occurs when the angle between the direction of the Moon and the direction of the Sun is 90°.

nebula A cloud of gas or dust in space.

nebula theory of planet formation The concept that planets grow out of rings of gas, dust, and ice surrounding a new-born star.

negative anomaly An area where the magnetic field strength is less than expected.

negative feedback Feedback that slows a process down or reverses it.

Neocrystallization The growth of new crystals, not in the protolith, during metamorphism.

Nevadan orogeny A convergent-margin mountain-building event that took place in western North America during the Late Jurassic Period.

nonconformity A type of unconformity at which sedimentary rocks overlie basement (older intrusive igneous rocks and/or metamorphic rocks).

nonflowing artesian well An artesian well in which water rises on its own up to a level that lies below the ground surface.

nonfoliated metamorphic rock Rock containing minerals that recrystallized during metamorphism, but which has no foliation.

nonmetallic mineral resources Mineral resources that do not contain metals; examples include building stone, gravel, sand, gypsum, phosphate, and salt.

nonplunging fold A fold with a horizontal hinge.

nonrenewable resource A resource that nature will take a long time (hundreds to millions of years) to replenish or may never replenish.

nonsystematic joints Short cracks in rocks that occur in a range of orientations and are randomly placed and oriented.

nor'easter A large, mid-latitude North American cyclone; when it reaches the East Coast, it produces strong winds that come out of the northeast.

normal fault A fault in which the hanging-wall block moves down the slope of the fault.

normal force The component of the gravitational force acting perpendicular to a slope.

normal polarity Polarity in which the paleomagnetic dipole has the same orientation as it does today.

normal stress The push or pull that is perpendicular to a surface.

North Atlantic deep-water mass The mass of cold, dense water that sinks in the north polar regions.

northeast tradewinds Surface winds that come out of the northeast and occur in the region between the equator and 30°N.

nuclear fuel Pellets of concentrated uranium oxide or a comparable radioactive material that can provide energy in a nuclear reactor.

nuclear fusion The process by which the nuclei of atoms fuse together, thereby creating new, larger atoms.

nuclear reactor The part of a nuclear power plant where the fission reactions occur.

nuée ardente *Pyroclastic flow.*

oasis A verdant region surrounded by desert, occurring at a place where natural springs provide water at the surface.

oblique-slip fault A fault in which sliding occurs diagonally along the fault plane.

obsidian An igneous rock consisting of a solid mass of volcanic glass.

occluded front A front that no longer intersects the ground surface.

oceanic crust The crust beneath the oceans; composed of gabbro and basalt, overlain by sediment.

oceanic lithosphere Lithosphere topped by oceanic crust; it reaches a thickness of 100 km.

Oil Age The period of human history, including our own, so named because the economy depends on oil.

oil field A region containing a significant amount of accessible oil underground.

oil reserve The known supply of oil held underground.

oil shale Shale containing kerogen.

oil trap A geologic configuration that keeps oil underground in the reservoir rock and prevents it from rising to the surface.

oil window The narrow range of temperatures under which oil can form in a source rock.

olistotrome A large, submarine slump block, buried and preserved.

ophiolite A slice of oceanic crust that has been thrust onto continental crust.

ordinary well A well whose base penetrates below the water table, and can thus provide water.

ore Rock containing native metals or a concentrated accumulation of ore minerals.

ore deposit An economically significant accumulation of ore.

ore minerals Minerals that have metal in high concentrations and in a form that can be easily extracted.

organic carbon Carbon that has been incorporated in an organism.

organic chemical A carbon-containing compound that occurs in living organisms, or that resembles such compounds; it consists of carbon atoms bonded to hydrogen atoms along with varying amounts of oxygen, nitrogen, and other chemicals.

organic coast A coast along which living organisms control landforms along the shore.

organic sedimentary rock Sedimentary rock (such as coal) formed from carbon-rich relicts of organisms.

organic shale Lithified, muddy, organic-rich ooze that contains the raw materials from which hydrocarbons eventually form.

orogen (or orogenic belt) A linear range of mountains.

orogenic collapse The process in which mountains begin to collapse under their own weight and spread out laterally.

orogeny A mountain-building event.

orographic barrier A landform that diverts air flow upward or laterally.

outcrop An exposure of bedrock.

outer core The section of the core, between 2,900 and 5,150 km deep, that consists of liquid iron alloy.

outwash plain A broad area of gravel and sandbars deposited by a braided stream network, fed by the meltwater of a glacier.

overburden The weight of overlying rock on rock buried deeper in the Earth's crust.

overriding plate (or slab) The plate at a subduction zone that overrides the downgoing plate.

oversaturated solution A solution that contains so much solute (dissolved ions) that precipitation begins.

oversized stream valley A large valley with a small stream running through it; the valley formed earlier when the flow was greater.

oxbow lake A meander that has been cut off yet remains filled with water.

oxidation reaction A reaction in which an element loses electrons; an example is the reaction of iron with air to form rust.

ozone O_3, an atmospheric gas that absorbs harmful ultraviolet radiation from the Sun.

ozone hole An area of the atmosphere, over polar regions, from which ozone has been depleted.

pahoehoe A lava flow with a surface texture of smooth, glassy, rope-like ridges.

paleoclimate The past climate of the Earth.

paleomagnetism The record of ancient magnetism preserved in rock.

paleopole The supposed position of the Earth's magnetic pole in the past, with respect to a particular continent.

paleosol Ancient soil preserved in the stratigraphic record.

Paleozoic The oldest era of the Phanerozoic Eon.

Pangaea A supercontinent that assembled at the end of the Paleozoic Era.

Pannotia A supercontinent that may have existed sometime between 800 Ma and 600 Ma.

parabolic dunes Dunes formed when strong winds break through transverse dunes to make new dunes whose ends point upwind.

parallax The apparent movement of an object seen from two different points not on a straight line from the object (e.g., from your two different eyes).

parallax method A trigonometric method used to determine the distance from the Earth to a nearby star.

parent isotope A radioactive isotope that undergoes decay.

partial melting The melting in a rock of the minerals with the lowest melting temperatures, while other minerals remain solid.

passive margin A continental margin that is not a plate boundary.

passive-margin basin A thick accumulation of sediment along a tectonically inactive coast, formed over crust that stretched and thinned when the margin first began.

patterned ground A polar landscape in which the ground splits into pentagonal or hexagonal shapes.

pause An elevation in the atmosphere where temperature stops decreasing and starts increasing, or vice versa.

peat Compacted and partially decayed vegetation accumulating beneath a swamp.

pedalfer soil A temperate-climate soil characterized by well-defined soil horizons and an organic A-horizon.

pediment The broad, nearly horizontal bedrock surface at the base of a retreating desert cliff.

pedocal soil Thin soil, formed in arid climates. It contains very little organic matter, but significant precipitated calcite.

pegmatite A coarse-grained igneous rock containing crystals of up to tens of centimeters across and occurring in dike-shaped intrusions.

pelagic sediment Microscopic plankton shells and fine flakes of clay that settle out and accumulate on the deep-ocean floor.

Pelé's hair Droplets of basaltic lava that mold into long glassy strands as they fall.

Pelé's tears Droplets of basaltic lava that mold into tear-shaped glassy beads as they fall.

peneplain A nearly flat surface that lies at an elevation close to sea level; thought to be the product of long-term erosion.

perched water table A quantity of groundwater that lies above the regional water table because an underlying lens of impermeable rock or sediment prevents the water from sinking down to the regional water table.

percolation The process by which groundwater meanders through tiny, crooked channels in the surrounding material.

peridotite A coarse-grained ultramafic rock.

periglacial environment A region with widespread permafrost but without a blanket of snow or ice.

period An interval of geologic time representing a subdivision of a geologic era.

permafrost Permanently frozen ground.

permanent magnet A special material that behaves magnetically for a long time all by itself.

permanent stream A stream that flows year-round because its bed lies below the water table, or because more water is supplied from upstream than can infiltrate the ground.

permeability The degree to which a material allows fluids to pass through it via an interconnected network of pores and cracks.

permineralization The fossilization process in which plant material becomes transformed into rock by the precipitation of silica from groundwater.

petrified A term used by geologists to describe plant material that has transformed into rock by permineralization.

petroglyph Drawings formed by chipping into the desert varnish of rocks to reveal the lighter rock beneath.

petroleum *Oil.*

phaneritic A textural term used to describe coarse-grained igneous rock.

Phanerozoic Eon The most recent eon, an interval of time from 542 Ma to the present.

phenocryst A large crystal surrounded by a finer-grained matrix in an igneous rock.

photochemical smog Brown haze that blankets a city when exhaust from cars and trucks reacts in the presence of sunlight.

photosynthesis The process during which chlorophyll-containing plants remove carbon dioxide from the atmosphere, form tissues, and expel oxygen back to the atmosphere.

phreatomagmatic eruption An explosive eruption that occurs when water enters the magma chamber and turns into steam.

phyllite A fine-grained metamorphic rock with a foliation caused by the preferred orientation of very fine-grained mica.

phyllitic luster A silk-like sheen characteristic of phyllite, a result of the rock's fine-grained mica.

phylogenetic tree A chart representing the ideas of paleontologists showing which groups of organisms radiated from which ancestors.

physical weathering The process in which intact rock breaks into smaller grains or chunks.

piedmont glacier A fan or lobe of ice that forms where a valley glacier emerges from a valley and spreads out into the adjacent plain.

pillow basalt Glass-encrusted basalt blobs that form when magma extrudes on the sea floor and cools very quickly.

placer deposit Concentrations of metal grains in stream sediment that develop when rocks containing native metals erode and create a mixture of sand grains and metal fragments; the moving water of the stream carries away lighter mineral grains.

planetesimal Tiny, solid pieces of rock and metal that collect in a planetary nebula and eventually accumulate to form a planet.

plankton Tiny plants and animals that float in sea or lake water.

plastic deformation The deformational process in which mineral grains behave like plastic and, when compressed or sheared, become flattened or elongate without cracking or breaking.

plate One of about twenty distinct pieces of the relatively rigid lithosphere.

plate boundary The border between two adjacent lithosphere plates.

plate-boundary earthquakes The earthquakes that occur along and define plate boundaries.

plate-boundary volcano A volcanic arc or mid-ocean ridge volcano, formed as a consequence of movement along a plate boundary.

plate interior A region away from the plate boundaries that consequently experiences few earthquakes.

plate tectonics *Theory of plate tectonics.*

playa The flat, typically salty lake bed that remains when all the water evaporates in drier times; forms in desert regions.

Pleistocene ice age(s) The period of time from about 2 Ma to 14,000 years ago, during which the Earth experienced an ice age.

plunge pool A depression at the base of a waterfall scoured by the energy of the falling water.

plunging fold A fold with a tilted hinge.

pluton An irregular or blob-shaped intrusion; can range in size from tens of m across to tens of km across.

pluvial lake A lake formed to the south of a continental glacier as a result of enhanced rainfall during an ice age.

point bar A wedge-shaped deposit of sediment on the inside bank of a meander.

polar cell A high-latitude convection cell in the atmosphere.

polar easterlies Prevailing winds that come from the east and flow from the polar high to the subpolar low.

polar front The convergence zone in the atmosphere at latitude 60°.

polar glacier *Dry-bottom glacier.*

polar high The zone of high pressure in polar regions created by the sinking of air in the polar cells.

polarity The orientation of a magnetic dipole.

polarity chron The time interval between polarity reversals of Earth's magnetic field.

polarity subchron The time interval between magnetic reversals if the interval is of short duration (less than 200,000 years long).

polarized light A beam of filtered light waves that all vibrate in the same plane.

polar wander The phenomenon of the progressive changing through time of the position of the Earth's magnetic poles relative to a location on a continent; significant polar wander probably doesn't occur—in fact, poles seem to remain fairly fixed, while continents move.

polar-wander path The curving line representing the apparent progressive change in the position of the Earth's magnetic pole, relative to a locality X, assuming that the position of X on Earth has been fixed through time (in fact, poles stay fixed while continents move).

pollen Tiny grains involved in plant reproduction.

polymorphs Two minerals that have the same chemical composition but a different crystal lattice structure.

pore A small open space within sediment or rock.

pore collapse The closer packing of grains that occurs when groundwater is extracted from pores, thus eliminating the support holding the grains apart.

porosity The total volume of empty space (pore space) in a material, usually expressed as a percentage.

porphyritic A textural term for igneous rock that has phenocrysts distributed throughout a finer matrix.

positive anomaly An area where the magnetic field strength is stronger than expected.

positive-feedback mechanism A mechanism that enhances the process that causes the mechanism in the first place.

potentiometric surface The elevation to which water in an artesian system would rise if unimpeded; where there are flowing artesian wells, the potentiometric surface lies above ground.

pothole A bowl-shaped depression carved into the floor of a stream by a long-lived whirlpool carrying sand or gravel.

Precambrian The interval of geologic time between Earth's formation about 4.57 Ga and the beginning of the Phanerozoic Eon 542 Ma.

precession The gradual conical path traced out by Earth's spinning axis; simply put, it is the "wobble" of the axis.

precious metals Metals (like gold, silver, and platinum) that have high value.

precipitation (1) The process by which atoms dissolved in a solution come together and form a solid; (2) rainfall or snow.

preferred mineral orientation The metamorphic texture that exists where platy grains lie parallel to one another and/or elongate grains align in the same direction.

pressure Force per unit area, or the "push" acting on a material in cases where the push is the same in all directions.

pressure gradient The rate of pressure change over a given horizontal distance.

pressure solution The process of dissolution at points of contact, between grains, where compression is greatest, producing ions that then precipitate elsewhere, where compression is less.

prevailing winds Surface winds that generally flow in the same direction for long time periods.

primary porosity The space that remains between solid grains or crystals immediately after sediment accumulates or rock forms.

principal aquifer The geologic unit that serves as the primary source of groundwater in a region.

principle of baked contacts When an igneous intrusion "bakes" (metamorphoses) surrounding rock, the rock that has been baked must be older than the intrusion.

principle of cross-cutting relations If one geologic feature cuts across another, the feature that has been cut is older.

principle of fossil succession In a stratigraphic sequence, different species of fossil organisms appear in a definite order; once a fossil species disappears in a sequence of strata, it never reappears higher in the sequence.

principle of inclusions If a rock contains fragments of another rock, the fragments must be older than the rock containing them.

principle of original continuity Sedimentary layers, before erosion, formed fairly continuous sheets over a region.

principle of original horizontality Layers of sediment, when originally deposited, are fairly horizontal.

principle of superposition In a sequence of sedimentary rock layers, each layer must be younger than the one below, for a layer of sediment cannot accumulate unless there is already a substrate on which it can collect.

principle of uniformitariansim The physical processes we observe today also operated in the past in the same way, and at comparable rates.

prograde metamorphism Metamorphism that occurs as temperatures and pressures are increasing.

Proterozoic The most recent of the Precambrian eons.

protocontinent A block of crust composed of volcanic arcs and hot-spot volcanoes sutured together.

protolith The original rock from which a metamorphic rock formed.

protoplanet A body that grows by the accumulation of planetesimals but has not yet become big enough to be called a planet.

protoplanetary nebula A ring of gas and dust that surrounded the newborn Sun, from which the planets were formed.

protostar A dense body of gas that is collapsing inward because of gravitational forces and that may eventually become a star.

pumice A glassy igneous rock that forms from felsic frothy lava and contains abundant (over 50%) pore space.

punctuated equilibrium The hypothesis that evolution takes place in fits and starts; evolution occurs very slowly for quite a while and then, during a relatively short period, takes place very rapidly.

P-waves Compressional seismic waves that move through the body of the Earth.

P-wave shadow zone A band between 103° and 143° from an earthquake epicenter, as measured along the circumference of the Earth, inside which P-waves do not arrive at seismograph stations.

pycnocline The boundary between layers of water of different densities.

pyroclastic debris Fragmented material that sprayed out of a volcano and landed on the ground or sea floor in solid form.

pyroclastic flow A fast-moving avalanche that occurs when hot volcanic ash and debris mix with air and flow down the side of a volcano.

pyroclastic rock Rock made from fragments blown out of a volcano during an explosion that were then packed or welded together.

quarry A site at which stone is extracted from the ground.

quartzite A metamorphic rock composed of quartz and transformed from a protolith of quartz sandstone.

quenching A sudden cooling of molten material to form a solid.

quick clay Clay that behaves like a solid when still (because of surface tension holding the water-coated clay flakes together), but that flows like a liquid when shaken.

radial network A drainage network in which the streams flow outward from a cone-shaped mountain, and define a pattern resembling spokes on a wheel.

radioactive decay The process by which a radioactive atom undergoes fission or releases particles thereby transforming into a new element.

radioactive isotope An unstable isotope of a given element.

radiometric dating The science of dating geologic events in years by measuring the ratio of parent radioactive atoms to daughter product atoms.

rain band A spiraling arm of a hurricane radiating outward from the eye.

rain shadow The inland side of a mountain range, which is arid because the mountains block rain clouds from reaching the area.

range (for fossils) The interval of a sequence of strata in which a specific fossil species appears.

rapids A reach of a stream in which water becomes particularly turbulent; as a consequence, waves develop on the surface of the stream.

reach A specified segment of a stream's path.

recessional moraine The end moraine that forms when a glacier stalls for a while as it recedes.

recharge area A location where water enters the ground and infiltrates down to the water table.

recrystallization The process in which ions or atoms in minerals rearrange to form new minerals.

rectangular network A drainage network in which the streams join each other at right angles because of a rectangular grid of fractures that breaks up the ground and localizes channels.

recurrence interval The average time between successive geologic events.

red giant A huge red star that forms when Sun-sized stars start to die and expand.

red shift The phenomenon in which a source of light moving away from you very rapidly shifts to a lower frequency; that is, toward the red end of the spectrum.

reef bleaching The death and loss of color of a coral reef.

reflected ray A ray that bounces off a boundary between two different materials.

refracted ray A ray that bends as it passes through a boundary between two different materials.

refraction The bending of a ray as it passes through a boundary between two different materials.

reg A vast stony plain in a desert.

regional metamorphism *Dynamothermal metamorphism;* metamorphism of a broad region, usually the result of deep burial during an orogeny.

regolith Any kind of unconsolidated debris that covers bedrock.

regression The seaward migration of a shoreline caused by a lowering of sea level.

relative age The age of one geologic feature with respect to another.

relative humidity The ratio between the measured water content of air and the maximum possible amount of water the air can hold at a given condition.

relative plate velocity The movement of one lithosphere plate with respect to another.

relief The difference in elevation between adjacent high and low regions on the land surface.

renewable resource A resource that can be replaced by nature within a short time span relative to a human life span.

reservoir rock Rock with high porosity and permeability, so it can contain an abundant amount of easily accessible oil.

residence time The average length of time that a substance stays in a particular reservoir.

residual mineral deposit Soils in which the residuum left behind after leaching by rainwater is so concentrated in metals that the soil itself becomes an ore deposit.

resurgent dome The new mound, or cone, of igneous rock that grows within a caldera as an eruption begins anew.

retrograde metamorphism Metamorphism that occurs as pressures and temperatures are decreasing; for retrograde metamorphism to occur, water must be added.

return stroke An upward-flowing electric current from the ground that carries positive charges up to a cloud during a lightning flash.

reversed polarity Polarity in which the paleomagnetic dipole points north.

reverse fault A steeply dipping fault on which the hanging-wall block slides up.

Richter magnitude scale A scale that defines earthquakes on the basis of the amplitude of the largest ground motion recorded on a seismogram.

ridge axis The crest of a mid-ocean ridge; the ridge axis defines the position of a divergent plate boundary.

right-lateral strike-slip fault A strike-slip fault in which the block on the opposite fault plane from a fixed spot moves to the right of that spot.

rip current A strong, localized seaward flow of water perpendicular to a beach.

riprap Loose boulders or concrete piled together along a beach to absorb wave energy before it strikes a cliff face.

roche moutonnée A glacially eroded hill that becomes elongate in the direction of flow and asymmetric; glacial rasping smoothes the upstream part of the hill into a gentle slope, while glacial plucking erodes the downstream edge into a steep slope.

rock A coherent, naturally occurring solid, consisting of an aggregate of minerals or a mass of glass.

rock burst A sudden explosion of rock off the ceiling or wall of an underground mine.

rock cycle The succession of events that results in the transformation of Earth materials from one rock type to another, then another, and so on.

rock flour Fine-grained sediment produced by glacial abrasion of the substrate over which a glacier flows.

rock glacier A slow-moving mixture of rock fragments and ice.

rock slide A sudden downslope movement of rock.

rocky coast An area of coast where bedrock rises directly from the sea, so beaches are absent.

Rodinia A proposed Precambrian supercontinent that existed around 1 billion years ago.

rotational axis The imaginary line through the center of the Earth around which the Earth spins.

R-waves (*Rayleigh waves*) Surface seismic waves that cause the ground to ripple up and down, like water waves in a pond.

sabkah A region of formerly flooded coastal desert in which stranded seawater has left a salt crust over a mire of mud that is rich in organic material.

salinity The degree of concentration of salt in water.

saltation The movement of a sediment in which grains bounce along their substrate, knocking other grains into the water column (or air) in the process.

salt dome A rising bulbous dome of salt that bends up the adjacent layers of sedimentary rock.

salt wedging The process in arid climates by which dissolved salt in groundwater crystallizes and grows in open pore spaces in rocks and pushes apart the surrounding grains.

sand spit An area where the beach stretches out into open water across the mouth of a bay or estuary.

sandstone Coarse-grained sedimentary rock consisting almost entirely of quartz.

sand volcano (or sand blow) A small mound of sand produced when sand layers below the ground surface liquify as a result of seismic shaking, causing the sand to erupt onto the Earth's surface through cracks or holes in overlying clay layers.

saprolite A layer of rotten rock created by chemical weathering in warm, wet climates.

Sargasso Sea The center of North Atlantic Gyre, named for the tropical seaweed sargassum, which accumulates in its relatively noncirculating waters.

saturated solution Water that carries as many dissolved ions as possible under given environmental conditions.

saturated zone The region below the water table where pore space is filled with water.

scattering The dispersal of energy that occurs when light interacts with particles in the atmosphere.

schist A medium-to-coarse-grained metamorphic rock that possesses schistosity.

schistosity Foliation caused by the preferred orientation of large mica flakes.

scientific method A sequence of steps for systematically analyzing scientific problems in a way that leads to verifiable results.

scoria A glassy mafic igneous rock containing abundant air-filled holes.

scouring A process by which running water removes loose fragments of sediment from a stream bed.

sea arch An arch of land protruding into the sea and connected to the mainland by a narrow bridge.

sea-floor spreading The gradual widening of an ocean basin as new oceanic crust forms at a mid-ocean ridge axis and then moves away from the axis.

sea ice Ice formed by the freezing of the surface of the sea.

seal A relatively impermeable rock, such as shale, salt, or unfractured limestone, that lies above a reservoir rock and stops the oil from rising further.

seam A sedimentary bed of coal interlayered with other sedimentary rocks.

seamount An isolated submarine mountain.

seasonal floods Floods that appear almost every year during seasons when rainfall is heavy or when winter snows start to melt.

seasonal well A well that provides water only during the rainy season when the water table rises below the base of the well.

sea stack An isolated tower of land just offshore, disconnected from the mainland by the collapse of a sea arch.

seawall A wall of riprap built on the landward side of a backshore zone in order to protect shore cliffs from erosion.

second The basic unit of time measurement, now defined as the time it takes for the magnetic field of a cesium atom to flip polarity 9,192,631,770 times, as measured by an atomic clock.

secondary enrichment The process by which a new ore deposit forms from metals that were dissolved and carried away from preexisting ore minerals.

secondary porosity New pore space in rocks, created some time after a rock first forms.

secondary recovery technique A process used to extract the quantities of oil that will not come out of a reservoir rock with just simple pumping.

sediment An accumulation of loose mineral grains, such as boulders, pebbles, sand, silt, or mud, that are not cemented together.

sedimentary basin A depression, created as a consequence of subsidence, that fills with sediment.

sedimentary rock Rock that forms either by the cementing together of fragments broken off preexisting rock or by the precipitation of mineral crystals out of water solutions at or near the Earth's surface.

sedimentary sequence A grouping of sedimentary units bounded on top and bottom by regional unconformities.

sediment budget The proportion of sand supplied to sand removed from a depositional setting.

sediment load The total volume of sediment carried by a stream.

sediment maturity The degree to which a sediment has evolved from a crushed-up version of the original rock into a sediment that has lost its easily weathered minerals and become well sorted and rounded.

sediment sorting The segregation of sediment by size.

seep A place where oil-filled reservoir rock intersects the ground surface, or where fractures connect a reservoir to the ground surface, so that oil flows out onto the ground on its own.

seiche Rhythmic movement in a body of water caused by ground motion.

seismic belts (or zones) The relatively narrow strips of crust on Earth under which most earthquakes occur.

seismicity Earthquake activity.

seismic-moment magnitude scale A scale that defines earthquake size on the basis of calculations involving the amount of slip, length of rupture, depth of rupture, and rock strength.

seismic ray The changing position of an imaginary point on a wave front as the front moves through rock.

seismic-reflection profile A cross-sectional view of the crust made by measuring the reflection of artificial seismic waves off boundaries between different layers of rock in the crust.

seismic tomography Analysis by sophisticated computers of global seismic data in order to create a three-dimensional image of variations in seismic-wave velocities within the Earth.

seismic velocity The speed at which seismic waves travel.

seismic-velocity discontinuity A boundary in the Earth at which seismic velocity changes abruptly.

seismic (earthquake) waves Waves of energy emitted at the focus of an earthquake.

seismogram The record of an earthquake produced by a seismograph.

seismograph (seismometer) An instrument that can record the ground motion from an earthquake.

semipermanent pressure cell A somewhat elliptical zone of high or low atmospheric pressure that lasts much of the year; it forms because high-pressure zones tend to be narrower over land than over sea.

Sevier orogeny A mountain-building event that affected western North America between about 150 Ma and 80 Ma, a result of convergent margin tectonism; a fold-thrust belt formed during this event.

shale Very fine-grained sedimentary rock that breaks into thin sheets.

shatter cones Small, cone-shaped fractures formed by the shock of a meteorite impact.

shear strain A change in shape of an object that involves the movement of one part of a rock body sideways past another part so that angular relationships within the body change.

shear stress A stress that moves one part of a material sideways past another part.

shear waves Seismic waves in which particles of material move back and forth perpendicular to the direction in which the wave itself moves.

shear zone A fault in which movement has occurred ductilely.

sheetwash A film of water less than a few mm thick that covers the ground surface during heavy rains.

shield An older, interior region of a continent.

shield volcano A subaerial volcano with a broad, gentle dome, formed either from low-viscosity basaltic lava or from large pyroclastic sheets.

shocked quartz Grains of quartz that have been subjected to intense pressure such as occurs during a meteorite impact.

shoreline The boundary between the water and land.

shortening The process during which a body of rock or a region of crust becomes shorter.

short-term climate change Climate change that takes place over hundreds to thousands of years.

Sierran arc A large continental volcanic arc along western North America that initiated at the end of the Jurassic Period and lasted until about 80 million years ago.

silica SiO_2.

silicate minerals Minerals composed of silicon-oxygen tetrahedra linked in various arrangements; most contain other elements too.

silicate rock Rock composed of silicate minerals.

siliceous sedimentary rock Sedimentary rock that contains abundant quartz.

silicic Rich in silica with relatively little iron and magnesium.

sill A nearly horizontal table-top-shaped tabular intrusion that occurs between the layers of country rock.

siltstone Fine-grained sedimentary rock generally composed of very small quartz grains.

sinkhole A circular depression in the land that forms when an underground cavern collapses.

slab-pull force The force that downgoing plates (or slabs) apply to oceanic lithosphere at a convergent margin.

slate Fine-grained, low-grade metamorphic rock, formed by the metamorphism of shale.

slaty cleavage The foliation typical of slate, and reflective of the preferred orientation of slate's clay minerals, that allows slate to be split into thin sheets.

slickensides The polished surface of a fault caused by slip on the fault; lineated slickensides also have groves that indicate the direction of fault movement.

slip face The lee side of a dune, which sand slides down.

slip lineations Linear marks on a fault surface created during movement on the fault; some slip lineations are defined by grooves, some by aligned mineral fibers.

slope failure The downslope movement of material on an unstable slope.

slumping Downslope movement in which a mass of regolith detaches from its substrate along a spoon-shaped sliding surface and slips downward semicoherently.

smelting The heating of a metal-containing rock to high temperatures in a fire so that the rock will decompose to yield metal plus a nonmetallic residue (slag).

snotite A long gob of bacteria that slowly drips from the ceiling of a cave.

snow line The boundary above which snow remains all year.

soda straw A hollow stalactite in which calcite precipitates around the outside of a drip.

soil Sediment that has undergone changes at the surface of the Earth, including reaction with rainwater and the addition of organic material.

soil erosion The removal of soil by wind and runoff.

soil horizon Distinct zones within a soil, distinguished from each other by factors such as chemical composition and organic content.

soil moisture Underground water that wets the surface of the mineral grains and organic material making up soil, but lies above the water table.

soil profile A vertical sequence of distinct zones of soil.

solar wind A stream of particles with enough energy to escape from the Sun's gravity and flow outward into space.

solid-state diffusion The slow movement of atoms or ions through a solid.

solifluction The type of creep characteristic of tundra regions; during the summer, the uppermost layer of permafrost melts, and the soggy, weak layer of ground then flows slowly downslope in overlapping sheets.

solstice A day on which the polar ends of the terminator (the boundary between the day hemisphere and the night hemisphere) lie 23.5° away from the associated geographic poles.

Sonoma orogeny A convergent-margin mountain-building event that took place on the western coast of North America in the Late Permian and Early Triassic periods.

sorting (1) The range of clast sizes in a collection of sediment; (2) the degree to which sediment has been separated by flowing currents into different-size fractions.

source rock A rock (organic-rich shale) containing the raw materials from which hydrocarbons eventually form.

southeast tradewinds Tradewinds in the Southern Hemisphere, which start flowing northward, deflect to the west, and end up flowing from southeast to northwest.

southern oscillation The oscillating of atmospheric pressure cells back and forth across the Pacific Ocean, in association with El Niño.

specific gravity A number representing the density of a mineral, as specified by the ratio between the weight of a volume of the mineral and the weight of an equal volume of water.

speleothem A formation that grows in a limestone cave by the accumulation of travertine precipitated from water solutions dripping in a cave or flowing down the wall of a cave.

sphericity The measure of the degree to which a clast approaches the shape of a sphere.

spreading boundary *Divergent plate boundary.*

spreading rate The rate at which sea floor moves away from a mid-ocean ridge axis, as measured with respect to the sea floor on the opposite side of the axis.

spring A natural outlet from which groundwater flows up onto the ground surface.

spring tide An especially high tide that occurs when the Sun is on the same side of the Earth as the Moon.

stable air Air that does not have a tendency to rise rapidly.

stable slope A slope on which downward sliding is unlikely.

stalactite An icicle-like cone that grows from the ceiling of a cave as dripping water precipitates limestone.

stalagmite An upward-pointing cone of limestone that grows when drips of water hit the floor of a cave.

standing wave A wave whose crest and trough remain in place as water moves through the wave.

star dune A constantly changing dune formed by frequent shifts in wind direction; it consists of overlapping crescent dunes pointing in many different directions.

stick-slip behavior Stop-start movement along a fault plane caused by friction, which prevents movement until stress builds up sufficiently.

stone rings Ridges of cobbles between adjacent bulges of permafrost ground.

stoping A process by which magma intrudes; blocks of wall rock break off and then sink into the magma.

storm An episode of severe weather in which winds, precipitation, and in some cases lightning become strong enough to be bothersome and even dangerous.

storm-center velocity A storm's (hurricane's) velocity along its track.

storm surge   Excess seawater driven landward by wind during a storm; the low atmospheric pressure beneath the storm allows sea level to rise locally, increasing the surge.

strain   The change in shape of an object in response to deformation (i.e., as a result of the application of a stress).

stratified drift   Glacial sediment that has been redistributed and stratified by flowing water.

stratigraphic column   A cross-section diagram of a sequence of strata summarizing information about the sequence.

stratigraphic formation   A recognizable layer of a specific sedimentary rock type or set of rock types, deposited during a certain time interval, that can be traced over a broad region.

stratigraphic sequence   An interval of strata deposited during periods of relatively high sea level, and bounded above and below by regional unconformities.

stratopause   The temperature pause that marks the top of the stratosphere.

stratosphere   The stable, stratified layer of atmosphere directly above the troposphere.

stratovolcano   A large, cone-shaped subaerial volcano consisting of alternating layers of lava and tephra.

stratus cloud   A thin, sheet-like, stable cloud.

streak   The color of the powder produced by pulverizing a mineral on an unglazed ceramic plate.

stream   A ribbon of water that flows in a channel.

stream bed   The floor of a stream.

stream capacity   The total quantity of sediment a stream carries.

stream capture (or piracy)   The situation in which headward erosion causes one stream to intersect the course of another, previously independent stream, so that the intersected stream starts to flow down the channel of the first stream.

stream competence   The maximum particle size that a stream can carry.

stream gradient   The slope of a stream's channel in the downstream direction.

stream rejuvination   The renewed downcutting of a stream into a floodplain or peneplain, caused by a relative drop of the base level.

stress   The push, pull, or shear that a material feels when subjected to a force; formally, the force applied per unit area over which the force acts.

stretching   The process during which a layer of rock or a region of crust becomes longer.

striations   Linear scratches in rock.

strike-slip fault   A fault in which one block slides horizontally past another (and therefore parallel to the strike line), so there is no relative vertical motion.

strip mining   The scraping off of all soil and sedimentary rock above a coal seam in order to gain access to the seam.

stromatolite   Layered mounds of sediment formed by cyanobacteria; cyanobacteria secrete a mucuous-like substance to which sediment sticks, and as each layer of cyanobacteria gets buried by sediment, it colonizes the surface of the new sediment, building a mound upward.

structural control   The condition in which geologic structures, such as faults, affect the distribution and drainage of water or the shape of the land surface.

subaerial   Pertaining to land regions above sea level (i.e., under air).

subduction   The process by which one oceanic plate bends and sinks down into the asthenosphere beneath another plate.

subduction zone   The region along a convergent boundary where one plate sinks beneath another.

sublimation   The evaporation of ice directly into vapor without first forming a liquid.

submarine canyon   A narrow, steep canyon that dissects a continental shelf and slope.

submarine fan   A wedge-shaped accumulation of sediment at the base of a submarine slope; fans usually accumulate at the mouth of a submarine canyon.

submarine slump   The underwater downslope movement of a semicoherent block of sediment along a weak mud detachment.

submergent coast   A coast at which the land is sinking relative to sea level.

subpolar low   The rise of air where the surface flow of a polar cell converges with the surface flow of a Ferrel cell, creating a low-pressure zone in the atmosphere.

subsidence   The vertical sinking of the Earth's surface in a region, relative to a reference plane.

substrate   A general term for material just below the ground surface.

subtropical high (subtropical divergence zone)   A belt of high pressure in the atmosphere at 30° latitude formed where the Hadley cell converges with the Ferrel cell, causing cool, dense air to sink.

subtropics   Desert climate regions that lie on either side of the equatorial tropics between the lines of 20° and 30° north or south of the equator.

summit eruption   An eruption that occurs in the summit crater of a volcano.

sunspot cycle   The cyclic appearance of large numbers of sunspots (black spots thought to be magnetic storms on the Sun's surface) every 9 to 11.5 years.

supercontinent cycle   The process of change during which supercontinents develop and later break apart, forming pieces that may merge once again in geologic time to make yet another supercontinent.

supernova   A short-lived, very bright object in space that results from the cataclysmic explosion marking the death of a very large star; the explosion ejects large quantities of matter into space to form new nebulae.

superplume   A huge mantle plume.

superposed stream   A stream whose geometry has been laid down on a rock structure and is not controlled by the structure.

surface current   An ocean current in the top 100 m of water.

surface waves   Seismic waves that travel along the Earth's surface.

surface westerlies   The prevailing surface winds in North America and Europe, which come out of the west or southwest.

surf zone   A region of the shore in which breakers crash onto the shore.

surge (glacial)   A pulse of rapid flow in a glacier.

suspended load   Tiny solid grains carried along by a stream without settling to the floor of the channel.

swamp   A wetland dominated by trees.

swash   The upward surge of water that flows up a beach slope when breakers crash onto the shore.

S-waves   Seismic shear waves that pass through the body of the Earth.

S-wave shadow zone   A band between 103° and 180° from the epicenter of an earthquake inside of which S-waves do not arrive at seismograph stations.

swelling clay   Clay possessing a mineral structure that allows it to absorb water between its layers and thus swell to several times its original size.

symmetry   The condition in which the shape of one part of an object is a mirror image of the other part.

syncline   A trough-shaped fold whose limbs dip toward the hinge.

systematic joints Long planar cracks that occur fairly regularly throughout a rock body.

tabular intrusions Sheet intrusions that are planar and of roughly uniform thickness.

Taconic orogeny A convergent mountain-building event that took place around 400 million years ago, in which a volcanic island arc collided with eastern North America.

tailings pile A pile of waste rock from a mine.

talus A sloping apron of fallen rock along the base of a cliff.

tar Hydrocarbons that exist in solid form at room temperature.

tarn A lake that forms at the base of a cirque on a glacially eroded mountain.

tar sand Sandstone reservoir rock in which less viscous oil and gas molecules have either escaped or been eaten by microbes, so that only tar remains.

taxonomy The study and classification of the relationships among different forms of life.

tension A stress that pulls on a material and could lead to stretching.

tephra Unconsolidated accumulations of pyroclastic grains.

terminal moraine The end moraine at the farthest limit of glaciation.

terminator The boundary between the half of the Earth that has daylight and the half experiencing night.

terrace The elevated surface of an older floodplain into which a younger floodplain had cut down.

terrestrial A term used to describe the inner, Earth-like planets.

thalweg The deepest part of a stream's channel.

theory A scientific idea supported by an abundance of evidence that has passed many tests and failed none.

theory of plate tectonics The theory that the outer layer of the Earth (the lithosphere) consists of separate plates that move with respect to one another.

thermal metamorphism Metamorphism caused by heat conducted into country rock from an igneous intrusion.

thermocline A boundary between layers of water with differing temperatures.

thermohaline circulation The rising and sinking of water driven by contrasts in water density, which is due in turn to differences in temperature and salinity; this circulation involves both surface and deep-water currents in the ocean.

thermosphere The outermost layer of the atmosphere containing very little gas.

thin section A 3/100-mm-thick slice of rock that can be examined with a petrographic microscope.

thin-skinned deformation A distinctive style of deformation characterized by displacement on faults that terminate at depth along a subhorizontal detachment fault.

thrust fault A gently dipping reverse fault; the hanging-wall block moves up the slope of the fault.

tidal bore A visible wall of water that moves toward shore with the rising tide in quiet waters.

tidal flat A broad, nearly horizontal plain of mud and silt, exposed or nearly exposed at low tide but totally submerged at high tide.

tidal reach The difference in sea level between high tide and low tide at a given point.

tide The daily rising or falling of sea level at a given point on the Earth.

tide-generating force The force, caused in part by the gravitational attraction of the Sun and Moon, and in part by the centrifugal force created by the Earth's spin, that generates tides.

till A mixture of unsorted mud, sand, pebbles, and larger rocks deposited by glaciers.

tillite A rock formed from hardened ancient glacial deposits and consisting of larger clasts distributed through a matrix of sandstone and mudstone.

toe (terminus) The leading edge or margin of a glacier.

tombolo A narrow ridge of sand that links a sea stack to the mainland.

topographical map A map that uses contour lines to represent variations in elevation.

topography Variations in elevation.

topsoil The top soil horizons, which are typically dark and nutrient-rich.

tornado A near-vertical, funnel-shaped cloud in which air rotates extremely rapidly around the axis of the funnel.

tornado swarm Dozens of tornadoes produced by the same storm.

tower karst A karst landscape in which steep-sided residual bedrock towers remain between sinkholes.

transform fault A fault marking a transform plate boundary; along mid-ocean ridges, transform faults are the actively slipping segment of a fracture zone between two ridge segments.

transform plate boundary A boundary at which one lithosphere plate slips laterally past another.

transgression The inland migration of shoreline resulting from a rise in sea level.

transition zone The middle portion of the mantle, from 400 to 670 km deep, in which there are several jumps in seismic velocity.

transpiration The release of moisture as a metabolic byproduct.

transverse dune A simple, wave-like dune that appears when enough sand accumulates for the ground surface to be completely buried, but only moderate winds blow.

travel-time curve A graph that plots the time since an earthquake began on the vertical axis, and the distance to the epicenter on the horizontal axis.

trellis network A drainage system that develops across a landscape of parallel valleys and ridges so that major tributaries flow down the valleys and join a trunk stream that cuts through the ridge; the resulting map pattern resembles a garden trellis.

trench A deep elongate trough bordering a volcanic arc; a trench defines the trace of a convergent plate boundary.

triangulation The method for determining the map location of a point from knowing the distance between that point and three other points; this method is used to locate earthquake epicenters.

tributary A smaller stream that flows into a larger stream.

triple junction A point where three lithosphere plate boundaries intersect.

tropical depression A tropical storm with winds reaching up to 61 km per hour; such storms develop from tropical disturbances, and may grow to become hurricanes.

tropical disturbance Cyclonic winds that develop in the tropics.

tropopause The temperature pause marking the top of the troposphere.

troposphere The lowest layer of the atmosphere, where air undergoes convection and where most wind and clouds develop.

truncated spur A spur (elongate ridge between two valleys) whose end was eroded off by a glacier.

trunk stream The single larger stream into which an array of tributaries flow.

tsunami A large wave along the sea surface triggered by an earthquake or large submarine slump.

tuff A pyroclastic igneous rock composed of volcanic ash and fragmented pumice, formed when accumulations of the debris cement together.

tundra   A cold, treeless region of land at high latitudes, supporting only species of shrubs, moss, and lichen capable of living on permafrost.

turbidite   A graded bed of sediment built up at the base of a submarine slope and deposited by turbidity currents.

turbidity current   A submarine avalanche of sediment and water that speeds down a submarine slope.

turbulence   The chaotic twisting, swirling motion in flowing fluid.

typhoon   The equivalent of a hurricane in the western Pacific Ocean.

ultimate base level   Sea level; the level below which a trunk stream cannot cut.

ultramafic   A term used to describe igneous rocks or magmas that are rich in iron and magnesium and very poor in silica.

unconfined aquifer   An aquifer that intersects the surface of the Earth.

unconformity   A boundary between two different rock sequences representing an interval of time during which new strata were not deposited and/or were eroded.

unconsolidated   Consisting of unattached grains.

undercutting   Excavation at the base of a slope that results in the formation of an overhang.

undersaturated   A term used to describe a solution capable of holding more dissolved ions.

unsaturated zone   The region of the subsurface above the water table.

unstable air   Air that is significantly warmer than air above and has a tendency to rise quickly.

unstable ground   Land capable of slumping or slipping downslope in the near future.

unstable slope   A slope on which sliding will likely happen.

updraft   Upward-moving air.

upper mantle   The uppermost section of the mantle, reaching down to a depth of 400 km.

upwelling zone   A place where deep water rises in the ocean, or hot magma rises in the asthenosphere.

U-shaped valley   A steep-walled valley shaped by glacial erosion into the form of a U.

vacuum   Space that contains very little matter in a given volume (e.g., a region in which air has been removed).

valley   A trough with sloping walls, cut into the land by a stream.

valley glacier   A river of ice that flows down a mountain valley.

Van Allen radiation belts   Belts of solar wind particles and cosmic rays that surround the Earth, trapped by Earth's magnetic field.

varve   A pair of thin layers of glacial lake-bed sediment, one consisting of silt brought in during the spring floods, and the other of clay deposited during the winter when the lake's surface freezes over and the water is still.

vascular plant   A plant with woody tissue and seeds and veins for transporting water and food.

vein   A seam of minerals that forms when dissolved ions carried by water solutions precipitate in cracks.

vein deposit   A hydrothermal deposit in which the ore minerals occur in veins that fill cracks in preexisting rocks.

velocity-versus-depth curve   A graph that shows the variation in the velocity of seismic waves with increasing depth in the Earth.

ventifact (faceted rock)   A desert rock whose surface has been faceted by the wind.

vesicles   Open holes in igneous rock formed by the preservation of bubbles in magma as the magma cools into solid rock.

viscosity   The resistance of material to flow.

volatiles   Elements or compounds such as H_2O and CO_2 that evaporate easily and can exist in gaseous forms at the Earth's surface.

volatility   A specification of the ease with which a material evaporates.

volcanic arc   A curving chain of active volcanoes formed adjacent to a convergent plate boundary.

volcanic ash   Tiny glass shards formed when a fine spray of exploded lava freezes instantly upon contact with the atmosphere.

volcanic bomb   A large piece of pyroclastic debris thrown into the atmosphere during a volcanic eruption.

volcanic-danger-assessment map   A map delineating areas that lie in the path of potential lava flows, lahars, debris flows, or pyroclastic flows of an active volcano.

volcanic gas   Elements or compounds that bubble out of magma or lava in gaseous form.

volcanic island arc   The volcanic island chain that forms on the edge of the overriding plate where one oceanic plate subducts beneath another oceanic plate.

volcano   (1) A vent from which melt from inside the Earth spews out onto the planet's surface; (2) a mountain formed by the accumulation of extrusive volcanic rock.

V-shaped valley   A valley whose cross-sectional shape resembles the shape of a V; the valley probably has a river running down the point of the V.

Wadati-Benioff zone   A sloping band of seismicity defined by intermediate- and deep-focus earthquakes that occur in the downgoing slab of a convergent plate boundary.

wadi   The name used in the Middle East and North Africa for a dry wash.

warm front   A front in which warm air rises slowly over cooler air in the atmosphere.

waste rock   Rock dislodged by mining activity yet containing no ore minerals.

waterfall   A place where water drops over an escarpment.

water gap   An opening in a resistant ridge where a trunk river has cut through the ridge.

watershed   The region that collects water that feeds into a given drainage network.

water table   The boundary, approximately parallel to the Earth's surface, that separates substrate in which groundwater fills the pores from substrate in which air fills the pores.

wave base   The depth, approximately equal in distance to half a wavelength in a body of water, beneath which there is no wave movement.

wave-cut bench   A platform of rock, cut by wave erosion, at the low-tide line that was left behind a retreating cliff.

wave-cut notch   A notch in a coastal cliff cut out by wave erosion.

wave erosion   The combined effects of the shattering, wedging, and abrading of a cliff face by waves and the sediment they carry.

wave front   The boundary between the region through which a wave has passed and the region through which it has not yet passed.

wavelength   The horizontal difference between two adjacent wave troughs or two adjacent crests.

wave refraction (ocean)   The bending of waves as they approach a shore so that their crests make no more than a 5° angle with the shoreline.

weather Local-scale conditions as defined by temperature, air pressure, relative humidity, and wind speed.

weathered rock Rock that has reacted with air and/or water at or near the Earth's surface.

weathering The processes that break up and corrode solid rock, eventually transforming it into sediment.

weather system A specific set of weather conditions, reflecting the configuration of air movement in the atmosphere, that affects a region for a period of time.

welded tuff Tuff formed by the welding together of hot volcanic glass shards at the base of pyroclastic flows.

well A hole in the ground dug or drilled in order to obtain water.

Western Interior Seaway A north-south-trending seaway that ran down the middle of North America during the Late Cretaceous Period.

wet-bottom (temperate) glacier A glacier with a thin layer of water at its base, over which the glacier slides.

wetted perimeter The area in which water touches a stream channel's walls.

wind abrasion The grinding away at surfaces in a desert by windblown sand and dust.

wind gap An opening through a high ridge that developed earlier in geologic history by stream erosion, but that is now dry.

xenolith A relict of wall rock surrounded by intrusive rock when the intrusive rock freezes.

yardang A mushroom-like column with a resistant rock perched on an eroding column of softer rock; created by wind abrasion in deserts where a resistant rock overlies softer layers of rock.

yazoo stream A small tributary that runs parallel to the main river in a floodplain because the tributary is blocked from entering the main river by levees.

Younger Dryas An interval of cooler temperatures that took place 4,500 years ago during a general warming/glacier-retreat period.

zeolite facies The metamorphic facies just above diagenetic conditions, under which zeolite minerals form.

zone of ablation The area of a glacier in which ablation (melting, sublimation, calving) subtracts from the glacier.

zone of accumulation (1) The layer of regolith in which new minerals precipitate out of water passing through, thus leaving behind a load of fine clay; (2) the area of a glacier in which snowfall adds to the glacier.

zone of aeration *Unsaturated zone.*

zone of leaching The layer of regolith in which water dissolves ions and picks up very fine clay; these materials are then carried downward by infiltrating water.

Credits

UNNUMBERED PHOTOS AND ART

NUMBERED PHOTOS AND ART

Drilling Program); 18.24: Stephen Marshak; **18.26A:** NASA; **18.26B:** Stephen Marshak; **18.29C:** © G.R. Roberts; **18.29D:** Stephen Marshak; **18.31:** Stephen Marshak; **18.32A:** David Muench/Corbis; **18.32B:** © Stephen McDaniel; **18.34B:** © G.R. Roberts; **18.35A:** © Hal Beral/Visuals Unlimited; **18.35B:** © David B. Fleetham/Visuals Unlimited; **18.37:** Adapted from Skinner and Porter, 1995; **18.39:** Adapted from Kraft, 1973; **18.40A:** Annie Griffiths Belt/Corbis; **18.40B:** Dave Saville/FEMA News Photo.

Chapter 19: 19.1: © GeoPhoto Publishing Company; **19.12A:** Vince Streano/Corbis; **19.14A:** Australian Picture Library/Corbis; **19.15H:** Stephen Marshak; **19.16A:** JPL/NASA; **19.16C:** Courtesy of the Food and Agriculture Organization of the United Nations; **19.17:** Stephen Marshak; **19.18A–B:** Stephen Marshak; **19.18C:** National Parks Service Photo; **19.18D:** Stephen Marshak; **19.19B:** Stephen Marshak; **19.20C–D:** Adapted from Galloway et al., 1999; **19.21G:** Photo by Richard O. Ireland, U.S. Geological Survey; **19.21H–I:** Stephen Marshak; **19.23D:** Courtesy of C Tech Development Corporation www.ctech.com; **19.27:** Stephen Marshak; **19.29A:** © G.R. Roberts; **19.29B:** Photo by David Parker, courtesy of the National Astronomy and Ionosphere Center–Arecibo Observatory, a facility of the NSF; **19.29C:** Lois Kent; **19.29D:** © G.R. Roberts; 19.31: Stephen Marshak; **19.32A:** © Kjell B. Sandyed/Visuals Unlimited; **19.32B:** Photo by Jim Pisarowicz.

Chapter 20: 20.1: Agence France Presse/Getty Images; **20.3:** Reuters NewMedia, Inc./Corbis; **20.7:** Bjorn Backe, Papilio/Corbis; **20.9A:** Corbis; **20.9B:** Lieutenant Mark Boland/NOAA Photo Library; **20.12A:** Adapted from Lutgens and Tarbuck, 1998; **20.12C:** Adapted from Lutgens and Tarbuck, 1998; **20.19:** Hubert Stadler/Corbis; **20.21:** Adapted from Lutgens and Tarbuck, 1998; **20.23:** Tim Thompson/Corbis; **20.26A:** Jim Zuckerman/Corbis; **20.26B:** Agence France Presse/Getty Images; **20.27:** NASA/GSFC/METI/ERSDAC/JAROS, and U.S./ Japan ASTER Science Team; **20.28:** Adapted from Coch, 1995; **20.29:** Adapted from NOAA; **20.30A:** Reuters NewMedia, Inc./Corbis; **20.30B:** Adapted from Coch, 1995; **20.30C:** Adapted from Lutgens and Tarbuck, 1998; **20.32:** NOAA Photo Library; **20.33:** Adapted from Getis et al., 1991; **20.34:** Adapted from Lutgens and Tarbuck, 1998; **20.35A:** FAO-SDRN Agrometeorolgy Group; **20.35B:** NASA/GSFC; **20.36:** Modified from Lutgens and Tarbuck, 1998.

Chapter 21: 21.1: Yann Arthus-Bertrand/Corbis; **21.2:** O. Alamany & E. Vicens/Corbis; **21.6C:** Charles O'Rear/Corbis; **21.7A:** Stephen Marshak; **21.7B:** Gordon Whitten/Corbis; **21.8:** Stephen Marshak; **21.9:** Liba Taylor/Corbis; **21.12A–B:** Stephen Marshak; **21.13D:** Stephen Marshak; **21.14:** O. Alamany & E. Vicens/Corbis; **21.15:** © Steve Mulligan Photography; **21.16:** JPL/NASA; **21.17:** Stephen Marshak; **21.18:** © Martin G. Miller/Visuals Unlimited; **21.19A:** © G.R. Roberts; **21.19B:** © Martin Miller; **21.20E–F:** Stephen Marshak; **21.24B:** Stephen Marshak; **21.26A:** J. Van Acker/FAO Photo; **21.26C–D:** Stephen Marshak; **21.27:** Stephen Marshak; **21.28:** Stephen Marshak; **21.29B:** Charles and Josette Lenars/Corbis; **21.29C:** Stephen Marshak; **21.29D:** David Turnley/Corbis; **21.29E:** Corbis; **21.30:** GSFC/NASA.

Chapter 22: 22.1: Stephen Marshak; **22.2A:** © 1996 Galen Rowell; **22.2B:** © 1986 Galen Rowell; **22.3A:** Photos courtesy Kenneth G. Libbrecht, www.snowcrystals.com; **22.3B:** Stephen Marshak; **22.3C:** Lowell Georgia/Corbis; **22.4B:** Courtesy Joel Harper, Institute of Arctic and Alpine Research, University of Colorado; **22.4C:** Stephen Marshak; **22.4D:** USGS/EROS Data Center Satellite Systems Branch; **22.5A:** Modified from Flint, 1971; **22.5B:** Modified from Radok, 1985; **22.9:** © Harry M. Walker;

22.11B: Modified from Raymond, 1971; **22.14B:** Antarctic Search for Meteorites (ANSMET) program/Dr. Larry Nittler, Carnegie Institute of Washington. Photo courtesy Case Western Reserve University, Department of Geological Sciences; **22.15A:** Stephen Marshak; **22.15B:** Courtesy Ted Scambos, National Snow and Ice Data Center, University of Colorado, Boulder; **22.17:** Ralph A. Clevenger/Corbis; **22.18:** Stephen Marshak; **22.19:** Stephen Marshak; **22.20A:** © 1986 Jack Olson; **22.20B:** © Art Wolfe; **22.20C:** Stephen Marshak; **22.21C–D:** Stephen Marshak; **22.22D:** © Gerald and Buff Corsi/Visuals Unlimited; **22.22E:** Ric Ergenbright/Corbis; **22.22F:** © 1986 Keith S. Walklet/Quietworks; **22.23:** © Martin G. Miller; **22.24:** Stephen Marshak; **22.25B:** Stephen Marshak; **22.26A–B:** Stephen Marshak; **22.26C:** Courtesy Duncan Heron; **22.26D:** Stephen Marshak; **22.28:** Adapted from Tarbuck and Lutgens, 1996; **22.29B:** John S. Shelton; **22.29C:** Modified from U.S. Geological Survey; **22.30C:** © Glenn Oliver/Visuals Unlimited; **22.31:** Tom Bean/Corbis; **22.33:** Giovanni Sella; **22.34C:** Stephen Marshak; **22.35B:** Adapted from Hamblin and Christiansen, 1998; **22.36A:** After Flint, 1971; **22.36B:** Scott T. Smith/Corbis; **22.37B:** Reproduced with permission of Dan Armstrong; **22.37C:** Mark A. Kessler, Complex Systems Laboratory, UCSD; **22.38:** Charles Merguerian; **22.39A:** Reproduced with permission of the Minister of Public Works and Government Services Canada, 2004 and Courtesy of Natural Resources Canada, Geological Survey of Canada; **22.39B:** After Flint, 1971; **22.40:** Modified from Skinner and Porter, 1995; **22.41:** detail of mural by Charles R. Knight, American Museum of Natural History, #4950(5). Photo by Denis Finnin; **22.42:** Adapted from Glacial Map of the United States, Geological Society of America; **22.43:** Stephen Marshak; **22.46:** After American Geophysical Union; **22.47:** Hendrick Averkamp, *Winter Scene with Ice Skaters*, ca. 1600. Courtesy Rijksmuseum, Amsterdam; **22.48:** Stephen Marshak.

Chapter 23: 23.2: NASA; **23.6A:** After Van Andel, 1994; **23.6B–C:** After Sloss, 1962; **23.6D:** After Boggs, 1995; 23.6E: After Van Andel,1994; **23.9A:** After Johnson, 1972; **23.9B–C:** After Van Andel, 1994; **23.10A:** Courtesy of the Climatic Research Unit, University of East Anglia; **23.10B:** © G.R. Roberts; **23.11A:** After Kump et al., 1999; **23.11B:** After Van Andel, 1994; **23.12A:** After McKenzie, 1998; **23.12B:** © Crown Copyright. Reproduced courtesy of Historic Scotland, Edinburgh; **23.12C–D:** Reprinted with permission from *Understanding Climate Change.* Copyright 1975 by the National Academy of Sciences. Courtesy of the National Academy Press, Washington, D.C.; **23.13A:** Courtesy of SOHO/MDI consortium. SOHO is a project of international cooperation between ESA and NASA; **23.13B:** Adapted from Chaisson and McMillan, 1998; **23.14A–B:** After Kump et al., 1999; **23.14C:** © William K. Hartmann/Planetary Science Institute; **23.15A:** Gianni Dagli Orti/Corbis; **23.16A:** Courtesy P&H Mining Equipment; **23.16B:** Richard Hamilton Smith/Corbis; **23.17A:** World Resources Institute, in collaboration with the World Conservation Monitoring Center and the World Wildlife Fund. In D. Bryant, et al., *The Last Frontier Forests: Ecosystems and Economies on the Edge* (World Resources Institute: Washington, D.C., 1997); **23.17B:** SeaWifs Project, NASA/Goddard Space Flight Center, and Orbimage; **23.17C:** Kennan Ward/Corbis; **23.18:** After Turco, 1997; **23.19A:** Courtesy of GSFC/NASA; **23.21C:** Adapted from Kattenberg et al., 1996; **23.22:** Adapted from McKenzie, 1998; **23.23:** Adapted from McKenzie, 1998.

Appendix A: a.1: Robert Glusic/PhotoDisc; **a.6:** Image courtesy of Arthur Smith and Randall Feenstra, Carnegie Mellon University; **a.13D:** Visuals Unlimited.

Appendix B: **b.2:** NGDC/NOAA; **b.3:** Dale Sawyer/Rice University; **b.4:** Lisa Gagagan/PLATES Project, University of Texas, Institute for Geophysics; **b.5:** National Resources Conservation Source/USDA; **b.6:** JPL/NASA; **b.7a:** AFP/Getty Images; **b.7b:** DigitalGlobe; **b.7c:** Earth and Space Sciences/University of Washington; **b.7d1:** Space Imaging; **b.7d2:** Space Imaging.

Index

NOTE: Italicized page numbers refer to pictures, tables, and figures. Bold page numbers refer to key words.